Lecture Notes in Computer Science   13246

More information about this series at https://link.springer.com/bookseries/558

Arnab Bhattacharya · Janice Lee Mong Li ·
Divyakant Agrawal · P. Krishna Reddy ·
Mukesh Mohania · Anirban Mondal ·
Vikram Goyal · Rage Uday Kiran (Eds.)

Database Systems for Advanced Applications

27th International Conference, DASFAA 2022
Virtual Event, April 11–14, 2022
Proceedings, Part II

Springer

Editors
Arnab Bhattacharya
Indian Institute of Technology Kanpur
Kanpur, India

Divyakant Agrawal
University of California, Santa Barbara
Santa Barbara, CA, USA

Mukesh Mohania
Indraprastha Institute of Information
Technology Delhi
New Delhi, India

Vikram Goyal
Indraprastha Institute of Information
Technology Delhi
New Delhi, India

Janice Lee Mong Li
National University of Singapore
Singapore, Singapore

P. Krishna Reddy 🆔
IIIT Hyderabad
Hyderabad, India

Anirban Mondal
Ashoka University
Sonepat, Haryana, India

Rage Uday Kiran
University of Aizu
Aizu, Japan

ISSN 0302-9743　　　　　ISSN 1611-3349 (electronic)
Lecture Notes in Computer Science
ISBN 978-3-031-00125-3　　　ISBN 978-3-031-00126-0 (eBook)
https://doi.org/10.1007/978-3-031-00126-0

This Springer imprint is published by the registered company Springer Nature Switzerland AG
The registered company address is: Gewerbestrasse 11, 6330 Cham, Switzerland

General Chairs' Preface

On behalf of the Organizing Committee, it is our great pleasure to welcome you to the proceedings of the 27th International Conference on Database Systems for Advanced Applications (DASFAA 2022), which was held during April 11–14, 2022, in Hyderabad, India. The conference has returned to India for the second time after a gap of 14 years, moving from New Delhi in 2008 to Hyderabad in 2022. DASFAA has long established itself as one of the leading international conferences in database systems. We were expecting to welcome you in person and give you a feel of our renowned Indian hospitality. However, unfortunately, given the Omicron wave of COVID-19 and the pandemic circumstances, we had to move the conference to a fully online mode.

Our gratitude goes first and foremost to the researchers, who submitted their work to the DASFAA 2022 main conference, workshops, and the data mining contest. We thank them for their efforts in submitting the papers, as well as in preparing high-quality online presentation videos. It is our distinct honor that five eminent keynote speakers graced the conference: Sunita Sarawagi of IIT Bombay, India, Guoliang Li of Tsinghua University, China, Gautam Das of the University of Texas at Arlington, Ioana Manolescu of Inria and Institut Polytechnique de Paris, and Tirthankar Lahiri of the Oracle Corporation. Each of them is a leader of international renown in their respective areas, and their participation significantly enhanced the conference. The conference program was further enriched with a panel, five high–quality tutorials, and six workshops on cutting-edge topics.

We would like to express our sincere gratitude to the contributions of the Senior Program Committee (SPC) members, Program Committee (PC) members, and anonymous reviewers, led by the PC chairs, Arnab Bhattacharya (IIT Kanpur), Lee Mong Li Janice (National University of Singapore), and Divyakant Agrawal (University of California, Santa Barbara). It is through their untiring efforts that the conference had an excellent technical program. We are also thankful to the other chairs and Organizing Committee members: industry track chairs, Prasad M. Deshpande (Google), Daxin Jiang (Microsoft), and Rajasekar Krishnamurthy (Adobe); demo track chairs, Rajeev Gupta (Microsoft), Koichi Takeda (Nagoya University), and Ladjel Bellatreche (ENSMA); workshop chairs, Maya Ramanath (IIT Delhi), Wookey Lee (Inha University), and Sanjay Kumar Madria (Missouri Institute of Technology); tutorial chairs, P. Sreenivasa Kumar (IIT Madras), Jixue Liu (University of South Australia), and Takahiro Hara (Osaka university); panel chairs, Jayant Haritsa (Indian Institute of Science), Reynold Cheng (University of Hong Kong), and Georgia Koutrika (Athena Research Center); Ph.D. consortium chairs, Vikram Pudi (IIIT Hyderabad), Srinath Srinivasa (IIIT Bangalore), and Philippe Fournier-Viger (Harbin Institute of Technology); publicity chairs, Raj Sharma (Goldman Sachs), Jamshid Bagherzadeh Mohasefi (Urmia University), and Nazha Selmaoui-Folcher (University of New Caledonia); publication chairs, Vikram Goyal (IIIT Delhi), and R. Uday Kiran (University of Aizu); and registration/local arrangement chairs, Lini Thomas (IIIT Hyderabad), Satish Narayana Srirama (University of Hyderabad), Manish Singh (IIT Hyderabad), P. Radha Krishna (NIT Warangal), Sonali Agrawal (IIIT Allahabad), and V. Ravi (IDRBT).

We appreciate the hosting organization IIIT Hyderabad, which is celebrating its silver jubilee in 2022. We thank the researchers at the Data Sciences and Analytics Center (DSAC) and the Kohli Center on Intelligent Systems (KCIS) at IIIT Hyderabad for their support. We also thank the administration and staff of IIIT Hyderabad for their help. We thank Google for the sponsorship. We feel indebted to the DASFAA Steering Committee for its continuing guidance.

Finally, our sincere thanks go to all the participants and volunteers. There would be no conference without them. We hope all of you enjoy these DASFAA 2022 proceedings.

February 2022

P. Krishna Reddy
Mukesh Mohania
Anirban Mondal

Program Chairs' Preface

It is our great pleasure to present the proceedings of the 27th International Conference on Database Systems for Advanced Applications (DASFAA 2022). DASFAA is a premier international forum for exchanging original research results and practical developments in the field of databases.

For the research track, we received 488 research submissions from across the world. We performed an initial screening of all submissions, leading to the desk rejection of 88 submissions due to violations of double-blind and page limit guidelines. For submissions entering the double-blind review process, each paper received at least three reviews from Program Committee (PC) members. Further, an assigned Senior Program Committee (SPC) member also led a discussion of the paper and reviews with the PC members. The PC co-chairs then considered the recommendations and meta-reviews from SPC members in making the final decisions. As a result, 72 submissions were accepted as full papers (acceptance ratio of 18%), and 76 submissions were accepted as short papers (acceptance ratio of 19%). For the industry track, 13 papers were accepted out of 36 submissions. Nine papers were accepted out of 16 submissions for the demo track. For the Ph.D. consortium, two papers were accepted out of three submissions. Four short research papers and one industry paper were withdrawn. The review process was supported by Microsoft's Conference Management Toolkit (CMT).

The conference was conducted in an online environment, with accepted papers presented via a pre-recorded video presentation with a live Q&A session. The conference program also featured five keynotes from distinguished researchers in the community, a panel, five high–quality tutorials, and six workshops on cutting-edge topics.

We wish to extend our sincere thanks to all SPC members, PC members, and external reviewers for their hard work in providing us with thoughtful and comprehensive reviews and recommendations. We especially thank the authors who submitted their papers to the conference. We hope that the readers of the proceedings find the content interesting, rewarding, and beneficial to their research.

March 2022

Arnab Bhattacharya
Janice Lee Mong Li
Divyakant Agrawal
Prasad M. Deshpande
Daxin Jiang
Rajasekar Krishnamurthy
Rajeev Gupta
Koichi Takeda
Ladjel Bellatreche
Vikram Pudi
Srinath Srinivasa
Philippe Fournier-Viger

Organization

DASFAA 2022 was organized by IIIT Hyderabad, Hyderabad, Telangana, India.

Steering Committee Chair

Lei Chen   Hong Kong University of Science and Technology, Hong Kong

Honorary Chairs

P. J. Narayanan   IIIT Hyderabad, India
S. Sudarshan   IIT Bombay, India
Masaru Kitsuregawa   University of Tokyo, Japan

Steering Committee Vice Chair

Stephane Bressan   National University of Singapore, Singapore

Steering Committee Treasurer

Yasushi Sakurai   Osaka University, Japan

Steering Committee Secretary

Kyuseok Shim   Seoul National University, South Korea

General Chairs

P. Krishna Reddy   IIIT Hyderabad, India
Mukesh Mohania   IIIT Delhi, India
Anirban Mondal   Ashoka University, India

Program Committee Chairs

Arnab Bhattacharya   IIT Kanpur, India
Lee Mong Li Janice   National University of Singapore, Singapore
Divyakant Agrawal   University of California, Santa Barbara, USA

Steering Committee

Zhiyong Peng	Wuhan University, China
Zhanhuai Li	Northwestern Polytechnical University, China
Krishna Reddy	IIIT Hyderabad, India
Yunmook Nah	Dankook University, South Korea
Wenjia Zhang	University of New South Wales, Australia
Zi Huang	University of Queensland, Australia
Guoliang Li	Tsinghua University, China
Sourav Bhowmick	Nanyang Technological University, Singapore
Atsuyuki Morishima	University of Tsukaba, Japan
Sang-Won Lee	Sungkyunkwan University, South Korea
Yang-Sae Moon	Kangwon National University, South Korea

Industry Track Chairs

Prasad M. Deshpande	Google, India
Daxin Jiang	Microsoft, China
Rajasekar Krishnamurthy	Adobe, USA

Demo Track Chairs

Rajeev Gupta	Microsoft, India
Koichi Takeda	Nagoya University, Japan
Ladjel Bellatreche	ENSMA, France

PhD Consortium Chairs

Vikram Pudi	IIIT Hyderabad, India
Srinath Srinivasa	IIIT Bangalore, India
Philippe Fournier-Viger	Harbin Institute of Technology, China

Panel Chairs

Jayant Haritsa	Indian Institute of Science, India
Reynold Cheng	University of Hong Kong, China
Georgia Koutrika	Athena Research Center, Greece

Sponsorship Chair

P. Krishna Reddy	IIIT Hyderabad, India

Publication Chairs

Vikram Goel IIIT Delhi, India
R. Uday Kiran University of Aizu, Japan

Workshop Chairs

Maya Ramanath IIT Delhi, India
Wookey Lee Inha University, South Korea
Sanjay Kumar Madria Missouri Institute of Technology, USA

Tutorial Chairs

P. Sreenivasa Kumar IIT Madras, India
Jixue Liu University of South Australia, Australia
Takahiro Hara Osaka University, Japan

Publicity Chairs

Raj Sharma Goldman Sachs, India
Jamshid Bagherzadeh Mohasefi Urmia University, Iran
Nazha Selmaoui-Folcher University of New Caledonia, New Caledonia

Organizing Committee

Lini Thomas IIIT Hyderabad, India
Satish Narayana Srirama University of Hyderabad, India
Manish Singh IIT Hyderabad, India
P. Radha Krishna NIT Warangal, India
Sonali Agrawal IIIT Allahabad, India
V. Ravi IDRBT, India

Senior Program Committee

Avigdor Gal Technion - Israel Institute of Technology, Israel
Baihua Zheng Singapore Management University, Singapore
Bin Cui Peking University, China
Bin Yang Aalborg University, Denmark
Bingsheng He National University of Singapore, Singapore
Chang-Tien Lu Virginia Tech, USA
Chee-Yong Chan National University of Singapore, Singapore
Gautam Shroff Tata Consultancy Services Ltd., India
Hong Gao Harbin Institute of Technology, China

Jeffrey Xu Yu	Chinese University of Hong Kong, China
Jianliang Xu	Hong Kong Baptist University, China
Jianyong Wang	Tsinghua University, China
Kamalakar Karlapalem	IIIT Hyderabad, India
Kian-Lee Tan	National University of Singapore, Singapore
Kyuseok Shim	Seoul National University, South Korea
Ling Liu	Georgia Institute of Technology, USA
Lipika Dey	Tata Consultancy Services Ltd., India
Mario Nascimento	University of Alberta, Canada
Maya Ramanath	IIT Delhi, India
Mohamed Mokbel	University of Minnesota, Twin Cities, USA
Niloy Ganguly	IIT Kharagpur, India
Sayan Ranu	IIT Delhi, India
Sourav S. Bhowmick	Nanyang Technological University, Singapore
Srikanta Bedathur	IIT Delhi, India
Srinath Srinivasa	IIIT Bangalore, India
Stephane Bressan	National University of Singapore, Singapore
Tok W. Ling	National University of Singapore, Singapore
Vana Kalogeraki	Athens University of Economics and Business, Greece
Vassilis J. Tsotras	University of California, Riverside, USA
Vikram Pudi	IIIT Hyderabad, India
Vincent Tseng	National Yang Ming Chiao Tung University, Taiwan
Wang-Chien Lee	Pennsylvania State University, USA
Wei-Shinn Ku	Auburn University, USA
Wenjie Zhang	University of New South Wales, Australia
Wynne Hsu	National University of Singapore, Singapore
Xiaofang Zhou	Hong Kong University of Science and Technology, China
Xiaokui Xiao	National University of Singapore, Singapore
Xiaoyong Du	Renmin University of China, China
Yoshiharu Ishikawa	Nagoya University, Japan
Yufei Tao	Chinese University of Hong Kong, China

Program Committee

Abhijnan Chakraborty	IIT Delhi, India
Ahmed Eldawy	University of California, Riverside, USA
Akshar Kaul	IBM Research, India
Alberto Abell	Universitat Politecnica de Catalunya, Spain
An Liu	Soochow University, China
Andrea Cali	Birkbeck, University of London, UK

Andreas Züfle	George Mason University, USA
Antonio Corral	University of Almeria, Spain
Atsuhiro Takasu	National Institute of Informatics, Japan
Bin Wang	Northeastern University, China
Bin Yao	Shanghai Jiao Tong University, China
Bo Jin	Dalian University of Technology, China
Bolong Zheng	Huazhong University of Science and Technology, China
Chandramani Chaudhary	National Institute of Technology, Trichy, India
Changdong Wang	Sun Yat-sen University, China
Chaokun Wang	Tsinghua University, China
Cheng Long	Nanyang Technological University, Singapore
Chenjuan Guo	Aalborg University, Denmark
Cheqing Jin	East China Normal University, China
Chih-Ya Shen	National Tsing Hua University, Taiwan
Chittaranjan Hota	BITS Pilani, India
Chi-Yin Chow	Social Mind Analytics (Research and Technology) Limited, Hong Kong
Chowdhury Farhan Ahmed	University of Dhaka, Bangladesh
Christos Doulkeridis	University of Pireaus, Greece
Chuan Xiao	Osaka University and Nagoya University, Japan
Cindy Chen	University of Massachusetts Lowell, USA
Cuiping Li	Renmin University of China, China
Dan He	University of Queensland, Australia
Demetrios Zeinalipour-Yazti	University of Cyprus, Cyprus
De-Nian Yang	Academia Sinica, Taiwan
Dhaval Patel	IBM TJ Watson Research Center, USA
Dieter Pfoser	George Mason University, USA
Dimitrios Kotzinos	University of Cergy-Pontoise, France
Fan Zhang	Guangzhou University, China
Ge Yu	Northeast University, China
Goce Trajcevski	Iowa State University, USA
Guoren Wang	Beijing Institute of Technology, China
Haibo Hu	Hong Kong Polytechnic University, China
Haruo Yokota	Tokyo Institute of Technology, Japan
Hiroaki Shiokawa	University of Tsukuba, Japan
Hongzhi Wang	Harbin Institute of Technology, China
Hongzhi Yin	University of Queensland, Australia
Hrishikesh R. Terdalkar	IIT Kanpur, India
Hua Lu	Roskilde University, Denmark
Hui Li	Xidian University, China
Ioannis Konstantinou	University of Thessaly, Greece

Iouliana Litou	Athens University of Economics and Business, Greece
Jagat Sesh Challa	BITS Pilani, India
Ja-Hwung Su	Cheng Shiu University, Taiwan
Jiali Mao	East China Normal University, China,
Jia-Ling Koh	National Taiwan Normal University, Taiwan
Jian Dai	Alibaba Group, China
Jianqiu Xu	Nanjing University of Aeronautics and Astronautics, China
Jianxin Li	Deakin University, Australia
Jiawei Jiang	ETH Zurich, Switzerland
Jilian Zhang	Jinan University, China
Jin Wang	Megagon Labs, USA
Jinfei Liu	Zhejiang University, China
Jing Tang	Hong Kong University of Science and Technology, China
Jinho Kim	Kangwon National University, South Korea
Jithin Vachery	National University of Singapore, Singapore
Ju Fan	Renmin University of China, China
Jun Miyazaki	Tokyo Institute of Technology, Japan
Junjie Yao	East China Normal University, China
Jun-Ki Min	Korea University of Technology and Education, South Korea
Kai Zeng	Alibaba Group, China
Karthik Ramachandra	Microsoft Azure SQL, India
Kento Sugiura	Nagoya University, Japan
Kesheng Wu	Lawrence Berkeley National Laboratory, USA
Kjetil Nørvåg	Norwegian University of Science and Technology, Norway
Kostas Stefanidis	Tempere University, Finland
Kripabandhu Ghosh	Indian Institute of Science Education and Research Kolkata, India
Kristian Torp	Aalborg University, Denmark
Kyoung-Sook Kim	Artificial Intelligence Research Center, Japan
Ladjel Bellatreche	ENSMA, France
Lars Dannecker	SAP, Germany
Lee Roy Ka Wei	Singapore University of Technology and Design, Singapore
Lei Cao	Massachusetts Institute of Technology, USA
Leong Hou U.	University of Macau, China
Lijun Chang	University of Sydney, Australia
Lina Yao	University of New South Wales Australia
Lini Thomas	IIIT Hyderabad, India

Liping Wang   East China Normal University, China
Long Yuan   Nanjing University of Science and Technology,
                          China
Lu-An Tang   NEC Labs America, USA
Makoto Onizuka   Osaka University, Japan
Manish Kesarwani   IBM Research, India
Manish Singh   IIT Hyderabad, India
Manolis Koubarakis   University of Athens, Greece
Marco Mesiti   University of Milan, Italy
Markus Schneider   University of Florida, USA
Meihui Zhang   Beijing Institute of Technology, China
Meng-Fen Chiang   University of Auckland, New Zealand
Mirella M. Moro   Universidade Federal de Minas Gerais, Brazil
Mizuho Iwaihara   Waseda University, Japan
Navneet Goyal   BITS Pilani, India
Neil Zhenqiang Gong   Iowa State University, USA
Nikos Ntarmos   Huawei Technologies R&D (UK) Ltd., UK
Nobutaka Suzuki   University of Tsukuba, Japan
Norio Katayama   National Institute of Informatics, Japan
Noseong Park   George Mason University, USA
Olivier Ruas   Inria, France
Oscar Romero   Universitat Politècnica de Catalunya, Spain
Oswald C.   IIT Kanpur, India
Panagiotis Bouros   Johannes Gutenberg University Mainz, Germany
Parth Nagarkar   New Mexico State University, USA
Peer Kroger   Christian-Albrecht University of Kiel, Germany
Peifeng Yin   Pinterest, USA
Peng Wang   Fudan University, China
Pengpeng Zhao   Soochow University, China
Ping Lu   Beihang University, China
Pinghui Wang   Xi'an Jiaotong University, China
Poonam Goyal   BITS Pilani, India
Qiang Yin   Shanghai Jiao Tong University, China
Qiang Zhu   University of Michigan – Dearborn, USA
Qingqing Ye   Hong Kong Polytechnic University, China
Rafael Berlanga Llavori   Universitat Jaume I, Spain
Rage Uday Kiran   University of Aizu, Japan
Raghava Mutharaju   IIIT Delhi, India
Ravindranath C. Jampani   Oracle Labs, India
Rui Chen   Samsung Research America, USA
Rui Zhou   Swinburne University of Technology, Australia
Ruiyuan Li   Xidian University, China

Sabrina De Capitani di Vimercati	Università degli Studi di Milano, Italy
Saiful Islam	Griffith University, Australia
Sanghyun Park	Yonsei University, South Korea
Sanjay Kumar Madria	Missouri University of Science and Technology, USA
Saptarshi Ghosh	IIT Kharagpur, India
Sebastian Link	University of Auckland, New Zealand
Shaoxu Song	Tsinghua University, China
Sharma Chakravarthy	University of Texas at Arlington, USA
Shiyu Yang	Guangzhou University, China
Shubhadip Mitra	Tata Consultancy Services Ltd., India
Shubhangi Agarwal	IIT Kanpur, India
Shuhao Zhang	Singapore University of Technology and Design, Singapore
Sibo Wang	Chinese University of Hong Kong, China
Silviu Maniu	Université Paris-Saclay, France
Sivaselvan B.	IIIT Kancheepuram, India
Stephane Bressan	National University of Singapore, Singapore
Subhajit Sidhanta	IIT Bhilai, India
Sungwon Jung	Sogang University, South Korea
Tanmoy Chakraborty	Indraprastha Institute of Information Technology Delhi, India
Theodoros Chondrogiannis	University of Konstanz, Germany
Tien Tuan Anh Dinh	Singapore University of Technology and Design, Singapore
Ting Deng	Beihang University, China
Tirtharaj Dash	BITS Pilani, India
Toshiyuki Amagasa	University of Tsukuba, Japan
Tsz Nam (Edison) Chan	Hong Kong Baptist University, China
Venkata M. Viswanath Gunturi	IIT Ropar, India
Verena Kantere	National Technical University of Athens, Greece
Vijaya Saradhi V.	IIT Guwahati, India
Vikram Goyal	IIIT Delhi, India
Wei Wang	Hong Kong University of Science and Technology (Guangzhou), China
Weiwei Sun	Fudan University, China
Weixiong Rao	Tongji University, China
Wen Hua	University of Queensland, Australia
Wenchao Zhou	Georgetown University, USA
Wentao Zhang	Peking University, China
Werner Nutt	Free University of Bozen-Bolzano, Italy
Wolf-Tilo Balke	TU Braunschweig, Germany

Wookey Lee	Inha University, South Korea
Woong-Kee Loh	Gacheon University, South Korea
Xiang Lian	Kent State University, USA
Xiang Zhao	National University of Defence Technology, China
Xiangmin Zhou	RMIT University, Australia
Xiao Pan	Shijiazhuang Tiedao University, China
Xiao Qin	Amazon Web Services, USA
Xiaochun Yang	Northeastern University, China
Xiaofei Zhang	University of Memphis, USA
Xiaofeng Gao	Shanghai Jiao Tong University, China
Xiaowang Zhang	Tianjin University, China
Xiaoyang Wang	Zhejiang Gongshang University, China
Xin Cao	University of New South Wales, Australia
Xin Huang	Hong Kong Baptist University, China
Xin Wang	Tianjin University, China
Xu Xie	Peking University, China
Xuequn Shang	Northwestern Polytechnical University, China
Xupeng Miao	Peking University, China
Yan Shi	Shanghai Jiao Tong University, China
Yan Zhang	Peking University, China
Yang Cao	Kyoto University, Japan
Yang Chen	Fudan University, China
Yanghua Xiao	Fudan University, China
Yang-Sae Moon	Kangwon National University, South Korea
Yannis Manolopoulos	Aristotle University of Thessaloniki, Greece
Yi Yu	National Institute of Informatics, Japan
Yingxia Shao	Beijing University of Posts and Telecommunication, China
Yixiang Fang	Chinese University of Hong Kong, China
Yong Tang	South China Normal University, China
Yongxin Tong	Beihang University, China
Yoshiharu Ishikawa	Nagoya University, Japan
Yu Huang	National Yang Ming Chiao Tung University, Taiwan
Yu Suzuki	Gifu University, Japan
Yu Yang	City University of Hong Kong, China
Yuanchun Zhou	Computer Network Information Center, China
Yuanyuan Zhu	Wuhan University, China
Yun Peng	Hong Kong Baptist University, China
Yuqing Zhu	California State University, Los Angeles, USA
Zeke Wang	Zhejiang University, China

Zhaojing Luo	National University of Singapore, Singapore
Zhenying He	Fudan University, China
Zhi Yang	Peking University, China
Zhixu Li	Soochow University, China
Zhiyong Peng	Wuhan University, China
Zhongnan Zhang	Xiamen University, China

Industry Track Program Committee

Karthik Ramachandra	Microsoft, India
Akshar Kaul	IBM Research, India
Sriram Lakshminarasimhan	Google Research, India
Rajat Venkatesh	LinkedIn, India
Prasan Roy	Sclera, India
Zhicheng Dou	Renmin University of China, China
Huang Hu	Microsoft, China
Shan Li	LinkedIn, USA
Bin Gao	Facebook, USA
Haocheng Wu	Facebook, USA
Shivakumar Vaithyanathan	Adobe, USA
Abdul Quamar	IBM Research, USA
Pedro Bizarro	Feedzai, Portugal
Xi Yin	International Digital Economy Academy, China
Xiangyu Niu	Facebook

Demo Track Program Committee

Ahmed Awad	University of Tartu, Estonia
Beethika Tripathi	Microsoft, India
Carlos Ordonez	University of Houston, USA
Djamal Benslimane	Université Claude Bernard Lyon 1, France
Nabila Berkani	Ecole Nationale Supérieure d'Informatique, Algeria
Philippe Fournier-Viger	Shenzhen University, China
Ranganath Kondapally	Microsoft, India
Soumia Benkrid	Ecole Nationale Supérieure d'Informatique, Algeria

Sponsoring Institutions

Google, India

INTERNATIONAL INSTITUTE OF
INFORMATION TECHNOLOGY

H Y D E R A B A D

IIIT Hyderabad, India

Contents – Part II

Applications of Machine Learning

Recommendation Systems

MDKE: Multi-level Disentangled Knowledge-Based Embedding for Recommender Systems

Haolin Zhou, Qingmin Liu, Xiaofeng Gao$^{(\boxtimes)}$, and Guihai Chen

MoE Key Lab of Artificial Intelligence, Department of Computer Science and Engineering,
Shanghai Jiao Tong University, Shanghai, China
koziello@sjtu.edu.cn, {gao-xf,gchen}@cs.sjtu.edu.cn

Abstract. Recommender systems aim to dig out the potential interests of users and find out items that might be connected with target users. Accuracy of the recommendation list is crucial for user-oriented applications. Many knowledge-based approaches combine graph neural networks with exploring node structural similarity, while paying little attention to semantically distinguishing potential user interests and item attributes. Therefore, personalized node embeddings fail to be captured after simply aggregating neighborhood information. In this work, we propose **M**ulti-level **D**isentangled **K**nowledge-based **E**mbedding (**MDKE**) for item recommendation. We divide the embedding learning of users and items into disentangled *semantic-level* and *structural-level* subspaces. Specifically, semantic-level embeddings correspond to item content vector, which is learned by knowledge graph embedding, and user profile vector that is obtained through a carefully designed preference propagation mechanism. For structural-level embeddings, a graph embedding-based module is proposed to capture local structure features on the redefined knowledge interaction graph. The latent vectors of users and items can be obtained by aggregating the aforementioned multi-level embeddings. Experiments on two real-world datasets verify the effectiveness of MDKE versus a series of state-of-the-art approaches.

Keywords: Recommender systems · Knowledge-based embedding ·
Representation learning · Graph neural network

1 Introduction

Recommender systems aim to predict how likely a user is going to buy an item. Some state-of-the-art models [16,23] combine collaborative filtering methods and graph embedding technology to dig into potential connections between users and items. Moreover, side information from knowledge graph (KG) [3,9,12,15] helps to explore a user's

This work was supported by the National Key R&D Program of China [2020YFB1707903]; the National Natural Science Foundation of China [61872238, 61972254]; Shanghai Municipal Science and Technology Major Project [2021SHZDZX0102]; the CCF-Tencent Open Fund [RAGR20200105] and the Tencent Marketing Solution Rhino-Bird Focused Research Program [FR202001].

A. Bhattacharya et al. (Eds.): DASFAA 2022, LNCS 13246, pp. 3–18, 2022.
https://doi.org/10.1007/978-3-031-00126-0_1

profound preference. Knowledge graph could be regarded as a semantic network that represents interlinked relationships between entities.

Graph neural networks have inherent advantages in processing graph data. In recommender system, there exists a variety of graph-structured data, such as user-item bipartite graph and knowledge graph. Therefore, some KG-aware models apply graph convolutional networks to entity representation learning. However, simply using graph embedding to aggregate information of nodes could eliminate part of node's personalized features, indicating that node embeddings become gradually indistinguishable after multi-layer graph convolutional operation due to the effect of Laplacian smoothing [22,25]. In the scenario of complex heterogeneous networks, the various connections of a given entity focus on reflecting different aspects. Aggregating neighboring nodes from a single embedding subspace might weaken the unique user profiles and item attributes contained in the node representation, thus leading to underutilization of rich relation-specific and semantic information. Besides, previous methods tend to treat users and items as equal nodes, failing to capture user interest in a user-centric way. For users who have clicked on the same item, the attribution behind their interacting behaviors, i.e., interest domain, is completely different. For example, a user may watch *Titanic* because he is a fan of *Leonardo DiCaprio* or simply because is interested in movies that belong to the genre of *Romance*. Therefore, it is necessary to explicitly propagate user preference information on KG to capture users' personal interests.

To overcome the above-mentioned challenges, we hereby propose *Multi-level Disentangled Knowledge-Based Embedding* (MDKE) for recommender systems to explore potential connection on the user-item bipartite graph and distinguish interest domains among different users. To deal with the first challenge of node convergence, the representation learning of users and items are separated into disentangled structural and semantic subspaces. The embeddings generated from structural subspace concentrate on capturing high-order connectivities between nodes, while the semantic embeddings are generated to preserve differences of user interests and item attributes. We define item embedding learned directly from KG as item content vector. Correspondingly, user embedding obtained by propagating entity messages according to connecting relationship on KG is called user profile vector. These two types of vectors are uniformly defined as semantic-level vector that represents an object's inherent preference or attribute. Besides, we combine KG and user-item bipartite graph by regarding user's clicking or purchasing behavior as a type of relation. The structural vectors of users and items are learned by applying graph convolutional network on the redefined knowledge interaction graph, reflecting a node's neighborhood information.

In terms of exploring users' personal interests explicitly, i.e., generating user profile vector from a user-centric view, we design a propagation mechanism on KG after obtaining item content vectors through knowledge graph embedding. For each user, we construct several subgraphs on the KG according to his directly interacted items, arguing that these subgraphs with included entities reflect users' interests towards different aspects, which in turn can help build user profiles. With the assistance of learned disentangled embeddings, possible pairwise interaction between user and item could be discovered by computing the prediction score.

In summary, our contributions in this paper are threefold:

(1) We highlight the importance of disentangling the learning process of semantic and structural-level features in KG-aware recommendation to better reflect the personalization and neighborhood features of users and items.
(2) We propose MDKE, an end-to-end framework that embeds users and items respectively from semantic and structural levels. The carefully designed propagation mechanism helps to capture user interests explicitly.
(3) Experiments on two real-world datasets and comparison with other state-of-the-art models show the effectiveness of MDKE. The proposed model increases the accuracy of recommendation results and also provides more interpretability.

2 Preliminaries

We denote the user set and item set as $\mathcal{U} = \{u_1, u_2, ..., u_n\}$ and $\mathcal{I} = \{i_1, i_2, ..., i_m\}$, respectively. The size of a set is represented by $|\cdot|$. The interaction matrix is defined as $R \in \{0,1\}^{|\mathcal{U}| \times |\mathcal{I}|}$ consisting of $y_{u,i}$ that indicates the interaction between user u and item i. If there exists any operation such as rating, clicking, or purchasing between them, $y_{u,i}$ is set to 1, and 0 otherwise.

For fully exploring side information of items, an item-based knowledge graph is extracted and we define it as $\mathcal{KG}$. The knowledge graph can be regarded as a heterogeneous network that consists of triplets holding the form of $[$entity: h, relation: r, entity: $t]$. The entity can be an item or a type of attribute. Therefore, $\mathcal{KG}$ is presented as $\{(h, r, t)|(h, t) \in \mathcal{E}^2, r \in \mathcal{R}\}$, where $\mathcal{E}$ is the set of entities, and $\mathcal{R}$ presents the set of relations.

The input data includes $\mathcal{U}$, $\mathcal{I}$, R, and $\mathcal{KG}$. The goal is to generate an item list with K elements for each user u. We denote the predicted likelihood between user u and item i as $\hat{y}_{u,i}$. The item list for a target user can be hereby regarded as the top K elements ranked by $\hat{y}_{u,i}$.

3 Methodology

As mentioned in Sect. 1, semantic and structural level information of users and items should be decoupled in separated embedding subspaces. Existing recommender systems based on graph neural network and knowledge graph implicitly model user interests and item attributes during the node aggregation procedure. Such approach leads nodes to lose their own semantic features to some extent. In order to distinguish different user interaction attributions, thus mining user profound interests and making full use of the auxiliary information in KG to extract item content, we design the framework of MDKE as shown in Fig. 1. We will elaborate on the technical details of the three main modules in the following subsections.

3.1 Item Content Extraction

To capture semantic-level features, we leverage the auxiliary information in KG, which consists of a bulk of triplets and textual information of entities. To concentrate on specific relation domains for each entity in different triplets [6], entities are projected into

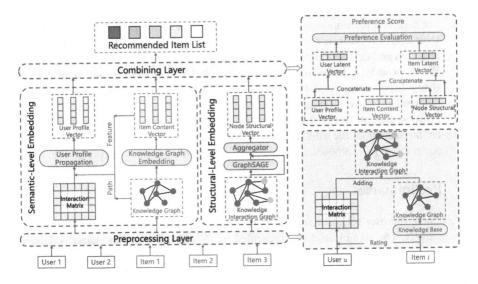

Fig. 1. Framework of MDKE for generating recommendation list

corresponding relation space through a set of projection matrix M_r by $v_e^r = v_e M_r$, where v_e is the embedding of either h or t in the entity space. For each triplet, a distance function between h and t in the relation-specific mapping space is computed by

$$f_r(\boldsymbol{h}, \boldsymbol{t}) = ||\boldsymbol{v}_h^r + \boldsymbol{r} - \boldsymbol{v}_t^r||_2^2, \tag{1}$$

reflecting closeness of relationship between different entities in the relation space. Furthermore, when defining the distance between two entities, the relevance of their semantics also needs to be considered [20]. Therefore, we project the entity vectors into a redefined semantic hyperplane. Given entity e, a semantic embedding w_e can be initialized with pre-trained word vectors [3,8]. The normal vector of semantic hyperplane for entities h and t is defined as

$$\mathbf{w}_\perp = \frac{\mathbf{w_h} + \mathbf{w_t}}{||\mathbf{w_h} + \mathbf{w_t}||_2^2} \tag{2}$$

The distance between h and t in the semantic hyperplane is thus obtained by

$$f_s(\boldsymbol{h}, \boldsymbol{t}) = \left|\left| \boldsymbol{v}_h - \mathbf{w}_\perp{}^T \boldsymbol{v}_h \mathbf{w}_\perp - (\boldsymbol{v}_t - \mathbf{w}_\perp{}^T \boldsymbol{v}_t \mathbf{w}_\perp) + \boldsymbol{r}_\perp \right|\right|_2^2, \tag{3}$$

where $(\boldsymbol{v}_e - \mathbf{w}_\perp{}^T \boldsymbol{v}_e \mathbf{w}_\perp)$ represents the projection vector of entity e, and $\boldsymbol{r}_\perp$ is the translation vector in the corresponding semantic hyperplane. Therefore, an objective function of KG can be defined as

$$\mathcal{L}_{KG} = \sum_{(h,r,t,t') \in \mathcal{KG} \times \mathcal{E}} [\gamma + f_r(\boldsymbol{h}, \boldsymbol{t}) - f_r(\boldsymbol{h}', \boldsymbol{t}') + \lambda(f_s(\boldsymbol{h}, \boldsymbol{t}) - f_s(\boldsymbol{h}', \boldsymbol{t}'))]_+, \tag{4}$$

where γ is the margin, λ denotes the contribution strength of text-aware regularization, t' is a negative entity sampled through modifying one of the original triplet (h, r, t)

randomly, and $[\cdot]_+ = \max(\cdot, 0)$ is the hinge loss. Item content vector, defined as v_i^c, is thereby extracted in an explainable way by optimizing Eq. (4).

3.2 User Preference Propagation

Many existing methods [24] obtain the potential interests of users through the interacted item list l_u. However, it should be noted that only relying on l_u is hard to capture user preference precisely. The connection between entities in KG reflects the similarity of inherent item attributes, which should be also considered in the preference exploration.

A toy example is illustrated in Fig. 2 to represent such implicit entity connections. The interacted movie *Inception* as well as its related entities and relations are extracted from KG. The user's interest towards a movie is actually attributed to the related genre or the involved actor. Therefore, propagating user preference through l_u to neighboring entities and aggregating these propagated embeddings reversely could have better representation ability, as such mechanism further digs into potential connection in KG.

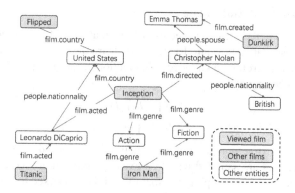

Fig. 2. Illustration of connections in KG

Specifically, user preference propagation starts from l_u, regarded as seed entities. We define l_u^z to denote the set of involved entities in the zth propagation hop:

$$l_u^z = \{h \mid \forall t \in l_u^{z-1}, (h, r, t) \in \mathcal{KG}\} \cup \{t \mid \forall h \in l_u^{z-1}, (h, r, t) \in \mathcal{KG}\}, \quad (5)$$

where $z \geq 1$, and l_u^0 represents the directly interacted items of user. The size of l_u^z is fixed to $|l|$, and the set of all the involved entities l_u^* during the preference propagation can be acquired by

$$I_u^* = \cup_{0 \leqslant z \leqslant H} l_u^z, \quad (6)$$

where H is the total hop number, i.e., the propagation distance. It should be underlined that in I_u^*, an entity may appear several times since there exists multiple paths between entities in KG. However, we don't have to make the elements in I_u^* mutually exclusive because when the user's interest spreads to a certain entity frequently, the entity is more likely to be in line with the user's preference. Finally, user profile vector, i.e., user semantic-level embedding, can be obtained by combining all the involved entities:

$$v_u^p = \frac{\sum_{e \in I_u^*} \alpha_{u,e} \cdot v_e}{||\sum_{e \in I_u^*} \alpha_{u,e} \cdot v_e||}, \tag{7}$$

where $\alpha_{u,e}$ represents the importance weight of entity e on the target u:

$$\alpha_{u,e} = \frac{1}{d_{u,e}^{1-\mu}}, \tag{8}$$

where $d_{u,e}$ stands for the distance between u and e. In other words, if $e \in l_u^{z-1}$, then $d_{u,e}$ equals to z. Intuitively, we argue that lower hop neighborhood is more important to reveal the user's real interest. For example, the user may be more interested in the director (1-hop) of the movie *Inception* rather than the spouse of the director (2-hop) as shown in Fig. 2. The parameter μ is defined as a discount factor less than 1 that distinguishes contribution strength of different hop neighbors.

3.3   Graph Structural Embedding

A variety of entities in KG are connected to each other and thus form a topological structure. In MDKE, users are added into KG to jointly learn the structural-level embeddings with other entities. Therefore, a knowledge interaction graph $\mathcal{G} = (V, E)$ is constructed before graph embedding by

$$\begin{aligned}
V &= \{h | (h, r, t) \in \mathcal{KG}\} \cup \{t | (h, r, t) \in \mathcal{KG}\} \cup \{u | u \in \mathcal{U}\} \\
E &= \{(h, t) | (h, r, t) \in \mathcal{KG}\} \cup \{(u, i) | y_{u,i} > 0\}
\end{aligned} \tag{9}$$

The process of extracting structure information is composed of two stages. In the first phase, neighbors are sampled based on existing connection records. Considering computation efficiency, the sampling number of each layer is fixed to a constant. In the second phase, different aggregators are built to integrate information propagated from neighbors. The structural-level embedding of entities is defined as v_e^s. The two optional graph aggregators we utilized in MDKE are the ones based on GraphSage [5] and graph attention mechanism [11] (GAT), which are defined as following:

– **GraphSage aggregator** generates the representation of nodes directly by

$$v_e^s \leftarrow \sigma(W \cdot MEAN(\{v_e^s\} \cup \{v_n^s, \forall n \in \mathcal{N}(e)\})), \tag{10}$$

where W is trainable variables, and $\mathcal{N}(e)$ represents the set of sampled neighbors.
– **GAT aggregator** considers that neighbor nodes have different effects on the target node when they are aggregated:

$$v_e^s \leftarrow \sigma(\sum_{n \in \mathcal{N}(e) \cup \{e\}} \frac{\exp\left(\pi_{e,n}^r\right)}{\sum_{n' \in \mathcal{N}(e) \cup \{e\}} \exp\left(\pi_{e,n'}^{r'}\right)} v_n^s) \tag{11}$$

$$\pi_{e,n}^r \leftarrow \text{LeakyReLU}(W_r^T(v_e^s \| v_n^s) + b_r), \tag{12}$$

where $\pi_{e,n}$ represents the computed attention score between target node e and its neighbor n; $W_r \in \mathbb{R}^{2d}$ and $b_r \in \mathbb{R}$ are relation-specific trainable parameters.

Note that we can iterate Eq. (10) or Eq. (11) several times to perform multi-layer node aggregation. The generated structural vectors v_e^s contain topological similarity that we obtain from structural embedding subspace.

3.4 Model Prediction and Training

We generate user latent embedding v_u according to user preference vector v_u^p and structural-level vector v_u^s. Similarly, item latent vector v_i is generated through item content vector v_i^c and item structure vector v_i^s. Therefore, the latent user and item embeddings to compute the prediction score could be obtained by

$$v_u = CONCAT(v_u^p, v_u^s), \quad v_i = CONCAT(v_i^c, v_i^s), \tag{13}$$

where the multi-layer vectors are directly concatenated together to retain the information learned from decoupled embedding subspaces as much as possible.

Algorithm 1: The training procedure of MDKE

Input: User set $\mathcal{U}$, item set $\mathcal{I}$, knowledge graph $\mathcal{KG}$, interaction matrix R, propagation distance H, detph of graph embedding layer D

1   Construct knowledge interaction matrix $\mathcal{G} = (V, E)$ following Eq. (9);
2   **while** *MDKE not converge* **do**
3   Sample (h, r, t, t') in $\mathcal{KG}$ randomly and generate semantic embeddings;
4   Compute the projetion distance of triplets and update trainable variables by $\nabla_\Theta L_{KG}$ following Eq. (4);
5   Sample $(u, i) \in \mathcal{U} \times \mathcal{I}$ to construct the training set $O \times R_u^+ \times R_u^-$;
6   Extract entity content embeddings v_e^c (v_i^c) from $\mathcal{KG}$;
7   **for** $u \in O$ **do**
8   Query seed entities l_u^0 according to R;
9   **for** $z \leftarrow 1$ *to* H **do**
10   Propagate user preference for the zth hop following Eq. (5);
11   **end**
12   Obtain the set of involved entites I_u^* following Eq. (6) and generate v_u^p by combining all the involved entity vectors in I_u^* following Eq. (7);
13   **end**
14   **for** $k \leftarrow 1$ *to* D **do**
15   **for** $e \in O \cup \mathcal{R}_u^+ \cup \mathcal{R}_u^-$ **do**
16   Aggregate neighborhood information of e by Eq. (10);
17   **end**
18   **end**
19   Obtain node structural embedding v_e^s (v_u^s and v_i^s);
20   Generate corresponding latent vectors v_u and v_i by Eq. (13);
21   Compute L_{CE} by Eq. (14) and update trainable variable by $\nabla_\Theta L_{CE}$;
22 **end**

The objective function is defined as cross-entropy loss:

$$\mathcal{L}_{CE} = -\sum_{u \in O} \sum_{i \in R_u^+ \times R_u^-} y_{u,i} \log \hat{y}_{u,i} + (1 - y_{u,i}) \log(1 - \hat{y}_{u,i}) + \beta \|\Theta\|_2^2, \quad (14)$$

where O denotes the sampled user set, R_u^+ is the interacted items set while R_u^- is generated from non-interacted items, $\hat{y}_{u,i}$ stands for the interaction possibility computed by performing dot product between v_u and v_i, Θ represents all the trainable variables, and β denotes the L_2 regularization coefficient to prevent overfitting. The whole training process of MDKE is shown in Algorithm 1, and we optimize $\mathcal{L}_{CE}$ and $\mathcal{L}_{KG}$ alternately.

In terms of the time complexity, we can see that the main cost comes from user propagation and structural information aggregation. For the former part, the cost is $\mathcal{O}(|O||l|^H d)$, where d denotes the size of single-level embedding. It can be found that the set of entities affected by the preference propagation increases exponentially with H. For structure information extraction, the computational complexity is $\mathcal{O}(|O||N|^D d^2)$, where D is the depth of graph convolutional layer and $|N|$ denotes the size of sampled neighbor set on the knowledge interaction graph.

3.5 Analysis of MDKE

The motivation for proposing MDKE is to take into account the node structural and semantic information jointly in the process of representation learning, exploring structure commonalities and feature differences simultaneously. The user and item embeddings for other recent KG-aware methods based on graph embedding like KGCN [14] and KGAT [15] are generated in an integrated manner from the aggregation stage. These methods claim that they have better node representation ability benefited from graph convolutional networks (GCN). However, GCN as a low-pass filters mainly retains the commonness of node features, while inevitably neglecting feature differences, making the learned embeddings of connected nodes become similar [2]. In recommender system, nodes with different connected neighbors or topologies also have a certain possibility to interact with each other, as users show different intents and items share various attributes. Therefore, aggregating semantic information obtained from KG through GCN directly will weaken the personalized features, and thus fail to capture user potential interests and item attributes. In the framework of MDKE, the semantic and structural-level modules are separated. Structural learning stage makes adjacent nodes to have similar embeddings, however, in semantic learning stage, the relevance of embeddings actually lies in the semantic hyperplane and relation-specific space. Therefore, these two levels of embeddings have explanatory factors from different sources, which should be modeled in a disentangled way. As shown in Fig. 3, *Contact* and *Rain Man* are two candidate films to be recommended in the test set for target user u_{146}. Although u_{146} and *Rain Man* share some common neighbors, indicating their similarity in the structural subspace, *Contact* is proved to be a more valid choice, as its overall correlation with u_{146} considering both embedding subspaces is stronger. In other words, the semantic attributes of *Contact* are more in line with u_{146}'s preference for watching movies. The toy example illustrates that disentangled representation learning in KG-aware recommendation is of great significance, as embeddings learned from decoupled subspaces originate in different relation sources.

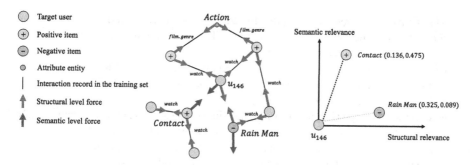

Fig. 3. Mechanism of decoupled representation learning in **MDKE**

4   Experiments

4.1   Experimental Settings

A various of experiments is conducted to verify the effectiveness of **MDKE**. We first introduce the statistics of two datasets on which the model performance is evaluated.

Datasets. Two classic real-world datasets, MovieLens-100K[1] and Last.FM[2], are chosen for conducting the experiment. Dataset information is shown in Table 1. For the task of *Top-k* ranking, we randomly divide the entire user-item interaction records into a training set and a test set at a ratio of 4:1. We query items in MovieLens from DBPedia[3] while Microsoft Satori[4] is queried for Last.FM to build an item-based knowledge graphs for each dataset. The triplets on the KG are extracted and recorded through IDs.

Table 1. Summary for two evaluation datasets

	MovieLens	Last.FM
# Users	943	23,566
# Items	1,098	48,123
# Interactions	70,166	1,289,003
# Entities	12,049	106,389
# Relationships	26	9
# KG Triplets	57,742	464,567

Baselines. We compare **MDKE** with the following state-of-the-art models:

– **DeepFM:** DeepFM [4] can learn low order and high order crossing features based on factorization machine, while leveraging the generalization ability of the deep part. We use user ID and item embeddings learned from TransR [6] as inputs.

[1] https://grouplens.org/datasets/movielens/.
[2] https://grouplens.org/datasets/hetrec-2011/.
[3] http://wiki.dbpedia.org/.
[4] https://searchengineland.com/library/bing/bing-satori.

- **NGCF:** Neural Graph Collaborative Filtering (NGCF) [16] captures the high-order connectivity in the user-item bipartite graph. The aggregation function considers the structural similarity between connecting nodes.
- **CFKG:** Collaborative Filtering based on Knowledge Graph (CFKG) [24] appends users into KG and jointly learn the vectors of users and items to incorporate user behaviors and entity knowledge.
- **KGCN:** Knowledge Graph Convolutional Networks (KGCN) [14] use spatial GCN approach to capture high-order connectivity between entities in the KG.
- **RippleNet:** RippleNet [12] is a path-based approach that makes full use of all the connected entities in the KG to propagate user profound interests.
- **KGAT:** Knowledge Graph Attention Network (KGAT) [15] aggregates node information on the collaborative knowledge graph and discriminate their contribution based on graph attention mechanism.

Parameter Settings. We set the semantic and structural-level feature embeddings to 64, and the depth of GraphSage layer is set as 2. The length of user preference propagation is selected as 4 and the size of $|l|$ is set to 8. We will discuss the impacts of key parameters on the performance of MDKE in Sec. 4.3. For the settings of other baseline methods, the embedding size is set as 128 equally. For NGCF and KGAT, the depth of convolution layer is set as 3. For KGCN, the depth of receptive field is set as 2. In terms of RippleNet, the hop number is 2 and the size of user ripple set is 32.

Evaluation Metrics. In our experiments, we use Precision@K, Recall@K, and MAP@K to evaluate the performance of methods, and the value of K indicates the length of recommendation list.

The experimental results focus on answering the following three research questions:

- **RQ1** How does MDKE perform compared with other state-of-the-art methods?
- **RQ2** How do the key hyper-parameters affect the performance of MDKE?
- **RQ3** Are the three modules: preference propagation, semantic embedding, and structural embedding indispensable to improve MDKE performance?

4.2 MDKE Performance

For **RQ1**, we conduct experiments through MoviesLens and Last.FM on MDKE and other recommendation-oriented baselines. The experimental results are shown in Table 2. We have the following observations:

First, empirical results shows that our model outperforms other methods by an obvious margin on the two evaluation datasets. Particularly, for the state-of-the-art approaches RippleNet and KGAT, which also leverage knowledge graph for recommendation. In RippleNet, a preference propagation mechanism is performed to model user deep interest towards entities, while in KGAT, the embeddings generated from KG are further fed into graph attention networks to aggregate neighbor node information. However, MDKE still outperforms these two models. The improvement mainly comes from two aspects. On the one hand, the semantic and structural modules in MDKE

Table 2. Experimental results of recommendation methods on evaluation datasets. Bold scores are the best in each column while underlined scores are the best among all baselines. Δ MDKE denotes the improvement versus best baseline results.

Metric (%)	MoviesLens						Last.FM					
	P@5	P@20	R@5	R@20	M@5	M@20	P@5	P@20	R@5	R@20	M@5	M@20
DeepFM	6.607	4.930	1.645	4.168	0.937	1.326	5.480	3.988	2.126	2.885	1.172	1.544
NGCF	10.876	6.912	2.722	6.332	1.967	2.364	5.617	4.013	2.176	2.984	1.246	1.734
CFKG	11.633	8.167	3.814	7.671	2.101	3.141	6.288	4.408	2.381	3.723	1.384	2.030
KGCN	11.234	7.632	3.542	7.472	2.013	2.996	6.113	4.102	2.232	3.012	1.301	1.912
RippleNet	11.684	8.736	3.823	7.810	2.144	3.460	6.654	4.588	2.543	3.962	1.433	2.127
KGAT	12.435	9.539	3.732	7.838	2.195	3.325	6.796	4.609	2.455	4.046	1.451	2.134
MDKE	**12.733**	**10.067**	**4.131**	**8.326**	**2.324**	**3.692**	**6.967**	**4.818**	**2.611**	**4.147**	**1.559**	**2.246**
Δ MDKE	2.4%	5.5%	8.1%	6.2%	5.9%	6.7%	2.5%	4.5%	2.7%	2.5%	7.4%	5.2%

are decoupled and processed separately, retaining as much heterogeneous information as possible that reflects node difference. During the process of node propagation in KGAT, semantic information can be generalized and weakened by aggregating node embeddings through attention network simultaneously. On the other hand, compared with RippleNet, the redesigned preference propagation mechanism helps to dig out users' deep interests more efficiently, as hop distance and semantic relevance are both taken into account to generate user and entity vectors, thus discriminating nodes and enhancing expressive ability.

Second, KG-aware models perform better than other baselines in general, which can be seen as a demonstration of the positive impacts that side information brings to recommender system. Since in KG, there exists plenty of auxiliary information that could be used to enrich item content and user preference representations. More potential connections are exploited with the addition of KG, thus improving the precision of results. We also observe that KGCN is slightly inferior to other KG-aware models in terms of performance. The reason is that user's embeddings are not fully exploited when propagating message simply on the KG. Besides, the performance improvement of CFKG compared to KGCN may come from the use of additional relation information.

Third, among non-KG-based approaches, we notice that there exists a performance gap between NGCF and DeepFM. We attribute it to the graph convolutional network used in NGCF that captures high-order connection relationship, thus discovering potential node interaction. In turn, the results also prove that manual feature engineering does not always capture effective feature combination.

4.3   Impacts of Hyperparameters

This part corresponds to **RQ2**. We conduct a series of experiments to capture the impact of hyperparameters on final recommendation qualities. Three parameters are tested:

Propagation Distance: The propagation distance denotes the hop number for user preference in KG, which varies from 1 to 7 in our experiments. In terms of overall

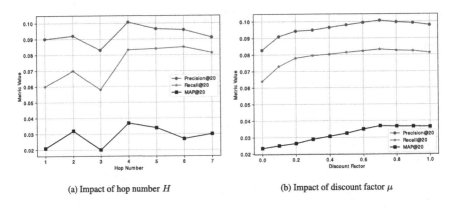

(a) Impact of hop number H (b) Impact of discount factor μ

Fig. 4. Impacts of hop number H and discount factor μ on MovieLens

tendency, a longer propagation distance improves the model performance. However, this improvement is not stable. Surprisingly, we find in Fig. 4a that recommendation quality becomes better when $H = 2$ and $H = 4$, compared with other lateral hop numbers. Considering the structure of KG, as shown in Fig. 2, when the current H is odd, there is a high possibility that the current propagated node represents an item while the current node tends to represent other attribute entities when H is even. Any two adjacent items are connected through holding the form as $user \xrightarrow{r_1} item \xrightarrow{r_2} attribute \xrightarrow{-r_2} item$. Such relation chain indicates that user preference is better learned by aggregating attribute vectors of items than by aggregating item vectors directly.

Discount Factor: The discount factor μ is utilized to control the impact of distance between target user and propagated entities as defined in Eq. (8). When μ equals to 0, it turns out that the nearby entity will have more impact on user embeddings, while all the involved entities in the set I_u^* contributes equally to the preference embedding of users when μ is set to 1. As shown in Fig. 4b, the model performance achieves the best when μ is around 0.7. Intuitively, the user should pay more attention to closer entities, leading μ to approach 0. However, when μ decreases, the main contribution of user embedding will concentrate on the interacted items, thus reducing the semantic information carried by the user preference vector. The trade-off between distance impact and semantic richness makes the optimal value of μ fall around 0.7.

Graph Node Aggregator: This parameter refers to the approach we aggregate neighborhood information while obtaining the node structural-level embedding. There are four optional aggregators including Max or Mean Pooling, GraphSage, and GAT. For the two pooling aggregators [5], an element-wise pooling operation is applied to aggregate information from sampled neighbor set after each neighbor's embedding comes through a trainable transformation matrix. The performance of the four aggregators are evaluated as shown in Fig. 5. A key observation is that two GNN-based aggregators, i.e., GraphSage and GAT, perform better than the other two pooling aggregators in general, indicating that the effect of pooling methods to remove redundant information is

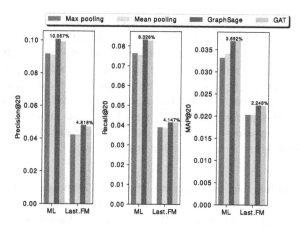

Fig. 5. Impacts of different graph nodes aggregators on two evaluation datasets (Displayed values are related to GraphSage)

not apparent in recommendation. The results also show that graph network is an expert in learning node neighborhood structure. Comparing the two graph embedding-based aggregators, GAT considers the importance of neighbor nodes and assign each node different weights. However, its modification does not show an advantage over Graph-Sage in MDKE. We argue that the disentangled embedding learning method helps to distinguish different nodes and is conducted in a more explicit and effective way. On the other hand, GAT aggregator takes nearly twice as long to generate a prediction for all the test instances in MovieLens as GraphSage. Therefore, we think GraphSage can best balance time efficiency and model performance among the four aggregators.

4.4 Ablation Study

We analyze the impacts of core components in MDKE by conducting ablation study. The results are presented in Table 3. In general, removing any part can have an obvious negative impact on the effectiveness of MDKE.

Table 3. Ablation study on two datasets.

Metric (%)	MovieLens			Last.FM		
	P@20	R@20	M@20	P@20	R@20	M@20
MDKE	**10.067**	**8.326**	**3.692**	**4.818**	**4.147**	**2.246**
- Preference prop	9.499	8.156	3.361	4.653	4.052	2.073
Δ Preference prop	−5.6%	−2.0%	−9.0%	−3.3%	−2.3%	−7.7%
- Semantic emb	9.043	7.683	3.144	4.409	3.941	1.991
Δ Semantic emb	−10.2%	−7.7%	−14.8%	−8.5%	−5.0%	−11.4%
- Structural emb	9.744	7.978	3.392	4.499	3.867	2.004
Δ Structural emb	−3.2%	−4.2%	−8.1%	−6.6%	−6.8%	−10.8%

Preference Propagation. In the first step, preference propagation is removed from the model, which means that user preference vectors are captured merely from directly interacted items. As shown in Table 3, a decline in model performance can be observed. For example, in terms of Precision@20, the decrease reaches 5.6%, which proves the positive impact of the propagation mechanism on explicit user interest modeling. As shown in the previous propagation distance experiments, learning user preference from other entities has better performance than directly from items. It proves once again that preference propagation helps to dig out deep interests of users and item attributes, which is crucial in generating recommendation for a target user.

Semantic Embedding. In this scenario, the entire semantic embedding part is removed. Results in Table 3 show a severe drop in performance, which is even larger than removing preference propagation. Without semantic knowledge, one can hardly distinguish user interests towards two items that share similar neighbors or in the same cluster. As shown in Fig. 3, the disentangled semantic and structural modules work together. When structural-level aggregation makes the embeddings of adjacent nodes converge, semantic information captures different attribute characteristics of nodes, helping the recommendation model to find possible interactions more accurately.

Structural Embedding. The third ablation study is conducted by removing structural module. The process of capturing high-order connectivity using GraphSage can be seen as an enhanced variant of traditional collaborative method, assuming that a user's buying behavior or an item's audience is similar to their second-order neighbors. The experiment results tell that even though the model merely with semantic embedding already achieves good performance, appending structural-level module from the perspective of connectivity can further improve the precision of recommendation.

5   Related Work

5.1   Graph-Based Methods for Recommendation

The main reason for using graph embedding technology in recommender system [18] is to exploit relations among interaction instances. The idea of message instruction and neighborhood aggregation could combine collaborative filtering with graph neural networks. GC-MC [19] aims to solve the task of predicting users' ratings of movies. The model uses a graph auto-encoder to generate embedded representations of users and items. Unlike the above two works, NGCF [16] believes that when information is disseminated on the graph, influence of different neighbor nodes on the target node should also be distinguished. Therefore, the model introduces node similarity mechanism, which can be seen as a variant of attention.

In recent years, KG-aware recommendation model has drawn significant focus. Using knowledge graph as side information can enrich embeddings by considering item attributes. KGCN [14] and KGNN-LS [13] connect the knowledge entities with the target item. The propagation stage only contains items. The user information is considered in the final prediction stage. Besides the knowledge graph, KGAT [15] also considers the user-item bipartite graph in the embedding propagation. Moreover, KARN [26] combines the session graph and knowledge graph to capture historical interest from user's click sequence.

5.2 Disentangled Representation Learning

Representation learning helps to extract useful information from data and is crucial for the performance of many other downstream tasks. Recently, disentangled representation learning is proposed to embed objects from multiple perspectives [1] through separating the explanatory factors from different sources. Such learning paradigm has been applied to many fields with rich data interaction, such as images [10], user behavior capturing, and knowledge graph embedding. In [7], a decoupled auto-encoder model is proposed to model user preference behind behaviors. DisenKGAT [17] considers micro and macro-disentanglement of entities to distinguish node embeddings and enhance the interpretability. [21] transfers disentangled embedding into CTR prediction. An attention-based network is designed to decouple the learning of unary and pairwise term. Considering that simply aggregating neighborhood information will filter out node's personalized features when generating embeddings, decoupling the subspaces of representation learning into semantic and structural-level is of great significance. Our work acquires item semantic information from content extraction stage and design a preference propagation mechanism with hop distance, which distinguishes MDKE from all the above-mentioned methods.

6 Conclusion

In this paper, we propose **M**ulti-level **D**isentangled **K**nowledge-Based **E**mbedding to generate user and item vectors for recommendation from different embedding subspaces, i.e., semantic and structural-level. To fully leverage the auxiliary information brought by KG, three main components are contained in MDKE. The semantic-level vectors of items are captured through the knowledge graph embedding approach, and user profile vectors are extracted through a carefully designed propagation mechanism. After appending users and constructing the knowledge interaction graph, a graph-based node aggregator is learned to generate node vectors to capture structural-level information. Finally, the forementioned multi-level embeddings are aggregated to form latent vectors, which are further used to estimate possible interaction between users and items. Experimental results show the effectiveness of our model compared with other baselines for recommendation, especially with the state-of-the-art models which also leverages knowledge graph. In terms of future work, we tend to explore information interaction between different embedding subspaces with a finer granularity.

References

1. Bengio, Y., Courville, A., Vincent, P.: Representation learning: a review and new perspectives. IEEE Trans. Pattern Anal. Mach. Intell. **35**(8), 1798–1828 (2013)

2. Bo, D., Wang, X., Shi, C., Shen, H.: Beyond low-frequency information in graph convolutional networks. In: AAAI. AAAI Press (2021)
3. Cao, X., et al.: DEKR: description enhanced knowledge graph for machine learning method recommendation. In: SIGIR, pp. 203–212. ACM (2021)
4. Guo, H., Tang, R., Ye, Y., Li, Z., He, X.: DeepFM: a factorization-machine based neural network for CTR prediction. In: IJCAI, pp. 1725–1731 (2017). ijcai.org
5. Hamilton, W., Ying, Z., Leskovec, J.: Inductive representation learning on large graphs. In: NIPS, pp. 1024–1034. MIT Press (2017)
6. Lin, Y., Liu, Z., Sun, M., Liu, Y., Zhu, X.: Learning entity and relation embeddings for knowledge graph completion. In: AAAI, pp. 2181–2187. AAAI Press (2015)
7. Ma, J., Zhou, C., Cui, P., Yang, H., Zhu, W.: Learning disentangled representations for recommendation. In: NIPS, pp. 5712–5723. MIT Press (2019)
8. Pennington, J., Socher, R., Manning, C.D.: Glove: Global vectors for word representation. In: EMNLP, pp. 1532–1543. ACL (2014)
9. Tai, C.Y., Wu, M.R., Chu, Y.W., Chu, S.Y., Ku, L.W.: MVIN: learning multiview items for recommendation. In: SIGIR, pp. 99–108. ACM (2020)
10. Tran, L., Yin, X., Liu, X.: Disentangled representation learning GAN for pose-invariant face recognition. In: CVPR, pp. 1415–1424. IEEE Computer Society (2017)
11. Veličković, P., Cucurull, G., Casanova, A., Romero, A., Lio, P., Bengio, Y.: Graph attention networks. In: ICLR (2017)
12. Wang, H., et al.: RippleNet: propagating user preferences on the knowledge graph for recommender systems. In: CIKM, pp. 417–426. ACM (2018)
13. Wang, H., et al.: Knowledge-aware graph neural networks with label smoothness regularization for recommender systems. In: SIGKDD, pp. 968–977. ACM (2019)
14. Wang, H., Zhao, M., Xie, X., Li, W., Guo, M.: Knowledge graph convolutional networks for recommender systems. In: WWW, pp. 3307–3313. ACM (2019)
15. Wang, X., He, X., Cao, Y., Liu, M., Chua, T.S.: KGAT: knowledge graph attention network for recommendation. In: SIGKDD, pp. 950–958. ACM (2019)
16. Wang, X., He, X., Wang, M., Feng, F., Chua, T.S.: Neural graph collaborative filtering. In: SIGIR, pp. 165–174. ACM (2019)
17. Wu, J., et al.: DisenKGAT: knowledge graph embedding with disentangled graph attention network. In: CIKM, pp. 2140–2149. ACM (2021)
18. Wu, S., Sun, F., Zhang, W., Cui, B.: Graph neural networks in recommender systems: a survey. arXiv preprint arXiv:2011.02260 (2020)
19. Wu, Y., Liu, H., Yang, Y.: Graph convolutional matrix completion for bipartite edge prediction. In: KDIR, pp. 49–58 (2018)
20. Xiao, H., Huang, M., Meng, L., Zhu, X.: SSP: semantic space projection for knowledge graph embedding with text descriptions. In: AAAI, pp. 3104–3110. AAAI Press (2017)
21. Xu, Y., Zhu, Y., Yu, F., Liu, Q., Wu, S.: Disentangled self-attentive neural networks for click-through rate prediction. In: CIKM, pp. 3553–3557. ACM (2021)
22. Yang, C., Wang, R., Yao, S., Liu, S., Abdelzaher, T.: Revisiting over-smoothing in deep GCNs. arXiv preprint arXiv:2003.13663 (2020)
23. Ying, R., He, R., Chen, K., Eksombatchai, P., Hamilton, W.L., Leskovec, J.: Graph convolutional neural networks for web-scale recommender systems. In: SIGKDD, pp. 974–983. ACM (2018)
24. Zhang, Y., Ai, Q., Chen, X., Wang, P.: Learning over knowledge-base embeddings for recommendation. arXiv preprint arXiv:1803.06540 (2018)
25. Zhao, L., Akoglu, L.: PairNorm: tackling oversmoothing in GNNs. arXiv preprint arXiv:1909.12223 (2019)
26. Zhu, Q., Zhou, X., Wu, J., Tan, J., Guo, L.: A knowledge-aware attentional reasoning network for recommendation. In: AAAI, pp. 6999–7006. AAAI Press (2020)

M^3-IB: A Memory-Augment Multi-modal Information Bottleneck Model for Next-Item Recommendation

Yingpeng Du[1], Hongzhi Liu[1(✉)], and Zhonghai Wu[2,3(✉)]

[1] School of Software and Microelectronics, Peking University,
Beijing, People's Republic of China
{dyp1993,liuhz}@pku.edu.cn
[2] National Engineering Research Center of Software Engineering,
Peking University, Beijing, People's Republic of China
[3] Key Lab of High Confidence Software Technologies (MOE),
Peking University, Beijing, People's Republic of China
wuzh@pku.edu.cn

Abstract. Modeling of users and items is essential for accurate recommendations. Traditional methods focused only on users' behavior data for recommendation. Several recent methods attempted to use multimodal data (e.g. items' attributes and visual features) to better model users and items. However, these methods fail to model users' dynamic and personalized preferences on different modalities. In addition, besides useful information for recommendation, the multi-modal data also contains a great deal of irrelevant and redundant information that may mislead the learning of recommendation models.

To solve these problems, we propose a Memory-augment Multi-Modal Information Bottleneck method, named M^3-IB, for next item recommendation. First, we design a memory network framework to maintain modality-specific knowledge and capture users' dynamic modality-specific preferences. Second, we propose to model and fuse users' personalized preferences on different modalities with a multi-modal probabilistic graph. Then, to filter out irrelevant and redundant information in multi-modal data, we extend the information bottleneck theory from single-modal to multi-modal scenario and design a multi-modal information bottleneck (M2IB) model. Finally, we provide a variational approximation and a flexible implementation of the M2IB model for next item recommendation. Experiential results on five real-world data sets demonstrate the promise of the proposed method.

Keywords: Next item recommendation · Multi-modal data · Memory networks · Information bottleneck · Variational approximation

---

Supplementary Information The online version contains supplementary material available at https://doi.org/10.1007/978-3-031-00126-0_2.

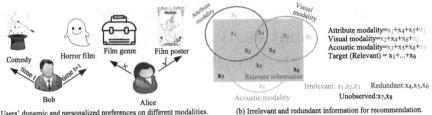

Fig. 1. Motivations and difficulties of exploiting multi-modal data for next item recommendation.

1   Introduction

Recommender systems gain popularity for many online applications such as e-commerce and online video sharing, which can enhance both users' satisfaction and platforms' profits. In real-world scenarios, users always engage items, such as click or purchase, in a sequential form. Therefore, predicting and recommending the next item that the target user may engage, known as next item recommendation, is a key part of recommender systems.

Multi-modal data, such as items' visual and acoustic features, carries rich content information of items, which can help to better model users' preferences and items' characteristic for next item recommendation. Several recent methods try to utilize multi-modal data by the constant feature-level combination for next item recommendation [2,9,12,15]. However, these methods ignore users' dynamic and personalized preferences on different modalities, which may restrict the recommendation performance. Figure 1 (a) shows a scenario of online film website. Bob preferred comedy films previously (at time t) but now he prefers horror films (at time $t+1$), which indicates users' dynamic preferences about the genre modality. Alice focused more on the poster than on the genre when selecting films, which indicates users' personalized preferences on different modalities.

Although the multi-modal data carries useful information for next item recommendation, it also contains a great deal of irrelevant and redundant information, which may mislead the learning of recommendation models. As shown in Fig. 1 (b), we have three modalities of data (circles) for the recommendation task (gray rectangle). It shows that multi-modal data may be overlapped and redundant (e.g. the repeated parts in x_4, x_5, and x_6), and not all information of multi-modal data is relevant for recommendation (e.g. z_1, z_2 and z_3) [23]. However, most existing methods adopt the combination strategies, such as feature-level aggregation or concatenation, for multi-modal data utilization, which exacerbates the misleading of irrelevant and redundant information for recommendation.

To address these problems, we propose a Memory-augment Multi-Modal Information Bottleneck model (M^3-IB) for next item recommendation. We first design a memory network based framework that consists of several knowledge memory regions and a lot of user-state memory regions. Each knowledge region stores the extracted item relevancies and similarities from a data modality as the modality-specific knowledge, and each private user-state region maintains the state of a specific user according to his/her dynamic behaviors. To learn users' dynamic preferences on different modalities, we extract user-state related knowledge in each modality-specific knowledge region for the preference modeling. To learn users' personalized preferences on different modalities, we propose a multi-modal probabilistic graph to discriminate and fuse the multi-modal information for different users. Furthermore, to mine useful information (e.g. $x_1, \cdots, x_6$) and filter out irrelevant information (e.g. z_1, z_2, and z_3) and redundant information (e.g. the repeated parts in x_4, x_5, x_6) in multi-modal data, we design a multi-modal information bottleneck (M2IB) model based on the multi-modal probabilistic graph. Finally, we provide a variational approximation and a flexible implementation of the M2IB model for next item recommendation. Extensive experiments on five real-world data sets show that the proposed model outperforms several state-of-the-art methods, and ablation experiments prove the effectiveness of the proposed model and verify our motivations.

2   Related Work

Multi-modal Recommendation. Several recent studies attempted to design personalized recommendations models based on static multi-modal data, which can be divided into collaborative filtering models and graph models. The collaborative filtering models tried to combine user-item interaction records and items' multi-modal content information for recommendation, i.e. treating structured attributes and users engaged items as factors of model [7]. For example, Chen et al. [3] proposed an attentive collaborative filtering method for multimedia recommendation, which can automatically assign weights to the item-level and content-level feedback in a distant supervised manner. As graph models can capture the collaborative or similarity relationship among the entities, they are widely used in recommender systems [22,27,28]. For example, Wei et al. [28] constructed a user-video bipartite graph, and utilized graph neural networks to learn modality-specific representations of users and micro-videos through message-passing. Although some of these methods [3,28] explored users' personalized preferences on different data modalities to some extent, they can not capture users' dynamic modality-specific preferences with their dynamic behaviors.

In real-world scenarios, users' behavior data always appears as a sequence and every behavior is linked with a timestamp. As a result, sequential recommendation, which aims at recommending the next item to users, has been becoming a hot research topic in recent years [5,12]. To fully exploit multi-modal information, most of existing methods adopted feature-level combination for next item recommendation. Some methods concatenated the features of different modalities in user's behavior order, and mined their transition patterns between adjacent behaviors [5,29]. For example, Zhang et al. [29] tried to model item- and attribute- level transition patterns from adjacent behaviors for next item recommendation. Some methods combined items' multi-modal content to enrich their representations, thereafter, such representations were fed into a sequential model such as RNN for next item recommendation [2,9,12,15]. For example, Huang et al. [9] combined unified multi-type actions and multi-modal content representation, and proposed a contextual self-attention network for sequential recommendation. However, these methods failed to explore users' dynamic modality-specific and personalized preferences on different modalities. More importantly, these feature-level combination strategies may exacerbate the misleading for the recommendation models by irrelevant and redundant information.

Memory Networks Recommendation. Recently, memory networks gain their popularity in recommender systems, which can capture users' historical behaviors or auxiliary information for recommendation [6,30]. For example, Zhu et al. [30] utilized the memory networks to store the categories of items in the last basket, and inferred the categories that users may need for next-basket prediction. However, these methods rely on specific data, which may be not applicable for dynamic and heterogeneous multi-modal data. In addition, these methods can not filter out the irrelevant and redundant information for recommendation.

Information Bottleneck. The information bottleneck (IB) principle [25] provides an information-theoretic method for representation learning. Compared to the traditional dimension reduction methods that fail to capture varying information amounts of different inputs (e.g. as in the encoder neural networks), the IB principle permits encodings to keep the amount of effective information varying with inputs in a high-dimensional space [13]. As the IB principle is computationally challenging, Alemi et al. [1] presented a variational approximation to it with a single-modality input. However, the existing IB models are not designed for recommendation tasks, which fails to fuse multi-modal information and discriminate the significance of individual modalities for each user.

We notice that the proposed model is related to some existing models such as variational auto-encoding (VAE) [11] based recommendation. For example, Liang et al. [14] proposed to combine MF and VAE to enforce the consistency of items embedding and their content (e.g. poster and plot) for a recommendation task. But it fails to utilize sequential information and remove irrelevant and redundant information for recommendation. Ma et al. [17] explored the macro and micro disentanglement for recommendation via the β-VAE scheme. However, it is impossible to learn disentangled representation with unsupervised learning [16]. Instead of learning disentangled representation of users, we attempt to mine the most meaningful and relevant information in multi-modal data with supervised learning.

3   Problem Definition

Let $\mathcal{U} = \{u_1, \cdots, u_M\}$ and $\mathcal{I} = \{i_1, \cdots, i_N\}$ denote the sets of M users and N items, respectively. Besides user-item interaction records, we suppose to know the order or timestamps of interactions. Therefore, we denote the sequential behaviors of each user $u \in U$ as $S_u(t) = [s_1, \cdots, s_t]$, where $s_j \in \mathcal{I}$ denotes the j-th item that user u engaged. In addition, we assume that multi-modal content of items is available, e.g. attributes, visual, acoustic, and textual features. For structured data $\mathcal{A} = \{a_1, \cdots, a_{|\mathcal{A}|}\}$, we suppose to know each item i's a_k-typed (e.g. Actor) attribute values $V_i(a_k) = \{v_1, \cdots, v_{|V_i(a_k)|}\}$ (e.g. Leonardo, Downey).

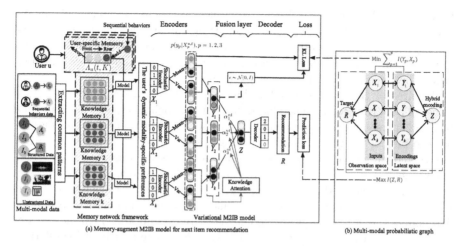

(a) Memory-augment M2IB model for next item recommendation

(b) Multi-modal probabilistic graph

Fig. 2. The architecture of the memory-augment M2IB model and the multi-modal probabilistic graph.

For unstructured data $\mathcal{C} = \{visual, acoustic, textual\}$, we can obtain their semantic representations $\{e_i^m \in \mathbb{R}^d | m \in \mathcal{C}\}$ of item i with feature extraction strategies, e.g. deep CNN visual features [18].

The goal of sequential recommendation is to learn a prediction function $f(\cdot)$ based on all users' sequential behaviors and multi-modal data to predict the next item that each user u may engage. In this paper, we take items' multi-modal content information into consideration and define the prediction function as $s(u,t) = \max_{i \in \mathcal{I}} f(u, i, S_u(t), \mathcal{A} \cup \mathcal{C})$.

4   The Proposed Method

The overall architecture of the proposed method is shown in Fig. 2. First to extract from multi-modal data, we design a memory network framework to capture users' dynamic modality-specific preferences according to their dynamic behaviors (Sect. 4.1). Then to fuse users' multi-modal preferences, we propose the M2IB model with a multi-modal probabilistic graph to learn users' personalized preferences on different modalities, and filter out irrelevant and redundant information (Sect. 4.2). Finally, we provide the detailed implementation of the variational M2IB model, which can encode, fuse, and decode users' multi-model preferences for next item recommendation (Sect. 4.3).

4.1   Memory Network Framework

To make use of multi-modal data, we propose a memory network framework called MemNet as shown in the left part of Fig. 2 (a), which consists of several knowledge memory regions and user-state memory regions.

Knowledge Memory Regions. To exploit the heterogeneous multi-modal data for recommendation, we organize multi-modal data as unified item-to-item similarity matrices. These matrices can be seen as the global-sense knowledge shared by all users, which measures the modality-specific relevancies and similarities between items with different modalities.

Inspired by the collaborative filtering models [21], we assume that there exist some collaborative correlations that describe users' behavior patterns. For example, some persons always buy pencils and erasers together, and some persons are used to buying the gum shell after buying the cigarette. To this end, we calculate the collaborative correlations between items based on user-item behavioral and sequential interactions respectively, which are denoted as $M^c, M^s \in \mathbb{R}^{N \times N}$. Specifically, co-engaged collaborative correlation M^c_{ij} counts the number of users who engaged both item i and item j, while sequential collaborative correlation M^s_{ij} counts the number of users who engaged item i and item j chronologically.

Inspired by the content based models [4], we assume that items' similarities measured by each content modality reflect users' inherent preference from a specific aspect. For example, some persons prefer to watch similar movies. For structured data such as items' a_k-typed attributes, we calculate the similarities M^{a_k} among items by counting the shared a_k-typed attributes by item i and item j, i.e. $M^{a_k}_{ij} = |V_i(a_k) \cap V_j(a_k)|$. For unstructured data such as items' visual content, we calculate their similarities M^m among items based on the extracted features e^m, e.g. inner product similarity.

To alleviate the differences between the measurement strategies, we only keep the F most frequent patterns for each modality inspired by the frequent pattern model [23] and the reassignment strategy [27]. We store the knowledge in P knowledge regions of the MemNet framework, which is denoted as $\mathcal{M} = [M^c, M^s, M^{a_1}, \cdots, M^{a_{|A|}}, M^{\text{visual}}, M^{\text{acoustic}}, M^{\text{textual}}]$.

Private User-Specific Memory Regions. To capture users' dynamic preferences, we propose to maintain the dynamic behaviors of each user in user-state memory regions. Let $\Lambda_u(t, K)$ denote the user-state memory for user u, where K is the memory capacity. After getting a new behavior of user u, i.e. user u interacting with item i at time t, we update the memory queue $\Lambda_u(t, K)$ to maintain its dynamic nature. We first drop the front item from the queue if the memory queue is full, and then add the new item i into the rear of the memory queue.

To model user u's dynamic modality-specific preference $X^{u,t}_p \in \mathbb{R}^N$ for different modalities, we aggregate the user-state related knowledge of each modality for the user's preference modeling at time t, i.e.

$$X^{u,t}_p = \frac{1}{|\Lambda_u(t, K)|} \sum_{i \in \Lambda_u(t,K)} \mathcal{M}^p_i \tag{1}$$

where $\mathcal{M}^p_i$ denotes the i-th row (related to item i) of the p-th knowledge region matrix $\mathcal{M}^p$.

4.2   Multi-modal Information Bottleneck Model

Multi-modal Probabilistic Graph with IB Principle. To learn the user's personalized preferences on different modalities, we propose a multi-modal probabilistic graph (M2PG) to discriminate and fuse his/her modality-specific preferences as shown in Fig. 2 (b). Specifically, the user's modality-specific preferences $[X_1, \cdots, X_P]$ are taken as the inputs, which are mapped into a latent space as the multi-modal preference encodings $[Y_1, \cdots, Y_P]$. Then, we aggregate these encodings into the hybrid encoding Z by attentional weights, which model users' personalized preference on different modalities. Finally, we map Z into the scores of all items to predict for the user's next behavior (target R).

To mine relevant information and filter out irrelevant and redundant information, we try to optimize two goals for next item recommendation. The first goal of M2IB model is to be expressive about the target R, which can be achieved by maximizing the mutual information between the hybrid encoding Z and the target R, i.e. $\max I(Z, R)$. The second goal of M2IB model is to minimize the information of the M2IB model retained from the inputs, which can be achieved by minimizing the mutual information between the encoding Y_p and the input X_p for different data modalities, i.e. $\min \sum_{p=1}^{P} I(Y_p, X_p)$. Combination of the above two goals means retaining as little as possible the most useful information for the target, i.e.

$$\max I(Z, R) - \lambda \sum_{p=1}^{P} I(Y_p, X_p) \tag{2}$$

where λ denotes a trade-off coefficient of the two goals.

Variational Approximation. To optimize objective function Eq. (2) which includes two kinds of intractable mutual information, we introduce two theorems for the mutual information variational approximation, whose proofs are shown in the supplementary materials[1]. Assuming we have Q samples $\{(r^q, x_1^q, .., x_P^q)|q = 1, \cdots, Q\}$ from a data distribution D that subjects to the empirical data distribution as follows:

$$p(r, x_1, .., x_P) = \begin{cases} 1/Q, & (r, x_1, .., x_P) \in D, \\ 0, & else. \end{cases} \tag{3}$$

where r denotes the target and $x_1, .., x_P$ denote the inputs.

Theorem 1. *With Q samples $\{(r^q, x_1^q, .., x_P^q)|q = 1, \cdots, Q\}$ and M2PG, the mutual information $I(Z, R)$ has an approximation lower bound, i.e.*

$$I(Z, R) \geq L_1 \approx \frac{1}{Q} \sum_{q=1}^{Q} E_{\prod_{p=1}^{P} p(y_p|x_p^q)p(z|y_1,..,y_P)} \log q(r^q|z) \tag{4}$$

where $q(r^q|z)$ denotes the variational inference of posterior $p(r^q|z)$.

---

[1] Link: https://pan.baidu.com/s/1dm0jIwes1DLIqNjIn3V1oA, code:gqc6.

Theorem 2. *With Q samples $\{(r^q, x_1^q, .., x_P^q)|q = 1, \cdots, Q\}$, the mutual information $\sum_{p=1}^{P} I(Y_p, X_p)$ has an approximation upper bound as follows:*

$$\sum_{p=1}^{P} I(Y_p, X_p) \leq L_2 \approx \frac{1}{Q} \sum_{q=1}^{Q} \sum_{p=1}^{P} KL(p(y_p|x_p^q)|\epsilon) \tag{5}$$

where $\epsilon \sim \mathcal{N}(0, I)$ and $KL(\cdot)$ denotes the Kullback-Leibler divergence.

Encoder-Fusion-Decoder Framework. According to the Theorems 1 and 2, we can obtain the variational M2IB model in an encoder-fusion-decoder framework.

$$I(Z, R) - \lambda \sum_{p=1}^{P} I(Y_p, X_p) \geq L = L_1 - \lambda L_2 \tag{6}$$

where L_1 in Eq.(4) consists of the $p(y_p|x_p^q)$ that denotes an **encoder** of input x_p, $p(z|y_1, .., y_P)$ that denotes the **fusion layer** for multi-modal information, and variational estimation $q(r^q|z)$ that denotes a **decoder** for the output.

4.3   Implementation for Next-Item Recommendation

In this subsection, we detailed the implementation of variational M2IB model as shown in the right part of Fig. 2 (a). It consists of multiple encoders, a fusion layer, a decoder, a prediction loss, and a KL divergence loss.

Encoder Layer: To exploit users' modality-specific preferences $[x_1, \cdots, x_P]$, we encode them into a latent space by encoders $\{p(y_p|x_p)|p = 1, \cdots, P\}$. For user u at time t, we encode his/her p-th modality preference $X_p^{u,t} \in \mathbb{R}^N$ as a random variable $p(y_p^{u,t}|X_p^{u,t}) \sim \mathcal{N}(\mu_p^{u,t}, diag(\sigma_p^{u,t}))$, where the mean $\mu_p^{u,t}$ and the standard deviation $diag(\sigma_p^{u,t})$ are parameterized by neural networks f_p^μ and f_p^σ as follows:

$$\mu_p^{u,t} = f_p^\mu(X_p^{u,t}) = W_p^\mu \cdot X_p^{u,t} + b_p^\mu \tag{7}$$

$$\sigma_p^{u,t} = f_p^\sigma(X_p^{u,t}) = \kappa(W_p^\sigma \cdot X_p^{u,t} + b_p^\sigma) \tag{8}$$

where $W_p^* \in \mathbb{R}^{d \times N}$ and $b_p^* \in \mathbb{R}^d$ denote the fully connected layers and the biases, respectively. d denotes the dimension of the latent space and the activation function $\kappa(x) = \exp(x/2)$ keeps the standard deviation to be positive.

Fusion Layer. To learn the user's personalized preference on different modalities in the fusion layer $p(z|y_1, \cdots, y_P)$, we adopt an attentional mechanism to aggregate his/her preference encodings into the hybrid preference encoding, i.e.

$$z^{u,t} = \sum_{p=1}^{P} \alpha_p^{ut} \cdot y_p^{u,t} \tag{9}$$

$$\alpha_p^{ut} = \frac{\exp(\text{l\_2}(y_p^{u,t})^\top \cdot \text{l\_2}(h_p))}{\sum_{p=1}^{P} \exp(\text{l\_2}(y_p^{u,t})^\top \cdot \text{l\_2}(h_p))} \tag{10}$$

where $L_2(\cdot)$ denotes the L2 normalization and $h_p \in \mathbb{R}^d$ denotes the encoding of the p-th modality to model its importance for recommendation, which represents the key for the p-th modality in the attentional mechanism. And $y_p^{u,t}$ represents both queries and values as in the self-attention mechanism [26].

Decoder Layer. To predict next behavior of user u at time t, we adopt the decoder $q(r|z)$ to map his/her hybrid preference encoding $z^{u,t}$ into the score or probability of all items that user u will engage:

$$\hat{r}^{u,t} = z^{u,t} \cdot H \tag{11}$$

where $H \in \mathbb{R}^{d \times N}$ is the decoding matrix which is the parameters of the decoder.

Loss Functions. To measure the effectiveness of the recommendation results, we adopt a pair-wise loss [19] to formulate the lower bound of $I(Z, R)$ in Eq. (4), i.e.

$$L_1 \approx \frac{1}{|D|} \sum_{(u,i_1,i_2,t) \in D} \log(\sigma(\hat{r}_{i_1}^{u,t} - \hat{r}_{i_2}^{u,t})) \tag{12}$$

where $\sigma(\cdot)$ denotes the sigmoid function. $D = \{(u, i_1, i_2, t)|u \in \mathcal{U}\}$ denotes the train set and (u, i_1, i_2, t) means user u has engaged item i_1 but not engaged item i_2 at time t. To limit the information retained from the inputs, we formulate the upper bound of L_2 term in Eq. (5) by calculating the KL divergence between the encoding $p(y_p^{u,t}|X_p^{u,t}) \sim \mathcal{N}(\mu_p^{u,t}, diag(\sigma_p^{u,t}))$ and $\epsilon \sim \mathcal{N}(0, I)$ for different modalities, i.e.

$$L_2 \approx \frac{1}{2|D|} \sum_{(u,t) \in D} \sum_{p=1}^{P} \sum_{l=1}^{d} (-\log((\sigma_{p,l}^{u,t})^2) + (\mu_{p,l}^{u,t})^2 + (\sigma_{p,l}^{u,t})^2)$$

4.4   Model Learning and Complexity Analysis

Parameters Learning. The overall objective function combines both the prediction loss L_1 and the KL divergence loss L_2 as follows:

$$\max_{\theta} L = L_1 - \lambda L_2 - \mu_\theta ||\theta||^2 \tag{13}$$

where λ denotes the trade-off coefficient. θ denotes all parameters need to be learned in the proposed model and μ_θ denotes the regularization coefficient of L2-norm $||\cdot||^2$. As the objective function is differentiable, we optimize it by stochastic gradient descent and adaptively adjust the learning rate by AdamGrad, which can be automatically implemented by TensorFlow[2].

Complexity. For the model learning process, updating one sample for the proposed model takes $O(P \cdot F \cdot d + p \cdot d)$ time. Specifically, P denotes the number of knowledge regions. $O(F \cdot d + p \cdot d)$ denotes the computational complexity of the encoder with sparse inputs and fusion layers, where F denotes the number

---

[2] https://www.tensorflow.org/.

of frequent patterns. Thus, one iteration takes $O(|D| \cdot P \cdot F \cdot d + |D| \cdot p \cdot d)$ time, where $|D|$ is the number of all users' interactions. The proposed model shares roughly equal time complexity to the self-attention method such as SASRec [10] and FDSA [29] $O(|D| \cdot K^2 \cdot d + |D| \cdot K \cdot d^2)$, where K denotes the window size of users' behaviors and $O(P \cdot F) \approx O(K^2)$ with appropriate hyper-parameters.

5   Experiment

In this section, we aim to evaluate the performance and effectiveness of the proposed method. Specifically, we conduct several experiments to study the following research questions:

RQ1: Whether the proposed method outperforms state-of-the-art methods for next item recommendation?

RQ2: Whether the proposed method benefits from the MemNet framework to capture users' dynamic modality-specific preferences?

RQ3: Whether the proposed M2IB model with M2PG outperforms the traditional IB model with single-modal probabilistic graph to model users' personalized preferences on different modalities?

RQ4: (a) Whether the proposed method benefits from removing irrelevant and redundant information compared with traditional dimension reduction? (b) To what extent the proposed method can filter out these irrelevant and redundant information?

5.1   Experimental Setup

Data Sets: We adopt five public data sets as the experimental data, including ML (Hetrec-MovieLens[3]), Dianping[4], Amazon[5]_Kindle (Amazon Kindle Store), Amazon_Game (Amazon Video Games), and Tiktok[6]. The ML data set contains users' ratings on movies with timestamps and the attributes of movies (e.g. Actors, Directors, Genres, and Countries). The Dianping data set consists of the users' ratings on restaurants in China with timestamps and attributes of restaurants (e.g. City, Business district, and Style). Amazon_Kindle and Amazon_Game data sets contain users' ratings on books and video games with timestamps and metadata of items such as genre and visual features, where the visual features are extracted from each product image using a deep CNN [18]. Tiktok data set is collected from a micro-video sharing platform. It consists of users, micro-videos, their interactions with timestamps, and visual and acoustic features which are extracted and published without providing the raw data. Due to the possible redundancy of visual and acoustic features, we reduce them to d-dimensions with principal component analysis (PCA) as in [9]. For these data sets, we filter users and items which have less than $[10, 25, 50, 50, 20]$ records for these data sets

---

[3] https://grouplens.org/datasets/movielens/.

[4] https://www.dianping.com/.

[5] http://jmcauley.ucsd.edu/data/amazon/links.html.

[6] http://ai-lab-challenge.bytedance.com/tce/vc/.

Table 1. Statistics of the experimental data sets

Data set	# User	# Item	# Interact	Attribute	Visual	Acoustic
ML	2,059	4,220	287,033	✓	✗	✗
Dianping	9,329	8,811	121,257	✓	✗	✗
Amazon_Kindle	5,862	7,951	190,573	✓	✓	✗
Amazon_Game	8,966	8,659	106,513	✓	✓	✗
Tiktok	10,037	3,617	190,144	✗	✓	✓

respectively according to their original scales, and consider ratings higher than 3.5 points as positive interactions as in [5]. The characteristics of these data sets are summarized in Table 1.

Evaluation Methodology and Metrics: We sort all interactions chronologically, then use the first 80% interactions as the train set and hold the last 20% for testing as in [24]. To stimulate the dynamic data sequence in a real-world scenario, we test the interactions from the hold-out data one by one correspondingly. Experimental results are recorded as the average of five runs with different random initializations of model parameters.

To evaluate the performances of the proposed method and the baseline methods, we adopt two widely used evaluation metrics for top-n recommendation as in [5], including hit ratio (hr) and normalized discounted cumulative gain (ndcg). Intuitively, hr@n considers all items ranked within the top-n list to be equally important, while ndcg@n uses a monotonically increasing discount to emphasize the importance of higher ranking positions versus lower ones.

Baselines: We take the following state-of-the-art methods as the baselines.
-**BPR** [19]. It proposes a pair-wise loss function to model the relative preferences of users.
-**VBPR** [8]. It integrates the visual features and ID embeddings of each item as its representation, and uses MF to reconstruct the historical interactions between users and items.
-**FPMC** [20]. It combines MF and MC to capture users' dynamic preferences with the pair-wise loss function.
-**Caser** [24]. It adopts the convolutional filters to learn users' union and skip patterns for sequential recommendation.
-**SASRec** [10]. It is a self-attention based sequential model, which utilizes item-level sequences for next item recommendation.
-**ANAM** [2]. It utilizes a hierarchical architecture to incorporate the attribute information by an attention mechanism for sequential recommendation.
-**FDSA** [29]. It models item and attribute transition patterns for next item recommendation based on the Transformer model [26].
-**HA-RNN** [15]. It combines the representation of items and their attributes, then fits them into an LSTM model for sequential recommendation.

Table 2. Performances of different methods. * indicates statistically significant improvement on a paired t-test ($p < 0.01$).

Method	ML		Dianping		Aamazon_Kindle		Aamazon_Game		Tiktok	
	hr@50	ndcg@50	hr@50	ndcg@50	hr@50	ndcg@50	hr@50	ndcg@50	hr@50	ndcg@50
BPR	0.1241	0.0357	0.1277	0.0359	0.1242	0.0364	0.1422	0.048	0.0849	0.0224
VBPR	-	-	-	-	0.1263	0.0367	0.1463	0.0503	0.0792	0.0213
FPMC	0.1767	0.0521	0.1391	0.0403	0.3110	0.1090	0.2165	0.0789	0.1284	0.0360
Caser	0.1438	0.0416	0.1177	0.0335	0.2643	0.0981	0.1829	0.0728	0.1393	0.0411
SASRec	0.1946	0.0540	0.1277	0.0352	0.2888	0.0889	0.2457	0.1052	0.1157	0.0337
ANAM	0.1173	0.0338	0.1285	0.0364	0.1096	0.0316	0.1352	0.0425	-	-
FDSA	0.1980	0.0552	0.1388	0.0382	0.3007	0.0948	0.2402	0.0905	0.1151	0.0340
ACF	0.1789	0.0537	0.1324	0.0372	0.2438	0.0885	0.2203	0.0962	0.1253	0.0378
HA-RNN	0.2047	0.0596	0.1476	0.0415	0.2808	0.0968	0.2194	0.0844	0.1296	0.0383
SNR	0.2241	0.0685	0.1501	0.0440	0.3029	0.1209	0.2487	0.1100	0.1188	0.0361
M^3-IB-noMM	0.1922	0.0571	0.1429	0.04	0.2666	0.1028	0.2019	0.0886	0.1350	0.0415
M^3-IB-noM2PG	0.2287	0.0694	0.1572	0.0451	0.3367	0.1220	0.2686	0.1062	0.1522	0.0458
M^3-IB-noBN	0.2306	0.0690	0.1536	0.0439	0.3440	0.1274	0.2544	0.1007	0.1577	0.0474
M^3-IB	**0.2399***	**0.0728***	**0.1637***	**0.0474***	**0.3479***	**0.1328***	**0.2758***	**0.1104***	**0.1600***	**0.0484***
Improve.	7.03%	6.22%	9.07%	7.73%	11.85%	9.84%	10.88%	0.35%	14.89%	17.81%

-**ACF** [3]. It consists of the component-level and item-level attention modules that learn to informative components of multimedia items and score the item preferences. We utilize users' recent engaged items as the neighborhood items in ACF to learn users' dynamic preferences.

-**SNR** [5]: It is a sequential network based recommendation method for next item recommendation, which tries to explore and combine multi-factor and multi-faceted preference to predict users' next behaviors.

Among these baseline methods, BPR and VBPR utilize static user-item interaction while others utilize users' sequential behaviors for recommendation. Besides, ANAM explores items' structured attributes information, while ACF, FDSA, SNR and HA-RNN explore both structured attributes and unstructured features for next item recommendation.

Implementation Details: We set the initial learn rating $\gamma = 1 \times 10^{-3}$, regularization coefficient $\mu_\theta = 1 \times 10^{-5}$, and the dimension $d = 64$ to all methods for a fair comparison. We set the frequent pattern number F as 20, the user-state memory capacity K as 5, and the trade-off parameter λ as 1.0 for the moderate hyper-parameter selection of the propose model. For baseline models, we tune their parameters to the best. The source code is available in https://github.com/kk97111/M3IB.

5.2   Model Comparison

Table 2 shows the performances of different methods for next item recommendation. To make the table more notable, we bold the best results and underline the best baseline results in each case. Due to ANAM is designed for the structured attribute features, it can not work with only unstructured information such as visual and acoustic features. VBPR is designed for visual features and can not work with structured attributes. From the Table 2, we can observe that: 1) The proposed method M^3-IB outperforms all baselines significantly in all cases. The relative improvement over the best baselines is 10.75% and 8.39% on average according to hr@50 and ndcg@50, respectively. It confirms the effectiveness of our proposed model (RQ1). 2) The multi-modal methods, e.g. HA-RNN and SNR, achieve better performance among baseline models on data sets that contain structured attribute information, which confirms the necessity of extracting the rich information from multi-modal data. But some multi-modal methods (e.g. ANAM, ACF, VBPR) show low accuracy in some cases, which indicates they may suffer from the misleading of irrelevant and redundant information for next item recommendation. 3) The methods without consideration of user-item dynamic relations, e.g. BPR and VBPR, perform badly among the baselines, which indicates it is necessary to capture users' dynamic preferences for next item recommendation.

5.3   Ablation Study

To evaluate the effectiveness of the module design of the proposed method and the motivations of this paper, we take some special cases of the proposed method as comparisons.

-M^3-**IB-noMM**: It is a variant of M^3-IB method without the MemNet framework. Following the idea in [15], we first enrich items' representations with multi-modal content, and then takes the representations of users' recent engaged (several) items as the multi-source input for the M2IB model.

-M^3-**IB-noM2PG**: It is a variant of M^3-IB method without the M2PG, which adopts the traditional IB model [1] that takes the concatenation of users' multi-modal preferences, i.e. $X_1 \oplus \cdots \oplus X_P$, as inputs.

-M^3-**IB-noBN**: It is a variant of M^3-IB without using the IB information constraints i.e. set $\lambda = 0$ in Eq. (2), which only uses dimension reduction for information processing.

The results of the ablation study are shown at the bottom part of Table 2. On all data sets, the proposed method M^3-IB with the MemNet framework significantly outperforms the variant M^3-IB-noMM that utilizes the feature-level combination strategy, which proves the necessity of capturing users' dynamic modality-specific preferences (RQ2). Besides, the proposed method M^3-IB outperforms the variant M^3-IB-noM2PG for multi-modal information fusion. It indicates the effectiveness of M2IB model to model users' personalized preferences on different modalities (RQ3). Finally, the proposed method M^3-IB outperforms the variant M^3-IB-noBN with traditional dimension reduction, which confirms the effectiveness of the proposed method M^3-IB to filter out the irrelevant and redundant information based on the IB principle (RQ4.(a)).

5.4   Hyper-Parameter Study

There are several key parameters for the proposed method M^3-IB, including the number of retained the most frequent patterns F for each modality, the capacity

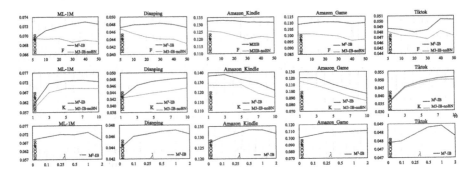

Fig. 3. The influence of number of most frequent patterns F and users' recent K behaviors to the method M^3-IB and the variant M^3-IB-noBN. And performances of the method M^3-IB varies with the trade-off parameter λ.

of each user memory region K, and the trade-off coefficient λ. Exploring them can help us with better understanding to what extent the M^3-IB method can extract useful information and filter out irrelevant and redundant information (RQ4.(b)). Figure 3 shows the performances of the method M^3-IB and the variant M^3-IB-noBN with various F and K. The larger F implies the more knowledge is retained from the data modality. It will contain more useful information while introducing more irrelevant and redundant information for next item recommendation. The larger K implies the more users' recent behaviors are memorized and utilized for next item recommendation. The method M^3-IB consistently outperforms the variant M^3-IB-noBN with sufficient information in all data sets, which confirms the necessity to remove irrelevant information. In addition, the method M^3-IB is more effective and stable than the variant M^3-IB-noBN when introducing massive noise, which shows IB model can capture useful information effectively and remove irrelevant and redundant information stably. Figure 3 also shows the performances of the method M^3-IB with various λ. The larger λ implies the more strict rule to remove the irrelevant and redundant information. It shows that too large or too small λ limits the performance of the method M^3-IB. It will remove some useful information when λ is too large, and introduce too much irrelevant and redundant information when λ is too small.

6   Conclusion

In this paper, we proposed a memory-augment multi-modal information bottleneck method to make use of information from multi-modal data for next item recommendation. The proposed method can model users' dynamic and personalized preferences on different modalities, and capture the most meaningful and relevant information with the IB principle. Extensive experiments show the proposed method outperforms state-of-the-art methods. Besides, the ablation experiments verify the validity of our research motivations: a) the proposed method benefits from exploring users' dynamic and personalized preferences on different modalities compared to existing content combination based strategies; b) the proposed method benefits from the M2IB model compared to traditional IB method for multi-modal fusion and representation learning; c) the proposed method shows effectiveness and stability to capture useful information effectively and remove irrelevant and redundant information for next item recommendation. In the future, we will study how to extend the proposed method to make use of unaligned data such as cross-domain information.

Acknowledgement. This work was supported by Peking University Education Big Data Project (Grant No. 2020YBC10).

References

1. Alemi, A.A., Fischer, I., Dillon, J.V., Murphy, K.: Deep variational information bottleneck. arXiv preprint arXiv:1612.00410 (2016)
2. Bai, T., Nie, J.Y., Zhao, W.X., Zhu, Y., Du, P., Wen, J.R.: An attribute-aware neural attentive model for next basket recommendation. In: Proceedings of SIGIR, pp. 1201–1204 (2018)
3. Chen, J., Zhang, H., He, X., Nie, L., Liu, W., Chua, T.S.: Attentive collaborative filtering: multimedia recommendation with item-and component-level attention. In: Proceedings of the 40th International ACM SIGIR Conference on Research and Development in Information Retrieval, pp. 335–344 (2017)
4. Di Noia, T., Mirizzi, R., Ostuni, V.C., Romito, D., Zanker, M.: Linked open data to support content-based recommender systems. In: Proceedings of the 8th International Conference on Semantic Systems, pp. 1–8 (2012)
5. Du, Y., Liu, H., Wu, Z.: Modeling multi-factor and multi-faceted preferences over sequential networks for next item recommendation. In: Oliver, N., Pérez-Cruz, F., Kramer, S., Read, J., Lozano, J.A. (eds.) ECML PKDD 2021. LNCS (LNAI), vol. 12976, pp. 516–531. Springer, Cham (2021). https://doi.org/10.1007/978-3-030-86520-7_32
6. Ebesu, T., Shen, B., Fang, Y.: Collaborative memory network for recommendation systems. In: The 41st International ACM SIGIR Conference On Research & Development in Information Retrieval, pp. 515–524 (2018)
7. Guo, H., Tang, R., Ye, Y., Li, Z., He, X.: DeepFM: a factorization-machine based neural network for CTR prediction. In: Proceedings of the 26th International Joint Conference on Artificial Intelligence, pp. 1725–1731 (2017)
8. He, R., McAuley, J.: VBPR: visual Bayesian personalized ranking from implicit feedback. In: Proceedings of the AAAI Conference on Artificial Intelligence, vol. 30 (2016)
9. Huang, X., Qian, S., Fang, Q., Sang, J., Xu, C.: CSAN: contextual self-attention network for user sequential recommendation. In: Proceedings of the 26th ACM international conference on Multimedia, pp. 447–455 (2018)
10. Kang, W.C., McAuley, J.: Self-attentive sequential recommendation. In: 2018 IEEE International Conference on Data Mining (ICDM), pp. 197–206. IEEE (2018)
11. Kingma, D.P., Welling, M.: Auto-encoding variational Bayes. arXiv preprint arXiv:1312.6114 (2013)
12. Li, X., et al.: Adversarial multimodal representation learning for click-through rate prediction. In: Proceedings of the Web Conference 2020, pp. 827–836 (2020)

13. Li, X.L., Eisner, J.: Specializing word embeddings (for parsing) by information bottleneck. In: Proceedings of the 2019 Conference on Empirical Methods in Natural Language Processing and the 9th International Joint Conference on Natural Language Processing (EMNLP-IJCNLP), pp. 2744–2754 (2019)

14. Liang, D., Krishnan, R.G., Hoffman, M.D., Jebara, T.: Variational autoencoders for collaborative filtering. In: Proceedings of the 2018 World Wide Web Conference, pp. 689–698 (2018)

15. Liu, K., Shi, X., Natarajan, P.: Sequential heterogeneous attribute embedding for item recommendation. In: Proceedings of ICDMW, pp. 773–780 (2017)

16. Locatello, F., et al.: Challenging common assumptions in the unsupervised learning of disentangled representations. In: International Conference on Machine Learning, pp. 4114–4124. PMLR (2019)

17. Ma, J., Zhou, C., Cui, P., Yang, H., Zhu, W.: Learning disentangled representations for recommendation. In: Advances in Neural Information Processing Systems, pp. 5711–5722 (2019)

18. McAuley, J., Targett, C., Shi, Q., Van Den Hengel, A.: Image-based recommendations on styles and substitutes. In: Proceedings of the 38th international ACM SIGIR Conference on Research and Development in Information Retrieval, pp. 43–52 (2015)

19. Rendle, S., Freudenthaler, C., Gantner, Z., Schmidt-Thieme, L.: BPR: Bayesian personalized ranking from implicit feedback. In: Proceedings of the Twenty-Fifth Conference on Uncertainty in Artificial Intelligence, pp. 452–461 (2009)

20. Rendle, S., Freudenthaler, C., Schmidt-Thieme, L.: Factorizing personalized Markov chains for next-basket recommendation. In: Proceedings of WWW, pp. 811–820 (2010)

21. Schafer, J.B., Frankowski, D., Herlocker, J., Sen, S.: Collaborative filtering recommender systems. In: Brusilovsky, P., Kobsa, A., Nejdl, W. (eds.) The Adaptive Web. LNCS, vol. 4321, pp. 291–324. Springer, Heidelberg (2007). https://doi.org/10.1007/978-3-540-72079-9_9

22. Sun, R., et al.: Multi-modal knowledge graphs for recommender systems. In: Proceedings of the 29th ACM International Conference on Information & Knowledge Management, pp. 1405–1414 (2020)

23. Swesi, I.M.A.O., Bakar, A.A., Kadir, A.S.A.: Mining positive and negative association rules from interesting frequent and infrequent itemsets. In: 2012 9th International Conference on Fuzzy Systems and Knowledge Discovery, pp. 650–655. IEEE (2012)

24. Tang, J., Wang, K.: Personalized top-n sequential recommendation via convolutional sequence embedding. In: Proceedings of WSDM, pp. 565–573 (2018)

25. Tishby, N., Pereira, F.C., Bialek, W.: The information bottleneck method, pp. 368–377 (1999)

26. Vaswani, A., et al.: Attention is all you need. In: Proceedings of NIPS, pp. 5998–6008 (2017)

27. Wang, Z., Liu, H., Du, Y., Wu, Z., Zhang, X.: Unified embedding model over heterogeneous information network for personalized recommendation. In: Proceedings of the 28th International Joint Conference on Artificial Intelligence, pp. 3813–3819. AAAI Press (2019)

28. Wei, Y., Wang, X., Nie, L., He, X., Hong, R., Chua, T.S.: MMGCN: multi-modal graph convolution network for personalized recommendation of micro-video. In: Proceedings of the 27th ACM International Conference on Multimedia, pp. 1437–1445 (2019)

29. Zhang, T., et al.: Feature-level deeper self-attention network for sequential recommendation. In: Proceedings of IJCAI, pp. 4320–4326 (2019)
30. Zhu, N., Cao, J., Liu, Y., Yang, Y., Ying, H., Xiong, H.: Sequential modeling of hierarchical user intention and preference for next-item recommendation. In: Proceedings of the 13th International Conference on Web Search and Data Mining, pp. 807–815 (2020)

Fully Utilizing Neighbors
for Session-Based Recommendation
with Graph Neural Networks

Xingyu Zhang[1,2](✉) and Chaofeng Sha[1,2]

[1] School of Computer Science, Fudan University, Shanghai, China
{xingyuzhang19,cfsha}@fudan.edu.cn
[2] Shanghai Key Laboratory of Data Science, Fudan University, Shanghai, China

Abstract. Session-based recommendation (SBR) has attracted many researchers due to its highly practical value in many online services. Recently, graph neural networks (GNN) are widely applied to SBR due to their superiority on learning better item and session embeddings. However, existing GNN-based SBR models mainly leverage direct neighbors, lacking efficient utilization of multi-hop neighbors information. To address this issue, we propose a multi-head graph attention diffusion layer to utilize multi-hop neighbors information. Furthermore, we spot the information loss of local contextual aggregation in existing GNN-based models. To handle this problem, we propose positional graph attention aggregation layer to exploit direct neighbors information. Combining these two designs, we propose a novel model named FUN-GNN (Fully Utilizing Neighbors with Graph Neural Networks) for session-based recommendation. Experiments on three real datasets demonstrate its superiority over existing state-of-the-art baselines in terms of Precision and Mean Reciprocal Rank metrics.

Keywords: Session-based recommendation · Graph neural network · Graph attention diffusion

1 Introduction

Session-based recommendation (SBR) systems are pervasive in various E-commerce and web applications where users' behaviors are recorded in anonymous sessions. Given current sessions, the SBR task aims at predicting the next item that users might be interested in. Due to the sequential or temporal patterns inherent in sessions, we could model users' behaviors based on Markov chain as FPMC did [15]. Inspired by the resurgence of deep learning, many neural SBR models have been proposed which learn sequential patterns of session prefix by employing recurrent neural network [3,8] or attention mechanism [11], thanks to the analogy between next-item prediction in SBR and next word prediction in language models.

Recently, following the successful applications of graph neural networks (GNN) on a broad spectrum, they are exploited to build SBR models [1,14,21–24]. In these GNN-based SBR models, firstly sessions are transformed into subgraphs or graphs which could model the relation and interaction between items

© The Author(s), under exclusive license to Springer Nature Switzerland AG 2022
A. Bhattacharya et al. (Eds.): DASFAA 2022, LNCS 13246, pp. 36–52, 2022.
https://doi.org/10.1007/978-3-031-00126-0_3

in sessions. Then GNNs such as gated graph neural networks (GGNN) [10] or graph attention networks (GAT) [18] are used to learn item and session embeddings. The outperformance of these GNN-based SBR models is demonstrated in the literatures [1,14,21–24].

However, there exist some limitations with these GNN-based SBR models. The first limitation is the ineffective utilization of multi-hop information. Simply stacking GNN layers and increasing iterations of message passing usually suffer from over-smoothing problem [9]. Previous GNN-based SBR models strive to capture multi-hop information via applying attention mechanism over all previous nodes [1,22,23] for each node. However, they do not distinguish the influence of items/nodes with different hops and simply cast the prefix of sessions as an item set, which might ignore sequential information within sessions (e.g. item transitions, item orders, relative positions).

The second limitation is the information loss of local contextual aggregation. Previous studies [2,22–24] apply GGNN [10] to aggregate neighbors information with a linear convolution, treating each neighbor the same. The works of [14,21] add a self-loop to the constructed graph and apply graph attention mechanism to aggregate neighbors information, but failed to utilize relative position information of neighbors. In LESSR [1], a EOPA layer is used to aggregate neighbors information via a GRU to capture relative position information, but fails to utilize semantic relationship (i.e. item similarity) between neighbors.

To solve above limitations, we propose a novel model named FUN-GNN (Fully Utilizing Neighbors with Graph Neural Networks) for session-based recommendation, which consists of a PGAA layer to utilize direct neighbors and a MGAD layer to utilize multi-hop neighbors. The design of Multi-head Graph Attention Diffusion (MGAD) layer is inspired by MAGNA [19], which attends multi-hop neighbors to capture the long-range patterns in sessions and distinguish the influence of different nodes. The Positional Graph Attention Aggregation (PGAA) layer uses the position embedding and attention mechanism to aggregate the local contextual information. The performance of FUN-GNN is demonstrated through extensive experiments on three public datasets, which are accompanied by ablation studies of each component. The main contributions of our paper are summarized as follows:

- We adopt the graph attention diffusion (GAD) to capture long-range dependencies between items in sessions and distinguish the influence of nodes with different hops. We propose an early stopping strategy for GAD, which is essential when diffusing attentions on session graphs.
- We employ position embedding to highlight the relative position between items and attention mechanism to aggregate contextual features for each node.
- We propose a novel SBR model FUN-GNN combining MGAD and PGAA layer. We conduct extensive experiments on three real datasets. The results demonstrate the superiority of FUN-GNN over existing state-of-the-art baselines.

2   Related Work

SBR (Session-based Recommendation) is a typical task in the field of recommendation system, which predicts next item based on previous sequence in a given session. Early SBR systems mainly adopt models from general recommendation systems, such as CF (Collaborative Filtering). For instance, Item-based KNN proposed by [16] can be adapted for SBR by simply recommending next item that is most similar with the last item in a given session. Session-based KNN proposed by [4] recommends items based on the cosine similarity between session binary vectors. However, these methods only consider co-occurrence patterns of items and sessions, ignoring the sequential information of items within sessions.

Towards this end, Markov chain-based models are adapted in SBR systems to capture sequential information. For instance, [15] propose FPMC which combines matrix factorization and Markov chain to learn user's general taste and sequential behavior respectively. Since the user profile is not available, FPMC is then adopted to SBR while ignoring the user latent representation.

Deep learning based models are popular in recent years with RNN's success on sequence modeling. [3] first applies GRU, a special form of RNN, to session-based recommendation task, which is further extended by [17] to boost recommendation performance via data augmentation. NARM [8] stacks GRU as the encoder and applies attention mechanism to capture both user's sequential behavior and main purpose. Besides RNN, STAMP [11] proposes a short-term attention priority model, which uses attention layer instead of GRU layer as the encoder to effectively capture user's long and current-term preferences.

More recently, graph neural networks (GNNs) have been widely used in many tasks to deal with unstructured data. SR-GNN [22] first applies GNN to session-based recommendation, which encodes session sequence to session graph with edges representing item transitions. SR-GNN then applies gated graph neural network (GGNN) [10] to learn item embeddings on the session graph. Many extensions of SR-GNN are also proposed [2,24]. TAGNN [24] extends SR-GNN with target-aware mechanism, which generates specific session representation for each target item. Based on SR-GNN, NISER+ [2] applies embedding normalization for both item and session embedding to avoid popularity bias on item predicting, and incorporates position embedding to improve recommendation performance. Besides GGNN, many other graph neural networks are proposed to deal with SBR task [1,14]. FGNN [14] proposes WGAT, which transforms session to weighted directed graph with edges representing the frequency of item transitions, and employs multi-layer graph attention network (GAT) [18] to extract item features. LESSR [1] proposes EOPA and SGAT: EOPA considers the edge orders of session graph, and captures neighbors information with a GRU aggregator; SGAT converts a session to a shortcut graph where each node is fully connected by its previous nodes, and captures long-range dependencies via attention mechanism. Moreover, many other graph construction methods are also explored for session sequence [13,21]. However, these GNN-based models have limitations to effectively utilize multi-hop neighbors due to the over-smoothing problem. Thus, attention-based readout function [2,13,14,22,24] or attention

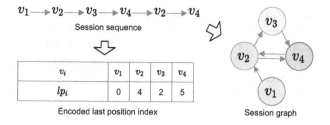

Fig. 1. An example of session graph construction with last position index encoded.

layer on complete graph [1,23] is employed to capture long-range dependencies. These attention-based methods simply cast all previous nodes as a node set, which does not consider the pattern of message passing in session graphs.

Our work focus on solving the limitations of both local and long-range dependencies utilization in existing methods. Our proposed MGAD is inspired by [7,19], which provide deep insights into graph diffusion. As will be demonstrated in ablation study of MGAD, the graph attention diffusion is also capable of improving recommendation performance when applied to session graphs.

3 Preliminaries

In this section, we first formulate the definition of session-based recommendation and give the notations (Sect. 3.1), and then introduce the construction of session graph (Sect. 3.2). Finally, we describe Graph Attention Diffusion (GAD) process which is used in the MGAD layer of our model (Sect. 3.3).

3.1 Problem Definition

Session-based recommendation aims to predict the next item on previous user interactions in an anonymous session. We formulate this problem as below.

Let $\mathcal{V} = \{v_1, v_2, v_3, ..., v_{|\mathcal{V}|}\}$ denote the item set consisting of all unique items involved in all sessions. The anonymous session is then defined as a sequence of items $S = [v_{s,1}, v_{s,2}, v_{s,3}, ..., v_{s,n}]$, where $v_{s,i}$ is ordered by timestamp and n is the length of S. The objective of this problem is to predict the next item $\hat{v}_{s,n+1}$ based on the previously recorded item sequence. Following [3,8,22], our model generates a probability distribution $\hat{y} = \{\hat{y}_1, ..., \hat{y}_{|\mathcal{V}|}\}$ over all items, where $\hat{y}_i = \hat{p}(v_{s,n+1} = v_i|S)$ denotes the probability of the next item v_i.

3.2 Session Graph Construction

Following the construction method of [22], we transform each session sequence S into a directed graph $G_s = (V_s, E_s)$. The node set V_s contains all the unique items in the corresponding session sequence S. Each node v_i in V_s is attributed by the initial item embedding $x_i^{(0)} \in \mathbb{R}^d$, where d denotes the embedding size.

Each edge $(v_i, v_j) \in E_s$ represents an transition $v_i \rightarrow v_j$ meaning that v_j is clicked after v_i. We use $\mathcal{N}_j = \{v_i | (v_i, v_j) \in E_s\}$ to denote the direct neighbor set of v_j. Moreover, the last position index (i.e. the position of last occurrence) is encoded as lp_i for each node v_i, which will be used in both PGAA layer (Sect. 4.1) and MGAD layer (Sect. 4.2). Figure 1 illustrates an example of the session graph construction and encoded last position index.

3.3   Graph Attention Diffusion

In this subsection we describe the Graph Attention Diffusion (GAD) [19], which is used in MGAD layer to incorporate multi-hop neighboring context information. GAD consists of three steps: (1) It first computes attention scores on all edges. (2) Then it uses diffusion strategy to compute diffused attention scores between any node pair based on the edge attention scores. (3) Finally GAD aggregates contextual features from multi-hop neighbors into each node representation via the diffused attention scores.

Graph Attention Computation on All Edges. Given the input features $H = [h_1, h_2, ..., h_{n_s}]$, where $h_i \in \mathbb{R}^d$ and n_s denotes number of unique nodes in the graph, we first apply a shared parameter W_h to transform input features of each node to higher-level features. Then we use a shared attentional vector a to obtain attention coefficients e_{ij} for each directed edge:

$$e_{ij} = a^\top [W_h h_i || W_h h_j] \quad \text{s.t.} \quad j \in \mathcal{N}_i \tag{1}$$

where $||$ denotes the concatenation operator and $W_h \in \mathbb{R}^{d \times d}, a \in \mathbb{R}^{2d}$ are learnable parameters. Then we use LeakyReLU unit with negative input slope 0.2 to activate e_{ij}, followed by a softmax function to generate attention matrix A:

$$A_{ij} = \begin{cases} \frac{\exp(\text{LeakyReLU}(e_{ij}))}{\sum_{k \in \mathcal{N}_i} \exp(\text{LeakyReLU}(e_{ik}))}, & \text{if } j \in \mathcal{N}_i \\ 0, & \text{otherwise} \end{cases} \tag{2}$$

Graph Attention Diffusion to Multi-hop Neighbors. The attention matrix A only considers 1-hop connected nodes, we now apply graph diffusion and generate a diffusion matrix $\mathcal{A}$, which is defined as follows:

$$\mathcal{A} = \sum_{i=0}^{\infty} \theta_i A^i \quad \text{s.t.} \quad \theta_i > 0, \ \theta_i > \theta_{i+1}, \text{ and } \sum_{i=0}^{\infty} \theta_i = 1. \tag{3}$$

where θ_i represents the weighting coefficient and A^i represents powers of attention matrix which gives the attention score between two nodes combining all the paths of length i. This diffusion process creates attention shortcut and is capable of increasing receptive field of attention mechanism [19]. In our implementation we use Personalized PageRank (PPR) procedure [12] and set $\theta_i = \alpha(1 - \alpha)^i$ where the hyper-parameter $\alpha \in (0, 1)$ indicates the decay rate of the diffusion weight θ_i.

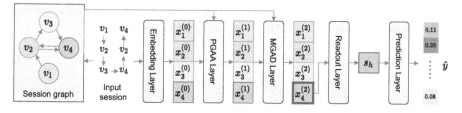

Fig. 2. The outline of the proposed model FUN-GNN. Given an input session, we first construct the session graph and generate initial node representation $x^{(0)}$ via embedding layer. Based on the graph, we apply PGAA layer to aggregate contextual information for each node and MGAD layer to capture long-range preferences. The readout layer takes the last node representation (with bold border) as input and generates the session embedding s_h. The prediction layer finally outputs the probability distribution of next item over all items $\hat{y}$.

Features Aggregation. Once obtained, the row normalized diffusion matrix $\mathcal{A}$ is used to aggregate features from multi-hop neighbors. Formally, the representation of node v_i is updated as follows:

$$\bar{h}_i = \sum_{j=1}^{n_s} \mathcal{A}_{ij} h_j \tag{4}$$

4   Methodology

In this section, we detail the proposed model FUN-GNN (Sects. 4.1 and 4.4). Figure 2 illustrates the architecture of FUN-GNN.

Based on the constructed session graph (Sect. 3.2) from the input session, FUN-GNN first uses a PGAA layer (Sect. 4.1) to capture rich local contextual information which considers the positional relationships between neighbors. FUN-GNN then employs a MGAD layer (Sect. 4.2) to capture long-range preferences behind multi-hop neighbors (item subsequences), which applies attention mechanism and diffusion strategy to utilize multi-hop neighbors information. The representation of the last item learnt by MGAD layer is fed to readout layer (Sect. 4.3) to generate the graph-level session embedding. Finally, the prediction layer (Sect. 4.4) makes predictions over all items based on the generated session embedding.

4.1   Positional Graph Attention Aggregation Layer

In this subsection, we describe the Positional Graph Attention Aggregation (PGAA) layer which takes the positional relationships between nodes into account. The PGAA layer consists of three steps:

- First, a shared position embedding matrix is used to utilize the position information. Inspired by the reversed position embedding used in [21], we use reversed last position index $n_s - lp_i$ for each node to obtain position embedding, which is then added with the initial item embedding:

$$x_i^p = x_i^{(0)} + p_{n_s - lp_i} \tag{5}$$

where $x_i^{(0)}$ and $p_{n_s - lp_i}$ denote the initial item embedding and the position embedding indexed by the reversed last position $n_s - lp_i$ respectively.

- Then, the attention mechanism is employed to obtain a weighted aggregation x_i^{agg} which could reflect the influence of previous items on target item v_i:

$$x_i^{agg} = \sum_{j \in \mathcal{N}_i} \delta_{ij} x_j^p \quad \text{where } \delta_{ij} = q^\top \sigma(W_{tar} x_i^p + W_{src} x_j^p + b) \tag{6}$$

where $W_{tar}, W_{src} \in \mathbb{R}^{d \times d}$ and $b, q \in \mathbb{R}^d$ are learnable parameters.

- Finally a GRU cell (i.e. GRU with only 1 recurrence step) is used to update each node based on the aggregated information:

$$x_i^{(1)} = \text{GRU}(x_i^{agg}, x_i^p) \tag{7}$$

4.2   Multi-head Graph Attention Diffusion Layer

In this subsection, we describe Multi-head Graph Attention Diffusion (MGAD) layer based on the graph attention diffusion process introduced in Sect. 3.3. This layer consists of layer normalization and residual connection as the design of MAGNA [19]. Usually the constructed session graphs is smaller than those considered in [19], we employ an Attention Diffusion with Early Stopping (ADES) strategy when training our GNN model. Figure 3 illustrates the architecture of MGAD layer with ADES strategy.

Attention Diffusion with Early Stopping. Since computing the powers of attention (Eq. 3) is extremely expensive [6] especially for large graphs, [19] proposes an approximate computation method for diffusion matrix $\mathcal{A}$ which successfully optimize the complexity of attention computation to $O(|E|)$ as defined in Eq. (8):

$$Z^{(0)} = H, \quad Z^{(k)} = (1 - \alpha)\mathcal{A}Z^{(k-1)} + \alpha H \tag{8}$$

which converges to the value of $\mathcal{A}H$ with the iterations $\kappa \to \infty$, i.e. $\lim_{\kappa \to \infty} Z^{(\kappa)} = \mathcal{A}H$. For more detailed analysis of this approximate computation, please refer to [19]. However, this approximate method conflicts with the condition $\theta_\kappa < \theta_{\kappa-1}$ in Eq. 3 when $0 < \alpha < 0.5$. Intuitively this error can be ignored when κ is large enough or $\alpha > 0.5$, but resulting in higher compute time or lower prediction performance respectively. To handle this error, we initialize the item embeddings with $Z^{(0)} = \alpha H$.

However, each node in session graph has different number of predecessors, i.e. its previous items in session sequence. Simply using the same iterations κ for all nodes is inappropriate. For example, in a session $\{v_1 \to v_2 \to v_3 \to v_4\}$,

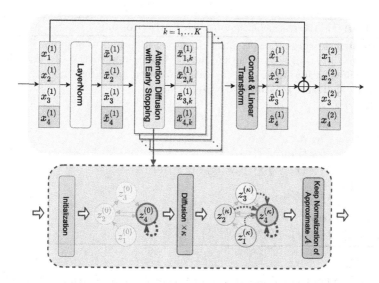

Fig. 3. The architecture of MGAD layer. The red dashed arrow in the bottom denotes the attention weights of $\mathcal{A}$ to v_4 through attention diffusion process.

after diffusion with $\kappa = 3$ and $\alpha = 0.2$, v_4 attends v_4, v_3, v_2, v_1 with weights of $0.2, 0.16, 0.128, 0.1024$, whereas v_3 attends v_3, v_2, v_1 with $0.2, 0.16, 0.2304$, which conflicts with the intuition that nearer neighbors should be more attended. This is because v_1 has no in-edges from other nodes, then the node representation is not updated during diffusion process. Thus, we modify the diffusion process with early stopping strategy as in Algorithm 1. The iterations for each node are limited by the minimum of its last position in current session and the hyperparameter κ. With the early stopping strategy, the diffusion iterations for v_3 decrease to 2, resulting in the attention weight 0.128 to v_1 instead of 0.2304. Finally, to keep the normalization of each row of diffusion matrix $\mathcal{A}$, we scale each z_i^κ as line 9 of Algorithm 1. The effect of diffusion with early stopping strategy is discussed in Sect. 5.5.

We define the whole process, including Eqs. 1, 2 and Algorithm 1, as Attention Diffusion with Early Stopping (ADES):

$$\bar{h}_i = \text{ADES}(h_i|G, \Theta) \quad \text{where } \Theta = \{W_h, a\} \tag{9}$$

Multi-head Graph Attention Diffusion. After obtaining $x_i^{(1)}$ from the PGAA layer, we use ADES to enhance item representation learning. Following [19], we propose a MGAD layer including three steps:

- we first apply layer normalization to the input representation, i.e. $\tilde{x}_i^{(1)} = \text{LayerNorm}(x_i^{(1)})$.

Algorithm 1: Attention Diffusion with Early Stopping

Input: item features $H = [h_1, ..., h_{n_s}]$; last position indices of items
$lp = [lp_1, ..., lp_{n_s}]$; attention matrix A computed by (Eq. 2); attention
diffusion iterations κ; decay weight of attention diffusion α

Output: aggregated features $\bar{H} = [\bar{h}_1, ..., \bar{h}_{n_s}]$.

1 $Z^{(0)} = \alpha H$
2 **for** $k = 1 : \kappa$ **do**
3 　　**for** $i = 1 : n_s$ **do**
4 　　　**if** $k \le lp_i$ **then**
5 　　　　$z_i^{(k)} = (1 - \alpha) \sum_{j=1}^{n_s} A_{ij} z_j^{(k-1)} + \alpha h_i$
6 　　　**else**
7 　　　　$z_i^{(k)} = z_i^{(k-1)}$
8 **for** $i = 1 : n_s$ **do**
9 　　$\bar{h}_i = z_i^{(\kappa)} / \left(1 - (1 - \alpha)^{\min(lp_i, \kappa)+1}\right)$

- Then, the ADES is extended to multi-head version, which can attend information jointly from different subspaces at different viewpoints:.

$$\hat{x}_i^{(1)} = W_m(\|_{k=1}^K \bar{x}_{i,k}^{(1)}) \quad \text{where} \quad \bar{x}_{i,k}^{(1)} = \text{ADES}(\tilde{x}_i^{(1)} | G, \Theta_k) \tag{10}$$

where Θ_k denotes the learnable parameters of k-th head as in Eq. 9, $\|$ denotes the concatenation, and $W_m \in \mathbb{R}^{d \times Kd}$ is used to aggregate all diffused features from K heads.

- Finally, the residual operation is applied to output the final representation, i.e. $x_i^{(2)} = \hat{x}_i^{(1)} + x_i^{(1)}$.

4.3 Session Embedding Readout

In our proposed model, we take the embedding of the last item $x_{last}^{(2)}$ to generate session embedding, i.e. $s_h = W_s x_{last}^{(2)}$, where $W_s \in \mathbb{R}^{d \times d}$ is a learnable parameter.

4.4 Prediction and Training

Following [2], we employ cosine similarity between item and session embeddings to obtain the final score, i.e. $\tilde{y}_i = x_i^{(0)} \cdot s_h / \|x_i^{(0)}\| \|s_h\|$. Since cosine similarity is restricted to $[-1, 1]$, we apply temperature β to the softmax function and generate probability distribution over all candidate items:

$$\hat{y}_i = \frac{\exp(\beta \tilde{y}_i)}{\sum_{j=1}^{|\mathcal{V}|} \exp(\beta \tilde{y}_j)} \tag{11}$$

In training process, the loss function is defined as the cross-entropy between the prediction distribution $\hat{y}$ and the ground truth y encoded by one-hot vectors:

$$\mathcal{L}(y, \hat{y}) = -\sum_{i=1}^{|\mathcal{V}|} y_i \log(\hat{y}_i) + (1 - y_i) \log(1 - \hat{y}_i) \tag{12}$$

5 Experiments

In this section, we first describe the experimental settings including datasets, baselines, evaluation metrics and implementation details. Then we make extensive analysis on experimental results. The research questions are summarized as follows:

- RQ1: Does FUN-GNN achieve the new state-of-the-art performance over the three datasets?
- RQ2: Does Positional Graph Attention Aggregation layer successfully capture direct neighbors information and improve the performance?
- RQ3: Does Multi-head Graph Attention Diffusion layer successfully capture multi-hop neighbors information and improve the performance?
- RQ4: How does the diffusion iterations κ, diffusion decay α and embedding size d affect the performance?

5.1 Datasets

We conduct the experiments on three benchmark datasets: *Yoochoose*[1], *Diginetica*[2], *Retailrocket*[3]. The detail statistics are shown in Table 1.

Yoochoose is a challenge dataset published by the RecSys Challenge 2015. Following [3,8,11], we use sessions at last day as test set. Since the training set is quite large and [17] shows training on more recent fractions performs better, We follow [8,11,17,22] and sort the training sessions by the time-stamp of the last event and keep the most recent fraction 1/64 as the training set.

Diginetica is a challenge dataset obtained from the CIKM Cup 2016, whose transaction data is suitable for session-based recommendation. Following [8,11], we use sessions in last week as test set.

Retailrocket is a dataset published by an e-commerce company. We split each user's click stream data with event type *view* into different sessions at the interval of 30 min, and use sessions in last 2 weeks as test set.

We follow the same preprocessing method as [17,22]: (1) We filter out all session with length less than 2 and items appearing less than 5 times; (2) Each session $S = [v_{s,1}, v_{s,2}, ..., v_{s,n}]$ is split into $([v_{s,1}], v_{s,2}), ..., ([v_{s,1}, ..., v_{s,n-1}], v_{s,n})$ for both training and test set.

5.2 Baselines and Evaluation Metrics

We compare our proposed FUN-GNN with the following representative models:

- **S-POP** recommends top-k frequent items in the current session.
- **Item-KNN** [16] is a non-parametric method which recommend next item based on the similarity with the last item in a given session.

---

[1] http://2015.recsyschallenge.com/challenge.html.
[2] https://competitions.codalab.org/competitions/11161.
[3] https://www.kaggle.com/retailrocket/ecommerce-dataset.

Table 1. Statistics of datasets

Dataset	*Yoochoose*	*Diginetica*	*Retailrocket*
# event	557,248	982,961	1,082,370
# train	369,859	719,470	715,496
# test	55,898	60,858	60,588
# item	16,766	43,097	48,727
Avg length	6.16	5.12	6.74

- **FPMC** [15] is a hybrid model on the next-basket recommendation based on Markov chain.
- **GRU4Rec** [3] applies GRUs to model session sequence and make recommendation.
- **NARM** [8] employs attention mechanism to model user interactions and capture the user's main purpose in current session to make recommendation.
- **STAMP** [11] proposes a short-term attention/memory priority model which can simultaneously capture long-term and short-term interests of a session.
- **SR-GNN** [22] converts sessions into weighted directed graphs, and employs GGNN to obtain item embeddings.
- **NISER+** [2] uses $L2$ normalization to both item and session representations to handle popularity bias problem for long-tail items.
- **TAGNN** [24] introduces target-aware attention module to generate item-specific session embeddings.
- **LESSR** [1] proposes a new edge encoding schema based on the graph constructed from SR-GNN with a GRU aggregator to preserve the information of edge order. Then it constructs a shortcut graph where each node is fully connected by its previous nodes to capture long-range dependencies.

Following [1,3,8,11,22] we use **P@20**[4] and **MRR@20** as evaluation metrics.

5.3   Implementation Details

Our implementation is based on Deep Graph Library [20] and PyTorch framework[5]. All the hyper-parameters are tuned via grid search using last 10% of training set as validation set, choosen from following ranges: embedding size $d \in \{50, 100, ..., 300\}$, diffusion iterations $\kappa \in \{1, 2, 3, ..., 10\}$, diffusion decay $\alpha \in \{0.01, 0.05, 0.1, 0.15, ..., 0.7\}$, heads of MGAD $K \in \{1, 2, 4, 8\}$, softmax temperature $\beta \in \{1, 2, 3, ..., 20\}$.

Section 5.6 discusses the effect of the hyper-parameters on the proposed model. In our implementation, we set $\beta = 13, d = 200, \kappa = 5$ for all three

---

[4] Note that in SBR task, the target item is unique for each session, so the metrics Precision (used in [11,22]), Recall (used in [3,8]), and Hit Rate (used in [1,23]) use the same formula to obtain the results.

[5] Our implementation is available at https://github.com/xingyuz233/FUN-GNN.

Table 2. Overall Performance Comparison with Baselines

Datasets	Yoochoose		Diginetica		Retailrocket	
Metrics	P@20	MRR@20	P@20	MRR@20	P@20	MRR@20
S-POP	31.04	18.59	21.08	13.68	37.84	30.78
Item-KNN	52.73	22.30	37.28	12.00	27.74	11.67
FPMC	44.83	14.90	25.10	9.06	29.28	15.10
GRU4Rec	60.84	25.69	34.78	10.92	48.32	26.92
NARM	68.32	28.63	49.70	16.17	54.46	33.87
STAMP	68.74	29.67	45.64	15.48	50.42	30.88
SR-GNN	70.57	30.94	50.73	17.59	58.07	32.84
LESSR	69.56	30.67	51.12	17.67	58.38	34.11
TAGNN	71.02	31.12	51.31	18.03	58.75	33.35
NISER+	<u>71.27</u>	<u>31.59</u>	<u>53.36</u>	<u>18.72</u>	<u>60.99</u>	<u>35.36</u>
FUN-GNN	**72.19**	**32.14**	**54.93**	**19.11**	**62.97**	**36.45**

datasets. α is set to $0.1, 0.3, 0.2$ for *Yoochoose, Diginetica, Retailrocket*. K is set to $1, 2, 4$ for *Yoochoose, Diginetica, Retailrocket*.

The Adam [5] optimizer is applied to optimize all the learnable parameters, with batch size of 512 and initialized learning rate of 0.001 which decays by 0.1 every 3 epochs. Furthermore, $L2$ penalty is set to 10^{-5} to avoid overfitting. For baselines, we use the optimal hyper-parameter settings in their original papers.

5.4 Overall Comparison (RQ1)

Results of the proposed FUN-GNN and other baselines on three datasets are shown in Table 2. We have the following observations:

Conventional baselines including S-POP, Item-KNN, FPMC under-perform all deep learning based baselines except GRU4Rec on the three datasets, which indicates the limitation of conventional methods to capture session information. S-POP performs the worst among all baselines, because it simply recommends top-N popular items in a given session which ignores any sequential information; FPMC performs better than S-POP, as it utilizes Markov chain and matrix factorization; Item-KNN achieves the best performance among all the convenient baselines, but is still not competitive compared to deep learning based methods. The reason might be that Item-KNN only considers the last item and ignores any global information of the given session.

Compared with conventional methods, deep learning based models achieve better performance on the three datasets, which demonstrates the effect of deep learning in session-based recommendation. Note that even GRU4Rec, the worst one amongst deep learning based models, can outperform conventional models on *Yoochoose* and *Diginetica* datasets, which shows the capability of RNN to model session sequence. NARM and STAMP outperform GRU4Rec, which shows the capability of attention mechanism to capture long-term preferences.

GNN-based models (SR-GNN, TAGNN, LESSR, NISER+) achieve better performance among all neural-network-based SBR models, which demonstrate graph structure is more suitable than sequence structure for modeling sessions. Specifically, TAGNN outperforms SR-GNN through enhancing target-aware session representation. NISER+ outperforms SR-GNN with extensions of position embedding and $L2$ normalization of items' and sessions' embedding, which could overcome the popular bias problem and gain much higher performance over SR-GNN. LESSR outperforms SR-GNN only on diginetica dataset which indicates the limitation of simply stacking different GNN layers.

Overall, FUN-GNN achieves sufficient gains over all baselines on all datasets. The out-performance of FUN-GNN is due to our designs: (1) applies attention mechanism to capture the relationship between directly connected items, (2) uses diffusion strategy to utilize information of multi-hop neighbors.

5.5   Ablation Study

Impact of PGAA Layer (RQ2). We examine the impact of PGAA layer compared with previous local context aggregation methods including (1) GGNN [10] used in [2,22–24], with one-step recurrence to capture the information of directly connected neighbors, (2) WGAT from [14] which extends the attention mechanism of GAT [18] by treating the frequency of item transitions as edge weight and applying it to computation of attention coefficients, (3) EOPA from [1], using a GRU to aggregate information from order-preserving neighboring nodes. To analyze the effectiveness of position embedding, we compare it with (1) forward PE: position embedding indexed by sequential order, where the position of each node v_i is simply indexed by lp_i. (2) w/o PE: without position embedding. RQ2 of Table 3 shows the results on three datasets.

WGAT performs the worst among all methods, which indicates its limitation to capture direct neighbors information. EOPA outperforms GGNN which might be due to the fact that EOPA takes the relative order of edges into account. The above two observations are also consistent with the findings of [1]. The proposed PGAA significantly outperforms both GGNN and EOPA, indicating the effectiveness of position embedding and attention aggregation. Note that even without position embedding, the model (i.e. w/o ES) also outperforms GGNN in most cases, which demonstrates the effectiveness of attention-based aggregation method. In contrast to reversed PE used in our model, forward PE does not gain significant improvement over w/o PE. This is because forward position embedding cannot capture the distance from each item to the target item, which also echoes the finding of [21].

Impact of MGAD Layer (RQ3). In this subsection, we study whether MGAD layer can capture global dependencies. We first examine the impact of attention diffusion early stopping strategy and residual connection. Then we compare MGAD with GNN layers from other state-of-the-art GNN-based SBR models, including: (1) SGAT from LESSR [1], which aggregates information from

Table 3. Ablation study of PGAA layer and MGAD layer

Datasets		*Yoochoose*		*Diginetica*		*Retailrocket*	
Metrics		P@20	MRR@20	P@20	MRR@20	P@20	MRR@20
Ours	PGAA+MGAD	**72.19**	**32.14**	**54.93**	**19.11**	**62.97**	**36.45**
RQ2	forward PE	71.75	31.82	54.61	18.93	62.55	35.90
	w/o PE	71.71	31.62	54.58	18.93	62.73	36.00
	GGNN+MGAD	71.71	31.40	54.39	18.94	62.44	35.85
	EOPA+MGAD	71.73	31.75	54.41	18.98	62.66	36.25
	WGAT+MGAD	70.92	30.41	53.42	18.97	61.86	36.06
	only MGAD	71.54	31.29	54.25	18.70	62.00	34.81
RQ3	w/o ES	71.82	31.58	54.63	19.03	62.59	36.07
	w/o residual	71.49	30.98	54.76	18.74	62.36	35.52
	PGAA+SGAT	71.57	31.47	54.27	18.97	62.20	35.89
	PGAA+SAN	71.60	31.45	54.30	18.80	62.10	35.70
	only PGAA	71.35	31.73	53.86	18.76	62.19	35.95

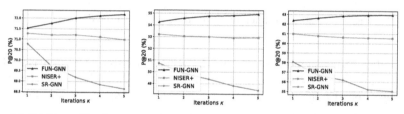

(a) P@20 on *Yoochoose*   (b) P@20 on *Diginetica*   (c) P@20 on *Retailrocket*

Fig. 4. P@20 with different iterations of attention diffusion on three datasets.

all its previous nodes for each node via attention mechanism, (2) SAN from GC-SAN [23], which aggregates information from all its previous nodes for each node by stacking self-attention layer and feed-forward layer with residual connection. RQ3 of Table 3 shows the results.

We can see that the combination of attention diffusion strategy with early stopping and residual connection achieves significant gains on the three datasets. MGAD consistently outperforms SGAT and SAN, demonstrating its effectiveness to capture long-range dependencies within the sessions.

5.6   Hyper-parameter Analysis (RQ4)

Analysis of Iterations κ.   Figure 4 reports the effect of iterations κ on the P@20 metric of FUN-GNN. We also compare it with SR-GNN and NISER+, with the iterations κ defined as the recurrence steps of GGNN. The result shows that FUN-GNN performs better when $\kappa > 1$, indicating its superiority to utilize multi-hop neighbors information. In contrast, the performance of GGNN-based

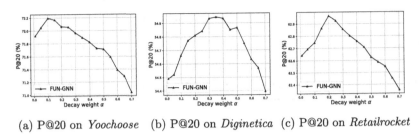

(a) P@20 on *Yoochoose*   (b) P@20 on *Diginetica*   (c) P@20 on *Retailrocket*

Fig. 5. P@20 with different decay weight of attention diffusion on three datasets

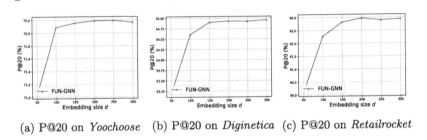

(a) P@20 on *Yoochoose*   (b) P@20 on *Diginetica*   (c) P@20 on *Retailrocket*

Fig. 6. P@20 with different embedding size on three datasets

SBR models drops significantly when the recurrence step size increases. The reason might be that these models suffer from over-smoothing problem with recurrence step size increased and fail to utilize multi-hop neighbors effectively.

Analysis of Decay Weight α. Figure 5 illustrates the effect of decay weight α on the P@20 metric of FUN-GNN. The optimal α is nearly at $\lambda = 0.2$, only with a slight variation among different datasets. The performance drops significantly with larger $\alpha > 0.5$. These two observations also echoes the findings of [7,19].

Analysis of Embedding Size d. Figure 6 shows the effect of embedding size d on the P@20 metric of FUN-GNN. We observe that increasing the embedding size generally improves the performance. This is because larger embedding size increases the model's capacity to learn item representations. Moreover, the P@20 of FUN-GNN eventually stabilizes with $d \geq 200$ on three datasets.

6   Conclusion

This paper mainly focuses on the limitations of neighbors utilizing in existing GNN-based models for session-based recommendation. We propose a novel model FUN-GNN, including a PGAA layer and a MGAD layer to utilize direct and multi-hop neighbors respectively. Experiments conducted on three benchmarks show that FUN-GNN achieves significant gains over state-of-the-art baselines.

Acknowledgements. This work is supported by National Natural Science Foundation of China under Grant No. 71772042.

References

1. Chen, T., Wong, R.C.: Handling information loss of graph neural networks for session-based recommendation. In: KDD, pp. 1172–1180. ACM (2020)
2. Gupta, P., Garg, D., Malhotra, P., Vig, L., Shroff, G.: NISER: normalized item and session representations with graph neural networks. CoRR abs/1909.04276 (2019)
3. Hidasi, B., Karatzoglou, A., Baltrunas, L., Tikk, D.: Session-based recommendations with recurrent neural networks. In: ICLR (Poster) (2016)
4. Jannach, D., Ludewig, M.: When recurrent neural networks meet the neighborhood for session-based recommendation. In: RecSys, pp. 306–310. ACM (2017)
5. Kingma, D.P., Ba, J.: Adam: a method for stochastic optimization. In: ICLR (Poster) (2015)
6. Klicpera, J., Bojchevski, A., Günnemann, S.: Predict then propagate: graph neural networks meet personalized PageRank. In: ICLR (Poster) (2019). OpenReview.net
7. Klicpera, J., Weißenberger, S., Günnemann, S.: Diffusion improves graph learning. In: NeurIPS, pp. 13333–13345 (2019)
8. Li, J., Ren, P., Chen, Z., Ren, Z., Lian, T., Ma, J.: Neural attentive session-based recommendation. In: CIKM, pp. 1419–1428. ACM (2017)
9. Li, Q., Han, Z., Wu, X.: Deeper insights into graph convolutional networks for semi-supervised learning. In: AAAI, pp. 3538–3545. AAAI Press (2018)
10. Li, Y., Tarlow, D., Brockschmidt, M., Zemel, R.S.: Gated graph sequence neural networks. In: ICLR (Poster) (2016)
11. Liu, Q., Zeng, Y., Mokhosi, R., Zhang, H.: STAMP: short-term attention/memory priority model for session-based recommendation. In: KDD, pp. 1831–1839. ACM (2018)
12. Page, L., Brin, S., Motwani, R., Winograd, T.: The PageRank citation ranking: bringing order to the web. Technical report, Stanford InfoLab (1999)
13. Pan, Z., Cai, F., Chen, W., Chen, H., de Rijke, M.: Star graph neural networks for session-based recommendation. In: CIKM, pp. 1195–1204. ACM (2020)
14. Qiu, R., Li, J., Huang, Z., Yin, H.: Rethinking the item order in session-based recommendation with graph neural networks. In: CIKM, pp. 579–588. ACM (2019)
15. Rendle, S., Freudenthaler, C., Schmidt-Thieme, L.: Factorizing personalized Markov chains for next-basket recommendation. In: WWW, pp. 811–820. ACM (2010)
16. Sarwar, B.M., Karypis, G., Konstan, J.A., Riedl, J.: Item-based collaborative filtering recommendation algorithms. In: WWW, pp. 285–295. ACM (2001)
17. Tan, Y.K., Xu, X., Liu, Y.: Improved recurrent neural networks for session-based recommendations. In: DLRS@RecSys, pp. 17–22. ACM (2016)
18. Velickovic, P., Cucurull, G., Casanova, A., Romero, A., Liò, P., Bengio, Y.: Graph attention networks. In: ICLR (Poster) (2018). OpenReview.net
19. Wang, G., Ying, R., Huang, J., Leskovec, J.: Multi-hop attention graph neural networks. In: IJCAI, pp. 3089–3096 (2021). ijcai.org
20. Wang, M., et al.: Deep graph library: a graph-centric, highly-performant package for graph neural networks. arXiv preprint arXiv:1909.01315 (2019)
21. Wang, Z., Wei, W., Cong, G., Li, X., Mao, X., Qiu, M.: Global context enhanced graph neural networks for session-based recommendation. In: SIGIR, pp. 169–178. ACM (2020)
22. Wu, S., Tang, Y., Zhu, Y., Wang, L., Xie, X., Tan, T.: Session-based recommendation with graph neural networks. In: AAAI, pp. 346–353. AAAI Press (2019)

23. Xu, C., et al.: Graph contextualized self-attention network for session-based rec-
    ommendation. In: IJCAI, pp. 3940–3946 (2019)
24. Yu, F., Zhu, Y., Liu, Q., Wu, S., Wang, L., Tan, T.: TAGNN: target attentive
    graph neural networks for session-based recommendation. In: SIGIR, pp. 1921–
    1924. ACM (2020)

Inter- and Intra-Domain Relation-Aware Heterogeneous Graph Convolutional Networks for Cross-Domain Recommendation

Ke Wang[1,2], Yanmin Zhu[1,2(✉)], Haobing Liu[1], Tianzi Zang[1,2],
Chunyang Wang[1,2], and Kuan Liu[1,2]

[1] Department of Computer Science and Engineering, Shanghai Jiao Tong University,
Shanghai, China
{onecall,yzhu,liuhaobing,zangtianzi,wangchy,lk_shjd}@sjtu.edu.cn
[2] State Key Laboratory of Satellite Ocean Environment Dynamics, Second Institute
of Oceanography, MNR, Hangzhou, China

Abstract. Cross-domain recommendation aims to model representations of users and items with the incorporation of additional knowledge from other domains, so as to alleviate the data sparsity issue. While recent studies demonstrate the effectiveness of cross-domain recommendation systems, there exist two unsolved challenges: (1) existing methods focus on transferring knowledge to generate shared factors implicitly, which fail to distill domain-shared features from explicit cross-domain correlations; (2) The majority of solutions are unable to effectively fuse domain-shared and domain-specific features. To this end, we propose Inter- and Intra-domain **R**elation-aware **C**ross-**D**omain **R**ecommendation framework (I^2RCDR) to explicitly learn domain-shared representations by capturing high-order inter-domain relations. Specifically, we first construct a cross-domain heterogeneous graph and two single-domain heterogeneous graphs from ratings and reviews to preserve inter- and intra-domain relations. Then, a relation-aware graph convolutional network is designed to simultaneously distill domain-shared and domain-specific features, by exploring the multi-hop heterogeneous connections across different graphs. Moreover, we introduce a gating fusion mechanism to combine domain-shared and domain-specific features to achieve dual-target recommendation. Experimental results on public datasets show that the effectiveness of the proposed framework against many strong state-of-the-art methods.

Keywords: Cross-domain recommendation · Inter-domain relations · Relation-aware graph convolutional network · Gating mechanism

1 Introduction

Recommendation systems learn representations from interactions between users and items. Many traditional methods [1,11] leverage historical feedbacks (e.g.,

A. Bhattacharya et al. (Eds.): DASFAA 2022, LNCS 13246, pp. 53–68, 2022.
https://doi.org/10.1007/978-3-031-00126-0_4

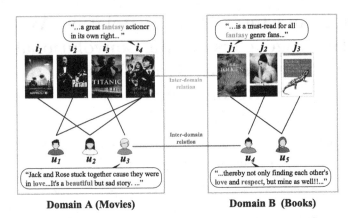

Fig. 1. An illustrative example of inter-domain relations for CDR systems. Both the reviews of items i_4 and j_1 are about the fantasy genre, while both the user u_3 and u_4 commented on the love theme in their reviews.

ratings) to capture users' preferences on items. However, the user-item interactions are usually sparse, which makes recommendation systems unable to generate optimal representations of users who give few ratings. Therefore, recent efforts [3,17,31] merge extra information from auxiliary domains into target domains to build Cross-Domain Recommendation (CDR).

Existing CDR approaches can be divided into two categories, transferring knowledge in different directions. The first category transfers in a unidirectional way, which focuses on learning the features of users and items from an auxiliary domain and then transferring learned features to a target domain. The most popular unidirectional technique is EMCDR [20], which learns a mapping function to execute target-domain recommendation utilizing the two-stage embedding-and-mapping paradigm [7,32]. However, the unidirectional method could easily accumulate noise in the intermediate steps and fall into sub-optimal learning [30]. The second category transfers in a bidirectional way. Considering that each domain is considerably richer in particular sorts of information, some recent works leverage bidirectional knowledge transfer to improve the performance on both domains simultaneously. This category selects overlapped users (or items) as a bridge to extract shared factors or patterns, which are transferred between two domains to achieve dual-target CDR [12,16,29].

Unfortunately, the aforementioned methods implicitly transfer knowledge to generate shared factors, which fail to distill domain-shared features from explicit correlations between cross-domain users (or items). In CDR systems, there are many inter-domain user-user and item-item relations, which are domain independent and could be used to explicitly portray domain-shared features. For example, as shown in Fig. 1, the reviews of movie i_4 in domain A and book j_1 in domain B which contain similar content (highlighted in boldface) show that the two items both belong to the fantasy genre, so we think the two items have the same property and there exists the inter-domain item-item relation. Similarly, the users u_3 and u_4 have the same preference and there exists the inter-domain user-user relation. We argue that such inter-domain relations are beneficial for learning domain-shared features, which are ignored by existing approaches.

Moreover, most models are incapable of fusing domain-shared and domain-specific features efficiently. In fact, user's (or item's) features are comprised of domain-shared part and domain-specific part, which reveal the cross-domain consistency and single-domain peculiarity, respectively [15]. Recently, emerged models simply conduct combination operations over domain-shared and domain-specific features, such as concatenation or addition [5,17]. However, the linear integration makes these models unable to fully explore the nonlinear interaction between the two kinds of features.

In this paper, we aim to capture inter-domain relations associating different domains to explicitly learn domain-shared features of users and items. However, there are a few challenges. The first challenge is how to explicitly establish the inter-domain relations between users (or items) from different domains. The most of existing CDR methods only model the cross-domain interactions from distinct domains, which are insufficient. Data from different domains may have similar semantic relations. Book and movie as an example, contents on both domains have some common topics, e.g., genre, plot, scene. As such, a good CDR model should have the ability to capture correlated information across domains. The second challenge is how to respectively distill domain-shared and domain-specific features. Capturing inter-domain relations results in a reconsideration for the current CDR methods that implicitly model intra-domain relations to generate domain-shared features. We should distinguish inter-domain and intra-domain relations to separately distill domain-shared and domain-specific features. Third, an effective strategy should be used to fuse domain-shared and domain-specific features, aiming to adaptively assign weights to balance cross-domain consistency and single-domain peculiarity.

To tackle these challenges, we propose a graph-structured CDR framework, namely I^2RCDR, which can collectively model the high-order **I**nter- and **I**ntra-domain **R**elations for dual-target **C**ross-**D**omain **R**ecommendation. First, we construct a cross-domain heterogeneous graph from interactions (ratings and reviews) to preserve inter-domain relations (including user-user and item-item relations) and user-item relations within each domain. Motivated by the landmark research BERT [6], we apply the SentenceBERT [21], which fine-tunes the BERT model and can transform reviews into semantic embeddings, to model the inter-domain relations. Meanwhile, a single-domain heterogeneous graph for each domain is also designed to represent intra-domain relations (including user-user and item-item relations) and user-item relation. Second, we introduce a relation-aware graph convolutional network (GCN) to extract domain-shared and domain-specific features, which can not only explore the multi-hop heterogeneous connections between users and items but also inject relation property into embedding propagation by encoding multi-typed relations. Finally, to achieve dual-target recommendation, we propose a gating fusion mechanism that can share partial parameters to seamlessly combine domain-shared and domain-specific features. The main contributions of this paper are as follows:

- We explore inter-domain relations to learn domain-shared representations of users and items explicitly. Three heterogeneous graphs are constructed from ratings and reviews to preserve not only cross-domain consistency and single-domain peculiarity but also multi-typed relations.
- We propose a novel graph-structured CDR framework to jointly model the inter- and intra-domain relations by the relation-aware graph convolutional networks. A gating fusion mechanism is designed to combine domain-shared and domain-specific features for dual-target recommendation.
- We perform extensive experiments on real-world datasets to show the effectiveness of our approaches. The experimental results demonstrate that our approaches significantly outperform the state-of-the-art approaches through detailed analysis.

2   Related Work

2.1   Graph-Based Recommendation

The core idea of graph neural networks (GNNs) is to iteratively aggregate neighbors' features to update the target node feature via propagation mechanism. The mainstream paradigm, such as GCN [14] and GraphSage [9], transforms interactions into a user-item graph to explore the multi-hop connections. GC-MC [1] designs one graph convolutional layer to construct user's and item's embeddings. NGCF [23] explicitly encodes collaborative signals in graph convolutional network (GCN). DGCF [24] disentangles user intents to generate disentangled representations. Furthermore, He et al. [10] propose LightGCN to systematically study several essential components in GCN. However, only considering singular user-item relation makes GNNs impossible to describe users' various preferences.

GNNs can also be used to model heterogeneous graphs constructed from other data. Recent efforts have exhibited incredible performance on many recommendation tasks. GHCF [2] and KHGT [26] design multi-behavior graphs to model multi-typed interactive patterns between users and items. To address the cold-start problem [28], Liu et al. [18] integrate social relation and semantic relation into the user-item graph. Xu et al. [27] utilize interaction sequences to construct a user-subsequence graph and an item-item graph to model user and item similarity on behaviors. Wang et al. [22] incorporate knowledge graphs into sequential recommendation model to enhance user representations.

2.2   Cross-Domain Recommendation

A popular path of CDR systems concentrates on learning a mapping function to transfer knowledge across domains in the same space. Man et al. [20] first introduce the embedding and mapping framework (EMCDR) to achieve cross-domain recommendation step by step. Based on the EMCDR model, CDLFM [25] incorporates users rating behaviors to generate precise latent factors. RC-DFM [7] extend stacked denoising autoencoders to deeply fuse textual contents and ratings. DCDCSR [32] and SSCDR [13] optimize the CDR model to obtain an accurate mapping function. However, the two-stage embedding-and-mapping strategy prevents CDR systems from accomplishing global optimization.

Recently, some methods aim to distill domain-shared features as a bridge between domains and apply symmetrical frameworks to accomplish dual-target CDR. CoNet [12] designs collaborative cross networks to select valuable features for dual knowledge transfer across domains. ACDN [16] integrates users aesthetic features into the CoNet model to transfer domain independent aesthetic preferences. DDTCDR [15] design a latent orthogonal mapping function to capture domain-shared preferences. PPGN [29] and BiTGCF [17] treat overlapped users as shared nodes to construct user-item graphs from ratings, then use GCN to learn cross-domain high-order representations. DA-GCN [8] and π-Net [19] utilize the recurrent neural network and GCN to accomplish the cross-domain sequential recommendation. Differing from existing dual-target CDR works, our method can explicitly transfer knowledge across domains and distill domain-shared features and domain-specific features in a direct way.

3 Preliminaries

Given two specific domains A and B, where the set of users U (of size $M = |U|$) are fully overlapped, the set of items in domains A and B are defined as I (of size $N^A = |I|$) and J (of size $N^B = |J|$), respectively. $u \in U$ indexes a user, $i \in I$ (or $j \in J$) indexes an item from A (or B). The user-item interaction matrices are defined as $\boldsymbol{R}_A \in \mathbb{R}^{M \times N^A}$ in domain A and $\boldsymbol{R}_B \in \mathbb{R}^{M \times N^B}$ in domain B.

Definition 1: Single-domain Heterogeneous Graph. The single-domain graphs $\boldsymbol{G}^A = (\mathcal{V}^U \cup \mathcal{V}^I, \mathcal{E}^A, \mathcal{R}^A)$ and $\boldsymbol{G}^B = (\mathcal{V}^U \cup \mathcal{V}^J, \mathcal{E}^B, \mathcal{R}^B)$ retain the intra-domain relations and user-item relation in A and B, respectively.

Definition 2: Cross-domain Heterogeneous Graph. The cross-domain graph $\boldsymbol{G}^C = (\mathcal{V}^U \cup \mathcal{V}^I \cup \mathcal{V}^J, \mathcal{E}^C, \mathcal{R}^C)$ retains the inter-domain relations and user-item relations within each domain, where $\mathcal{V}^U$, $\mathcal{V}^I$, and $\mathcal{V}^J$ are sets of nodes indicating users, items in A and B, respectively.

Definition 3: Relations. Each edge $e \in \{\mathcal{E}^C, \mathcal{E}^A, \mathcal{E}^B\}$ is associated with the function $\psi(e) : \mathcal{E} \to \mathcal{R}$. $\mathcal{R}$ defines the relations indicating edge types. Formally, we define the type-specific relation $\mathcal{R}_r \in \mathcal{R}$, $r = 1, 2, ..., 8$. $\mathcal{R}_1 = \{\text{inter-domain user-user relation}\}$, $\mathcal{R}_2 = \{\text{inter-domain item-item relation}\}$, $\mathcal{R}_3 = \{\text{intra-domain user-user relation in A}\}$, $\mathcal{R}_4 = \{\text{intra-domain user-user relation in B}\}$, $\mathcal{R}_5 = \{\text{intra-domain item-item relation in A}\}$, $\mathcal{R}_6 = \{\text{intra-domain item-item relation in B}\}$, $\mathcal{R}_7 = \{\text{user-item relation in A}\}$, $\mathcal{R}_8 = \{\text{user-item relation in B}\}$. Particularly, $\mathcal{R}^C = \{\mathcal{R}_1, \mathcal{R}_2, \mathcal{R}_7, \mathcal{R}_8\}$, $\mathcal{R}^A = \{\mathcal{R}_3, \mathcal{R}_5, \mathcal{R}_7\}$, and $\mathcal{R}^B = \{\mathcal{R}_4, \mathcal{R}_6, \mathcal{R}_8\}$.

Problem Formulation. Our task is to simultaneously predict the probability $\hat{y}_{ui}^A$ and $\hat{y}_{uj}^B$ of unseen interaction between user u, item i in domain A, and item j in domain B, to enhance the accuracy for dual-target recommendation.

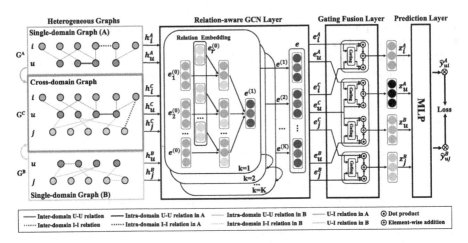

Fig. 2. An illustration of I^2RCDR structure. The U-I relation, U-U relation, I-I relation are short for user-item relation, user-user relation, and item-item relation, respectively.

4   Proposed Framework

Our proposed **Inter-** and **Intra-**domain **R**elation-aware heterogeneous graph convolutional networks for **C**ross-**D**omain **R**ecommendation (I^2RCDR) is an end-to-end learning framework. The main structure of our proposed framework is illustrated in Fig. 2.

4.1   Graph Construction and Embedding

Inter-domain Relation and Cross-Domain Graph. We first establish the inter-domain user-user relation. For user u, we concatenate all the user's reviews to generate a document d_u. We introduce SentenceBERT [21] to convert the documents d_u into a fixed-size text vector $\boldsymbol{D}_u$, which is formulated as:

$$\boldsymbol{D}_u = SentenceBERT(d_u). \tag{1}$$

To capture the user-user relation between two users across different domains, we compute the cosine similarity to generate the inter-domain semantic link. Particularly, for user u in domain A and u' in domain B, the existing probability score $Pr(u, u')$ of the link between u and u' is as follows:

$$Pr(u, u') = \varphi(\frac{\boldsymbol{D}_u \boldsymbol{D}_{u'}}{\|\boldsymbol{D}_u\| \|\boldsymbol{D}_{u'}\|}), \tag{2}$$

where $Pr(u, u')$ indicates the weight of an edge between u and u'. $\varphi(x) = max(0, x)$ is the ReLU function that normalizes the cosine similarities.

We calculate all the similarities between users from different domains and obtain the inter-domain user-user matrix $\boldsymbol{C}_U \in \mathbb{R}^{M \times M}$. Similarly, we can generate the inter-domain item-item relation matrix $\boldsymbol{C}_I \in \mathbb{R}^{N^A \times N^B}$. With all the matrices formally defined, we then describe the cross-domain graph G^C as shown below:

$$\tilde{A}^C = \begin{bmatrix} C_U & R_A^\top & R_B^\top \\ R_A & 0 & C_I^\top \\ R_B & C_I & 0 \end{bmatrix}, \tag{3}$$

where $\tilde{A}^C \in \mathbb{R}^{(M+N^A+N^B)\times(M+N^A+N^B)}$ is the adjacency matrix. R_A and R_B are user-item interaction matrices in domains A and B, respectively. $R_A^\top$, $R_B^\top$, and $C_I^\top$ are the transposed matrices.

Intra-domain Relation and Single-Domain Graph. We explore the user-user (or item-item) relation from domain A to establish the intra-domain user-user relation matrix A_U (or item-item relation matrix A_I), which computes the cosine similarities between users (or items) in domain A and gains the probability scores as shown in Eq. (2). We describe the single-domain graph G^A as shown below:

$$\tilde{A}^A = \begin{bmatrix} A_U & R_A^\top \\ R_A & A_I \end{bmatrix}, \tag{4}$$

where $\tilde{A}^A \in \mathbb{R}^{(M+N^A)\times(M+N^A)}$ is the adjacency matrix. Similarly, we obtain the adjacency matrix $\tilde{A}^B$ of graph G^B.

Dense Embedding. For user u, item i in domain A, and item j in domain B, we define them using one-hot encodings, namely $x_u \in \mathbb{R}^M$, $x_i \in \mathbb{R}^{N^A}$, and $x_j \in \mathbb{R}^{N^B}$. For relation $\mathcal{R}_r$ which indexes edge type, we also define $x_r \in \mathbb{R}^8$. Then, we map the one-hot encodings into dense embeddings as follows:

$$h_u = P_u x_u, h_i = P_i x_i, h_j = P_j x_j, h_r = P_r x_r, \tag{5}$$

where $P_u = \{P_u^C, P_u^A, P_u^B\} \in \mathbb{R}^{(M+N^A+N^B)\times d}$, $P_i = \{P_i^C, P_i^A\} \in \mathbb{R}^{(M+N^A)\times d}$, $P_j = \{P_j^C, P_j^B\} \in \mathbb{R}^{(M+N^B)\times d}$ and $P_r \in \mathbb{R}^{8\times d}$ are transformation matrices. d denotes the embedding size. $h_u = \{h_u^C, h_u^A, h_u^B\}$ indicate embeddings of u in G^C, G^A, and G^B, respectively. $h_i = \{h_i^C, h_i^A\}$ indicate embeddings of i in G^C and G^A. $h_j = \{h_j^C, h_j^B\}$ indicate embeddings of j in G^C and G^B.

4.2 Relation-Aware GCN Layer

Next, we introduce GCN to capture relation-aware multi-hop heterogeneous connections between users and items in different graphs. Considering different types of relations between nodes, we compose a neighboring node with respect to its relation to model a relation-aware target node. Specifically, we incorporate the relation embeddings into the propagation process via element-wise addition. We take user u as an example and formulate relation-aware GCN as follows:

$$e_u^{(k+1)} = \sigma \left(\sum_{(v_r)\in\mathcal{N}_u} \frac{1}{\sqrt{|\mathcal{N}_u||\mathcal{N}_{v_r}|}} W^{(k)} \left(e_{v_r}^{(k)} \oplus e_r^{(k)} \right) \right), \tag{6}$$

where $e_{v_r}^{(k)}$ and $e_r^{(k)}$ respectively denote the embeddings of node v_r and relation r after k layers propagation. σ is the nonlinear activate function and $\boldsymbol{W}^{(k)}$ is the weight matrix. $\frac{1}{\sqrt{|\mathcal{N}_u||\mathcal{N}_{v_r}|}}$ is a symmetric normalization constant. $\mathcal{N}_u$ and $\mathcal{N}_{v_r}$ denote the neighbors of u and v_r. v_r are the neighbors of user u under relation type $\mathcal{R}_r$. For example, in graph $\boldsymbol{G}^C$, the neighbors $v_r \in \{v_1, v_7, v_8\}$ correspond to three type-specific relations $\{\mathcal{R}_1, \mathcal{R}_7, \mathcal{R}_8\}$. $\oplus$ denotes the element-wise addition. Note that other composition ways like element-wise product can also be used, we leave it for future research. $e_r^{(k)}$ defines the relation embeddings, which is updated as follows:

$$e_r^{(k+1)} = \boldsymbol{W}_r^{(k)} e_r^{(k)}, \tag{7}$$

where $\boldsymbol{W}_r^{(k)}$ is a weight matrix which maps relations to the same space as nodes. Note that h_u, h_i, h_j and h_r are defined as initial embeddings $e_u^{(0)}$, $e_i^{(0)}$, $e_j^{(0)}$, and $e_r^{(0)}$ for node u, i, j, and relation $\mathcal{R}_r$, respectively.

To offer a easy-to-understand perspective of propagation mechanism, we use matrix form to describe this propagation process (equivalent to Eq. (6)):

$$E_u^{(k+1)} = \sigma \left(\hat{A}(E_{v_r}^{(k)} \oplus E_r^{(k)}) \boldsymbol{W}^{(k)} \right), \tag{8}$$

where $E_{v_r}^{(k)}$ and $E_r^{(k)}$ are the embeddings of v_r and corresponding relation $\mathcal{R}_r$ obtained after k steps of propagation. $\hat{A} = D^{-\frac{1}{2}} \tilde{A} D^{-\frac{1}{2}}$ denotes the symmetrically normalized matrix. $\tilde{A} \in \{\tilde{A}^C, \tilde{A}^A, \tilde{A}^B\}$ and D is the diagonal degree matrix of $\tilde{A}$.

After aggregating and propagating with K steps, we generate and stack multiple embeddings of user u, namely $\left\{ e_u^{(0)}, e_u^{(1)}, \cdots, e_u^{(K)} \right\}$. The combination of embeddings refined from different order neighbors can better indicate the features of user u. As such, we concatenate different embeddings with the following formula and get the final embeddings of user u.

$$e_u = e_u^{(0)} \| \cdots \| e_u^{(k)} \| \cdots \| e_u^{(K)}, \tag{9}$$

where $\|$ denotes concatenation operation. $e_u^{(k)}$ is the embeddings of user u with k steps, $k = 0, 1, 2, ..., K$. $e_u \in \mathbb{R}^{d'}$ is the final embeddings of user u, d' denotes the embedding size after K steps of propagation and concatenation.

In this layer, we can utilize the relation-aware GCN to generate not only domain-shared features e_u^C of user u in $\boldsymbol{G}^C$ but also domain-specific features e_u^A (or e_u^B) of user u in $\boldsymbol{G}^A$ (or $\boldsymbol{G}^B$). Similarly, we obtain domain-shared features e_i^C of item i in $\boldsymbol{G}^C$, domain-shared features e_j^C of item j in $\boldsymbol{G}^C$, domain-specific features e_i^A of item i in $\boldsymbol{G}^A$, and domain-specific features e_j^B of item j in $\boldsymbol{G}^B$.

4.3   Gating Fusion Layer

The relation-aware GCN can simultaneously model the inter- and intra-domain relations to distill domain-shared and domain-specific features. Therefore, the combination of two types of features needs to balance the cross-domain consistency and single-domain peculiarity. Motivated by the milestone work Gated Recurrent Unit (GRU) [4], we introduce the gating mechanism to differentiate the importance of domain-shared and domain-specific features for dual-target recommendation. To combine user's features e_u^C, e_u^A, and e_u^B, we use two neural gating units that share partial parameters to generate two combination features of domains A and B. The process is computed as:

$$G_u^A = \text{sigmoid}\left(V_u^C e_u^C + V_u^A e_u^A\right), \tag{10}$$

$$G_u^B = \text{sigmoid}\left(V_u^C e_u^C + V_u^B e_u^B\right), \tag{11}$$

$$z_u^A = G_u^A \odot e_u^C + (1 - G_u^A) \odot e_u^A, \tag{12}$$

$$z_u^B = G_u^B \odot e_u^C + (1 - G_u^B) \odot e_u^B, \tag{13}$$

where $V_u^C \in \mathbb{R}^{d' \times d'}$ is a shared weight matrix for the two gating units, V_u^A and $V_u^B \in \mathbb{R}^{d' \times d}$ are weight matrices. z_u^A and z_u^B indicate the combination features of user u in A and B, respectively. The shared parameters V_u^C make two combination features assigned to the same weight for bidirectional knowledge transfer, ensuring the consistency of domain-shared part.

For the features e_i^C and e_i^A of item i in domain A, we apply a standard gating unit to combine them adaptively with the following formula:

$$G_i^A = \text{sigmoid}\left(V_i^C e_i^C + V_i^A e_i^A\right), \tag{14}$$

$$z_i^A = G_i^A \odot e_i^C + (1 - G_i^A) \odot e_i^A, \tag{15}$$

where V_i^C and $V_i^A \in \mathbb{R}^{d' \times d'}$ are weight matrices. Similarly, we can gain the combined feature z_j^B of item j in domain B.

4.4 Prediction Layer

After the combination of features, we generate user features (z_u^A and z_u^B) and item features (z_i^A and z_j^B). To endow the CDR systems with non-linearity, we apply the multi-layer perception (MLP) to model the user-item interactions. The formula in domain A is as follows:

$$\phi^1 = a^1(S^1 \begin{bmatrix} z_u^A \\ z_i^A \end{bmatrix} + b^1), \tag{16}$$

$$\cdots\cdots \tag{17}$$

$$\phi^L = a^L(S^L \phi^{L-1} + b^L), \tag{18}$$

$$\hat{y}_{ui}^A = f(\phi^L), \tag{19}$$

where S^l and b^l denote the trainable matrix and bias term for the l-th layer, respectively. a^l denotes the activation function such as sigmoid, ReLU, and hyperbolic tangent (tanh). ϕ^l denotes the output result for the l-th layer. $f(\cdot)$ is the prediction function, which maps ϕ^L to the probability $\hat{y}_{ui}^A$. Analogously, we generate the probability score $\hat{y}_{uj}^B$ in domain B.

4.5   Model Training

For CDR systems, an appropriate loss function can make the model achieve global optimization and speed up the model convergence. Considering the nature of implicit feedback, we select cross-entropy as the loss function which is defined as follows:

$$\mathcal{L}\left(\hat{y}_{uv}, y_{uv}\right) = - \sum_{(u,v)\in\mathcal{P}^+\cup\mathcal{P}^-} y_{uv}\log\hat{y}_{uv} + (1 - y_{uv})\log\left(1 - \hat{y}_{uv}\right), \qquad (20)$$

where y_{uv} defines an observed interaction and $\hat{y}_{uv}$ defines its corresponding predicted interaction. $\mathcal{P}^+$ is the set of observed interactions, $\mathcal{P}^-$ is a certain number of negative instances that can be randomly sampled from unobserved interaction to prevent over-fitting.

We aim to simultaneously enhance the performance of recommendation in both domains. Hence, the joint loss function to be minimized for domain A ($\mathcal{L}_A$) and domain B ($\mathcal{L}_B$) is defined as:

$$\mathcal{L}_{joint} = \alpha\mathcal{L}_A + \beta\mathcal{L}_B + \mathcal{L}_{reg} = \alpha\mathcal{L}(\hat{y}_{ui}^A, y_{ui}^A) + \beta\mathcal{L}(\hat{y}_{uj}^B, y_{uj}^B) + \gamma\|\Theta\|_2^2, \qquad (21)$$

where $\mathcal{L}(\hat{y}_{ui}^A, y_{ui}^A)$ and $\mathcal{L}(\hat{y}_{uj}^B, y_{uj}^B)$ define the loss function in domains A and B, respectively. Considering that the sparseness of interactions in the two domains is inconsistent, we leverage α and β to control sample balance. Here, we set $\alpha = \beta = 1$, considering two recommendation tasks for domains A and B are of equal importance. $\mathcal{L}_{reg}$ is a regularization term, in which γ is a hyper-parameter that controls the importance of $L2$ regularization and Θ are network parameters.

5   Experiments

This section answers the following questions:

RQ1: How does I^2RCDR perform compared with baselines?
RQ2: How do different designed modules (i.e., relation-aware GCN, gating fusion mechanism, and MLP) contribute to the model performance?
RQ3: Do inter-domain relations provide valuable information?
RQ4: How does I^2RCDR perform with different parameter settings?

5.1   Experimental Settings

Dataset. We examine the performance of our CDR framework on the real-world and well-known Amazon dataset[1], which includes abundant rating and review data and is widely used in CDR systems. We choose three pairs of datasets to organize our experiments. We first transform the ratings into implicit data, where each interaction is marked as 0 or 1, indicating whether the user has rated the item. We then select overlapped users from each pair of datasets and filter out non-overlapped users. Table 1 summarizes the detailed statistics of datasets.

---

[1] http://jmcauley.ucsd.edu/data/amazon/.

Table 1. Statistics of datasets.

Dataset	# Users	# Items	# Interactions	Density
Toys and Games (Toy)	1,380	6,773	19,831	0.212%
Video Games (Video)	1,380	6,667	21,359	0.232%
Sports and Outdoors (Sport)	3,908	13,057	43,996	0.086%
Clothing Shoes and Jewelry (Cloth)	3,908	13,044	35,115	0.069%
Home and Kitchen (Home)	14,059	25,995	179,543	0.049%
Health and Personal Care (Health)	14,059	17,663	174,998	0.071%

Evaluation Metric. We employ the *leave-one-out* strategy to conduct the evaluation. Since we focus on the top-N recommendation tasks, we apply two widely adopted metrics, namely Hit Ratio (HR) and Normalized Discounted Cumulative Gain (NDCG) [13], to efficiently estimate the performance of our method and baselines. For each user, we randomly sample 99 negative items that have not been rated with the user and combine them with the positive instance the user has been rated as the list waiting to be sorted in the ranking procedure. We repeat this process 5 times and show the average ranking results.

Comparison Methods. We compare our proposed method with state-of-the-art approaches and categorize the baselines into three groups: Single-Domain Recommendation (SDR), MLP-based CDR, and GNN-based CDR.

- **SDR.** GC-MC [1] applies GNNs to recommendation tasks and converts rating user-item interaction into a bipartite graph. NGCF [23] explicitly injects collaborative signals into the user-item graph. LightGCN [10] researches three different components in GCN and demonstrates neighbor aggregation is the most essential factor. HGNR [18] uses ratings, reviews, and social network data to construct a heterogeneous graph.
- **MLP-based CDR.** MLP-based CDR mainly introduces MLP to learn hidden features. CoNet [12] designs cross-connection networks to achieve dual knowledge transfer based on the cross-stitch network model. DTCDR [31] proposes a dual-target CDR framework to integrate the domain-shared features of overlapped users from two domains.
- **GNN-based CDR.** This group applies GNN to learn high-order features from graphs. PPGN [29] fuses user-item graphs from two domains into a holistic graph. BiTGCF [17] establishes a user-item graph for each domain and uses the overlapped user as the bridge to fuse users domain-shared and domain-specific features. GA-DTCDR [33] designs two independent heterogeneous graphs which are constructed from ratings and reviews.

Parameter Settings. We utilize Tensorflow to implement our framework and all the baselines. In our model learning stage, we choose *Adam* as the optimizer

Table 2. Recommendation performance of compared methods in terms of HR and NDCG. The best performance is in boldface and the best baseline is underlined.

Dataset	Toy & Video				Home & Hearth				Sport & Cloth			
	Toy		Video		Home		Hearth		Sport		Cloth	
Method	H@10	N@10	H@10	N@10	H@10	N@10	H@10	N@10	H@10	N@10	H@10	N@10
GC-MC	0.276	0.135	0.317	0.154	0.284	0.138	0.275	0.136	0.371	0.194	0.385	0.201
NGCF	0.295	0.156	0.370	0.190	0.326	0.169	0.312	0.162	0.393	0.212	0.412	0.221
LightGCN	0.313	0.159	0.397	0.205	0.300	0.146	0.289	0.135	0.372	0.189	0.386	0.191
HGNR	0.323	0.170	0.405	0.224	0.356	0.202	0.311	0.164	0.404	0.234	0.410	0.233
CoNet	0.375	0.229	0.476	0.275	0.366	0.213	0.318	0.193	0.417	0.224	0.420	0.196
DTCDR	0.455	0.254	0.518	0.294	0.394	0.210	0.382	0.195	0.444	0.253	0.471	0.268
PPGN	0.457	0.265	0.498	0.285	0.396	0.174	0.386	0.197	0.446	0.234	0.460	0.228
BiTGCF	0.443	0.231	_0.524_	0.297	0.397	0.229	0.346	0.181	0.456	0.264	0.474	0.274
GA-DTCDR	_0.464_	_0.254_	0.521	_0.302_	_0.425_	_0.226_	_0.398_	_0.212_	_0.459_	_0.266_	_0.481_	_0.286_
I^2RCDR	**0.496**	**0.285**	**0.558**	**0.342**	**0.441**	**0.252**	**0.434**	**0.246**	**0.479**	**0.281**	**0.524**	**0.297**

for all models to update model parameters and set the initial learning rate as 0.001. We sample four negative instances for each positive instance to generate the training dataset. The batch size is set to 512. we set $d = 32$ for the embedding size of all the methods. Furthermore, we use dropout techniques to further prevent over-fitting and fix the dropout rate as 0.1. For our proposed method and GNN-based baselines, we set $k = 4$. For HGNR, we use reviews to construct user-user graph instead of the social network graph. For DTCDR and GA-DTCDR, we only consider the set of overlapped uses and model the review information to ensure fairness.

5.2 RQ1: Performance Comparison

Table 2 reports the summarized results of our experiments on three pairs of datasets in terms of HR@10 (H@10) and NDCG@10 (N@10). It can be seen that I^2RCDR consistently achieves the best performance compared with all the baselines, which reveals the superiority of modeling inter-domain and intra-domain relations collectively by relation-aware GCN. Compared with SDR approaches (GC-MC, NGCF, LightGCN, and HGNR), CDR models usually obtain better performance, benefit from fusing more useful knowledge from both two domains during the transfer learning phase. In CDR methods, GNN-based models (PPGN, BiTGCF, and GA-DTCDR) outperform MLP-based models (CoNet and DTCDR) by 6.28% H@10 and 4.72% N@10 on average, respectively. This observation justifies that GNNs can achieve better recommendation performance by modeling the high-order representations. Furthermore, DTCDR and GA-DTCDR relying on integrating review information into CDR systems, are indeed better than other CDR models which only use rating information, but are still weaker than our method which fuses reviews and ratings to learn domain-shared and domain-specific features explicitly.

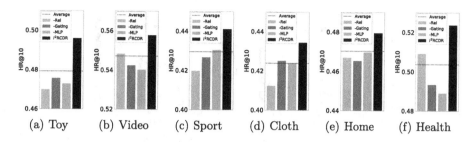

Fig. 3. Performance of I^2RCDR compared with different variants in terms of HR@10

5.3 RQ2: Ablation Study

We attempt to validate whether I^2RCDR benefits from the influence of different modules. Therefore, we compare I^2RCDR with the following variant versions:

- **-Rel.** This method performs a typical GCN model [14] instead of our relation-aware GCN.
- **-Gating.** This variant employs an element-wise attention mechanism to replace our gating fusion mechanism.
- **-MLP.** This method removes MLP in the prediction layer and defines an inner product as the interaction function.

As presented in Fig. 3, we observe that our I^2RCDR is better than all the variants in terms of HR@10. We overlook the performance of NDCG which follows the similar trend due to the space limitation.

5.4 RQ3: Effect of Inter-domain Relations

To verify the effectiveness of inter-domain user-user and item-item relations, we compare the performance of our method leveraging one or two kinds of relations in terms of HR@10 and NDCG@10. Figure 4 illustrates the comparison results concerning I^2RCDR-UI (without considering inter-domain relations), I^2RCDR-I (only considering user-user relation), I^2RCDR-U (only considering item-item relation), I^2RCDR (simultaneously considering two kinds of relations). I^2RCDR-I and I^2RCDR-U both have better performance by adding inter-domain user-user or item-item relation than I^2RCDR. This proves the effectiveness of integrating inter-domain relations in CDR tasks. In addition, we ignore the results about the effect of intra-domain relations due to space limitations, which exhibit a similar pattern to inter-domain relations.

5.5 RQ4: Parameter Analysis

We vary the layer numbers which are in the range of {2, 3, 4, 5, 6}, to verify whether I^2RCDR benefits from multiple propagation layers. Figure 5 shows the results about different layers on three pairs of datasets in terms of HR@10. When

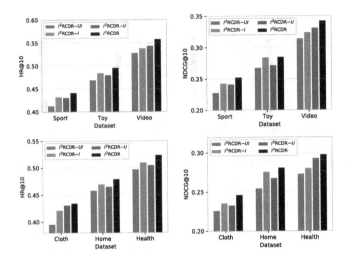

Fig. 4. Performance of I^2RCDR with different inter-domain relations

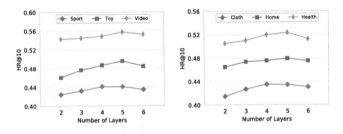

Fig. 5. Effect of propagation layer numbers

the number of layers is 4 or 5, we obtain the finest results. The performance significantly improves as the number of layers increases. We think this is because more potential relations have been mined with the increasing of model depth. However, the results also show that longer layers (e.g., 6-hop) maybe make a lot of noise, which affects the recommendation accuracy.

6   Conclusion and Future Work

In this paper, we propose I^2RCDR, an end-to-end graph-structured framework that naturally incorporates inter-domain relations into CDR systems. I^2RCDR designs the relation-aware GCNs to encode heterogeneous graphs and jointly model inter- and intra-domain relations. To balance cross-domain consistency and single-domain peculiarity, we design a gating fusion mechanism to fuse domain-shared and domain-specific features for dual-target recommendation. Extensive experiments are carried out on three pairs of datasets and the results demonstrate the effectiveness of the proposed framework. Currently, we only deal with the holistic user-user and item-item relations produced from reviews. In future work, we will consider disentangling semantic relations to extract multifaceted features and model users' fine-grained preferences.

Acknowledgments. This research is supported in part by the 2030 National Key AI Program of China 2018AAA0100503 (2018AAA0100500), National Science Foundation of China (No. 62072304, No. 61772341, No. 61832013), Shanghai Municipal Science and Technology Commission (No. 19510760500, No. 19511101500, No. 19511120300), the Oceanic Interdisciplinary Program of Shanghai Jiao Tong University (No. SL2020MS032), Scientific Research Fund of Second Institute of Oceanography, the open fund of State Key Laboratory of Satellite Ocean Environment Dynamics, Second Institute of Oceanography, MNR, and Zhejiang Aoxin Co. Ltd.

References

1. van den Berg, R., Kipf, T.N., Welling, M.: Graph convolutional matrix completion. arXiv preprint arXiv:1706.02263 (2017)
2. Chen, C., et al.: Graph heterogeneous multi-relational recommendation. In: AAAI, vol. 35, pp. 3958–3966 (2021)
3. Chen, C., et al.: An efficient adaptive transfer neural network for social-aware recommendation. In: SIGIR, pp. 225–234 (2019)
4. Cho, K., et al.: Learning phrase representations using RNN encoder-decoder for statistical machine translation. arXiv preprint arXiv:1406.1078 (2014)
5. Cui, Q., Wei, T., Zhang, Y., Zhang, Q.: HeroGRAPH: a heterogeneous graph framework for multi-target cross-domain recommendation. In: ORSUM@ RecSys (2020)
6. Devlin, J., Chang, M.W., Lee, K., Toutanova, K.: BERT: pre-training of deep bidirectional transformers for language understanding. arXiv preprint arXiv:1810.04805 (2018)
7. Fu, W., Peng, Z., Wang, S., Xu, Y., Li, J.: Deeply fusing reviews and contents for cold start users in cross-domain recommendation systems. In: AAAI, vol. 33, pp. 94–101 (2019)
8. Guo, L., Tang, L., Chen, T., Zhu, L., Nguyen, Q.V.H., Yin, H.: DA-GCN: a domain-aware attentive graph convolution network for shared-account cross-domain sequential recommendation. arXiv preprint arXiv:2105.03300 (2021)
9. Hamilton, W.L., Ying, R., Leskovec, J.: Inductive representation learning on large graphs. In: NIPS, pp. 1025–1035 (2017)
10. He, X., Deng, K., Wang, X., Li, Y., Zhang, Y., Wang, M.: LightGCN: simplifying and powering graph convolution network for recommendation. arXiv preprint arXiv:2002.02126 (2020)
11. He, X., Liao, L., Zhang, H., Nie, L., Hu, X., Chua, T.S.: Neural collaborative filtering. In: WWW, pp. 173–182 (2017)
12. Hu, G., Zhang, Y., Yang, Q.: CoNET: collaborative cross networks for cross-domain recommendation. In: CIKM, pp. 667–676 (2018)
13. Kang, S., Hwang, J., Lee, D., Yu, H.: Semi-supervised learning for cross-domain recommendation to cold-start users. In: CIKM, pp. 1563–1572 (2019)
14. Kipf, T.N., Welling, M.: Semi-supervised classification with graph convolutional networks. arXiv preprint arXiv:1609.02907 (2016)

15. Li, P., Tuzhilin, A.: DDTCDR: deep dual transfer cross domain recommendation. In: WSDM, pp. 331–339 (2020)
16. Liu, J., et al.: Exploiting aesthetic preference in deep cross networks for cross-domain recommendation. In: WWW, pp. 2768–2774 (2020)
17. Liu, M., Li, J., Li, G., Pan, P.: Cross domain recommendation via bi-directional transfer graph collaborative filtering networks. In: CIKM, pp. 885–894 (2020)
18. Liu, S., Ounis, I., Macdonald, C., Meng, Z.: A heterogeneous graph neural model for cold-start recommendation. In: SIGIR, pp. 2029–2032 (2020)
19. Ma, M., Ren, P., Lin, Y., Chen, Z., Ma, J., de Rijke, M.: π-Net: a parallel information-sharing network for shared-account cross-domain sequential recommendations. In: SIGIR, pp. 685–694 (2019)
20. Man, T., Shen, H., Jin, X., Cheng, X.: Cross-domain recommendation: an embedding and mapping approach. In: IJCAI, vol. 17, pp. 2464–2470 (2017)
21. Reimers, N., Gurevych, I.: Sentence-BERT: sentence embeddings using Siamese BERT-networks. arXiv preprint arXiv:1908.10084 (2019)
22. Wang, C., Zhu, Y., Liu, H., Ma, W., Zang, T., Yu, J.: Enhancing user interest modeling with knowledge-enriched itemsets for sequential recommendation. In: CIKM, pp. 1889–1898 (2021)
23. Wang, X., He, X., Wang, M., Feng, F., Chua, T.S.: Neural graph collaborative filtering. In: SIGIR, pp. 165–174 (2019)
24. Wang, X., Jin, H., Zhang, A., He, X., Xu, T., Chua, T.S.: Disentangled graph collaborative filtering. In: SIGIR, pp. 1001–1010 (2020)
25. Wang, X., Peng, Z., Wang, S., Yu, P.S., Fu, W., Hong, X.: Cross-domain recommendation for cold-start users via neighborhood based feature mapping. In: Pei, J., Manolopoulos, Y., Sadiq, S., Li, J. (eds.) DASFAA 2018. LNCS, vol. 10827, pp. 158–165. Springer, Cham (2018). https://doi.org/10.1007/978-3-319-91452-7_11
26. Xia, L., et al.: Knowledge-enhanced hierarchical graph transformer network for multi-behavior recommendation. In: AAAI, vol. 35, pp. 4486–4493 (2021)
27. Xu, Y., Zhu, Y., Shen, Y., Yu, J.: Learning shared vertex representation in heterogeneous graphs with convolutional networks for recommendation. In: IJCAI, pp. 4620–4626 (2019)
28. Zang, T., Zhu, Y., Liu, H., Zhang, R., Yu, J.: A survey on cross-domain recommendation: taxonomies, methods, and future directions. arXiv preprint arXiv:2108.03357 (2021)
29. Zhao, C., Li, C., Fu, C.: Cross-domain recommendation via preference propagation GraphNet. In: CIKM, pp. 2165–2168 (2019)
30. Zhao, C., Li, C., Xiao, R., Deng, H., Sun, A.: CATN: cross-domain recommendation for cold-start users via aspect transfer network. In: SIGIR, pp. 229–238 (2020)
31. Zhu, F., Chen, C., Wang, Y., Liu, G., Zheng, X.: DTCDR: a framework for dual-target cross-domain recommendation. In: CIKM, pp. 1533–1542 (2019)
32. Zhu, F., Wang, Y., Chen, C., Liu, G., Orgun, M., Wu, J.: A deep framework for cross-domain and cross-system recommendations. arXiv preprint arXiv:2009.06215 (2020)
33. Zhu, F., Wang, Y., Chen, C., Liu, G., Zheng, X.: A graphical and attentional framework for dual-target cross-domain recommendation. In: IJCAI, pp. 3001–3008 (2020)

Enhancing Graph Convolution Network for Novel Recommendation

Xuan Ma, Tieyun Qian$^{(\boxtimes)}$, Yile Liang, Ke Sun, Hang Yun, and Mi Zhang

School of Computer Science, Wuhan University, Hubei, China
{yijunma0721,qty,liangyile,sunke1995,yunhgg,mizhanggd}@whu.edu.cn

Abstract. Graph convolution network based recommendation methods have achieved great success. However, existing graph based methods tend to recommend popular items yet neglect tail ones, which are actually the focus of novel recommendation since they can provide more surprises for users and more profits for enterprises. Furthermore, current novelty oriented methods treat all users equally without considering their personal preference on popular or tail items.

In this paper, we enhance graph convolution network with novelty-boosted masking mechanism and personalized negative sampling strategy for novel recommendation. Firstly, we alleviate the popularity bias in graph based methods by obliging the learning process to pay more attention to tail items which are assigned to a larger masking probability. Secondly, we empower the novel recommendation methods with users' personal preference by selecting true negative popular samples. Extensive experimental results on three datasets demonstrate that our method outperforms both graph based and novelty oriented baselines by a large margin in terms of the overall F-measure.

Keywords: Recommender systems · Novel recommendation · Masking mechanism · Negative sampling

1 Introduction

Recommender systems have been a fundamental component in many web services such as e-commerce and social networks [3,8,15,25,26]. Early methods utilize collaborative filtering (CF) technique to leverage the user-item interaction data for recommendation. Recently, deep learning methods become the mainstream. Among which, the graph based methods, especially those with graph convolution network as the main framework, have achieved promising performance in conventional recommendation, where the accuracy serves as the dominant or even the sole target.

Classical graph based models aim to learn meaningful representations for graph data by iteratively aggregating messages passed by neighbors. Meanwhile, the node information and topological structure can be encoded during the process of transformation and aggregation. Graph based models have been demonstrated powerful in representation learning [5,6,12]. As the users' interaction behaviors can be represented as a user-item bipartite graph, the graph based models also become prevalent in the field of recommendation [2,7,9,21,31]. Though effective, existing graph based methods tend to recommend popular items to users since they help yield high accuracy. However, there are only a few popular items in the real world and most items are long-tail ones, i.e., the majority of interactions are occupied by a small number of popular items. Recommending tail items can provide more surprises for users and more profits for enterprises. Consequently, novel recommendation has aroused great research interests in recent years, but it is largely neglected in graph based methods.

Most novel recommendation studies adopt a post-processing strategy [11, 14,19,34]. They first achieve high accurate recommendation results, and then re-rank the results based on the novelty metric. The loss functions in these methods are not optimized for both accuracy and novelty at the same time and their performance is limited to the base models. To address these issues, the end-to-end frameworks are proposed [14,16,17]. Although accuracy and novelty can be simultaneously optimized, the accuracy of the recommendation results is relatively low. More importantly, current novel recommendation methods do not take the users' personal preferences into consideration. Indeed, some users prefer popular items while some others prefer tail items. Recommending niche items to users having popular preference will significantly decrease the customer satisfaction, and vice versa.

In this paper, we propose to enhance the graph convolution network for novel recommendation. On one hand, we aim to *alleviate the popularity bias in graph based methods*. On the other hand, we aim to *deepen novel recommendation towards a more personalized way*. To achieve these two goals, we first oblige the learning procedure to pay more attention to tail items by assigning them a larger masking probability. In this way, the representation of tail items can be enhanced since they build more relations to their neighbors. We then classify the users' preference into normal preference and niche preference, and differentiate "true" negative samples from "false" negative ones to realize personalized novel recommendation. We conduct extensive experiments on three real-world datasets. The results demonstrate that our proposed model yields significant improvements over the state-of-the-art graph based and novelty oriented baselines in terms of the trade-off among accuracy, coverage, and novelty.

2   Related Work

2.1   Graph Based Methods for Recommendation

The graph based recommendation models learn the user and item embeddings from the user-item bipartite graph structure, and have achieved the state-of-the-art recommendation performance in terms of accuracy [2,9,27,29]. Motivated by the strength of graph convolution to capture signals in high-hop neighbors, graph convolution network (GCN) has been widely adopted in graph based recommendation models. For example, Monti et al. [18] build the user-user or item-item graphs and adopt GCN framework to aggregate information from these graphs. Berg et al. [2] construct a bipartite graph for rating matrix completion. Ying et al. [31] combine random walks and graph convolutions to incorporate graph structure and node feature. Zhang et al. [32] design a stacked and reconstructed GCN to improve the prediction performance. Wu et al. [28] deploy graph attention networks for modeling social effects for recommendation tasks. Wang et al. [23] encode the high-order connectivity for social recommendation. He et al. [9] discard feature transformation and nonlinear activation for construct light-weighted GCN. Yang et al. [29] propose an enhanced graph learning network to make the graph learning module and the node embedding module iteratively learned from each other.

In summary, these studies have shown promising results by learning user or item embeddings to facilitate downstream tasks. However, all these models ignore the long-tail phenomenon in user-item interactions. This makes the models biased towards popular items, and thus is unfair to the representation learning of tail items. In view of this, we propose to alleviate the popularity bias when learning item embeddings for graph based recommendation models.

2.2 Novel Recommendation

Novel recommendation, also known as novelty oriented recommendation or long-tail recommendation, plays a vital role in promoting the customer experience and increasing sales profit [10]. Existing work in novel recommendation can be categorized into three groups. The first group adopts a two-stage strategy that re-ranking the results from the raw model to enhance novelty [11,34]. For example, Jugovac et al. [11] ensure the accuracy and novelty by re-ranking items in the first top-N and the next M candidate items. The second group uses the category as supplement information through clustering [20,22]. The third group adopts an end-to-end framework [16,17]. For example, Li et al. [13] decompose the overall interested items into a low-rank part for head items and a sparse part for tail items, and independently reveal two parts in the training stage.

Overall, existing novelty oriented recommendation models ignore the users' personal preference in head or tail items. In contrast, we empower novel recommendation with personalized negative sampling strategy by differentiating "true" negative samples from "false" negative ones which are popular yet not interacted by the user and her neighbors.

3   Our Proposed Model

In this section, we introduce our *enhanced masking and negative sampling* (*EMNS*) model for graph convolution network by first formulating the problem and then presenting the details.

3.1   Problem Formulation

Let $U = \{u_1, u_2, \ldots, u_{|U|}\}$ denote the set of users and $I = \{i_1, i_2, \ldots, i_{|I|}\}$ denote the set of items, where $|U|$ and $|I|$ are the number of users and items, respectively.

Let $R = \{r_{ui}\} \in \mathbb{R}^{|U| \times |I|}$ denote the user-item interaction matrix. R consists of 0 and 1 entries, where $r_{ui} = 1$ if the user u has clicked the item i and $r_{ui} = 0$ otherwise. We then construct a bipartite graph, whose nodes are users and items in U and V in two disjoint sets, and the edges are interactions between users and items.

Given the above interaction graph, we aim to recommend the top-N unrated items as the conventional recommendation. In real-world recommendation scenarios, the users often purchase popular items, which is known as the conformity effect in psychology. As a result, we can get an accurate recommendation list by recommending hot items. However, only providing popular items and ignoring the user's own preference for niche items will reduce the user's satisfaction on recommendation results. This arouses the research in novel recommendation, and we also take both novelty and accuracy into account at the same time in this study.

3.2   Model Overview

Graph convolution network has shown promising performance in conventional recommendation. Beyond that, we aim at introducing more novelty oriented techniques into graph learning such that head or tail items can be properly recommended to the users. To this end, we propose *a novelty-boosted masking mechanism* and *a personalized negative sampling strategy*.

Due to the lack of interactions, the long-tail items cannot obtain enough information in classical graph based models, and thus the model is biased towards popular items. To solve this problem, we first make adjustments to the process of classical graph based models. Inspired by the excellent context fusion capability of the masked language model, we oblige the learning procedure to pay more attention on tail items by assigning them a larger masking probability, such that we can strengthen the learning process for novel items. This is our *novelty-boosted masking mechanism*.

Furthermore, recommending niche items to users having popular preference will sacrifice accuracy, and vice versa. So we propose to deepen the model towards a more personalized way. We first classify the user's preference into normal and niche one, and then enlarge the probability of selecting tail items for a user with niche item preference by selecting "true" negative samples. This is our *personalized negative sampling strategy*.

Through these two special designs, we hope to alleviate the popularity bias and develop a more personalized novel recommendation model. Figure 1 illustrates the overall structure of our proposed EMNS model. It mainly consists of six layers, i.e., an embedding layer, a dual branch masking layer, a stacked graph convolution layers, a negative sampling layer, a reconstruction layer, and a gated fusion layer. In the following subsections, we will present the details for these layers.

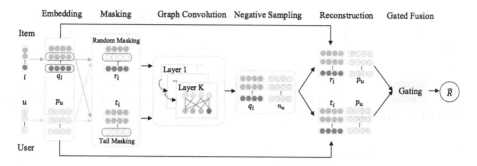

Fig. 1. Overview of our proposed EMNS model. The dark blue nodes in the red rectangle represent the long-tail items and the light blue ones indicate the masked items. (Color figure online)

3.3 Embedding Layer

Following existing embedding based recommendation models, we use $P \in \mathbb{R}^{D \times |U|}$ and $Q \in \mathbb{R}^{D \times |I|}$ to represent the free embeddings of users and items. By performing an index operation, the lookup table P transforms the one-hot representations of a user id into the user's latent vector $p_u \in \mathbb{R}^D$. Similarly, we can obtain the item's latent vector $q_i \in \mathbb{R}^D$ through Q.

3.4 Masking Layer

As mentioned above, the majority of interactions are occupied by a small number of popular items in the real world. Due to the sparse interaction of tail items, it is hard to learn good representations for them. To address this problem, we design a novelty-boosted masking strategy in this layer to force the learning procedure to pay more attention to tail items.

We first put the head items and tail items into two learning branches. In the "novelty learning branch", we aim to strengthen the representation learning of tail items by enlarging their probability of being masked. In the "normal learning branch", we perform the normal masking strategy for all items except those in the novelty learning branch. We then mask $p\%$ of the whole item nodes in two branches respectively. We use r_i to represent that an item i belongs to the normal random learning branch and use t_i for the novelty learning branch.

In the novelty learning branch, we take the long-tail distribution phenomena of items in the interaction into consideration by applying tail masking mechanism. We set the masking possibility of an item node proportional to its novelty, i.e., the less popular the items are, the higher the probability they are masked. Following [1], we denote the popularity of an item i as its frequency in the training set, i.e., $p_i = |U_i|$. We denote the maximum number of interactions in the whole item set as p_{max}. We then calculate the masking possibility of an item i in this branch as:

$$M_i = \frac{w_i}{\sum_{j=1}^{N} w_j}, \tag{1}$$

where $w_i = p_i/p_{max}$. We can see that the sampling probability is inversely proportional to its popularity. Next, we sample an item i according to the value of M_i. In this way, we force the model to adaptively pay more attention to the embedding learning process of long-tail items to alleviate the popularity bias.

In the normal learning branch, we apply the random masking strategy. We select each item with equal probability, and thus this branch retains the original distribution information of items.

3.5   Graph Convolutional Layer

After the masking layer, we employ a conventional stacked graph convolution layer for aggregating messages from the neighborhood. Here we choose the state-of-the-art LightGCN [9] which has four layers as the backbone for this layer. Given the user's representation p_u^{l-1} and the item's representation q_i^{l-1} at the $(l$-1$)$-th layer, the output of graph convolution operation at the l-th layer is as follows,

$$p_u^l = \sum_{i \in I_u} \frac{1}{\sqrt{|I_u|}\sqrt{|U_i|}} q_i^{l-1}, \quad q_i^l = \sum_{u \in U_i} \frac{1}{\sqrt{|U_i|}\sqrt{|I_u|}} p_u^{l-1}, \tag{2}$$

where I_u represents the set of items that are interacted by the user u and U_i represents the set of users that interact with item i.

3.6   Negative Sampling Layer

The selection of positive and negative samples plays a crucial role in the recommendation process. When training the model, the embedding of positive items and negative items learn in different directions through optimizing the loss function. Specifically, the positive ones become closer to the target user embedding while the negative ones become far away from it. Therefore, selecting high-quality negative samples is beneficial for model optimization.

All existing methods take the random sampling strategy. However, this strategy tends to select "false" negative samples. For example, selecting niche items as negative samples for a user with niche preference is unreasonable and will compromise the accuracy. As a result, it is necessary to choose "true" negative samples in contrast to the user's preference and make personalized recommendations for different users.

With this in mind, we propose to choose the negative samples based on the user preference patterns. We select the "true" negative samples for users with niche preference to improve novelty, which are popular yet not interacted by the target user and her neighbors.

The detailed negative sampling procedure is summarized in Algorithm 1. We first calculate the preference patterns based on users' statistics. The popularity of a user is relevant to items which she interacts with. For example, a user with niche preference will buy more items with low popularity scores. To model the user's preference patterns, a natural way is using the mean and standard deviation:

$$\theta_u = \frac{1}{|I_u|} \sum_{i \in I_u} p_i, \quad \sigma_{\theta_u} = \sqrt{\frac{\sum_{i \in I_u}(\theta_u - \theta_i)^2}{|I_u|}}, \tag{3}$$

Algorithm 1. Personalized Negative Sampling Procedure

Input: Positive samples P; Item set I; User set U; Negative sample rate T; Popular item set I_p

Output: The user's negative sampling pool Φ; The item's negative sampling pool Ω

1: $\theta \leftarrow$ the mean of θ_u in a batch
2: $\sigma \leftarrow$ the mean of σ_{θ_u} in a batch
3: **for** each positive sample $(u,i,\text{True}) \in P$ **do**
4:      **for** $t \leftarrow 1$ to T **do**
5:          **if** $\theta_u < \theta$ and $\sigma_{\theta_u} < \sigma$ **then**
6:              $f_u \leftarrow$ Find_the_most_similar_user $\cup \{u\}$
7:              $\Phi_u \leftarrow \Phi_u + \{I_p \setminus I_{f_u}\}$
8:          **else**
9:              $\Phi_u \leftarrow \Phi_u + \{I \setminus I_u\}$
10:         $f_i \leftarrow$ Find_the_most_similar_item $\cup \{i\}$
11:         $\Omega_i \leftarrow \Omega_i + \{U \setminus U_{f_i}\}$
12: **return** Φ, Ω

---

where U_i represents the sets of users that interact with the item i and I_u represents the sets of items that are interacted by the user u.

For users with niche preference, they may have low mean scores due to the low popularity scores of the items they interact with. Furthermore, since their interests are concentrated in niche items, their range of interests fluctuates less and thus they have a low standard deviation score. By calculating the value of mean and standard deviation, we can divide user preference into niche and normal one.

The Find_the_most_similar_user/item function in Algorithm 1 is used to choose the most similar user/item. We realize this based on the user's/item's behavior. Here we take the user side as an example. We first calculate the similarity between a user u and a user v as:

$$sim(u,v) = \frac{|I_u \cap I_v|}{\sqrt{|I_u \cap I_v|}} \tag{4}$$

If the neighbor user has not clicked on one popular item as well, it confirms that such item is unattractive to this type of users. Thus, we choose this popular item into the negative sampling pool of the target niche user to increase the reliability of negative samples.

For users with normal preference, we take the same random sampling strategy as previous studies to increase dynamic selection because these users usually have a wide range of interests.

At the same time, we perform item negative sampling for positive items. We treat items with the highest similarity as the neighbor of the target item. Then we choose the users who have not clicked on both the target item and its neighbor into the negative sampling pool for the target item.

After preparing the negative sampling pools of both users and items, we randomly select negative samples from them. By adopting personalized negative sampling strategy, the recommendation model is optimized to distinguish the user's preferences on popular or tail items. For users with niche preference, the popular items which no similar users click on are less likely to be retrieved. Consequently, this increases the possibility to recommend tail items.

3.7 Reconstruction Layer

As we mask items according to different strategies in the masking layer, in this layer we need to reconstruct them to maximize the consistency between the node convolutional representation induced from the neighborhood and its initial embedding.

To be specific, we first use a two-layer feedforward neural network as the decoder, i.e., $\hat{q}_i = FFN(q_i^l)$. Then we minimize the distance between the initial embedding of an item q_i and the decoded representation $\hat{q}_i$ as follows,

$$L_{recon} = \frac{1}{2|I|} \sum_{i \in I} ||q_i - \hat{q}_i||^2 \tag{5}$$

Through the reconstruction process, we make messages delivered to the node to be consistent with the preference of the node itself, and we also strengthen the restriction on the node representation.

3.8 Gated Fusion Layer

We have obtained the hidden representation of items and users from the normal learning and the novelty learning branches. We then combine these two kinds of hidden representation together to promote forecasting users' preferences for unrated items. Specifically, the fusion operation for an item i is computed as:

$$\alpha = \frac{exp(\sigma(W_1 r_i))}{exp(\sigma(W_1 r_i)) + exp(\sigma(W_2 t_i))}, \quad \tilde{q}_i = \alpha \odot r_i + (1 - \alpha) \odot t_i, \tag{6}$$

where $\sigma(\cdot)$ is the sigmoid function, $W_1 \in \mathbb{R}^{D \times D}$ and $W_2 \in \mathbb{R}^{D \times D}$ are weight matrices and $\odot$ denotes the element-wise product. We can obtain $\tilde{p}_u$ in a similar way.

Given a user $u's$ final representation $\tilde{p}_u$ and an item $i's$ final representation $\tilde{q}_i$ after the gating operation, we can get the predicted rating of the user u to the item i as:

$$\hat{r}_{ui} = <\tilde{p}_u, \tilde{q}_i>, \tag{7}$$

where $<,>$ denotes the inner product operation between two vectors.

3.9 Training

In our model, we employ the pairwise ranking based BPR loss, which assumes that the prediction values for rated items should be higher than those for unrated ones. As we perform negative sampling for both the user and the item, there are two pairs in every training process of each interaction between a user u and an item i, i.e., (u, i, i_n) and (i, u_p, u_n), where i_n and u_n stand for negative samples and u_p stands for positive sample for the item i. Thus the objective function for the main prediction task can be formulated as:

$$L_{main} = -\sum_{u=1}^{|U|}\sum_{i\in I_u}\sum_{i_n\in\Phi_u} ln\sigma(\hat{r}_{ui} - \hat{r}_{ui_n}) - \lambda_1\sum_{i=1}^{|I|}\sum_{u_p\in U_i}\sum_{u_n\in\Omega_i} ln\sigma(\hat{r}_{iu_p} - \hat{r}_{iu_n}), \quad (8)$$

where Φ_u and is the negative sampling pool of the user u and the item i, respectively, and λ_1 is the hyper-parameter to control the impact of item negative sampling.

Furthermore, since the users in this two pairs are connected by the same positive item i, there should be restrictions that the users with similar interests have similar representations. To this end, we first extract the users from these two pairs to form a new triple (u, u_p, u_n). We then let the correlation between the latent representation of the target user u and the positive user sample u_p be larger than that between u and the negative user sample u_n. We finally define the mutual information (MI) loss function for (u, u_p, u_n) as:

$$L_{mi} = -\sum_{(u,u_p,u_n)\in T} log\sigma(f(u, u_p) - f(u, u_n)), \quad (9)$$

where $f(\cdot)$ is the Euclidean distance and T denotes constructed training triples.

Given the main prediction loss L_{main}, the reconstruction loss L_{recon}, and the mutual information loss L_{mi}, we combine them into the final optimization loss. We add two hyper-parameters λ_2 and λ_3 to balance the three losses. The overall objective function can be formulated as follows:

$$L = L_{main} + \lambda_2 L_{recon} + \lambda_3 L_{mi} \quad (10)$$

The training of our model is performed for optimizing this overall loss function.

4   Experiments

In this section, we present our experimental settings and the main results.

4.1   Experimental Settings

Datasets To evaluate the performance of our proposed method, we use three real-world datasets. One dataset is **MovieLens**[1], and two others are **Digital Music** and **Instant Video** from Amazon[2]. We take the 100K version from MovieLens which has 100K interactions and the 5-core version for the Amazon datasets where each user or item has at least five interactions. The statistics of the datasets are summarized in Table 1. We randomly select 20% of historical interactions for each user as the testing set, and the remaining 80% historical interactions are treated as training set.

---

[1] https://grouplens.org/datasets/movielens/.
[2] http://jmcauley.ucsd.edu/data/amazon/links.html.

Table 1. Statistics of the datasets.

Dataset	Users	Items	Interactions	Sparsity
MovieLens	943	1682	100000	93.70%
Digital Music	5541	3568	64706	99.67%
Amazon instant video	5130	1685	37126	99.57%

Baselines. We choose the following ten state-of-the-art methods as our baselines and categorize them into three groups: conventional neural graph models, long-tail recommendation methods, and popularity bias eliminating methods.

(1) Conventional graph based recommendation methods.

- **STAR-GCN** [32] employs a stacked GCN structure with the masking mechanism. Some nodes' embeddings are randomly masked and reconstructed during the training phase.
- **LightGCN** [9] only keeps the neighborhood aggregation component in GCN to make the model easier to implement and train.
- **SGL** [27] introduces self-supervised learning into graph based recommendation. We adopt SGL-ED as our baseline since it performs the best among all variants.
- **EGLN** [29] enhances graph learning by iteratively learning node embeddings and the graph structure.

(2) Long-tail recommendation methods.

- **GANC** [34] integrates user's niche preference into a re-ranking framework that customizes the balance between accuracy, coverage, and novelty.
- **PPNW** [17] proposes a personalized pairwise novelty weighting for the BPR loss function to introduce the novelty of users and items.
- **TailNet** [16] determines the user's preference on popular or niche items in session-based recommendation. We remove its GRU module which is tailed for session recommendation.

(3) Latest popularity bias eliminating methods.

- **MF-PC** [33] presents a post-processing approach by adding compensation to less popular items.
- **MACR** [24] eliminates the popularity bias from the cause-effect perspective by analyzing the contribution of each cause and then performing counterfactual inference to remove the effect of item popularity.

Evaluation Metrics. We use accuracy, coverage, novelty, and the trade-off metrics to validate the accuracy and novelty of the methods. Recall@N(Rec@N) [16,34] is used as the accuracy which measures how an algorithm can rank items for each user. For coverage, we use Coverage@N(Cov@N) [16,30,34] as the metric. For novelty, we extend LTAccuracy@N [34] to measure how many novel items are in each top-N recommendation list, which we call Novelty@N(Nov@N). Note that the head items are the 20% most popular items, and the rest are tail items according to the Pareto principle [30]. To measure the trade-off, we employ F-score [4] to calculate the harmonic mean of conflicting accuracy, novelty, and coverage, i.e., $F1@N = \frac{3*Rec@N*Cov@N*Nov@N}{Rec@N+Cov@N+Nov@N}$.

Settings. For fair comparison, we set the embedding size to 64 and the mini-batch size to 2048. Since we use the pair-wise ranking based loss, for each observed user-item interaction, we randomly select one unobserved item as the candidate negative sample for baselines. All the models are optimized by the Adam optimizer with the learning rate of 0.01. For model-agnostic baselines including GACN, PPNW, IDS4NR, and MACR, we use the same LightGCN backbone as our EMNS and fine-tune the hyper-parameters.

4.2 Main Results

We compare our model with the baseline methods on three datasets. The results are shown in Table 2. It can be seen that our EMNS model always achieves the best performance in terms of the trade-off F1 metric on all three datasets. This clearly demonstrates the superiority of our model for novel recommendation. Compared with graph based methods, our model gets much better novelty and coverage scores since we alleviate the popularity bias with the specialized masking strategy technique. Compared with novel recommendation methods, our model reaches better accuracy by differentiating users' preference on popular or niche items.

From these results, we have some other observations.

Firstly, the graph based recommendation methods often outperform novelty oriented ones in terms of accuracy. In particular, LightGCN achieves the best Acc@5 on three datasets. However, its novelty score is low as it does not consider the recommendation of tail items. In real applications, it is necessary to recommend niche items that match users' preference because recommending the most popular items may give the user a negative impression of not being treated seriously. Moreover, STAR-GCN is the second best in terms of Cov@5 on Music and Video owing to its random masking strategy. However, without the novelty-boosted masking and personalized sampling strategies, its novelty and accuracy is much inferior to our model.

Secondly, we find that most long-tail recommendation baselines sacrifice too much accuracy to improve novelty. For example, GANC reaches an extremely high Nov@5 score 0.9807 on music, but the corresponding Rec@5 score 0.0134 is very poor. This may seriously damage the users' experience on recommendation results since there are too many irrelevant items. In contrast, our EMNS achieves a better balance between accuracy and novelty which can better improve customer satisfaction.

Thirdly, for both popularity bias eliminating baselines, their scores are relatively balanced, i.e., their accuracy, coverage, and novelty scores are all in the middle. However, their process focuses on debiasing from the perspective of matching score and does not learn accurate representation for both users and items, so they are still inferior to our model.

Table 2. The overall performance comparison on three datasets. The best performance among all is in bold while the second best one is underlined.

Dataset	Model	Accuracy		Coverage		Novelty		Trade-off	
		Rec@5	Rec@10	Cov@5	Cov@10	Nov@5	Nov@10	F1@5	F1@10
MovieLens	STAR-GCN	0.0973	0.1704	0.3538	0.4393	0.2229	0.2315	0.0341	0.0617
	LightGCN	**0.1289**	**0.2115**	0.3406	0.4315	0.0979	0.1201	0.0227	0.0430
	SGL	0.0996	0.1626	0.2802	0.3519	0.2911	0.3016	0.0352	0.0622
	EGLN	0.0969	<u>0.1707</u>	0.3886	0.4772	0.1739	0.1870	0.0296	0.0546
	GANC	0.0286	0.0551	0.1938	0.2941	**0.5923**	<u>0.5582</u>	0.0119	0.0297
	PPNW	0.0742	0.1287	**0.6015**	<u>0.6845</u>	<u>0.5745</u>	**0.5764**	<u>0.0609</u>	<u>0.1089</u>
	Tailnet	0.0567	0.1073	0.3224	0.3858	0.3152	0.2951	0.0230	0.0429
	MF-PC	0.0752	0.1272	0.4625	0.5642	0.2508	0.2697	0.0331	0.0603
	MACR	0.0966	0.1282	0.2538	0.3314	0.3898	0.5348	0.0386	0.0683
	EMNS	<u>0.1033</u>	0.1669	<u>0.5889</u>	**0.7051**	0.4769	0.4830	**0.0744**	**0.1258**
Music	STAR-GCN	0.1049	0.1542	<u>0.8469</u>	<u>0.9310</u>	0.3689	0.3790	0.0744	0.1115
	LightGCN	**0.1531**	0.2121	0.7421	0.8730	0.2877	0.3155	0.0829	0.1251
	SGL	0.1465	0.2058	0.7988	0.8926	0.4276	0.4434	<u>0.1093</u>	<u>0.1585</u>
	EGLN	<u>0.1530</u>	**0.2194**	0.7264	0.8657	0.2707	0.2981	0.0785	0.1228
	GANC	0.0134	0.0216	0.2830	0.4237	**0.9807**	**0.9768**	0.0087	0.0189
	PPNW	0.1383	0.1967	0.7544	0.8469	0.4293	0.4370	0.1016	0.1475
	Tailnet	0.0283	0.0700	0.2407	0.6491	<u>0.6463</u>	0.4914	0.0144	0.0553
	MF-PC	0.0413	0.0700	0.8318	0.9150	0.6075	0.5420	0.0423	0.0682
	MACR	0.1500	<u>0.2136</u>	0.7497	0.9100	0.3305	0.3795	0.0906	0.1472
	EMNS	0.1373	0.1898	**0.8589**	**0.9398**	0.5213	<u>0.5472</u>	**0.1215**	**0.1746**
Video	STAR-GCN	0.1183	0.1730	<u>0.9151</u>	0.9697	0.3376	0.3313	0.0800	0.1131
	LightGCN	**0.2166**	<u>0.2842</u>	0.8569	0.9525	0.2195	0.2547	0.0945	0.1387
	SGL	0.2083	0.2722	0.8770	0.9490	0.3704	0.3880	<u>0.1394</u>	0.1869
	EGLN	0.2122	**0.2857**	0.8528	0.9483	0.2213	0.2490	0.0934	0.1365
	GANC	0.0503	0.0861	0.6830	0.8261	**0.8707**	**0.8528**	0.0560	0.1031
	PPNW	0.2035	0.2634	0.8516	0.9198	0.3589	0.3669	0.1319	0.1720
	Tailnet	0.0761	0.1217	0.4848	0.6207	0.2147	0.1512	0.0306	0.0383
	MF-PC	0.1110	0.1482	0.8320	0.9412	0.2837	0.3306	0.0640	0.0974
	MACR	<u>0.2134</u>	0.2809	0.8676	**0.9839**	0.3394	0.3735	0.1327	<u>0.1890</u>
	EMNS	0.1876	0.2349	**0.9337**	<u>0.9766</u>	<u>0.5367</u>	<u>0.5736</u>	**0.1700**	**0.2211**

5   Detailed Study

5.1   Ablation Study

The core merit of our EMNS lies in the novelty-boosted masking for niche items and the negative sampling to recognize the users' real preference on popular items. We also strengthen the correlation between the user and her similar users by adding a mutual information loss. We perform a series of ablation studies on each of our proposed components to examine their effects.

- EMNS$_{w/o\text{-}BBM}$: This variant removes the novelty-boosted masking mechanism and performs convolution directly.
- EMNS$_{w/o\text{-}MI}$: This variant removes the mutual information loss hence there is no restrictions on a user and her similar users.
- EMNS$_{w/o\text{-}NS}$: This variant replaces novelty-boosted negative sampling strategy with a random one. It is based on EMNS$_{w/o\text{-}MI}$ and further removes all proposed sampling process.

Table 3. Results for ablation study.

Dataset	Model	Accuracy		Coverage		Novelty		Trade-off	
		Rec@5	Rec@10	Cov@5	Cov@10	Nov@5	Nov@10	F1@5	F1@10
MovieLens	EMNS	0.1033	0.1669	0.5889	0.7051	0.4769	0.4830	**0.0744**	**0.1258**
	EMNS$_{w/o\text{-}BBM}$	0.0806	0.1333	0.4980	0.5897	0.5206	0.5313	0.0571	0.0999
	EMNS$_{w/o\text{-}MI}$	0.0983	0.1607	0.6076	0.7217	0.4863	0.4909	0.0731	0.1244
	EMNS$_{w/o\text{-}NS}$	0.1229	0.2012	0.4001	0.4904	0.1713	0.1940	0.0364	0.0648
Music	EMNS	0.1373	0.1898	0.8589	0.9398	0.5213	0.5472	**0.1215**	**0.1746**
	EMNS$_{w/o\text{-}BBM}$	0.1290	0.1779	0.8520	0.9363	0.5624	0.5885	0.1201	0.1727
	EMNS$_{w/o\text{-}MI}$	0.1362	0.1880	0.8570	0.9403	0.5181	0.5447	0.1200	0.1726
	EMNS$_{w/o\text{-}NS}$	0.1469	0.2062	0.7584	0.8963	0.2972	0.3287	0.0826	0.1273
Video	EMNS	0.1876	0.2349	0.9337	0.9766	0.5367	0.5736	**0.1700**	**0.2211**
	EMNS$_{w/o\text{-}BBM}$	0.1870	0.2329	0.9192	0.9643	0.5402	0.5811	0.1692	0.2202
	EMNS$_{w/o\text{-}MI}$	0.1858	0.2304	0.9388	0.9838	0.5316	0.5623	0.1680	0.2152
	EMNS$_{w/o\text{-}NS}$	0.2069	0.2682	0.8807	0.9697	0.2650	0.3069	0.1071	0.1550

From the results in Table 3, we can find that each component alone indeed contributes to the overall trade-off metric, and combining them together achieves the most balanced novel recommendation.

Firstly, a single GCN module EMNS$_{w/o\text{-}BBM}$ without the novelty-boosted masking is inferior to the complete EMNS in terms of accuracy. This phenomenon indicates that our specialized masking mechanism drives the model to recover the representations of tail items from its neighbor items, hence it can learn better representations for tail items.

Secondly, the accuracy performance for EMNS$_{w/o\text{-}MI}$ declines on all three datasets. This suggests that making restrictions on the representations of users with the same interests has positive impacts.

Finally, the selection of negative samples is also very important. After removing the personalized negative sampling component, EMNS$_{w/o\text{-}NS}$ degrades to a naive sampling method, and it is hard to find true negative samples.

5.2   Parameter Study

There are two types of hyperparameters in our method: λ in the loss function and p for masking percentage. We only show the results of parameter study on Movielens due to the space limitation.

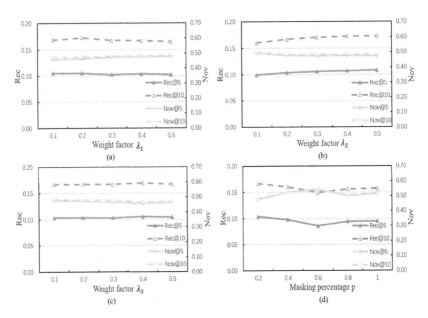

Fig. 2. Impacts of hyperparameters $\lambda_1, \lambda_2, \lambda_3, p$ on MovieLens dataset.

We can find that the accuracy and novelty show the opposite trends. This is consistent with the problem setting and the results in previous studies. Our EMNS performs the user and item negative sampling at the same time, and the weighting factor λ_1 controls the proportion of item negative sampling. As we can see in Fig. 2(a), with the increase of λ_1, the accuracy gradually decreases and the novelty increases. This indicates that choosing a suitable negative sample for the item and making personalized recommendation do help improve novelty. We set λ_1 to 0.4 for MovieLens and Music datasets and 0.1 for Video dataset.

The weight factor λ_2 controls the proportion of masking and reconstructing task. When λ_2 is small, the novelty is high but the accuracy is relatively low in Fig. 2(b). This illustrates that the proper value of λ_2 can alleviate the popularity bias and improve the recommendation performance. We set λ_2 to 0.2 for MovieLens dataset, 0.05 for Music dataset, and 0.01 for Video dataset.

The weight factor λ_3 also has an impact on accuracy according to Fig. 2(c). It controls the strength of the representations of users with similar interests. When λ_3 becomes large, the restrictions also increase. We set λ_3 to 0.1 for MovieLens dataset and 0.01 for two Amazon datasets.

From Fig. 2(d), we can find that with the increase of masking percentage, the representations of tail items are obliged to be learned from their neighbors, thus the model reaches a better novelty, but its accuracy declines at the same time. We set p to 0.2 for MovieLens dataset and 0.4 for two Amazon datasets for a better balance.

6   Conclusion

In this work, we propose to enhance graph based recommendation methods with novelty-enhanced masking technique and deepen novel recommendation methods with personalized negative sampling. Firstly, we force the classical graph learning based models to pay more attention to tail items by assigning them larger masking probability. In this way, we alleviate the popularity bias when learning the item representations. We further classify users' preference into niche preference and normal preference when learning the user representations. Then, we debias the user's preference on popular items by differentiating the "true" or "false" negative samples for her. We conduct extensive experiments on three real-world datasets. The results demonstrate that our proposed model yields significant improvements over the state-of-the-art graph based and novelty oriented baselines.

Acknowledgments. This work has been supported in part by the NSFC Project (61572376).

References

1. Adomavicius, G., Kwon, Y.: Improving aggregate recommendation diversity using ranking-based techniques. TKDE **24**(5), 896–911 (2011)
2. van den Berg, R., Kipf, T.N., Welling, M.: Graph convolutional matrix completion. arXiv:1706.02263 (2017)
3. Chen, J., Zhang, H., He, X., Nie, L., Liu, W., Chua, T.S.: Attentive collaborative filtering: multimedia recommendation with item-and component-level attention. In: SIGIR, pp. 335–344 (2017)
4. Cheng, P., Wang, S., Ma, J., Sun, J., Xiong, H.: Learning to recommend accurate and diverse items. In: WWW, pp. 183–192 (2017)
5. Defferrard, M., Bresson, X., Vandergheynst, P.: Convolutional neural networks on graphs with fast localized spectral filtering. In: NIPS, pp. 3844–3852 (2016)
6. Derr, T., Ma, Y., Tang, J.: Signed graph convolutional networks. In: ICDM (2018)
7. Fan, W., et al.: Graph neural networks for social recommendation. In: WWW, pp. 417–426 (2019)
8. He, R., McAuley, J.: VBPR: visual Bayesian personalized ranking from implicit feedback. In: AAAI (2016)
9. He, X., Deng, K., Wang, X., Li, Y., Zhang, Y., Wang, M.: LightGCN: simplifying and powering graph convolution network for recommendation. In: SIGIR (2020)
10. Hurley, N., Zhang, M.: Novelty and diversity in top-n recommendation-analysis and evaluation. TOIT **10**(4), 1–30 (2011)
11. Jugovac, M., Jannach, D., Lerche, L.: Efficient optimization of multiple recommendation quality factors according to individual user tendencies. ESWA **81**, 321–331 (2017)
12. Kipf, T.N., Welling, M.: Semi-supervised classification with graph convolutional networks. arXiv:1609.02907 (2016)
13. Li, J., Lu, K., Huang, Z., Shen, H.T.: On both cold-start and long-tail recommendation with social data. TKDE **33**(1), 194–208 (2019)

14. Liang, Y., Qian, T., Li, Q., Yin, H.: Enhancing domain-level and user-level adaptivity in diversified recommendation. In: SIGIR, pp. 747–756 (2021)

15. Linden, G., Smith, B., York, J.: Amazon.com recommendations: item-to-item collaborative filtering. IEEE Internet Comput. **7**(1), 76–80 (2003)

16. Liu, S., Zheng, Y.: Long-tail session-based recommendation. In: RecSys (2020)

17. Lo, K., Ishigaki, T.: Matching novelty while training: novel recommendation based on personalized pairwise loss weighting. In: ICDM, pp. 468–477 (2019)

18. Monti, F., Bronstein, M.M., Bresson, X.: Geometric matrix completion with recurrent multi-graph neural networks. In: NIPS, pp. 3700–3710 (2017)

19. Oh, J., Park, S., Yu, H., Song, M., Park, S.T.: Novel recommendation based on personal popularity tendency. In: ICDM, pp. 507–516 (2011)

20. Park, Y.J.: The adaptive clustering method for the long tail problem of recommender systems. TKDE **25**(8), 1904–1915 (2012)

21. Qian, T., Liang, Y., Li, Q., Xiong, H.: Attribute graph neural networks for strict cold start recommendation. TKDE (2020)

22. de Sousa Silva, D.V., Durão, F.A.: Dynamic clustering personalization for recommending long tail items. In: FedCSIS, pp. 417–425 (2020)

23. Wang, X., He, X., Wang, M., Feng, F., Chua, T.S.: Neural graph collaborative filtering. In: SIGIR, pp. 165–174 (2019)

24. Wei, T., Feng, F., Chen, J., Wu, Z., Yi, J., He, X.: Model-agnostic counterfactual reasoning for eliminating popularity bias in recommender system. In: SIGKDD (2021)

25. Wu, C., Wu, F., An, M., Huang, J., Huang, Y., Xie, X.: NPA: neural news recommendation with personalized attention. In: SIGKDD, pp. 2576–2584 (2019)

26. Wu, C., Wu, F., An, M., Huang, Y., Xie, X.: Neural news recommendation with topic-aware news representation. In: ACL. pp. 1154–1159 (2019)

27. Wu, J., et al.: Self-supervised graph learning for recommendation. In: SIGIR, pp. 726–735 (2021)

28. Wu, Q., et al.: Dual graph attention networks for deep latent representation of multifaceted social effects in recommender systems. In: WWW, pp. 2091–2102 (2019)

29. Yang, Y., Wu, L., Hong, R., Zhang, K., Wang, M.: Enhanced graph learning for collaborative filtering via mutual information maximization. In: SIGIR (2021)

30. Yin, H., Cui, B., Li, J., Yao, J., Chen, C.: Challenging the long tail recommendation. arXiv:1205.6700 (2012)

31. Ying, R., He, R., Chen, K., Eksombatchai, P., Hamilton, W.L., Leskovec, J.: Graph convolutional neural networks for web-scale recommender systems. In: SIGKDD (2018)

32. Zhang, J., Shi, X., Zhao, S., King, I.: Star-GCN: stacked and reconstructed graph convolutional networks for recommender systems. arXiv:1905.13129 (2019)

33. Zhu, Z., He, Y., Zhao, X., Zhang, Y., Wang, J., Caverlee, J.: Popularity-opportunity bias in collaborative filtering. In: WSDM, pp. 85–93 (2021)

34. Zolaktaf, Z., Babanezhad, R., Pottinger, R.: A generic top-n recommendation framework for trading-off accuracy, novelty, and coverage. In: ICDE (2018)

Knowledge-Enhanced Multi-task Learning for Course Recommendation

Qimin Ban[1], Wen Wu[1,2(✉)], Wenxin Hu[3], Hui Lin[4], Wei Zheng[5], and Liang He[1]

[1] School of Computer Science and Technology,
East China Normal University, Shanghai, China
51194506001@stu.ecnu.edu.cn, wwu@cc.ecnu.edu.cn, lhe@cs.ecnu.edu.cn
[2] Shanghai Key Laboratory of Mental Health and Psychological Crisis Intervention,
School of Psychology and Cognitive Science, East China Normal University,
Shanghai, China
[3] School of Data Science and Engineering,
East China Normal University, Shanghai, China
wxhu@cc.ecnu.edu.cn
[4] Shanghai Liulishuo Information Technology Co., Ltd., Shanghai, China
hui.lin@liulishuo.com
[5] Information Technology Services, East China Normal University, Shanghai, China
wzheng@admin.ecnu.edu.cn

Abstract. Knowledge tracing (KT) aims to model learners' knowledge level and predict future performance given their past interactions in learning applications. Adaptive learning systems mainly generate course recommendations based on learner's knowledge level acquired by KT. However, for KT tasks, learners' forgetting has not been well modeled. In addition, learner's individual differences also influence the accuracy of knowledge level prediction. While for recommendation tasks, most of methods are conducted separately from KT tasks, ignoring the deep connection between them. In this paper, we are motivated to propose a **K**nowledge-Enhanced **M**ulti-task Learning model for **C**ourse **R**ecommendation (KMCR), which regards the improved knowledge tracing task (IKTT) as an auxiliary task to assist the primary course recommendation task (CRT). Specifically, in IKTT, for assessing dynamic evolving knowledge level, we not only design a personalized controller to enhance the deep knowledge tracing model for modeling learner's forgetting behavior, but also use personality to model the individual differences based on the theory of cognitive psychology. In CRT, we adaptively combine learner's knowledge level obtained by IKTT with their sequential behavior to generate learners' representation. The experimental results on real-world datasets demonstrate that our approach outperforms related methods in terms of recommendation accuracy.

Keywords: Knowledge tracing · Course recommendation · Multi-task learning · Personality-based individual differences

1   Introduction

In response to the recent Covid19 pandemic, adaptive learning plays a more and more important role in online learning systems such as Khan Academy and Coursera. Compared with traditional classroom instruction by providing the same learning items for all learners, adaptive learning pays more attention to the differences between individuals. One of the common approaches to implement adaptive learning is to recommend a list of tailored learning items based on learners' needs, ability, and preferences, with the aim of facilitating learners to master new learning items efficiently [2].

Existing recommendations for learning items mainly depend on learners' knowledge level captured by knowledge tracing (KT) [11,23,25], which reflects learner's degree of mastery for learning items. As a typical work, context extended deep knowledge tracing is used to capture learners' knowledge level for all learning items and then recommend the learning items that the learner has not mastered yet [11]. Although these methods have made a great success in adaptive learning, there are still some limitations. For KT tasks, forgetting is a constant phenomenon in the learning process according to the Ebbinghaus forgetting curve theory. However, it is hard to handle the issue by using the widely used Long Short-Term Memory (LSTM) [10] based deep knowledge tracing model, because LSTM does the same operation for each state while the forgetting phenomenon is time-sensitive and uneven variational. In addition, the individual differences existed among learners are normally ignored, but it would influence the prediction performance of knowledge level. It has been proven that individual differences make impacts on learners' ability to automatically detect and extract the input information (i.e., course) [14]. For recommendation tasks, current studies considering the tasks of knowledge tracing and recommendation in separate steps, and they provide recommendation based on the output of the knowledge tracing task, which neglects the deep connection between two tasks as well as the rich information embedded inside the recommendation models such as learner's behavior.

To cope with the aforementioned issues, we propose a **K**nowledge-Enhanced **M**ulti-task Learning model for **C**ourse **R**ecommendation (KMCR). By sharing information between the improved knowledge tracing task (IKTT) and course recommendation task (CRT), one learner's knowledge level captured by IKTT is used as hints to guide the improvement of CRT. The full exploitation of the correlations between the IKTT and the CRT potentially enables the CRT task to be better off. Specifically, in IKTT, for capturing dynamic knowledge level more accurate, on one hand, we design a personalized controller, which consists three time-related features and the enhanced LSTM, for improving deep knowledge tracing model and deal with learners' complex forgetting problem more effective. On the other hand, we consider the influence caused by individual differences. We utilize personality to model the individual differences, which is supported by both cognitive psychology theory [14] and practical verification. Specifically, from the perspective of theory, in recent years, personality has been recognized as a valuable personal factor for personalized learning, as educational psychologists prove that the individual differences reflected by personality is important for understanding how learners perform in courses [4,5,14,24]. Besides, it is proved that personality can moderate language learning and comprehension ability [14], in particular in the area of second language learning [4]. Regarding practical verification, a preliminary experiment on our dataset shows that the personality trait of Neuroticism has significant positive correlation with learner's course accuracy in our dataset ($p < 0.01$). When integrating personality into IKTT to take individual differences into consideration, we map personality into a difference common space, which represents that different learner's personality have specific processing and understanding for courses. In CRT, for constructing learners' profiles, we adaptively fuse learners' previously acquired knowledge level, and sequential behavior.

The main contributions of this paper are as follows:

- We propose an end-to-end **K**nowledge-Enhanced **M**ulti-task Learning model for **C**ourse **R**ecommendation (KMCR), which regards the improved knowledge tracing task (IKTT) as an auxiliary task to assist the primary course recommendation task (CRT).
- In IKTT, we design a personalized controller to enhance the deep knowledge tracing model by incorporating three time-related features and the enhanced LSTM with the aim of assessing dynamic evolving knowledge level more accurate. In addition, inspired by the theory of cognitive psychology, we integrate personality into IKTT to take individual differences into consideration.
- In CRT, we adaptively fuse learners' knowledge level and sequential behavior to model learners' profile.
- We conduct experiments on real-world datasets, and the results show that our model outperforms the related models in terms of recommendation accuracy.

2 Related Work

2.1 Knowledge Tracing

Knowledge tracing is used to capture learners' knowledge level for the purpose of predicting learners' performance about the next interacted learning item [3]. The deep knowledge tracing (DKT) model [19] utilizes deep learning to predict learners' knowledge level. In the learning process, forgetting is a constant phenomenon. Some studies w.r.t. knowledge tracing have attempted to solve the forgetting problem. For example, [17] extended the DKT to consider forgetting by incorporating the lag time from the previous interaction, and the number of past trials on a question. [6,18] used position information to indicate time context and modeled learner's forgetting by exponential time interval for knowledge tracing. However, these studies neglect the priors and dependencies among learning items. Besides, they did not take individual differences among learners into consideration.

2.2   Personality and Learning

In knowledge tracing, individual differences cannot be ignored because it undoubtedly influences learner's performance from the aspect of processing and understanding information [14]. We utilize personality to model the individual differences which is inspired by cognitive psychology. Personality is considered one of the primary factors that influence human behavior, attitudes, interests, and taste [13]. A widely used personality model is the Big-Five model [7], which defines personality into five traits: *Openness to Experience (O), Conscientiousness (C), Extraversion (E), Agreeableness (A), Neuroticism (N)*. Personality has been regarded as a valuable component for personalized learning recently. Concretely, [5] observed that personality can affect students' academic performance. For example, those who are more self-disciplined (with a high C value) or emotionally stable (with a low N value) tend to achieve better academic outcomes. [24] found the significant correlations between personality and learners' communication behavior in Web-based Learning Systems. [14] indicated that the effect of statistical learning ability on second-language comprehension is moderated by personality traits. [4] showed that personality has an effect on second language learning. Although it has been proven that personality plays an important role of in learning, to our knowledge, few studies have incorporated the individual differences caused by personality into practical applications especially for knowledge tracing.

2.3   Course Recommendation

Existing work of course recommendation mainly depends on knowledge tracing. For instance, [23] obtained knowledge level by performance factors analysis and utilized the cosine similarity to recommend the most similar texts with the knowledge concepts of questions incorrectly answered. [11,25] used DKT to predict learners' knowledge level. Regarding the recommendations, [11] recommended the learning items that the learner has not mastered yet. [25] regarded the knowledge level as difficulty to filter the exercises. However, when capturing the learners' knowledge level, the aforementioned methods did not consider the factors that may influence learner's performance such as their forgetting behavior as well as their individual differences comprehensively. Besides, they directly generated recommendation items based on the output of the knowledge tracing task, which not only ignored the deep connection between them but also neglected learners' preference for the type of course.

When it comes to model learners' preference, sequential recommendation has achieved great success to model dynamic users' preference. For example, [28] used interest extractor layer to capture temporal interests from history behavior sequence. [26] utilized a time-aware controller and a content-aware controller to model users' long and short-term preference.

Regarding the deep connection between different tasks, multi-task learning (MTL) can use the learning results of one task as hints to guide another task to learn better [27]. MTL is widely used for recommendation systems due to its superiority. For example, [8] simultaneously learned parameters of the ranking task and rating task by a multi-task framework. [16] utilized knowledge graph embedding task to assist the recommendation task.

In adaptive learning, recommendation task is closely related to the knowledge tracing task. We are thus motivated to explore whether optimizing recommendation and knowledge tracing tasks in a joint and unified framework is beneficial. Regarding the recommendation task, in order to integrate learners' knowledge level and their preference to generate tailored course, we further research whether modeling the course recommendation task as a sequence recommendation task with a multi-task learning framework is effective. Besides, we investigate the effects of utilizing time context to model forgetting behavior and exploit personality-based individual differences in knowledge tracing task.

3 Model Description

3.1 Problem Definition

Let Learners $= \{u_1, u_2, \ldots, u_m\}$ be a set of learners, and Courses $= \{c_1, c_2, \ldots, c_n\}$ be a set of courses, where $|\text{Learners}| = m$, $|\text{Courses}| = n$. For each learner u, we have their personality traits and behavior history. Learner's personality traits contain five dimension which are represented by $u_{po}, u_{pc}, u_{pe}, u_{pa}, u_{pn}$. A learner behavior history is represented by the course ID sequence which the learner learned. For each interactive course in learner's learning sequence, we have its time context ($tContext$) (i.e. the time when the interaction happened) and accuracy (y_{aux})(i.e. the degree of mastery about the learner for the interactive course). Each course is denoted by its ID c. Our goal is to generate a top N recommendation list based on the sorted recommendation score $\hat{y}_{pri}$. Specifically, the score $\hat{y}_{pri}$ is obtained by learner's profile U and course description C, and U is represented by adaptively fusing learner's knowledge level $\hat{y}_{aux_{c_{k+1}}}$ and sequential behavior u_b. When accessing learner's knowledge level, we use time context $tContext$ to enhance deep knowledge tracing and integrate learners' personality-based individual differences.

3.2 Our Framework

In this section, we describe our proposed Knowledge-Enhanced Multi-task Learning model for Course Recommendation (KMCR). As shown in Fig. 1, it includes four parts: 1) **Input Layer**, where contains the features we used; 2) **Embedding Layer**, where the features used in our model are embedded into low-dimensional dense vector; 3) **Improved Knowledge Tracing Task (IKTT)**, where learners' dynamic knowledge level $\hat{y}_{aux}$ is captured; 4) **Course Recommendation Task (CRT)**, where the recommendation score $\hat{y}_{pri}$ of courses are figured out. A top-N recommended list is finally presented based on the sorted $\hat{y}_{pri}$.

Next, we will introduce the details of each part and explain how they are optimized jointly in an end-to-end framework.

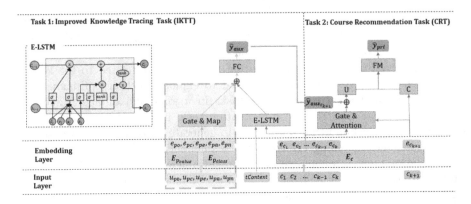

Fig. 1. Framework of the KMCR model

Input Layer. We use four types of features: 1) *learner's personality represen-tation*, which contains $u_{po}, u_{pc}, u_{pe}, u_{pa}, u_{pn}$[1]; 2) *learner's behavior*, composed of course ID that appears in learning sequence; 3) *course's description*, including course's ID c; 4) *interaction's context*, made up of the time t for each interaction in the learning sequence, named *tContext*.

Embedding Layer. $e_{po}, e_{pc}, e_{pe}, e_{pa}, e_{pn}$, e_c are embedding vector of the feature $u_{po}, u_{pc}, u_{pe}, u_{pa}, u_{pn}$ and c respectively. $e_{po}, e_{pc}, e_{pe}, e_{pa}, e_{pn}$ are from parameter matrix $\mathbf{E}_{\mathbf{Pvalue}} \in R^{13 \times dp}$ and $\mathbf{E}_{\mathbf{Pclass}} \in R^{5 \times dp}$,[2] and e_c is from $\mathbf{E_c} \in R^{n \times d}$ respectively, where dp and d denote the embedding size of personality and course ID. n is the total number of courses.

Improved Deep Knowledge Tracing Task (IKTT). In IKTT, we design a personalized controller to enhance the DKT model by considering both learn-ers' behavior sequence and time context of each interaction. We also integrate personality into IKTT to consider individual differences. Specifically, for per-sonalized controller, inspired by [26], we propose three time-related features as shown in Fig. 2:

Time interval feature δ_{tk} (Eq. 1). We consider this feature because the interval learning phenomenon is common in learners' behavior sequence. In IKTT, we think the longer the interval time from the previous interaction, the greater probability the learner will forget something. Suppose learner a's learning record is $B_a = \{(i_a^1; Jan\ 1^{st}); (i_a^2; Jan\ 2^{nd}); (i_a^3; Feb\ 2^{nd}); (i_a^4; Feb\ 3^{rd})\}$, it is reasonable to forget more information from interaction i_a^2 to i_a^3 than from i_a^3 to i_a^4, because

---

[1] The value of each personality dimension (on a [1–7] scale) is the average of two related problems (i.e., the positive score and 8 minus the corresponding reverse score).

[2] 13 indicates all the possible value (1,1.5,...,6.5,7), 5 denotes the class of personality (O,C,E,A,N). The embedding e_p* is the concatenation of personality value embed-ding and personality class embedding.

Fig. 2. The three time-related features, each circle denotes an interaction (i.e., a course in learner's learning sequence) and each semicircle corresponds to the knowledge skill which is contained in the course. A course may include one or more knowledge skills. The features of time interval, time span, and time lag are represented by the blue, green, and red bidirectional arrows respectively. (Color figure online)

interaction i_a^3 is learned one month after interaction i_a^2 while interaction i_a^4 is learned just one day after learning interaction i_a^3.

$$\delta_{tk} = tanh(w_\delta log(t_{i+1} - t_i) + b_\delta) \tag{1}$$

where w_δ and b_δ are trainable parameters. t_{i+1} and t_i denote the time of $(i+1)$-th and i-th interaction occurred in learner's behavior sequence. The time interval feature δ_{tk} encodes the temporal distance of two consecutive interactions (the blue bidirectional arrows).

Time span feature s_{tk} (Eq. 2). This feature is considered because there exist precedence and dependency relationships between courses and they will influence each other. For the aforementioned learner a's learning record B_a, the corresponding course of interaction i_a^1 may be the prerequisite course for the corresponding course of interaction i_a^4, and the time span between interaction i_a^1 and interaction i_a^4 influences the prediction accuracy of interaction i_a^4.

$$s_{tk} = tanh(w_s log(t_{k+1} - t_i) + b_s) \tag{2}$$

where w_s and b_s are trainable parameters. t_{k+1} indicates the occurrence time of predict $(k+1)$-th interaction. The time span feature s_{tk} encodes the temporal distance between each course and the predicted target course (c_{k+1}, which occurs in the $(k+1)$-th interaction) (the green bidirectional arrows).

Time lag feature ϵ_{tk} (Eq. 3). We consider this feature because there also exist a strong effect brought by the previously learned same course when predicting the interactive accuracy. For B_a, interaction i_a^2 does a lot for the accuracy of interaction i_a^4 if they correspond to the same course.

$$\epsilon_{tk} = tanh(w_\epsilon log(t_{k+1} - t_{k+1_{sc}}) + b_\epsilon) \tag{3}$$

where w_ϵ and b_ϵ are trainable parameters. $t_{k+1_{sc}}$ is the time of the latest interaction whose course is the same with the predicted course. The time lag feature ϵ_{tk} encodes temporal distance between the predicted target course (c_{k+1}) and the previous same course ($c_{k+1_{sc}}$) (the red bidirectional arrows).

Furtherly, we convert the time-related features (δ_{tk}, s_{tk}) into dense vectors by a fully connected layer as [1], and compute time gates (T_δ and T_s) accordingly. To better imitate human natural forgetting on learning, we exploit the three time-related features to enhance the forgetting gate and corresponding cell states of LSTM (i.e., E-LSTM in Fig. 1):

$$T_\delta = \sigma(x_k w_{x\delta} + \delta_{tk} w_{t\delta} + b_{t\delta}) \tag{4}$$

$$T_s = \sigma(x_k w_{xs} + s_{tk} w_{ts} + b_{ts}) \tag{5}$$

$$f_k = \sigma(x_k w_{xf} + \delta_{tk} w_{\delta f} + s_{tk} w_{sf} + \epsilon_{tk} w_{ef} + h_{k-1} w_{hf} + b_f) \tag{6}$$

$$c_k = f_k T_\delta c_{k-1} + i_k T_s tanh(x_k w_c + h_{k-1} w_{kc} + b_c) \tag{7}$$

The learners' knowledge hidden state h_k at the k-th input step is updated:

$$h_k = o_k tanh(c_k) \tag{8}$$

where w_* and b_* are trainable parameters. σ and tanh are sigmoid and tanh activate function respectively, and x_k, f_k, c_k, h_k, o_k are the input, forget gate, memory cell, hidden state and output gate of LSTM respectively.

For personality-based individual differences, considering different dimension may have different priority, a gating mechanism that accepts input and controls how much information passes is exploited, which is similar to [22]:

$$u_{p(o,...,n)} = \sigma(w_p e_{p(o,...,n)} + b_p) \odot \tanh(w_{ep} e_{p(o,...,n)} + b_{ep}) \tag{9}$$

where w_* and b_* are trainable parameters. $(o, ..., n)$ represents the abbreviation of (o, c, e, a, n).

Then the personality-based individual difference is accessed by using personality to map individual difference into a difference common space:

$$c_{diff} = w_d u_{p(o,...,n)} + b_d \tag{10}$$

where w_d and b_d are learned weight matric and bias. Finally, by using a fully connected with sigmoid activation ,we jointly exploit the learner's historical performance h_k and the personality-based individual difference c_{diff} by concatnation to predict learner's knowledge level $\hat{y}_{aux}$ for all courses :

$$\hat{y}_{aux} = \sigma(w_a(h_k \oplus c_{diff})) + b_a \tag{11}$$

where w_a and b_a are trainable parameters. $\oplus$ denotes the concatenation.

Regarding the loss function, we use it depends on the formation of dataset. Specifically, we employ the cross entropy loss when the ground truth of knowledge level is category data, which only contain 0 and 1. Otherwise, we apply the mean squared error loss if the ground truth of knowledge level is continuous value in 0–1 rather than discrete value of 0 and 1. In this case, our model could obtain the learners' mastery of specific knowledge more precisely.

Formally, the cross entropy loss is:

$$L_K = - \sum_{(u,c)\in\Omega} y_{aux_{c_{k+1}}} log(\hat{y}_{aux_{c_{k+1}}}) + (1 - y_{aux_{c_{k+1}}}) log(1 - \hat{y}_{aux_{c_{k+1}}}) \tag{12}$$

The mean squared error loss function is:

$$L_K = \frac{1}{|\Omega|} \sum_{(u,c)\in\Omega} (y_{aux_{c_{k+1}}} - \hat{y}_{aux_{c_{k+1}}})^2 \tag{13}$$

where Ω is the training set and $|\,|$ denotes its size. $y_{aux_{c_{k+1}}}$ and $\hat{y}_{aux_{c_{k+1}}}$ are the accuracy of ground-truth and prediction for the course of $(k+1)$-th interaction, i.e. c_{k+1}, respectively.

Course Recommendation Task (CRT). We generate the score of recommendation $\hat{y}_{pri}$ based on learner's profile U and course's representation C. Concerning learner's profile U, we adaptively fuse two related features based on the specific context: 1) *learner's current knowledge level of target course* $\hat{y}_{aux_{c_{k+1}}}$, which is acquired by IKTT. 2) *learner's sequential behavior representation* u_b, which is included as it may reflect learning preference. Intuitively, not all courses that a learner learned is important, so we firstly design a gating mechanism that accepts each course as an input e_{c_i} and controls how much information passes through to the next level, and then use a attention mechanism to adaptively calculate the representation vector of learner interests by taking into consideration the relevance of historical behaviors. As for course's representation C, it is denoted by the embedding of target course, i.e. $e_{c_{k+1}}$.

$$\bar{e}_{c_i} = \sigma(w_c e_{c_i} + b_c) \odot \tanh(w_{cc} e_{c_i} + b_{cc} e_{c_i}) \tag{14}$$

$$\alpha_j = \frac{exp(\bar{e}_{c_j} W e_{c_{k+1}})}{\sum_{p=1}^{k} exp(\bar{e}_{c_p} W e_{c_{k+1}})} \tag{15}$$

$$u_b = \sum_{j=1}^{k} \alpha_j \bar{e}_j \tag{16}$$

$$U = \beta * \hat{y}_{aux_{c_{k+1}}} + (1 - \beta) * u_b \tag{17}$$

We apply the factorization machine (FM) [20] to compute the predicted recommendation score $\hat{y}_{pri}$:

$$\hat{y}_{pri} = \sum_{i=1}^{d} w_i x_i^r + \sum_{i=1}^{d} \sum_{j=i+1}^{d} < z_i, z_j > x_i^r x_j^r + b \tag{18}$$

where $x^r = U \oplus C$. b denotes bias term. w is the weight for linear regression and $\{z_i\}_{i=1}^{d}$ is the weight of the pairwise interaction between x_i^r and x_j^r. $< \cdot, \cdot >$ indicates the inner product.

We exploit the pairwise Bayesian personalized ranking (BPR) loss [21] to optimize model parameters. It assumes that the observed interactions should be assigned higher prediction scores than unobserved ones:

$$L_R = \sum_{(u,i,j) \in O} -ln\sigma(\hat{y}_{pri_i} - \hat{y}_{pri_j}) \tag{19}$$

where $O = \{(u,i,j)|(u,i) \in R^+, (u,j) \in R^-\}$ indicates the pairwise training data. R^+ and R^- denote the observed interactions and the unobserved interactions respectively.

The Objective of Multi-task Learning. Different types of loss functions are linearly combined to jointly learn two tasks in an end-to-end manner:

$$L = L_R + \lambda L_K \tag{20}$$

where λ is control parameters.

4 Experiments

In this section, we conduct experiments to evaluate our method, aiming to answer the following research questions:

- **RQ1**: In general, would our proposed model perform better than the classical recommendation methods?
- **RQ2**: Would the auxiliary improved deep knowledge tracing task (IKTT) assist the main course recommendation task (CRT)?
- **RQ3**: Whether the components of personalized controller and personality-based individual differences affect the final recommendation performance?

4.1 Dataset

To answer our research questions, we evaluate the proposed model on a public dataset (POJ) and our own industrial dataset.

POJ Dataset. It is a public dataset which is collected from RKT [18]. [18] obtained it from Peking online platform of coding practices, which consists of computer programming questions. We processed the data based on RKT and then we found that there exist some anomalous learners who do the exercise more than 50000 times, so we further excluded 11 learners who solve exercise more than 8000 times. The period of data is between 2019-07-27 and 2020-04-11. It contains 13,289 learners, 2,030 exercise and 424,004 interactions. Each interaction has its time context and accuracy of exercise. The accuracy of exercise only contains 0 and 1, which denotes the wrong and correct result about learner's submit respectively. Therefore, we apply the cross entropy loss in IKTT task on this dataset.

Industrial Dataset. We collected the real data from a Chinese AI-driven education technology company (Liulishuo). It contains learners' records including their learning sequence, corresponding time context, and accuracy of courses. To obtain each learner's Big-Five personality, we exploited the TIPI questionnaire [7], where the value of each dimension (on a [1–7] scale) is the average of two related problems. After shifting out the survey with invalid answers,[3] we

---

[3] To clean the data, we first excluded learners whose average learning time shorter than 30 min and total learning days no more than a week based on their behavior and corresponding time context. We then filtered out all of the contradictory records (e.g., a user rated both 1 or 7 on two opposite statements "I think I am extraverted, enthusiastic" and "I think I am reserved, quiet") by analyzing their answer to questionnaire.

ultimately retain 2,063 learners enrolled in 1,198 courses with 312,379 interactions. The mean value of each personality trait is: *Openness to Experience* (M = 4.87, SD = 1.01), *Conscientiousness* (M = 4.75, SD = 1.18), *Extraversion* (M = 4.14, SD = 1.41), *Agreeableness* (M = 5.20, SD = 0.88), and *Neuroticism* (M = 3.59, SD = 1.22).[4] The period of data is between 2019-12-31 and 2020-02-29. Besides, the accuracy of courses is continuous value in 0–1, which denotes the learner's master degree about course. For this reason, on this dataset, we employ mean squared error loss in IKTT task.

For each learner, we reserved the last two interactions to validation and testing sets, while the rest interactions were used for training. Moreover, in the training process, we treated each observed interaction as a positive instance and pair a negative course with it and we supervised each step like [28]. In the test set, according to the scale of our dataset, we matched each observed interaction with 25 negative courses by random sampling, which is similar to [9], and ultimately generated the predicted scores for the 26 instances (1 positive and 25 negatives).

4.2   Evaluation Metrics

We leverage the widely used metrics to evaluate all the methods, which contain Normalized Discounted Cumulative Gain of Top-N items (NDCG@N, the larger the better) [12] and Hit Ratio of Top-N items (HR@N, the larger the better). NDCG@N accounts for the predicted position of the ground truth instance, and HR@N measures the successfully recommended percentage of ground truth instances among Top-N recommendations:

$$NDCG@N = \frac{1}{Z} DCG@N = \frac{1}{Z} \sum_{i=1}^{N} \frac{2^{I(|T_u \cap \{r_u^i\}|)} - 1}{log_2(i+1)} \tag{21}$$

$$HR@N = \frac{1}{m} \sum_{u} I(|R_u \cap T_u|) \tag{22}$$

where Z indicates a normalization constant of the maximum possible value of DCG@N. T_u is the interaction course set in test for learner u, and r_u^i denotes the i-th course in recommend list R_u based on predict score. $I(x)$ denotes an indicator function and the value is 1 if $x > 0$ else 0. m is the number of learners in our test set.

4.3   Compared Methods

To evaluate the performance on the recommendation task, we compared our model with the following classical recommendation methods:

– **BPRMF** [21] is matrix factorization optimized by the BPR loss, which exploits the learner-course ID alone.

---

[4] M and SD refer to Mean and Standard Deviation respectively.

Table 1. Performance comparison on POJ dataset. The best performing result for each metric is boldfaced, and the best baseline method is underlined.

Model	NDCG@1	NDCG@5	NDCG@10	NDCG@20	HR@5	HR@10	HR@20
BPRMF	0.0381	0.1159	0.1776	0.2786	0.1981	0.3921	0.7934
FM	0.0388	0.1167	0.1784	0.2799	0.1993	0.3985	0.7955
LSTM	0.0392	0.1186	0.1866	0.2805	0.2061	0.4132	0.7959
DIEN	0.0399	0.1189	0.1870	0.2813	0.2072	0.4147	0.7974
SLI-REC	<u>0.0436</u>	<u>0.1207</u>	<u>0.1872</u>	<u>0.2857</u>	<u>0.2075</u>	<u>0.4183</u>	<u>0.7986</u>
KMCR-K	0.0445	0.1211	0.1883	0.2859	0.2096	0.4195	0.799
KMCR-T	0.0461	0.1223	0.1894	0.2867	0.2109	0.4211	0.7993
KMCR-P	**0.0473**	**0.1246**	**0.1942**	**0.2883**	**0.2141**	**0.4255**	**0.8015**
$\triangle^a$ SLI-REC(%)	8.39	3.27	3.73	0.92	3.19	1.73	0.36

[a] $\triangle$ means the gain of the method which is proposed by us on the previous line over the methods which followed by $\triangle$. Gain% $= \frac{Value_{OurModel}-Value_{Baseline}}{Value_{Baseline}} \times 100\%$, where $Value_{OurModel}$ and $Value_{Baseline}$ denote the performance of our approach and the baseline model respectively

- **FM** [20] uses the learner-course ID to present recommendation via factorization machine.
- **LSTM** [10] uses learner's sequential behaviors to model learner's preference for sequential prediction by incorporating long short-term memory.
- **DIEN** [28] uses two layers of GRU to model the learners' sequential behaviors. In the second layer, GRU with attentional update gate is leveraged.
- **SLI-REC** [26] is based on LSTM and exploits attention mechanism to adaptively combine learners' long and short-term preferences. Time and semantic context are considered to model short-term preferences.

To investigate the effective of each component of our model, we further compared our model with the following variants methods:

- **KMCR-P** removes the part of individual differences in knowledge tracing task (the part which denoted by yellow rectangle in Fig. 1).
- **KMCR-K** removes the auxiliary task of IKTT (i.e., set L_K in Eq. 20 to equal 0) from KMCR-P.
- **KMCR-T** removes the personalized controller which is introduced in IKTT from KMCR-P (i.e., the IKTT is made up of the original DKT).

We implemented models by using Tensorflow.[5] We initialized the hyperparameters for the baselines by following related work and carefully tuned them to ensure that they achieve the optimal performance. For our model, the learning rate was tuned amongst {0.0001, 0.0003, 0.0005, 0.001, 0.003, 0.005}. As a result, the optimal setting for learning rate is 0.001. Dropout ratio is 0.8 to reduce overfitting. The embedding size of dp for personality and d for other parameters are fixed to 5 and 18. λ was searched in {0.1,0.2,...,0.9,1}, and we finally set it to 1. We optimized all models with the Adam optimizer, where the batch size was fixed at 128.

---

[5] https://www.tensorflow.org.

Table 2. Performance comparison on industrial dataset. The best performing result for each metric is boldfaced, and the best baseline method is underlined.

Model	NDCG@1	NDCG@5	NDCG@10	NDCG@20	HR@5	HR@10	HR@20
BPRMF	0.0383	0.1163	0.1782	0.2793	0.1992	0.3957	0.7962
FM	0.0391	0.1172	0.1788	0.2837	0.1999	0.3998	0.7987
LSTM	0.0394	0.1219	0.1871	0.2811	0.2138	0.4188	0.7990
DIEN	0.0401	0.1253	0.1899	0.2823	0.2184	0.4190	0.8003
SLI-REC	<u>0.0408</u>	<u>0.1258</u>	<u>0.1904</u>	<u>0.2852</u>	<u>0.2186</u>	<u>0.4197</u>	<u>0.8019</u>
KMCR-K	0.0462	0.1264	0.1913	0.2865	0.2189	0.421	0.8025
KMCR-T	0.0486	0.1285	0.1926	0.2894	0.2195	0.4223	0.8094
KMCR-P	**0.0494**	**0.1304**	**0.1933**	**0.2932**	**0.2207**	0.4230	**0.8110**
$\Delta^a{}_{SLI-REC(\%)}$	21.08	3.66	1.52	2.81	0.96	0.79	1.13
KMCR	**0.0524**	**0.1358**	**0.2030**	**0.3006**	**0.2230**	**0.4319**	**0.8236**
$\Delta^a{}_{SLI-REC(\%)}$	28.43	7.95	6.62	5.40	2.01	2.91	2.71
$\Delta^a{}_{KMCR-P(\%)}$	6.07	4.14	5.02	2.52	1.04	2.10	1.55

[a] $\triangle$ means the gain of the method which is proposed by us on the previous line over the methods which followed by $\triangle$. Gain% = $\frac{Value_{OurModel} - Value_{Baseline}}{Value_{Baseline}} \times 100\%$, where $Value_{OurModel}$ and $Value_{Baseline}$ denote the performance of our approach and the baseline model respectively

4.4   Performance Comparison

Overall Comparison (RQ1). To answer RQ1, we compared *KMCR-P*, the variant of our model, with baseline methods on two datasets because POJ dataset does not have learner's personality.

Tables 1 and 2 report the overall performance of different methods on POJ and industrial dataset w.r.t. NDCG and HR respectively. We observed that sequential behavior based recommendation methods (including our method and *SLI-REC*, *DIEN* and *LSTM*) are better than traditional collaborative filtering models (e.g., *BPRMF* and *FM*). On the one hand, these methods utilize deep learning and it is more effective for learner and course representation. On the other hand, this demonstrates that the sequential behavior which reflects learners' recent learning intent is important. Among them, our proposed *KMCR-P* gets the best performance. Specifically, the average gain[5] of *KMCR-P* over the strongest baseline (*SLI-REC*) w.r.t. NDCG and HR are 4.08% and 3.42% on POJ dataset, while the values on industrial dataset are 7.27% and 5.99%. Compared with *SLI-REC*, *KMCR-P* further integrates learner's knowledge level when generating recommendation. Therefore, the experiment results indicate that learner's knowledge level can help generate the personalized course recommendation. Moreover, the improvement in terms of NDCG is larger than HR. It denotes that *KMCR-P* can generate better sorting results with the help of knowledge level.

Effect of Multi-task Learning (RQ2). Tables 1 and 2 show the performances of multi-task *KMCR-P* and single-task *KMCR-K* on POJ and industrial dataset respectively. It shows that our multi-task learning model (*KMCR-P*) achieves

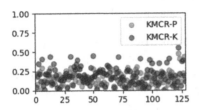

Fig. 3. The predicted absolute error on industrial dataset. The green and purple points denote the absolute error between the ground-truth and the predicted knowledge level by *KMCR-P*, *KMCR-K*.

Table 3. The ACC and AUC on POJ dataset.

	ACC	AUC
KMCR-K	0.6518	0.5537
KMCR-P	0.6691	0.5781

Table 4. The predicted absolute error on industrial dataset.

	Mean	SD
KMCR-K	0.1784	0.1052
KMCR-P	0.1268	0.0764

higher recommendation accuracy than the single-task model (*KMCR-K*) in all evaluation metrics (e.g., the average gain[5] of *KMCR-P* over the *KMCR-K* w.r.t. NDCG and HR are 3.27% and 2.53% on POJ dataset, while the values on industrial dataset are 3.37% and 2.32%.). The results are possibly because we regard IKTT as an optimizer target other than the recommendation task, and it makes the predicted knowledge level more accurate in this case.

To explicitly verify it, we make a comparison of the predicted knowledge level, which is generated by *KMCR-P* and *KMCR-K*. On the classificatory POJ dataset, we use the accuracy (ACC) and the Area Under Curve (AUC) to evaluate. When comparing on industrial dataset, we use the mean and standard deviation (SD) of the predicted absolute error[6], which is the differences between the ground truth and the predicted knowledge level by *KMCR-P* and *KMCR-K* respectively. The results on POJ and industrial dataset are shown in Table 3 and Table 4 respectively. It denotes that *KMCR-P* can get better predicted knowledge level compared with *KMCR-K* (The metric on POJ dataset are ACC: 0.6691 vs. 0.6518, AUC: 0.5781 vs. 0.5537. The larger ACC and AUC are better. The results on industrial dataset are Mean: 0.1268 vs. 0.1784, SD: 0.0764 vs. 0.1052. Lower mean and standard deviation of predicted absolute error are better). Furthermore, we made a case study on industrial dataset and visualized the predicted error results in the test set shown in Fig. 3. It indicates that the predicted absolute error by *KMCR-K* is bigger (i.e., the top of all points are the purple points in most cases), which in accordance with what we suspected.

Effect of Each Component (RQ3). Tables 1 and 2 also reflect the effectiveness of the personalized controller on POJ and industrial dataset. Concretely,

---

[6] The predicted absolute error = |ground_truth - *predict_value*|, and the mean of predicted absolute error = average(predicted absolute error), the standard deviation of predicted absolute error = standard deviation(predicted absolute error). The ground_truth is learner's knowledge level y_{aux} (i.e., the accuracy of course), and the *predict_value* is the predicted knowledge level $\hat{y}_{aux}$ by *KMCR-P* or *KMCR-K*.

the performance of *KMCR-T*, which removes the personalized controller introduced in IKTT (i.e., the IKTT is made up of the original DKT), achieves poor
performance than *KMCR-P* (the average of decrease in term of NDCG and HR
are 1.88% and 1.34% on POJ dataset, while the values on industrial dataset are
1.2% and 0.64%.). It validates that the proposed IKTT is superior to DKT and
using time context to model learner's forgetting behavior is important.

As for the effectiveness of the personality-based individual difference, it can
be seen from Table 2 that *KMCR*, which contains the personality-based individual difference is the best. Specifically, compared with *KMCR-P*, *KMCR* gets
4.44% and 2.69% improvement on average w.r.t. NDCG and HR. It demonstrates that the personality-based individual differences would improve the personalized recommendation by lifting the prediction performance of knowledge
level. Besides, it is worth noting that *KMCR* has a greater improvement than
KMCR-P on top 1 (6.07%), and the result is valuable because learners usually
choose top-ranked recommended courses in real-life learning scenario.

5 Conclusion

In this paper, we propose a Knowledge-Enhanced Multi-task Learning model for
Course Recommendation (KMCR), which regards the improved knowledge tracing task (IKTT) as auxiliary task to assist the primary course recommendation
task. As for IKTT, in order to precisely capture learners' dynamic knowledge
level, a personalized controller is designed to model learner's forgetting behavior
and a personality-based individual difference is innovatively integrated inspired
by cognitive psychology. When generating recommendation, we adaptively combine the knowledge level captured by IKTT with learner's behavior for modeling
the learner's profile. Experiments on two real-world datasets show that our approach outperforms related work in terms of recommendation accuracy. When it
comes to implementation, although we collect learner's personality by questionnaire, the acquisition of it would not impede the practical application of our
proposed method to generalize to other scenarios. It is because implicit acquisition of personality has been studied in recent years and personality has been
proven to be accurately predicted by multi-source heterogeneous data [15].

Acknowledgements. We thank editors and reviewers for their suggestions and comments. We also thank National Natural Science Foundation of China (under project
No. 61907016) and Science and Technology Commission of Shanghai Municipality
(under projects No. 21511100302, No. 19511120200 and 20dz2260300) for sponsoring
the research work.

References

1. Beutel, A., et al.: Latent cross: making use of context in recurrent recommender systems. In: WSDM, pp. 46–54 (2018)
2. Carbonell, J.R.: AI in CAI: an artificial-intelligence approach to computer-assisted instruction. IEEE Trans. Human-Mach. Syst. **11**(4), 190–202 (1970)
3. Corbett, A., et al.: Knowledge tracing: modeling the acquisition of procedural knowledge. User Model. User Adapt. Interact. **4**(4), 253–278 (1994)
4. Dewaele, J.M.: Personality in second language acquisition. In: The Encyclopedia of Applied Linguistics, pp. 4382–4389
5. Duff, A., et al.: The relationship between personality, approach to learning and academic performance. Personal. Individ. Diff. **36**(8), 1907–1920 (2004)
6. Gan, W., et al.: Modeling learner's dynamic knowledge construction procedure and cognitive item difficulty for knowledge tracing. Appl. Intell. **50**(11), 3894–3912 (2020)
7. Gosling, S.D., et al.: A very brief measure of the big-five personality domains. J. Res. Personal. **37**(6), 504–528 (2003)
8. Hadash, G., et al.: Rank and rate: multi-task learning for recommender systems. In: RecSys, pp. 451–454 (2018)
9. He, X., et al.: Neural collaborative filtering. In: WWW, pp. 173–182 (2017)
10. Hochreiter, S., Schmidhuber, J.: Long short-term memory. Neural Comput. **9**(8), 1735–1780 (1997)
11. Huo, Y., et al.: Knowledge modeling via contextualized representations for ISTM-based personalized exercise recommendation. Inf. Sci. **523**, 266–278 (2020)
12. Järvelin, K., et al.: IR evaluation methods for retrieving highly relevant documents. In: SIGIR, vol. 51, pp. 243–250. ACM, New York (2017)
13. John, O.P., Srivastava, S., et al.: The big five trait taxonomy: history, measurement, and theoretical perspectives. Handbook Personal. Theory Res. **2**, 102–138 (1999)
14. Kerz, E., Wiechmann, D., Silkens, T.: Personality traits moderate the relationship between statistical learning ability and second-language learners' sentence comprehension. In: CogSci (2020)
15. Mehta, Y., Majumder, N., Gelbukh, A., Cambria, E.: Recent trends in deep learning based personality detection. Artificial Intelligence Review **53**(4), 2313–2339 (2019). https://doi.org/10.1007/s10462-019-09770-z
16. Meng, W., et al.: Incorporating user micro-behaviors and item knowledge into multi-task learning for session-based recommendation. In: SIGIR, pp. 1091–1100 (2020)
17. Nagatani, K., et al.: Augmenting knowledge tracing by considering forgetting behavior. In: WWW, pp. 3101–3107 (2019)
18. Pandey, S., Srivastava, J.: RKT: Relation-aware self-attention for knowledge tracing. In: CIKM, pp. 1205–1214 (2020)
19. Piech, C., et al.: Deep knowledge tracing. Adv. Neural Inf. Process. Syst. **28**, 505–513 (2015)
20. Rendle, S.: Factorization machines with libfm. TIST **3**(3), 1–22 (2012)
21. Rendle, S., et al.: BPR: Bayesian personalized ranking from implicit feedback. arXiv preprint arXiv:1205.2618 (2012)
22. Tay, Y., et al.: Multi-pointer co-attention networks for recommendation. In: SIGKDD, pp. 2309–2318 (2018)
23. Thaker, K., et al.: Recommending remedial readings using student knowledge state. In: EDM, pp. 233–244. ERIC (2020)

24. Wu, W., et al.: Inferring students' personality from their communication behavior in web-based learning systems. Int. J. Artif. Intell. Educ. **29**(2), 189–216 (2019)
25. Wu, Z., et al.: Exercise recommendation based on knowledge concept prediction. Knowl. Based Syst. **210**, 106481 (2020)
26. Yu, Z., et al.: Adaptive user modeling with long and short-term preferences for personalized recommendation. In: IJCAI, pp. 4213–4219 (2019)
27. Zhang, Y., Yang, Q.: A survey on multi-task learning. IEEE Trans. Knowl. Data Eng. (2021)
28. Zhou, G., et al.: Deep interest evolution network for click-through rate prediction. In: AAAI, vol. 33, pp. 5941–5948 (2019)

Learning Social Influence from Network Structure for Recommender Systems

Ting Bai[1,2], Yanlong Huang[1], and Bin Wu[1,2(✉)]

[1] Beijing University of Posts and Telecommunications, Beijing, China
{baiting,HuangYL-0619,wubin}@bupt.edu.cn
[2] Key Laboratory of Trustworthy Distributed Computing and Service (BUPT),
Ministry of Education, Beijing, China

Abstract. The purchase decision of users is influenced by their basic preference of items, as well as the social influence of peers. Most social recommendation methods focus on incorporating the semantic collaborative information of social friends. In this paper, we argue that the semantic strength of their friends is also influenced by the subnetwork structure of friendship groups, which had not been well addressed in social recommendation literature. We propose a deep adversarial social model (SoGAN) that can automatically integrate the subnetwork structure of social groups and their semantic information into a unified recommendation framework. Specifically, we first align users in two different views, i.e., the "social-friend" view and "co-purchase" view. Then a generative adversarial network is used to learn the structure information of social groups to enhance the performance of recommender systems. We utilize the structural similarity between two views to produce true samples in SoGAN, and generate the mimic data based on the similarity between the semantic representations of users in two views. By discriminating the true instances based on structure similarity, we naturally inject the structure information into semantic learning of users. Extensive experiments on three real-world datasets, show the superiority of incorporating the social structure impact in recommender systems.

Keywords: Social recommendation · Generative adversarial networks · Multi-view graph · Network structure

1 Introduction

The social activities among friends in recommender systems enhance the activeness and retention of users [6], and had attracted increasing attention in the research area [12,28]. The social recommendation aims to incorporate the collaborative influence from social friends, which had been proved to be effective

This work is supported by the National Natural Science Foundation of China under Grant No. 62102038; the National Natural Science Foundation of China under Grant No. 61972047, the NSFC-General Technology Basic Research Joint Funds under Grant U1936220.

in modeling users' preference of items, as well as improving the performance of recommendation systems. Most existing social recommendation methods mainly focus on incorporating the different semantic collaborative information of social friends in different ways. For example, SocialMF [15] integrates the influences of users' friends into the matrix factorization model, based on trust relationships between users. SocialReg [16] incorporates social information into recommender systems by regularizing users' latent factors with latent factors of their connected users, and they assume that similar users would share similar latent preferences. Recently, some deep neural networks based methods [8,12,27,28] mainly focus on learning the strength of social ties by assigning different attention weights or finding more reliable social friends by using generative adversarial neural networks. For example, the hand-craft structure properties, like motifs, had been used to select the reliable neighborhood information of a user to enhance social recommendation [27]. But few of them had addressed the structural collaborative information of social groups, as well as its impact on the social influence of users.

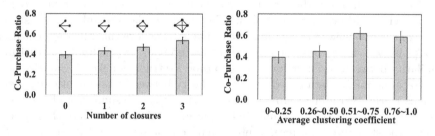

Fig. 1. The correlation between structure properties of social groups, i.e., triadic closure pattern and clustering coefficient, and their influence on user behavior, i.e., co-purchase of items.

In this paper, we argue that the semantic strength of their friends is also influenced by the subnetwork structure of friendship groups, which had not been well addressed in social recommendation literature. As shown in Fig. 1, we explore the subnetwork structure of social groups and their influence on a user's purchase behavior. We use two representative structure properties to characterize the structure of subnetwork, i.e., triadic closure pattern: the most basic unit of social network structure for studying group phenomena [14] and clustering coefficient: measuring how connected a vertex's neighbors are to one another, and use co-purchase ratio of items to represent the social influence of the friends on a user, that means the higher co-purchase ratio of items, the more influence of his friends. We can see that, there are some correlations between subnetwork structure and user's purchase behavior, for example, as the number of closures increases, the social influence will be higher, so is the clustering coefficient that is grouped by different threshold values (from 0–0.75).

However, the correlations are not consistent when measured by different structure properties, and the correlations within certain structure properties are also nonlinear, making it very difficult to find out the relevance between subnetwork structure of social groups and their social impact on users' interests. What's more, it is also impossible to enumerate all the hand-craft features to characterize subnetwork structure of friendship groups. In order to automatically learn the structure information of social groups, in this paper, we propose a deep adversarial social model (SoGAN) that can automatically integrate the subnetwork structure of social groups and their semantic collaborative information into a unified recommendation framework. As shown in Fig. 2, we decouple the different relationships into three views, "user-item" view (user-item relation), "co-purchase" view (user-user relation) and "social-friend" view (user-user relation). Different from other generative adversarial social recommendation models [25,27], we firstly select the useful social information in homogeneous "co-purchase" and "social-friend" views then integrate it into "user-item" view for the final prediction of user interests, which avoids the negative transfer of information across the heterogeneous views.

To make full use of the structure information, we align the users in "social-friend" view and "co-purchase" view. The more similar the subnetwork structures of the two views are, the more reasonable to believe that the different user groups in two views share the same purchase interests. We utilize the structural similarity between two user-user views to produce true instances in SoGAN, and generate the mimic data based on the similarity between the semantic representations of the two user groups. By forcing the semantic similarity of user representations to approach the inherent structure similarity, our model naturally injects the structure information into semantic representation learning of users. Finally, we update the representation of users in "user-item" view by incorporating the semantic information from both social-friends view and co-purchase view for the final prediction. Our contributions of this paper are summarized as follows:

- We propose a novel generative adversarial model SoGAN. By forcing the semantic similarity of user representation to approach the inherent structure similarity, SoGAN has the ability to take full use of the structure information of social groups and can automatically incorporate its impact into the collaborative social influence learning from other users.
- We decouple two types of relationships, user-item and user-user relations, into three different views. The useful social information of users is only selected in homogeneous "co-purchase" and "social-friend" views with user-user relation, and it is further integrated into heterogeneous "user-item" view for the final prediction of user interests, which avoids the negative transfer of information across the heterogeneous view.
- We conduct extensive experiments on three real-world datasets, showing the effectiveness of incorporating the social structure impact in social recommender systems.

2   Related Works

Our work is mostly related to social recommendation and adversarial learning. The related work is summarized as follows.

2.1   Social Recommendation

Traditional recommendation methods [17,18] use the basic interaction information of users and items, which suffer from the data sparsity problem. Leveraging social network information provides an effective approach to alleviate data sparsity and improve the model performance. The social information has been used in several studies [7,16]. For example, SocialMF [15] integrates the influences of users' friends into the matrix factorization model, based on trust relationships between users. SocialReg [16] incorporates social information into recommender systems by regularizing users' latent factors with latent factors of their connected users, and it assumes that similar users would share similar latent preferences. SBPR [29] assumes that users tend to assign higher ratings to the items which their friends prefer and incorporates this assumption into the pair-wise ranking loss. As friends at different degrees of closeness have different social influences, many methods also measure the strength of social ties. The attention mechanism is widely adopted in assigning different weights on the users' friends when modeling users' preferences in social recommendations [5,8].

Some graph-based recommendation models are proposed to capture the high order neighborhood relations [1,8,20,20,22,23]. For example, GraphRec [8] is a graph neural network based model for rating prediction in a social recommendation setting, and it aggregates representations for items and users from their connected neighbors. Diffnet [22] designs a layer-wise influence propagation structure to model how users' preferences evolve as the social influence propagates recursively. MHCN [28] is a social recommendation system based on hypergraph convolution to capture users' high-level information to improve recommendation performance. SEPT [26] argues that the supervision signals from other nodes are also likely to be beneficial to the representation learning, and proposes a general social perception self-supervision framework. Few of them had addressed the structural collaborative information of social groups, as well as its impact on the social influence of users.

2.2   Generative Adversarial Learning in RS

As the successful usages of generative adversarial networks (GAN) in many areas. Some GAN-based recommendation models have been proposed [2-4,19,21,30]. CFGAN [4] is a GAN-based collaborative filtering framework, in which the generator generates realistic purchase vectors instead of discrete item indexes for a given user. Instead of relying on a static and fixed rating score prediction function, an adversarial framework for collaborative ranking is proposed in [21] to approximate a continuous score function with pairwise comparisons. Some GAN-based models are proposed for social recommendation [9,13,25]. RSGAN [25] utilizes adversarial training to generate implicit friends, which are further used to generate social feedbacks to improve recommendation performance. DASO [9] proposes a deep adversarial social recommendation framework composed of two adversarial learning components in the social domain and item domain respectively, which adopts a bidirectional mapping method to transfer users' information between two domains. APR [13] adds perturbations to the latent factors of recommendation models as adversarial attacks, which enhances the performance and robustness of Bayesian personalized ranking.

3   Preliminary

3.1   Problem Statement

Assume we have a set of users and items, denoted by U and I respectively. $\mathbf{R} \in \mathbb{R}^{|U| \times |I|}$ is a binary adjacent matrix that records user-item interactions. If a user u consumed/clicked an item i, $r_{ui} = 1$, otherwise $r_{ui} = 0$. In social recommendation scenario, the behavior of a user is also influenced by their friends. We define the users who produce social influence on a user u as S_u and use $\mathbf{R}_S \in \mathbb{R}^{|U| \times |U|}$ to denote the social adjacency matrix, which is binary and symmetric because we work on undirected social networks.

Given the purchase history I^u of a user u and his/her social connected peers S_u, the recommender system aims to predict the interests of user u in the next purchase, defined as:

$$P(i_{next}^u) = \mathcal{F}(i \in I | u, I^u, S_u),\tag{1}$$

where $P(i_{next}^u)$ is the probability of item $i \in I$ being purchased by user u at the next time, and $\mathcal{F}$ is the prediction function. The prediction problem can also be formulated as a ranking problem so that the top K items to user u are recommended.

3.2   Construction of Multi-views

To avoid the negative transfer of information across the heterogeneous views, we decoupling the different relationships into three views, "user-item" view (user-item relation), "co-purchase" view (user-user relation) and "social-friend" view (user-user relation). Different views are aligned by the users that appear in the views at the same time. We firstly select the useful social information in homogeneous "co-purchase" and "social-friend" views, then integrate it into "user-item" view for the final prediction of user interests. The details are as follows:

- User-item view: consists of the interacted user and item nodes, as well as the edges of their interaction relations.
- Co-purchase view: the edge in co-purchase view is built between two users if they had purchased the same items. It intuitively reflects the same purchasing habit between users [6]. Considering the strength of co-purchase relation is

low and may introduce noise to the social impact learning, we filter the edges between two users by adding the restriction of social-friend relations. The adjacency matrix in co-purchase view is $\mathbf{M}_C = (\mathbf{R} \cdot \mathbf{R}^T) \odot \mathbf{R}_S$, where $\mathbf{R}$ and $\mathbf{R}_S$ are the adjacency matrix of user-item interaction and user-user social relationship, $\odot$ is the Hadamard product.

– Social-friend view: the edge in social-friend view is constructed according to the social friendship relations. Considering some friends may establish relationships occasionally [24,27], and it will not play a positive role in the modeling of users' social information. According to the stability of relationship mentioned in [14,26]: the structure of friendships in ternary closures is stable, we keep the edges with ternary closure social structure in social-friend view. The adjacency matrix of social-friend view $\mathbf{M}_S$ is computed as $\mathbf{M}_S = (\mathbf{R}_S \cdot \mathbf{R}_S) \odot \mathbf{R}_S$.

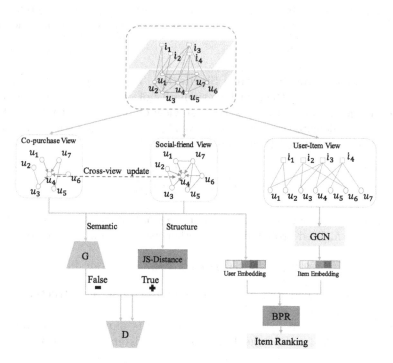

Fig. 2. The architecture of SoGAN model, including network structure learning component and item prediction component. The network structure is learned by generative adversarial networks, in which the mimic data is generated by computing the similarity between semantic representations of user in different views, while the true instance is computed based on the structure similarity of social groups. We update the representation of u_4 in "user-item" view by incorporating the semantic information from both "social-friend" views and "co-purchase" view for the final item prediction.

4   Our Proposed Model

The overview of our proposed model SoGAN is illustrated in Fig. 2. SoGAN is consist of two components, the subnetwork structure learning component by generative adversarial networks (GAN) and the prediction component for item ranking.

4.1   Network Structure Learning Component

The more similar the subnetwork structures of the two views are, the more reasonable to believe that the different user groups in two views share the similar user representations. We utilize the structural similarity between two user-user views to produce true instances and generate the mimic data based on the similarity between the semantic representations. By forcing the semantic similarity of user representation to approach the inherent structure similarity, our model naturally injects the structure information into semantic representation learning of the user.

True Instances from Structure Similarity. We generate the true training instances by computing the structure similarity of social groups in co-purchase and social-friend views. Given co-purchase view C and social-friend view S, for a user u_a, we calculate the normalized probability of his neighbors u_b according to the adjacency matrix $\mathbf{M}_C$ in co-purchase view.

$$p(u_b|u_a, \mathbf{M}_C) = \frac{\mathbf{M}_C(u_a, u_b)}{\sum_{u_k \in U} \mathbf{M}_C(u_a, u_k)}. \tag{2}$$

We formula the topological structure of user u_a as a vector $\mathbf{t}(u_a, C)$ by concatenating the normalized probability of all other users in view C, as follows:

$$\mathbf{t}(u_a, C) = p(u_1|u_a, \mathbf{M}_C) \oplus p(u_2|u_a, \mathbf{M}_C), ..., \oplus p(u_{|U|}|u_a, \mathbf{M}_C). \tag{3}$$

For the social-friend view S, we can obtain the topological structure $\mathbf{t}(u_a, S)$ of user u_a by the same way as in view C.

Then the locally topological structural similarity of user u_a between views C and S can be calculated by the Jensen-Shannon distance between $\mathbf{t}(u_a, C)$ and $\mathbf{t}(u_a, S)$ as:

$$D_{JS}(\mathbf{t}(u_a, C)||\mathbf{t}(u_a, S)) = \frac{1}{2}[D_{KL}(\mathbf{t}(u_a, C)||M) + D_{KL}(\mathbf{t}(u_a, S)||M)], \tag{4}$$

where $M = \frac{\mathbf{t}(u_a, C) + \mathbf{t}(u_a, S)}{2}$, and D_{KL} denotes the Kullback-Leibler divergence:

$$D_{KL}(P||Q) = \sum_x P(x)log\frac{P(x)}{Q(x)}. \tag{5}$$

Given a user u_a, we get the structure similarity of two views C and S by :

$$S_{structure}(C, S|u_a) = 1 - D_{JS}(\mathbf{t}(u_a, C)||\mathbf{t}(u_a, S)). \tag{6}$$

Finally, given a user u_a and a specific view C, we can estimate the true instance $P_{true}(S|C, u_a)$ for discriminator according to:

$$P_{true}(S|C, u_a) = \frac{S_{structure}(S, C|u_a)}{\sum_{r \in \{C,S\}} S_{structure}(r, C, u_a)}. \tag{7}$$

Mimic Instances Based on Semantic Similarity. The mimic data is generated based on the similarity between the semantic representations of a user in two views. Given a specific view C, we use graph neural networks [11] to capture the semantic information of user u_a by aggregating the neighborhood information in the view, termed as "inner-view aggregation", formulated as:

$$\mathbf{v}_{u_a,C}^k = \text{aggregation}(\mathbf{v}_{u_b,C}^{k-1}|u_b \in \mathcal{N}_{u_a,C}), \tag{8}$$

where $\mathcal{N}_{u_a,C}$ is the set of neighbors (include user u_a, i.e., self-loop) associated with user u_a in view C, and k is a hyper parameter to represent the depth for inner-view aggregation.

Following GraphSAGE [11], the aggregation operation can have many forms, we use mean aggregator, formulated as:

$$\mathbf{v}_{u_a,C}^k = \sigma(\mathbf{W}_C^k \cdot \text{mean}(\mathbf{v}_{u_b,C}^{k-1}|u_b \in \mathcal{N}_{u_a,C})), \tag{9}$$

where $\sigma(x) = \frac{1}{1+exp(-x)}$ is the sigmoid activation function.

The semantic representation of user u_a in the social-friend view can be computed by the same way. Given the semantic vectors $\mathbf{v}_{u_a,C}^k$ and $\mathbf{v}_{u_a,S}^k$ in copurchase view and social-friend view respectively, we compute the semantic similarity of u_a in two views as:

$$S_{semantic}(C, S|u_a) = \cos \left\langle \mathbf{v}_{u_a,C}^k, \mathbf{v}_{u_a,S}^k \right\rangle. \tag{10}$$

The mimic data generated based on the semantic similarity of user in two views is:

$$\mathcal{G}(S|C, u_a) = \frac{\exp(S_{semantic}(S, C|u_a))}{\sum_{r \in \{C,S\}} \exp(S_{semantic}(r, C, u_a))}, \tag{11}$$

where exp is the exponential function to make the similarity value be positive.

Based on the semantic similarity of two views, we update the representation of user u_a in co-purchase view C (same for u_a in social-friend view S) by incorporating the information from other view, termed as "cross-view aggregation", defined as:

$$\mathbf{v}_{u_a,C}^k = \mathbf{v}_{u_a,C}^k + \mathcal{G}(S|C, u_a) \cdot \mathbf{v}_{u_a,S}^k. \tag{12}$$

4.2   The Generative and Adversarial Process

We have generated the true and mimic data based on the structure and semantic similarity respectively. By discriminating the true and mimic data, our model will force the semantic similarity of user representations to approach the inherent structure similarity, so as to inject the structure information into semantic representation learning of users. The discriminator is defined as:

$$\mathcal{D}(u; \theta_{\mathcal{D}}) = \sigma(\mathbf{W}_{\mathcal{D}} \cdot \mathbf{e}_u + \mathbf{b}), \tag{13}$$

where $\theta_{\mathcal{D}}$ is the parameters optimized in discriminator, $\mathbf{e}_u$ is the input data from structure similarity or semantic similarity, and $\mathbf{W}_{\mathcal{D}}$ and $\mathbf{b}$ are the translation matrix and bias vector.

Following the optimization of GAN [10], we maximize the output log-probability when the similarity is computed by sub-structure of social group (see Eq. 7), and minimize the output log-probability when the similarity is computed based on the semantic representations of user in two views (see Eq. 11). The parameters are optimized by:

$$\min_{\mathcal{G}} \max_{\mathcal{D}} V(\mathcal{G}, \mathcal{D}) = \sum_{u_a \in U} \left\{ \mathbb{E}_{X \sim P_{true}(S|C, u_a)}[log\mathcal{D}(X; \theta_{\mathcal{D}})] \right.$$
$$\left. + \mathbb{E}_{Z \sim \mathcal{G}(S|C, u_a)}[log(1 - \mathcal{D}(Z; \theta_{\mathcal{D}}))] \right\}. \tag{14}$$

4.3   The Optimized Loss Function

We update the representation of users in user-item view by incorporating the semantic information from both social-friend view and co-purchase view for the final prediction, formulated as:

$$\mathbf{v}_{u_a, P} = \mathbf{v}_{u_a}^P + \mathbf{W}_C \cdot \mathbf{v}_{u_a, C} + \mathbf{W}_S \cdot \mathbf{v}_{u_a, S}, \tag{15}$$

where $\mathbf{v}_{u_a}^P$ is the user representation generated by GCN in user-item view. $\mathbf{v}_{u_a, C}$ and $\mathbf{v}_{u_a, S}$ are the user representations in the generator of GAN in co-purchase view and social-friend view.

The generator of GAN is optimized by:

$$\mathcal{L}_1 = \mathbb{E}_{Z \sim \mathcal{G}(S|C, u_a)}[log(1 - \mathcal{D}(Z; \theta_{\mathcal{D}}))]]. \tag{16}$$

Given a user u and item i, the predicted interaction value r_{ui} is defined as:

$$\hat{r}_{ui} = \mathbf{v}_{u_a, P} \cdot \mathbf{v}_i, \tag{17}$$

where $\mathbf{v}_i$ is the learned vector of item i.

We use Bayesian Personalized Ranking (BPR) loss [17] to optimize our model, defined as:

$$\mathcal{L}_2 = \sum_{i, j \in I^a, u \in U} -log\sigma(\hat{r}_{ui} - \hat{r}_{uj}). \tag{18}$$

Finally, we integrate the GAN training loss function Eq. (16) and BPR loss Eq. (18) by weight β as the final optimization function:

$$\mathcal{L} = \beta\mathcal{L}_1 + \mathcal{L}_2. \tag{19}$$

The training process of our model is summarized in Algorithm 1.

Algorithm 1: SoGAN Algorithm

Input: User-Item interactions $(U, I, \mathbf{R})$, Social relations $\mathbf{S}$.
1 Initialize all parameters for $\mathcal{G}$ and $\mathcal{D}$
2 Sample true training instances by Eq. (7)
3 **while** *not converge* **do**
4   **for** *Generator_steps* **do**
5   Generate the similarity of views for each user a in view C by Eq. (11)
6   Incorporate the information from generated view by Eq. (12)
7   Update $\mathcal{G}$ by minimizing Eq. (19)
8   **for** *Discriminator_steps* **do**
9   Sample true similarity based on structure for each user a in view C by Eq. (7)
10   Generate the mimic similarity for each user a in view C by Eq. (11)
11   Update $\mathcal{D}$ by maximizing the discriminator in Eq. (14)
12 **return** the representations of user and item in Eq.(15)

---

5   Experiments

5.1   Experimental Settings

Datasets. We experiment with three representative real-world datasets Ciao,[1] Douban Book[2] and Yelp.[3] Ciao is a review website where users give ratings and opinions on various products, and have trust relationships with other users. Douban is a popular site on which users can review movies, music, and books. Yelp comes from the Yelp challenge 2019. It contains users' reviews and check-in information of restaurants, businesses and so on. All the datasets contain both the user-item interactions and the user-user social relations, such as trust and following. Table 1 summarizes the statistics of the datasets.

Baseline Methods. The primary goal of our work is to employ the user's local structure information in different views to improve the accuracy of recommendation. We compare our model with the state-of-the-art baseline methods including MF-based, GAN-based, and GCN-based methods. The detailed introduction of the baselines are as follows:

---

[1] https://www.librec.net/datasets.html.
[2] https://www.dropbox.com/s/u2ejjezjk08lz1o/Douban.tar.gz?dl=0.
[3] https://www.yelp.com/dataset/challenge.

Table 1. Statistics of the datasets.

Datasets	Ciao	Douban Book	Yelp
# Users	3,787	23,585	84,528
# Items	16,121	23,386	44,691
# Interactions	35,310	809,697	1,018,438
# Social Links	9,215	22,863	7,766

- **BPR** [17]: It only uses the user-item interaction information. It optimizes the matrix factorization model with a pairwise ranking loss. This is a state-of-the-art model for traditional item recommendation.
- **SBPR** [29]: It's a social Bayesian ranking model that considers social relationships in the learning process. It assumes that users tend to assign higher ranks to items that their friends prefer.
- **SocialReg** [16]: It's a typical social-based model that regularizes and weighs friends' latent factors with social network information.
- **SoicalMF** [15]: It's a social-based model by reformulating the contributions of trusted users to the information of activating user's user-specific vector.
- **IRGAN** [19]: It's a GAN-based framework that unifies the generative and discriminative information retrieval models. It's a pioneering method that demonstrates the potential of GAN in information matching.
- **CFGAN** [4]: It's a GAN-based collaborative filtering framework, in which the generator generates realistic purchase vectors instead of discrete item indexes for a given user.
- **RSGAN** [25]: It's a state-of-the-art GAN-based social recommendation framework, where the generator generates reliable implicit friends for each user and the discriminator ranks items according to each user's own preference and her generated friends' preference.
- **Diffnet** [22]: It models the recursive dynamic social diffusion in social recommendation with a layer-wise propagation structure.
- **LightGCN** [12]: It is a graph-based model, consisted of two basic components: light graph convolution and layer combination for recommendation.
- **MHCN** [28]: It proposes a multi-channel hypergraph convolutional network, which works on multiple motif-induced hypergraphs integrating self-supervised learning into the training to improve social recommendation.
- **SEPT** [26]: It's a socially-aware self-supervised tri-training framework to improve recommendation by discovering self-supervision signals from two complementary views of the raw data.

Evaluation Metrics. Given a user, we infer the item that the user will interact with. Each candidate method will produce an ordered list of items for the recommendation. We adopt two widely used ranking-based metrics to evaluate the performance: Hit ratio at rank k (Hit Ratio@k) and Normalized Discounted Cumulative Gain at rank k (NDCG@k). We report the top K ($K = 5$ and $K = 10$) items in the ranking list as the recommended set.

Parameter Settings. We use different settings on validation data to obtain the best results. For the embedding size, we test the values of $[8, 16, 32, 64, 128, 256]$. The batch size and learning rate are searched in $[32, 64, 128, 256, 512]$ and $[0.001, 0.005, 0.01, 0.05, 0.1]$, respectively. The parameters for the baselines algorithms are carefully tuned to achieve optimal performance. We use Adam for optimization. The depth k for inner-view aggregation in set to 1 for model efficiency. The embedding size is set to $\{32, 64, 128, 256\}$ for the three datasets, respectively. The learning rate is set to $\{0.0001, 0.0005, 0.0005, 0.001\}$. We stop training if the evaluation metrics on the validation set decreased for 10 successive epochs. The dropout is set to 0.5 to avoid overfitting. The weight β is set to 0.5 in Eq. (19).

Table 2. Performance comparison of SoGAN and other methods.

Dataset	Ciao				Douban Book				Yelp			
Metric	H@5	H@10	N@5	N@10	H@5	H@10	N@5	N@10	H@5	H@10	N@5	N@10
BPR	0.219	0.329	0.145	0.180	0.418	0.541	0.306	0.346	0.572	0.741	0.405	0.460
SBPR	0.241	0.359	0.157	0.192	0.443	0.562	0.332	0.370	0.609	0.772	0.437	0.491
SocialReg	0.240	0.356	0.157	0.195	0.431	0.561	0.336	0.374	0.604	0.772	0.438	0.492
SocialMF	0.241	0.356	0.157	0.194	0.431	0.562	0.335	0.374	0.601	0.771	0.435	0.490
IRGAN	0.245	0.363	0.161	0.196	0.459	0.565	0.337	0.375	0.615	0.774	0.445	0.503
CFGAN	0.247	0.368	0.167	0.203	0.464	0.579	0.345	0.380	0.629	0.778	0.451	0.516
RSGAN	0.252	0.372	0.169	0.208	0.478	0.588	0.359	0.395	0.638	0.791	0.463	0.522
Diffnet	0.255	0.373	0.163	0.201	0.460	0.583	0.343	0.382	0.642	0.791	0.472	0.521
LightGCN	0.247	0.373	0.167	0.207	0.480	0.596	0.360	0.398	0.634	0.780	0.468	0.516
MHCN	0.242	0.360	0.163	0.201	0.482	<u>0.604</u>	0.361	0.400	<u>0.674</u>	<u>0.818</u>	<u>0.502</u>	<u>0.549</u>
SEPT	<u>0.251</u>	<u>0.381</u>	<u>0.170</u>	<u>0.212</u>	<u>0.486</u>	0.603	<u>0.367</u>	<u>0.405</u>	0.671	0.816	0.497	0.544
SoGAN	**0.294**	**0.432**	**0.199**	**0.244**	**0.665**	**0.781**	**0.517**	**0.554**	**0.717**	**0.853**	**0.544**	**0.597**
Gain[%]	16.83↑	13.38↑	17.06↑	15.09↑	36.81↑	29.49↑	40.58 ↑	36.74↑	6.87↑	4.52↑	9.41↑	9.59↑

5.2 Main Results

The performance of different models is shown in Table 2. We can see that:

(1) Our proposed model SoGAN outperforms all the baselines, including MF-based, GAN-based, or GCN-based models. This indicates that the local structure information of social groups plays an important role in improving the accuracy of the recommender systems, and our model can effectively learn the network structure information of social groups.

(2) GCN-based and GAN-based models perform better than MF-Based methods. The collaborative social information can be well captured by aggregating the information from neighbors in graph neural networks. SEPT and MHCN achieve the best performance in all the GCN-based baselines. The self-supervised signals in SEPT and MHCN are used to learn information from different views of the original data, which improves the model performance.

(3) GAN-based models perform better than MF-Based models, i.e., the GAN-based social method RSGAN achieves better performance than SocialReg and SBPR. IRGAN and CFGAN have comparable results and outperform

the classical method BPR. This indicates the usefulness of using generative adversarial training process in social recommender systems.

(4) The models that leverage social relations perform much better than the general models, i.e., SocialReg and SBPR outperform BPR, RSGAN outperforms IRGAN and CFGAN, SEPT and MHCN outperform LightGCN. This shows that the social information can be well utilized to improve the model performance.

5.3   Experimental Analysis

The Effectiveness of Generative Adversarial Learning. We verify the effectiveness of generative adversarial learning and evaluate its ability to distinguish the complementary information from social structure. We remove the GAN training process, and the variant model is termed as "SoGAN(-GAN)", in which the semantic information of users in two views is obtained by simply aggregating their neighborhood information and optimized by Eq. (18). The comparisons of SoGAN and its variant model SoGAN(-GAN) are shown in Fig. 3. We can see that: by using generative adversarial training process, our model can learn from the local structure information of social groups. Such structure information is useful to enhance the performance of social recommender systems.

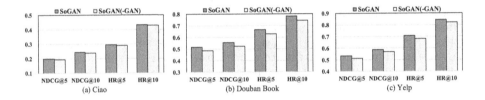

Fig. 3. The effectiveness of generative adversarial learning.

The Effectiveness of Inner-view and Cross-view Aggregation. Two aggregation methods are used in SoGAN: inner-view aggregation (see Eq. 8) and cross-view aggregation (see Eq. 12). To better understand and verify the effect of each aggregation operation, we make ablation analysises by removing one of the aggregation operation from our model. The comparisons of SoGAN with two degenerated variants: SoGAN(-Inner) and SoGAN(-Cross), by removing inner-view aggregation and cross-view updating respectively, are shown in Fig. 4. We can see that: SoGAN(-Inner) and SoGAN(-Cross) perform worse than SoGAN, indicating that the collaborative neighborhood information in the view itself (captured by inner-view aggregation) and complementary collaborative information from another view (learned by cross-view updating) are both critical for learning a better user representation. The complementary information from other view is incorporated based on the structural similarity of two views, showing the usefulness of using the network structure information of social groups.

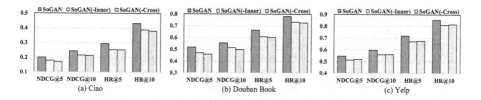

Fig. 4. The usefulness of inner-view and cross-view aggregations.

Model Convergence. The GAN framework encounters the problem of slow model convergence and long training time during the training process, especially when it is applied to the model with discrete data sampling. We show the learning curves of NDCG@10 and HR@10 about each GAN-based model and our model on datasets Ciao in Fig. 5, from which we can see that our model converges faster than other GAN-based models, and can gain superior experimental results. The results are consistent in other datasets with larger data amount, showing that our model can achieve better recommendation performance, and require less time for model training.

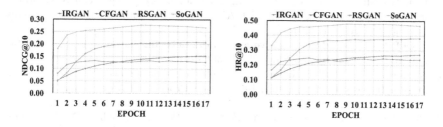

Fig. 5. The convergence curve of the GAN-based models on Ciao.

6   Conclusion

In this paper, we verify that the influence of social friends for a user is also influenced by the subnetwork structure of friendship groups. We propose a deep generative adversarial model SoGAN to learn the structural information of users from different views, and integrate the subnetwork structure of social groups and their semantic collaborative information into a unified recommendation framework. By learning the social influence from the network structure, our model achieves a better performance in social recommendation task.

References

1. Berg, R.V.D., Kipf, T.N., Welling, M.: Graph convolutional matrix completion. arXiv preprint arXiv:1706.02263 (2017)

2. Bharadhwaj, H., Park, H., Lim, B.Y.: Recgan: recurrent generative adversarial networks for recommendation systems. In: Proceedings of the 12th ACM Conference on Recommender Systems, pp. 372–376. ACM (2018)

3. Cai, X., Han, J., Yang, L.: Generative adversarial network based heterogeneous bibliographic network representation for personalized citation recommendation. In: Thirty-Second AAAI Conference on Artificial Intelligence (2018)

4. Chae, D.K., Kang, J.S., Kim, S.W., Lee, J.T.: CFGAN: a generic collaborative filtering framework based on generative adversarial networks. In: Proceedings of the 27th ACM International Conference on Information and Knowledge Management, pp. 137–146. ACM (2018)

5. Chen, C., Zhang, M., Liu, Y., Ma, S.: Social attentional memory network: modeling aspect-and friend-level differences in recommendation. In: Proceedings of the Twelfth ACM International Conference on Web Search and Data Mining, pp. 177–185. ACM (2019)

6. Cialdini, R.B., Goldstein, N.J.: Social influence: compliance and conformity. Annu. Rev. Psychol. **55**, 591–621 (2004)

7. Dai, B.R., Lee, C.Y., Chung, C.H.: A framework of recommendation system based on both network structure and messages. In: 2011 International Conference on Advances in Social Networks Analysis and Mining, pp. 709–714. IEEE (2011)

8. Fan, W., et al.: Graph neural networks for social recommendation. In: The World Wide Web Conference, pp. 417–426. ACM (2019)

9. Fan, W., Ma, Y., Yin, D., Wang, J., Tang, J., Li, Q.: Deep social collaborative filtering. In: Proceedings of the 13th ACM Conference on Recommender Systems, pp. 305–313. ACM (2019)

10. Goodfellow, I., et al.: Generative adversarial nets. In: Advances in Neural Information Processing Systems, pp. 2672–2680 (2014)

11. Hamilton, W., Ying, Z., Leskovec, J.: Inductive representation learning on large graphs. In: Advances in Neural Information Processing Systems, pp. 1024–1034 (2017)

12. He, X., Deng, K., Wang, X., Li, Y., Zhang, Y., Wang, M.: LightGCN: Simplifying and powering graph convolution network for recommendation. In: Proceedings of the 43rd International ACM SIGIR conference on research and development in Information Retrieval, pp. 639–648 (2020)

13. He, X., He, Z., Du, X., Chua, T.S.: Adversarial personalized ranking for recommendation. In: The 41st International ACM SIGIR Conference on Research & Development in Information Retrieval, pp. 355–364. ACM (2018)

14. Huang, H., Tang, J., Wu, S., Liu, L., Fu, X.: Mining triadic closure patterns in social networks. In: Proceedings of the 23rd International Conference on World Wide Web, pp. 499–504 (2014)

15. Jamali, M., Ester, M.: A matrix factorization technique with trust propagation for recommendation in social networks. In: Proceedings of the Fourth ACM Conference on Recommender Systems, pp. 135–142. ACM (2010)

16. Ma, H., Zhou, D., Liu, C., Lyu, M.R., King, I.: Recommender systems with social regularization. In: Proceedings of the Fourth ACM International Conference on Web Search and Data Mining, pp. 287–296. ACM (2011)

17. Rendle, S., Freudenthaler, C., Gantner, Z., Schmidt-Thieme, L.: BPR: Bayesian personalized ranking from implicit feedback. In: Proceedings of the Twenty-Fifth Conference on Uncertainty in Artificial Intelligence, pp. 452–461. AUAI Press (2009)

18. Rendle, S., Freudenthaler, C., Schmidt-Thieme, L.: Factorizing personalized Markov chains for next-basket recommendation. In: Proceedings of the 19th International Conference on World Wide Web, pp. 811–820. ACM (2010)
19. Wang, J., et al.: IRGAN: a minimax game for unifying generative and discriminative information retrieval models. In: Proceedings of the 40th International ACM SIGIR Conference on Research and Development in Information Retrieval, pp. 515–524. ACM (2017)
20. Wang, X., He, X., Wang, M., Feng, F., Chua, T.: Neural graph collaborative filtering. In: Proceedings of the 42nd International ACM SIGIR Conference on Research and Development in Information Retrieval, SIGIR 2019, Paris, 21–25 July 2019, pp. 165–174 (2019)
21. Wang, Z., Xu, Q., Ma, K., Jiang, Y., Cao, X., Huang, Q.: Adversarial preference learning with pairwise comparisons. In: Proceedings of the 27th ACM International Conference on Multimedia, pp. 656–664 (2019)
22. Wu, L., Sun, P., Fu, Y., Hong, R., Wang, X., Wang, M.: A neural influence diffusion model for social recommendation. In: Proceedings of the 42nd International ACM SIGIR Conference on Research and Development in Information Retrieval, pp. 235–244 (2019)
23. Ying, R., He, R., Chen, K., Eksombatchai, P., Hamilton, W.L., Leskovec, J.: Graph convolutional neural networks for web-scale recommender systems. In: Proceedings of the 24th ACM SIGKDD International Conference on Knowledge Discovery & Data Mining, pp. 974–983. ACM (2018)
24. Yu, J., Gao, M., Li, J., Yin, H., Liu, H.: Adaptive implicit friends identification over heterogeneous network for social recommendation. In: Proceedings of the 27th ACM International Conference on Information and Knowledge Management, pp. 357–366 (2018)
25. Yu, J., Gao, M., Yin, H., Li, J., Gao, C., Wang, Q.: Generating reliable friends via adversarial training to improve social recommendation. arXiv preprint arXiv:1909.03529 (2019)
26. Yu, J., Yin, H., Gao, M., Xia, X., Zhang, X., Hung, N.Q.V.: Socially-aware self-supervised tri-training for recommendation. arXiv preprint arXiv:2106.03569 (2021)
27. Yu, J., Yin, H., Li, J., Gao, M., Huang, Z., Cui, L.: Enhance social recommendation with adversarial graph convolutional networks. IEEE Trans. Knowl. Data Eng. (2020)
28. Yu, J., Yin, H., Li, J., Wang, Q., Hung, N.Q.V., Zhang, X.: Self-supervised multi-channel hypergraph convolutional network for social recommendation. In: Proceedings of the Web Conference 2021, pp. 413–424 (2021)
29. Zhao, T., McAuley, J., King, I.: Leveraging social connections to improve personalized ranking for collaborative filtering. In: Proceedings of the 23rd ACM International Conference on Conference on Information and Knowledge Management, pp. 261–270. ACM (2014)
30. Zhou, F., Yin, R., Zhang, K., Trajcevski, G., Zhong, T., Wu, J.: Adversarial point-of-interest recommendation. In: The World Wide Web Conference, pp. 3462–34618. ACM (2019)

PMAR: Multi-aspect Recommendation Based on Psychological Gap

Liye Shi[1], Wen Wu[1,2(✉)], Yu Ji[1], Luping Feng[3], and Liang He[1]

[1] School of Computer Science and Technology, East China Normal University, Shanghai, China
lyshi@ica.stc.sh.cn, 52205901009@stu.ecnu.edu.cn, lhe@cs.ecnu.edu.cn
[2] Shanghai Key Laboratory of Mental Health and Psychological Crisis Intervention, School of Psychology and Cognitive Science, East China Normal University, Shanghai, China
wwu@cc.ecnu.edu.cn
[3] School of Data Science and Engineering, East China Normal University, Shanghai, China

Abstract. Review-based recommendations mainly explore reviews that provide actual attributes of items for recommendation. In fact, besides user reviews, merchants have their descriptions of the items. The inconsistency between the descriptions and the actual attributes of items will bring users psychological gap caused by the Expectation Effect. Compared with the recommendation without merchant's description, users may feel more unsatisfied with the items (below expectation) or be more impulsive to produce unreasonable consuming (above expectation), both of which may lead to inaccurate recommendation results. In addition, as users attach distinct degrees of importance to different aspects of the item, the personalized psychological gap also needs to be considered. In this work, we are motivated to propose a novel Multi-Aspect recommendation based on Psychological Gap (PMAR) by modelling both user's overall and personalized psychological gaps. Specifically, we first design a gap logit unit for learning the user's overall psychological gap towards items derived from textual review and merchant's description. We then integrate a user-item co-attention mechanism to calculate the user's personalized psychological gap. Finally, we adopt Latent Factor Model to accomplish the recommendation task. The experimental results demonstrate that our model significantly outperforms the related approaches w.r.t. rating prediction accuracy on Amazon datasets.

Keywords: Review-based recommendation · Collaborative filtering · Psychological gap · Deep learning

1 Introduction

A variety of review-based recommendations have been proposed, which incorporate the valuable information in user-generated textual reviews into the recommendation process [2,14,15]. Recently, deep learning methods have achieved

A. Bhattacharya et al. (Eds.): DASFAA 2022, LNCS 13246, pp. 118–133, 2022.
https://doi.org/10.1007/978-3-031-00126-0_8

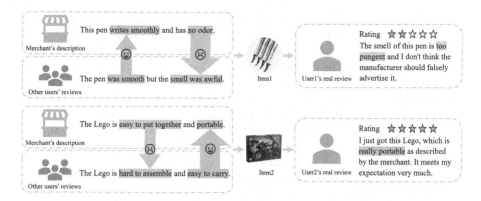

Fig. 1. Illsutrates the pshycholgical gap produced by the inconsistency between merchant's description and textual review (Note: red arrows indicate the user feels unsatisfied because the item's actual attributes are not as good as described by the merchant; in contrast, green arrows denote that the user has an impulse to consume the item. The size or thickness of an arrow represents the degree to which users value the corresponding aspect of the item.) (Color figure online)

good performance for review-based recommendations. Although these methods perform well, there are still some limitations that may influence the performance.

Firstly, existing studies mainly model a user's preference for items based on textual reviews written by users to provide actual attributes of items [7,8]. In fact, in addition to reviews, the merchants have their descriptions of the items. The inconsistency of descriptions and actual attributes of items will bring users psychological gap, which is a phenomenon caused by the Expectation Effect [1]. Compared with the recommendations without merchant's description, if an item's actual attributes are lower than expected, the user may be more unsatisfied with the item [17]. Conversely, if the actual attributes of an item are higher than expected, users are more likely to produce unreasonable consumption [18]. Both situations may lead to inaccurate recommendation results. In our work, we consider both user's overall psychological gap and personalized psychological gap in the recommendation process. Generally speaking, we model the overall psychological gap based on the merchant's description and actual attributes reflected by other users' reviews of the item. As for the personalized psychological gap, because users pay attention to different aspects of the item, we assign distinct importance to each aspect's psychological gap.

Figure 1 illustrates one user's overall psychological gap and personalized psychological gap towards different aspects of items in our review-based recommendation process. For item1, we observe that actual attributes (such as smell) reflected by other users' reviews are not as good as the merchant's description, which will cause user1 to be more unsatisfied with item1, and even feel cheated by the merchant. Under this circumstance, the recommendation should reduce the probability of recommending item1 to user1. While for item2, the actual attributes are more than expected relative to the merchant's description (i.e., more than expected), which will make the user more impulsive to buy. In this case, our model would increase the probability of recommending item2 to user2. Regarding personalized psychological gap, take item2 which contains two aspects (i.e., "easy to assemble" and "portable") as an example. Our model will predict that user2 pays the most attention to the "portable" aspect of item2. In this case, user2's psychological gap in the "portable" aspect is considered to be more essential to user2's preference for item2. As for the "easy to assemble" aspect, even if the description of the merchant is inconsistent with the actual attributes of item2 reflected by other user reviews, the psychological gap of user2 based on this aspect has little impact on user2's preference towards item2. In this way, we can infer that item2 still meets user2's preference.

In this paper, we are motivated to propose a Multi-Aspect recommendation based on Psychological Gap (shorten as PMAR). Based on textual review and merchant's description, PMAR models a user's overall psychological gap and personalized psychological gap for better recommendation. We first design an Overall Psychological Gap Module which adopts a gap logit unit to model the user's overall psychological gap based on merchants' descriptions and item reviews. We then propose a Personalized Psychological Gap Module which uses an attention mechanism to select relevant user history reviews based on the semantic information of the review and corresponding item intrinsic information. Besides, a user-item co-attention mechanism is used to obtain the user's personalized psychological gap towards the item. Lastly, we get the user's and item's final representations and apply the Latent Factor Model (LFM) [3] model to complete the rating prediction task. We conduct experiments on several real-world datasets to verify the effectiveness of our model. The experimental results show that our model achieves significantly higher rating prediction accuracy than state-of-the-art models.

In the following sections, we first introduce related work in Sect. 2. We then describe our proposed PMAR model in Sect. 3. We further present the details of our experimental design and results analysis in Sect. 4. Finally, we conclude the paper and indicate some future directions in Sect. 5.

2   Related Work

In recent years, how to integrate the rich information embedded in reviews into the recommendation generation has attracted increasing attention [12]. The review-based recommendations can help derive user preferences and item attributes by considering the semantic information reflected in reviews. With the breakthrough of deep learning technology, many works have been proposed to model contextual information from reviews for better recommendation. For example, DeepCoNN [19] adopted two parallel CNNs [4] in the last layer in order to generate potential representations of users and items at the same time. It placed a shared layer on the top to couple these two CNNs together. Neural Attentional Regression model with Review-level Explanations (NARRE) [2] utilized an attention mechanism to select essential reviews when modelling users

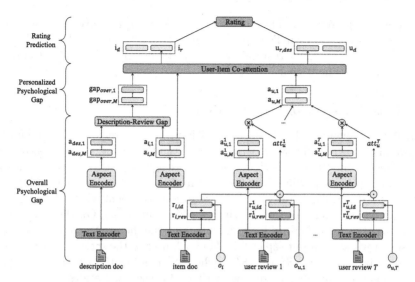

Fig. 2. The architecture of PMAR.

and items. Dual Attention Mutual Learning between Ratings and Reviews for Item Recommendation (DAML) [7] utilized local and mutual attention of the CNN to learn the features of reviews jointly. Very recently, MRCP [9] designed a three-tier attention network to select the informative words and reviews for users and items in a dynamic manner. In fact, in addition to reviews, merchants' descriptions of items are also important. The inconsistency of the merchant's descriptions and the item's actual attributes will arise psychological gaps of users [17,18]. For instance, as stated in [17], users would feel a tremendous psychological gap and even worse cheated when the quality of items did not match what the merchant advertised. Besides, [18] argued that when a user bought an item, the user's experiences exceeding expectations towards the item would increase his/her satisfaction.

3   Methodology

In this section, we will introduce the details of our PMAR model. The architecture is shown in Fig. 2 including three major modules. The first module is called *Overall Psychological Gap Module*. A Text Encoder layer is integrated to capture the contextual information of words in reviews and descriptions. Then an Aspect Encoder layer is designed to obtain the aspect representations of different aspects. In addition, a Description-Review Gap layer is used to model the user's overall psychological gap on the item based on merchant's description and item reviews. The second module is the *Personalized Psychological Gap Module*, where a User Review Selection layer is adopted to select relevant user history reviews based on the semantic information of reviews and the item information corresponding to reviews. Then a User-Item Co-attention layer is designed to calculate the user's personalized psychological gap. The last module is the *Rating Prediction Module*, where a Latent Factor Model (LFM) is utilized to predict users' ratings for the items.

3.1   Overall Psychological Gap Module

As an item's reviews are homogeneous which means each review is written for the same item, we merge all reviews written by distinct users for the item as the item review document. Item i's review document is represented as a word sequence $D_i = \{w_1, w_2, ..., w_{l_i}\}$ where l_i is the length of the review document. Similarly, we treat the description of item i as a description document which is represented as D_{des}, and the length of the description document is l_{des}. On the other hand, reviews written by a user are for different items (i.e. heterogeneous), so we take the user's reviews as input separately. For each user u, the set of his/her reviews is denoted by $S_u = \{S_u^1, S_u^2, \cdots, S_u^T\}$, where S_u^t denotes the t-th review of user u, T denotes the number of reviews written by user u. Besides, user u's t-th review can be represented as a word sequence $S_u^t = \{w_1^t, w_2^t, ..., w_{l_u}^t\}$, where l_u is the length of each review. Furthermore, the design of this module includes three layers: text encoder layer, aspect encoder layer, and description-review gap layer (see Fig. 2).

- **Text Encoder**

Given item i's review document $D_i = \{w_1, w_2, ..., w_{l_i}\}$, we first project each word to its embedding representation $\mathbf{D}_i = [\mathbf{w}_1, \mathbf{w}_2, ..., \mathbf{w}_{l_i}]$ where $\mathbf{w}_j \in R^{d_w}$ is the embedding vector for the j-th word, d_w is the dimension of word embedding, and l_i is the length of item review document. Then CNNs [4] are adopted to capture the context information around each word. Specifically, f_c convolution filters with the sliding window of size s are applied over matrix $\mathbf{D}_i$ to obtain contextual features of each word. The feature matrix $\mathbf{C}_i = [\mathbf{c}_{i,1}, \mathbf{c}_{i,2}, ..., \mathbf{c}_{i,l_i}]$ is the resultant feature, where $\mathbf{c}_{i,j}$ is the latent contextual feature vector for word w_j in item i's review document.

- **Aspect Encoder**

As mentioned before, an item has many aspects, such as quality, price and service. Each word in the review plays a distinct role on different aspects, so our goal is to derive a set of aspect-level item representations. Specifically, we first design M different aspect projection matrices over words which are denoted as $\mathbf{AM} = [\mathbf{W}_1, ..., \mathbf{W}_M]$, where M is the number of aspect, $\mathbf{W}_m \in R^{d_w \times d_w}$ represents the m-th aspect projection matrix. To ensure our model is end to end, the aspect in our model is defined as an implicit feature, and there is no need to pre-process to extract explicit aspects with other tools. The aspect-specific document embedding matrix $\mathbf{G}_{i,m}$ for item i's review document and the m-th aspect can be calculated as:

$$\mathbf{G}_{i,m} = [\mathbf{g}_{i,1,m}, \mathbf{g}_{i,2,m}, \cdots, \mathbf{g}_{i,l_i,m}] \tag{1}$$

$$\mathbf{g}_{i,j,m} = \mathbf{W}_m \mathbf{c}_{i,j} + b_m \tag{2}$$

where $\mathbf{g}_{i,j,m}$ is the word w_j's embedding towards the m-th aspect, b_m is the bias. Since we need to select important words under different aspects, we generate M aspect query vectors which are denoted as $\mathbf{V} = [\mathbf{v}_1, \ldots, \mathbf{v}_M]$, where $\mathbf{v}_m$ represents the m-th aspect query vector. Then the m-th aspect representation of item i's review document $\mathbf{a}_{i,m}$ can be represented as:

$$\mathbf{a}_{i,m} = \sum_{j=1}^{l_i} \beta_{i,j,m} \mathbf{g}_{i,j,m} \tag{3}$$

$$\beta_{i,j,m} = \frac{\exp\left(\mathbf{v}_m^\top \mathbf{g}_{i,j,m}/\tau\right)}{\sum_{k=1}^{l_i} \exp\left(\mathbf{v}_m^\top \mathbf{g}_{i,k,m}/\tau\right)} \tag{4}$$

Here, τ is the temperature parameter that is used to sharpen weights of important words, and $\beta_{i,j,m}$ is the importance of word w_j towards the m-th aspect. Similarly, we can get the m-th aspect representation of item i's description document $\mathbf{a}_{des,m}$. As mentioned above, the input of user reviews is a series of reviews. So we regard each review of user u's review set as a piece of document and repeat the steps similar to item document. Then $\mathbf{C}_u^t = \left[\mathbf{c}_{u,1}^t, \mathbf{c}_{u,2}^t, \ldots, \mathbf{c}_{u,l_u}^t\right]$ is the feature matrix of the t-th review of user u after convolution process. The aspect-specific review embedding matrix $\mathbf{G}_{u,m}^t$ for user u's t-th review and the m-th aspect can be calculated as:

$$\mathbf{G}_{u,m}^t = \left[\mathbf{g}_{u,1,m}^t, \mathbf{g}_{u,2,m}^t, \ldots, \mathbf{g}_{u,l_u,m}^t\right] \tag{5}$$

$$\mathbf{g}_{u,j,m}^t = \mathbf{W}_m \mathbf{c}_{u,j}^t + b_m \tag{6}$$

where $\mathbf{g}_{u,j,m}^t$ is the embedding of word w_j^t towards the m-th aspect, and b_m is the bias. Then we get the m-th aspect representation of user u's t-th review $\mathbf{a}_{u,m}^t$ as:

$$\mathbf{a}_{u,m}^t = \sum_{j=1}^{l_u} \beta_{u,j,m}^t \mathbf{g}_{u,j,m} \tag{7}$$

$$\beta_{u,j,m}^t = \frac{\exp\left(\mathbf{v}_m^\top \mathbf{g}_{u,j,m}^t/\tau\right)}{\sum_{k=1}^{l_u} \exp\left(\mathbf{v}_m^\top \mathbf{g}_{u,k,m}^t/\tau\right)} \tag{8}$$

where τ is the temperature parameter, and $\beta_{u,j,m}^t$ is the importance of word w_j^t towards the m-th aspect.

- **Description-Review Gap**

In reality, in addition to the reviews, the merchants have their own descriptions of the items. In this layer, we propose a Description-Review Gap layer to model the overall psychological gap between users and items based on item descriptions and item reviews. According to the Aspect Encoder layer, we have obtained the representation of each aspect of item's description document and review document. In order to calculate the difference between the description and the review in each aspect, inspired by CARP [6], we design a gap logit unit to represent the overall psychological gap of the m-th aspect between item description document and reviews document as follows:

$$\mathbf{gap}_{over,m} = [(\mathbf{a}_{i,m} - \mathbf{a}_{des,m}) \oplus (\mathbf{a}_{i,m} \odot \mathbf{a}_{des,m})] \tag{9}$$

3.2 Personalized Psychological Gap Module

As users focus on different aspects of items when purchasing items, they will have a personalized psychological gap towards different aspects of items. In this module, we first use a User Review Selection layer to select important user reviews based on textual message and corresponding item intrinsic message. Then we design a User-Item Co-attention layer to model the user's personalized psychological gap.

- **User Review Selection**

In this layer, we use an attention mechanism to extract important user reviews according to the current item. We average each word in the item review to get the semantic representation of the document review. Similarly, we average the words in each user review to get the semantic representation of the current review. Unlike the previous method of extracting essential user reviews based on semantic similarity alone, we consider both the semantic information of the reviews and the item ID information corresponding to each user's review. We add two information as the representation of each review. Similarly, for item document reviews, as the document reviews of items are all written to the current item, we add the ID information of the current item. Specifically, user u's fusion representation $\mathbf{r}_u^t$ towards the t-th review can be represented as follows:

$$\mathbf{r}_u^t = \mathbf{r}_{u,rev}^t + \mathbf{r}_{u,id}^t \tag{10}$$

$$\mathbf{r}_{u,rev}^t = avg(\mathbf{c}_{u,1}^t, \mathbf{c}_{u,2}^t, ..., \mathbf{c}_{u,l_u}^t), \quad \mathbf{r}_{u,id}^t = \mathbf{o}_{u,t} \tag{11}$$

where $\mathbf{o}_{u,t}$ is the corresponding item's ID embedding about t-th review of user u, and avg is the average pooling method. Similarly, we represent the item document review as follows:

$$\mathbf{r}_i = \mathbf{r}_{i,rev} + \mathbf{r}_{i,id} \tag{12}$$

$$\mathbf{r}_{i,rev} = avg(\mathbf{c}_{i,1}, \mathbf{c}_{i,2}, ..., \mathbf{c}_{i,l_i}), \quad \mathbf{r}_{i,id} = \mathbf{o}_i \tag{13}$$

where $\mathbf{r}_i$ is the item i's fusion representation based on its review and ID embedding, and $\mathbf{o}_i$ is the item i's ID embedding. Then we use an attention mechanism to select important user reviews according to item i as follows:

$$att_u^t = \frac{\exp\left(\mathbf{r}_i^\top \mathbf{r}_u^t\right)}{\sum_{k=1}^{T} \exp\left(\mathbf{r}_i^\top \mathbf{r}_u^k\right)} \tag{14}$$

where att_u^t represents the importance of user review S_u^t towards item i. We then get different aspects' weighted sum representation of user reviews as:

$$\mathbf{a}_{u,m} = \sum_{t=1}^{T} att_u^t \mathbf{a}_{u,m}^t \tag{15}$$

- **User-Item Co-attention**

As stated above, users value different aspects when facing different items. The psychological gaps caused by different aspects of descriptions and comments will also be different. We design a user-item co-attention mechanism to model the importance of different aspects. Formally,

$$weight_m = \mathbf{W}_{uv,m}[\mathbf{a}_{u,m}, \mathbf{a}_{i,m}, \mathbf{a}_{u,m} \odot \mathbf{a}_{i,m}] + b_{uv,m} \tag{16}$$

$$att_m = \frac{\exp{(weight_m)}}{\sum_{k=1}^{M} \exp{(weight_k)}} \tag{17}$$

where att_m represents the importance of the m-th aspect when user u values the item i, $\mathbf{W}_{uv,m} \in R^{3d_w \times 1}$ is the weight parameter, and $b_{uv,m}$ is the bias. Then we get the user's personalized psychology gap:

$$\mathbf{gap}_{per} = \sum_{m=1}^{M} att_m \mathbf{gap}_{over,m} \tag{18}$$

Similarly, user u's and item i's representation from reviews can be denoted as:

$$\mathbf{u}_r = \sum_{m=1}^{M} att_m \mathbf{a}_{u,m}, \quad \mathbf{i}_r = \sum_{m=1}^{M} att_m \mathbf{a}_{i,m} \tag{19}$$

We then concat user u's review representation and personalized psychology gap representation, and pass the obtained vector to the Multilayer Perception to get the user u's representation $\mathbf{u}_{r,des}$:

$$\mathbf{u}_{r,des} = \mathbf{W}_p[\mathbf{u}_r, \mathbf{gap}_{per}] + b_p \tag{20}$$

where $\mathbf{W}_p \in R^{3d_w \times d}$ is the weight parameter, d is the dimension of latent factors, and b_p is the bias.

3.3 Rating Prediction Module

Although the user representation $\mathbf{u}_{r,des}$ learned from reviews contains rich aspect information of users, there are some latent characteristics of users which can not be captured in reviews but can be inferred from the rating patterns. Thus, we also represent users according to their ids to capture the latent factors of users. The final representation $\mathbf{u}_f$ of user u is the concatenation of the user representation $\mathbf{u}_{r,des}$ learned from reviews and the user embedding $\mathbf{u}_d \in R^d$ from user ID.

Similarly, we can get the final representation $\mathbf{i}_f$ of item i. Formally, it can be represented as follows:

$$\mathbf{u}_f = [\mathbf{u}_d, \mathbf{u}_{r,des}], \quad \mathbf{i}_f = [\mathbf{i}_d, \mathbf{i}_r] \tag{21}$$

We concatenate two representations as $\mathbf{x}_{u,i} = [\mathbf{u}_f, \mathbf{i}_f]$, and then pass it into a Latent Factor Model (LFM) [3]. The LFM function is defined as follows:

$$\hat{y}_{u,i} = \mathbf{W}_x (\mathbf{x}_{u,i}) + b_u + b_i + \mu \tag{22}$$

where $\hat{y}_{u,i}$ denotes the predicted rating, $\mathbf{W}_x$ denotes the parameter matrix of the LFM model, b_u denotes the user bias, b_i denotes the item bias, and μ denotes the global bias.

3.4   Objective Function

We optimize the model parameters to minimize recommendation task loss $\mathcal{L}$. Concretely, the mean squared error is used as the loss function of recommendation task loss, which measures the divergences between the rating score that the model predicts and the gold rating score that a user gives to an item. The recommendation task loss can be calculated as:

$$\mathcal{L} = \frac{1}{|\Omega|} \sum_{u,i \in \Omega} (\hat{y}_{u,i} - y_{u,i})^2 \tag{23}$$

where Ω denotes the set of instances for training, $y_{u,i}$ is the gold rating score, and $\hat{y}_{u,i}$ is the predicted rating score of the user u to the item i separately.

4   Experiments

4.1   Experiment Setup

Datasets and Evaluation Metric. We used four publicly accessible datasets from Amazon 5-core[1] (i.e., *Automotive* (shorten as "Auto"), *Office Products* (Office), *Grocery and Gourmet Food* (Food) and *Toys and Games* (Toys)), which included the review information that users wrote for items they had purchased. We filtered items that did not have descriptions. Following NARRE [2], we randomly split the dataset into training (80%), validation (10%), and testing (10%) sets. At least one interaction per user/item was included in the training set.

The statistics of the datasets are summarized in Table 1. As for the evaluation, we used Mean Square Error (MSE) [16] to measure the rating prediction. Concretely, we computed the square error between the predicted rating $\hat{y}_{u,i}$ and the ground truth $y_{u,i}$, where Ω indicated the set of the user-item pairs in the testing set (see Eq. 24).

$$MSE = \frac{1}{|\Omega|} \sum_{u,i \in \Omega} (\hat{y}_{u,i} - y_{u,i})^2 \tag{24}$$

---

[1] http://jmcauley.ucsd.edu/data/amazon.

Table 1. Statistics of datasets used for rating prediction task

Dataset	Auto	Office	Food	Toys
Number of users	996	2,370	10,446	15,096
Number of items	683	973	6,305	9,814
Number of ratings	6,553	24,120	100,980	125,944
Words per user	279.27	834.24	454.57	408.57
Words per item	368.49	453.93	395.37	406.09
Density of ratings	0.96%	1.05%	0.15%	0.09%

In addition, to make a more intuitive comparison, we used Accuracy Improvement Percentage (AIP) to measure the accuracy improvement percentage of our proposed model against other compared methods (see Eq. 25).

$$AIP = \frac{MSE_{comparedmethod} - MSE_{ourmethod}}{MSE_{comparedmethod}} \tag{25}$$

For each dataset, we performed fivefold cross-validation to avoid any biases. As the data were not normally distributed, we adopted permutation test [11] for significance tests. In our experiments, pre-trained word embeddings were adopted from Google News [10], where the word embedding size was set to 100. We kept the number and the length of reviews covering pe percent users, where pe was set to 0.85 for all four datasets. For item doc and description doc, we kept the length of doc 500 and 100 respectively. Adam was used to updating parameters when training. The learning rate was determined by grid search amongst $\{0.0001, 0.0002, 0.001, 0.002\}$, the dropout ratio was explored amonst $\{0.0, 0.1, ..., 0.9\}$, and the batch size was set amonst $\{32, 64, 128, 256\}$. The window size s was empirically set as 3 and τ was set as 0.5. All hyperparameters were tuned according to the validation set.

Compared Methods. We compared our approach with eight related methods. These algorithms can be classified into three categories: *rating-based* (NMF and SVD), *review-based* (e.g. DeepCoNN, D-ATTn, NARRE, DAML, NRMA and MRCP) and *variations of our method* (e.g. PMAR-O, PMAR-P and PMAR-ID). The detailed description of each method is given below.

- **NMF** [5]: Non-negative Matrix Factorization is a traditional model which casts the MF model within a neural framework and combines the output with multi-layered perceptrons.
- **SVD** [3]: Singular Value Decomposition is a matrix factorization model which reduces the dimension of a rating matrix and eliminates its sparsity.
- **DeepCoNN** [19]: Deep Cooperative Neural Networks is a neural network which learns the representations of users and items from reviews by using convolutional neural networks.
- **D-ATTn** [13]: Dual Attention CNN Model is an interpretable, dual attention-based CNN model which combines reviews and ratings for product rating prediction.

- **NARRE** [2]: Neural Attentional Rating Regression with Review-level Explanations is a model which uses attention mechanism to explore the usefulness of reviews.
- **DAML** [7]: Dual Attention Mutual Learning is a model which utilizes local and mutual attention of CNN to learn the features of reviews.
- **NRMA** [8]: Neural Recommendation Model with Hierarchical Multi-view Attention is a model which designs a review encoder with multiview attention to learn representations of reviews from words.
- **MRCP** [9]: Multi-aspect Neural Recommendation Model with Context-aware Personalized Attention is a most recently model which designs three encoders to extract hierarchical features of reviews, aspects and users/items.
- **PMAR-O**: This method removes the Description-Review Gap Layer from our PMAR model. That is, the model does not consider the psychological gap.
- **PMAR-P**: As the second variation of our method, this method removes the User-Item Co-attention Layer from our PMAR model so that the weight of the psychological gap of each aspect is the same. That is, it only considers the overall psychological gap rather than the personalized psychological gap.
- **PMAR-ID**: As another variation of our method, this method uses textual message alone to model user's and item's fusion representations in Eq. 10 and Eq. 12.

4.2   Overall Performance Comparison

The overall comparison results are shown in Table 2. We observe that our proposed PMAR method is the best in terms of MSE, and the improvements against the best baselines are significant (p-value < 0.05 via permutation test). Concretely, the advantages against MRCP in Office and Toys are 2.64% and 2.06% respectively (refer to the value of ΔMRCP). The possible reason is that our method not only considers the multi-aspect information in reviews which can construct diverse user preferences and item characteristics, but also the personalized psychological gap brought by the inconsistency of merchant's description and the item's actual attributes to the user. Another interesting observation from Table 2 is that our PMAR model achieves the largest improvement against other baselines in Office datasets (e.g., Δ MRCP = 2.64% in Office vs. 1.89% in Food). It is reasonable because the density of Office dataset is higher relative to others (see Table 1). In this case, our PMAR model is able to learn user preference and item characteristic more comprehensively, leading to the more accurate personalized psychological gap construction.

Furthermore, when compared with three variations (i.e., PMAR-O, PMAR-P and PMAR-ID), our complete model PMAR also performs better in all four datasets. Among the variations, compared with our PMAR model, PMAR-O which does not consider the psychological gap has the largest drop in the results of all datasets. Specifically, in Office, our PMAR model's advantage against PMAR-O is 2.61% while against PMAR-P is 1.72%. It is reasonable because PMAR-O models neither overall psychological gap nor personalized psychological gap.

Table 2. Overall comparison results of rating prediction are measured by MSE (Note: * denotes the statistical significance for p-value < 0.05 compared to the best baseline, the boldface indicates the best model result of the dataset, and the underline indicates the best baseline result of the dataset.)

Method		Auto	Office	Food	Toys
Rating-based	NMF	1.0083	0.8388	1.2707	1.0486
	SVD	0.8276	0.7626	1.0293	0.8931
Review-based	DeepCoNN	0.7473	0.7338	0.9881	0.8501
	D-ATT	0.7532	0.7397	0.9815	0.8355
	NARRE	0.7685	0.7124	0.9919	0.8217
	DAML	<u>0.7320</u>	0.7033	0.9908	0.8711
	NRMA	0.7658	0.7118	0.9891	0.8229
	MRCP	0.7565	<u>0.6967</u>	<u>0.9729</u>	<u>0.8191</u>
Variations of our method	PMAR-O	0.7347	0.6965	0.9657	0.8112
	PMAR-P	0.7201	0.6902	0.9561	0.8062
	PMAR-ID	0.7302	0.6816	0.9599	0.8066
Our Method	PMAR	**0.7137***	**0.6783***	**0.9545***	**0.8022***
AIP of PMAR	ΔNMF	29.22%	19.13%	24.88%	23.50%
	ΔSVD	13.76%	11.05%	7.27%	10.18%
	ΔDeepCoNN	4.50%	7.56%	3.40%	5.63%
	ΔD-ATT	5.24%	8.30%	2.75%	3.99%
	ΔNARRE	7.13%	4.79%	3.77%	2.37%
	ΔDAML	<u>2.50%</u>	3.55%	3.66%	7.91%
	ΔNRMA	6.80%	4.71%	3.50%	2.52%
	ΔMRCP	5.66%	<u>2.64%</u>	<u>1.89%</u>	<u>2.06%</u>
	ΔPMAR-O	2.86%	2.61%	1.16%	1.11%
	ΔPMAR-P	0.89%	1.72%	0.17%	0.50%
	ΔPMAR-ID	2.26%	0.48%	0.56%	0.55%

In this way, when there is a great gap between the item's actual attributes and the merchant's description, the psychological gap of the user will have a large impact on the accuracy for recommendation. In addition, the suboptimal results of PMAR-P model (e.g., Δ PMAR-P $= 0.89\%$ in Auto vs. 1.72% in Office) validate the importance of the personalized psychological gap. It is likely because the variation ignores that users pay distinct attentions to different aspects, which will cause users' personalized psychological gaps, resulting in a great impact on accuracy for the recommendation. Besides, the degraded MSEs of PMAR-ID demonstrate that combining both ID and review messages can capture textual message as well as intrinsic attributes of the reviewed item (e.g., Δ PMAR-ID $= 0.56\%$ in Food vs. 0.55% in Toys). Although PMAR-ID uses both the overall psychological gap and personality psychological gap, it uses textual message alone which can not capture intrinsic attributes of the reviewed item.

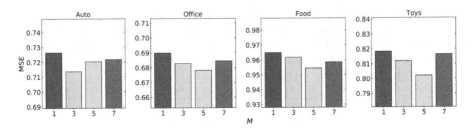

Fig. 3. Experimental results with the change of Aspects' Number.

4.3 Influence of the Number of Aspects

To investigate whether different number of aspects will influence the performance of our PMAR model, we change M (refer to Sect. 3.1). Following MRCP, the aspects' number M is set amongst $\{1, 3, 5, 7\}$. We can observe from Fig. 3 that the multi-aspect model (i.e., $M > 1$) always performs better than that with only one aspect (i.e., $M = 1$), indicating the effectiveness of multi-aspect diverse representations of users and items. Another finding is that when M equals to 5, our PMAR model performs the best in three datasets (i.e., Office, Food and Toys). In contrast, our PMAR achieves the best result in the Auto dataset when M equals to 3. We argued that the value of the optimal aspects' number might be related to the average number of words per user and each item (e.g., Words Per User: 279.27 in Auto vs 834.24 in Office; Words Per Item: 368.49 in Auto vs 453.93 in Office). That is, more words may contain richer aspects of users and items. In addition, setting the aspects' number to 5 would be an optimal choice for most recommendations.

4.4 Case Study

To better understand how our model facilitates the recommendation system based on the user's psychological gap caused by the inconsistency of the merchant's description and the item's actual attributes, we conducted a case study in Fig. 4. We randomly sampled two user-item pairs from Auto and Office datasets. For each user-item pair, we set M as 5 and extracted the most valued aspect by the current user (i.e., the highest attention weight according to att_m in Eq. 17) and another aspect that the user less valued. Similar to CARP [6], to better visualize an aspect, we retrieved the top-K phrases whose weight is the sum of the weights (i.e., $\beta_{m,j,i}$ in Eq. 4 and $\beta_{u,j,m}^t$ in Eq. 8) of the constituent words in the convolutional window. We then selected the most informative sentences containing these phrases to represent the corresponding aspect. Here, we choose $K = 30$ which is the same setting as CARP [6]. In Fig. 4, red and green which are predicted by our PMAR model indicate the phrases in reviews and descriptions of the user's most valued and less valued aspects respectively. As a reference, we manually displayed the parts matched well by the most valued aspects in the target user-item review with yellow color. In addition, $y_{u,i}$ represents the user's

user1 – item1 (Heavy-Duty Dual Propane Tank Cover): $y_{u,i} = 3.0$ $\hat{y}_{u,i} = 3.3$ $\hat{y}'_{u,i} = 3.9$			aspect importance
aspect 5	item description:	... is made of heavy-duty polypropylene and protects propane gas bottles.	0.287
	item history reviews:	It's not as heavy duty as I thought.	
aspect 4	item description:	Easy to assemble in about 15 minutes with supplied hardware.	0.100
	item history reviews:	The cover was easy to assemble.	
user history reviews:		... hang very secure to under sink door.	
target review:		I guess I expected 'Heavy duty' to be really heavy duty. I felt the product was somewhat flimsy. I think the product should not advertize it to be more than what it is.	

user2 – item2 (EnduraGlide Dry-Erase Markers): $y_{u,i} = 5.0$ $\hat{y}_{u,i} = 4.8$ $\hat{y}'_{u,i} = 4.1$			aspect importance
aspect 2	item description:	... write smoother, and erase cleanly.	0.396
	item history reviews:	... and erase really well. I think this is a really nice set of dry-erase markers.	
aspect 1	item description:	... that delivers bold, continuous color on dry-erase boards.	0.028
	item history reviews:	The colors (assorted option) aren't as bold as I may have liked them.	
user history reviews:		Of course does not leave any residue either.	
target review:		I was looking to a inexpensive dry-erase marker set when a few of my oldmarkers dried out. These work perfectly. The print erases as expected.	

Fig. 4. Example study of two user-item paris from Auto and Office.

gold rating of the current item, $\hat{y}_{u,i}$ represents the rating that our PMAR model predicts, and $\hat{y}'_{u,i}$ denotes the rating predicted by PMAR-O (i.e., w/o Overall Psychological Gap) model which is a variation of our PMAR model.

Example 1: the first user-item pair is a user who gave a tank cover "Heavy-Duty Dual Propane Tank Cover" a low score (i.e. $y_{u,i} = 3.0$). We can observe that our PMAR model predicts that user1 assigns the highest weight on aspect 5 of item1 (e.g., aspect importance: 0.287 of aspect 5 vs. 0.100 of aspect 4) which is regarding the heavy-duty aspect and s/he likes sturdy items according to his/her history reviews. This indicates aspect 5 brings a greater psychological gap to user1 than other aspects. Besides, item1 in aspect 5 is not as heavy-duty as described by the merchant. Although item1 is easy to assemble as advertised by the merchant in aspect 4 (see Fig. 4), the recommendation system should still reduce the probability of recommending item1 to user1. In the rating prediction stage, our model PMAR predicts that user1's rating for item1 is 3.3 points while PMAR-O model predicts 3.9 points. It is reasonable because the variant does not consider the impact of user1's psychological gap which causes user1 to be more disappointed with item1, leading to the inaccuracy for recommendation. Finally, it can be seen that user1's target review denotes that user1 is not satisfied with item1 and considers that the merchant falsely advertises the heavy-duty aspect s/he values most, which is in accordance with our PMAR model's prediction.

Example 2: compared with the first pair, the second one illustrates a user-item pair towards a pen "EnduraGlide Dry-Erase Markers" by user2 with a high rating (i.e., $y_{u,i} = 5.0$). It can be seen that our PMAR model predicts that user2 values aspect 2 of item2 most (e.g., aspect importance: 0.396 of aspect 2 vs. 0.028 of aspect 1) which is about the erase-cleanly aspect. Besides, user2 likes items that do not leave any residue according to his/her history reviews. This implies that item2 erases cleanly as the merchant's description in aspect 2 (see Fig. 4), leading to a greater psychological gap to user2 than other aspects. Although regarding the less valued aspect 1 in which item2 is not as bold as advertised by the merchant, the recommendation system should still increase the probability of introducing item2 to user2. As for rating prediction, our model PMAR and PMAR-O models predict that user2's ratings for item2 are 4.8 and 4.1 points respectively. The possible reason is that the variant does not consider the impact of user2's psychological gap which causes user2 to be more satisfied with item2. Finally, we can observe that user2's target review denotes that item2 satisfies user2's expectation of the erase-cleanly aspect, which also meets the prediction of our PMAR model.

5   Conclusions and Future Work

In this work, we are motivated by the Expectation Effect to propose a novel PMAR model which learns a user's overall and personalized psychological gap. Concretely, we first construct an Overall Psychological Gap Module which uses a logit unit to calculate the overall psychological gap of merchants description and item reviews. Secondly, we design a Personalized Psychological Gap Module which uses an attention mechanism to select relevant user reviews and item reviews based on review semantic information and the corresponding item intrinsic information. A user-item co-attention mechanism is adopted to calculate the user's personalized psychological gap. We evaluate our model on four real-world datasets (i.e., Auto, Office, Food and Toys). The experimental results show that our model significantly outperforms other state-of-the-art methods in terms of rating prediction accuracy.

Acknowledgements. We thank editors and reviewers for their suggestions and comments. We also thank National Natural Science Foundation of China (under project No. 61907016) and Science and Technology Commission of Shanghai Municipality (under projects No. 21511100302, No. 19511120200 and 20dz2260300) for sponsoring the research work.

References

1. Anderson, R.E.: Consumer dissatisfaction: the effect of disconfirmed expectancy on perceived product performance. J. Mark. Res. **10**(1), 38–44 (1973)
2. Chen, C., Zhang, M., Liu, Y., Ma, S.: Neural attentional rating regression with review-level explanations. In: WWW, pp. 1583–1592 (2018)
3. Koren, Y., Bell, R., Volinsky, C.: Matrix factorization techniques for recommender systems. Computer **42**(8), 30–37 (2009)
4. Lawrence, S., Giles, C.L., Tsoi, A.C., Back, A.D.: Face recognition: a convolutional neural-network approach. IEEE Trans. Neural Netw. **8**(1), 98–113 (1997)
5. Lee, D., Sebastian Seung, H.: Algorithms for non-negative matrix factorization. In: NIPS, pp. 556–562 (2001)

6. Li, C., et al.: A capsule network for recommendation and explaining what you like and dislike. In: SIGIR, pp. 275–284 (2019)
7. Liu, D., Li, J., Du, B., Chang, J., Gao, R.: DAML: dual attention mutual learning between ratings and reviews for item recommendation. In: KDD, pp. 344–352 (2019)
8. Liu, H., et al.: Hierarchical multi-view attention for neural review-based recommendation. In: NLPCC, pp. 267–278 (2020)
9. Liu, H., et al.: Toward comprehensive user and item representations via three-tier attention network. ACM Trans. Inf. Syst. **39**(3), 1–22 (2021)
10. Mikolov, T., Sutskever, I., Chen, K., Corrado, G.S., Dean, J.: Distributed representations of words and phrases and their compositionality. In: NIPS, pp. 3111–3119 (2013)
11. Nichols, T.E., Holmes, A.P.: Nonparametric permutation tests for functional neuroimaging: a primer with examples. Hum. Brain Mapp. **15**(1), 1–25 (2002)
12. Peng, Q., et al.: Mutual self attention recommendation with gated fusion between ratings and reviews. In: DASFAA, pp. 540–556 (2020)
13. Seo, S., Huang, J., Yang, H., Liu, Y.: Interpretable convolutional neural networks with dual local and global attention for review rating prediction. In: RecSys, pp. 297–305 (2017)
14. Wang, X., et al.: Neural review rating prediction with hierarchical attentions and latent factors. In: DASFAA, pp. 363–367 (2019)
15. Wang, X., Xiao, T., Tan, J., Ouyang, D., Shao, J.: MRMRP: multi-source review-based model for rating prediction. In: DASFAA, pp. 20–35 (2020)
16. Willmott, C.J., Matsuura, K.: Advantages of the mean absolute error (MAE) over the root mean square error (RMSE) in assessing average model performance. Clim. Res. **30**(1), 79–82 (2005)
17. Wu, X.F., Zhou, J., Yuan, X.: The research of factors which effect B2C E-commerce trust-based on the mechanism of process. In: AMR, pp. 2583–2586 (2012)
18. Xiao, B., Benbasat, I.: E-commerce product recommendation agents: use, characteristics, and impact. MIS Q. **31**, 137–209 (2007)
19. Zheng, L., Noroozi, V., Yu, P.S.: Joint deep modeling of users and items using reviews for recommendation. In: WSDM, pp. 425–434 (2017)

Meta-path Enhanced Lightweight Graph Neural Network for Social Recommendation

Hang Miao[1,2], Anchen Li[1,2], and Bo Yang[1,2(✉)]

[1] Key Laboratory of Symbolic Computation and Knowledge Engineering of Ministry of Education, Jilin University, Changchun, China
[2] College of Computer Science and Technology, Jilin University, Changchun, China
{miaohang19,liac20}@mails.jlu.edu.cn, ybo@jlu.edu.cn

Abstract. Social information is widely used in recommender systems to alleviate data sparsity. Since users play a central role in both user-user social graphs and user-item interaction graphs, many previous social recommender systems model the information diffusion process in both graphs to obtain high-order information. We argue that this approach does not explicitly encode high-order connectivity, resulting in potential collaborative signals between user and item not being captured. Moreover, direct modeling of explicit interactions may introduce noises into the model and we expect users to pay more attention to reliable links. In this work, we propose a new recommendation framework named Meta-path Enhanced Lightweight Graph Neural Network (ME-LGNN), which fuses social graphs and interaction graphs into a unified heterogeneous graph to encode high-order collaborative signals explicitly. We consider using a lightweight GCN to model collaborative signals. To enable users to capture reliable information more efficiently, we design meta-paths to further enhance the embedding learning by calculating meta-path dependency probabilities. Empirically, we conduct extensive experiments on three public datasets to demonstrate the effectiveness of our model.

Keywords: Social recommendation · Recommender systems · Graph convolutional network · Collaborative filtering

1 Introduction

The volume of data is growing with the rapid development of society, leading to information overload. In response, recommender systems have been proposed for personalized information filtering, which centers on predicting whether a

Supported by the National Key R&D Program of China under Grant Nos. 2021ZD0112501 and 2021ZD0112502; the National Natural Science Foundation of China under Grant Nos. 62172185 and 61876069; Jilin Province Key Scientific and Technological Research and Development Project under Grant Nos. 20180201067GX and 20180201044GX; and Jilin Province Natural Science Foundation under Grant No. 20200201036JC.

user will interact with a certain item. Collaborative filtering (CF) [4] has been extensively studied in recommender systems, which uses user-item interaction data to learn the user and item embeddings [9–13,16,17,19,23]. However, the data sparsity that accompanies CF hinders the development of recommender systems. With the rapid increase in social platforms, users are willing to make friends and express their preferences on social platforms. Inspired by this, the social recommendation was proposed to alleviate the data sparsity issue through abundant social relationships.

As users take on an important role in both user-user social networks and user-item interaction graphs, the key to social recommendation is that the final embedding contains information from two interaction graphs. Traditional social recommendation methods [5,10,20,26] aim at adding the influence of social neighbors to the user embedding representation via a user-user interaction matrix with the matrix factorization. These improvements can be seen as utilizing first-order neighbors of the graph structure to make improvements. In the following research, Wu et al. proposed Diffnet [18] and Diffnet++ [21] to further enhance the embedding learning by incorporating high-order neighborhood information into the embedding learning process. Although the performance of the recommendation models of these social networks has improved, we believe that there is still room for improvement in the present social recommendation models. Specifically, these methods are not explicitly coded between the user and the long-distance item. For example, there is such an association $\mathcal{U}_4 \to \mathcal{U}_1 \to \mathcal{I}_1$ in Fig. 1, then $\mathcal{U}_4$ and $\mathcal{I}_1$ are intrinsically related, but Diffnet does not explicitly model these latent collaborative signals, we hope to encode such collaboration signals in an explicit way. As well, during the training of the previous models, all interactions between users and users (items) are coded uniformly, regardless of the reliability of the connections between them. Some unreliable connections may distract the nodes and lead to diminishing embedding representation capabilities. Therefore, we expect users to focus on those users and items that are more relevant to them in the modeling.

In this work, we propose a new recommendation framework named Meta-path Enhanced Lightweight Graph Neural Network (ME-LGNN), which models high-order collaborative signals explicitly. Specifically, we merge the user-user social network and the user-item interaction graph into a unified social recommendation HIN, then we utilize a lightweight graph convolution operation to aggregate neighbors' information. By stacking multiple layers, users and items can obtain the characteristics of their high-order neighbors. In the aggregation process, we use the attention network to adaptively aggregate the embedding representation of different neighbors. In addition, to enable users to capture collaborative signals more efficiently, we devise a series of interpretable meta-paths in HIN. To make the model focus more on reliable connections during training, we reduce noises by constraining the dependency probability of meta-paths, making the embedding representation more capable. The two training processes are alternated to improve the overall recommendation performance.

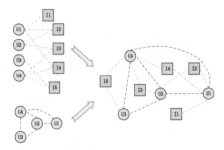

Fig. 1. Heterogeneous information network (HIN) constructed by the user-item inter-action graph and the social network. The blue node represents the user, the green node represents the item, the orange dashed line indicates the interaction between the user and the item, the purple dashed line indicates the trust relationship between the users. (Color figure online)

In summary, the contributions of this work are as follows:

- We emphasize the significance of explicitly incorporating high-order collabo-rative signals into embedding in social recommendation models.
- We fuse the user-user social network and the user-item interaction graph into a unified heterogeneous network. On the basis of the network, we propose a new social recommendation framework called ME-LGNN, which can model high-order collaborative signals explicitly.
- To enable users to focus more on reliable connections, we designed a series of meaningful meta-paths, which further improve embedding learning by con-straining meta-path dependency probabilities.
- We conduct extensive experiments on three public datasets. Results demon-strate competitive performance of ME-LGNN and the effectiveness of explicit modelling of unified HINs.

2   Related Work

In this section, we briefly review the work related to social recommendation.

SoRec [15] is an early social recommendation model based on matrix factor-ization, which proposed a factor analysis method using the probability matrix. To incorporate the preferences of friends trusted by the user, Ma et al. pro-posed a recommendation model RSTE [14] with weighted social information on top of SoRec. However, this model does not propagate information about users' preferences in the network, so SocialMF [10] was proposed, which is based on the trust propagation mechanism of social networks to obtain the embedding representation of users. Observing that users tend to give higher scores to items that their friends like, Zhao et al. proposed the SBPR [26] model, which uses social information to select training examples and incorporates social relation-ships into the model. TBPR [20] was built on it by incorporating strong and weak ties in social relationships into social recommendations. These models proposed different methods to address the sparse issue of collaborative filtering.

The recommender systems based deep learning aim to capture non-linear features from the interaction graph. NeuMF [8] combines traditional matrix factorization and Multilayer Perceptron (MLP) to extract both low and high dimensional features with promising recommendation performance. DeepSoR [2] proposed to learn embedding representations from social relationships and then integrated the user's embedding representation into a probability matrix factorization for evaluating prediction. However, these methods do not encode the interaction of information in an explicit way.

In recent years, graph neural networks are becoming well known for their powerful performance in learning graph data. Recommendation tasks also can be well represented as graph structures, so GNNs provide great potential for the development of recommendation tasks. GC-MC [1] was proposed to construct embedding of users and items by passing messages on the user-item interaction graph, but this passing only operates on a layer of links and does not incorporate high order interaction information. The NGCF [19] proposed by Wang et al. designs a graph convolution operation on the user-item interaction graph to capture collaborative filtering signals in high-order connections. LightGCN [7] makes the model more suitable for collaborative filtering tasks by removing feature transformations and non-linear variations from the GCN of the NGCF. To incorporate social relationships, GraphRec [3] learns an adequate embedding representation from the rich social relationships by fusing first-order interactions on social networks and user-item interaction graphs with neural network processes. Diffnet [18] leverages convolutional operations to perform recursive diffusion in social networks to obtain high-order collaborative signals from users. With these existing models, our work performs differently in that we fuse the social network and user-item interaction graphs into a unified heterogeneous information graph that explicitly encodes potential collaborative signals by propagating information over the heterogeneous graph.

Considered from the perspective of heterogeneous graphs, a number of meta-path-related models have been proposed to solve the recommendation task. The recommendation model based on the meta-path can perform interpretable analysis of the recommendation results. Yu et al. proposed HteRec [25] which targets implicit feedback and obtains different user preference matrices based on different meta-paths, and then goes through a matrix factorization model to implement recommendation tasks. Han et al. proposed the NeuACF [6] model to extract different dimensions of information through multiple meta-paths in an attempt to fuse different aspects of information. The IF-BPR [24] model also learns the user's embedding representation through a meta-path approach and then discovers the user's potential friends through an adaptive approach. On the contrary, our work aims to enhance representation by calculating meta-path dependency probabilities in such a way that users can focus on connections that are more closely related to themselves and ignore connections that are more dissimilar to themselves in a heterogeneous graph.

3   Methodology

In general, ME-LGNN is composed of four parts: (1) Embedding Layer: providing the initial embedding of users and items. (2) Aggregation layer: learning embedding representations of users and items through a lightweight GCN with the attention mechanism; (3) Enhancement layer: designing some reasonable meta-paths, through meta-path constraints to further enhance the embedding representation capabilities. (4) Predicting Layer. We show the overall neural architecture of ME-LGNN in Fig. 2. We first describe the problem formulation, then introduce the four components of our model, and finally discuss the training optimization process.

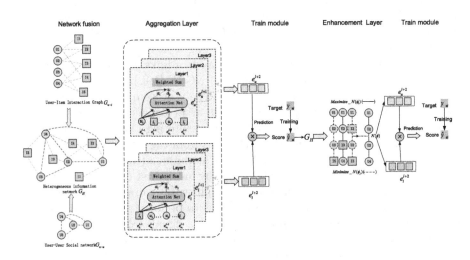

Fig. 2. Illustration of the ME-LGNN model architecture (arrow lines indicate information flow). In the aggregation layer user (blue) and item (green) are aggregated through multiple propagation layers for neighborhood aggregation, the output is subjected to the first round of training, then further enhanced by a meta-path enhancement layer for embedding learning, and its output is subjected to the final prediction. (Color figure online)

3.1   Notation and Problem Formulation

We introduce necessary notations and definitions by describing the graph-based collaborative filtering problem. In graph-based social recommendation, there are two types of entities, user set U ($\mid U \mid = M$) and item set I ($\mid I \mid = N$). Two interaction graphs are formed from U and I: (1) User-item interaction graph G_{u-i}: we can use the adjacency matrix to represent the interactions in the graph. $R \in \mathbb{R}^{M \times N}$ is the interaction matrix between users and items. If there is an interaction between the user u and the item i, y_{ui} is 1, otherwise it is 0. (2) User-user social network G_{u-u}: $S \in \mathbb{R}^{M \times M}$ is the interaction matrix between users. If the user u_1 and the user u_2 trust each other, $y_{u_1 u_2}$ is 1, otherwise it is 0. The type of relationship between entities a and b is r_{ab}. Relationship types include social relationships between users and user ratings of items. We merge the two networks to obtain a unified heterogeneous information network G_H. For social recommendation, given the evaluation matrix R of users and items and the social network interaction matrix S, our goal is to predict the user's preference for unknown items. Next, we define the problems investigated in this work as follows:

Input: user-user social network G_{u-u}, user-item interaction graph G_{u-i}, user set U and item set I.

Output: a personalized ranking function that maps an item to a real value for each user: $f_u : I \rightarrow \mathbb{R}$.

In this process, we first merge the two input interaction graphs to obtain a new heterogeneous information network G_H. The rest of the training is carried out on G_H.

3.2   Embedding Layer

For social recommendation, we consider it as a representation learning problem. Following some mainstream models [7,18,19,21], we use the embedding vector $e_u \in \mathbb{R}^d$ ($e_i \in \mathbb{R}^d$) to describe the user (item) and use the embedding vector $r \in \mathbb{R}^d$ to describe the relationship type, d represents the dimension of the embedding. We construct two parameter matrices as embedding lookup tables:

$$E = [\overbrace{e_{u_1}, \cdots, e_{u_N}}^{\text{user embeddings}}, \overbrace{e_{i_1}, \cdots, e_{i_M}}^{\text{item embeddings}}], R = [\quad \overbrace{r_{uu}, r_{ui}}^{\text{relationship type embeddings}} \quad]. \tag{1}$$

3.3   Aggregation Layer

At the aggregation layer, we employ the attention network structure to aggregate adaptively the learning embeddings of different neighbors in the heterogeneous graph G_H. Through iterative aggregation, information can be diffused and propagated in the network. Some high-order cooperative information can also be captured, so the model can model high-order collaborative signals. Recent work [7] has found that the most common feature transformations and non-linear activations in GCN contribute little to collaborative filtering and instead make training more difficult and reduce recommendation performance. In our work, we remove the two steps of feature transformation and nonlinear activation to reduce model complexity. We aggregate all nodes in the heterogeneous information network, the graph convolution operation in the ME-LGNN is defined as:

$$e_a^* = AGG\left(e_a, \mathcal{N}(a)\right) = e_a + \sum_{b \in \mathcal{N}(a)} \tilde{\pi}_{ab} \cdot e_b, \tag{2}$$

where e_a^* represents the embedding representation of node a, $\mathcal{N}(a)$ represents the set of neighbors of a, and $\tilde{\pi}_{ab}$ is the attention value, which indicates that how strongly node a is influenced by node b. The attention value is specifically defined as follows:

$$\tilde{\pi}_{ab} = \frac{exp(\pi_{ab})}{\sum_{b' \in \mathcal{N}(a)} exp(\pi_{ab'})}, \tag{3}$$

$$\pi_{ab} = (e_a \odot e_b)^T tanh(r_{ab} \odot e_b), \tag{4}$$

where r_{ab} is the interaction type between node a and node b. High-order correlation is crucial for encoding collaborative signals and estimating correlation scores between users and items. We iteratively execute the above aggregation process and encode high-order collaborative signals to explore high-order connectivity information.

3.4 Enhancement Layer

In the training process of the previous model, Fig. 1 shows that the interaction information between all users and users (items) is uniformly coded, regardless of the user's high-order information relevance. Some unreliable links may distract nodes. To alleviate the above problems, we do the following work to enhance the representation learning ability.

Design Meta-path. Referring to [24], the interaction in this paper is undirected. Here we have designed some reasonable meta-paths as shown in Table 1. We generate these meta-paths by means of a random walk method on the heterogeneous graph G_H. Then, a new path set P is obtained.

Table 1. Meta-paths designed for social recommendation.

Path	Schema	Description
P_1	U–I–U	Users who have consumed the same item are similar
P_2	U–U–U	Users may trust their friends' friends
P_3	U–U–I	Users may have similar preferences to their friends
P_4	U–U–U–U	Users who share the same friends are similar with each other
P_5	U–U–U–I	Users may have similar preferences as their friends' friends
P_6	U–U–I–U	Users' preferences may be similar to those of their friends who have similar preferences
P_7	U–I–U–U	Users who have consumed the same items may have similar preferences

Calculate Dependent Path Probability. We reconstruct a dependency path. For the path head and tail nodes, if there are interactions between the two nodes in the ground-truth, the path will have a higher score, otherwise, it will have a lower score. The goal is to enable users to pay more attention to the nodes that are more likely to interact with themselves when learning embedding, so as to further enhance the embedding learning.

Inspired by [22], we calculate the dependency probabilities of paths with the following approach. Unlike [22] which proposed to model reconstructed sequences as sequence generation, the sequence nature of paths in our study is not obvious. Specifically, since there is a relationship between the two nodes in the path, we use self-attention to calculate the hidden state $\boldsymbol{p}_{b_l}$ of each node $\boldsymbol{q}_{b_{l-1}}$ on the path $\phi(n)$, as follows:

$$\boldsymbol{p}_{b_l} = \textit{self-attention}(\boldsymbol{p}_{b_l-1}, \boldsymbol{q}_{b_{l-1}}), \tag{5}$$

where $\boldsymbol{p}_{b_l}$ is the embedding of the node, and the initialization of $\boldsymbol{p}_{b_0}$ is the output of the aggregation layer. Then $\boldsymbol{p}_{b_l}$ is given to the softmax layer to calculate the probability of node v_{b_l} in the path:

$$P\left(v_{b_l} \mid v_{<b_l}\right) = \frac{exp\left(\boldsymbol{p}_{b_l} \boldsymbol{W}_r \boldsymbol{q}_{b_l}\right)}{\sum_n exp\left(\boldsymbol{p}_{b_l} \boldsymbol{W}_r \boldsymbol{q}_{b_n}\right)}, \tag{6}$$

where $\boldsymbol{W}_r \in \mathbb{R}^{d \times d}$ is the parameter matrix. Then we get the set of node probabilities for path $\phi(n)$ as $\{P(v_{b_1} \mid v_{<b_1}), P(v_{b_2} \mid v_{<b_2}), \cdots, P(v_{b_L} \mid v_{<b_L})\}$, finally, the probability of path $\phi(n)$ is calculated as:

$$N(\phi(n)) = \prod_{l=1}^{L} P\left(v_{b_l} \mid v_{<b_l}\right). \tag{7}$$

3.5 Predicting Layer

After propagation through k layers, we obtain a representation of the user embedding $\{\boldsymbol{e}_u^1, \boldsymbol{e}_u^2, \cdots, \boldsymbol{e}_u^k\}$ and item embedding $\{\boldsymbol{e}_i^1, \boldsymbol{e}_i^2, \cdots, \boldsymbol{e}_i^k\}$ for each layer. Since each layer may reflect different dimensions of user preferences, we concatenate each layer of user representation as the final representation: $\boldsymbol{e}_u^* = [\boldsymbol{e}_u^0 \parallel \boldsymbol{e}_u^1 \parallel \cdots \parallel \boldsymbol{e}_u^k]$. we use a similar approach to get the final representation of item: $\boldsymbol{e}_i^* = [\boldsymbol{e}_i^0 \parallel \boldsymbol{e}_i^1 \parallel \cdots \parallel \boldsymbol{e}_i^k]$. Ultimately we calculate the user's preference $\hat{y}_{ui}$ for a certain item via inner product:

$$\hat{y}_{ui} = \boldsymbol{e}_u^{*T} \boldsymbol{e}_i^*. \tag{8}$$

3.6 Model Training

To learn the model parameters, for the aggregation layer, we adopt cross-entropy loss as the loss function:

$$\mathcal{L}_A = \sum_{(u,i,j) \in \mathcal{O}} -\ln \sigma(\hat{y}_{ui} - \hat{y}_{uj}), \tag{9}$$

where $\mathcal{O} = \{(u, i, j) \mid (u, i) \in \mathcal{R}^+, (u, j) \in \mathcal{R}^-\}$ denotes the pairwise training data, $\mathcal{R}^+$ denotes positive samples, $\mathcal{R}^-$ denotes negative samples.

For the enhancement layer, the $\hat{y}_{ui}$ was calculated using the same method as in Eq. (8). We calculate the additional reconstruction loss for the generated training set of reconstructed paths. We get fewer positive samples by random walk through the meta-path, so we give a larger weight to the positive samples during training. The difference with Eq. (9) is that here the training is performed through a cross-entropy loss function with weights:

$$\mathcal{L}_E = \sum_{(u,i,j)\in\mathcal{O}} -\ln\sigma(\mu\hat{y}_{ui} - \hat{y}_{uj}), \tag{10}$$

$$\mu = \sqrt{\frac{\|\mathcal{R}^+\| + \|\mathcal{R}^-\|}{\|\mathcal{R}^+\|}}, \tag{11}$$

where μ is the weight value of the positive sample.

The complete loss function of ME-LGNN is as follows:

$$\mathcal{L} = \mathcal{L}_A + \mathcal{L}_E + \lambda\|\Theta\|^2, \tag{12}$$

where $\Theta = \{E, R, W_r\}$ represents all the trainable model parameters. It should be noted here that the only training for our aggregation layer is the embedding of the 0th layer. Compared with the traditional graph convolution operation, the amount of parameters is reduced. Our enhancement layer model training parameters and the conversion matrix in self-attention. λ is used to balance the power of L_2 regularization and prevent overfitting. $\mathcal{L}_A$ and $\mathcal{L}_E$ are trained alternately to jointly improve the performance of the model. We adopt Adam optimizer to optimize the aggregation layer and enhancement layer to update the parameters. Adam is a commonly used optimizer which can adaptively adjust the learning rate.

The specific process of ME-LGNN is presented in Algorithm 1. A training epoch involves two stages: the aggregation layer (line 3–5) and the enhancement layer (line 6–8). In each iteration, we execute the aggregation layer and the enhancement layer in turn to enhance embedding learning.

3.7   Complexity Analysis

In this section, we analyse the complexity of our model.

Model Size. We adopt the alternative optimization strategy, the training matrix of our model consists of three parts: user and item embeddings, the parameter matrix, transformation matrixes. For the aggregation layer, we only need to learn the 0th layer user embeddings $U^{(0)} \in \mathbb{R}^{N \times d}$ and item embeddings $I^{(0)} \in \mathbb{R}^{M \times d}$. For the enhancement layer, model training matrix W^r and three conversion matrixes in self-attention need to be learned. The number of parameters in each matrix is $d \times d$. In total, the overall model size is approximately $(N + M + 4d)d$. It can be seen that our model is very lightweight.

Algorithm 1. ME-LGNN

Input: User-item interaction matrix R, user-user interaction matrix S, heterogeneous information network G_H, user set U and item set I
Output: Prediction function $f_u : I \to \mathbb{R}$
 1: Initialize model parameters
 2: **for** number of training iteration **do**
 3:     //the aggregation layer
 4:     Sample minibatch of positivite and negative interactions from R;
 5:     Update parameters by gradient descent on Equation (2)-(4), (8), (12);
 6:     //the enhancement layer
 7:     Obtain meta-paths by meta-path-based random walk method according to Table 1 from G_H;
 8:     Update parameters by gradient descent on Equation (5)-(8), (12);
 9: **end for**
10: Get the trained model parameters and calculate the prediction score
---

Time Complexity. Time consumption comes in five main parts: graph convolution, aggregation attention, self-attention, dependent path probability, and prediction. We first calculate the number of interactions $t = \mid R^+ \mid + \mid S^+ \mid$ in the adjacency matrix, where $\mid R^+ \mid$ and $\mid S^+ \mid$ denote the number of nonzero elements in R and S. For the graph convolution and attention through k layers, time consumption both are $O(tdk)$, d is the dimension of embedding. The self-attention and dependent path probability have computational complexity $O(Nd)$. For the prediction layer, only inner product calculations were carried out, taking time $O(NMd^2)$. Since our model removes the feature transformations and non-linear transformations from the GCN according to [7], our model is also more efficient in the training process than traditional GNN-based social recommendation models.

4   Experiments

In this section, we conduct experiments on three real datasets, Yelp, Douban and Lastfm-2k, to answer the following three questions:

*RQ*1 Can ME-LGNN perform better than other competitive methods?

*RQ*2 Does our proposed meta-path enhancement module improve recommendation performance?

*RQ*3 Is ME-LGNN effective in mitigating data sparsity problems?

4.1   Experimental Settings

Dataset. To validate the performance of the model proposed in this paper, we conduct experiments on three public datasets, Yelp, Douban and Lastfm-2k. The three datasets are described in detail as follows.

Table 2. The statistics of the three datasets.

Dataset	Yelp	Douban	Lastfm-2k
Users	10580	12748	1892
Items	13870	22347	17632
Total links	79295	84575	84845
Ratings	102662	555739	55625
Link density	0.07%	0.05%	2.37%
Rating density	0.07%	0.19%	0.17%

- **Yelp:** The dataset is an online location-based social network where users rate and interact with each other on top of the site.
- **Douban:** The dataset is a crawl of Douban reading data, which describes the interaction behaviour of users on Douban.
- **Lastfm-2k:** The dataset contains the social networks of the 2K users set from the Last.fm online music system, as well as information about the users' listening.

With the dataset described above, we randomly selected 60% of the user interaction dataset as the training set, 20% as the test set and the remaining 20% as the validation set to tune the hyperparameters. Next, we perform a negative sampling strategy on the dataset, sampling the items that have not been consumed by the users as negative samples. The information of the dataset after pre-processing is shown in Table 2.

Evaluation Indicators. In this article, with an aim of evaluating the top-N, We use two benchmarks which are commonly used in recommender systems, Recall and Normalized Discounted Cumulative Gain (NDCG). Recall measures how many positive examples are judged to be positive, NDCG considers not only the ranking but also the relevance of the top positive examples.

Baselines. To verify the performance of our model, we compare ME-LGNN with the following methods.

- **TBPR** [20]: The approach incorporates the important concept of strong and weak ties into social recommendation, combining the BPR [16] model to discriminate between strong and weak ties. The authors propose an EM model-based algorithm for discriminating strong and weak ties in social networks, which learns the potential feature vectors of users and items based on the optimal recommendation accuracy.
- **SocialMF** [10]: The method constrains that a user's preferences should be as similar as possible to the average preferences of the social neighbors to which the user is connected, and introduces trust propagation in the matrix factorization so that users are represented as being close to the users they trust.

- **NeuMF** [8]: The method is a typical recommendation algorithm based on deep learning. It combines traditional matrix factorization and a multi-layer perceptron to extract both low and high dimensional features with impressive recommendation results.
- **Diffnet** [18]: This approach uses GCN to model the user's social network to obtain the user's embeddings then leverages the SVD++ [11] framework to implement the recommendation task.
- **LightGCN** [7]: This model is also a graph-based model, where it explores recommendation tasks on a user-item interaction graph. In this work, the authors' goal is to simplify the design of the GCN to make it cleaner and more suitable for collaborative filtering tasks.

Parameter Settings. We use tensorflow to implement our model. For all models relying on gradient descent-based approaches in the model learning process, we utilise Adam as the optimization method. For all comparison models, we tune the learning rate between [0.001, 0.005, 0.01.0.02, 0.05] to get the best results. The regularization factor was tuned between $[10^{-6}, 10^{-5}, \cdots, 10^1, 10^2]$ to obtain the best results. The training batch size is set to 1024. For NeuMF, we set the hidden layers as suggested in [8]. For Diffnet, LightGCN, we tune the GCN layers between [1, 2, 3]. For our model, the number of GCN layers is set to 2. Finally, we set the embedding dimension to 32 and 64 respectively to compare the recommended performance.

4.2   Overall Comparison (RQ1)

In Table 3 and Table 4, we show the overall performance of the top-10 recommendations for all models with different dimensional embeddings in the three datasets. It can be observed that almost all the performances are improved accordingly with increasing embedding dimension d. Both TBPR [20] and SocialMF [10] leverage the social connections of users to mitigate the problem of sparsity, and while TBPR [20] incorporates the strength of ties, SocialMF [10] introduces the propagation of trust. NeuMF [8] introduce high-dimensional features on the basis of traditional methods and achieve considerable results. Graph convolution models Diffnet [18] and LightGCN [7] show a significant improvement, as reflected in the great potential of graph-based recommendation models for recommendation tasks. Our ME-LGNN is superior on almost all data sets, which shows the effectiveness of explicit modeling for high-order collaborative signals. We carry out further experiments on this aspect, the results are depicted in Table 5 and Table 6. They present the experimental performance under different Top-N when $d = 64$, which are consistent with our previous analysis, further verifying the effectiveness of our model. Such results and explanations prove that our model is effective through specific experiments. As can be seen from the results of the experiment, the results perform best when the dimension is 64, so we will use $d = 64$ for model analysis in the following experiments.

Table 3. Recall@10 comparisons for different dimension size d.

Models	Yelp		Douban		Lastfm-2k	
	d = 32	d = 64	d = 32	d = 64	d = 32	d = 64
TBPR	0.0247	0.0256	0.0246	0.0358	0.0621	0.0605
SocialMF	0.0182	0.0245	0.0410	0.0438	0.0766	0.0725
NeuMF	0.0258	0.0283	0.0473	0.0456	0.0934	0.0936
Diffnet	0.0223	0.0218	0.0309	0.0453	0.0834	0.0865
LightGCN	0.0344	0.0361	0.0422	0.0512	0.1118	0.1033
ME-LGNN	**0.0407**	**0.0412**	**0.0497**	**0.0514**	**0.1159**	**0.1200**

Table 4. NDCG@10 comparisons for different dimension size d.

Models	Yelp		Douban		Lastfm-2k	
	d = 32	d = 64	d = 32	d = 64	d = 32	d = 64
TBPR	0.0201	0.0198	0.0324	0.0631	0.0754	0.0734
SocialMF	0.0159	0.0218	0.0542	0.0605	0.0882	0.0873
NeuMF	0.0216	0.0226	0.0617	0.0636	0.1028	0.1171
Diffnet	0.0195	0.0207	0.0397	0.0585	0.0988	0.1068
LightGCN	0.0316	0.0334	0.0657	**0.0696**	0.1271	0.1282
ME-LGNN	**0.0346**	**0.0353**	**0.0670**	0.0691	**0.1396**	**0.1415**

Table 5. Recall@N comparisons for different top-N values.

Models	Yelp		Douban		Lastfm-2k	
	N = 10	N = 20	N = 10	N = 20	N = 10	N = 20
TBPR	0.0256	0.0353	0.0358	0.0546	0.0605	0.0898
SocialMF	0.0245	0.0462	0.0439	0.0727	0.0724	0.1263
NeuMF	0.0283	0.0499	0.0456	0.0720	0.0937	0.1461
Diffnet	0.0219	0.0338	0.0453	0.0648	0.0866	0.1354
LightGCN	0.0361	0.0582	0.0512	0.0735	0.1033	0.1596
ME-LGNN	**0.0412**	**0.0653**	**0.0514**	**0.0815**	**0.1200**	**0.1728**

Table 6. NDCG@N comparisons for different top-N values.

Models	Yelp		Douban		Lastfm-2k	
	N = 10	N = 20	N = 10	N = 20	N = 10	N = 20
TBPR	0.0198	0.0219	0.0631	0.0623	0.0733	0.0847
SocialMF	0.0218	0.0281	0.0605	0.0661	0.0874	0.1113
NeuMF	0.0226	0.0283	0.0636	0.0689	0.1171	0.1385
Diffnet	0.0207	0.0240	0.0585	0.0607	0.1060	0.1279
LightGCN	0.0334	0.0383	**0.0696**	0.0722	0.1282	0.1521
ME-LGNN	**0.0353**	**0.0421**	0.0691	**0.0740**	**0.1415**	**0.1609**

4.3   Ablation Experiments (RQ2)

In this subsection, we examine whether our proposed meta-path enhancement module is effective through ablation experiments. We remove the meta-path enhancement module to train the model and obtain experimental results for the model LGNN. The results of the experiment are shown in Table 7 and Table 8, it can be observed that the addition of the meta-path enhancement module to the LGNN yields the ME-LGNN, which further improves performance and demonstrates the effectiveness of valuing reliable connections for modeling. Moreover, after removing the meta-path enhancement module, our model still exhibits impressive performance, further demonstrating its effectiveness in explicit modeling of HINs.

Table 7. Recall@N comparisons for different top-N values.

Models	Yelp		Douban		Lastfm-2k	
	N = 10	N = 20	N = 10	N = 20	N = 10	N = 20
LGNN	0.0407	0.0646	0.0509	0.0821	0.1163	0.1686
ME-LGNN	**0.0412**	**0.0653**	**0.0514**	**0.0815**	**0.1200**	**0.1728**

Table 8. NDCG@N comparisons for different top-N values.

Models	Yelp		Douban		Lastfm-2k	
	N = 10	N = 20	N = 10	N = 20	N = 10	N = 20
LGNN	0.0345	0.0409	0.0685	0.0737	0.1390	0.1598
ME-LGNN	**0.0353**	**0.0421**	**0.0691**	**0.0740**	**0.1415**	**0.1609**

4.4   Performance Comparison Under Different Sparsity (RQ3)

To verify whether our model can mitigate the data sparsity problem, we conduct sparsity experiments on the Yelp and Douban datasets. For the users in the training data, we first group them according to the quantity of interactions. For example, [16, 32) means that the user interacts with the item at least 16 times and at most 32 times in the training set. The experimental results are presented in Fig. 3, from which we are able to observe a general trend of improved performance with increasing numbers of interactions. The result is cognitive, since the more interactions there are, the more information that can be captured. It can also be noted that our model exhibits respectable performance in most cases, with a significant improvement on the Yelp dataset in particular, indicating that our model can mitigate the data sparsity problem.

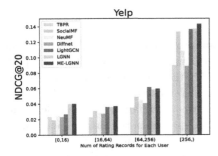
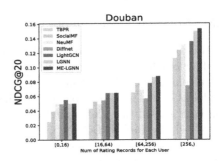

Fig. 3. Performance under different rating sparsity on two datasets.

5 Conclusion

In this work, we propose a new recommendation framework called Meta-path Enhanced Lightweight Graph Neural Network (ME-LGNN) to explicitly model high-order collaborative signals. In order to make users pay more attention to reliable links, we design a series of meaningful meta-paths, random walk based on the meta-paths, and constrain the links by calculating the dependency probabilities of the meta-paths. Extensive experimental analysis on three public datasets shows the effectiveness of our proposed model.

Currently, our approach shows promising performance in handling simple heterogeneous information networks. In future, we consider improving our approach to enable the handling of complex network structures with more attribute information. We will also consider how to make more rational use of meta-paths to improve the interpretability of recommendation results.

References

1. van den Berg, R., Kipf, T.N., Welling, M.: Graph convolutional matrix completion. arXiv preprint arXiv:1706.02263 (2017)
2. Fan, W., Li, Q., Cheng, M.: Deep modeling of social relations for recommendation. In: AAAI (2018)
3. Fan, W., et al.: Graph neural networks for social recommendation. In: WWW, pp. 417–426 (2019)
4. Goldberg, D., Nichols, D., Oki, B.M., Terry, D.: Using collaborative filtering to weave an information tapestry. Commun. ACM **35**(12), 61–70 (1992)
5. Guo, G., Zhang, J., Yorke-Smith, N.: TrustSVD: collaborative filtering with both the explicit and implicit influence of user trust and of item ratings. In: AAAI, vol. 29 (2015)
6. Han, X., Shi, C., Wang, S., Philip, S.Y., Song, L.: Aspect-level deep collaborative filtering via heterogeneous information networks. In: IJCAI, pp. 3393–3399 (2018)
7. He, X., et al.: LightGCN: simplifying and powering graph convolution network for recommendation. In: SIGIR, pp. 639–648 (2020)
8. He, X., et al.: Neural collaborative filtering. In: WWW, pp. 173–182 (2017)

9. Hu, Y., Koren, Y., Volinsky, C.: Collaborative filtering for implicit feedback datasets. In: 2008 Eighth IEEE International Conference on Data Mining, pp. 263–272. IEEE (2008)
10. Jamali, M., Ester, M.: A matrix factorization technique with trust propagation for recommendation in social networks. In: RecSys, pp. 135–142 (2010)
11. Koren, Y.: Factorization meets the neighborhood: a multifaceted collaborative filtering model. In: SIGKDD, pp. 426–434 (2008)
12. Koren, Y.: Collaborative filtering with temporal dynamics. In: SIGKDD, pp. 447–456 (2009)
13. Koren, Y., Bell, R., Volinsky, C.: Matrix factorization techniques for recommender systems. Computer **42**(8), 30–37 (2009)
14. Ma, H., King, I., Lyu, M.R.: Learning to recommend with social trust ensemble. In: SIGIR, pp. 203–210 (2009)
15. Ma, H., Yang, H., Lyu, M.R., King, I.: SoRec: social recommendation using probabilistic matrix factorization. In: CIKM, pp. 931–940 (2008)
16. Rendle, S., Freudenthaler, C., Gantner, Z., Schmidt-Thieme, L.: BPR: Bayesian personalized ranking from implicit feedback. arXiv preprint arXiv:1205.2618 (2012)
17. Schafer, J.B., Frankowski, D., Herlocker, J., Sen, S.: Collaborative filtering recommender systems. In: Brusilovsky, P., Kobsa, A., Nejdl, W. (eds.) The Adaptive Web. LNCS, vol. 4321, pp. 291–324. Springer, Heidelberg (2007). https://doi.org/10.1007/978-3-540-72079-9_9
18. Seah, B.S., Bhowmick, S.S., Dewey Jr., C.F.: DiffNet: automatic differential functional summarization of de-map networks. Methods **69**(3), 247–256 (2014)
19. Wang, X., He, X., Wang, M., Feng, F., Chua, T.S.: Neural graph collaborative filtering. In: SIGIR, pp. 165–174 (2019)
20. Wang, X., Lu, W., Ester, M., Wang, C., Chen, C.: Social recommendation with strong and weak ties. In: CIKM, pp. 5–14 (2016)
21. Wu, L., et al.: DiffNet++: a neural influence and interest diffusion network for social recommendation. TKDE (2020)
22. Xu, W., Chen, K., Zhao, T.: Document-level relation extraction with reconstruction. In: AAAI (2021)
23. Xue, H.J., Dai, X., Zhang, J., Huang, S., Chen, J.: Deep matrix factorization models for recommender systems. In: IJCAI, vol. 17, Melbourne, Australia, pp. 3203–3209 (2017)
24. Yu, J., Gao, M., Li, J., Yin, H., Liu, H.: Adaptive implicit friends identification over heterogeneous network for social recommendation. In: CIKM, pp. 357–366 (2018)
25. Yu, X., et al.: Personalized entity recommendation: a heterogeneous information network approach. In: WSDM, pp. 283–292 (2014)
26. Zhao, T., McAuley, J., King, I.: Leveraging social connections to improve personalized ranking for collaborative filtering. In: CIKM, pp. 261–270 (2014)

Intention Adaptive Graph Neural Network for Category-Aware Session-Based Recommendation

Chuan Cui, Qi Shen, Shixuan Zhu, Yitong Pang, Yiming Zhang,
Hanning Gao, and Zhihua Wei[✉]

Tongji University, Shanghai, China
{2033065,2130777,2130768,1930796,2030796,2030795,
zhihua_wei}@tongji.edu.cn

Abstract. Session-based recommendation (SBR) is proposed to recommend items within short sessions given that user profiles are invisible in various scenarios nowadays, such as e-commerce and short video recommendation. There is a common scenario that user specifies a target category of items as a global filter, however previous SBR settings mainly consider the item sequence and overlook the rich target category information. Therefore, we define a new task called Category-aware Session-Based Recommendation (CSBR), focusing on the above scenario, in which the user-specified category can be efficiently utilized by the recommendation system. To address the challenges of the proposed task, we develop a novel method called Intention Adaptive Graph Neural Network (IAGNN), which takes advantage of relationship between items and their categories to achieve an accurate recommendation result. Specifically, we construct a category-aware graph with both item and category nodes to represent the complex transition information in the session. An intention-adaptive graph neural network on the category-aware graph is utilized to capture user intention by transferring the historical interaction information to the user-specified category domain. Extensive experiments on three real-world datasets are conducted to show our IAGNN outperforms the state-of-the-art baselines in the new task.

Keywords: Session-based recommendation · Graph neural network

1 Introduction

Recommender systems (RS) have become indispensable nowadays in scenarios such as online shopping or social media, to provide users with accurate information in an effective way. Thanks to the development in deep neural networks, more and more powerful RS methods have been proposed. Most of them assume the user profile and historical interactions are well recorded. Nevertheless, many services allow user interaction without user identification, therefore *session-based recommendation* (SBR) was proposed specially. Most of them are based on *recurrent neural networks* (RNNs) [8,12] or *graph neural networks* (GNNs) [18,25].

C. Cui and Q. Shen—Contributed equally to this research.

© The Author(s), under exclusive license to Springer Nature Switzerland AG 2022
A. Bhattacharya et al. (Eds.): DASFAA 2022, LNCS 13246, pp. 150–165, 2022.
https://doi.org/10.1007/978-3-031-00126-0_10

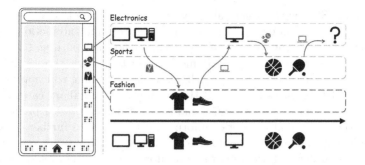

Fig. 1. A toy example of Category-aware Session-based Recommendation.

Previous SBR settings usually focus on the interacted item sequence without consideration of users' other types of behavior, e.g., clicking the specific category button to filter out items. However, this kind of behavior truely exists in a real-world SBR scenario. For instance, Fig. 1 illustrates a regular page that allows specifying a category to filter items, with the selection buttons on the navigation bar. During an online shopping trip, after several browsing among different categories, user may wonder if there is any interesting item in the category "Electronics". Therefore he taps the "Electronics" button and interacts with items in this target category. This implies the following possible user demands: (i) The recommendation direction deviates from the user's intention, therefore user wants to adjust the recommendation results via additional feedback in time, e.g., specifying the target category in accord with his intention. (ii) Sometimes the recommendation results could not keep up with user's dynamic and abruptly shifting intention, therefore he may want to filter the items with target category to locate the right item efficiently. (iii) During the interaction, user may tend to concentrate on items in one category than the others, and an autonomous category selection by the user becomes necessary. In summary, these scenarios require the system to model the dynamic and rich interaction sequence (including interactions with item and category), to make more accurate and user-controllable recommendations based on user-specified target category, rather than general recommendations under all categories. When facing above scenarios, previous SBR settings cannot perceive the specific category which user intends to view next, and the system fails in reacting to the shifted interest promptly, leading to the failure of meeting user expectations.

Due to above limitations, we propose a new task for SBR: In an ongoing session, with user-specified target category and each item's corresponding category, the system recommends item specifically in the target category. The category selection can be commonly seen in the navigation bar of online applications, e.g., the page of an e-commerce platform in Fig. 1. This task, namely Category-aware Session-Based Recommendation (CSBR), extends the interaction mode of users, *w.r.t.* specifying target category of next items, to accomplish a user-controllable SBR.

As for the SBR with target category, the key challenge is how to efficiently leverage the category information in addition to the original interaction sequence. A straightforward adaption method is to recommend the next item based on the interaction history via existing SBR models and then filter the results by the target category. However, this approach overlooks something which are the challenges for CSBR listed as follows: (i) How to inject the auxiliary category information into session representation dynamically. Compared with previous SBR methods, the additional categories information needs to be further considered, including category-level user interaction transition and item-category relations. (ii) How to transfer historical interaction information to the user-specified category domain efficiently. Intuitively, the interaction in one category might be helpful to improve recommendation effectiveness in other categories, for the idea that user behavior in different categories might reveal a particular user characteristics or interest. For example, as shown in Fig. 1, a user might firstly interact with several items in the order of categories "Electornics", "Fashion", "Electronics", "Sports" and then he clicks the "Electronics" category button to get the recommendation results only in "Electronics" category. During this process, some fashion item like "T-shirt with Super Mario print" may be helpful to predict the next item "Nintendo console" in the "Electronics" category because they are all related to the video game "Mario" series. However, adopting user interaction information from other categories will incorporate helpful but also noisy information at the same time. For instance, a "Swimming google" in "Sports" category might be irrelevant to a game console. Therefore, it is crucial to distill intention-adaptive information for matching the current interest of user.

To effectively address the aforementioned challenges, we propose a novel model named Intention Adaptive Graph Neural Network (IAGNN) for CSBR. In detail, we start by converting the complex interaction session into an Category-aware Graph, which is composed of item-level transitions and category-level transitions, and item-category relations. Besides, to transfer historical interaction information to the user-specified category domain, we instantiate the message path between target category node and item nodes with explicit links in the graph. Based on the lossless graph, we employ a position-aware graph neural network to learn the item-level and category-level representations. By stacking multiple layers, the complicated historical interaction information is aggregated to refine the user intention representation in target category iteratively. Finally, the user's intention is characterized by the target category and last item representations for CSBR.

Our main contributions of this work are summarized below:

- We consider a flexible session-based recommendation scenario when user's autonomous navigation in possible item category is perceived, which is introduced as the task **CSBR**, concentrating on leveraging the signals of user-specified target category and item interaction sequence, for more precise recommendation results.

- To address the challenges of CSBR, we propose a novel model named Intention Adaptive Graph Neural Network (**IAGNN**). Firstly, we construct a category-aware graph not only explicitly models the historical transitions and item-category relations, but also covers the connections of user-specified target category and historical interacted items. Moreover, we conduct a position-aware intention-adaptive graph neural network on the category-aware graph to capture user intention by transferring the historical interaction information to the user-specified category domain.
- Extensive experiments on three datasets demonstrate that our model is superior compared with state-of-the-art models for CSBR.

2   Related Works

In this section, we review some related works of session-based recommendation and category-based recommendation.

2.1   Session-Based Recommendation

Various neural network based approaches have been proposed for session-based recommendation(SBR) with the development of deep learning recently.

RNN-Based SBR. Recurrent neural networks (RNNs) are powerful sequence models which have been widely adopted for SBR tasks [8,12,14,21,23,27]. GRU4Rec [8] by Hidasi et al. was the first RNN-based method to capture information in user-item interaction sequences by simply utilizing several GRU layers. NARM [12] takes advantage of attention mechanism beyond GRU4Rec, by referring the last interacted item, and captures user's preference representations both from global and local perspective of current session. However, these sequential methods are unable to capture the items transition relationship efficiently.

GNN-Based SBR. Recently, graph neural networks (GNNs) has been proved to be competent to extract complex relationships between objects, so there emerges quite a few GNN-based methods for SBR [19,20,22,24–26]. SR-GNN [25] is the first work which leverages gated GNN (GGNN) on a directed graph constructed from the interaction sequence to learn item embeddings. Nevertheless, SR-GNN only propagates messages between adjacent items, which would fail to take long-distance item relations into consideration. LESSR [4] proposes a better architecture, including EOPA layer and SGAT layer to solve the lossy session encoding problem and propagate information along shortcut connections, which leads to lossless information presentation during graph construction and better performance for SBR. Pan et al. proposed StarGNN [18] with a star graph neural network to model the complex transition relationship between items with an additional star node connected to every item in an on going session, and applies a highway network to handle the over-fitting issue in GNNs.

2.2 Category Information in Recommendation

As a significant auxiliary information for items, category information has been explored in other recommendation areas.

Cross-Domain Recommendation. Cross-domain recommendation (CDR) [9,10] can be considered as general multi-category recommendation, which utilizes data from multiple domains to deal with issues like cold start [1] and data sparsity [17] in target domain (category). Recently, Ma et al. proposed π-net [15] to generate recommendation scores for every item in two domains by integrating information from both. DA-GCN [6] constructs a cross-domain sequence graph which explicitly links account-sharing users and items from two domains to learn expressive representations for recommendation.

Category-Aware Recommendation. Category-aware recommendation utilizes category information to enhance the item representation for better user preference modeling. There are several traditional category-aware recommendation methods. HRPCA [2] proposes a hybrid recommendation method to handle the customer preferences varieties in different product categories. Choi et al. [5] design a recommendation algorithm based on the category correlations to a user with certain preferences. Recently, LBPR [7] conducts list-wise bayesian ranking for next category and category-based location recommendation. CoCoRec [3] utilizes self-attention to model item transition patterns in category-specific action sub-sequences, and recommends items for user by collaborating neighbors' in-category preferences.

As outlined above, the rich category information has not been comprehensively explored by previous SBR methods. Also for previous category-related methods, they rarely focused on recommendation for a session with anonymous users. What is more, none of these task settings explicitly consider the user-specified category information, and there are huge gaps in this scenario.

3 Preliminary

Given the entire item set $\mathcal{V}$ and category set $\mathcal{C}$, we first define a category-augmented item session as $s = \{(v_1, c_{v_1}), (v_2, c_{v_2}), ..., (v_n, c_{v_n})\}$, in which (v_i, c_{v_i}) represents the user interacted item $v_i \in \mathcal{V}$ in category $c_{v_i} \in \mathcal{C}$, and n is the session length. Note that v_i and c_{v_i} can be repetitive in the sequence, since there can be repeated items in the session. Given session s, the user-specified target category c_t and the corresponding item set $\mathcal{V}_{c_t} = \{v_i | c_{v_i} = c_t, v_i \in \mathcal{V}\}$, the goal of CSBR is to predict a probability score for any item $v \in \mathcal{V}_{c_t}$ such that an item with higher score is more likely to be interacted next.

4 Methodology

4.1 Overall Architecture

Figure 2 illustrates the overall architecture of IAGNN under the context of user's target category specified. First, the Embedding Layer will initialize id embeddings for all items and categories. Second, we construct the Category-aware

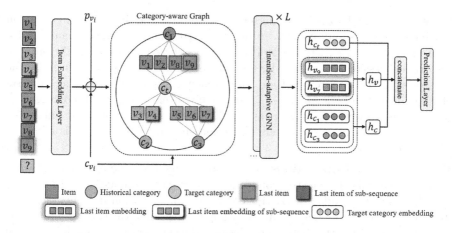

Fig. 2. The overview of IAGNN.

Graph to explicitly keep the transitions of in-category items and different categories, along with the relation between items and their corresponding categories. Third, a Heterogeneous Graph Attention Neural Network is introduced to propagate embeddings through attentive graph convolution. Finally, by leveraging a Embedding Fusion Layer and a Prediction Layer, items in user-specified target category will be recommended.

4.2   Category-Aware Graph Construction

According to previous SBR methods, category information is not explicitly integrated in the recommendation process. In our proposal, we take advantage of graph neural network to model the items and categories transition. The graph $\mathcal{G}_s = (\mathcal{V}_s, \mathcal{E}_s)$ is a directed heterogeneous graph, in which $\mathcal{V}_s = (\{v_1, v_2, v_3, ..., v_n\}, \{c_1, c_2, c_3, ..., c_h\}, c_t)$ indicates n item nodes, h category nodes and the user-specified target node c_t in the graph. $\mathcal{E}_s$ is comprised of the edges representing item transitions, category transitions, item-category connections and target category connections. Therefore, the semantics of item transition and category transition can be represented by these nodes and their relations. Generally, there are four sub-graphs with independent edge relations, and we will detail them as follows.

Item Transitions. At first, we split the original interaction sequence into several category-specific item sub-sequences. For instance, items in category c will be connected with their original interaction order in the session, and the same for other categories. Therefore, the in-category item-to-item transition patterns can be modeled in the sub-sequence.

In detail, for each item node v_i in the category c_{v_i}, we create a link starting from v_i to its next item v_j, representing the item transition in the category subsequence. And we also create self-loop for each item node to include the node itself when aggregating.

Category Transitions. Considering each item node v_i in the original session sequence, we can simultaneously initialize a corresponding category node c_{v_i} for it. And a category transition graph can be constructed according to the order of the original session sequence order. For instance, in Fig. 2, c_1, c_2, c_3 indicate the categories in colors 'Green', 'Blue' and 'Yellow'. The session sequence consists of items in category $[c_1, c_2, c_3, c_1]$ respectively, then a sub-graph composed of connections between categories can be constructed by the directed link $c_1 \to c_2 \to c_3 \to c_1$. Similar to the item transition sub-graph, a self-loop is added to every category node c_i.

Item-Category Connections. For representing the inherent item-category relations, we build bi-directed connections between item node v_i and its corresponding category node c_{v_i}. By this connection, information from non-adjacent items nodes can be propagated in a two-hop way through the corresponding category node as an intermediate node. Besides, the category transition information which implies the user interest changes across different categories can be propagated back to all the related item nodes. Finally, this approach can enhance the representation of both items and categories.

Target Category Connections. Moreover, the user-specified target category is a given information in our task, to better leverage this information, we add a particular node c_t to represent the target category, which is connected to all the item nodes (In Fig. 2, c_t represents the same category with c_3). Therefore, it can not only solve the long-range information propagation issue between non-adjacent items, but also build the transfer from all the historical information to the target category domain as the the process of message passing from items to target category node. Meanwhile, as a pseudo node for the next item, the representation of c_t can be an intermediate variable to bridge the gap of all the interacted items and the actual next item in the session.

4.3   Intention-Adaptive Graph Neural Network

Next, we present how intention-adaptive graph neural networks propagate messages between different nodes.

Embeddings. For each item v_i, it is projected by the item embedding layer $E_i \in \mathbb{R}^{|V| \times d_v}$ into dense embedding e_{v_i}. In which d_v denotes the item embedding size. Meanwhile the category embedding layer $E_c \in \mathbb{R}^{|C| \times d_c}$ transform category unique identifications to category embedding e_{c_i}. Note that since c_t represents the target category, it can also be initialized in the same way as a regular category node.

Also for a session with no more than n items, we take advantage of the reversed position embeddings $\mathbf{P}_{-i} \in \mathbb{R}^{n \times d_P}$ by following GCE-GNN [24], which shows superiority performance because the session length is versatile. Moreover, we conduct a positive-going position embedding in our experiments. Also the category embeddings $\mathbf{e}_{c_{v_i}}$ are added for each item v_i to enhance the item representation by concatenation and projection:

$$\mathbf{e}_{v_i} = \mathbf{W}' \left[\mathbf{e}_{v_i} \parallel \mathbf{e}_{c_{v_i}} \parallel \mathbf{p}_{v_i} \right] \tag{1}$$

where $\|$ is the concatenation operation, $\mathbf{W}' \in \mathbb{R}^{d_v \times (d_v + d_c + d_P)}$ is the projection matrix to keep the size of e_{v_i}, $\mathbf{p}_{v_i} \in \mathbf{P}_{-i}$ is the reversed position embedding for v_i.

Message Propagation. Our method uses the same message propagation mechanism for the different edge types in the graph. We use item-item message propagation as an example here. As mentioned above, the item node v_i is linked to its previous item in the in-category sequence along with the self-loop connection. Then the item message that node v_i received can be represented as:

$$\mathbf{h}_{v_i \leftarrow v}^{(l+1)} = \sigma\Big(\sum_{j \in \mathcal{N}_{v_i}(v)} \alpha_{ij}^{(l)} \mathbf{W}_v \mathbf{h}_{v_j}^{(l)} \Big) \tag{2}$$

$$\alpha_{ij} = \frac{\exp(\text{LeakyReLU}(\mathbf{a}_{l_v} \mathbf{W}_v \mathbf{h}_i + \mathbf{a}_{r_v} \mathbf{W}_v \mathbf{h}_j))}{\sum_{k \in \mathcal{N}_{v_i}(v)} \exp(\text{LeakyReLU}(\mathbf{a}_{l_v} \mathbf{W}_v \mathbf{h}_i + \mathbf{a}_{r_v} \mathbf{W}_v \mathbf{h}_k))} \tag{3}$$

where $\mathbf{h}_{v_i \leftarrow v}^{(l+1)} \in \mathbb{R}^{d_v}$ is the item-source message representation of node v_i at layer $(l+1)$ and $\mathcal{N}_{v_i}(v)$ stands for the set of neighbor item nodes. α_{ij} is a scalar representing the attention coefficient. $\mathbf{W}_v \in \mathbb{R}^{d_v \times d_v}$ and $\mathbf{a}_{l_v}, \mathbf{a}_{r_v} \in \mathbb{R}^{1 \times d_v}$ are the shared linear transformation parameters for the source and target node.

Graph Nodes Aggregation. Similar to the item-item message aggregation (Eqs. 2 and 3), we can get other types of messages from the adjacent nodes. Therefore, the final representation for an item node at layer $(l+1)$ can be represented as:

$$\mathbf{h}_{v_i}^{(l+1)} = \mathbf{h}_{v_i \leftarrow v}^{(l+1)} + \mathbf{h}_{v_i \leftarrow c}^{(l+1)} + \mathbf{h}_{v_i \leftarrow c_t}^{(l+1)} \tag{4}$$

where $\mathbf{h}_{v_i \leftarrow v}$ denotes the messages passed from other item nodes, $\mathbf{h}_{v_i \leftarrow c}$ represents the messages from its corresponding category node, and $\mathbf{h}_{v_i \leftarrow c_t}$ stands for the messages from target category.

As for the category nodes, its representation at layer $(l+1)$ is:

$$\mathbf{h}_{c_i}^{(l+1)} = \mathbf{h}_{c_i \leftarrow v}^{(l+1)} + \mathbf{h}_{c_i \leftarrow c}^{(l+1)} \tag{5}$$

In which $\mathbf{h}_{c_i \leftarrow v}^{(l+1)}$ is the messages propagated from adjacent item nodes, and $\mathbf{h}_{c_i \leftarrow c}^{(l+1)}$ represents the messages from adjacent category nodes.

The same for the target category node c_t, its $(l+1)$ layer representation is:

$$\mathbf{h}_{c_t}^{(l+1)} = \mathbf{h}_{c_t \leftarrow v}^{(l+1)} \tag{6}$$

Since the target category node connects to all the item nodes, only $\mathbf{h}_{c_t \leftarrow v}^{(l+1)}$, the messages aggregated from item nodes are used for its representation.

As we employs an L-layer IAGNN, the embedding of each node before and after the IAGNN can be represented as $\mathbf{h}^{(0)}$ and $\mathbf{h}^{(L)}$ respectively. In order to fuse the semantics before and after the network and avoid possible overfitting, we utilize the *gate* mechanism to get the final embeddings for each kind of node:

$$\mathbf{h}^L = \mathbf{g} \odot \mathbf{h}^{(0)} + (1 - \mathbf{g}) \odot \mathbf{h}^{(L)} \tag{7}$$

$$\mathbf{g} = \sigma(\mathbf{W}[\mathbf{h}^{(0)} \| \mathbf{h}^{(L)}]) \tag{8}$$

where the **g** is the gate factor to control information contributed from $\mathbf{h}^{(0)}$ and $\mathbf{h}^{(L)}$. $\mathbf{W} \in \mathbb{R}^{(d_h \times 2) \times d_h}$ is the transform matrix to get the gate. σ denotes the *Sigmoid* activation function. Here we use a function *Gated* to simplify Eqs. 7 and 8:

$$\mathbf{h}^L = \text{Gated}\left(\mathbf{h}^{(0)}, \mathbf{h}^{(L)}\right) \tag{9}$$

4.4   Embeddings Fusion and Prediction

In a SBR scenario, the last item user interacted contributes a lot to the next item user will interact with. In our case, there are two last items, one is the last item v_l of the original interaction sequence, the other is the last item v_{l_c} of the in-category sub-sequence. Therefore, after going through the IAGNN, we fuse these two item representations to a session item representation:

$$\mathbf{h}_v = \text{Gated}\left(\mathbf{h}_{v_l}^L, \mathbf{h}_{v_{l_c}}^L\right) \tag{10}$$

Accordingly, we fuse the category node embeddings c_l and c_{l_c} which is related to the sequence last item v_l and last item v_{l_c} of in-category sub-sequence as the session category representation:

$$\mathbf{h}_c = \text{Gated}\left(\mathbf{h}_{c_l}^L, \mathbf{h}_{c_{l_c}}^L\right) \tag{11}$$

After obtaining the item and category session representation h_v and h_c, we can also get the embedding of target category node:

$$\mathbf{h}_{c_t} = \mathbf{h}_{c_t}^L \tag{12}$$

Then, we combine them into one embedding for the session:

$$\mathbf{h}_s = \mathbf{W}_s \left[\mathbf{h}_v \parallel \mathbf{h}_c \parallel \mathbf{h}_{c_t}\right] \tag{13}$$

where $\mathbf{W}_s \in \mathbb{R}^{d_v \times (d_v + d_c \times 2)}$ is used to project the concatenation result to an embedding of size d_v. Note that we did not introduce additional attention mechanism for all the items as a readout step, because regarding the target category node, the attention mechanism is equivalently done after message propagation through multi-layer GNN.

By multiplying the session representation with all item embeddings, we can get the prediction score as follows:

$$\hat{y}_i = \mathbf{h}_s^{\mathsf{T}} \tilde{\mathbf{e}}_{v_i} \tag{14}$$

where $\tilde{e}_{v_i}$ is the projected candidate item embeddings through item embedding layer $E_i \in \mathbb{R}^{|V| \times d_v}$.

To train our network, a cross-entropy loss function is employed to optimize the model parameters:

$$L(\hat{\mathbf{y}}) = -\sum_{i=1}^{|V|} y_i \log(\hat{y}_i) + (1 - y_i)\log(\hat{y}_i) \tag{15}$$

where y_i denotes the ground-truth item one-hot encoding vector.

5   Experiments

We conduct extensive experiments to evaluate our method in comparison with other state-of-the-art methods[1]. The goal in this section is to answer the following research questions:

- **RQ1:** What is the performance difference between SOTA baselines and ours?
- **RQ2:** Can the category information of the sequence items contribute to the recommendation performance?
- **RQ3:** Will the graph construction benefit the model performance?
- **RQ4:** How does our model perform on the task CSBR with different hyperparameters setup?

Table 1. Statistics of datasets used in experiments.

Statistic	Diginetica*	Yoochoose*	Jdata*
No. of items	33,596	13,317	79,356
No. of sessions	482,743	400,613	959,824
Avg. of session length	6.48	10.27	6.90
No. of categories	982	12	79
Avg. of categories per session	2.28	2.07	2.35

5.1   Experimental Setup

Dataset. We performed the evaluation on three public datasets: Diginetica*, Yoochoose* and Jdata*, which are widely used in the session-based recommendation research [4,16,24]. These datasets contain additional category information which can support our work for CSBR.

- Diginetica*[2] includes user sessions extracted from e-commerce search logs, with desensitized user ids, hashed queries, hashed query terms, hashed product descriptions and meta-data, log-scaled prices, clicks, and purchases.
- Yoochoose*[3] is the dataset for RecSys Challenge 2015, which contains user clicks and purchases of an online retailer within several months. In our case, we use the fractions 1/4 of Yoochoose data.
- Jdata*[4] is also a dataset of a challenge hosted by JD.com. The data was filtered by 1 h duration to extract the session data.

---

[1] Source code and data will be released in the future for research purpose.
[2] https://competitions.codalab.org/competitions/11161.
[3] https://www.kaggle.com/chadgostopp/recsys-challenge-2015.
[4] https://jdata.jd.com/html/detail.html?id=8.

To better apply the datasets to our task, we pre-process them before training. At first, as described in [12,14,25], we applied a data augmentation by regarding the ith item as the label and items before i as the input sequence. Moreover, the category of the label item is considered as the user-specified target category. Then, since there could be short sessions or infrequent items, we removed all sessions of length ≤ 2 and items which have an occurrence less than 5 times in all datasets. In the end, for the given specific category in one session, we filtered the candidate items by keeping the category of candidate item the same as the target category. We summarized the preprocessed dataset in Table 1.

Baseline Models. To evaluate the performance of our method, we compare it with several baselines. Note that the candidate items in the experiments are all filtered with the target category, which means we evaluate the precision and rank performance only on the candidate items in the target category for all models.

- **GRU4Rec** [8] captures patterns in user-item interaction sequences by simply utilizing several GRU layers.
- **NARM** [12] combines the global and local representations to generate the session embedding through attention mechanism and RNN model.
- **SR-GNN** [25] transforms user interaction sequences into directed graphs and utilizes the GGNN layer [13] to learn the embeddings of sessions and items.
- **LESSR** [4] uses a lossless GRU and a shortcut graph attention layer to capture long-range dependencies and lossless embeddings.
- **StarGNN** [18] employs a additional star graph neural network to model the complex transition in sessions.
- **DA-GCN**† [6] uses a domain-aware GNN and two novel attention mechanisms to learn the sequence representation. Here we change the domain concept of the original DA-GCN to category in DA-GCN† by employing in-category sub-sequences, along with the target category integrated.

Nevertheless, we adopt the target category information to NARM, LESSR, StarGNN by fusing the target category embedding to the query part of their attention mechanism, named NARM†, LESSR† and StarGNN†.

Evaluation Metrics. We employ two widely used metrics: *Precision* (P@k) and *Mean reciprocal rank* (mrr@k) following [4,25], where $k = \{10, 20\}$. They respectively represent the correct proportion of the top-k result items and the order of recommendation ranking.

Implementation Details. By leveraging the frameworks PyTorch and Deep Graph Library, we implement our method. We fix the embedding size of items and categories to 128. The model parameters are initialized with a Gaussian distribution with $\mu = 0$ and $\sigma = 0.1$, where μ and σ are the statistical mean and standard deviation. To help the model converge, We employ the *Adam* [11] optimizer with the mini-batch size of 256. A grid search for the hyper-parameters is the following: learning rate η in $\{0.001, 0.005, 0.01, 0.05, 0.1\}$, learning rate decay step in $\{2, 3, 4\}$, number of graph neural network layers L in $\{1, 2, 3, 4, 5\}$. Furthermore, we split the dataset to train, test and validation set by ratio 8:1:1.

Table 2. Experimental results (%) of different models in terms of P@{10, 20}, and mrr@{10, 20} on three datasets. The * means the best results on baseline methods. *Improv.* means improvement over the state-of-art methods. The bold number indicates the improvements over the strongest baseline are statistically significant ($p < 0.01$) with paired t-tests.

Models	Diginetica*				Yoochoose*				Jdata*			
	P@10	P@20	mrr@10	mrr@20	P@10	P@20	mrr@10	mrr@20	P@10	P@20	mrr@10	mrr@20
GRU4Rec	27.47	36.52	15.02	15.89	10.25	13.93	3.92	4.38	31.20	43.79	13.91	14.97
NARM	28.86	37.84	16.89	17.53	11.86	15.25	5.98	6.21	33.92	45.01	15.52	16.29
SR-GNN	30.71	41.86	16.27	17.17	11.46	15.39	4.84	5.11	35.00	46.30	16.07	16.88
LESSR	30.49	41.42	16.52	17.27	11.90	15.67	5.85	6.24	35.95	46.92	16.95	17.71
StarGNN	29.86	40.96	15.33	16.09	12.46	17.30	5.47	5.80	36.91	48.40	17.15	17.94
NARM[†]	28.79	38.48	16.95*	17.61	11.95	15.76	6.28	6.54	33.98	44.97	15.61	16.38
LESSR[†]	30.55	41.39	16.77	17.51	12.95	17.24	6.20	6.52	36.05	47.22	17.01	17.79
StarGNN[†]	30.67	41.34	16.93	17.66*	11.73	16.27	5.24	5.55	37.21*	48.75*	17.53*	18.33*
DA-GCN[†]	31.49*	42.53*	16.86	17.62	15.53*	20.15*	7.09*	7.41*	36.42	48.30	16.52	17.29
IAGNN	**32.63**	**43.80**	**17.35**	**18.11**	**17.03**	**21.42**	**8.18**	**8.49**	**39.86**	**51.79**	**18.89**	**19.72**
Improv.	3.62%	2.99%	2.36%	2.55%	9.66%	6.30%	15.37%	14.57%	7.12%	6.24%	7.76%	7.58%

Regarding the baseline methods, we either directly make minimal changes to their original source code for supporting our dataset, or implement by ourselves according to their papers.

5.2 Experimental Results (RQ1)

Table 2 shows the comparison results of IAGNN over other baselines on the preprocessed datasets Diginetica*, Yoochoose* and Jdata*.

Comparison of Different Baselines. The performances of two RNN-based methods, GRU4Rec and NARM, are not so competitive. Nevertheless, NARM performs much better than GRU4Rec because of its usage of attention mechanism to capture the user interest. And for the GNN-based methods SR-GNN, LESSR and StarGNN, they performs significantly better than the RNN-based methods, which proved that the GNN has promising advantage for SBR. Moreover, we can discover that LESSR and StarGNN outperform SR-GNN in most cases because they use shortcut graph or the star node to capture global dependencies between distant items, showing the effectiveness of explicitly considering long-range dependencies.

Significance of Target Category. Furthermore, for the methods with target category information adopted, NARM[†], LESSR[†] and StarGNN[†] generally outperform their original implementation, which indicates that integrating user-specified target category information can improve the performance significantly for CSBR. We can notice that DA-GCN[†] performs better than the other modified baselines. This means the graph construction method including in-category sub-sequence can benefit our task. Note that for dataset *Diginetica**, the overall improvements are not obvious because it has too many categories which can be noisy to the category enhanced models.

Model Effectiveness. Our model outperforms all the other GNN-based models from Table 2, which shows our model has a better effectiveness. We can elaborate the advantage of our method in three aspects. First, we integrate the user-specified target category into the graph as a dedicated node, such that the model can be aware of the target category as the session context. Second, we inherit the in-category sub-sequence graph construction from the CDR methods, which can model the item transition per category. Third, we introduce the nodes and transitions for categories, and along with the in-category sub-sequence, we keep the original item sequence which can lead to less information loss.

Table 3. Performance comparison for ablation study.

Model setting	Yoochoose*		Jdata*	
	P@20	mrr@20	P@20	mrr@20
w/o Category nodes	20.89	8.12	50.84	19.32
w/o Target category information	20.91	8.04	50.87	19.08
w/o Category transition	21.36	8.13	51.14	19.47
Add Original item transition	21.22	8.41	51.70	19.68
Add Attention mechanism	20.76	7.68	50.14	18.97
Positive position information	21.80	8.04	51.49	19.62
IAGNN	21.42	8.49	51.79	19.72

5.3   Ablation Study (RQ2&3)

In this section, we performed some ablation studies to demonstrate the effectiveness of our model designs.

Category Information. In our proposed method, we created category nodes for in-category sub-sequences during the graph construction. Therefore, we compared our model with the version without sub-sequence category nodes ("w/o Category nodes", by removing $\mathbf{h}_{v_i \leftarrow c}$, $\mathbf{h}_c$ in Eqs. 4, 5, 11 and 13) in the graph in order to show the effectiveness of introducing these nodes. On the other hand, to show the importance of introducing user-specified target category information, we used mean pooling of all item representations as the representation of the target category node instead of the target category ("w/o Target category information") embedding to diminish the impact of target category for comparison:

$$\mathbf{e}_{c_t} = \text{MeanPool}\left(\mathbf{e}_{v_i}\right) \tag{16}$$

Moreover, we compared the case when transitions between each category node were removed ("w/o Category transition", by removing $\mathbf{h}_{c_i \leftarrow c}$ in Eq. 5). As illustrated in Table 3, the performance dropped on both datasets while detaching category information from the model in different mechanisms. This proved the graph construction details aforementioned in Sect. 4 help improving the model performance.

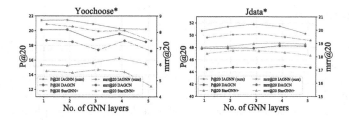

Fig. 3. Model comparison w.r.t. different depths of GNN.

Additional Model Operations. As we know, the items in session has its inherent order, and we added the reversed position embedding for every item node to keep the original item transition information. Thus we tried to add the original item transition as directed links for all the item nodes to check if an additional lossless item order can improve the performance. Meanwhile, following [4,12,25], an additional attention mechanism which gets each item contribution was added to our model. Also we want to see if a positive position embedding is better than the negative. Corresponding to above three conditions, we respectively conduct three alternative models: "Add Original item transition", "Add Attention mechanism" and "Positive position information" for comparison. Following the result in Table 3, The performance drops a little by adding original item transition links to the method. The reason is the model already takes advantage of the negative position embedding and category transition which persist the original item transition order, adding these links will be redundant. Meanwhile, by introducing attention mechanism, the performance was impaired because the target category information is much more significant than attention on every item from different categories. About the position information, albeit positive position embedding improves for P@20 on dataset Yoochoose*, overall the performance drops comparing to our method with negative position embedding.

5.4 Hyper-parameters Study (RQ4)

GNN Depths. We conducted experiments on different number of GNN layers for different GNN-based methods to check the performance impact. As illustrated in Fig. 3, our method keeps beyond the others. Also, we notice that the performance goes up when multi-layer GNN was conducted than that of only one single layer, because more GNN layers are able to capture the high-level and complex semantic information. Furthermore, the metrics of all methods decline approximately after 4 GNN layers, which indicates the over-smoothing is a common issue for GNN-based methods. Without exception, the performance of our method drops after about 3 layers. Nevertheless, as the number of GNN layers goes up, the performance of almost every method increases at the beginning, because GNN is competent for capturing the node embedding transitions, showing it is suitable for our task which involves item and category transitions.

6 Conclusion

This paper proposes a novel category-aware session-based recommendation task, which includes the rich category information both resided in the interaction sequence and the user's intention. Accordingly, a novel model Intention Adaptive Graph Neural Network is introduced to take advantage of these category information, and achieves a more effective performance by explicitly transferring the historical interaction information to the user-specified category domain. Extensive experiments on three real-world datasets are conducted to show our IAGNN outperforms the state-of-the-art baselines in the new task.

As to future work, we prefer to involve other types of attribute information besides category information which would extend the task to an attribute aware session-based recommendation task.

Acknowledgements. The work is partially supported by the National Nature Science Foundation of China (No. 61976160, 61976158, 61906137) and Technology research plan project of Ministry of Public and Security (No. 2020JSYJD01) and Shanghai Science and Technology Plan Project (No. 21DZ1204800).

References

1. Abel, F., Herder, E., Houben, G.J., Henze, N., Krause, D.: Cross-system user modeling and personalization on the social web. User Model. User-Adap. Inter. **23**(2), 169–209 (2013)
2. Albadvi, A., Shahbazi, M.: A hybrid recommendation technique based on product category attributes. Expert Syst. Appl. **36**(9), 11480–11488 (2009)
3. Cai, R., Wu, J., San, A., Wang, C., Wang, H.: Category-aware collaborative sequential recommendation. In: SIGIR, pp. 388–397 (2021)
4. Chen, T., Wong, R.C.W.: Handling information loss of graph neural networks for session-based recommendation. In: SIGKDD, pp. 1172–1180 (2020)
5. Choi, S.M., Han, Y.S.: A content recommendation system based on category correlations. In: 2010 Fifth International Multi-conference on Computing in the Global Information Technology, pp. 66–70. IEEE (2010)
6. Guo, L., Tang, L., Chen, T., Zhu, L., Nguyen, Q.V.H., Yin, H.: DA-GCN: a domain-aware attentive graph convolution network for shared-account cross-domain sequential recommendation. arXiv preprint arXiv:2105.03300 (2021)
7. He, J., Li, X., Liao, L.: Category-aware next point-of-interest recommendation via listwise Bayesian personalized ranking. In: IJCAI, vol. 17, pp. 1837–1843 (2017)
8. Hidasi, B., Karatzoglou, A., Baltrunas, L., Tikk, D.: Session-based recommendations with recurrent neural networks. arXiv preprint arXiv:1511.06939 (2015)
9. Hu, G., Zhang, Y., Yang, Q.: CONET: collaborative cross networks for cross-domain recommendation. In: CIKM, pp. 667–676 (2018)
10. Kang, S., Hwang, J., Lee, D., Yu, H.: Semi-supervised learning for cross-domain recommendation to cold-start users. In: CIKM, pp. 1563–1572 (2019)
11. Kingma, D.P., Ba, J.: Adam: a method for stochastic optimization. arXiv preprint arXiv:1412.6980 (2014)
12. Li, J., Ren, P., Chen, Z., Ren, Z., Lian, T., Ma, J.: Neural attentive session-based recommendation. In: CIKM, pp. 1419–1428 (2017)

13. Li, Y., Tarlow, D., Brockschmidt, M., Zemel, R.: Gated graph sequence neural networks. arXiv preprint arXiv:1511.05493 (2015)
14. Liu, Q., Zeng, Y., Mokhosi, R., Zhang, H.: Stamp: short-term attention/memory priority model for session-based recommendation. In: SIGKDD, pp. 1831–1839 (2018)
15. Ma, M., Ren, P., Lin, Y., Chen, Z., Ma, J., Rijke, M.D.: π-net: a parallel information-sharing network for shared-account cross-domain sequential recommendations. In: SIGIR, pp. 685–694 (2019)
16. Meng, W., Yang, D., Xiao, Y.: Incorporating user micro-behaviors and item knowledge into multi-task learning for session-based recommendation. In: SIGIR, pp. 1091–1100 (2020)
17. Pan, W., Xiang, E., Liu, N., Yang, Q.: Transfer learning in collaborative filtering for sparsity reduction. In: AAAI, vol. 24 (2010)
18. Pan, Z., Cai, F., Chen, W., Chen, H., de Rijke, M.: Star graph neural networks for session-based recommendation. In: CIKM, pp. 1195–1204 (2020)
19. Pang, Y., et al.: Heterogeneous global graph neural networks for personalized session-based recommendation. arXiv preprint arXiv:2107.03813 (2021)
20. Qiu, R., Li, J., Huang, Z., Yin, H.: Rethinking the item order in session-based recommendation with graph neural networks. In: CIKM, pp. 579–588 (2019)
21. Ren, P., Chen, Z., Li, J., Ren, Z., Ma, J., De Rijke, M.: RepeatNet: a repeat aware neural recommendation machine for session-based recommendation. In: AAAI, vol. 33, pp. 4806–4813 (2019)
22. Shen, Q., et al.: Multi-behavior graph contextual aware network for session-based recommendation. arXiv preprint arXiv:2109.11903 (2021)
23. Song, J., Shen, H., Ou, Z., Zhang, J., Xiao, T., Liang, S.: ISLF: interest shift and latent factors combination model for session-based recommendation. In: IJCAI, pp. 5765–5771 (2019)
24. Wang, Z., Wei, W., Cong, G., Li, X.L., Mao, X.L., Qiu, M.: Global context enhanced graph neural networks for session-based recommendation. In: SIGIR, pp. 169–178 (2020)
25. Wu, S., Tang, Y., Zhu, Y., Wang, L., Xie, X., Tan, T.: Session-based recommendation with graph neural networks. In: AAAI, vol. 33, pp. 346–353 (2019)
26. Xu, C., et al.: Graph contextualized self-attention network for session-based recommendation. In: IJCAI, vol. 19, pp. 3940–3946 (2019)
27. Zhou, K., et al.: s^3 - rec: self-supervised learning for sequential recommendation with mutual information maximization. In: CIKM (2020)

Multi-view Multi-behavior Contrastive Learning in Recommendation

Yiqing Wu[1,2,3], Ruobing Xie[3], Yongchun Zhu[1,2], Xiang Ao[1,2], Xin Chen[3],
Xu Zhang[3], Fuzhen Zhuang[4,5(✉)], Leyu Lin[3], and Qing He[1,2(✉)]

[1] Key Lab of Intelligent Information Processing of Chinese Academy of Sciences
(CAS), Institute of Computing Technology, CAS, Beijing 100190, China
{wuyiqing20s,zhuyongchun18s,aoxiang,heqing}@ict.ac.cn
[2] University of Chinese Academy of Sciences, Beijing 100049, China
[3] WeChat Search Application Department, Tencent, China
{ruobingxie,andrewxchen,xuonezhang,goshawklin}@tencent.com
[4] Institute of Artificial Intelligence, Beihang University, Beijing 100191, China
zhuangfuzhen@buaa.edu.cn
[5] Xiamen Institute of Data Intelligence, Xiamen, China

Abstract. Multi-behavior recommendation (MBR) aims to jointly consider multiple behaviors to improve the target behavior's performance. We argue that MBR models should: (1) model the coarse-grained commonalities between different behaviors of a user, (2) consider both individual sequence view and global graph view in multi-behavior modeling, and (3) capture the fine-grained differences between multiple behaviors of a user. In this work, we propose a novel Multi-behavior Multi-view Contrastive Learning Recommendation (MMCLR) framework, including three new CL tasks to solve the above challenges, respectively. The multi-behavior CL aims to make different user single-behavior representations of the same user in each view to be similar. The multi-view CL attempts to bridge the gap between a user's sequence-view and graph-view representations. The behavior distinction CL focuses on modeling fine-grained differences of different behaviors. In experiments, we conduct extensive evaluations and ablation tests to verify the effectiveness of MMCLR and various CL tasks on two real-world datasets, achieving SOTA performance over existing baselines. Our code will be available on https://github.com/wyqing20/MMCLR.

Keywords: Multi-behavior recommendation · Contrastive learning

1 Introduction

Personalized recommendation aims to provide appropriate items for users according to their preferences. The core problem of personalized recommendation is how to accurately capture user preferences from user behaviors. In

---

Y. Wu and R. Xie—Equal contributions.

A. Bhattacharya et al. (Eds.): DASFAA 2022, LNCS 13246, pp. 166–182, 2022.
https://doi.org/10.1007/978-3-031-00126-0_11

real-world scenarios, users usually have different types of behaviors to interact with recommender systems. For example, users can *click, add to cart, purchase,* and *write reviews* for items in E-commerce systems (e.g., Amazon, Taobao). Some conventional recommendation models [15] often rely on a single behavior for recommendation. However, it may suffer from severe data sparsity [11,14,38] and cold-start problems [10,26,37,39] in practical systems, especially for some high-cost and low-frequency behaviors. In this case, other behaviors (e.g., *click* and *add to cart*) can provide additional information for user understanding, which reflect user diverse and multi-grained preferences from different aspects.

Multi-behavior recommendation (MBR), which jointly considers different types of behaviors to learn user preferences better, has been widely explored and verified in practice [1,2,19]. ATRank [34] uses self-attention to model feature interactions between different behaviors of a user in sequence-based recommendation with focusing on the **individual sequence view** of a single user's historical behaviors. In contrast, MBGCN [9] considers different behaviors in graph-based recommendation, focusing on the **global graph view** of all users' interactions. However, there are still three challenges in MBR:

(1) *How to model the coarse-grained commonality between different behaviors of a user?* All types of behaviors of a user reflect this user's preferences from certain aspects, and thus these behaviors naturally share some commonalities. Considering the commonalities between different behaviors could help to learn better user representations to fight against the data sparsity issues. However, it is challenging to extract informative commonalities between different behaviors for recommendation, which is often ignored in existing MBR models.

(2) *How to jointly consider both individual and global views of multi-behavior modeling?* Conventional MBR models are often implemented on either sequence-based or graph-based models separately based on different views. The sequence-based MBR focuses more on the individual view of a user's multiple sequential behaviors to learn user representations [34]. In contrast, the graph-based MBR often concentrates on the global view of all users' behaviors, with multiple behaviors regarded as edges [9]. Different views (individual/global) and modeling methods (sequence/graph-based) build up different sides of users, which are complementary to each other and are helpful in MBR.

(3) *How to learn the fine-grained differences between multiple behaviors of a user?* Besides the coarse-grained commonalities, users' multiple behaviors also have fine-grained differences. There are preference priorities even among the target and other behaviors (e.g., *purchase > click*). In real-world E-commerce datasets, the average number of *click* is often more than 7 times that of the average number of *purchase* [9]. The large numbers of clicked but not purchased items, viewed as hard negative samples, may reflect essential latent disadvantages that prevent users to purchase items. Existing works seldom consider the differences between multiple behaviors, and we attempt to encode this fine-grained information into users' multi-behavior representations.

Recently, contrastive learning (CL) has shown its magic in recommendation, which greatly alleviates the data sparsity and popularity bias issues [36]. We find that CL is naturally suitable for modeling coarse-grained commonalities and fine-grained differences between multi-behavior and multi-view user representations. To solve above challenges, we propose a novel **Multi-behavior Multi-view Contrastive Learning Recommendation (MMCLR)** framework. Specifically, MMCLR contains a sequence module and a graph module to jointly capture multiple behaviors' relationships, learning multiple user representations from different views and behaviors. We design three contrastive learning tasks for existing challenges, including the multi-behavior CL, the multi-view CL, and the behavior distinction CL. (1) The *multi-behavior CL* is conducted between multiple behaviors in both sequence and graph views. It assumes that user representations learned from different behaviors of the same user should be closer to each other compared to other users' representations, which focuses on extracting the commonalities between different types of behaviors. (2) The *multi-view CL* is a harder CL conducted between user representations in two views. It highlights the commonalities between the local sequence-based and the global graph-based user representations after behavior-level aggregations, and thus improves both views' modeling qualities. (3) The *behavior distinction CL*, unlike the multi-behavior CL, concentrates on the fine-grained differences rather than the coarse-grained commonalities between different types of behaviors. It is specially designed to capture users' fine-grained preferences on the target behavior's prediction task (e.g., purchase). The combination of CL tasks can multiply the additional information brought by multiple behaviors in the target recommendation task. Through the MMCLR framework assisted with three types of auxiliary CL losses, MBR models can better understand the informative commonalities and differences between different user behaviors and modeling views, and thus improve the overall performances.

In experiments, we evaluate MMCLR on real-world MBR datasets. The significant improvements over competitive baselines and ablation versions demonstrate the effectiveness of MMCLR and its different CL tasks and components. The contributions of this work are summarized as follows:

- We systematically consider multiple contrastive learning tasks in MBR. To the best of our knowledge, this is the first attempt to bring in contrastive learning in multi-behavior recommendation.
- We propose a multi-behavior CL task and a multi-view CL task, which model the coarse-grained commonalities between different behaviors and (individual sequence/global graph) views for better representation learning.
- We also design a behavior distinction CL task, which creatively highlights the fine-grained differences and behavior priorities between multiple behaviors via a contrastive learning framework.
- MMCLR outperforms SOTA baselines on all datasets and metrics. All proposed CL tasks and the capability on cold-start scenarios are also verified.

2   Related Work

Sequence-Based and Graph-Based Recommendation. *Sequence-based recommendation* mainly leverages users' sequential behavior to mine users' interests, which focuses on individual information. Recently, various deep neural networks have been employed for sequence-based recommendation, e.g., RNN [7], memory networks [3], attention mechanisms [15,23,30,35] and mixed models [20,29]. *Graph-based recommendation* aims to use high-order interaction information contained in the graph, which is able to model the global information of user preferences. Existing works have proved the effectiveness of GNNs in learning user and item representations [17,27]. In this work, we exploit both individual sequence view and global graph view in MBR.

Multi-behavior Recommendation. Inspired by transfer learning [40–42], multi-behavior recommendation takes advantage of other behavior of user to help the prediction of target behavior. Ajit et al. [14] take multi-behavior into consideration via a collective matrix factorization. Recent works often model MBR via sequence or graph-based models [19,25]. MRIG [16] builds sequential graphs on users' behavior sequences. MBGCN [9] learns user-item and item-item similarities on the designed user-item graph and different co-behavior graphs. Other works combine MBR with meta-learning [22] and external knowledge [21]. However, these methods do not make full use of the correlations between behaviors via CL. In this paper, we propose a universal framework that utilizes contrastive learning to model the relations of different behaviors.

Self-supervised Learning. Self-supervised learning (SSL) aims at training a network by pretext tasks, which are designed according to the characteristics of raw data. Recently, self-supervised learning has been shown its superior ability in CV [5,31], NLP [4], and Graph [12] fields. Some works also adopt self-supervised learning in recommender systems [18,24,28,36].

However, most of them fall into single-behavior methods. In this paper, we focus on modeling the commonalities and differences between multiple behaviors and views of users with CL.

3   Methodology

3.1   Preliminaries

MMCLR aims to make full use of multi-behavior and multi-view information to learn better representations for recommendation. We first give detailed definitions of the key notions in our multi-behavior recommendation as follows:

Multi-behavior Modeling. In MBR, the most important and profitable behavior (e.g., *purchase* in E-commerce) is regarded as the target behavior. While it suffers from data sparsity issues. Specifically, we denote the user and item as $u \in U$ and $v \in V$, where U and V are user set and item set. We suppose that users have B types of behaviors $\{b_1, \cdots, b_B\}$ in a system, where b_t is the target behavior.

Multi-view Modeling. Users' multiple behaviors can be modeled with different views, highlighting different aspects of user preferences. In this work, we construct two views, including the sequence vie and the graph view. For the *sequence view*, we represent the multi-behavior historical sequence of user u as $S_u = \{s_u^{b_1}, s_u^{b_2}, ..., s_u^{b_B}\}$, where s_u^b is the behavior sequence of user u under behavior b. For each behavior, we have the item sequence $s_u^b = \{v_1, v_2, ..., v_{|s_u^b|}\}$. For the *graph view*, we build a global multi-relation user-item graph $G = (\mathcal{V}, \mathcal{E})$, where $\mathcal{V}$ is the node set, and $\mathcal{E}$ is the edge set. If user u and item v have an interaction under a certain behavior b, there is a edge $e = (u, v, b) \in \mathcal{E}$ in graph G. We use $\boldsymbol{u}_i^0$ and $\boldsymbol{v}_j^0$ to represent the corresponding raw feature of u_i and v_j.

Problem Definition. Given a user's multi-behavior sequences S_u and the global multi-relation user-item graph G, MMCLR should predict the most appropriate item v that the user will interact under the target behavior b_t.

3.2 Framework of Multi-view Multi-behavior Recommendation

Overview. The model structure of MMCLR is illustrated in Fig. 1. Our model mainly has three parts: multi-view encoder, multi-behavior fusion, and multi-view fusion. Three types of contrastive learning tasks are proposed to capture the multi-behavior and multi-view feature interactions. Specifically, for a user u, the global user-item graph G and the user's multi-behavior sequence S_u are first fed to the sequence-view encoder and the graph-view encoder as inputs. In both sequence and graph encoders, we build B user single-behavior representations according to each behavior, respectively. Second, these single-behavior representations under the same view are fused by the multi-behavior fusion module, with sequence/graph-based multi-behavior CL and behavior distinction CL tasks as auxiliary losses. Then, we combine the sequence-view and graph-view user representations by the multi-view fusion module with the multi-view CL, jointly considering individual and global preferences. Finally, the similarity between the fused user and item representations is viewed as the ranking score.

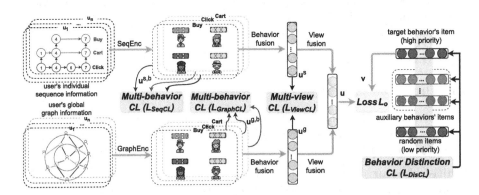

Fig. 1. Overall architecture of MMCLR with our proposed contrastive learning tasks.

Multi-view Encoder. Conventional sequence-based recommendation models [15,34] often focus on the individual historical behaviors of a user, which aims to precisely capture the local sequential information of a user. In contrast, graph-based recommendation models [9,33] are often conducted on the whole user-item graph built by all users' behaviors, which can benefit from the global interactions. We argue that both individual sequence and global graph views are beneficial in multi-behavior recommendation.

Specifically, we implement an individual sequence-based encoder SeqEnc($\cdot$) and a global graph-based encoder GraphEnc($\cdot$) to learn users' and items' single-behavior representations separately. Formally, for the behavior b:

$$\boldsymbol{u}^{s,b} = \text{SeqEnc}^b(\boldsymbol{s}_u^b), \quad \boldsymbol{u}^{g,b} = \text{GraphEnc}^b(G, u, b), \qquad (1)$$

where $\boldsymbol{s}_u^b$ is the user's historical behavior sequence of b, and G is the global user-item graph. $\boldsymbol{u}^{s,b}$ and $\boldsymbol{u}^{g,b}$ indicate the user sequence-view and graph-view single-behavior representation of b. Finally, we learn $2B$ single-behavior representations in two views for the next multi-behavior and multi-view fusions. Note that we can flexibly select appropriate sequence and graph models for SeqEnc$^b(\cdot)$ and GraphEnc$^b(\cdot)$. Specifically, We adopt Bert4rec and lightGCN as sequence encoder and graph encoder. For lightGCN we replace the original aggregator with meaning aggregator.

Multi-behavior Fusion. Single-behavior representations may suffer from data sparsity issues, especially for some high-cost and low-frequent target behaviors (e.g., *purchase*). In this case, other auxiliary behaviors (e.g., *click, add to cart*) could provide essential information to infer user preferences on the target behaviors. Hence, we build a multi-behavior fusion module to fuse user single-behavior representations in each view to get the integrated sequence-view representation $\boldsymbol{u}^s$ and the integrated graph-view representation $\boldsymbol{u}^g$, which is noted as:

$$\boldsymbol{u}^s = \text{MLP}^s(\boldsymbol{u}^{s,b_1}||,\cdots,||\boldsymbol{u}^{s,b_B}), \quad \boldsymbol{u}^g = \text{MLP}^g(\boldsymbol{u}^0||\boldsymbol{u}^{g,b_1}||,\cdots,||\boldsymbol{u}^{g,b_B}). \quad (2)$$

$\boldsymbol{u}^0$ is the raw user embedding in the graph view. MLP^s and MLP^g are two-layer MLPs with ReLU as activation. We also build the graph-view item representation $\boldsymbol{v}^g$ similar to $\boldsymbol{u}^g$, where $\boldsymbol{v}^0$ is also used as the raw behavior features in Eq. (1).

Multi-view Fusion. To take advantage of representations in both views, we apply a multi-view fusion to learn the final user and item representations, which contain both individual and global information. We formalize the integrated user representation $\boldsymbol{u}$ and item representation $\boldsymbol{v}$ as follows:

$$\boldsymbol{u} = \text{MLP}^U(\boldsymbol{u}^s||\boldsymbol{u}^g), \quad \boldsymbol{v} = \text{MLP}^V(\boldsymbol{v}^0||\boldsymbol{v}^g). \qquad (3)$$

Following the classical ranking model [13], the inner product of $\boldsymbol{u}$ and $\boldsymbol{v}$ is used to calculate the ranking scores of user-item pairs, trained under L_o as:

$$L_o = - \sum_{(u,v_i)\in S^+} \sum_{(u,v_j)\in S^-} \log \sigma(\boldsymbol{u}^\top \boldsymbol{v}_i - \boldsymbol{u}^\top \boldsymbol{v}_j), \qquad (4)$$

where $(u, v_i) \in S^+$ indicates the positive set of the target behavior, and $(u, v_j) \in S^-$ indicates the randomly-sampled negative set.

Multi-view Multi-behavior Contrastive Learning. The above architecture is a straightforward combination of multi-view multi-behavior representations. To better capture the coarse-grained commonalities and fine-grained differences between different behaviors and views to learn better user representations in different views and behaviors, we design three types of CL tasks. Next we will introduce details of them.

3.3   Multi-behavior Contrastive Learning

A user's single-behavior representations reflect user preferences on the corresponding behaviors, which also share certain commonalities to reflect the user itself. We build two multi-behavior CL tasks in the sequence and graph views respectively as auxiliary losses to better use multi-behavior information.

Sequential Multi-behavior CL. We adopt a sequential multi-behavior CL, which attempts to minimize the differences between different single-behavior representations of the same user and maximize the differences between different users. In this case, we naturally regard different single-behavior representations of a user as certain kinds of (behavior-level) user augmentations.

Precisely, considering a mini-batch of N users $\{u_1, \cdots, u_N\}$, we randomly select two single-behavior representations $(\boldsymbol{u}_i^{s,b_1}, \boldsymbol{u}_i^{s,b_2})$ of behavior b_1 and b_2 for each u_i as the positive pair in CL. And we consider $(\boldsymbol{u}_i^{s,b_1}, \boldsymbol{u}_j^{s,b_2})$ as the negative pair. Following [2], we also conduct a projector $\mathrm{MLP}_{p_1}(\cdot)$ to map all user single-behavior representations into the same sequential semantic space. We have:

$$\boldsymbol{u}_{i,p_1}^{s,b_1} = \mathrm{MLP}_{p_1}(\boldsymbol{u}_i^{s,b_1}), \quad \boldsymbol{u}_{j,p_1}^{s,b_2} = \mathrm{MLP}_{p_1}(\boldsymbol{u}_j^{s,b_2}). \tag{5}$$

The sequential multi-behavior CL loss L_{SeqCL} is defined as follows:

$$L_{SeqCL} = -\sum_{i=1}^{N} \sum_{u_j \neq u_i} f(\boldsymbol{u}_{i,p_1}^{s,b_1}, \boldsymbol{u}_{i,p_1}^{s,b_2}, \boldsymbol{u}_{j,p_1}^{s,b_2}),$$

$$f(\boldsymbol{x}, \boldsymbol{y}, \boldsymbol{z}) = \log(\sigma(\boldsymbol{x}^\top \boldsymbol{y} - \boldsymbol{x}^\top \boldsymbol{z})). \tag{6}$$

$f(\boldsymbol{x}, \boldsymbol{y}, \boldsymbol{z})$ denotes our pair-wise distance function, $\sigma(\cdot)$ is the sigmoid activation.

Graphic Multi-behavior CL. Similar with the sequential multi-behavior CL, we also build a graphic multi-behavior CL for the graph-view representations. For $\boldsymbol{u}_i^{g,b_1}$, we consider $\boldsymbol{u}_i^{g,b_2}$ as the positive sample and $\boldsymbol{u}_j^{g,b_2}$ as the negative sample in this CL. We also have $\boldsymbol{u}_{i,p_2}^{g,b_1} = \mathrm{MLP}_{p_2}(\boldsymbol{u}_i^{g,b_1})$ and $\boldsymbol{u}_{j,p_2}^{g,b_2} = \mathrm{MLP}_{p_2}(\boldsymbol{u}_j^{g,b_2})$ as Eq. (5). We define the graphic multi-behavior CL loss $L_{GraphCL}$ as follows:

$$L_{GraphCL} = -\sum_{i=1}^{N} \sum_{u_j \neq u_i} f(\boldsymbol{u}_{i,p_2}^{g,b_1}, \boldsymbol{u}_{i,p_2}^{g,b_2}, \boldsymbol{u}_{j,p_2}^{g,b_2}), \qquad (7)$$

in which $f(\boldsymbol{x}, \boldsymbol{y}, \boldsymbol{z})$ is the same as Eq. (6). Through the sequential and graphic multi-behavior CL tasks, MMCLR can learn better and more robust single-behavior representations, which is the fundamental of user diverse preferences. It functions well, especially when the target behaviors are sparse.

3.4   Multi-view Contrastive Learning

The multi-view CL aims to highlight the relationships between the individual sequence and global graph views. It is natural that the sequence-view and graph-view user representations of the same user should be closer than others, since they reflect the same user's preferences (though learned from different information). Hence, we propose the multi-view CL task on the integrated sequence-view and graph-view user representations in Eq. (2). We regard $(\boldsymbol{u}_i^s, \boldsymbol{u}_i^g)$ of the same user u_i as the positive pair, considering $\boldsymbol{u}_i^s$ and $\boldsymbol{u}_i^g$ as different view-level user augmentations of u_i, and regard $(\boldsymbol{u}_i^s, \boldsymbol{u}_j^g)$ and $(\boldsymbol{u}_i^g, \boldsymbol{u}_j^s)$ as the in-batch negative pairs of two views. After the projector, we have $\boldsymbol{u}_{i,p_3}^s = \mathrm{MLP}_{p_3}(\boldsymbol{u}_i^s)$ and $\boldsymbol{u}_{j,p_3}^g = \mathrm{MLP}_{p_3}(\boldsymbol{u}_j^g)$. The multi-view CL loss L_{ViewCL} is noted as follows:

$$L_{ViewCL} = -\sum_{i=1}^{N} \sum_{u_j \neq u_i} f(\boldsymbol{u}_{i,p_3}^s, \boldsymbol{u}_{i,p_3}^g, \boldsymbol{u}_{j,p_3}^g). \qquad (8)$$

We are the first to propose the notion of multi-view CL. Through this CL, individual sequence and global graph views can cooperate well in MBR.

3.5   Behavior Distinction Contrastive Learning

The above two CL tasks highlight the commonalities between a user's multiple behaviors and views compared to other users' representations. However, the fine-grained differences between different behaviors of a user are also essential. For example, in E-commerce, the low-frequent high-cost *purchase* behaviors reflect the user's high-priority preferences, comparing with other low-cost auxiliary behaviors like *click* and *add to cart*. To some extent, these auxiliary behaviors (viewed as positive pair instances in multi-behavior CL) could be even regarded as certain hard negative samples of the high-cost target behaviors [8]. Considering the fine-grained differences and behavior priorities can further improve the target behavior's (e.g., purchase) performances, especially when distinguishing "the good but negative" candidates (e.g., clicked but not purchased items), which are challenging interference terms in practical ranking systems. Hence, we propose a novel **behavior distinction CL** for the first time in MBR.

Specifically, we define the behavior priority in MBR as follows: *items of the target behavior* v_i > *items of auxiliary behaviors* v_j >> *other random in-batch items* v_k. In the target behavior prediction task, the integrated user representation $\boldsymbol{u}$ should firstly be close to $\boldsymbol{v}_i$, and then the hard negative samples of auxiliary behaviors $\boldsymbol{v}_j$, and finally be distinct with the random negative items $\boldsymbol{v}_k$. Similarly, we conduct a projector MLP_{p_4} to get $\boldsymbol{u}_{p_4}$, $\boldsymbol{v}_{i,p_4}$, $\boldsymbol{v}_{j,p_4}$, and $\boldsymbol{v}_{k,p_4}$, and then learn the item-aspect behavior distinction CL L_{DisCL} as follows:

$$L_{DisCL} = -\sum_{u} \sum_{(v_i,v_j)} \sum_{v_k} (f(\boldsymbol{u}_{p_4}, \boldsymbol{v}_{i,p_4}, \boldsymbol{v}_{j,p_4}) + \beta f(\boldsymbol{u}_{p_4}, \boldsymbol{v}_{j,p_4}, \boldsymbol{v}_{k,p_4})). \tag{9}$$

β is a loss weight, v_i and v_j are one of the target/auxiliary behaviors of u.

The multi-behavior CL (i.e., Eq. (6 and 7)) aims to narrow the distances between different behaviors of a user from the global perspective, thus distinguishing them from other items. In contrast, the behavior distinction CL explores to capture the fine-grained differences between different types of behaviors of a user, achieving deeper and more precise understandings of user's target-behavior preferences.

3.6   Optimization

Overall Loss. The overall loss L is defined with hyper-parameters λ as:

$$L = \lambda_o L_o + \lambda_1 L_{SeqCL} + \lambda_2 L_{GraphCL} + \lambda_3 L_{ViewCL} + \lambda_4 L_{DisCL}. \tag{10}$$

Model Analysis. For complexity, the graph and sequential encoders can run parallel, so the encoder complexity is decided by the more complex model. Hence, MMCLR does not produce extra encoding time. For contrastive tasks, the training complexity of the MLP layer is $O(|U|d^2)$, and the complexity of CL is $O(|U|Nd)$, where $|U|$ is the number of users and N is the batch size. The complexity is equal with existing CL models [18,36] and can be computed in parallel with fusion operations. Moreover, the CL losses are only calculated in offline, which means our model has equal online serving complexity as others.

4   Experiments

In this section, we aim at answering the following research questions: **(RQ1)** How does MMCLR perform compared with other SOTA baselines in MBR on various evaluation metrics? **(RQ2)** What are the effects of different contrastive learning tasks in our proposed MMCLR? **(RQ3)** How does MMCLR perform on cold-start scenarios compared to baselines and ablation versions? **(RQ4)** How do different hyper-parameters affect the final performance?

4.1   Datasets

We evaluate MMCLR on two real-world MBR datasets on E-commerce, including the Tmall and CIKM2019 EComm AI dataset. *Tmall*[1]*:* It is collected by Tmall, which is one of the largest E-commerce platforms in China. We process this dataset following [2]. After processing, our Tmall dataset contains 22,014 users and 27,155 items. We consider three behaviors (i.e., click, add-to-cart, purchase), collecting 83,778 purchase behaviors, 44,717 add-to-cart behaviors, and 485,483 click behaviors. *CIKM2019 EComm AI:* It is provided by the CIKM2019 EComm AI challenge. In this dataset, each instance is made up by an item, a user and a behavior label (i.e., click, add-to-cart, purchase). We process this dataset following [2] as well. Finally, this dataset includes 23,032 users, 25,054 items, 100,529 purchase behaviors, 38,347 add-to-cart behaviors, and 276,750 click behaviors.

4.2   Competitors

We compare MMCLR against several state-of-the-art baselines. For baselines not designed for MBR, we adopt our MMCLR's fusion function to jointly consider multi-behavior data. All baselines exploit data of multiple behaviors.

- **BERT4Rec^{MB}.** BERT4Rec [15] is a self-attention-based sequential recommendation model. We conduct separate Transformer encoders on all behaviors, and fuse them via MMCLR's fusion function, denoted as BERT4Rec^{MB}.
- **LightGCN^{MB}.** lightGCN [6] is a widely-used GNN model. Similarly, we construct multiple user-item graphs for all behaviors, encode them by it.
- **MRIG.** MRIG [16] is one of the SOTA sequence-based models for MBR. It adopts user's individual behavior sequence to build a sequential graph, which regards two items having an edge if they are adjacent in a sequence.
- **MBGCN.** MBGCN [9] is a recent graph-based MBR model. It integrates multi-behavior information by user-item and item-item propagations.
- **MBGMN.** MBGMN [22] is one of the SOTA graph-based models for MBR. MBGMN first models the behavior heterogeneity and interaction diversity jointly with the meta-learning paradigm.
- **MGNN.** MGNN [32] is one of the SOTA multiplex-graph-based models for MBR. It builds users' multi-behavior to a multiplex-graph and learns shared graph embedding and behavior-specific embedding for recommendation.

We also compare with MMCLR's ablation versions for further comparisons:

- **BERT4Rec^{CL}.** We add the sequential multi-behavior CL L_{SeqCL} to the BERT4Rec^{MB}, which is noted as BERT4Rec^{CL}.
- **LightGCN^{CL}.** Similarly, We also add the graphic multi-behavior CL to the LightGCN^{MB}, which is denoted as LightGCN^{CL}.

---

[1] https://tianchi.aliyun.com/competition/entrance/231721/introduction.

- **MMR.** MMR is an ablation version of MMCLR without all CL tasks. It can be viewed as a simple multi-view multi-behavior model, which combines BERT4Rec^{MB} with LightGCN^{MB} via embedding concatenation and MLP.

4.3   Experimental Settings

Parameter Settings. The embedding sizes of users and items are 64 and batch size is 256 for all methods. We optimize all models by Adam optimizer. For BERT4Rec, we stack two-layer transformers and each transformer with two attention heads. The depth of our graph encoder is set to 2. The learning rate and L2 normalization coefficient of MMCLR are set as $1e^{-3}$ and $1e^{-4}$, respectively. The weights of supervised loss L_o and four CL losses (i.e., L_{SeqCL}, $L_{GraphCL}$, L_{ViewCL}, L_{DisCL}) are set as 1.0, 0.2, 0.2, 0.2, and 0.05, respectively. For all baselines, We conduct a grid search for parameter selections.

Evaluation Protocols. Following [28,36], We adopt the leave-one-out strategy to evaluate the models' performance; We also employ the top-K hit rate (HIT), top-K Normalized Discounted Cumulative Gain (NDCG), Mean Reciprocal Rank (MRR), and AUC (Area Under the Curve). For HIT and NDCG, we report top 5 and 10; For each ground truth, we randomly sample 99 items that user did not interact with under the target behavior as negative samples.

4.4   Results of Multi-behavior Recommendation (RQ1)

The main MBR results are shown in Table 1, from which we find that:

(1) MMCLR performs the best among all baselines and ablation versions of MMCLR on all metrics in two datasets. It achieves 4%–11.8% improvements over the best baselines on most metrics, with the significance level as $p < 0.05$ (paired t-test of MMCLR V.S. baselines). It indicates that MMCLR can well capture the commonalities and differences between different behaviors and views, and thus can better take advantage of all multi-view and multi-behavior information in MBR. (2) BERT4Rec^{CL} and LightGCN^{CL} perform much better than their original models without CL. It verifies the importance of modeling relations between different types of behaviors when jointly learning user representations. It also implies that our multi-behavior CL can help to capture the behavior-level commonalities. Nevertheless, MMCLR still performs better than single-view models, which verifies the significance of jointly modeling multi-view information. (3) We notice that MMR performs comparably with BERT4Rec^{MB}. It reflects that the simple fusion of individual sequence-based and global graph-based models may not make full use of the multi-view information.

Table 1. Results on multi-behavior recommendation. * indicates significance ($p < 0.05$).

Database	Model	MRR	AUC	HIT@5	NDCG@5	HIT@10	NDCG@10
Tmall	BERT4Rec^{MB}	0.1568	0.6671	0.2138	0.1448	0.3133	0.1769
	LightGCN^{MB}	0.1449	0.6542	0.1983	0.1318	0.3020	0.1651
	MRIG	0.1545	0.6823	0.2084	0.1401	0.3207	0.1762
	MBGCN	0.1534	0.6912	0.2100	0.1396	0.3208	0.1751
	MBGMN	0.1673	0.6808	0.2273	0.1559	0.3308	0.1892
	MGNN	<u>0.1782</u>	<u>0.6955</u>	<u>0.2332</u>	<u>0.1651</u>	<u>0.3389</u>	<u>0.1991</u>
	LightGCN^{CL}	0.1609	0.6863	0.2201	0.1483	0.3293	0.1835
	BERT4Rec^{CL}	0.1754	0.6971	0.2385	0.1641	0.3467	0.1990
	MMR	0.1576	0.6606	0.2152	0.1466	0.3108	0.1773
	MMCLR	**0.1861***	**0.7237***	**0.2608***	**0.1770***	**0.3751***	**0.2138***
	Improvement	**4.4%**	**4.1%**	**11.8%**	**7.3%**	**10.7%**	**7.4%**
CIKM	BERT4Rec^{MB}	0.1792	0.6990	0.2451	0.1687	0.3552	0.2042
	LightGCN^{MB}	0.1705	0.6979	0.2332	0.1584	0.3466	0.1949
	MRIG	0.1795	0.7026	0.2489	0.1696	0.3649	0.2068
	MBGCN	0.1850	0.6897	0.2479	0.1751	0.3492	0.2077
	MBGMN	0.1887	0.7035	0.2575	0.1795	0.3648	0.2140
	MGNN	<u>0.1973</u>	<u>0.7116</u>	<u>0.2616</u>	<u>0.1866</u>	<u>0.3718</u>	<u>0.2222</u>
	LightGCN^{CL}	0.1746	0.7031	0.2398	0.1633	0.3530	0.1998
	BERT4Rec^{CL}	0.1984	0.7282	0.2728	0.1912	0.3929	0.2281
	MMR	0.1788	0.6941	0.2506	0.1700	0.3627	0.2061
	MMCLR	**0.2046***	**0.7313***	**0.2878***	**0.1981***	**0.4049***	**0.2358***
	Improvement	**3.7%**	**2.9%**	**10.0%**	**6.2%**	**8.9%**	**6.1%**

4.5   Ablation Study (RQ2)

In this section, we aim to prove that MMCLR can solve the three challenges mentioned in the introduction section via three CL tasks. We build seven ablation versions of MMCLR, which are different combinations of CL tasks and the multi-view fusion, to show the effectiveness of different components. Specifically, we regard the basic sequence-based model of MMCLR with multi-behavior information as *seq* (i.e., BERT4Rec^{MB}), and the basic graph-based model of enhanced LightGCN with multi-behavior information as *graph* (i.e., LightGCN^{MB}). We set *seq+graph* as the simple multi-view fusion version (i.e., MMR). Moreover, we represent the multi-behavior CL, multi-view CL, and behavior distinction CL as BCL, VCL, and DCL, respectively. The final MMCLR is noted as seq+graph +BCL+VCL+DCL. From Table 2, we can observe that:

(1) Comparing ablation versions with and without BCL, we find that both sequential and graphic multi-behavior CL tasks are beneficial. BCL tasks even function well on the seq+graph model. The improvements of BCL are impressive, which have over 2% improvements in most metrics. It is because that multiple behaviors produced by the same user should reflect related preferences of the

user. Modeling the coarse-grained commonalities of different behaviors helps to learn better representations to fight against the data sparsity issues. Moreover, through BCL, we can learn better user representations that are more precise and distinguishable from other users'. It reconfirms the effectiveness of the multi-behavior CL in modeling such coarse-grained commonality. (2) Comparing models with and without VCL, we know that the multi-view CL is also essential in multi-view fusion (getting nearly 1% improvements on most metrics). We also implement a simple fusion model with seq and graph models, whose improvements over single-view models are marginal. The multi-view CL smartly aligns sequence-view and graph-view representations via the CL-based learning, which well captures useful information from both individual and global aspects. These improvements verify the significance of multi-view CL. (3) Comparing with the last two versions, we can observe that the behavior distinction CL further improves the performances on all metrics. The 0.6–1.4% improvements are significant. It verifies that jointly considering both coarse-grained commonalities and fine-grained differences are essential in MMCLR.

Table 2. Ablation tests on CL tasks and multi-view fusion in MMCLR.

Ablation	HIT@5	NDCG@5	HIT@10	NDCG@10
seq	0.2138	0.1448	0.3133	0.1769
graph	0.2108	0.1442	0.3136	0.1773
seq+graph	0.2152	0.1466	0.3108	0.1773
seq+BCL	0.2385	0.1641	0.3467	0.1990
graph+BCL	0.2380	0.1620	0.3456	0.1966
seq+graph+BCL	0.2418	0.1632	0.3527	0.1988
seq+graph+BCL+VCL	0.2521	0.1722	0.3614	0.2074
MMCLR (final)	**0.2608***	**0.1770***	**0.3751***	**0.2138***

4.6 Results on Cold-Start Scenarios (RQ3)

Real-world multi-behavior recommendation systems usually suffer from cold-start issues (e.g., cold-start users that have few historical behaviors), especially for the high-cost purchase behaviors in MBR of E-commerce. Hence, we further conduct an evaluation on the cold-start (user) scenario to verify the effectiveness of MMCLR on more challenging tasks. Without loss of generality, we regard all users that have less than 3 target behaviors in the train set as our cold-start users and select these cold-start users' test instances in the overall Tmall dataset as the test set of the cold-start scenario. To comprehensively display the effectiveness of MMCLR and its multiple CL tasks on the cold-start scenario, we draw three figures in Fig. 2 from different aspects. Precisely, we can observe that: (1) Figure 2(a) shows different models' NDCG performances in both overall and cold-start users. We can know that: (a) All models perform better on the overall

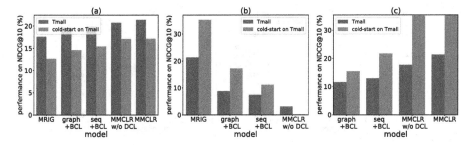

Fig. 2. Results of different models and ablation versions on the overall and cold-start scenarios. (a) NDCG@10 on the overall and cold-start datasets. (b) MMCLR's relative improvements of NDCG@10 on different baselines. (c) Different MMCLR's ablation versions' relative improvements of NDCG@10 on the baseline MRIG.

users than the cold-start users. (b) Results on both overall and cold-start users have consistent improvements from graph+BCL to MMCLR. (2) Figure 2(b) shows MMCLR's relative improvements on other models. We find that: (a) Comparing with different models and ablation versions (except MMCLR w/o DCL), MMCLR has higher improvements on cold-start scenarios (e.g., nearly 35% astonishing improvements on MRIG). It is because that MMCLR can make full use of the multi-behavior and multi-view information via CL tasks, which can alleviate the data sparsity in cold-start users. (b) We notice that DCL brings in a slight improvement on cold-start users. It is natural since cold-start users usually have very few target behaviors, and rely more on auxiliary behaviors via the commonality-led CL tasks as supplements. (3) Figure 2(c) gives the relative improvements of different MMCLR's ablation versions on MRIG. We observe that: (a) Both sequential and graphic multi-behavior CL, multi-view CL, and behavior distinction CL has improvements on cold-start scenarios. (b) Relatively, the multi-behavior CL contributes more on the overall dataset, while the multi-view CL focuses more on the cold-start users. It may be because that a different view can bring in more information for cold-start users thanks to the global graph view and its multi-view CL task.

4.7 Parameter Analyses (RQ4)

Loss Weight. We start the experiment with different main-task loss weights on the Tmall dataset to explore its influence. We change the weight of supervised L_o among $\{0.2, 1, 2, 4, 8\}$. From Fig. 3(a) we can find that: (1) Both HIT@10 and NDCG@10 first increase and then decrease from 0.2 to 8, and MMCLR achieves the best results when $\lambda_o = 1.0$ (here CL loss weights are 0.2, 0.2, 0.2, and 0.05). It indicates that the supervised loss is the fundamental of model training, and a proper loss weight helps to balance the supervised and self-supervised learning. (2) MMCLR consistently outperforms baselines with different weights. It shows the effectiveness and robustness of our model with different loss weights.

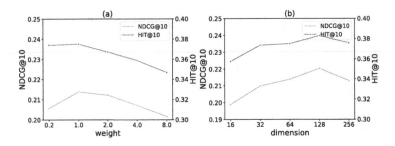

Fig. 3. Parameter analyses on (a) loss weights, and (b) embedding dimensions.

Embedding Dimension. We also test different input embedding dimensions on the Tmall dataset. We vary the embedding dimensions in $\{16, 32, 64, 128, 256\}$, and keep other optimal hyper-parameters unchanged. The results of different dimensions are shown in Fig. 3. We observe that the model achieves better performance with bigger dimension, while dimension from 16 to 128. It shows that enough embedding dimension helps to increase model capacity. In contrast, the model with 256 dimensions has worse performance than 128 dimensions. The performance may be suffered from overfitting. It also suggests that too large an embedding dimension is not necessary.

5   Conclusion

In this work, We study the multi-behavior recommendation problem. Specifically, to alleviate the sparsity problem of target behaviors existing in recommender systems, we propose a novel MMCLR framework to jointly consider the commonalities and differences between different behaviors and views in MBR via three CL tasks. Extensive experimental results verify the effectiveness of our MMCLR and its CL tasks. The performance of MMCLR on cold-start users further demonstrates the superiority of MMCLR on the cold-start problem.

Acknowledgments. The research work supported by the National Natural Science Foundation of China under Grant No. 61976204, U1811461, U1836206. Xiang Ao is also supported by the Project of Youth Innovation Promotion Association CAS, Beijing Nova Program Z201100006820062.

References

1. Chen, C., et al.: Graph heterogeneous multi-relational recommendation. In: Proceedings of AAAI (2021)
2. Chen, C., Zhang, M., Zhang, Y., Ma, W., Liu, Y., Ma, S.: Efficient heterogeneous collaborative filtering without negative sampling for recommendation. In: Proceedings of AAAI (2020)
3. Chen, X., et al: Sequential recommendation with user memory networks. In: Proceedings of WSDM (2018)

4. Devlin, J., Chang, M.W., Lee, K., Toutanova, K.: BERT: pre-training of deep bidirectional transformers for language understanding. arXiv preprint (2018)
5. Doersch, C., Gupta, A., Efros, A.A.: Unsupervised visual representation learning by context prediction. In: Proceedings of ICCV (2015)
6. He, X., Deng, K., Wang, X., Li, Y., Zhang, Y., Wang, M.: LightGCN: simplifying and powering graph convolution network for recommendation. In: Proceedings of SIGIR (2020)
7. Hidasi, B., Karatzoglou, A., Baltrunas, L., Tikk, D.: Session-based recommendations with recurrent neural networks. In: ICLR (2016)
8. Huang, J.T., et al.: Embedding-based retrieval in Facebook search. In: Proceedings of KDD (2020)
9. Jin, B., Gao, C., He, X., Jin, D., Li, Y.: Multi-behavior recommendation with graph convolutional networks. In: Proceedings of SIGIR (2020)
10. Pan, F., Li, S., Ao, X., Tang, P., He, Q.: Warm up cold-start advertisements: improving CTR predictions via learning to learn ID embeddings. In: Proceedings of SIGIR, pp. 695–704 (2019)
11. Pan, W., Xiang, E., Liu, N., Yang, Q.: Transfer learning in collaborative filtering for sparsity reduction. In: Proceedings of AAAI, vol. 24 (2010)
12. Perozzi, B., Al-Rfou, R., Skiena, S.: DeepWalk: online learning of social representations. In: Proceedings of KDD (2014)
13. Rendle, S., Freudenthaler, C., Gantner, Z., Schmidt-Thieme, L.: BPR: Bayesian personalized ranking from implicit feedback. arXiv preprint (2012)
14. Singh, A.P., Gordon, G.J.: Relational learning via collective matrix factorization. In: Proceedings of KDD, pp. 650–658 (2008)
15. Sun, F., et al.: BERT4Rec: sequential recommendation with bidirectional encoder representations from transformer. In: Proceedings of CIKM (2019)
16. Wang, W., et al.: Beyond clicks: modeling multi-relational item graph for session-based target behavior prediction. In: Proceedings of WWW (2020)
17. Wang, X., He, X., Wang, M., Feng, F., Chua, T.S.: Neural graph collaborative filtering. In: Proceedings of SIGIR (2019)
18. Wu, J., et al.: Self-supervised graph learning for recommendation. In: Proceedings of SIGIR (2021)
19. Xi, D., et al.: Modeling the sequential dependence among audience multi-step conversions with multi-task learning in targeted display advertising. In: Proceedings of KDD (2021)
20. Xi, D., et al.: Neural hierarchical factorization machines for user's event sequence analysis. In: Proceedings of SIGIR, pp. 1893–1896 (2020)
21. Xia, L., et al.: Knowledge-enhanced hierarchical graph transformer network for multi-behavior recommendation. In: Proceedings of AAAI (2021)
22. Xia, L., Xu, Y., Huang, C., Dai, P., Bo, L.: Graph meta network for multi-behavior recommendation. In: Proceedings of SIGIR (2021)
23. Xiao, C., et al.: UPRec: user-aware pre-training for recommender systems. arXiv preprint (2021)
24. Xie, R., Liu, Q., Wang, L., Liu, S., Zhang, B., Lin, L.: Contrastive cross-domain recommendation in matching (2021)
25. Xie, R., Liu, Y., Zhang, S., Wang, R., Xia, F., Lin, L.: Personalized approximate pareto-efficient recommendation. In: Proceedings of the Web Conference 2021, pp. 3839–3849 (2021)
26. Xie, R., Qiu, Z., Rao, J., Liu, Y., Zhang, B., Lin, L.: Internal and contextual attention network for cold-start multi-channel matching in recommendation. In: Proceedings of IJCAI, pp. 2732–2738 (2020)

27. Xie, R., et al.: Long short-term temporal meta-learning in online recommendation. In: Proceedings of WSDM (2022)
28. Xie, X., et al.: Contrastive learning for sequential recommendation. arXiv preprint (2020)
29. Ying, H., et al.: Sequential recommender system based on hierarchical attention network. In: Proceedings of IJCAI (2018)
30. Zeng, Z., et al.: Knowledge transfer via pre-training for recommendation: a review and prospect. Front. Big Data 4, 602071 (2021)
31. Zhang, R., Isola, P., Efros, A.A.: Colorful image colorization. In: Leibe, B., Matas, J., Sebe, N., Welling, M. (eds.) ECCV 2016. LNCS, vol. 9907, pp. 649–666. Springer, Cham (2016). https://doi.org/10.1007/978-3-319-46487-9_40
32. Zhang, W., Mao, J., Cao, Y., Xu, C.: Multiplex graph neural networks for multi-behavior recommendation. In: Proceedings of CIKM (2020)
33. Zheng, Y., Gao, C., He, X., Li, Y., Jin, D.: Price-aware recommendation with graph convolutional networks. In: Proceedings of ICDE (2020)
34. Zhou, C., et al.: ATRank: an attention-based user behavior modeling framework for recommendation. In: Proceedings of AAAI (2018)
35. Zhou, G., et al.: Deep interest network for click-through rate prediction. In: Proceedings of KDD (2018)
36. Zhou, K., et al.: S3-Rec: self-supervised learning for sequential recommendation with mutual information maximization. In: Proceedings of CIKM (2020)
37. Zhu, Y., et al.: Transfer-meta framework for cross-domain recommendation to cold-start users. In: Proceedings of SIGIR (2021)
38. Zhu, Y., et al.: Personalized transfer of user preferences for cross-domain recommendation. In: Proceedings of WSDM (2021)
39. Zhu, Y., et al.: Learning to warm up cold item embeddings for cold-start recommendation with meta scaling and shifting networks. In: Proceedings of SIGIR (2021)
40. Zhu, Y.: Multi-representation adaptation network for cross-domain image classification. Neural Netw. 119, 214–221 (2019)
41. Zhu, Y., et al.: Deep subdomain adaptation network for image classification. IEEE Trans. Neural Netw. Learn. Syst. 32, 1713–1722 (2020)
42. Zhuang, F., et al.: A comprehensive survey on transfer learning. In: Proceedings of the IEEE (2020)

Joint Locality Preservation and Adaptive Combination for Graph Collaborative Filtering

Zhiqiang Guo[1], Chaoyang Wang[1], Zhi Li[1(✉)], Jianjun Li[1], and Guohui Li[2]

[1] School of Computer Science and Technology,
Huazhong University of Science and Technology, Wuhan, China
{zhiqiangguo,sunwardtree,leoric,jianjunli}@hust.edu.cn
[2] School of Software Engineering,
Huazhong University of Science and Technology, Wuhan, China
guohuili@hust.edu.cn

Abstract. Due to its powerful representation ability, Graph Convolutional Network (GCN) based collaborative filtering (CF), which treats the interaction of user-items as a bipartite graph, has become the upstart in recommender systems. Nevertheless, existing GCNs based recommendation model only compromisingly exploits the shallow relationship (generally less than 4 layers) to represent the user and item with different number of interactions, which limits their performance. To address this problem, we propose a novel recommendation framework named *joint Locality preservation and Adaptive combination for Graph Collaborative Filtering* (LaGCF), which contains two components: locality preservation and adaptive combination, where locality preservation explicitly integrates local features with high-order features at each propagation layer for obtaining better performance faster, while adaptive combination adds adaptive weight and identity matrix in aggregation to enhance the representing power of deep-level GCN. Finally, extensive experiments are conducted on four publicly available datasets, and the results demonstrate the effectiveness and superior performance of our model from both analytical and empirical perspectives.

Keywords: Collaborative filtering · Graph neural network · Locality preservation · Adaptive combination

1 Introduction

Recommender systems (RSs) has greatly alleviated the *information overload* problem by performing personalized information filtering [5,8,26]. Among various recommendation methods, collaborative filtering (CF) [10,18,23,31], which focuses on modeling the historical user-item interactions without domain knowledge, has made substantial progress towards personalized recommendation.

Supported by the National Natural Science Foundation of China under Grant No. 61672252.

The common paradigm for CF models is to learn the latent features (embeddings) of users and items and then perform prediction based on the embedding vectors. Traditional CF methods, represented by matrix factorization (MF), obtain the latent features of users and items by factorizing the user-item interactive matrix. Recently, inspired by the success of graph convolutional networks (GCNs) in effectively extracting features in non-Euclidean spaces, some researchers try to exploit the user-item bipartite graph structure by propagating embeddings on it, aiming at achieving more effective latent features. For example, Wang et al. [31] proposed NGCF, which adopts the same propagation rules as in standard GCN (including neighborhood aggregation, feature transformation and non-linear activation) to capture the high-order connectivity between users and items by stacking multiple feature propagation layers, and achieves promising results. Later, Chen et al. [3] found that the feature transformation and non-linear activation in standard GCN (used in NGCF) are actually unnecessary, and proposed LR-GCCF by removing these two components to achieve better recommendation performance. Recently, He et al. [17] further proposed LightGCN to abandon the self-connection operation and utilize layer combination to achieve state-of-the-art performance for graph-based CF.

Though achieving promising results, existing GCN-based recommendation models still suffer from the notorious *over-smoothing* problem inherited from GCN, i.e., with the increase of the number of propagation layers, the embeddings of all the nodes learned by GCN tend to converge to a constant value, which makes them hard to distinguish from each other. Since different users and items usually have different number of interactions, existing GCN-based recommendation models can only aggregate the shallow relationship to achieve the peak performance with at most three or four layers [4], which makes them fail to utilize deep-level high-order relationships for better performance. We analyze that the major factor causing such a problem is **local interactivity dilution**. The interactive feature (first-order feature in GCNs) between user and item is always important for achieving good performance in CF-based models. However, existing GCN-based models all simply and directly propagate high-order features on the user-item bipartite graph. With the increase of the layer number, the high-order neighbors that are aggregated to represent a user (or an item) are farther and farther away from the initial node, resulting in local interactive features to be gradually forgotten. Such a local interactivity dilution phenomenon has negative impact on the uniqueness of features, and eventually will deteriorate the recommendation performance. Extensive ablation studies in LightGCN demonstrate our thoughts about the significance of local interactivity for recommendation. It is noteworthy that the layer combination after aggregation in LightGCN is a compensatory technique to alleviate local interactivity dilution. However, the layer combination after aggregation only uses **fixed coefficient** for combination and cannot balance the local interactive feature and the high-order feature of users well. Therefore, an adaptive combination method is desirable to strengthen the representing ability of deep-level GCNs and further alleviate the *over-smoothing* problem in recommendation.

To solve the above problems, we propose a novel recommendation model named LaGCF, which contains two components: locality preservation and adaptive combination. Specifically, we first utilize locality preservation to deal with the local interactivity dilution problem when deepening propagation layers. Instead of directly aggregating the features of higher-order neighbors, we explicitly integrate interactive features with high-order features at each propagation layer. Such an operation ensures that the deep high-order representation retains at least a fraction of locality. Secondly, we introduce adaptive combination that adds adaptive weight and identity matrix in aggregation, to enhance the representing power of deep-level GCNs by implementing layer combination with arbitrary coefficients. Extensive experiments on four real-word datasets verify the effectiveness of our model. The results clearly show that our model not only outperforms the state-of-the-art methods within four layers, but also can keep achieving promising performance with more layers. To summarize, the main contributions of this work are as follows:

- We empirically show that existing GCN-based CF methods face serious *over-smoothing* problem when deepening propagation layers. Further analysis show that the local interactive features in LightGCN play an important role for enhancing the recommendation performance when propagating on more layers.
- We propose LaGCF, which contains locality preservation and adaptive combination, to address the local interactivity dilution problem and give full play to the representation power of deep-level GCNs in recommendation.
- Extensive experiments are conducted on four benchmark datasets, and the results verify the superior performance of LaGCF over several state-of-the-art methods. Moreover, the experimental results also show that LaGCF can consistently achieve promising performance when stacking more layers.

2   Related Work

2.1   Collaborative Filtering

Collaborative filtering [12] is the most influential and widely used model for recommendation, which focuses on modeling the historical user-item interactions. Most CF-based models are based on learning latent representations of users and items [18,19,22,30,33]. Matrix factorization (MF) [23] is the classical model-based CF, which directly encodes the single ID of a user (an item) as a feature vector and employs the inner product to model the user-item interactions. FISM [19] further learns the item similarity matrix and exhibits improvements in predicting item ranking. Recently, based on the powerful neural networks, the recommendation methods have made a great process. DMF [33] maps the users and items into a low-dimensional space with non-linear projections, and then utilizes cosine similarity as the matching function to calculate predictive scores. By replacing the inner product with non-linear neural networks as the matching function, NCF [18] learns the non-linear representations between users and items to improve recommendation. MultiVAE [25] extends variational autoencoders to

CF for latent representation and implicit feedback, and uses Bayesian inference for parameter estimation. Deng et al. [9] categorize CF models into representation learning-based CF and matching function learning-based CF, and propose a deep collaborative filtering model, which combines the strengths of these two types of methods to achieve better performance.

From the perspective of user-item interaction graph, even the neural CF models can only exploit the one-hop neighbors of a user to learn the embedding, which limits their performance.

2.2 GCN-Based Recommendation

Graph convolutional network is designed to handle the graph structure and has made a great success in tasks such as node classification [2,21], transportation network [35], and social network analysis [13,32]. Recently, it has been proved to be an efficient way to improve recommendation performance by learning representations of users and items from the graph structure [6,17,27]. Some early efforts take a personalized random walk approach for recommendation [7,11]. However, most of these models rely on carefully constructing the random walk process and face the problem of high time complexity. Later, Pinsage [34] applies a sampling technique GraphSAGE [14] for graph convolution aggregation and achieves high scalability. GC-MC [1] adopts the structure of autoencoder and applies GCN to make link prediction. However, GC-MC only considers one-hop neighbor, which makes it fail to capture high-order collaborative signals. To exploit high-order connectivity, NGCF [31] redesigns the embedding communication layer to strengthen the connection between multi-hop nodes in user-item interaction graph. Most recently, LR-GCCF [3] finds that removing feature transformation and non-linear activation in NGCF can exhibit a substantial improvement of recommendation accuracy. Meanwhile, LightGCN [17] adopts the linear structure and further abandons the self-connection and employs layer combination to achieve state-of-the-art performance. UltraGCN [27] resorts to directly approximate the limit of infinite-layer graph convolutions via a constraint loss to get an ultra-simplified formulation of GCNs.

However, the aforementioned methods do not pay much attention to the *oversmoothing* problem in recommendation. Different from them, we try to utilize locality preservation and adaptive combination to alleviate this problem.

3 Preliminaries

3.1 Recap

We consider a recommender system with m users $\mathcal{U}$ and n items $\mathcal{I}$. User's implicit feedback on items is represented by a sparse interaction matrix $\mathbf{R} \in \mathbb{R}^{m \times n}$, where an element r_{ui} is 1 if $u \in \mathcal{U}$ has interacted with $i \in \mathcal{I}$ before, and 0 otherwise. Based on the interaction matrix $\mathbf{R}$, the user-item bipartite graph is constructed as $\mathcal{G} = <\mathcal{U} \cup \mathcal{I}, \mathbf{A}>$, where the node set of $\mathcal{G}$ consists of the two types of user nodes and item nodes, and $\mathbf{A} \in \mathbb{R}^{(m+n) \times (m+n)}$ denotes the adjacency matrix constructed from the interactive matrix $\mathbf{R}$ as,

$$\mathbf{A} = \begin{pmatrix} \mathbf{0} & \mathbf{R} \\ \mathbf{R}^T & \mathbf{0} \end{pmatrix}, \tag{1}$$

in which $\mathbf{0} \in \mathbb{R}^{m \times n}$ is a null matrix. Note that if self-connection is considered, the adjacency matrix $\tilde{\mathbf{A}}$ can be obtained by $\tilde{\mathbf{A}} = \mathbf{A} + \mathbf{I}$, where $\mathbf{I}$ is an identity matrix. A nonzero r_{ui} matches an edge between user u and item i on $\mathcal{G}$.

The matrix form is benefit for providing a holistic view of representation propagation [28]. Therefore, we use the matrix form to describe the follow-up work. Let $\mathbf{E}^{(0)}$ denote the 0-th layer embedding matrix for user and item as an embedding look-up table:

$$\mathbf{E}^{(0)} = [\ \overbrace{\mathbf{e}_{u_1}^{(0)}, \cdots, \mathbf{e}_{u_m}^{(0)}}^{users\ embeddings},\ \overbrace{\mathbf{e}_{i_1}^{(0)}, \cdots, \mathbf{e}_{i_n}^{(0)}}^{items\ embeddings}\ ]. \tag{2}$$

where $\mathbf{e}_u^{(0)} \in \mathbb{R}^{d_o}$ and $\mathbf{e}_i^{(0)} \in \mathbb{R}^{d_o}$ respectively denote the initial embedding of user u and item i, and d_0 is the initial embedding size. Based on the bipartite graph, the k-th layer of aggregation can be formalized,

$$\mathbf{E}^{(k+1)} = AGG(\mathbf{E}^{(k)}, \mathcal{G}) \tag{3}$$

where $\mathbf{E}^k \in \mathbb{R}^{(m+n) \times d_k}$ is the representation of user or item in the k-th layer, d_k is the embedding size of the k-th layer, and $AGG(\cdot)$ is the aggregation function. After the user and item embeddings are obtained, a dot-product is usually used to predict the preference of user to item.

NGCF. As a pioneer GCN-based CF method, NGCF [31] adopts the modified aggregation method to generate the high-order representation of user and item,

$$\mathbf{E}^{(k+1)} = \sigma \left(\tilde{\mathcal{L}} \mathbf{E}^{(k)} \mathbf{W}_1^{(k)} + \left(\tilde{\mathcal{L}} \mathbf{E}^{(k)} \odot \mathbf{E}^{(k)} \right) \mathbf{W}_2^{(k)} \right) \tag{4}$$

where $\mathbf{W}_1^{(k)}, \mathbf{W}_2^{(k)} \in \mathbb{R}^{d_k \times d_{k+1}}$ are trainable weight matrices to distill useful information for propagation, $\odot$ denotes the element-wise product to encode the interaction between user and its neighbors, σ is the $ReLU$ activation function, and $\tilde{\mathcal{L}} = \mathbf{D}^{-\frac{1}{2}} \tilde{\mathbf{A}} \mathbf{D}^{-\frac{1}{2}}$ is the graph Laplacian matrix for the user-item bipartite graph, where $\mathbf{D} \in \mathbb{R}^{(m+n) \times (m+n)}$ is a diagonal matrix, in which the j-th diagonal element $D_{jj} = |\mathcal{N}_j|$ represents the number of user or item interactions corresponding to the j-th row of $\mathbf{E}^{(0)}$. Thereby, we can get $\mathcal{L}_{ui} = \frac{1}{\sqrt{|\mathcal{N}_u||\mathcal{N}_i|}}$, where $\mathcal{N}_u$ and $\mathcal{N}_i$ respectively denote the number of user u's and item i's first-hop neighbors. Finally, NGCF obtains the K-th layer representation of user u by layer concatenation,

$$\mathbf{E}_u = [\mathbf{E}_u^{(0)}; \mathbf{E}_u^{(1)}; \cdots; \mathbf{E}_u^{(K-1)}; \mathbf{E}_u^{(K)}] \tag{5}$$

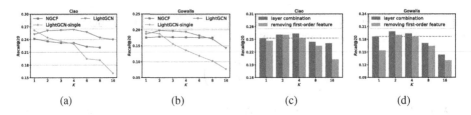

Fig. 1. Empirical explorations on Ciao and Gowalla. (a) and (b) show the trend of Recall@20 as deepening the number of layers. (c) and (d) show the performance comparison of LightGCN before and after removing the first-order features.

LightGCN. LightGCN [17] deserts the feature transformation and non-linear activation, and only utilizes the weighted sum aggregator to achieve state-of-the-art CF performance. The graph convolution operation in LightGCN is as follows:

$$\mathbf{E}^{(k+1)} = \mathcal{L}\mathbf{E}^{(k)} \tag{6}$$

Note LightGCN removes the self-connection operation. Hence, the embeddings of users and items obtained at each layer are independent. After K-layer propagation, LightGCN forms the representations of user u by layer combination:

$$\mathbf{E}_u = \sum_{k=0}^{K} \alpha_k \mathbf{E}_u^{(k)} \tag{7}$$

where $\alpha_k \geq 0$ is a hyper-parameter (set as $\frac{1}{K+1}$ in LightGCN) that denotes the significance of the k-th order feature in constituting the final embedding.

3.2   Empirical Explorations

To show the limitation of existing GCN-based CF methods, we conduct an empirical study on NGCF, LightGCN-single[1] and LightGCN. We use the codes released by the authors of NGCF and LightGCN to conduct this experiment on Ciao and Gowalla.[2] Meanwhile, we keep all hyper parameters (e.g., embedding size, learning rate, regularization coefficient) the same. Figure 1(a) and (b) shows the trend of Recall@20 as deepening the number of layer on Ciao and Gowalla respectively, based on which we have the following observations:

- NGCF, LightGCN-single and LightGCN all achieve promising performance in a shallow number of layers (less than 4). As the number of layers increased, their performance declines to varying degrees.
- Compared with LightGCN-single and LightGCN, NGCF exhibits the smallest performance degradation, which may be due to its use of learnable parameters to alleviate the negative effects of deepening layers.

---

[1] LightGCN-single denote a simplified version of LightGCN that only uses the embedding of the K-th layer to represent users and items.

[2] The two datasets are consistent with the ones stated in Sect. 5.1.

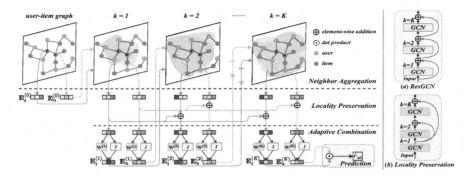

Fig. 2. The left shows the particular propagation of LaGCF (the orange and blue lines express the data flow of users and items, respectively). The right shows the (a) simplified illustration of ResGCN and (b) Locality preservation. (Color figure online)

- LightGCN achieves better performance than LightGCN-single, which proves the effect of layer combination. Nevertheless, LightGCN cannot maintain a competitive performance at deeper layers.

Though the layer combination after aggregation considers the fusion of feature from different layers, it uses fixed coefficients and treats the features from different layer equally, hence fails to recognize the importance of local interactivity. We further study the effect of local interactivity by removing the first-order feature from layer combination, the results are shown in Fig. 1(c) and (d), from which we have two major observations,

- The performance of the "removing first-order feature" model is always worse than the one with whole layer combination. Besides, the performance gap continues to increase with the deepening of layers. When $K = 16$, the performance gap reaches 15.17% and 9.05% on Ciao and Gowalla, respectively.
- Compared with the only two-layers combination (denoted by gray horizontal line), the mean combination of multi-layers high-order features will introduce negative impact, resulting in a significant performance drop with deep layers.

4 Methodology

4.1 LaGCF

Based on the analysis in Sect. 3.2, we propose a novel model named *Joint Locality Preservation and Adaptive Combination for Graph Collaborative Filtering* (LaGCF), as illustrated in Fig. 2. which mainly contains three components: neighbor aggregation, locality preservation and adaptive combination.

Neighbor Aggregation. Existing GCN-based methods only use a kind of Laplacian norm to perform neighbor aggregation. In order to take advantage of different aggregators, we combine two different aggregation methods to make full use of the aggregation features of the neighborhood of users and items. Based on LightGCN, we describe our aggregating process as,

$$\mathbf{E}^{(k+1)} = \mathcal{L}_l \mathbf{E}^{(k)} + \mathcal{L}_s \mathbf{E}^{(k)} = (\mathcal{L}_l + \mathcal{L}_s)\mathbf{E}^{(k)} = \hat{\mathcal{L}}\mathbf{E}^{(k)} \qquad (8)$$

where $\hat{\mathcal{L}} = \mathcal{L}_l + \mathcal{L}_s$ is the Laplacian matrix after combining different aggregators. Specifically, $\mathcal{L}_s = \mathbf{D}^{-\frac{1}{2}}\mathbf{A}\mathbf{D}^{-\frac{1}{2}}$ is the commonly used Laplacian matrix in GCN-based CF methods, which considers the degree of both current user and its neighbors, while $\mathcal{L}_l = \mathbf{D}^{-1}\mathbf{A}$ is a Laplacian matrix that only considers the degree of the current user.

Locality Preservation. Motivated by the significance of local interactivity analyzed in Sect. 3.2, we propose locality preservation to construct a connection between the interactive representation and the current layer's representation for fusing the local interactivity during aggregating. The graph convolution operation of locality preservation is defined as,

$$\mathbf{E}^{(k+1)} = \hat{\mathcal{L}}\mathbf{E}^{(k)} + \mathbf{E}^{(1)} \qquad (9)$$

where $\mathbf{E}^{(1)}$ signifies the local interactive feature. Note that when $k = 0$, we can obtain $\mathbf{E}^{(1)} = \hat{\mathcal{L}}\mathbf{E}^{(0)}$. Thus, we can get another form of locality preservation as,

$$\mathbf{E}^{(k+1)} = \hat{\mathcal{L}}(\mathbf{E}^{(k)} + \mathbf{E}^{(0)}) \qquad (10)$$

Figure 2(a) shows the traditional residual connection in GCN [24]. Instead of using a residual connection to propagate information from previous layer, our model directly constructs a connection to the local interactive feature $\mathbf{E}^{(1)}$ at each layer. Such an operation ensures that the deep high-order representation reserves local interactivity. Beyond that, we further analyze the effectiveness of locality preservation by expanding the Eq. (10) as,

$$\begin{aligned} \mathbf{E}^{(k+1)} &= \hat{\mathcal{L}}(\mathbf{E}^{(k)} + \mathbf{E}^{(0)}) = \hat{\mathcal{L}}\mathbf{E}^{(k)} + \hat{\mathcal{L}}\mathbf{E}^{(0)} \\ &= \hat{\mathcal{L}}(\hat{\mathcal{L}}\mathbf{E}^{(k-1)} + \hat{\mathcal{L}}\mathbf{E}^{(0)}) + \hat{\mathcal{L}}\mathbf{E}^{(0)} \\ &= \hat{\mathcal{L}}^2\mathbf{E}^{(k-1)} + \hat{\mathcal{L}}^2\mathbf{E}^{(0)} + \hat{\mathcal{L}}\mathbf{E}^{(0)} \\ &\cdots \\ &= \hat{\mathcal{L}}^k\mathbf{E}^{(0)} + \hat{\mathcal{L}}^{k-1}\mathbf{E}^{(0)} + \cdots + \hat{\mathcal{L}}^2\mathbf{E}^{(0)} + \hat{\mathcal{L}}\mathbf{E}^{(0)} \end{aligned} \qquad (11)$$

From Eq. (11), it can be observed that the locality preservation actually realizes the combination of feature from different layers with fixed coefficient during the aggregation. This operation not only can achieve better performance, but also exhibits faster training speed than the layer combination used in LightGCN. We will demonstrate it on real-world datasets in Sect. 5.2.

Nevertheless, it has also been shown in [21] that such locality preservation only partially alleviates the *over-smoothing* problem in recommendation. This suggests that the idea of locality preservation alone is not sufficient to relief the *over-smoothing* problem when deepening propagation layers, which motivates us to propose an adaptive combination strategy below.

Adaptive Combination. According to the analysis in Sect. 3.2, the weight matrix in NGCF helps to alleviate the *over-smoothing* problem in recommendation. Therefore, to amend the deficiency faced by locality preservation, we reintroduce the weight parameters in the linear aggregation. However, as proved by LightGCN, the weight parameters will increase the complexity of training, leading to performance degradation. Thus, we redesign the aggregation form after adding weight parameters. We first set the weight parameters to $\mathbf{W}_{re}^{(k)} \in \mathbb{R}^1$, which reduces the amount of parameters and makes different users (or items) features obtained from the same layer share a same weight,

$$\mathbf{E}^{(k+1)} = \hat{\mathcal{L}}(\mathbf{E}^{(k)} + \mathbf{E}^{(0)})\mathbf{W}_{re}^{(k)} \tag{12}$$

Then, inspired by ResNet [15], we introduce the identity mapping to further constrain the influence of weight parameters as follows,

$$\mathbf{E}^{(k+1)} = \hat{\mathcal{L}}\left(\mathbf{E}^{(k)} + \mathbf{E}^{(0)}\right)\left(\frac{\alpha}{k}\mathbf{W}_{re}^{(k)} + \left(1 - \frac{\alpha}{k}\right)\mathbf{I}^{(k)}\right). \tag{13}$$

where $\mathbf{I}^{(k)}$ is an identity matrix, $\frac{\alpha}{k}$ ensures the decay of the weight adaptively increases when deepening layers, and α is a hyper-parameter. We can set a sufficiently small α to weaken the influence of the weight parameters. Meanwhile, identity matrix is benefit for retaining the current layer feature to final representation. By setting $\mathbf{P}^{(k)} = \frac{\alpha}{k}\mathbf{W}^{(k)} + (1 - \frac{\alpha}{k})\mathbf{I}^{(k)} \in \mathbb{R}^1$, we can further expand the Eq. (13) as,

$$\begin{aligned}
\mathbf{E}^{(k+1)} &= \hat{\mathcal{L}}(\mathbf{E}^{(k)} + \mathbf{E}^{(0)})\mathbf{P}^{(k)} = \hat{\mathcal{L}}\mathbf{E}^{(k)}\mathbf{P}^{(k)} + \hat{\mathcal{L}}\mathbf{E}^{(0)}\mathbf{P}^{(k)} \\
&= \hat{\mathcal{L}}(\hat{\mathcal{L}}\mathbf{E}^{(k-1)}\mathbf{P}^{(k-1)} + \hat{\mathcal{L}}\mathbf{E}^{(0)}\mathbf{P}^{(k-1)})\mathbf{P}^{(k)} + \hat{\mathcal{L}}\mathbf{E}^{(0)}\mathbf{P}^{(k)} \\
&= \hat{\mathcal{L}}^2\mathbf{E}^{(k-1)}\mathbf{P}^{(k-1)} + \hat{\mathcal{L}}^2\mathbf{E}^{(0)}\mathbf{P}^{(k-1)}\mathbf{P}^{(k)} + \hat{\mathcal{L}}\mathbf{E}^{(0)}\mathbf{P}^{(k)} \\
&\quad \cdots \\
&= \hat{\mathcal{L}}^k\mathbf{E}^{(0)}\boldsymbol{\Gamma}^{(0)} + \hat{\mathcal{L}}^{k-1}\mathbf{E}^{(0)}\boldsymbol{\Gamma}^{(1)} + \cdots + \hat{\mathcal{L}}^2\mathbf{E}^{(0)}\boldsymbol{\Gamma}^{(k-1)} + \hat{\mathcal{L}}\mathbf{E}^{(0)}\boldsymbol{\Gamma}^{(k)}
\end{aligned} \tag{14}$$

where $\boldsymbol{\Gamma}^{(k)} = \prod_{l=k}^K \mathbf{P}^{(l)} \in \mathbb{R}^1$ is the adaptive parameters for layer combination during aggregation, which breaks through the limitation of fixed coefficients.

Model Prediction and Optimization. In our model, we use the final K-th layer embeddings of user u and item i to predict the preference of u to i by inner product as,

$$\hat{r}_{ui} = \mathbf{E}_u^{(K)^\mathsf{T}}\mathbf{E}_i^{(K)}, \tag{15}$$

To learn the parameters of our model, we utilize the Bayesian Personalized Ranking (BPR) loss [29], which is a pairwise loss that has been widely used in previous top-n recommendation [3,31]. The objective function is defined as,

$$L_{BPR} = \sum_{(u,i^+,i^-)\in\mathcal{O}} -\ln\sigma\left(\hat{r}_{ui^+} - \hat{r}_{ui^-}\right) + \lambda\|\Theta\|_2^2, \tag{16}$$

Table 1. Statistics of the four datasets

Dataset	#User	#Item	#Interaction	Sparsity
Ciao	442	534	9,822	95.839%
Yelp	7,062	7,488	167,456	99.683%
Gowalla	29,858	40,988	1027,464	99.916%
Abooks	35,736	38,121	1960,674	99.856%

where $\mathcal{O} = \{(u, i^+, i^-) \mid (u, i^+) \in \mathcal{R}^+, (u, i^-) \in \mathcal{R}^-\}$ indicates the pairwise training set, $\mathcal{R}^+$ denotes the observed user-item interactions in the training dataset, $\mathcal{R}^-$ is the sampled unobserved interaction, $\Theta = \{\mathbf{E}^{(0)}, \{\mathbf{W}_{re}^{(k)}\}_{k=1}^K\}$ are trainable parameters of the model, and λ represents the L_2 regularization coefficient. We employ the mini-batch Adam [20] to optimize the prediction model and update the model parameters.

5 Experiment

5.1 Experimental Setup

Dataset and Metrics. We evaluate the performance of LaGCF on four publicly available datasets: **Ciao DVD**[3] (Ciao for short), **Yelp**,[4] **Gowalla**,[5] and **Amazon-Books**[6] (Abooks for short). To ensure the quality of datasets, we filter the first three datasets to retain each user and item with at least 10 interactions. For Abooks, we ensure that each user and item have at least 20 interactions. After preprocessing, the statistics of the datasets are summarized in Table 1. For each dataset, we randomly hold 80% items in each user's interaction to constitute the training set, and the remaining items are treated as the testing set. To conduct pairwise learning, we treat each observed user-item interaction as positive instance, and randomly select an unobserved interaction as negative instance. We employ Recall@20 and Normalized Discounted Cumulative Gain [16] (NDCG@20) to evaluate the performance for top-n recommendation. After that, the evaluated recommendation methods can generate a ranked top-n list from all items to evaluate the metrics mentioned above.

Parameter Settings. We implement our model in Pytorch.[7] The embedding size is fixed to 64 for all models on four datasets. The Adam is used to optimized our model with a default learning rate of 0.001 and mini-batch size of 2048. The L_2 regularization coefficient λ and the identity mapping parameter α are important for our model, hence we search λ in the range of $\{1e^{-5}, 1e^{-4}, \ldots, 1e^{-1}\}$, and

---

[3] https://guoguibing.github.io/librec/datasets.html.

[4] This dataset is employed from the Yelp challenge, and we only choose the subset about Yelp's tip from https://www.yelp.com/dataset.

[5] https://snap.stanford.edu/data/loc-gowalla.html.

[6] http://jmcauley.ucsd.edu/data/amazon/.

[7] https://github.com/georgeguo-cn/LaGCF.

Table 2. Comparison of overall performance. Best performance in existing methods and our models is in **boldface**. The percent of improvement is about the best results between ours and the compared methods.

Dataset	Ciao		Yelp		Gowalla		Abooks	
Methods	Recall	NDCG	Recall	NDCG	Recall	NDCG	Recall	NDCG
MF	0.2467	0.1600	0.0833	0.0575	0.1536	0.1224	0.0895	0.0778
MLP	0.2185	0.1387	0.0764	0.0530	0.1150	0.0893	0.0589	0.0499
NeuMF	0.2419	0.1539	0.0803	0.0556	0.1266	0.1007	0.0786	0.0653
GC-MC	0.2168	0.1300	0.0860	0.0585	0.1758	0.1378	0.1015	0.0864
PinSage	0.2344	0.1510	0.0907	0.0606	0.1771	0.1391	0.1050	0.0904
NGCF	0.2419	0.1489	0.0928	0.0623	0.1785	0.1401	0.1072	0.0923
LR-GCCF	0.2537	0.1692	0.1020	0.0692	0.1485	0.1164	0.0802	0.0685
LightGCN	**0.2635**	**0.1743**	**0.1112**	**0.0760**	**0.1984**	**0.1607**	**0.1273**	**0.1125**
LaGCF	**0.2699**	**0.1756**	**0.1184**	**0.0805**	**0.2094**	**0.1705**	**0.1409**	**0.1279**
Improv.	2.42%	0.78%	6.42%	5.94%	5.53%	6.12%	10.67%	13.65%

α in the range of $\{0.0001, 0.001, 0.005, 0.01, 0.5, 1.0\}$. Unless otherwise specified, we set L_2 as $1e^{-4}$, and α as 0.001. We employ an early-stop strategy, in which the training process will stop if the performance do not increase after 50 steps.

5.2   Performance Comparison with SOTA Methods

Compared Methods. We compare our proposed model LaGCF with the following competitive methods:

- **MF** [23] directly encodes the single ID of a user (an item) as a feature vector and employs the inner product to model the user-item interactions.
- **MLP** [18] utilizes the multi-layer perceptron to replace inner product to learn the non-liner interactions between users and items. We set the layer number of MLP as 3.
- **NeuMF** [18] is a state-of-the-art neural CF method, which captures the feature on user-item interactions by combining multiple hidden layers and a generalized matrix factorization.
- **GC-MC** [1] adopts GCN to learn the representations of user and item, and only considers the convolutional layer on the first-order neighbors.
- **PinSage** [34] exploits GraphSAGE [14] on item-item graph. We revise PinSage to adapt it to user-item interaction graph and use two graph convolution layers to conduct graph convolution.
- **NGCF** [31] performs embedding propagation in user-item interaction graph to encodes the collaborative signal in the form of high-order connectivities.
- **LR-GCCF** [3] removes the feature transformation and non-linear activation in NGCF to construct a simple linear GCN-based recommendation model.
- **LightGCN** [17] makes a step further to remove the self-connection operation and employ layer combination to achieve the state-of-the-art CF performance.

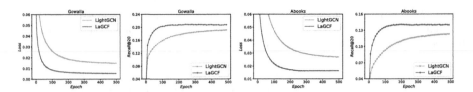

Fig. 3. Training curves of LaGCF and LightGCN, which are evaluated by training loss and testing recall per 10 epochs on Gowalla and Abooks.

Overall Comparison. Table 2 shows the overall performance comparison for all the compared methods in terms of Recall@20 and NGCF@20. We choose the best result of NGCF, LR-GCCF, LightGCN and our models by running them in shallow layer (less than 6). We further plot the training curves of the training loss and testing recall in Fig. 3 to reveal the advantages of LaGCF. From the results, we have the following observations:

- The GCN-based methods in general performs better than non-GCN based ones, indicating the advantages of utilizing graph structure for embedding learning. Among them, the performance of NGCF is consistently better than that of GC-MC and PinSage, which verifies that incorporating high-order connectivities in the embeddings can promote representation learning.
- LR-GCCF simplifies NGCF by removing feature transformation and non-linear activation, and performs better on Ciao and Yelp. However, on Gowalla and Abooks, LR-GCCF under-performs NGCF. The reason might be that concatenating different order of features but without feature transformation cannot bring its superiority into full play, especially on high sparsity datasets. LightGCN achieves substantial improvement over LR-GCCF and NGCF, due to that it further removes self-connection and introduces layer combination.
- The proposed LaGCF exhibits the best performance consistently on all the datasets. In particular, LaGCF reaches a relative improvement over the best baseline in terms of NDCG@20 by 0.78%, 5.94%, 6.12%, 13.65% on Ciao, Yelp, Gowalla, Abooks, respectively, which indicates the effectiveness of introducing locality preservation and adaptive combination. Obviously, LaGCF is more suitable for scenes with high sparsity interactions in practice, since LaGCF can utilize the deep high-order relationships when facing few interaction.
- During the training process, LaGCF consistently obtains lower training loss, indicating that LaGCF fits the training data better than LightGCN. More-over, the lower training loss successfully transfers to better testing accuracy, indicating the strong generalization power of LaGCF.

Performance Comparison w.r.t. Propagation Layers. Due to the well-known *over-smoothing* problem, increasing the number of propagation layers usually limits the expressiveness of GCN-based models. We investigate whether introducing the locality preservation and adaptive combination helps alleviate

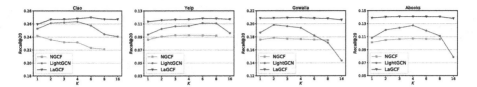

Fig. 4. The trend of Recall@20 of NGCF, LightGCN, LaGCF with increased number of layers on four datasets.

Table 3. The effectiveness of different components in LaGCF.

Dataset		Yelp		Gowalla	
Layer#	Methods	Recall	NDCG	Recall	NDCG
4-layers	LaGCF w/o $lp+ac$	0.1071	0.0745	0.1938	0.1542
	LaGCF w/o lp	0.1085	0.0757	0.2008	0.1621
	LaGCF w/o ac	0.1141	0.0784	0.2074	0.1673
	LaGCF w/o $\mathcal{L}_s$	0.0977	0.0629	0.1177	0.0763
	LaGCF w/o $\mathcal{L}_l$	0.1156	0.0787	0.1993	0.1590
	$LaGCFw/\mathrm{E}^{(0)}$	0.0963	0.0655	0.1749	0.1403
	LaGCF	**0.1169**	**0.0799**	**0.2094**	**0.1696**
16-layers	LaGCF w/o $lp+ac$	0.0978	0.0664	0.1535	0.1230
	LaGCF w/o lp	0.1081	0.0736	0.1950	0.1575
	LaGCF w/o ac	0.1075	0.0680	0.1874	0.1459
	LaGCF	**0.1175**	**0.0800**	**0.2060**	**0.1645**

such a problem. To this end, we perform experiments for GCN-based methods with different number of layers (K). Figure 4 presents the results regarding Recall@20 on four datasets. From the results, we have the following observations:

- Although NGCF performs not good, its performance fluctuation on all datasets is not significant, possibly due to that it keeps the feature transformation operation. However, when the layer number exceeds 8, it cannot be trained to obtain results, due to the vanishing gradient issue. After achieving the best result, LightGCN's performance declines rapidly and becomes the worst.
- LaGCF consistently achieves the best results with the increase of K. Especially when the layer number exceeds 4, the performance gap between our model and other methods becomes more significant. This further demonstrates the effectiveness of our design on introducing locality preservation and adaptive combination to improve the expressiveness of GCN-based models.

5.3   Ablation Study

To investigate how locality preservation (lp for short) and adaptive combination (ac for short) affect the performance, we consider a series of variants of LaGCF to conduct the ablation study, as shown in Table 3. Due to space limitation, we

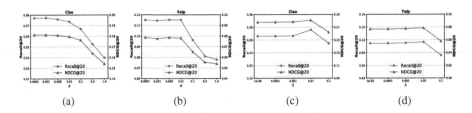

Fig. 5. Varying α and λ with 4-layer LaGCF on Ciao and Yelp

only report the results of LaGCF on Yelp and Gowalla, since LaGCF exhibits similar performance trend on the other two datasets. From the results, we find,

– With the same number of layers, the method without both lp and ac perform worse than the one that includes only one, while the one with both, i.e., LaGCF, always achieves the best result. Meanwhile, when $K = 4$, the method using only locality preservation is more outstanding than the one using only adaptive combination. An opposite result can be observed when $K = 16$. Thus, we can conclude that locality preservation plays a critical role to the performance improvement in shallow layers, while adaptive combination help to keep the representation ability of deep-level GCN-based CF method.
– Intuitively, using $\mathcal{L}_s$ to aggregate the neighbor will perform better than using $\mathcal{L}_l$, while LaGCN, which considers both $\mathcal{L}_s$ and $\mathcal{L}_l$, performs the best. Note that we only simply add up $\mathcal{L}_s$ and $\mathcal{L}_l$, it is possible to adjust their weights to obtain even better result.
– We study the effect of initial feature $\mathbf{E}^{(0)}$ for our model by replacing $\mathbf{E}^{(1)}$ in locality preservation. It can be observed the performance decreases sharply if $\mathbf{E}^{(0)}$ is applied, explaining why we use local interactivity in our model.

5.4   Hyper-parameter Sensitivity

We further conduct the hyper-parameter sensitivity analysis for key parameter, including α, L_2 regularization and coefficient λ, to investigate the performance of 4-layer LaGCF. Figure 5 shows the results with different α and λ. Due to space concern, we only present the results on Ciao and Yelp.

– α is the core parameter of adaptive combination to balance the weight parameter and identity matrix. The optimal values of α for Ciao and Yelp are 0.001 and 0.01, respectively. From the results in Fig. 5(a) and (b), when α is larger than 0.01, the performance declines quickly, which indicates that too much relaxation on the weight parameter will negatively affect model ability and hence is not favorable.
– As shown in Fig. 5(c) and (d), LaGCN is relatively insensitive to λ. The best result is obtained when $\lambda = 0.01$. Similar to α, the performance drops when λ is larger than 0.01. This shows that slightly stronger regularization will positively affect model performance and is advisable.

6 Conclusion

In this work, we first perform empirical studies to analyze the limitation of exacting GCN-based recommendation models and reveal the importance of local interactive features. Then, we propose a novel *joint Locality preservation and Adaptive combination for Graph Collaborative Filtering* (LaGCF) recommendation model. Specifically, locality preservation fuses local interactive features and deep high-order features in aggregation, which alleviates the local interactivity dilution. In adaptive combination, an identity matrix is added into the simplified weight parameter to ensure that the current layer feature is adaptive kept to the final representation. Extensive experiments on four real-world datasets demonstrate the rationality and effectiveness of our model on exploiting locality preservation and adaptive combination.

References

1. Berg, R., Kipf, T.N., Welling, M.: Graph convolutional matrix completion. arXiv preprint arXiv:1706.02263 (2017)
2. Camps-Valls, G., Marsheva, T.V.B., Zhou, D.: Semi-supervised graph-based hyperspectral image classification. IEEE Trans. Geosci. Remote Sens. **45**(10), 3044–3054 (2007)
3. Chen, L., Wu, L., Hong, R., Zhang, K., Wang, M.: Revisiting graph based collaborative filtering: a linear residual graph convolutional network approach. In: Proceedings of AAAI, vol. 34, pp. 27–34 (2020)
4. Chen, M., Wei, Z., Huang, Z., Ding, B., Li, Y.: Simple and deep graph convolutional networks. In: Proceedings of ICML, pp. 1725–1735 (2020)
5. Cheng, H.T., et al.: Wide & deep learning for recommender systems. In: Proceedings of the 1st Workshop on Deep Learning for Recommender Systems, pp. 7–10 (2016)
6. Choi, J., Jeon, J., Park, N.: LT-OCF: learnable-time ODE-based collaborative filtering. In: Proceedings of CIKM, pp. 251–260 (2021)
7. Christoffel, F., Paudel, B., Newell, C., Bernstein, A.: Blockbusters and wallflowers: accurate, diverse, and scalable recommendations with random walks. In: Proceedings of ACM RecSys, pp. 163–170 (2015)
8. Covington, P., Adams, J., Sargin, E.: Deep neural networks for Youtube recommendations. In: Proceedings of ACM RecSys, pp. 191–198 (2016)
9. Deng, Z.H., Huang, L., Wang, C.D., Lai, J.H., Philip, S.Y.: DeepCF: a unified framework of representation learning and matching function learning in recommender system. In: Proceedings of AAAI, vol. 33, pp. 61–68 (2019)
10. Ebesu, T., Shen, B., Fang, Y.: Collaborative memory network for recommendation systems. In: Proceedings of SIGIR, pp. 515–524 (2018)
11. Fouss, F., Pirotte, A., Renders, J.M., Saerens, M.: Random-walk computation of similarities between nodes of a graph with application to collaborative recommendation. IEEE Trans. Knowl. Data Eng. **19**(3), 355–369 (2007)
12. Goldberg, D., Nichols, D., Oki, B., Terry, D.: Using collaborative filtering to weave an information tapestry. Commun. ACM **35**, 61–70 (1992)
13. Grover, A., Leskovec, J.: node2vec: scalable feature learning for networks. In: Proceedings of KDD, pp. 855–864 (2016)

14. Hamilton, W.L., Ying, R., Leskovec, J.: Inductive representation learning on large graphs. arXiv preprint arXiv:1706.02216 (2017)
15. He, K., Zhang, X., Ren, S., Sun, J.: Deep residual learning for image recognition. In: Proceedings of CVPR, pp. 770–778 (2016)
16. He, X., Chen, T., Kan, M.Y., Chen, X.: TriRank: review-aware explainable recommendation by modeling aspects. In: Proceedings of CIKM, pp. 1661–1670 (2015)
17. He, X., Deng, K., Wang, X., Li, Y., Zhang, Y., Wang, M.: LightGCN: simplifying and powering graph convolution network for recommendation. In: Proceedings of SIGIR, pp. 639–648 (2020)
18. He, X., Liao, L., Zhang, H., Nie, L., Hu, X., Chua, T.S.: Neural collaborative filtering. In: Proceedings of WWW, pp. 173–182 (2017)
19. Kabbur, S., Ning, X., Karypis, G.: FISM: factored item similarity models for Top-N recommender systems. In: Proceedings of KDD, pp. 659–667 (2013)
20. Kingma, D.P., Ba, J.: Adam: a method for stochastic optimization. arXiv preprint arXiv:1412.6980 (2014)
21. Kipf, T.N., Welling, M.: Semi-supervised classification with graph convolutional networks. In: Proceedings of ICLR (2017)
22. Koren, Y.: Factorization meets the neighborhood: a multifaceted collaborative filtering model. In: Proceedings of KDD, pp. 426–434 (2008)
23. Koren, Y., Bell, R., Volinsky, C.: Matrix factorization techniques for recommender systems. Computer $42(8)$, 30–37 (2009)
24. Li, G., Muller, M., Thabet, A., Ghanem, B.: DeepGCNs: can GCNs go as deep as CNNs? In: Proceedings of ICCV, pp. 9267–9276 (2019)
25. Liang, D., Krishnan, R.G., Hoffman, M.D., Jebara, T.: Variational autoencoders for collaborative filtering. In: Proceedings of WWW, pp. 689–698 (2018)
26. Liu, F., Cheng, Z., Zhu, L., Gao, Z., Nie, L.: Interest-aware message-passing GCN for recommendation. arXiv preprint arXiv:2102.10044 (2021)
27. Mao, K., Zhu, J., Xiao, X., Lu, B., Wang, Z., He, X.: UltraGCN: ultra simplification of graph convolutional networks for recommendation. In: Proceedings of CIKM, pp. 1253–1262 (2021)
28. Qiu, J., Tang, J., Ma, H., Dong, Y., Wang, K., Tang, J.: DeepInf: social influence prediction with deep learning. In: Proceedings of KDD, pp. 2110–2119 (2018)
29. Rendle, S., Freudenthaler, C., Gantner, Z., Schmidt-Thieme, L.: BPR: Bayesian personalized ranking from implicit feedback. arXiv preprint arXiv:1205.2618 (2012)
30. Shah, S.T.U., Li, J., Guo, Z., Li, G., Zhou, Q.: DDFL: a deep dual function learning-based model for recommender systems. In: Nah, Y., Cui, B., Lee, S.-W., Yu, J.X., Moon, Y.-S., Whang, S.E. (eds.) DASFAA 2020. LNCS, vol. 12114, pp. 590–606. Springer, Cham (2020). https://doi.org/10.1007/978-3-030-59419-0_36
31. Wang, X., He, X., Wang, M., Feng, F., Chua, T.S.: Neural graph collaborative filtering. In: Proceedings of SIGIR, pp. 165–174 (2019)
32. Xu, F., Lian, J., Han, Z., Li, Y., Xu, Y., Xie, X.: Relation-aware graph convolutional networks for agent-initiated social e-commerce recommendation. In: Proceedings of CIKM, pp. 529–538 (2019)
33. Xue, H.J., Dai, X., Zhang, J., Huang, S., Chen, J.: Deep matrix factorization models for recommender systems. In: Proceedings of IJCAI, pp. 3203–3209 (2017)
34. Ying, R., He, R., Chen, K., Eksombatchai, P., Hamilton, W.L., Leskovec, J.: Graph convolutional neural networks for web-scale recommender systems. In: Proceedings of KDD, pp. 974–983 (2018)
35. Yu, B., Yin, H., Zhu, Z.: Spatio-temporal graph convolutional networks: a deep learning framework for traffic forecasting. arXiv preprint arXiv:1709.04875 (2017)

Gated Hypergraph Neural Network
for Scene-Aware Recommendation

Tianchi Yang[1], Luhao Zhang[2], Chuan Shi[1(✉)], Cheng Yang[1], Siyong Xu[1],
Ruiyu Fang[2], Maodi Hu[2], Huaijun Liu[2], Tao Li[2], and Dong Wang[2]

[1] Beijing University of Posts and Telecommunications, Beijing, China
{yangtianchi,shichuan,xusiyong}@bupt.edu.cn
[2] Meituan, Beijing, China
{zhangluhao,fangruiyu,humaodi,liuhuaijun,litao19,wangdong07}@meituan.com

Abstract. To improve e-commercial recommender systems, researchers
have never stopped exploring the interactions between users and items.
Unfortunately, most existing methods only explore one or some certain
components of the entire interactions. In fact, the entire interaction process is much richer and more complex, including but not limited to "who
purchases what items in which merchant under what interaction environments". Furthermore, many interactions have common features, thus
forming a scene, a kind of prior knowledge for predicting user interactions. In this paper, we make the first attempt to study the scene-aware
recommendation, which provides better recommendations with the entire
interaction modeling and the scene prior knowledge. To this end, we propose a novel gated hypergraph neural network for Scene-aware Recommendation (SREC). Particularly, we first construct a heterogeneous scene
hypergraph to model the entire interactions and scene prior knowledge.
Then we propose a novel scene-aware gate mechanism-based hypergraph
neural network to enrich their representations. Finally, we design a separable score function to predict the matching scores among user, scene,
merchant and interaction environments for training and inference procedures. Extensive experiments demonstrate that our SREC can fully leverage the scene prior knowledge and outperforms state-of-the-art methods
on real industrial datasets.

Keywords: Recommendation Systems · Hypergraph neural network

1 Introduction

With the rapid development of online social media and e-commerce, the benefits
of Recommendation Systems (RSs) are well recognized as a basic service of e-
commerce platforms [13,24], based on which users could filter out numerous
uninformative messages and facilitate decision-making [27].

Since the birth of the recommendation system, researchers have never
stopped exploring the interactions between users and items to accurately predict

A. Bhattacharya et al. (Eds.): DASFAA 2022, LNCS 13246, pp. 199–215, 2022.
https://doi.org/10.1007/978-3-031-00126-0_13

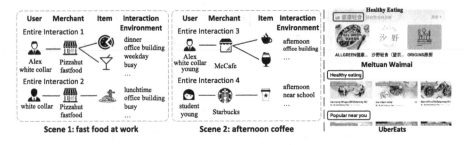

Fig. 1. (Left) Examples of entire interactions and scenes. The key features related to the scene are marked in red. (Right) Applications of scenes in e-commercial platforms. (Color figure online)

user preferences and suggest items. For example, classic recommendation methods, e.g., matrix factorization [18], mainly model users' preferences towards items using only simple historical user-item interaction records such as ratings [6]. Since these methods suffer from cold start and data sparsity problems [24], one or some kinds of interaction information are introduced into RSs. Social recommendation [10,17] and location-aware recommendation [2,22] respectively leverage social friendships and location relationships to enrich the interaction information. Similarly, context-aware recommendation [3–5,21] incorporates contextual information, e.g., weather, location, etc., alongside the core data (users and items) for better recommendations. Moreover, heterogeneous information network [12,15,24] is employed to fuse more variety of rich interaction attributes to learn better user/item embeddings.

Despite the great success of these methods by exploring the interaction, in practical applications, the interaction process is much richer and more complex, and these methods only explore some certain components of the entire interactions. For instance, in e-commercial applications, especially food delivery industry such as Meituan Waimai, UberEats, GrubHub, etc., an entire interaction includes but is not limited to "who purchases what items in which merchant under what interaction environments[1]", which involves multiple entities (e.g., users, merchants, items) and interaction environments (e.g., time period, location, season). As Fig. 1 illustrates, a white-collar, Alex, bought fast food, pizza and juice, for dinner at PizzaHut near his office building on a weekday to save time, etc. Most previous studies only consider the user and item, e.g., "Alex, pizza and juice", while some further introduce one or some more interaction information, e.g., "office building" in location-aware recommendation, "dinner, weekday" in context-aware recommendation. They still fail to formally and explicitly model the entire interactions.

Furthermore, we can find that many interactions have common features, which form a *scene*. As Fig. 1 illustrates, there are a series of interactions (e.g., entire interactions 1 and 2) containing the same features "white-collar, fast-food, office

---

[1] The term interaction environment refers to the properties specific to the interaction itself, e.g., spatial and temporal properties.

building, weekday" respectively from the user profiles, merchant/item attributes and interaction environments. Then we can summarize a scene named "fast food at work", representing a common prior knowledge that white-collars likely prefer to buy fast food when being busy at work. Similarly, we can conclude another scene named "afternoon coffee", representing that young people tend to buy coffee in the afternoon to keep awake. Scenes have begun to be employed to e-commercial applications, especially food delivery applications, such as Meituan Waimai as shown in Fig. 1, and these scene information have been summarized based on the entire interaction records through manual rules or statistical methods. Scene is totally different from context in context-aware recommendation. Although the meaning of the context is extended from temporal-spatial properties in early definitions [21] to click sequence [7], social relationships [17], etc., it is still a supporting component of interaction. Differently, a scene is the abstraction of a group of entire interactions involving users, merchants, items, and interaction environments. Note that, in this work, we focus on the usage of the scene prior knowledge tagged beforehand, and leave the automatically extracting scene from interactions for future work.

With the help of scenes, the recommendation system will benefit from the following advantages: (1) the entire interaction between user and items can be effectively characterized by scenes, since scene is a high-level abstraction of not only user and item, but also merchant and interaction environments; (2) the scenes bring more comprehensive prior knowledge about the entire interactions, so that avoid improper recommendation due to incomplete priors (e.g., recommend coffee to a coffee lover even late at night); (3) the scenes can make the recommendation list more interpretable and well-organized as shown in Fig. 1. In a word, we can provide better recommendations with the entire interaction modeling and the scene prior knowledge. We name this recommendation setting that introduces scene prior knowledge as scene-aware recommendation. Different from existing studies, which are limited on modeling interactions and hence failing to leverage the scene prior knowledge, we model the entire interactions from a more comprehensive perspective and successfully make full use of the scenes.

In this paper, we make the first attempt to study the scene-aware recommendation. However, this is challenging due to the following reasons: Firstly, each entire interaction involves multiple entities (user, merchant, item) and interaction environments (time period, location, season, etc.) as well as complex relations among them. How to model such complex entire interactions? Secondly, each specific entity is involved in multiple scenes since user preferences have always been changing (e.g., "Alex" is involved in two interactions of different scenes in Fig. 1). How to correctly extract the scene-specific information from each entity and make full use of it? Thirdly, the scene is already known in training samples, but in real applications, it is unknown to which scene the user's potential behavior will belong. How to bridge this gap and correctly infer the possible scenes for users while predicting user preference?

To tackle the aforementioned challenges, we propose a novel gated hypergraph neural network for **S**cene-aware **REC**ommendation, named SREC. Particularly, we first construct a heterogeneous scene hypergraph to model the entire interactions among users, merchants, items and interaction environments together with the scenes, which both comprehensively models the complex relationships among them all and appropriately incorporates the scene prior knowledge. Then we propose a novel gated hypergraph neural network to learn the representations of user, merchant and scene with a scene-aware gate mechanism designed to effectively discern different scene-specific information. Finally, we propose an effective separable score function to support a two-stage inference in line with practical requirements that firstly infers the proper scene before the final recommendation. The main contributions are summarized as follows:

(1) To the best of our knowledge, *this is the first attempt* to study scene-aware recommendation which provides better recommendation with the entire interaction modeling and the scene prior knowledge.
(2) We propose a novel gated hypergraph neural network for scene-aware recommendation. It first constructs a heterogeneous scene hypergraph to model the entire interactions and the scenes, then designs a gated hypergraph neural network followed by a separable score function to predict user preference with the help of scenes.
(3) Extensive experiments verify that our SREC can make full use of the entire interaction modeling and scene prior knowledge, thus greatly outperforming state-of-the-art (SOTA) methods for two settings on real industrial datasets.

2 Preliminary

As mentioned above, in e-commercial applications, an interaction usually forms "who purchases what items in which merchant under what interaction environments", commonly involving a user, a merchant, several items and the interaction environments. In order to model the interaction comprehensively, we formalize it as an *entire interaction*.

Definition 1. *Entire Interaction.* *Given a 4-tuple $<\mathcal{U}, \mathcal{M}, \mathcal{I}, \mathcal{C}>$ (denoted as Γ for short, $\mathcal{U}, \mathcal{M}, \mathcal{I}$ and $\mathcal{C}$ denoting the set of users, merchants, items and interaction environments, respectively), an entire interaction $\tau \in \Gamma$ is formulated as $\tau = <u, m, i, c>$, representing a user $u \in \mathcal{U}$ purchased some items $i \subset \mathcal{I}$ in the merchant $m \in \mathcal{M}$ under the interaction environments $c \subset \mathcal{C}$.*

Note that U, M, I may contain attribute information, which is default to make the definition clearer. Different from the traditional recommendation methods that only users and items are explicitly modeled, we also explicitly model merchants and interaction environments. They are somehow underestimated as implicit auxiliary attributes in traditional e-commercial apps, but have essential influence for location-based services, e.g., Meituan Waimai. Considering the interaction environment "afternoon", for instance, people are likely to drink coffee in the afternoon but hardly at night. Furthermore, we find that many interactions have common features, thus forming a *scene*, which is formalized as follows.

Definition 2. Scene. *A scene $s \in S$ is defined as a set of entire interactions that have common features. Therefore, each entire interaction can be labeled with a scene by the scene function $\psi : \Gamma \to S$.*

As Fig. 1 shows, each scene indicates a kind of common purchase patterns, e.g., "scene 1" represents "young white-collars often choose fast food at work on weekdays". The scenes are usually obtained by summarizing the entire interaction records through manual rules or statistical methods in industrial applications. Therefore, each entire interaction τ can be tagged with one of the predefined scenes by the manual-defined scene function $\psi(\tau)$. Therefore, we can provide better recommendations with the entire interaction modeling and scene prior knowledge. We name this recommendation setting *scene-aware recommendation*. In this work, we only focus on recommending merchants, an urgent practical task for industrial applications like food delivery, while the item recommendation is left as future work.

Definition 3. Scene-aware Recommendation. *Given the entire interactions labeled with $|S|$ scenes, i.e., $<\Gamma; S>$, for a user u under interaction environments c, scene-aware recommendation aims to predict a merchant m with the help of scenes S, i.e., $P(m|c, u; S)$.*

Note that scene-aware recommendation is different from existing recommendation. Previous recommendation settings, e.g., session-based recommendation, are mostly meant to predict m based on a given u, i.e., $P(m|u)$, while some others, e.g., context-aware recommendation, only further consider the interaction environments, i.e., $P(m|c, u)$. Our scene-aware recommendation will leverage the scene prior knowledge, thus forming $P(m|c, u; S)$.

3 Methodology

In this section, we first present an overview for the proposed SREC. The basic idea is that, as illustrated in Fig. 2, with comprehensively modeling entire interactions together with the scenes, we design a novel gated hypergraph neural network to enrich the representations of users, merchants and scenes, followed by a score function to make the scene bridge the given user and corresponding interaction environments to the recommended merchants.

In detail, to model the entire interactions among users, merchants, items, interaction environments and scenes, we construct a heterogeneous scene hypergraph based on the entire interaction records, where each hyperedge and its type represent an entire interaction and the corresponding scene. Next, a novel gated hypergraph neural network enriches the representations of users, merchants and scenes by aggregating the hypergraph-based neighboring information. During aggregation, a scene-aware gate mechanism is designed to effectively discern different scene-specific information and make full use of them correctly. Noting that we only explicitly enrich the embeddings of users, merchants and scenes in this work to reduce computing costs, since our task focuses on recommending

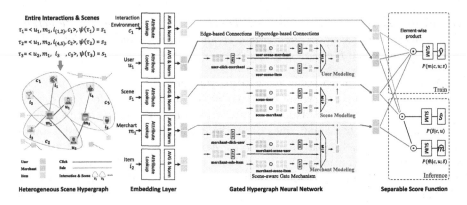

Fig. 2. Illustration of the proposed model SREC.

merchant to user. Finally, only after an interaction occurs can it be confirmed which scene this interaction will belong to, i.e., the scene is not available that the user's following purchase will belong to. Therefore, we design a separable score function to firstly infer proper scenes and then recommend merchants.

3.1 Heterogeneous Scene Hypergraph

Since each entire interaction involves multiple entities and interaction environments, we model them and scenes via a heterogeneous scene hypergraph.

Specifically, we first build a graph with three types of nodes, i.e., user, merchant and item nodes. To model the historical interaction records, we add a hyperedge for each entire interaction, e.g., a hyperedge $e = \{u_1, m_1, i_1, i_2\}$ is built for entire interaction τ_1 in Fig. 2. Naturally, we attach the entity attributes to the node features, and can attach the interaction environments to hyperedge features. Then, the scene tagged on each entire interaction $\psi(\tau_1)$ is attached to the hyperedge type (refer to the hyperedge color in Fig. 2). Therefore, each hyperedge and its connecting nodes together with their features can represent an entire interaction instance. Furthermore, two types of edges are added to enrich the information of the heterogeneous scene hypergraph. Particularly, we establish edges between a certain user and his/her clicking merchants to model the user's short-term historical behaviors within a session. Edges are built between a certain merchant and some items, representing sales relationship. In summary, the heterogeneous scene hypergraph has three types of nodes (users $\mathcal{U}$, merchants $\mathcal{M}$, items $\mathcal{I}$), two types of undirected edges for click and sale, and $|\mathcal{S}|$ types of attributed hyperedges.

3.2 Embedding Layer

We propose to initialize the representations of users, merchants, items, interaction environments and scenes with their attributes, since the attribute-based representations can alleviate the cold start problem for both entities and scenes.

Taking a user $u \in \mathcal{U}$ as an example, we first embed each feature field into a low-dimensional space and then fuse them into node embedding. Formally,

$$e_i = 1/q \cdot x_i \cdot F_i, \quad i = 1, 2, \cdots, F, \tag{1}$$

$$u = h(e_1, e_2, \cdots, e_F), \tag{2}$$

where F is number of feature fields, x_i is the one-/multi-hot encoding of the i-th field, F_i is attribute embedding matrix, q is the number of non-zero elements of x_i. Following traditional settings [7], average function is adopted as the field aggregation function. Besides, we further add a layer normalization for fast convergence, formally, $h = \text{LayerNorm} \circ \text{AVG}$.

Similarly, we can embed merchant, item, interaction environment and scene as dense embeddings m, i, c and s based on their attributes,[2] respectively.

3.3   Gated Hypergraph Neural Network

To show the intuition of the gated hypergraph neural network, we illustrate three observations based on the e-commercial data, being described in accordance with the left part in Fig. 2: (1) Apparently, the entities belonging to a certain interaction are related to each other, e.g., a node u_1 and its neighbors m_1, i_1, i_2 under the hyperedge τ_1 are related. (2) Each user (merchant, or item) will have interactions of different scenes, but the reasons why it belongs to one specific scene may be completely different from another. For instance, u_1 is included in both τ_1 and τ_2), but the reason why it belongs to the scene s_1 could be completely different from s_2. (3) Different users (merchants, or items) who have the interactions of the same scene are also related, i.e., two unconnected nodes are related by the same hyperedge type, e.g., u_1 and u_2 are related under scene s_1. Therefore, the scenes are exactly required by the gate mechanism, namely scene-aware gate mechanism, which could filter out irrelevant information to a specific scene while strengthen the relevant information during neighboring information propagation. In the following, we will introduce the propagation rule for user, merchant and scene, respectively.

User Modeling. In our hypergraph, if the central node u is a user node, it has three types of connections: user-click-merchant, user-hyperedge-merchant[3] and user-hyperedge-item. For user-click-merchant connections, following previous work [16], we apply Transformer to capture the user behavior information:

$$u_b = \text{Transformer}\,(m_1, m_2, \cdots, m_n)\,, \tag{3}$$

where n is the number of sampled neighbors. For the other two types of hyperedge-based connections, we design a scene-aware gate mechanism to filter

---

[2] We take the scene attributes from the common features of its containing entire interactions.

[3] User-hyperedge-merchant represents a series of connection forms: user-s_1-merchant, user-s_2-merchant, $\cdots$, and similarly for user-hyperedge-item.

out irrelevant information to a specific scene and allow the propagation of scene-aware information. Next, following [7], we use the average function to aggregate for reducing the computational complexity. Formally,

$$u_{sm} = \text{AVG}(\{m_j \odot s_j | j = 1, 2, \cdots, n\}), \tag{4}$$

$$u_{si} = \text{AVG}(\{i_j \odot s_j | j = 1, 2, \cdots, n\}), \tag{5}$$

where $\odot$ denotes element-wise product, m_j and i_j represent the neighbors based on user-hyperedge-merchant and user-hyperedge-item connections, respectively, and s_j is the representation of corresponding hyperedge type, i.e., scene. Hence, the scene-specific information will be amplified by the gating s_j, while irrelevant information will be blocked. Now, we have three type-level embeddings u_b, u_{sm} and u_{si} for user node u. Then we utilize a Multi-Layer Perception (MLP) to fuse them together, and update it to the raw user embedding u, formally,

$$u' = u + MLP(u_b \oplus u_{sm} \oplus u_{si}), \tag{6}$$

where $\oplus$ denotes the operation of concatenation.

Merchant Modeling. If the central node is a merchant node, it has four types of connections: merchant-clicked-user, merchant-sell-item, merchant-hyperedge-user and merchant-hyperedge-item. For merchant-clicked-user and merchant-sell-item connections, we also use average function to obtain the type-level embeddings m_u and m_i. For merchant-hyperedge-user and merchant-hyperedge-item connections, we similarly apply the scene-aware gate mechanism followed by average function to obtain the hyperedge-based type-level embeddings m_{su} and m_{si}. Finally, we apply another MLP to the four type embeddings and update its output to the raw merchant embedding, formally,

$$m' = m + MLP(m_u \oplus m_i \oplus m_{su} \oplus m_{si}). \tag{7}$$

Scene Modeling. According to observation (3), even two unconnected users (or merchants) will have some common characteristics because some of their belonging interactions are of the same scene. A straightforward solution is to design a cross-hyperedge propagation, which allows information flowing along with the type of hyperedges. However, this will lead to an inefficient computing process. Since each user/merchant is not bounded to a certain scene, we need to calculate the user/merchant representations based on all possible scenes. If this process is computed dynamically, it will extremely reduce the online efficiency. If we pre-calculate the user/merchant embeddings of all candidate scenes, it will require large storage space, i.e., each user/merchant needs to store $|S|$ embeddings.

Therefore, we explore another solution instead: since the common characteristics among the user/merchants are scene-specific, we could just incorporate them into the scene representation. Similarly, denoting u_j and m_j as the users/merchants included into the hyperedge assigned with scene $s \in S$, we have

$$\boldsymbol{s}_u = \mathrm{AVG}(\{\boldsymbol{u}_j \odot \boldsymbol{s} | j = 1, 2, \cdots, n\}), \tag{8}$$

$$\boldsymbol{s}_m = \mathrm{AVG}(\{\boldsymbol{m}_j \odot \boldsymbol{s} | j = 1, 2, \cdots, n\}), \tag{9}$$

$$\boldsymbol{s}' = \boldsymbol{s} + MLP(\boldsymbol{s}_u \oplus \boldsymbol{s}_m). \tag{10}$$

3.4   Separable Score Function

After obtaining the enriched embeddings of users, merchants and scenes, here we focus on how to evaluate the match score among the triple <user, scene, merchant> and the corresponding interaction environments. To avoid searching for the best from all the possible triples, we need to design a score function $f(m|c, u; s) = P(m|c, u; s)$ which meets the following requirements:

Separability. f should be separable: it can be split into a two-way sub-function $g(s|c, u)$ to predict match score between the user and candidate scenes, while f further predicts match score among user, scene and merchant.

Reusability. f and g should be reusable: when calculating f, some calculation results of g should be available and useful to improve online efficiency.

Consistency. f and g should be consistent: if f returns a large score, g should also return a relatively large score. This ensures that the "correct" merchant will not be directly filtered out when g selects the related scenes.

Here we discuss the two most widely used score functions: MLP and inner product. The former usually obtains a satisfactory performance due to strong fitting capability, but it is not separable due to its deep structure. Noticing the latter is actually the sum of element-wise multiplication of two vectors, it is easy to expand for multiple inputs and meanwhile keep reusable. Formally, we have

$$g(s|c, u) = \mathrm{sum}(\boldsymbol{c} \odot \boldsymbol{u} \odot \boldsymbol{s}), \qquad f(m|c, u; s) = \mathrm{sum}(\boldsymbol{c} \odot \boldsymbol{u} \odot \boldsymbol{s} \odot \boldsymbol{m}), \tag{11}$$

where each element of $\boldsymbol{c}, \boldsymbol{u}, \boldsymbol{m}$ and $\boldsymbol{s}$ is constrained to be non-negative real numbers to meet consistency requirement. In our work, we apply hard sigmoid function before the element-wise product to satisfy the non-negative condition. In the following, we present a brief proof for the above claim.

Theorem 1. *If there exists δ subject to $f \geq \delta$, g will have a low bound with the non-negative constraint.*

Proof. Considering the j-th dimension of the d-dimensional vectors $\boldsymbol{c}, \boldsymbol{u}, \boldsymbol{s}$ and $\boldsymbol{m}$, suppose the upper bound of m_j is σ_j. Then, we have $c_j, u_j, t_j, m_j \geq 0$ and $m_j \leq \sigma_j$. Suppose there exists a positive real number δ_j subject to $c_j \cdot u_j \cdot s_j \cdot m_j \geq \delta_j > 0$. Then we get $c_j \cdot u_j \cdot s_j \geq \frac{\delta_j}{m_j} \geq \frac{\delta_j}{\sigma_j}$. Therefore, for the entire vector, we have $g = \sum_{j=1}^d c_j \cdot u_j \cdot t_j \geq \sum_{j=1}^d \frac{\delta_j}{\sigma_j} \geq \frac{\sum_{j=1}^d \delta_j}{\max_j(\sigma_j)} = \frac{\delta}{\sigma}$. Here, $\delta = \sum_{j=1}^d \delta_j$ and $\sigma = \max_j(\sigma_j)$. That is, if we have $f \geq \delta$, g will have a low bound δ/σ.

Classification Setting and Model Training. Through the above modules, we can obtain the match score among user, scene and merchant under interaction

Table 1. Statistics of datasets

Dataset	#Train	#Test	Train user/Merchant/Item/Scene	Test user/Merchant/Item/Scene
1-day	1.13M	1.19M	153K/32K/409K/263	160K/33K/427K/263
3-day	3.51M	1.25M	381K/41K/892K/283	168K/33K/453K/265
5-day	6.13M	1.32M	589K/45K/1.27M/288	172K/33K/479K/257
7-day	8.55M	1.15M	746K/47K/1.54M/294	155K/33K/419K/258

environments by function f. In other words, in this case, the entire interaction and the scene are already given, hence we can directly predict their matching scores. We name this setting "classification", which is, following existing methods [7], also applied for model training. Formally, given a sample $<m, c, u; s>$:

$$\hat{y} = \text{sigmoid}(f(m|c, u; s)), \tag{12}$$

$$\mathcal{L} = \sum_{j \in \mathcal{Y}^+ \cup \mathcal{Y}^-} (y_j \log \hat{y}_j + (1 - y_j) \log(1 - \hat{y}_j)), \tag{13}$$

where y_j and $\hat{y}_j$ are the true label and prediction of the sample j. $\mathcal{Y}^+$ and $\mathcal{Y}^-$ are the positive and negative instance sets, respectively. The set of negative instances is composed of training triples with either the user, scene or merchant replaced. We will discuss in detail our negative sampling strategy in Sect. 4.2.

Inference Setting. As mentioned before, in practical applications, we will first recall highly related k_s scenes based on a given user and interaction environments with g. Then the most related k_m merchants based on each selected scene above can be recommended with function f. Finally, we will recommend $k_s \cdot k_m$ merchants in all. We name this procedure "inference", formally,

$$\{\hat{s}_j | j = 1, 2, \cdots, k_s\} = \text{Top}_{k_s} \ g(s|c, u), \tag{14}$$

$$\{\hat{m}_j | j = 1, 2, \cdots, k_m\} = \cup_{\hat{s} \in \{\hat{s}_j\}} \text{Top}_{k_m} \ f(m|c, u; \hat{s}) \tag{15}$$

4 Experiments

4.1 Experimental Setup

Datasets. A real-world large-scale dataset is built from the food delivery industry, i.e., Meituan Waimai platform. We collect 8-day orders of user purchases of foods and the corresponding click records before the purchase in Beijing District. Each order is an interaction instance, mostly containing a user, a merchant, several items and corresponding interaction environments, and has been already tagged with a scene based on some hand-craft rules and manual efforts. Then we use these orders to build a heterogeneous scene hypergraph as described in Sect. 3.1. For better validation, we split the whole data into several different scales of data: we use different periods (from 1 to 7 days) as the training data

Table 2. AUC comparisons of different methods. The last row indicates the improvements (%) compared to the best baseline (underlined).

Method	1-day			3-day			5-day			7-day		
	50%	75%	100%	50%	75%	100%	50%	75%	100%	50%	75%	100%
AutoInt	.5785	.5868	.5891	.5841	.5838	.5846	.5884	.5904	.5895	.5839	.5834	.5839
NIRec	.6589	.6712	.6877	.7075	.7295	.7433	.7319	.7534	.7719	.7502	.7664	.7742
AutoInt$_S$	.6781	.6803	.6892	.6842	.6913	.6987	.7135	.7198	.7203	.6906	.6939	.7001
MEIRec	.7497	.7565	.7634	.7393	.7567	.7658	.7814	.7761	.7956	.7956	.7851	.7891
NIRec$_S$	.7118	.7597	.7660	.7880	_.7975_	_.8072_	.8138	.8151	.8213	.8342	_.8398_	.8390
HAN	.7459	.7569	.7652	.7138	.7230	.7310	.8103	_.8193_	_.8258_	.8132	.8137	.8144
HGAT	.7471	.7584	_.7675_	.7627	.7667	.7697	.7846	.7839	.7923	.8255	.8336	_.8390_
Hyper-SAGNN	.7061	.7130	.7133	.7270	.7327	.7378	.7318	.7350	.7360	.7560	.7591	.7583
Hyper-SAGNN$_S$	_.7579_	_.7660_	.7649	_.7882_	.7944	.7980	_.8147_	.8186	.8176	_.8350_	.8367	.8340
SREC	**.8083**	**.8087**	**.8097**	**.8695**	**.8702**	**.8787**	**.8944**	**.8915**	**.8970**	**.8992**	**.9032**	**.8995**
Improvement	6.64	5.57	5.50	10.32	9.11	8.85	9.79	8.80	8.62	7.68	7.55	7.21

(about 10% data from the training set is extracted for validation) and predict the next one day. Therefore, we have four datasets marked as **1-day**, **3-day**, **5-day** and **7-day**. To get robust results, we vary the size of each training set from 50% to 100%. The detailed statistics of the data are reported in Table 1.

Baselines. We compare SREC with the following four groups of methods and their variants: recommendation methods without scenes: AutoInt [19], NIRec [12]; variant recommendation methods with scenes: AutoInt$_S$, MEIRec [7], NIRec$_S$; graph embedding methods: HAN [25], HGAT [14]; hypergraph-based methods: Hyper-SAGNN [31], Hyper-SAGNN$_S$. They are detailed as follows: AutoInt is SOTA feature-based Click-Through-Rate model. MEIRec and NIRec are meta-path-guided heterogeneous graph neural network based approaches for context-aware recommendation, while HAN and HGAT are both SOTA heterogeneous graph embedding models. Hyper-SAGNN is a SOTA hypergraph-based model for link prediction. For fair comparisons, we further modify the above methods to be accessible to the same information as ours (such as interaction environments, scenes, etc.) since the absence of any information harms the performance. Particularly, we add a channel of scene information to AutoInt and Hyper-SAGNN, denoting as AutoInt$_S$ and Hyper-SAGNN$_S$. For graph-based models, we transform the hypergraph into a heterogeneous graph: a summary node is introduced for each interaction instance, whose type and attributes depend on the corresponding scene and interaction environments, i.e., we transform the hyperedges into summary nodes, and link the summary nodes with the nodes related the corresponding interaction instance. For meta-path based methods, we select the following meta-paths based on experiments: UMU, USU, UMIMU, MUM, MIM, MSM, SUS, SMS, SIS.

Settings and Metrics. We evaluate model performance for both settings. For classification setting, following previous work [7], we use Area Under receiver operator characteristic Curve (AUC) for evaluation. We use HR@K (Hit Ratio) to evaluate under the inference setting. Without loss of generality, we set K as 10 (k_s) for scenes and 100 ($k_s \cdot k_m$) for merchants in this work, denoting HR@10-S and HR@100-M, respectively. This is computed only on the positive sample set.

Detailed Implementation. For our method, we set the dimension of attribute embeddings, linear transformations and 2-layer MLPs all as 64. We use 8 hidden neurons and 8 heads in the transformer. For all baselines, we also set the hidden dimensions as 64 for fair comparison. We set the batch size as 2048, and set the learning rate as 0.0001 with Adam optimizer.

4.2   Experimental Results

Main Results. We evaluate the AUC performance based on the constructed negative samples in Table 2. We can conclude as follows: (1) SREC significantly outperforms all the competitive baselines. Compared to the best performance of baselines, SREC gains 5.50%–10.32% improvement in the four datasets. Moreover, when fewer training instances (50%, 75%) are provided, our SREC achieves a higher improvement. It indicates the effectiveness of our completely modeling of interactions and integrating scene prior knowledge, which makes our model more robust. (2) The scene prior knowledge is greatly useful. Compared with the methods access to scene prior knowledge, those without scene prior knowledge exhibit an obvious performance drop. It demonstrates *the superiority to considerate scenes than to direct recommend merchants and the significance of this scene-aware recommendation problem.* (3) The more accurate and comprehensive the interaction modeling is, the better the performance is. In detail, the graph-based models are better than the traditional feature-based model due to the consideration of the interactive relations between entities. Next, the hypergraph-based model HYPER-SAGNN$_S$ further performs better in most cases, which verifies the better modeling of entire interactions within scenes as hyperedges, since introducing a summary node cannot correctly model the entire interaction. Moreover, the huge performance gap between HYPER-SAGNN$_S$ and SREC verifies the necessity of our scene-aware gate mechanism to distinguish the mixed scene-related information.

Evaluation of Inference. In practice, we cannot search for the best from all the possible <user, scene, merchant> triples in terms of efficiency, thus requiring the inference setting. Here we measure the inference performances for scenes and merchants, i.e., HR@10-S and HR@100-M. Since only HAN and HGAT support the inference procedure, which can export relatively fixed embeddings for users, merchants and scenes, we compare our SREC with these two baselines on the four datasets. As reported in Table 3, our method consistently outperforms greatly in terms of both scenes and merchants, while HAN and HGAT only achieve limited and unrobust performance for inference. We believe this huge performance gap is caused by the following reasons: (1) The three requirements of our separable score function play a vital role to ensure the effective inference procedure. (2) It is essential to comprehensively model the entire interactions by

Table 3. HR@K comparisons for the recall of scenes and merchants.

Method		1-day			3-day			5-day			7-day		
		50%	75%	100%	50%	75%	100%	50%	75%	100%	50%	75%	100%
HAN	HR@10-S	.3910	.3678	.4148	.3652	.3608	.3809	.1455	.1904	.2655	.2604	.2858	.2493
	HR@100-M	.0123	.0505	.1048	.0392	.0728	.0949	.0107	.0307	.1678	.0614	.0881	.1079
HGAT	HR@10-S	.3696	.3851	.4284	.3598	.3324	.3910	.1123	.1398	.1609	.2690	.2542	.2821
	HR@100-M	.0796	.0549	.0950	.0336	.0325	.0581	.0235	.0613	.0411	.0643	.1274	.0919
SREC	HR@10-S	**.9588**	**.9756**	**.9720**	**.9884**	**.9922**	**.9913**	**.9927**	**.9920**	**.9936**	**.9926**	**.9913**	**.9936**
	HR@100-M	**.2576**	**.2615**	**.2849**	**.5234**	**.5439**	**.5886**	**.6560**	**.6916**	**.6912**	**.7113**	**.7503**	**.7249**

hyperedges. Introducing summary node for each interaction will forcibly divide the indivisible whole interaction relation into several pair-wise sub-relations, resulting in information loss. (3) The information relevant to different scenes is mixed in every single entity. For a particular scene, the information relevant to other scenes becomes noise on the contrary.

Table 4. AUC and HR@100-M comparisons of our variants.

Variant	1-day		3-day		5-day		7-day	
	AUC	HR@100	AUC	HR@100	AUC	HR@100	AUC	HR@100
SREC	<u>.8097</u>	**.2849**	<u>.8787</u>	**.5886**	<u>.8970</u>	**.6912**	<u>.8995</u>	**.7249**
SREC\GNN	.7649	.1981	.7691	.2266	.7844	.3171	.7726	.3929
SREC\Scene	.7493	.2071	.8072	.2757	.8656	.4375	.8811	.4984
SREC\Gate	.8002	<u>.2113</u>	.8122	.3388	.8664	<u>.6164</u>	.8888	<u>.6062</u>
SREC-ReLU	.5107	.0019	.5010	.0016	.5134	.0018	.5029	.0010
SREC-None	**.8960**	.0034	**.9213**	.0057	**.9219**	.0112	**.9246**	.0054

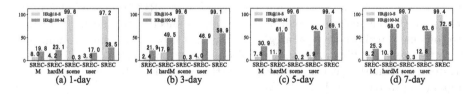

Fig. 3. HR@K (%) of SREC with different negative sampling strategies.

Comparison of Variants. We compare SREC with 2 groups of variants to validate the design of its modules: One group is used to verify the effectiveness of our propagation rule. The other group aims to verify the design of constraints in our separable score function. As reported in Table 4, we can draw the following conclusions. Firstly, the performances of SREC\GNN, SREC\Scene and

SREC\Gate, which are removed any neighboring information, scene type-specific information or scene-aware gate mechanism respectively, are limited in terms of both AUC and HR metrics, thus verifying the design of our propagation rule in gated hypergraph neural network. Secondly, ReLU, as a straightforward solution to for non-negative constraint, cannot improve the performance if replace hard-sigmoid activation in the separable score function with it, i.e., SREC-ReLU. Even worse, ReLU will cause large-scale death of neurons and thus the model cannot be trained, because its gradient in the negative range is 0 and the operation of element-wise product among multiple vectors will worsen this phenomenon. Moreover, the AUC metric gets better if the activation function hardsigmoid is directly removed, i.e., SREC-None, but the HR@K performance obtains a severe drop. Because the embedding space is limited by the hardsigmoid function, thereby removing it will strengthen the ability of fitting data (refer to the improvement on AUC). However, this constraint is indispensable to ensure the successful recall (refer to the performance drop on HR@K).

Impact of Negative Sampling Strategy. In our task, the model need learn based on triples <user, scene, merchant>, thus requiring a best negative sampling strategy. We have explored four negative sampling strategies: For each triple, SREC-M randomly replaces the merchant without any constraint; SREC-hardM replaces the merchant under the constraints of the same interaction environments; SREC-scene randomly replaces the scene; SREC-user replaces the user also under interaction environment constraints. The AUC metric is meaningless since the negative instances are different. Therefore, we choose HR@K metric on the positive instances in the test set for evaluation. As depicted in Fig. 3, the poor performance of SREC-M indicates that without any constraint is too simple to generate negative samples, while the performance of SREC-hardM is much better. Moreover, the result of SREC-scene shows that it is helpful for improving the HR@10-S to directly replace the scene, but it causes a lot of outrageous scene-merchant combinations, such as nutritious breakfast is paired with a barbecue restaurant. In our testing, we find SREC-user can effectively help correct scene-merchant pairing, although its effect is not ideal. Therefore, we combine the above three strategies to construct negative samples, to improve both the scene recalling and merchant recommendation. As shown in Fig. 3, this strategy combination at the expense of a little loss of HR@10-S, in exchange for a great improvement in HR@100-M.

5   Related Work

We first introduce some related recommendation methods, and then discuss the recent hypergraph-based methods.

Researchers have always been exploring the interactions between users and items for better recommendation. Since the classic methods such as matrix factorization [18] suffer from cold start and data sparsity problems [24], many kinds of interaction information are studied, e.g., social friendships [10,17], location relationships [2,22,26], contextual information [3–5,21], etc. HIN-based recommendation is then proposed to integrate any type of interaction information [12,24], attracting more research interests recently. However, the interaction process in real applications is still much richer and more complex, and these methods only explore some certain components of the entire interactions.

Hypergraph expands the concept of edge in graph into hyperedge that connects multiple nodes, thus having a strong ability to model a complex interaction among multiple entities [9]. HGNN [8] and HyperGCN [29] were the first to expand graph convolution to hypergraph, thereby inspiring researchers' enthusiasm for hypergraph neural network [1,20,31]. Recently, some researchers have begun to explore hypergraph-based methods for recommendation. For example, DHCF [11] developed a dual-channel learning strategy based on hypergraph that explicitly defined the hybrid high-order correlations, while [30] further integrates self-supervised learning into the training of the hypergraph convolutional network. There are also studies focusing on finer-grained recommendations, such as HyperRec for next-item recommendation [23], DHCN for session-based recommendation [28], etc. However, these methods still fail to comprehensively model the user's behaviors, thereby they can neither leverage the scene prior knowledge, limiting the performance of recommendation. Consequently, we are the first to fill this gap: A novel gated hypergraph neural network is proposed in this paper for scene-aware recommendation.

6   Conclusion

In this paper, we make the first attempt to study the scene-aware recommendation, where the entire interactions are modeled comprehensively and the scenes are introduced to guide the recommendation. Specifically, we propose a novel gated hypergraph neural network for scene-aware recommendation (SREC). It first constructs a heterogeneous scene hypergraph to comprehensively model the entire interactions and incorporate the scene prior knowledge, followed by a novel gated hypergraph neural network to learn the representations of users, merchants and scenes. Finally, it designs a separable score function to predict the match score, thus recommending the merchants. Extensive experiments demonstrate our SREC outperforms SOTA methods for both classification and inference on the real industrial datasets. In the future, we will explore clustering approaches for automatically discovering and tagging scenes in an unsupervised manner to reduce the costs of expert annotations.

Acknowledgments. This work is supported in part by the National Natural Science Foundation of China (No. U20B2045, 62192784, 62172052, 61772082, 62002029) and also supported by Meituan.

References

1. Bandyopadhyay, S., Das, K., Murty, M.N.: Line hypergraph convolution network: applying graph convolution for hypergraphs. arXiv (2020)

2. Chang, B., Jang, G., Kim, S., Kang, J.: Learning graph-based geographical latent representation for point-of-interest recommendation. In: CIKM (2020)
3. Chen, C., Zhang, M., Ma, W., Liu, Y., Ma, S.: Efficient non-sampling factorization machines for optimal context-aware recommendation. In: WWW (2020)
4. Chen, H., Li, J.: Adversarial tensor factorization for context-aware recommendation. In: RecSys, pp. 363–367 (2019)
5. Chen, L., Xia, M.: A context-aware recommendation approach based on feature selection. Appl. Intel. **51**(2), 865–875 (2020). https://doi.org/10.1007/s10489-020-01835-9
6. Da'u, A., Salim, N., Idris, R.: Multi-level attentive deep user-item representation learning for recommendation system. Neurocomputing **433**, 119–130 (2021)
7. Fan, S., et al.: Metapath-guided heterogeneous graph neural network for intent recommendation. In: KDD (2019)
8. Feng, Y., You, H., Zhang, Z., Ji, R., Gao, Y.: Hypergraph neural networks. In: AAAI, pp. 3558–3565 (2019)
9. Gao, Y., Zhang, Z., Lin, H., Zhao, X., Du, S., Zou, C.: Hypergraph Learning: Methods and Practices. TPAMI (2020)
10. Huang, C., et al.: Knowledge-aware coupled graph neural network for social recommendation. In: AAAI (2021)
11. Ji, S., Feng, Y., Ji, R., Zhao, X., Tang, W., Gao, Y.: Dual channel hypergraph collaborative filtering. In: KDD, pp. 2020–2029 (2020)
12. Jin, J., et al.: An efficient neighborhood-based interaction model for recommendation on heterogeneous graph. In: KDD, pp. 75–84 (2020)
13. Kou, F., Du, J., He, Y., Ye, L.: Social network search based on semantic analysis and learning. CAAI Trans. Intel. Technol. **1**(4), 293–302 (2016)
14. Linmei, H., Yang, T., Shi, C., Ji, H., Li, X.: Heterogeneous graph attention networks for semi-supervised short text classification. In: EMNLP (2019)
15. Luo, X., et al.: AliCoCo: Alibaba e-commerce cognitive concept net. In: SIGMOD (2020)
16. Luo, X., Yang, Y., Zhu, K.Q., Gong, Y., Yang, K.: Conceptualize and infer user needs in e-commerce. In: CIKM, pp. 2517–2525 (2019)
17. Ma, H., Zhou, T.C., Lyu, M.R., King, I.: Improving recommender systems by incorporating social contextual information. TOIS **29**(2), 1–23 (2011)
18. Rendle, S., Freudenthaler, C., Gantner, Z., Schmidt-Thieme, L.: BPR: Bayesian personalized ranking from implicit feedback. In: UAI, pp. 452–461 (2009)
19. Song, W., et al.: AutoInt: automatic feature interaction learning via self-attentive neural networks. In: CIKM, pp. 1161–1170 (2019)
20. Tu, K., Cui, P., Wang, X., Wang, F., Zhu, W.: Structural deep embedding for hyper-networks. In: AAAI, pp. 426–433 (2018)
21. Agagu, T., Tran, T.: Context-aware recommendation methods. IJISA **10**(9), 1–12 (2018)
22. Wang, H., Li, P., Liu, Y., Shao, J.: Towards real-time demand-aware sequential POI recommendation. Inf. Sci. **547**, 482–497 (2021)
23. Wang, J., Ding, K., Hong, L., Liu, H., Caverlee, J.: Next-item recommendation with sequential hypergraphs. In: SIGIR, pp. 1101–1110 (2020)
24. Wang, X., Bo, D., Shi, C., Fan, S., Ye, Y., Yu, P.S.: A survey on heterogeneous graph embedding: methods, techniques, applications and sources. arXiv (2020)
25. Wang, X., et al.: Heterogeneous graph attention network. In: WWW (2019)
26. Werneck, H., Silva, N., Viana, M.C., Mourão, F., Pereira, A.C.M., Rocha, L.: A survey on point-of-interest recommendation in location-based social networks. In: WebMedia, pp. 185–192 (2020)

27. Wu, S., Zhang, W., Sun, F., Cui, B.: Graph neural networks in recommender systems: a survey. arXiv (2020)
28. Xia, X., Yin, H., Yu, J., Wang, Q., Cui, L., Zhang, X.: Self-supervised hypergraph convolutional networks for session-based recommendation. In: AAAI (2021)
29. Yadati, N., Nimishakavi, M., Yadav, P., Nitin, V., Louis, A., Talukdar, P.: Hyper-GCN: a new method for training graph convolutional networks on hypergraphs. In: NeurIPS, pp. 1511–1522 (2019)
30. Yu, J., Yin, H., Li, J., Wang, Q., Hung, N.Q.V., Zhang, X.: Self-supervised multi-channel hypergraph convolutional network for social recommendation. In: WWW, pp. 413–424 (2021)
31. Zhang, R., Zou, Y., Ma, J.: Hyper-SAGNN: a self-attention based graph neural network for hypergraphs. In: ICLR (2020)

Hyperbolic Personalized Tag Recommendation

Weibin Zhao[1]([⊠]), Aoran Zhang[2], Lin Shang[1], Yonghong Yu[2], Li Zhang[3],
Can Wang[4], Jiajun Chen[1], and Hongzhi Yin[5]

[1] State Key Lab for Novel Software Technology, Nanjing University, Nanjing, China
njzhaowb@gmail.com, {shanglin,chenjj}@nju.edu.cn
[2] Tongda College, Nanjing University of Posts and Telecommunications,
Yangzhou, China
yuyh@njupt.edu.cn
[3] Department of Computer Science, Royal Holloway, University of London,
Surrey, UK
li.zhang@rhul.ac.uk
[4] School of Information and Communication Technology, Griffith University,
Brisbane, Australia
can.wang@griffith.edu.au
[5] School of Information Technology and Electrical Engineering,
The University of Queensland, Brisbane, Australia

Abstract. Personalized Tag Recommendation (PTR) aims to automatically generate a list of tags for users to annotate web resources, the so-called items, according to users' tagging preferences. The main challenge of PTR is to learn representations of involved entities (i.e., users, items, and tags) from interaction data without loss of structural properties in original data. To this end, various PTR models have been developed to conduct representation learning by embedding historical tagging information into low-dimensional Euclidean space. Although such methods are effective to some extent, their ability to model hierarchy, which lies in the core of tagging information structures, is restricted by Euclidean space's polynomial expansion property. Since hyperbolic space has recently shown its competitive capability to learn hierarchical data with lower distortion than Euclidean space, we propose a novel PTR model that operates on hyperbolic space, namely HPTR. HPTR learns the representations of entities by modeling their interactive relationships in hyperbolic space and utilizes hyperbolic distance to measure semantic relevance between entities. Specially, we adopt tangent space optimization to update model parameters. Extensive experiments on real-world datasets have shown the superiority of HPTR over state-of-the-art baselines.

Keywords: Hyperbolic spaces · Personalized · Tag recommendation · Embedding

A. Bhattacharya et al. (Eds.): DASFAA 2022, LNCS 13246, pp. 216–231, 2022.
https://doi.org/10.1007/978-3-031-00126-0_14

1 Introduction

Tagging systems have become essential in many web applications, such as Last.FM, Flickr, and YouTube. In the process of tagging, users are allowed to freely add metadata to the songs, videos, products, and other web resources(called items) in the form of keywords, the so-called tags. Besides annotating items, tags are beneficial to the systems and users for efficiently organizing, searching, and sharing the related items. With the increasing availability of tags in various domains, tag recommendation has become a popular service to help users acquire their desired tags more conveniently.

As a subtask of top-N ranking recommendation, the tag recommendation aims to assist users' tagging process by automatically suggesting a ranked list of tags. According to whether users' personalized preferences are considered, the tag recommendation tasks can be divided into the non-personalized and the personalized. The non-personalized tag recommendation (NPTR) [20,38,40,47] aims to generate a list of candidate tags ranked merely by their semantic relevance to the target item, this kind of recommendation will suggest the same tag list to all involved users. On the other hand, the personalized tag recommendation (PTR) [14,15,37,39] takes user's tagging preferences into account, its ultimate goal is to suggest tags that are relevant to both target item and target user, such goal make PTR more complex than NPTR, and as a result, PTR will recommend different tag lists to different users for the same target item. Due to users' diverse intentions and interests, PTR is more meaningful and practical in real scenarios of tag recommendation. Moreover, as indicated in [35], the PTR could outperform the theoretical upper bound of any NPTR.

Serving as the source data for PTR tasks, the users' historical tagging information implies a complex structure that involves three kinds of entities (i.e., users, items, and tags) and multiple interactive relationships, so it is a challenge for PTR to accurately learn latent representations of entities with preservation of their real semantic relevance in such data. To this end, various learning methods have been proposed to boost the performance of the PTR model. The core of tagging information is the ternary interaction, i.e., user-item-tag, which can be naturally represented by a three-order tensor. Thus tensor factorization techniques are widely adopted [3,9,35,37,39,49] to boost the performances of PTR models. Although tensor factorization-based PTR models are effective to a certain extent, all of them are conducted in Euclidean spaces. That is, they learn latent representations of entities in a low-dimensional embedding space and adopt matching functions that only cover the scope of Euclidean space, such as inner product, Euclidean distance, and neural networks, to compute the semantic relevance between embeddings [42,43] because Euclidean spaces are the natural generalization of our intuition.

On the other hand, recent research [8,11,25,29,30] has shown that Euclidean embedding have distortion for many real-world data, which follows power-law distribution or exhibits scale-free. It means that the actual logical patterns and semantic relevance can not be well preserved when embedding such data in Euclidean spaces. As mentioned in [29,30], the data with power-law distribution tend to have tree-like structures. For a tree, the number of its nodes grows exponentially with the tree depth, but the volume of Euclidean spaces grows polynomially with distance from the origin point. In terms of tagging information, the interactive relationships it contains have been found to follow power-law distribution [12,23,31,34], e.g., a small portion of tags (hot tags) are frequently used to annotate a certain kind of items while massive tags are seldom adopted. As power-law distributions and tree-like structures can be explained by assuming an underlying hierarchy in the data [29,30], it is therefore crucial for PTR to leverage this insight to develop an optimal learning model with lower embedding distortions [8]. Moreover, through our observation, this phenomenon does exist: For real users, some prefer to use tags with general semantics, while others prefer tags with descriptive semantics. Such personal preference divides users into different hierarchies with respect to the semantics. As shown in Fig. 1(a), in Last.FM, a popular track like Led Zeppelin's "Stairway to Heaven" has dozens of unique tags applied hundreds of times. According to our knowledge, some latent semantic hierarchies (Fig. 1(b)) may exist among these annotated tags.

(a)   (b)

Fig. 1. Hierarchies implied in tags

Notably, hyperbolic space has shown promise in modeling hierarchical data [5,8,29,30] in recent years. Hyperbolic space is a kind of non-Euclidean space with constant negative curvature. If we embed a disk into a two-dimensional hyperbolic space with curvature $K = -1$, its corresponding circumference $(2\pi \sinh r)$ and area $(2\pi(\cosh r - 1))$ both grow exponentially with the radius r, as opposed to the two-dimensional Euclidean space where the corresponding circumference $(2\pi r)$ and area (πr^2) grows linearly and quadratically respectively. Thus hyperbolic space can be viewed as the continuous version of a tree, and it is well-suited for embedding hierarchical data with lower dimensions than the Euclidean space.

Motivated by the above merits of hyperbolic space, we develop a novel PTR model with hyperbolic embedding, namely, HPTR. Our idea is to learn representations of entities by embedding historical tagging information into the Poincaré ball, which is an isometric hyperbolic space model and feasible to perform a gradient-descent step. As the PTR is an implicit feedback recommendation task, we build an objective function based on Bayesian Personalized Ranking (BPR) optimization criterion. Besides, there exist multiple relations between entities, which more or less affect the performance of PTR, so the main difficulty of HPTR is how to discover users' tagging preferences by dealing with these relations in hyperbolic space. Unlike the traditional item recommendation, in PTR scenarios, the users' preferences for items make no sense for learning and predicting. Thus we take two relations into consideration: $user-tag$ and $item-tag$, which have been proven effective in many PTR models [9,37,44,49]. Finally, we can utilize the hyperbolic distance between involved entities to reflect one user's tagging preference for a target item.

The main contributions of our work are summarized as follows:

- We bridge the gap between PTR and hyperbolic geometry by discovering common structural properties between tagging information and hyperbolic space. With the expectation for achieving better performance in the PTR task, we propose a novel PTR model conducted in hyperbolic space, namely HPTR, which learns the representations of entities on Poincaré ball and measures users' tagging preferences by hyperbolic distance. To the best of our knowledge, this is the first work to integrate PTR with hyperbolic space.
- We conduct extensive experiments on three real-world datasets to verify the efficiency of the proposed model, and experimental results show that our HPTR can outperform the state-of-the-art PTR model, especially with lower embedding dimensions.

2   Related Work

Our work is related to the following research directions.

2.1   Personalized Tag Recommendation Methods

With the popularization of tagging systems in various web applications, personalized tag recommendation (PTR) is becoming more attractive in the field of recommender systems. Considering the core of users' historical tagging information is the ternary interaction between entities, i.e., user, item, and tag, which can be represented by a three-order tensor naturally, so most early studies utilized tensor factorization techniques, especially the tucker decomposition (TD) to learn representations of involved entities in PTR tasks [3,35,39].

Due to the model equation of TD resulting in a cubic runtime in the factorization dimension, the computation cost of TD makes it infeasible for large-scale PTR tasks. Rendle et al. [37] proposed the pairwise interaction tensor factorization (PITF) model to tackle this problem, which explicitly models the pairwise interactions between entities and results in linear runtime. PITF has been extensively studied for its outperformance, and more learning methods derived from it [9,16] have been proposed to fit new problem scenarios. Moreover, to take advantage of the end-to-end learning ability of deep neural networks(DNN), several learning frameworks based on DNN [27,28,49] are developed to improve the performance of traditional PTR models further.

Note that all the above models are conducted in Euclidean spaces. As we mentioned before, their capabilities of learning the representations of hierarchical data are restricted by the polynomial expansion property of Euclidean space.

2.2 Hyperbolic Embedding

In the field of representation learning, hyperbolic spaces have started to get attention from the studies on how to discover suitable embedding spaces to model complex networks [21]. As scale-free and strong clustering are typical properties of complex networks, and such properties can be traced back to hierarchical structure within them, hyperbolic spaces have become a better choice for their capability of modeling hierarchical structures. Since then, there has been an increasing interest in utilizing hyperbolic embedding to learn representations of data with explicit or implicit hierarchies, representative research or applications of hyperbolic embedding include but are not limited to: Natural Language Processing [8,25,29], Knowledge Graph embedding [1,5,19], Heterogeneous Information Network embedding [46], Neural Networks based representation learning [6,11,24] and Computer Vision tasks [17,32]. In the domain of recommender systems, the original work exploring the use of hyperbolic space for the recommender systems is [42]. Subsequently, a number of models enhanced by hyperbolic embedding [4,10,22,26,43] have been proposed in order to get better performance in traditional recommendation tasks or cope with new tasks, via a series of corresponding experiments, these models have demonstrated the superiority over their Euclidean counterparts and state-of-the-art baselines. Nevertheless, the existing studies of hyperbolic recommender systems have not covered the scope of PTR.

3 Preliminaries

3.1 Problem Description

Unlike the item recommendation systems containing two types of entities, i.e., users and items, PTR consists of three types of entities: the set of users U, the set of items I, and the set of tags T. The historical tagging information between users, items, and tags is represented as $S \subseteq U \times I \times T$. A ternary $(u, i, t) \in S$ indicates that the user u has annotated the item i with the tag t. From the ternary relation set S, personalized tag recommendation methods usually deduce a three-order tensor $Y \in \mathbb{R}^{|U| \times |I| \times |T|}$, whose element $y_{u,i,t}$ is defined as follows:

$$y_{u,i,t} = \begin{cases} 1, & (u,i,t) \in S \\ 0, & otherwise, \end{cases} \tag{1}$$

where $y_{u,i,t} = 1$ indicates a positive instance, and the remaining data are the mixture of negative instances and missing values. In addition, the tagging information for the user-item pair (u, i) is defined as $\mathbf{y}_{u,i} = \{y_{u,i,t} | y_{u,i,t}, t \in T\}$.

PTR aims to recommend a ranked list of tags to a certain user for annotating a certain item. Usually, a score function $\widehat{Y} : U \times I \times T \longrightarrow \mathbb{R}$ is employed to measure users' preferences on tags for their target items. The entry $\widehat{y}_{u,i,t}$ of $\widehat{Y}$ indicates the degree to which a user u prefers to annotate the item i with the tag t. After predicting the score $\widehat{y}_{u,i,t}$ for all candidate tag t given a $user - item$ pair (u, i), the personalized tag recommender system returns a ranked list of Top-N tags in terms of the obtained scores. Formally, the ranked list of Top-N tags given to the $user - item$ pair (u, i) is defined as follows:

$$Top(u, i, N) = arg\underset{t \in T}{\overset{N}{max}}\ \widehat{y}_{u,i,t},\tag{2}$$

where N denotes the number of recommended tags.

3.2 Hyperbolic Embedding

Hyperbolic space is a smooth Riemannian manifold with constant negative curvature, and five isometric models can describe it [33], which are the Lorentz (hyperboloid) model, the Poincaré ball model, the Poincaré half space model, the Klein model, and the hemishpere model. Our work chooses the Poincaré ball to describe the embedding space, for it is relatively suitable for modeling a tree.

Let $\mathcal{B}^d = \{x \in \mathbb{R}^d \mid \|x\| < 1\}$ be the an open d-dimensional unit ball, where $\|\cdot\|$ denotes the Euclidean norm. The Poincaré ball can be defined by the Riemannian manifold $(\mathcal{B}^d, g_x^B)$, in which g_x^B is the Riemannian metric tensor given as:

$$g_x^B = \left(\frac{2}{1 - \|x\|^2}\right)^2 g^E\tag{3}$$

where $x \in \mathcal{B}^d$ and $g^E = \mathbf{I}$ denotes the Euclidean metric tensor. Furthermore, the distance between points $x, y \in \mathcal{B}^d$ is given as:

$$d_B(x, y) = \text{arcosh}\left(1 + 2\frac{\|x - y\|^2}{(1 - \|x\|^2)(1 - \|y\|^2)}\right)\tag{4}$$

It is worth noting that the Poincaré ball model is conformal: the angles of embedded vectors are equal to their angles in the Euclidean space, thus making Poincaré ball suitable for the gradient-based learning method.

4 HPTR Model

Due to the core of tagging information is the ternary interaction among entities, i.e., $user - item - tag$, our HPTR is committed to learn three embedding matrices: $\mathbf{U} \in \mathbb{R}^{|U| \times d}, \mathbf{I} \in \mathbb{R}^{|I| \times d}, \mathbf{T} \in \mathbb{R}^{|T| \times d}$ (d is the embedding dimension). In addition, there exist multiple relations between entities, which more or less affect the performance of PTR, so the main difficulty of HPTR is how to discover users' tagging preferences by dealing with these relations in hyperbolic space.

Inspired by the work [35], which claims that the relation between user and item is meaningless for modeling users' tagging preference. So we take two factors into consideration: $user - tag$ relation and $item - tag$ relation.

Furthermore, as indicated in Sect. 3, the final output of PTR is a list of Top-N tags, which ranked by the predicting score $\widehat{y}_{u,i,t}$ for all candidate tag t with respect to a certain pair (u, i), so we merely need to calculate the semantic relevance between (u, i) and t in embedding space, instead of u and t, i and t, respectively. Hence we add embedding of $user$ and $item$ together, i.e., $u + i$ to represent each (u, i) pair. Meanwhile, in the embedding space, if a user u prefers to annotate the item i with the tag t, the distance d_B between point $u + i$ and point t on $\mathcal{B}^d$ should be relatively shorter, and vice versa. Consequently, our HPTR can measure users' tagging preferences by using the score function $\widehat{y}_{u,i,t}$ defined as:

$$\widehat{y}_{u,i,t} = p \left(d_B \left((\mathbf{U_u} + \mathbf{I_i}), \quad \mathbf{T_t} \right) \right) \tag{5}$$

where $\mathbf{U}_u = \mathbf{U}.onehot(u)$, $\mathbf{I}_i = \mathbf{I}.onehot(i)$, $\mathbf{T}_t = \mathbf{T}.onehot(t)$ are embeddings of a given triple (u, i, t); $.onehot()$ indicates the operation of lookup in embedding table according to one-hot id encoding; $p(\cdot)$ is the transformation function for converting hyperbolic distances d_B to users' tagging preference, here we take it as $p(x) = \beta x + c$ with $\beta \in \mathbb{R}$ and $c \in \mathbb{R}$ similar to [42].

4.1   Objective Function

In this work, we agree with the assumption in [37]: When we observe a certain pair (u, i) in tagging information S, we believe that the user u should prefer tag t over tag t' iff the triple (u, i, t) can be observed from historical tagging information and (u, i, t') can not be observed. Based on this assumption, the training set D_S (i.e., the set of quadruple (u, i, t, t')) with the pairwise constraint is defined as:

$$D_S = \{(u, i, t, t') \mid (u, i, t) \in S \land (u, i, t') \notin S\} \tag{6}$$

The objective of model training is to maximize the margin between the scores $\widehat{y}_{u,i,t}$ of the positive triple (u, i, t) and negative triple (u, i, t'), so we adopt the Bayesian Personalized Ranking (BPR) optimization criterion [36] to learn model parameters $\Theta = \{\mathbf{U}, \mathbf{I}, \mathbf{T}, \beta, \mathbf{c}\}$, and build the objective function of HPTR as:

$$\mathcal{L}^{HPTR} = \min_{\Theta} \sum_{(u,i,t,t') \in D_S} -\ln \sigma \left(\widehat{y}_{u,i,t,t'} \right) + \lambda_\Theta \|\Theta\|_F^2 \tag{7}$$

4.2   Optimization

As the Poincaré ball is a Riemannian manifold with constant negative curvature, the parameters lies in the ball should be updated by Riemannian gradient, so the Riemannian stochastic gradient descent(RSGD) [2] has been applied to optimize most of Poincaré embedding based models [10, 29, 42, 46]. In terms of HPTR, the model parameters consist of embedded parameters(i.e., $\{\mathbf{U}, \mathbf{I}, \mathbf{T}\} \in \mathcal{B}^d$) and non-embedded parameters(i.e., $\{\beta, \mathbf{c}\} \notin \mathcal{B}^d$), therefore, we update the two types of parameters together via tangent space optimization [5,6] to avoid using two corresponding optimizers.

$$\boldsymbol{\theta}_{t+1} = \text{proj} \left(\boldsymbol{\theta}_t - \eta \frac{\left(1 - \|\boldsymbol{\theta}_t\|^2\right)^2}{4} \nabla_E \right) \tag{8}$$

We recall that a d-dimensional hyperbolic space is a Riemannian manifold $\mathcal{M}$ with a constant negative curvature $-c(c > 0)$, the tangent space $\mathcal{T}_x\mathcal{M}$ at point x on $\mathcal{M}$ is a d-dimensional flat space that best approximates $\mathcal{M}$ around x, and the elements $\mathbf{v}$ of $\mathcal{T}_x\mathcal{M}$ are referred to as tangent vectors. In our work, We define all of parameters in the tangent space of the Poincaré ball so that we can learn them via powerful Euclidean optimizers(e.g., Adam). In particular, for the calculation of $\mathbf{U_u} + \mathbf{I_i}$ in Eq. 5, we also do it in tangent space beforehand.

When it comes to calculate the hyperbolic distance d_B, we use the exponential map $\exp_x^c(\mathbf{v})$ to recover the corresponding parameters (map $\mathbf{v}$ of tangent space back to $\mathcal{B}^d$) as following:

$$\exp_x^c(\mathbf{v}) = x \oplus_c \left(\tanh \left(\sqrt{c}\frac{\lambda_x^c \|\mathbf{v}\|}{2} \right) \frac{\mathbf{v}}{\sqrt{c}\|\mathbf{v}\|} \right) \tag{9}$$

where $\oplus^c$ denotes the *Möbius* addition operator [11] that provides an analogue to Euclidean addition for hyperbolic space.

5   Experiments and Analysis

In this section, we conduct several groups of experiments on two real-world datasets to compare the performance of HPTR with other state-of-the-art PTR models.

5.1   Datasets and Evaluation Metrics

In our experiments, we choose two public available datasets,[1] i.e., LastFM and ML10M, to evaluate the performance of all compared methods. Similar to [35, 37], we preprocess each dataset to obtain their corresponding p-core, which is the largest subset where each user, item, and tag has to occur at least p times. In our experiments, every datasets is 5-core or 10-core. The general statistics of datasets are summarized in Table 1.

To evaluate the recommendation performance of all compared methods, we adopt the $leave - one - out$ evaluation protocol, which has been widely used in related studies. Specifically, for each pair (u, i), we select the last triple (u, i, t) according to the tagging time and remove it from S to S_{test}. The remaining observed $user - item - tag$ triples are the training set $S_{train} = S - S_{test}$. Similar

---

[1] https://grouplens.org/datasets/hetrec-2011/.

Table 1. Description of datasets.

Dataset	Users	Items	Tags	Tag assignments	Density
LastFM-core5	1348	6927	2132	162047	8.13989E−06
LastFM-core10	966	3870	1024	133945	3.49896E−05
ML10M-core5	990	3247	2566	61688	7.47871E−06
ML10M-core10	469	1524	1017	37414	5.14701E−05

to the item recommendation problem, the PTR provides a top-N highest ranked list of tags for a pair (u, i). We employ two typical ranking metrics to measure the performance of all compared methods, i.e., Precision@N and Recall@N. For both metrics, we set $N = 3, 5, 10$.

5.2 Experiment Settings

We choose the following traditional tag recommendation algorithms as baselines:

- PITF: PITF [37] explicitly models the pairwise interactions among users, items and tags by inner product, it is a strong competitor in the field of personalized tag recommendation.
- NLTF [9] is a non-linear tensor factorization model, which enhances PITF by exploiting the Gaussian radial basis function to capture the nonlinear interaction relations among users, items and tags.
- ABNT: ABNT [49] utilizes the multi-layer perception to model the nonlinearities of the interactions among users, items and tags.

We empirically set the parameters of compared models according to their corresponding literature in order to recover their optimal performance: the dimension of embedding d is set to 64. In addition, for the ABNT model, the number of hidden layers is set to 2. For HPTR, we set curvature $-c = -1$, and map tangent space $\mathcal{T}_x\mathcal{M}$ at origin point $x = 0$ on the Poincaré ball, the dimension of embedding d is tuned amongst $\{8, 16, 32, 64\}$. We choose Adam [18] as the optimizer for all involved models.

5.3 Performance Comparison

Tables 2, 3, 4 and 5 present the tag recommendation quality of all compared method on the selected four datasets.

From the inspection of Tables 2, 3, 4 and 5, we have observed the following experimental results:

- PITF is superior to NTLF and ABNT with respect to all evaluation metrics, which indicates that, for the adoption of Euclidean matching functions, the traditional inner product might be a better choice for the PTR model to measure the semantic relevance between entities.

Table 2. Recommendation quality comparisons on LastFM-core5.

Model	PITF	NLTF	ABNT	HPTR-8	HPTR-16	HPTR-32	HPTR-64
Precision@3	0.2127	0.1949	0.1563	0.1944	0.2425	0.2634	**0.2813**
Precision@5	0.1789	0.1678	0.1353	0.1590	0.1925	0.2112	**0.2229**
Precision@10	0.1274	0.1191	0.1018	0.1095	0.1275	0.1413	**0.1424**
Recall@3	0.2571	0.2275	0.1569	0.2477	0.3127	0.3432	**0.3600**
Recall@5	0.3479	0.3239	0.2194	0.3201	0.3857	0.4061	**0.4382**
Recall@10	0.4814	0.4523	0.3298	0.4065	0.4722	0.5129	**0.5191**

Table 3. Recommendation quality comparisons on LastFM-core10.

Model	PITF	NLTF	ABNT	HPTR-8	HPTR-16	HPTR-32	HPTR-64
Precision@3	0.2513	0.2443	0.1641	0.2343	0.2797	0.3094	**0.3162**
Precision@5	0.2088	0.2062	0.1367	0.1875	0.2153	0.2431	**0.2555**
Precision@10	0.1458	0.1249	0.0941	0.1261	0.1486	0.1631	**0.1675**
Recall@3	0.3204	0.2845	0.1579	0.2961	0.3552	0.3894	**0.4001**
Recall@5	0.4158	0.4017	0.2190	0.3740	0.4346	0.4762	**0.4914**
Recall@10	0.5654	0.5541	0.3034	0.4704	0.5262	0.5607	**0.5696**

Table 4. Recommendation quality comparisons on ML10M-core5.

Model	PITF	NLTF	ABNT	HPTR-8	HPTR-16	HPTR-32	HPTR-64
Precision@3	0.1398	0.1323	0.0822	0.0693	0.1218	0.1610	**0.1711**
Precision@5	0.1021	0.0972	0.0628	0.0523	0.0881	0.1116	**0.1206**
Precision@10	0.0641	0.0596	0.0400	0.0359	0.0547	0.0661	**0.0707**
Recall@3	0.3208	0.2974	0.2089	0.1574	0.2723	0.3480	**0.3717**
Recall@5	0.3910	0.3560	0.2538	0.1949	0.3206	0.3938	**0.4200**
Recall@10	0.4623	0.4270	0.3039	0.2622	0.3840	0.4507	**0.4766**

Table 5. Recommendation quality comparisons on ML10M-core10.

Model	PITF	NLTF	ABNT	HPTR-8	HPTR-16	HPTR-32	HPTR-64
Precision@3	0.1699	0.1436	0.0896	0.1259	0.1761	0.2075	**0.2189**
Precision@5	0.1173	0.1143	0.0759	0.0928	0.1251	0.1454	**0.1484**
Precision@10	0.0744	0.0714	0.0501	0.0590	0.0730	0.0806	**0.0825**
Recall@3	0.3770	0.3388	0.2210	0.3051	0.4078	0.4742	**0.4970**
Recall@5	0.4523	0.4334	0.3015	0.3605	0.4743	0.5395	**0.5486**
Recall@10	0.5205	0.5341	0.3858	0.4406	0.5358	0.5877	**0.5960**

- Compared against the most competitive PITF with respect to the same embedding dimension $d = 64$, HPTR improves the Precision@3 of PITF by 32.2%, 25.8%, 22.3%, and 28.9% on Lastfm-core5, Lastfm-core10, ML10M-core5, and ML10M-core10, respectively. For Recall@3, the improvements of HPTR over PITF are 40.0%, 24.9%, 15.9%, and 31.8% on the above four datasets, respectively. It implies that Hyperbolic space can provide a more suitable inductive-bias for modeling interactive relationships in tagging information.
- For each compared method, its recommendation performance is better on the 10-core datasets than that on the corresponding 5-core datasets. This observation indicates that HPTR may result in better recommendation performance on datasets with higher density.
- The HPTR has achieved the best recommendation performance over all evaluation metrics. Notably, its performance with the lower hyperbolic embedding dimension: on LastFM-core5, LastFM-core10, and ML10M-core10 with $d = 16$, and on ML10M-core5 with $d = 32$, HPTR has met or exceeded the performance of other models with higher Euclidean embedding dimension ($d = 64$), which confirms the prominent advantage of hyperbolic space in the representation capacity.

5.4 Parameters Sensitivity Analysis

In our proposed HPTR, the dimension of embeddings d is the most important parameter since it controls the capacity of the whole model, so we conduct additional experiments to study the sensitivity of d to the performance of PTR by tuning it within $\{8, 16, 32, 64, 128, 256, 512, 1024\}$. We also take Precision@N and Recall@N, and set $N = 3, 5, 10$, to give an insight of impact on performance with respect to parameter d.

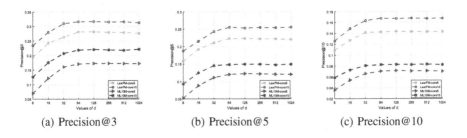

| (a) Precision@3 | (b) Precision@5 | (c) Precision@10 |

Fig. 2. Impact of d on Precision@N

From Fig. 2 and Fig. 3, we can observe that the curves of Precision@N and Recall@N show similar changing trends on four datasets. In the beginning, the values of Precision@N and Recall@N both increase stably with the growth of d, when d exceeds 64, most of Precision@N and Recall@N are no longer in an uptrend, which indicates that merely increasing the dimension is not conducive

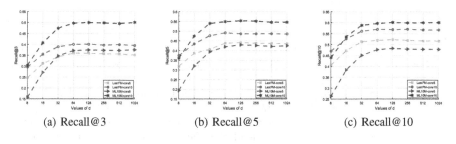

(a) Recall@3   (b) Recall@5   (c) Recall@10

Fig. 3. Impact of d on Recall@N

to sustained improvement of recommendation. One possible reason is that HPTR will obtain sufficient learning ability when d reaches a certain threshold, and after that, the higher dimension of embeddings will lead to the over-fitting problem.

6   Conclusion

In this paper, we take the initiative to conduct PTR task in hyperbolic space. Being aware of the common structural properties of hyperbolic space and tagging information, we propose the HPTR model to pursue better recommendation performances. HPTR is committed to learning optimal representation of users, items, and tags in the hyperbolic space. By embedding training data into Poincaré ball and adopting hyperbolic distance as the matching function, our HPTR is expected to exhibit better performance in the recommendation for its capability of capturing the hierarchical relationship within the training data. We conduct extensive experiments to verify the validity of HPTR, and the experimental results have shown its superiority over state-of-the-art baselines. Furthermore, HPTR with a lower hyperbolic embedding dimension can outperform baselines with higher Euclidean embeddings in the experiments.

It should be noted that, in our work, we have not made the most of side information within the tagging information and have overlooked the graph structure among entities. Recently, there appears some work [7,13,41,45,48,50] that has been made to improve the performance of recommendation models by inducing side information and graph learning methods. This will motivate us to consider whether we can utilize such information and methods to further boost the performance of PTR. Furthermore, since hyperbolic space can be described by more than one isometric model, our future work will mainly focus on exploring the performance of the other isometric models in PTR tasks.

Acknowledgements. The authors would like to acknowledge the support for this work from the Natural Science Foundation of the Higher Education Institutions of Jiangsu Province (17KJB520028), Tongda College of Nanjing University of Posts and Telecommunications (XK203XZ21001), and Future Network Scientific Research Fund Project (FNSRFP-2021-YB-54)

References

1. Balažević, I., Allen, C., Hospedales, T.: Multi-relational poincaré graph embeddings. In: Proceedings of the 33rd International Conference on Neural Information Processing Systems, pp. 4463–4473 (2019)
2. Bonnabel, S.: Stochastic gradient descent on Riemannian manifolds. IEEE Trans. Autom. Control **58**(9), 2217–2229 (2013)
3. Cai, Y., Zhang, M., Luo, D., Ding, C., Chakravarthy, S.: Low-order tensor decompositions for social tagging recommendation. In: Proceedings of the 4th ACM International Conference on Web Search and Data Mining, pp. 695–704 (2011)
4. Chamberlain, B.P., Hardwick, S.R., Wardrope, D.R., Dzogang, F., Daolio, F., Vargas, S.: Scalable hyperbolic recommender systems. arXiv preprint arXiv:1902.08648 (2019)
5. Chami, I., Wolf, A., Juan, D.C., Sala, F., Ravi, S., Ré, C.: Low-dimensional hyperbolic knowledge graph embeddings. In: Proceedings of the 58th Annual Meeting of the Association for Computational Linguistics, pp. 6901–6914 (2020)
6. Chami, I., Ying, R., Re, C., Leskovec, J.: Hyperbolic graph convolutional neural networks. In: Proceedings of the 33rd International Conference on Neural Information Processing Systems, pp. 4868–4879 (2019)
7. Chen, L., Wu, L., Hong, R., Zhang, K., Wang, M.: Revisiting graph based collaborative filtering: a linear residual graph convolutional network approach. In: Proceedings of the AAAI Conference on Artificial Intelligence, vol. 34, pp. 27–34 (2020)
8. Dhingra, B., Shallue, C., Norouzi, M., Dai, A., Dahl, G.: Embedding text in hyperbolic spaces. In: Proceedings of the 12th Workshop on Graph-Based Methods for Natural Language Processing, TextGraphs-12, pp. 59–69 (2018)
9. Fang, X., Pan, R., Cao, G., He, X., Dai, W.: Personalized tag recommendation through nonlinear tensor factorization using gaussian kernel. In: Proceedings of the 29th AAAI Conference on Artificial Intelligence, pp. 439–445 (2015)
10. Feng, S., Tran, L.V., Cong, G., Chen, L., Li, J., Li, F.: HME: a hyperbolic metric embedding approach for next-poi recommendation. In: Proceedings of the 43rd International ACM SIGIR Conference on Research and Development in Information Retrieval, pp. 1429–1438 (2020)
11. Ganea, O.E., Bécigneul, G., Hofmann, T.: Hyperbolic neural networks. In: Proceedings of the 32nd International Conference on Neural Information Processing Systems, pp. 5350–5360 (2018)
12. Halpin, H., Robu, V., Shepherd, H.: The complex dynamics of collaborative tagging. In: Proceedings of the 16th International Conference on World Wide Web, pp. 211–220 (2007)
13. Han, J.: Adaptive deep modeling of users and items using side information for recommendation. IEEE Trans. Neural Netw. Learn. Syst. **31**(3), 737–748 (2019)
14. Hotho, A., Jäschke, R., Schmitz, C., Stumme, G.: Information retrieval in folksonomies: search and ranking. In: Sure, Y., Domingue, J. (eds.) ESWC 2006. LNCS, vol. 4011, pp. 411–426. Springer, Heidelberg (2006). https://doi.org/10.1007/11762256_31
15. Jäschke, R., Marinho, L., Hotho, A., Schmidt-Thieme, L., Stumme, G.: Tag recommendations in folksonomies. In: Kok, J.N., Koronacki, J., Lopez de Mantaras, R., Matwin, S., Mladenič, D., Skowron, A. (eds.) PKDD 2007. LNCS (LNAI), vol. 4702, pp. 506–514. Springer, Heidelberg (2007). https://doi.org/10.1007/978-3-540-74976-9_52

16. Jiang, F., et al.: Personalized tag recommendation via adversarial learning. In: Developments of Artificial Intelligence Technologies in Computation and Robotics: Proceedings of the 14th International FLINS Conference, FLINS 2020, pp. 923–930. World Scientific (2020)

17. Khrulkov, V., Mirvakhabova, L., Ustinova, E., Oseledets, I., Lempitsky, V.: Hyperbolic image embeddings. In: Proceedings of the IEEE/CVF Conference on Computer Vision and Pattern Recognition, pp. 6418–6428 (2020)

18. Kingma, D.P., Ba, J.: Adam: a method for stochastic optimization. arXiv preprint arXiv:1412.6980 (2014)

19. Kolyvakis, P., Kalousis, A., Kiritsis, D.: HyperKG: hyperbolic knowledge graph embeddings for knowledge base completion. arXiv preprint arXiv:1908.04895 (2019)

20. Krestel, R., Fankhauser, P., Nejdl, W.: Latent dirichlet allocation for tag recommendation. In: Proceedings of the 3rd ACM Conference on Recommender Systems, pp. 61–68 (2009)

21. Krioukov, D., Papadopoulos, F., Kitsak, M., Vahdat, A., Boguná, M.: Hyperbolic geometry of complex networks. Phys. Rev. E **82**(3), 036106 (2010)

22. Li, A., Yang, B., Chen, H., Xu, G.: Hyperbolic neural collaborative recommender. arXiv preprint arXiv:2104.07414 (2021)

23. Li, X., Guo, L., Zhao, Y.E.: Tag-based social interest discovery. In: Proceedings of the 17th International Conference on World Wide Web, pp. 675–684 (2008)

24. Liu, Q., Nickel, M., Kiela, D.: Hyperbolic graph neural networks. In: Proceedings of the 33rd International Conference on Neural Information Processing Systems, pp. 8230–8241 (2019)

25. López, F., Heinzerling, B., Strube, M.: Fine-grained entity typing in hyperbolic space. In: Proceedings of the 4th Workshop on Representation Learning for NLP, RepL4NLP-2019, pp. 169–180 (2019)

26. Ma, C., Ma, L., Zhang, Y., Wu, H., Liu, X., Coates, M.: Knowledge-enhanced Top-K recommendation in poincaré ball. In: Proceedings of the AAAI Conference on Artificial Intelligence, vol. 35, pp. 4285–4293 (2021)

27. Nguyen, H.T.H., Wistuba, M., Grabocka, J., Drumond, L.R., Schmidt-Thieme, L.: Personalized deep learning for tag recommendation. In: Kim, J., Shim, K., Cao, L., Lee, J.-G., Lin, X., Moon, Y.-S. (eds.) PAKDD 2017. LNCS (LNAI), vol. 10234, pp. 186–197. Springer, Cham (2017). https://doi.org/10.1007/978-3-319-57454-7_15

28. Nguyen, H.T.H., Wistuba, M., Schmidt-Thieme, L.: Personalized tag recommendation for images using deep transfer learning. In: Ceci, M., Hollmén, J., Todorovski, L., Vens, C., Džeroski, S. (eds.) ECML PKDD 2017. LNCS (LNAI), vol. 10535, pp. 705–720. Springer, Cham (2017). https://doi.org/10.1007/978-3-319-71246-8_43

29. Nickel, M., Kiela, D.: Poincaré embeddings for learning hierarchical representations. In: Proceedings of the 31st International Conference on Neural Information Processing Systems, pp. 6341–6350 (2017)

30. Nickel, M., Kiela, D.: Learning continuous hierarchies in the Lorentz model of hyperbolic geometry. In: International Conference on Machine Learning, pp. 3779–3788. PMLR (2018)

31. Pan, Y., Huo, Y., Tang, J., Zeng, Y., Chen, B.: Exploiting relational tag expansion for dynamic user profile in a tag-aware ranking recommender system. Inf. Sci. **545**, 448–464 (2021)

32. Peng, W., Shi, J., Xia, Z., Zhao, G.: Mix dimension in poincaré geometry for 3d skeleton-based action recognition. In: Proceedings of the 28th ACM International Conference on Multimedia, pp. 1432–1440 (2020)

33. Peng, W., Varanka, T., Mostafa, A., Shi, H., Zhao, G.: Hyperbolic deep neural networks: a survey. IEEE Trans. Pattern Anal. Mach. Intel. (2021)

34. Rader, E., Wash, R.: Influences on tag choices in del.icio.us. In: Proceedings of the 2008 ACM Conference on Computer Supported Cooperative Work, pp. 239–248 (2008)

35. Rendle, S., Balby Marinho, L., Nanopoulos, A., Schmidt-Thieme, L.: Learning optimal ranking with tensor factorization for tag recommendation. In: Proceedings of the 15th ACM SIGKDD International Conference on Knowledge Discovery and Data Mining, pp. 727–736 (2009)

36. Rendle, S., Freudenthaler, C., Gantner, Z., Schmidt-Thieme, L.: BPR: Bayesian personalized ranking from implicit feedback. In: Proceedings of the 25th Conference on Uncertainty in Artificial Intelligence, pp. 452–461 (2009)

37. Rendle, S., Schmidt-Thieme, L.: Pairwise interaction tensor factorization for personalized tag recommendation. In: Proceedings of the 3rd ACM International Conference on Web Search and Data Mining, pp. 81–90 (2010)

38. Sun, B., Zhu, Y., Xiao, Y., Xiao, R., Wei, Y.: Automatic question tagging with deep neural networks. IEEE Trans. Learn. Technol. **12**(1), 29–43 (2018)

39. Symeonidis, P., Nanopoulos, A., Manolopoulos, Y.: Tag recommendations based on tensor dimensionality reduction. In: Proceedings of the 2008 ACM Conference on Recommender Systems, pp. 43–50 (2008)

40. Tang, S., et al.: An integral tag recommendation model for textual content. In: Proceedings of the AAAI Conference on Artificial Intelligence, vol. 33, pp. 5109–5116 (2019)

41. Vasile, F., Smirnova, E., Conneau, A.: Meta-Prod2vec: product embeddings using side-information for recommendation. In: Proceedings of the 10th ACM Conference on Recommender Systems, pp. 225–232 (2016)

42. Vinh, T.D.Q., Tay, Y., Zhang, S., Cong, G., Li, X.L.: Hyperbolic recommender systems. arXiv preprint arXiv:1809.01703 (2018)

43. Vinh Tran, L., Tay, Y., Zhang, S., Cong, G., Li, X.: HyperML: a boosting metric learning approach in hyperbolic space for recommender systems. In: Proceedings of the 13th International Conference on Web Search and Data Mining, pp. 609–617 (2020)

44. Wang, K., Jin, Y., Wang, H., Peng, H., Wang, X.: Personalized time-aware tag recommendation. In: 32nd AAAI Conference on Artificial Intelligence (2018)

45. Wang, Q., Yin, H., Wang, H., Nguyen, Q.V.H., Huang, Z., Cui, L.: Enhancing collaborative filtering with generative augmentation. In: Proceedings of the 25th ACM SIGKDD International Conference on Knowledge Discovery & Data Mining, pp. 548–556 (2019)

46. Wang, X., Zhang, Y., Shi, C.: Hyperbolic heterogeneous information network embedding. In: Proceedings of the AAAI Conference on Artificial Intelligence, vol. 33, pp. 5337–5344 (2019)

47. Wu, Y., Yao, Y., Xu, F., Tong, H., Lu, J.: Tag2Word: using tags to generate words for content based tag recommendation. In: Proceedings of the 25th ACM International on Conference on Information and Knowledge Management, pp. 2287–2292 (2016)

48. Ying, R., He, R., Chen, K., Eksombatchai, P., Hamilton, W.L., Leskovec, J.: Graph convolutional neural networks for web-scale recommender systems. In: Proceedings of the 24th ACM SIGKDD International Conference on Knowledge Discovery & Data Mining, pp. 974–983 (2018)

49. Yuan, J., Jin, Y., Liu, W., Wang, X.: Attention-based neural tag recommendation. In: Li, G., Yang, J., Gama, J., Natwichai, J., Tong, Y. (eds.) DASFAA 2019. LNCS, vol. 11447, pp. 350–365. Springer, Cham (2019). https://doi.org/10.1007/978-3-030-18579-4_21

50. Zhang, S., Yin, H., Chen, T., Hung, Q.V.N., Huang, Z., Cui, L.: GCN-based user representation learning for unifying robust recommendation and fraudster detection. In: Proceedings of the 43rd International ACM SIGIR Conference on Research and Development in Information Retrieval, pp. 689–698 (2020)

Diffusion-Based Graph Contrastive Learning for Recommendation with Implicit Feedback

Lingzi Zhang[1,2,4], Yong Liu[2,3], Xin Zhou[2], Chunyan Miao[1,2,3(✉)],
Guoxin Wang[4], and Haihong Tang[4]

[1] School of Computer Science and Engineering,
Nanyang Technological University (NTU), Singapore, Singapore
lingzi001@e.ntu.edu.sg
[2] Alibaba-NTU Singapore Joint Research Institute, NTU, Singapore, Singapore
{stephenliu,xin.zhou}@ntu.edu.sg
[3] Joint NTU-UBC Research Centre of Excellence in Active Living for the Elderly,
NTU, Singapore, Singapore
ASCYMiao@ntu.edu.sg
[4] Alibaba Group, Hangzhou, China
{xiaogong.wgx,piaoxue}@taobao.com

Abstract. Recent studies on self-supervised learning with graph-based recommendation models have achieved outstanding performance. They usually introduce auxiliary learning tasks that maximize the mutual information between representations of the original graph and its augmented views. However, most of these models adopt random dropout to construct the additional graph view, failing to differentiate the importance of edges. The insufficiency of these methods in capturing structural properties of the user-item interaction graph leads to suboptimal recommendation performance. In this paper, we propose a Graph Diffusion Contrastive Learning (GDCL) framework for recommendation to close this gap. Specifically, we perform graph diffusion on the user-item interaction graph. Then, the diffusion graph is encoded to preserve its heterogeneity by learning a dedicated representation for every type of relations. A symmetric contrastive learning objective is used to contrast local node representations of the diffusion graph with those of the user-item interaction graph for learning better user and item representations. Extensive experiments on real datasets demonstrate that GDCL consistently outperforms state-of-the-art recommendation methods.

1 Introduction

The recent development of deep learning motivates the emergence of various neural network-based recommendation models [14,28]. One representative group of studies leverages graph neural networks (GNN). With the advantage of exploiting the high-order connectivity of users and items through iterative message propagation, GNN-based models have shown prominent performance in recommendation tasks. However, these methods still suffer from insufficient supervision

A. Bhattacharya et al. (Eds.): DASFAA 2022, LNCS 13246, pp. 232–247, 2022.
https://doi.org/10.1007/978-3-031-00126-0_15

signals from observed sparse user-item interactions, unbalanced distributions of user (or item) node degree, and unavoidable noises (*e.g.*, accidentally clicking) from users' implicit feedback for most datasets [26].

Self-supervised learning (SSL) has recently become a promising technique to address these limitations for a more robust and generalized representation learning of GNNs [9]. Researchers have begun to actively explore SSL in GNN-based recommendation models [3,18,24,26]. Most of them concentrate on contrastive learning, which maximizes the agreement between representations of augmented views and the original graph. The key part of contrastive learning is how to design the graph augmentation strategy. The most commonly endorsed approach is randomly removing edges, either on the whole graph [26] or on an h-hop enclosed subgraph [3,18]. However, this method cannot differentiate the importance of edges. Ideally, we want to obtain an augmented graph by keeping informative edges while removing irrelevant or noisy edges.

To build a more effective augmentation view of the user-item interaction graph for contrastive learning, we propose to replace the random dropout with graph diffusion, which reconciles the spatial message passing by smoothing out the neighborhood over the graph [12]. The diffusion process defines a weighted graph exchanged from the original unweighted graph. The weights measure the relative importance of edges based on the graph structure. Hence, we can utilize these importance scores to design different sparsification methods so as to preserve a more effective neighborhood for each node in the diffusion graph. Meanwhile, most GNN models incorporate high-order information by increasing the number of convolutional layers. The iterative expansion not only includes more nodes for learning better representations but also introduces more noisy edges, which can deteriorate the recommendation performance. Graph diffusion does not have this constraint as it can extend connections in the graph from one-hop to multi-hops. We retrieve the information of a larger neighborhood by one layer of aggregation rather than stacking multiple layers of GNNs. Hence, the problem of having noises in real graphs can be further mitigated.

In this paper, we propose a simple yet effective Graph Diffusion Contrastive Learning (GDCL) framework for item recommendation with users' implicit feedback. Existing graph diffusion algorithms focus on homogeneous graphs, which include a single node type. In GDCL, we first devise the diffusion algorithm to consider different types of nodes in a heterogeneous graph (*i.e.*, user-item interaction graph). The derived diffusion graph consists of multiple types of relations between nodes. Specifically, besides the user-item relation, user-user, and item-item relations are also introduced. These heterogeneous relations are not fully captured if we simply apply graph convolutional network (GCN) based encoders to treat them uniformly as previous works [7,26]. Hence, we extend GCN to model the heterogeneity of the diffusion graph by maintaining a dedicated representation for every type of relations, and then fuse them using a mean aggregator. To train the overall model end-to-end, we leverage a multi-task training paradigm to jointly optimize the recommendation task and self-supervised learning task. For the recommendation task, previous SSL-based recommendation models [3,18,24,26] only rely on the user-item graph for user preference prediction. Differing from these works, we utilize representations learned from the auxiliary view (*i.e.*, diffusion graph) together with the user-item graph to improve the representation learning for users and items. For the self-supervised task, we contrast node representations encoded from two views by a symmetric mutual information maximization objective function. Experimental results on four publicly available datasets demonstrate that the proposed GDCL model consistently outperforms state-of-the-art recommendation methods.

2 Related Work

2.1 Graph-Based Recommendation

As the data for the recommendation task naturally forms a user-item bipartite graph, graph-based models have been widely applied for solving recommendation problems [15–17]. Earlier works are mainly based on random walk approaches. For example, [8] utilizes a generalized random walk with restart model to learn user preferences. Recent works leverage graph embedding approaches, which encode graph structure information into node embeddings. For instance, [5] proposes to use a heterogeneous information network to model objects and relations in the recommender system, and extract aspect-level latent factors for users and items. Most recent studies target at GNN-based approaches. [25] proposes the neural graph collaborative filtering (NGCF) model, which explicitly models higher-order connectivity in the user-item graph and effectively injects collaborative signals into the embedding learning process. [7] proposes to simplify NGCF by removing unnecessary components of GCN, *e.g.*, non-linear activation and feature transformation.

2.2 Graph Contrastive Learning

Contrastive learning has recently been extended to process graph data. Some works maximize the mutual information between local node and global graph representations. [6] employs graph diffusion to generate an additional structural view of a sampled graph, and uses a discriminator to contrast node representations from one view with the graph representation of another view and vice versa. Other studies emphasize the contrast between local node representations. [27] designs four types of graph augmentations to incorporate various priors, including node dropping, edge perturbation, attribute masking, and subgraph sampling. The contrastive loss is designed to distinguish representations of the same node from those of other nodes under different views.

As the research of self-supervised learning is still in its infancy, only a few works explore it for the recommendation task [3,18,24,26]. For example, [18] proposes to contrast the graph embedding with sampled subgraph embedding via a global-local infoMax objective. [26] is the most relevant existing work, which introduces a self-supervised graph learning (SGL) framework. SGL generates two views for each node via node dropping, edge dropping, and sampling subgraphs.

An auxiliary task is constructed to maximize the consistency between different views of the same node and enforce discrepancies with other nodes. Although these works achieve promising results, they simply construct the auxiliary graph view via stochastic sampling of the user-item interaction graph without differentiating the relative importance of connected edges. In addition, most works aggregate higher-order nodes through layer-to-layer propagation where the detrimental effect brought by noisy edges is inevitable. Instead, we employ graph diffusion for contrastive learning, which alleviates the problem of noisy edges and leads to improved performance.

3   The Proposed Recommendation Model

This work focuses on top-K item recommendations based on users' implicit feedback. Let $\mathcal{U} = \{u_1, u_2, \cdots, u_m\}$ and $\mathcal{V} = \{v_1, v_2, \cdots, v_n\}$ be the set of users and items, where m and n denote the number of users and items, respectively. Observed users' interactions with items can be described by a bipartite graph $\mathcal{G} = \{\mathcal{U}, \mathcal{V}, \mathcal{E}\}$, where $\mathcal{E}$ denotes the set of edges that represents all interactions between users and items. If a user u has interacted with an item v, we build an edge e_{uv} between the corresponding user node and item node. For each node t in $\mathcal{G}$, we denote the set of its first-hop neighbors in $\mathcal{G}$ by $\mathcal{N}_t$. We denote the adjacency matrix of $\mathcal{G}$ by $\mathbf{A} \in \mathbb{R}^{(m+n) \times (m+n)}$. In this work, we consider the user-item interaction graph as an undirected graph. Then, we set $A_{ij} = A_{ji} = 1$, if there exists an edge connecting two nodes t_i and t_j in $\mathcal{G}$; Otherwise, we set $A_{ij} = A_{ji} = 0$. The degree matrix $\mathbf{D} \in \mathbb{R}^{(m+n) \times (m+n)}$ of $\mathcal{G}$ is a diagonal matrix, where the diagonal element $D_{ii} = \sum_{j=1}^{m+n} A_{ij}$. Given the interaction graph $\mathcal{G}$ between users and items, our objective is to predict the probability that a user u would like to interact with a candidate item v, and then recommend K top-ranked candidate items to u. Figure 1 shows the overall framework of the proposed GDCL recommendation model. It consists of three main components: 1) diffusion-based graph augmentation, 2) graph encoders, and 3) self-supervised contrastive learning. Next, we introduce details of each component.

3.1   Diffusion-Based Graph Augmentation

Graph Diffusion Approximation. For a homogeneous graph, its diffusion matrix can be formulated as [6,12],

$$\mathbf{\Pi} = \sum_{k=0}^{\infty} \theta_k \mathbf{T}^k, \tag{1}$$

where $\mathbf{T}$ is the generalized transition matrix that can be defined by the symmetrically normalized adjacency matrix as $\mathbf{T} = \mathbf{D}^{-1/2} \mathbf{A} \mathbf{D}^{1/2}$, and θ_k is weighting coefficients of $\mathbf{T}^k$. In Eq. (1), $\mathbf{\Pi}$ has closed-form solutions when considering two special cases, *i.e.*, Personalized PageRank (PPR) [20] and heat kernel [23].

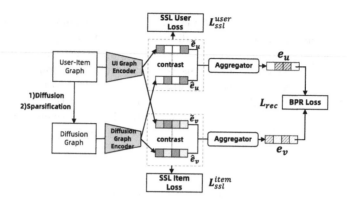

Fig. 1. The overall framework of the proposed GDCL recommendation model.

However, these two solutions involve matrix inverse and matrix exponential operations, which are computationally infeasible for processing large-scale graphs.

Inspired by [2], we resort to an efficient algorithm to approximate the diffusion matrix $\mathbf{\Pi}$. Specifically, we choose the following adaption of personalized PageRank [11] to instantiate $\mathbf{\Pi}$,

$$\mathbf{\Pi}_{PPR} = \alpha(\mathbf{I}_n - (1 - \alpha)\mathbf{D}^{-1}\mathbf{A})^{-1}, \tag{2}$$

where $\mathbf{I}_n$ is the identity matrix, and α is the teleport probability in a random walk. A larger teleport probability means higher chances to return back to the root node, and hence we can preserve more locality information. A smaller teleport probability allows us to reach out to a larger neighborhood. We can tune α to adjust the size of the neighborhood for different datasets. In [2], the push-flow algorithm [1] is used to obtain a sparse approximation for each row of $\mathbf{\Pi}_{PPR}$. Note that every row of this approximated matrix can be pre-computed in parallel using a distributed batch data processing pipeline. However, this algorithm is designed for homogeneous graphs, where all nodes are of the same type. We adapt it for the user-item interaction graph $\mathcal{G}$ which contains two different types of nodes. Precisely, we define teleport probabilities α_u and α_v for user nodes and item nodes, respectively. For the i-th row π_i in the approximated graph diffusion matrix $\mathbf{\Pi}_{PPR}$, we first determine the node type and then use its corresponding teleport probability for computation. As a result, we can control the amount of information being diffused to the neighborhood for different types of nodes. It has practical benefits for a recommender system. For example, real-world datasets usually have more long-tail items than long-tail users. In such cases, decreasing the teleport probability of items can help find more possibly relevant users and items, and hence learn better representations. Details of the diffusion matrix approximation algorithm are summarized in Algorithm 1.

Diffusion Matrix Sparsification. The diffusion matrix $\mathbf{\Pi}_{PPR}$ is a dense matrix, where each element reflects the relevance between two nodes based on

Algorithm 1. Approximate graph diffusion for the user-item interaction graph

1: **Inputs:** Graph $\mathcal{G}$, teleport probability α_u and α_v, target node $z \in \mathcal{U} \cup \mathcal{V}$, max.residual ϵ, node degree vector d
2: Initialize the estimate-vector $\pi = 0$ and the residual-vector r. r is a zero vector with only position z equal to α_u if z is a user node or α_v if z is an item node.
3: **while** $\exists\ t\ s.t.\ (r_t > \alpha_u \epsilon d_t$ if $t \in \mathcal{U})$ or $(r_t > \alpha_v \epsilon d_t$ if $t \in \mathcal{V})$ **do**
4:    **if** $t \in \mathcal{U}$ **then**
5:       $\alpha = \alpha_u$
6:    **else if** $t \in \mathcal{V}$ **then**
7:       $\alpha = \alpha_v$
8:    **end if**
9:    # approximate personalized PageRank [2]
10:    $\pi_t \mathrel{+}= r_t$
11:    $m = (1 - \alpha) \cdot r_t / d_t$
12:    **for** $u \in \mathcal{N}_t$ **do**
13:       $r_u \mathrel{+}= m$
14:    **end for**
15:    $r_t = 0$
16: **end while**
17: **Return** π

the graph structure. As suggested by [1], weights of personalized PageRank vectors are usually concentrated in a small subset of nodes. Thus, we can truncate small weights but still obtain a good approximation. In this work, we propose following three methods to sparsify the diffusion matrix,

– **Topk**: For each row π_i in $\mathbf{\Pi}_{PPR}$, we retain k entries with highest weights from user nodes and item nodes respectively, and set other entries to zero. Namely, we will keep $2k$ "neighbors" for each node in the diffusion graph.
– **Topk-rand**: We firstly select k user nodes and k item nodes with highest weights in each row π_i, following the Topk method. Then, we randomly drop selected nodes with a dropout ratio ρ (an adjustable hyper-parameter).
– **Topk-prob**: This method is similar to the Topk-rand method. The only difference is that, in this method, the probability for dropping a selected node is proportional to its weight in the weight vector π_i.

As the Topk sparsification method is deterministic, we train the GDCL model with the fixed sparsified diffusion matrix. The other two sparsification methods Topk-rand and Topk-prob are stochastic. When using Topk-rand and Topk-prob methods to train the proposed model, we perform diffusion matrix sparsification at each training epoch. We denote the sparsified diffusion graph by $\widetilde{\mathcal{G}}$.

3.2 Graph Encoders

The original user-item interaction graph $\mathcal{G}$ and the diffusion graph $\widetilde{\mathcal{G}}$ are treated as two congruent views for contrastive learning. Two different graph encoders are designed to capture the information in $\mathcal{G}$ and $\widetilde{\mathcal{G}}$. To begin with, we randomly initialize embeddings of a user u and an item v by $\mathbf{e}_u^{(0)}$ and $\mathbf{e}_v^{(0)}$ respectively, which are shared by both graph encoders.

User-Item Interaction Graph Encoder. In this work, we use LightGCN [7] to encode the user-item interaction graph $\mathcal{G}$. As shown in [7], LightGCN only keeps the neighborhood aggregation of GCN when propagating node embeddings. At the ℓ-th layer, the graph convolution operation on a user node u is defined as,

$$\widetilde{\mathbf{e}}_u^{(\ell)} = \sum_{v \in \mathcal{N}_u} \frac{1}{\sqrt{|\mathcal{N}_u|}\sqrt{|\mathcal{N}_v|}} \widetilde{\mathbf{e}}_v^{(\ell-1)}, \tag{3}$$

where $\widetilde{\mathbf{e}}_u^{(\ell)}$ denotes the representation of the user u at the ℓ-th layer, and $|\cdot|$ denotes the cardinality of a set. Then, we sum multiple representations from different layers to obtain the final user embedding derived from the user-item interaction graph as follows,

$$\widetilde{\mathbf{e}}_u = \mathbf{e}_u^{(0)} + \widetilde{\mathbf{e}}_u^{(1)} + \cdots + \widetilde{\mathbf{e}}_u^{(L)}, \tag{4}$$

where L denotes the number of GCN layers. Similarly, we can obtain the final representation $\widetilde{\mathbf{e}}_v$ of an item v based on the interaction graph $\mathcal{G}$.

Diffusion Graph Encoder. The diffusion graph $\widetilde{\mathcal{G}}$ built in Sect. 3.1 is derived from the user-item interaction graph, which includes two types of nodes. Thus, three types of relations are established in $\widetilde{\mathcal{G}}$, including user-item relation, user-user relation, and item-item relation. To effectively capture the heterogeneous structure of $\widetilde{\mathcal{G}}$, we propose a graph diffusion encoder to model these three relations separately and maintain a dedicated representation for every type of relations. As shown in Fig. 2, each node's adjacent nodes can either be users or items in $\widetilde{\mathcal{G}}$. We segregate them into two groups based on the node type and perform feature aggregation within each group. Hence, two representations are generated for every user (or item). One is derived from the user-item relation and the other one is derived from the user-user (or item-item) relation.

Traditional message passing neural networks, *e.g.*, GCN, aggregate their first-hop neighbors at each layer. Higher-order neighbors are only accessible through layer-to-layer propagation. The graph diffusion process breaks this constraint by creating connections to multi-hop nodes, and hence aggregation can be performed on a larger neighborhood without stacking multiple GNN layers [12]. Specifically, the derivation of user embeddings are as follows,

$$\widehat{\mathbf{e}}_{u1} = \sum_{u' \in \mathcal{N}_u^{(1)}} \pi_u(u')\mathbf{e}_{u'}^{(0)}, \quad \widehat{\mathbf{e}}_{u2} = \sum_{v \in \mathcal{N}_u^{(2)}} \pi_u(v)\mathbf{e}_v^{(0)}, \tag{5}$$

where $\widehat{\mathbf{e}}_{u1}$ and $\widehat{\mathbf{e}}_{u2}$ are embeddings of u obtained from the user-user diffusion graph and user-item diffusion graph, respectively. $\mathcal{N}_u^{(1)}$ and $\mathcal{N}_u^{(2)}$ denote sets of first-hop nodes of the user u on the user-user diffusion graph and user-item

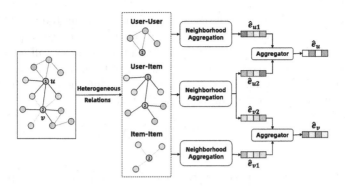

Fig. 2. The diffusion graph encoder. Blue and yellow colors denote user and item nodes. Edges in black, blue, and yellow denote user-item, user-user, and item-item relations. (Color figure online)

diffusion graph, respectively. $\boldsymbol{\pi}_u$ represents the diffusion vector of u, and $\pi_u(v)$ is the weight of the node v in the diffusion vector $\boldsymbol{\pi}_u$. Then, we combine these two embeddings by the MEAN operation,

$$\widehat{\mathbf{e}}_u = \mathrm{MEAN}\big(\widehat{\mathbf{e}}_{u1}, \widehat{\mathbf{e}}_{u2}\big), \tag{6}$$

which takes the average of user embeddings from the user-user diffusion graph and the user-item diffusion graph to be the final user representation $\widehat{\mathbf{e}}_u$ learned from the diffusion graph. Similarly, we can obtain the representation $\widehat{\mathbf{e}}_v$ of an item v learned from the diffusion graph.

3.3 Self-supervised Contrastive Leaning

After obtaining user and item representations from two graph views, we maximize the consistency between positive pairs and the inconsistency between negative pairs via contrastive learning. A positive pair is formed by representations of the same node from different views, whereas representations of different nodes form negative pairs. The GNN model is enforced to distinguish positive pairs from negative pairs.

Following [29], we use the InfoNCE_Sym mutual information estimator to define contrastive learning losses in this work. Specifically, negative pairs are categorized into inter-view negative pairs and intra-view negative pairs. An inter-view negative pair consists of two nodes, which are generated from different views. If both nodes of a negative pair are from the same view, they are called an intra-view negative pair. Let $\mathcal{B}$ denote the current mini-batch of observed user-item interaction pairs, $\mathcal{U}_\mathcal{B}$ and $\mathcal{V}_\mathcal{B}$ denote the list of users and items in $\mathcal{B}$, respectively. For a user $u \in \mathcal{U}_\mathcal{B}$, its contrastive loss is defined as,

$$\ell(\widetilde{\mathbf{e}}_u, \widehat{\mathbf{e}}_u) = \log\big(\frac{e^{cos(\widetilde{\mathbf{e}}_u, \widehat{\mathbf{e}}_u)/\tau}}{e^{cos(\widetilde{\mathbf{e}}_u, \widehat{\mathbf{e}}_u)/\tau} + \sum_{u' \neq u} e^{cos(\widetilde{\mathbf{e}}_u, \widehat{\mathbf{e}}_{u'})/\tau} + \sum_{u' \neq u} e^{cos(\widetilde{\mathbf{e}}_u, \widetilde{\mathbf{e}}_{u'})/\tau}}\big), \tag{7}$$

where $cos(\cdot)$ is the cosine similarity function and τ is the temperature hyper-parameter. Three terms in the denominator are the score of the positive pair, total scores of all inter-view negative pairs, and total scores of all intra-negative pairs. Note that we only consider negative instances from the current mini-batch. As two graph views are interchangeable, we can define the contrastive learning loss from the user perspective in a symmetric way as follows,

$$L_{ssl}^{user} = -\frac{1}{2|\mathcal{U_B}|} \sum_{u \in \mathcal{U_B}} [\ell(\widetilde{\mathbf{e}}_u, \widehat{\mathbf{e}}_u) + \ell(\widehat{\mathbf{e}}_u, \widetilde{\mathbf{e}}_u)]. \tag{8}$$

Similarly, we can define the contrastive learning loss from item perspective as,

$$L_{ssl}^{item} = -\frac{1}{2|\mathcal{V_B}|} \sum_{v \in \mathcal{V_B}} [\ell(\widetilde{\mathbf{e}}_v, \widehat{\mathbf{e}}_v) + \ell(\widehat{\mathbf{e}}_v, \widetilde{\mathbf{e}}_v)]. \tag{9}$$

3.4 Multi-task Training

To learn model parameters, we leverage multi-task training paradigm to jointly optimize the recommendation task and self-supervised task. For the recommendation task, we first aggregate representations from the user-item graph encoder and the diffusion graph encoder by element-wise summation,

$$\mathbf{e}_u = \widetilde{\mathbf{e}}_u + \widehat{\mathbf{e}}_u, \quad \mathbf{e}_v = \widetilde{\mathbf{e}}_v + \widehat{\mathbf{e}}_v, \tag{10}$$

where $\mathbf{e}_u$ and $\mathbf{e}_v$ are final representations of user u and item v. Then, the prediction of u's preference on the item v can be defined as,

$$\hat{y}_{uv} = \mathbf{e}_u^\top \mathbf{e}_v. \tag{11}$$

The Bayesian Pairwise Ranking (BPR) loss [22] is employed as the loss function for the recommendation task. Specifically, for each user-item pair $(u, v) \in \mathcal{B}$, we randomly sample an item w that has no interaction with u to form a triplet (u, v, w). Then, the loss function is defined as follows,

$$L_{rec} = \sum_{(u,v,w) \in \mathcal{D_B}} -\log \sigma (\hat{y}_{uv} - \hat{y}_{uw}), \tag{12}$$

where $\mathcal{D_B}$ denotes the set of triplets for all observed user-item pairs in $\mathcal{B}$. The total loss for learning the proposed recommendation model is as follows,

$$L = L_{rec} + \lambda_1 (L_{ssl}^{user} + L_{ssl}^{item}) + \lambda_2 \|\Theta\|_2^2, \tag{13}$$

where λ_1 balances the primary recommendation loss and auxiliary self-supervised loss, λ_2 controls the L2 regularization, and Θ denotes model parameters.

4 Experiments

4.1 Experimental Settings

Experimental Datasets. To examine the capability of the proposed model, we conduct experiments on three datasets: Amazon review [19], MovieLens-1M,[1] and Yelp2018.[2] For Amazon review, we choose subsets "Video Games" and "Arts" for evaluation. On each dataset, we only keep users and items that have at least 5 interactions. The statistics of datasets are shown in Table 1.

Table 1. Statistics of the experimental datasets.

Dataset	# Users	# Items	# Interactions	Sparsity
Amazon-Games	45,950	16,171	363,590	99.95%
Amazon-Arts	42,137	20,942	317,109	99.96%
MovieLens-1M	5,400	3,662	904,616	95.43%
Yelp2018	34,518	22,918	380,632	99.95%

Baseline Methods. We compare the GDCL model with following baselines: 1) **BPR** [22]: This is a classical collaborative filtering method based on matrix factorization; 2) **LightGCN** [7]: This is the state-of-the-art GNN based recommendation model. It simplifies the design of GCNs for collaborative filtering by discarding feature transformation and non-linear activation; 3) **BUIR** [13]: This work proposes a recommendation framework that does not require negative sampling. It utilizes two distinct encoder networks to learn from each other; 4) **BiGI** [3]: This work generates the graph-level representation and contrasts it with sampled edges' representations via a global-local infoMax objective. 5) **SGL** [26]: This work exploits self-supervised learning on the user-item graph. They devise three types of data augmentation operations on graph structure from different aspects to construct auxiliary contrastive task.

Evaluation Protocols. For each dataset, we sort observed user-item interactions in chronological order based on interaction timestamps. Then, first 80% of interactions are chosen for training. The next 10% and the last 10% of interactions are used for validation and testing. The performance of a recommendation model is measured by three widely used ranking-based metrics: Recall@K, NDCG@K, and Hit Ratio@K (denoted by R@K, N@K, and HR@K), where K is set to 5, 10, and 20. For each metric, we compute the performance of each user in the testing data and report the averaged performance over all users.

---

[1] https://grouplens.org/datasets/movielens/1m/.
[2] https://www.yelp.com/dataset.

Implementation Details. We implement GDCL by PyTorch [21]. Model parameters are initialized by the Xavier method [4] and learned by the Adam optimizer [10]. We continue to train the model until the performance is not increased for 50 consecutive epochs. All baselines are trained from scratch for fair comparison. The embedding dimension is fixed to 64, the batch size is set to 2048, and the learning rate is set to 0.001. The number of GCN layers is chosen from $\{1, 2, 3\}$. Empirically, we set teleport probability $\alpha_u = \alpha_v = 0.2$ and max.residual $\epsilon = 0.001$ for all datasets except MovieLens-1M dataset. On MovieLens-1M dataset, α_u and α_v are set to 0.1 and ϵ is set to 0.0001. We tune the temperature τ in contrastive learning in $\{0.1, 0.2\}$. The SSL regularization λ_1 is chosen from $\{0.1, 0.01, 0.001, 0.0001\}$, and the weight decay λ_2 is chosen from $\{1e-2, 1e-3, 1e-4, 1e-5\}$. The dropout ratio ρ in diffusion matrix sparsification is set to 0.1. For each method, we use grid-search to choose optimal hyper-parameters based on its performance on validation data.

4.2 Performance Comparison

Table 2 summarizes the performance of different models. GDCL and SGL outperform purely graph-based models by leveraging self-supervised learning that contrasts node-level representations from different graph views. This result demonstrates the effectiveness of incorporating self-supervised learning into the recommendation task. The proposed GDCL model consistently achieves the best performance on all datasets. Compared with the best baseline method SGL, GDCL contrasts representations learned from the user-item graph with those learned from the diffusion graph. Thus, more useful graph structure information can be preserved, and negative impacts of noisy edges can be reduced. Moreover, GDCL employs representations learned from both graph encoders to enhance user and item representations for better recommendation performance.

On MovieLens-1M, graph-based baseline models are inferior to BPR for most evaluation metrics. Compared with other datasets, MovieLens-1M has a higher density. The training data has enough supervision signals from interactions; thus simple BPR is adequate. On such datasets, GDCL can still achieve the best performance. It can be attributed to the sparsifed diffusion graph which helps to learn informative representations. BiGI performs worst on most datasets, which implies that contrasting the local sampled subgraph representations with global graph representations does not benefit much on the recommendation performance. In most cases, the non-negative sampling method BUIR achieves better results than LightGCN. However, it cannot compete with models (*e.g.*, SGL and GDCL) that use a joint learning framework with self-supervised contrastive loss.

4.3 Ablation Study

To study the impact of each component in GDCL, we consider following variants of GDCL: 1) **GDCL$_{Adj}$**: we replace the diffusion matrix with the adjacency matrix to evaluate the effectiveness of graph diffusion; 2) **GDCL$_{GCN}$**: we replace

Table 2. Overall performance comparison. Best results are in **boldface** and the second best is underlined. "%Improv" refers to the relative improvement of GDCL over the best baseline. * indicates the improvements are statistically significant with $p < 0.05$.

Dataset	Model	R@5	N@5	HR@5	R@10	N@10	HR@10	R@20	N@20	HR@20
Amazon Games	BPR	0.0111	0.0096	0.0256	0.0208	0.0127	0.0440	0.0336	0.0165	0.0672
	LightGCN	0.0157	0.0134	0.0335	0.0269	0.0173	0.0545	0.0429	0.0220	0.0823
	BUIR	0.0148	0.0125	0.0320	0.0265	0.0166	0.0538	0.0456	0.0224	0.0888
	BiGI	0.0134	0.0106	0.027	0.0228	0.0139	0.0462	0.0383	0.0185	0.0728
	SGL	0.0157	0.0130	0.0331	0.0269	0.0170	0.0552	0.0451	0.0225	0.0860
	GDCL	**0.0174***	**0.0137***	**0.0367***	**0.0307***	**0.0184***	**0.0604**	**0.0469***	**0.0234***	**0.0907***
	%Improv	10.82%	2.23%	9.55%	14.12%	6.35%	9.42%	2.85%	4.00%	2.13%
Amazon Arts	BPR	0.0111	0.0092	0.0237	0.0175	0.0115	0.0363	0.0292	0.0151	0.0575
	LightGCN	0.0139	0.0114	0.0288	0.0234	0.0149	0.0463	0.0374	0.0193	0.0733
	BUIR	0.0129	0.0113	0.0268	0.0233	0.0150	0.046	0.0375	0.0195	0.0726
	BiGI	0.0098	0.0083	0.0204	0.0177	0.0113	0.0360	0.0287	0.0148	0.0570
	SGL	0.0141	0.0116	0.0281	0.0241	0.0153	0.0473	0.0381	0.0197	0.0730
	GDCL	**0.0144***	**0.0117***	**0.0291***	**0.0251***	**0.0157***	**0.0499***	**0.0391***	**0.0201***	**0.0765***
	%Improv	2.12%	0.86%	1.04%	4.14%	2.61%	5.49%	2.62%	2.03%	4.36%
MovieLens 1M	BPR	0.0299	**0.2900**	0.6012	0.0512	0.2691	0.6974	0.0859	0.2505	0.7837
	LightGCN	0.0294	0.2820	0.6012	0.0480	0.2621	0.6815	0.0836	0.2442	0.7619
	BUIR	0.0237	0.2673	0.5665	0.0387	0.2483	0.6587	0.0696	0.2322	0.7391
	BiGI	0.0274	0.2532	0.5774	0.0473	0.2441	0.6885	0.0797	0.2310	0.7837
	SGL	0.0286	0.2814	0.5972	0.0512	0.2628	0.6984	0.0848	0.2482	0.7728
	GDCL	**0.0320***	0.2877	**0.6121***	**0.0533***	0.2696	**0.7202***	**0.0872***	0.2540	**0.7917***
	%Improv	7.02%	−0.79%	1.81%	4.10%	0.18%	3.12%	1.51%	1.39%	1.02%
Yelp2018	BPR	0.0161	0.0136	0.0393	0.0290	0.0182	0.0689	0.0499	0.0248	0.1121
	LightGCN	0.0193	0.0159	0.0465	0.0328	0.0208	0.0771	0.0543	0.0275	0.1202
	BUIR	0.0196	0.0167	0.0473	0.0338	0.0217	0.0783	0.0572	0.0291	0.1271
	BiGI	0.0143	0.0113	0.0335	0.0237	0.0147	0.0556	0.0415	0.0204	0.0960
	SGL	0.0191	0.0164	0.0465	0.0346	0.0220	0.0801	**0.0608**	0.0301	0.1323
	GDCL	**0.0213***	**0.0184***	**0.0510***	**0.0356***	**0.0235***	**0.0840***	0.0606	**0.0313***	**0.1340**
	%Improv	8.67%	10.17%	7.82%	2.89%	6.81%	4.86%	−0.32%	3.98%	1.28%

the diffusion graph encoder with one layer of LightGCN to evaluate the effectiveness of modeling heterogeneous relations in graph diffusion. It performs neighborhood aggregation without differentiating various relations between nodes; 3) **GDCL$_{w/o\,Diff}$**: we predict the user preferences only using adjacency matrix to evaluate the effectiveness of incorporating graph diffusion embeddings during inference. The performance achieved by different GDCL variants is summarized in Table 3. We find that the combination of all components consistently improves the model performance on all datasets. Three GDCL variants perform differently on different datasets. Graph diffusion benefits more on Amazon-Arts and MovieLens-1M, as GDCL$_{GCN}$ and GDCL$_{w/oDiff}$ perform better than GDCL$_{Adj}$. On Amazon-Games, simply adopting graph diffusion for contrastive learning (GDCL$_{GCN}$) is worse than using adjacency matrix (GDCL$_{Adj}$). With the designed diffusion graph encoder and its generated embeddings for recommendation prediction, the performance is improved significantly.

We also conduct experiments to investigate impacts of different matrix sparsification methods. From Table 4, Topk-rand and Topk-prob settings outperform Topk. We conjecture that introducing randomness during training enhances

Table 3. Performance achieved by different variants of GDCL.

Method	Amazon-Games		Amazon-Arts		MovieLens-1M	
	Recall@10	NDCG@10	Recall@10	NDCG@10	Recall@10	NDCG@10
$GDCL_{Adj}$	0.0293	**0.0188**	0.0237	0.0151	0.0514	0.2669
$GDCL_{GCN}$	0.0284	0.0176	0.0242	0.0154	0.0513	0.2653
$GDCL_{w/oDiff}$	0.0285	0.0181	0.0245	0.0154	0.0519	**0.2708**
GDCL	**0.0307**	0.0184	**0.0251**	**0.0157**	**0.0533**	0.2696

the generalization capability of the model. Topk-rand works better on Amazon-Games and MovieLens-1M, while Topk-prob is better on Amazon-Arts.

Table 4. Impacts of different sparsification methods

Method	Amazon-Games		Amazon-Arts		MovieLens-1M	
	Recall@10	NDCG@10	Recall@10	NDCG@10	Recall@10	NDCG@10
Topk	0.0293	0.0181	0.0242	0.0153	0.0480	0.2652
Topk-rand	**0.0307**	**0.0184**	0.0241	0.0151	**0.0533**	**0.2696**
Topk-prob	0.0294	0.0178	**0.0251**	**0.0157**	0.0485	0.2656

To study the model performance on different users' popularity groups, we split testing users into two groups. Group 1 has 25% of users who have the least number of interactions, while Group 2 has the remaining 75% of users with more interactions. We compare three models, including LightGCN, SGL, and GDCL. As shown in Fig. 3, GDCL achieves more significant improvement for users with more interaction data. We speculate that utilizing graph diffusion with sparsification can help alleviate the problem of noisy edges.

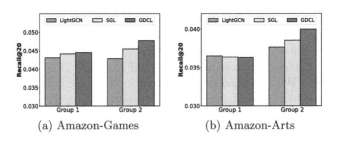

(a) Amazon-Games   (b) Amazon-Arts

Fig. 3. Performance comparison over different user groups.

4.4   Hyper-parameter Study

To study impacts of the GCN depth of user-item interaction graph encoder, we vary the number of layers L in $\{1, 2, 3\}$. As shown in Table 5, GDCL outperforms

Table 5. Performance of SGL and GDCL with different numbers of GCN layers

Method	Amazon-Games		Amazon-Arts		Yelp2018	
	Recall@10	NDCG@10	Recall@10	NDCG@10	Recall@10	NDCG@10
SGL (L=1)	0.0262	0.0170	0.0228	0.0146	0.0327	0.0207
GDCL (L=1)	**0.0285**	**0.0180**	**0.0232**	**0.0148**	**0.0343**	**0.0224**
SGL (L=2)	0.0269	0.0170	**0.0242**	0.0150	0.0338	0.0216
GDCL (L=2)	**0.0290**	**0.0180**	0.0241	**0.0153**	**0.0356**	**0.0235**
SGL (L=3)	0.0252	0.0162	0.0241	0.0153	**0.0346**	0.0220
GDCL (L=3)	**0.0307**	**0.0184**	**0.0251**	**0.0157**	0.0345	**0.0225**

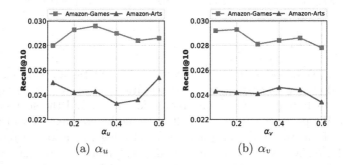

(a) α_u (b) α_v

Fig. 4. Performance trends of GDCL w.r.t α_u/α_v.

SGL on a majority of metrics. Furthermore, GDCL with one GCN layer achieves comparable performance with SGL with three GCN layers.

The teleport probability α is introduced to decide how much nearest neighborhood information is kept. If α is closer to 1, it is more likely to return to the starting node thus more information of first hop nodes is retained. Otherwise, if α is closer to 0, more weights will be diffused to multi-hop nodes. We define teleport probabilities α_u and α_v for user and item nodes. Figure 4 shows the model performance versus different settings of α_u and α_v in $\{0.1, 0.2, ..., 0.6\}$. The plots reveal that recommendation performance is sensitive to α. For Amazon-Games, taking smaller $\alpha = 0.2$ for both users and items can achieve higher accuracy. The performance drops with larger α. For Amazon-Arts, the performance is the best with $\alpha_u = 0.6$. As for α_v, the performance remains relatively steady when it is smaller than 0.6. When α_v is increased to 0.6, the performance declines.

We also study how the number of selected top user and item nodes in diffusion graph sparsification affects the model performance. We fix *Topk* of items to 32 and vary *Topk* of users in $\{8, 16, 32, 64\}$, and vice versa. The results are shown in Fig. 5. For Amazon-Games, the most effective neighborhood size is 32. The performance will decline when the size is too large or too small. For Amazon-Arts, including a larger number of users and items will gain better performance.

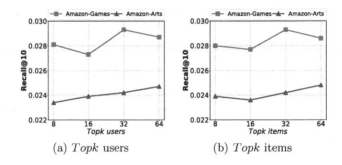

Fig. 5. Performance trends of GDCL w.r.t topk users/items.

5 Conclusion

In this paper, we introduce a self-supervised auxiliary task for learning user and item embeddings by contrasting two representations generated from graph structural views, including the user-item interaction graph and the sparsified diffusion graph. We propose a simple and effective model architecture to learn node embeddings from the diffusion graph by identifying its various heterogeneous relations. Extensive experiments on four datasets demonstrate that our model has achieved competitive performance compared with state-of-the-art baselines.

Acknowledgments. This research is supported by Alibaba Group through Alibaba Innovative Research (AIR) Program and Alibaba-NTU Singapore Joint Research Institute (JRI), Nanyang Technological University, Singapore.

References

1. Andersen, R., Chung, F., Lang, K.: Local graph partitioning using PageRank vectors. In: FOCS 2006, pp. 475–486 (2006)
2. Bojchevski, A., et al.: Scaling graph neural networks with approximate PageRank. In: KDD 2020, pp. 2464–2473 (2020)
3. Cao, J., Lin, X., Guo, S., Liu, L., Liu, T., Wang, B.: Bipartite graph embedding via mutual information maximization. In: WSDM 2021, pp. 635–643 (2021)
4. Glorot, X., Bengio, Y.: Understanding the difficulty of training deep feedforward neural networks. In: AISTATS 2010, pp. 249–256 (2010)
5. Han, X., Shi, C., Wang, S., Yu, P.S., Song, L.: Aspect-level deep collaborative filtering via heterogeneous information networks. In: IJCAI 2018, pp. 3393–3399 (2018)
6. Hassani, K., Khasahmadi, A.H.: Contrastive multi-view representation learning on graphs. In: ICML 2020, pp. 4116–4126 (2020)
7. He, X., Deng, K., Wang, X., Li, Y., Zhang, Y., Wang, M.: LightGCN: simplifying and powering graph convolution network for recommendation. In: SIGIR 2020 (2020)
8. Jiang, Z., Liu, H., Fu, B., Wu, Z., Zhang, T.: Recommendation in heterogeneous information networks based on generalized random walk model and Bayesian personalized ranking. In: WSDM 2018, pp. 288–296 (2018)

9. Jin, W., et al.: Self-supervised learning on graphs: deep insights and new direction. arXiv preprint arXiv:2006.10141 (2020)
10. Kingma, D.P., Ba, J.: Adam: a method for stochastic optimization. arXiv preprint arXiv:1412.6980 (2014)
11. Klicpera, J., Bojchevski, A., Günnemann, S.: Predict then propagate: graph neural networks meet personalized PageRank. arXiv:1810.05997 (2018)
12. Klicpera, J., Weißenberger, S., Günnemann, S.: Diffusion improves graph learning. In: NeurIPS 2019 (2019)
13. Lee, D., Kang, S., Ju, H., Park, C., Yu, H.: Bootstrapping user and item representations for one-class collaborative filtering. In: SIGIR 2021, pp. 317–326 (2021)
14. Lei, C., et al.: SEMI: a sequential multi-modal information transfer network for e-commerce micro-video recommendations. In: KDD 2021, pp. 3161–3171 (2021)
15. Liu, Y., et al.: Pre-training graph transformer with multimodal side information for recommendation. In: MM 2021, pp. 2853–2861 (2021)
16. Liu, Y., Yang, S., Xu, Y., Miao, C., Wu, M., Zhang, J.: Contextualized graph attention network for recommendation with item knowledge graph. TKDE (2021)
17. Liu, Y., Yang, S., Zhang, Y., Miao, C., Nie, Z., Zhang, J.: Learning hierarchical review graph representations for recommendation. TKDE (2021)
18. Liu, Z., Ma, Y., Ouyang, Y., Xiong, Z.: Contrastive learning for recommender system. arXiv:2101.01317 (2021)
19. Ni, J., Li, J., McAuley, J.: Justifying recommendations using distantly-labeled reviews and fine-grained aspects. In: EMNLP-IJCNLP 2019, pp. 188–197 (2019)
20. Page, L., Brin, S., Motwani, R., Winograd, T.: The PageRank citation ranking: Bringing order to the web. Technical report 1999-66 (1999)
21. Paszke, A., Gross, S., Massa, F., et al.: PyTorch: an imperative style, high-performance deep learning library. In: Advances in Neural Information Processing Systems, vol. 32, pp. 8024–8035 (2019)
22. Rendle, S., Freudenthaler, C., Gantner, Z., Schmidt-Thieme, L.: BPR: Bayesian personalized ranking from implicit feedback. arXiv:1205.2618 (2012)
23. Kondor, R.I., Lafferty, J.D.: Diffusion kernels on graphs and other discrete input spaces. In: ICML 2002 (2002)
24. Tang, H., Zhao, G., Wu, Y., Qian, X.: Multi-sample based contrastive loss for top-k recommendation. arXiv:2109.00217 (2021)
25. Wang, X., He, X., Cao, Y., Liu, M., Chua, T.S.: KGAT: knowledge graph attention network for recommendation. In: KDD 2019, pp. 950–958 (2019)
26. Wu, J., et al.: Self-supervised graph learning for recommendation. arXiv:2010.10783 (2021)
27. You, Y., Chen, T., Sui, Y., Chen, T., Wang, Z., Shen, Y.: Graph contrastive learning with augmentations. In: NeurIPS 2020, vol. 33, pp. 5812–5823 (2020)
28. Zhang, Y., Li, B., Liu, Y., Miao, C.: Initialization matters: regularizing manifold-informed initialization for neural recommendation systems. In: KDD 2021, pp. 2263–2273 (2021)
29. Zhu, Y., Xu, Y., Yu, F., Liu, Q., Wu, S., Wang, L.: Graph contrastive learning with adaptive augmentation. In: WWW 2021, pp. 2069–2080 (2021)

Multi-behavior Recommendation with Two-Level Graph Attentional Networks

Yunhe Wei[1], Huifang Ma[1,2,3(✉)], Yike Wang[1], Zhixin Li[2], and Liang Chang[3]

[1] College of Computer Science and Engineering, Northwest Normal University,
Lanzhou 730070, Gansu, China
mahuifang@yeah.net
[2] Guangxi Key Lab of Multi-source Information Mining and Security, Guangxi
Normal University, Guilin 541004, Guangxi, China
[3] Guangxi Key Lab of Trusted Software, Guilin University of Electronic Technology,
Guilin 541004, Guangxi, China

Abstract. Multi-behavior recommendation learns accurate embeddings of users and items with multiple types of interactions. Although existing multi-behavior recommendation methods have been proven effective, the following two insights are often neglected. First, the semantic strength of different types of behaviors is ignored. Second, these methods only consider the static preferences of users and the static feature of items. These limitations motivate us to propose a novel recommendation model AMR (**A**ttentional **M**ulti-behavior **R**ecommendation) in this paper, which captures hidden relations in user-item interaction network by constructing multi-relation graphs with different behavior types. Specifically, the node-level attention aims to learn the importance of neighbors under specific behavior, while the behavior-level attention is able to learn the semantic strength of different behaviors. In addition, we learn the dynamic feature of target users and target items by modeling the dependency relation between them. The results show that our model achieves great improvement for recommendation accuracy compared with other state-of-the-art recommendation methods.

Keywords: Multi-behavior recommendation · Node-level attention ·
Behavior-level attention · Dynamic feature

1 Introduction

In recent years, Graph Neural Networks (GNNs) [3,5], which can naturally integrate node information and topological structure, have been demonstrated to be useful in learning on graph data. Hence, GNNs techniques have been applied in recommendation systems [4,10]. Although these methods have shown promising results, they only model singular type of user-item interaction behavior, which makes them insufficient to extract complex preference information of users from multiple types of behaviors. In order to make effective use of multi-relational

A. Bhattacharya et al. (Eds.): DASFAA 2022, LNCS 13246, pp. 248–255, 2022.
https://doi.org/10.1007/978-3-031-00126-0_16

interaction data, several efforts [1, 2, 6, 8] on multi-behavior recommendation systems have been made, showing the superior performance in terms of learning users' preference. We categorize the existing multi-behavior recommendation methods related to our work into two types. The first category is to extend the sampling method by using multiple types of interaction data [7, 8]. For example, MC-BPR [7] utilizes multiple types of feedback data to pairwise learning-to-rank algorithm. The key to the method is to map different feedback channels to different levels that reflect the contribution that each type of feedback can have in the training phase. This is further extended by BPRH[8], designing a more complex training-pair sampling method based on multi-behavior data. The second category is to use a unified framework to model multi-relational interaction data in multi-behavior recommendations coherently[2, 6]. For instance, MBGCN [6] constructs a unified model to represent multi-behavior data, which captures behavior semantics through item propagation layer.

Although these multi-behavior recommendation models have been proven effective, there still exist some issues as follows. First, existing models ignore the semantic strength of multiple types of behaviors. Second, existing models statically model user and item characteristics in the form of fixed weights or fixed constraints.

To address the challenges above, we propose a novel Attentional Multi-behavior Recommendation method. AMR can effectively model and fuse multiple types of behavior factors and item characteristics from different behaviors. First, node-level attention is computed to aggregate the neighbors of users and items. Then, we utilize different weights to aggregate multiple types of behaviors based on behavior-level attention. In particular, we learn the dynamic preferences and characteristics of users and items by capturing the dependencies between the target user and the target item. In particular, the users dynamic preference don't change with time, but with the change of target items.

2   Preliminaries

We aim to recommend items for users on the target behavior by utilizing multiple types of behaviors. Let $U = \{u_1, u_2, \cdots, u_m\}$ and $V = \{v_1, v_2, \cdots v_n\}$ are the sets of users and items respectively, where m is the number of users, and n is the number of items.

User-Item Multi-behavior Interaction Graph G^l. The interaction data between users and items are defined as L interaction bipartite graphs $\{G^l = (U, V, E^l)\}_{l=1}^L$, where L is the number of behavior types. We assume that the L-th behavior is the target behavior and the others are auxiliary behaviors. Each edge in E represents an interaction between a user u and an item v.

User-User Multi-relation Graph G_u^l and Item-Item Multi-relation Graph G_v^l. To capture the user's and item's multiple types of relations, we define graphs $\{G_u^l = (U, E_u^l)\}_{l=1}^L$ and $\{G_v^l = (V, E_v^l)\}_{l=1}^L$ to represent multiple dependencies across users and items, respectively. In G_u^l, edge $e_{u,u'}^l$ linked between user u and u' with their meta relations. Similarly, we can define the item-item multi-relation graph G_v^l based on the meta relations of items. In particular, the user-user meta relation is defined as $u_i - v_j - u_{i'}$, similarly, the item-item meta relation is defined as $v_j - u_i - v_{j'}$.

3   Methodology

In this section, we show the framework of our proposed model AMR. The overall framework of our proposed architecture is depicted in Fig. 1.

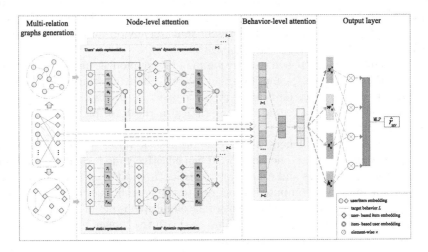

Fig. 1. The overall framework of the proposed AMR. The model requires user-item interaction graph as raw input. Firstly, user and item multi-relation graphs are captured via meta relations. Secondly, we use node-level attention and behavior-level attention to aggregate neighbor information and behavior information. Finally, we utilize MLP to predict the target user preference score for the target item.

3.1   Embedding Layer

The users' initial embedding and the items' initial embedding can be expressed by two embedding matrix U and V, where the i-th row of U and j-th row of V is the embedding of user i and item j. We use $Y_{n \times d}$ to represent the users' static preference matrix, where y_i denotes the static embedding vector for user i. In addition, we define another matrix $Z_{n \times d}$ to represent item-based users' dynamic embedding, where z_i denotes the dynamic embedding vector for user i. Similar to user presentation, the static and dynamic characteristic embedding matrices of items are denoted as $P_{m \times d}$ and $Q_{m \times d}$, where the p_j and the q_j represent the static and dynamic embedding of item j respectively.

$$Y = W_1 U; Z = W_2 U; P = W_1 V; Q = W_2 V \tag{1}$$

3.2   Attention Based Graph Convolution Layer

User Static Preference Learning. For each type of behavior, we have x_u as static representation of user u via node-level attention. We generate user static embedding by aggregating neighbor nodes in user-user multi-relation graph.

$$x_u = \sigma\left(\sum_{u' \in N_u} \alpha_{uu'} y_{u'}\right) \tag{2}$$

where σ is an activation function, *i.e.*, LeakyReLU, and x_u is the updated representation of user u, which incorporates user impact by using the attention weights $\alpha'_{uu'}$ obtained from G_u^l. The importance of user pair (u, u') can be formulated as follows:

$$\alpha'_{uu'} = \sigma(a_y^T[W y_u \| W y_{u'}]) \tag{3}$$

where $\|$ represents the concatenation operation and W is weight matrix. To make coefficients easily comparable across different nodes, we normalize them across all choices of u' using the softmax function:

$$\begin{aligned}
\alpha_{uu'} = softmax_{u'}(\alpha'_{uu'}) &= \frac{\exp(\alpha'_{uu'})}{\sum_{k \in N_u} \exp(\alpha'_{uk})} \\
&= \frac{\exp(\sigma(a_y^T[W y_u \| W y_{u'}]))}{\sum_{k \in N_u} \exp(\sigma(a_y^T[W y_u \| W y_{u'}]))}
\end{aligned} \tag{4}$$

After the node-level attention aggregation for each type of behavior, we generate L static embeddings for each user $\{x_u^1, x_u^2, \cdots x_u^L\}$.

To address the challenge of behavior selection and semantic fusion in a user-user multi-relation graph, we propose a novel behavior-level attention to automatically learn the importance of different behaviors to the target behavior.

$$x_u^* = \sigma\left(\sum_{l=1}^L \beta_{Ll} x_u^l\right) \tag{5}$$

where σ is activation function, and x_u^* is the representation of user u aggregating multiple behaviors, which incorporates behavior impact by using the behavior-level attention weights β'_{Ll}. The importance of auxiliary behavior l to target behavior L can be formulated as follows:

$$\beta'_{Ll} = \sigma(b_y^T[W x_u^L \| W x_u^l]) \tag{6}$$

After obtaining the importance of different behaviors, we normalize them to get the weight coefficient β_{Ll} via softmax function:

$$\beta_{Ll} = softmax_l(\beta'_{Ll}) = \frac{\exp(\beta'_{Ll})}{\sum_{l'=1}^L \exp(\beta'_{Ll'})} \tag{7}$$

User Dynamic Preference Learning. Through the embedding layer, we set $R_V(u)$ as items that user u has interacted with on the target behavior. Then we let each item in $R_V(u)$ interact with the target item v^+.

We define the item-based user embedding, which depends on target item v^+ with max pooling to select the most dominating features for D dimensions:

$$m_{u(v^+)d} = \max_{v \in R_V(u)} \{q_{v^+d} \cdot q_{vd}\} \qquad \forall d = 1, \dots D \tag{8}$$

where $m_{u(v^+)d}$, p_{v^+d}, and q_{vd} are the d-th feature of $\boldsymbol{m}_{u(v^+)}$, $\boldsymbol{q}_{v^+}$, $\boldsymbol{q}_v$ respectively. In order to define the user dynamic preference factor $\boldsymbol{w}_{u(v^+)}$, we proceed to incorporate the multi-relation information from user-user graph:

$$\boldsymbol{w}_{u(v^+)} = \sigma\Big(\sum_{u' \in N_u} \eta_{uu'} \boldsymbol{m}_{u'(v^+)} \Big) \tag{9}$$

$$\eta'_{uu'} = \sigma(\boldsymbol{a}_q^T [\boldsymbol{W}\boldsymbol{m}_{u(v^+)} || \boldsymbol{W}\boldsymbol{m}_{u'(v^+)}]) \tag{10}$$

$$\eta_{uu'} = \frac{\exp(\sigma(\boldsymbol{a}_q^T [\boldsymbol{W}\boldsymbol{m}_{u(v^+)} || \boldsymbol{W}\boldsymbol{m}_{u'(v^+)}]))}{\sum_{k \in N_u} \exp(\sigma(\boldsymbol{a}_q^T [\boldsymbol{W}\boldsymbol{m}_{u(v^+)} || \boldsymbol{W}\boldsymbol{m}_{k(v^+)}]))} \tag{11}$$

Note that the above attention weight $\eta'_{uu'}$ depends on the user's history of rated items as well as specific candidate item v^+, which indicates that factor embedding $\boldsymbol{w}_{u(v^+)}$ would change dynamically with different contexts. This design conforms to the intuition of context aware effect of user preference, so we term it as dynamic preference factor.

Through node-level attention, we can get dynamic embedding of users in different behaviors. In order to generate users' unified dynamic preferences, we aggregate the dynamic embedding of users on different behaviors:

$$\boldsymbol{w}^*_{u(i^+)} = \sigma\Big(\sum_{l=1}^{L} \mu_{Ll} \boldsymbol{w}^l_{u(i^+)} \Big) \tag{12}$$

$$\mu'_{Ll} = \sigma(\boldsymbol{b}_q^T [\boldsymbol{W}\boldsymbol{w}^L_{u(i^+)} || \boldsymbol{W}\boldsymbol{w}^l_{u(i^+)}]) \tag{13}$$

$$\mu_{Ll} = softmax_l(\mu_{Ll}) = \frac{\exp(\mu'_{Ll})}{\sum_{l'=1}^{L} \exp(\mu'_{Ll'})} \tag{14}$$

Based on the above process, we can obtain the static and dynamic embedding of each user. Similarly, we can obtain the static embedding $\boldsymbol{h}^*_v$ and dynamic embedding $\boldsymbol{s}^*_v$ of the item. We elaborately describe the user representation learning process here. Because the item representation learning is a dual process, we omit it for brevity.

3.3   Output Layer

We combine four interaction vectors into a unified representation and use a MLP network to implement the recommendation prediction. In particular, the input of MLP is the element-wise product of the user and the item vector.

$$\hat{r}_{uv} = MLP[\boldsymbol{x}_u^* \otimes \boldsymbol{h}_v^*; \quad \boldsymbol{x}_u^* \otimes \boldsymbol{s}_v^*; \quad \boldsymbol{w}_u^* \otimes \boldsymbol{h}_v^*; \quad \boldsymbol{w}_u^* \otimes \boldsymbol{s}_v^* \ ] \tag{15}$$

3.4   Model Training

To learn model parameters, we optimize the pairwise BPR [9] loss, which has been intensively used in recommendation systems.

$$Loss = \sum_{(u,i,j)\in O} -\ln \sigma(y(u,i) - y(u,j)) + \Psi||\Theta||^2 \tag{16}$$

where $O = \{(u,i,j)|(u,i) \in R^+, (u,j) \in R^-\}$ denotes the set of pairwise target behavior training data. R^+ represents the user-item pairs that have interacted under the target behavior. On the contrary, R^- represents user-item pairs that haven't interacted under the target project. σ is the sigmoid function, Θ denotes all trainable parameters and Ψ is the L2 normalization coefficient which controls the strength of the L2 normalization to prevent overfitting.

4   Experiments

4.1   Experimental Settings

Dataset. To evaluate the effectiveness of our method, we conduct extensive experiments on two real-world e-commerce datasets: Taobao and Beibei. For all experiments, we evaluate our model and baselines in terms of *Recall@K* and *NDCG@K*. We compare our AMR with two kinds of representative methods, including one-behavior model and multi-behavior model.

Table 1. Summary of the datasets used in the experiments

Dataset	Users	Items	View	Cart	Purchase
Taobao	48,749	39,493	1,548,126	193,747	259,747
Beibei	21,716	7,977	2,412,586	642,622	304,576

4.2   Overall Performance

The comparison results of our framework and the baselines are summarized in Table 2. In general, we can find that our AMR shows significant improvement compared with all baselines on all *Recall@K* and *NDCG@K* metrics. This indicates that the AMR is sufficient to capture the multi-types behavior between users and items.

Table 2. Recommendation results. Numbers in bold face are the best results for corresponding metrics.

Method	Taobao				Beibei			
	K = 10		K = 20		K = 10		K = 20	
	Recall	NDCG	Recall	NDCG	Recall	NDCG	Recall	NDCG
MF-BPR	0.0265	0.0155	0.0342	0.0179	0.0385	0.0215	0.0621	0.0263
NCF	0.0269	0.0157	0.0356	0.0189	0.0433	0.0242	0.0667	0.0294
NGCF	0.0286	0.0169	0.0373	0.0201	0.0449	0.0249	0.0684	0.0305
GraphSAGE	0.0225	0.0134	0.0289	0.0149	0.0313	0.0181	0.0587	0.0246
MC-BPR	0.0237	0.0143	0.0322	0.0168	0.0364	0.0207	0.0639	0.0274
NMTR	0.0319	0.0182	0.0393	0.0211	0.0397	0.0221	0.0648	0.0278
NGCF-MB	0.0297	0.0175	0.0429	0.0224	0.0426	0.0238	0.0679	0.0304
MBGCN	0.0384	0.0198	0.0516	0.0263	0.0479	0.0268	0.0718	0.0326
AMR	**0.0417**	**0.0211**	**0.0582**	**0.0295**	**0.0508**	**0.0281**	**0.0753**	**0.0341**
Improv	8.6%	6.6%	12.8%	12.2%	6.1%	4.9%	4.9%	4.6%
p-value	3.52e−8	5.83e−5	2.78e−4	6.25e−7	3.95-6	5.48e−7	3.63e−4	7.25e−3

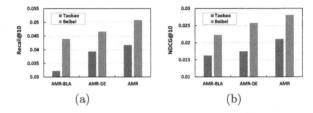

Fig. 2. Effect of behavior-level attention and dynamic embedding.

4.3 Ablation Study

- **Effect of behavior-level attention (AMR-BLA).** AMR-BLA is a variant of AMR, which removes the behavior-level attention mechanism. To examine the effectiveness of behavior-level attention mechanism, we compare the performance of this variant with AMR. It is clear that the performance of AMR degrades without the behavior-level attention mechanism on both datasets. This result demonstrates the benefits of the behavior-level attention mechanism on item aggregation and user aggregation.
- **Effect of dynamic embedding (AMR-DE).** AMR-DE is a variant of AMR, which removes the dynamic embedding mechanism. We can see that without dynamic embedding mechanism, the performance of interactive prediction is deteriorated significantly. It justifies our assumption that dynamic embedding mechanism have informative information that can help to learn user or item latent factors and improve the performance of recommendation.

5   Conclusion

In this work, we focus on the problem of enhancing recommendation based on multi-behavior. We devise a novel multi-behavior recommendation framework AMR, which can model multi-behavior information and dynamic information coherently. To overcome the sparsity and limitations of the target behavior data, we utilize auxiliary behavior to capture user preference in multi-behavior network. We conduct extensive experiments on two real-world datasets, demonstrating that our proposed AMR could boost the multi-behavior recommendation performance over existing methods.

Acknowledgements. This work is supported by the Gansu Natural Science Foundation Project (21JR7RA114), the National Natural Science Foundation of China (61762078, 61363058,U1811264, 61966004) and Northwest Normal University Young Teachers Research Capacity Promotion Plan (NWNU-LKQN2019-2).

References

1. Ding, J., et al.: Improving implicit recommender systems with view data. In: IJCAI, pp. 3343–3349 (2018)
2. Gao, C., et al.: Learning to recommend with multiple cascading behaviors. IEEE Trans. Knowl. Data Eng. **33**, 2588–2601 (2019)
3. Hamilton, W.L., Ying, R., Leskovec, J.: Inductive representation learning on large graphs. arXiv preprint arXiv:1706.02216 (2017)
4. He, X., Liao, L., Zhang, H., Nie, L., Hu, X., Chua, T.S.: Neural collaborative filtering. In: Proceedings of the 26th International Conference on World Wide Web, pp. 173–182 (2017)
5. Jiang, Y., Ma, H., Liu, Y., Li, Z., Chang, L.: Enhancing social recommendation via two-level graph attentional networks. Neurocomputing **449**, 71–84 (2021)
6. Jin, B., Gao, C., He, X., Jin, D., Li, Y.: Multi-behavior recommendation with graph convolutional networks. In: Proceedings of the 43rd International ACM SIGIR Conference on Research and Development in Information Retrieval, pp. 659–668 (2020)
7. Loni, B., Pagano, R., Larson, M., Hanjalic, A.: Bayesian personalized ranking with multi-channel user feedback. In: Proceedings of the 10th ACM Conference on Recommender Systems, pp. 361–364 (2016)
8. Qiu, H., Liu, Y., Guo, G., Sun, Z., Zhang, J., Nguyen, H.T.: BPRH: Bayesian personalized ranking for heterogeneous implicit feedback. Inf. Sci. **453**, 80–98 (2018)
9. Rendle, S., Freudenthaler, C., Gantner, Z., Schmidt-Thieme, L.: BPR: Bayesian personalized ranking from implicit feedback. arXiv preprint arXiv:1205.2618 (2012)
10. Wang, X., He, X., Wang, M., Feng, F., Chua, T.S.: Neural graph collaborative filtering. In: Proceedings of the 42nd International ACM SIGIR Conference on Research and Development in Information Retrieval, pp. 165–174 (2019)

Collaborative Filtering
for Recommendation in Geometric
Algebra

Longcan Wu[1], Daling Wang[1(✉)], Shi Feng[1], Kaisong Song[1,2], Yifei Zhang[1],
and Ge Yu[1]

[1] Northeastern University, Shenyang, China
{wangdaling,fengshi,zhangyifei,yuge}@cse.neu.edu.cn
[2] Alibaba Group, Hangzhou, China
kaisong.sks@alibaba-inc.com

Abstract. At present, recommender system plays an important role in many practical applications. Many recommendation models are based on representation learning, in which users and items are embedded into a low-dimensional vector space, and then historical interactions are used to train the models. We find that almost all of these methods model users and items in real-valued embedding space, which neglect the potential value of other non-real spaces. In this paper, we propose a **Geometric Algebra-based Collaborative Filtering (GACF)** model for recommendation. Specifically, GACF firstly uses multivectors to represent users and items. Then GACF uses geometric product and inner product to model the historical interaction between users and items. By using geometric product, the model prediction can obtain inter-dependencies between components of multivectors, which enable complex interactions between users and items to be captured. Extensive experiments on two real datasets demonstrate the effectiveness of GACF.

Keywords: Recommendation · Collaborative filtering · Geometric algebra

1 Introduction

Information overload is a common problem in today's Internet world. As an important mean to solve this problem, recommender system has been a great success. As a classic idea in the recommendation field, collaborative filtering (CF) has received a lot of attention [4,12]. The key idea of CF is that similar users will show similar preferences for items.

So far, many model-based collaborative filtering methods in real-valued embedding space have been proposed [15]. For example, matrix factorization (MF) [7] first randomly initializes the users and items, and then models the history interaction using inner product. Based on MF, NeuMF [4] models user-item interaction using a nonlinear neural network. Although using the nonlinear neural network can obtain the higher order information, related studies show that

A. Bhattacharya et al. (Eds.): DASFAA 2022, LNCS 13246, pp. 256–263, 2022.
https://doi.org/10.1007/978-3-031-00126-0_17

using simple inner product in MF still can achieve better results [3,8,9]. At the same time, the simple inner product in real space makes the modeling ability of MF still need to be improved.

Recently, Zhang et al. [14] proposed models CCF and QCF, which respectively represented users and items in complex and quaternion spaces [1,6] with CF. Thanks to the Hamilton product and inner product in quaternion space, inter-dependencies between components of users and items can be captured, which enhance the expressiveness of model. CCF and QCF show the potential value of inner product in other non-real spaces in recommendation [5,10,14]. In fact, real, complex number and quaternion are special cases of multivector, which is the element of geometric algebra (GA) space [11]. Compared with complex number and quaternion, multivector has more degree of freedom, good generalization capacity and excellent representation ability. Thanks to excellent properties of multivector, GA has been successfully applied to many fields, such as physics and engineering [2].

Considering the potential value of non-real spaces and successful application of GA in various fields, we first propose a **Geometric Algebra-based Collaborative Filtering (GACF)** model for recommendation. Specifically, GACF first uses N-grade multivector of geometric algebra to represent users and items, which makes the embeddings of users and items have higher degree of freedom as well richer representations. Then GACF reconstructs the historical interaction between users and items by inner product and asymmetrical geometric product [14], which can obtain inter-dependencies between components of multivector of users and items. Finally, bayesian personalized ranking (BPR) loss [7] is used to train the model in an end-to-end fashion. Extensive experiments on two real datasets demonstrate the effectiveness of GACF.

The main contributions of this paper are concluded as follows: (1) As far as we know, we are the first to model the recommender system in geometric algebra, which extends the study of recommender systems in non-real space. (2) We propose a novel recommendation model GACF, which uses multivectors and geometric product and inner product to model the historical interactions between users and items. (3) We conduct extensive experiments on two commonly used real-world datasets. Experiment results show that GACF achieves the state-of-the-art performance.

2 Methodology

In this section, we introduce the proposed recommendation model GACF. As shown in Fig. 1, GACF includes three parts: embedding layer, interaction modeling layer and output layer. In the embedding layer, we randomly initialize users and items with vectors and represent users and items in the form of multivectors. In the interaction modeling layer, we use inner product and asymmetric geometric products to model the interaction between users and items. Finally, we obtain the model prediction score in the output layer. In GACF model, we need to represent users and items with multivectors in the specific N-grade geometric algebraic space $\mathbb{G}^N$. So, in this section we give more details about two variants of GACF: GACF2 and GACF3, which are based on 2-grade multivector and 3-grade multivector respectively in geometric algebra.

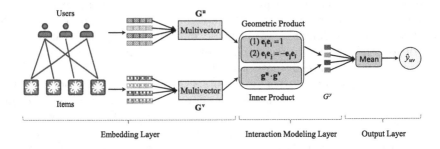

Fig. 1. Overall architecture of the proposed model GACF.

2.1 GACF2 and GACF3

GACF2. In the embedding of GACF2, we represent users and items in the form of 2-grade multivector with four random vectors:

$$\mathbf{G^u} = \mathbf{g_0^u} + \mathbf{g_1^u e_1} + \mathbf{g_2^u e_2} + \mathbf{g_{12}^u e_1 e_2}$$
$$\mathbf{G^v} = \mathbf{g_0^v} + \mathbf{g_1^v e_1} + \mathbf{g_2^v e_2} + \mathbf{g_{12}^v e_1 e_2} \tag{1}$$

where $\mathbf{g_0^u, g_1^u, g_2^u, g_{12}^u, g_0^v, g_1^v, g_2^v, g_{12}^v} \in \mathbb{R}^d$. As we can see, the embedding of users and items have four components, which make embeddings have richer representation capability. Then in the interaction modeling layer of GACF2, we use the geometric product and inner product to model the interaction between users and items. For GACF2, the geometric product G^y between user and item is shown as follows:

$$\begin{aligned}
G^y &= g_0^y + g_1^y \mathbf{e_1} + g_2^y \mathbf{e_2} + g_{12}^y \mathbf{e_1 e_2} \\
&= \mathbf{G^u} \otimes \mathbf{G^v} \\
&= \mathbf{g_0^u \cdot g_0^v} + \mathbf{g_1^u \cdot g_1^v} + \mathbf{g_2^u \cdot g_2^v} - \mathbf{g_{12}^u \cdot g_{12}^v} \\
&\quad + (\mathbf{g_0^u \cdot g_1^v} + \mathbf{g_1^u \cdot g_0^v} - \mathbf{g_2^u \cdot g_{12}^v} + \mathbf{g_{12}^u \cdot g_2^v})\mathbf{e_1} \\
&\quad + (\mathbf{g_0^u \cdot g_2^v} + \mathbf{g_1^u \cdot g_{12}^v} + \mathbf{g_2^u \cdot g_0^v} - \mathbf{g_{12}^u \cdot g_1^v})\mathbf{e_2} \\
&\quad + (\mathbf{g_0^u \cdot g_{12}^v} + \mathbf{g_1^u \cdot g_2^v} - \mathbf{g_2^u \cdot g_1^v} + \mathbf{g_{12}^u \cdot g_0^v})\mathbf{e_1 e_2}
\end{aligned} \tag{2}$$

where $g_0^y, g_1^y, g_2^y, g_{12}^y \in \mathbb{R}$, "$\cdot$" stands for inner product between vectors. Finally, in the output layer we respectively take the mean of all components of G^y in formula (2) as the model prediction score $\hat{y}_{uv}$ of GACF2:

$$\hat{y}_{uv} = (g_0^y + g_1^y + g_2^y + g_{12}^y)/4 \tag{3}$$

GACF3. In the embedding of GACF3, we represent users and items in the form of 3-grade multivector with eight random vectors:

$$G^u = g_0^u + g_1^u e_1 + g_2^u e_2 + g_3^u e_3 + g_{12}^u e_1 e_2 + g_{23}^u e_2 e_3 + g_{13}^u e_1 e_3 + g_{123}^u e_1 e_2 e_3$$
$$G^v = g_0^v + g_1^v e_1 + g_2^v e_2 + g_3^v e_3 + g_{12}^v e_1 e_2 + g_{23}^v e_2 e_3 + g_{13}^v e_1 e_3 + g_{123}^v e_1 e_2 e_3$$
$$\tag{4}$$

where $g_0^u, g_1^u, g_2^u, g_3^u, g_{12}^u, g_{23}^u, g_{13}^u, g_{123}^u, g_0^v, g_1^v, g_2^v, g_3^v, g_{12}^v, g_{23}^v, g_{13}^v, g_{123}^v \in \mathbb{R}^d$. The embedding of users and items have eight components, which make GACF3 have better representation capability than GACF2. Then in the interaction modeling layer of GACF3, we use geometric product and inner product to model the interaction between users and items:

$$G^y = g_0^y + g_1^y e_1 + g_2^y e_2 + g_3^y e_3 + g_{12}^y e_1 e_2 + g_{23}^y e_2 e_3 + g_{13}^y e_1 e_3 + g_{123}^y e_1 e_2 e_3$$
$$= G^u \otimes G^v \tag{5}$$

where $g_0^y, g_1^y, g_2^y, g_3^y, g_{12}^y, g_{23}^y, g_{13}^y, g_{123}^y \in \mathbb{R}$, "·" stands for inner product between vectors. Finally, in the output layer we respectively take the mean of all components of G^y in formula (5) as the model prediction score $\hat{y}_{uv}$ of GACF3:

$$\hat{y}_{uv} = (g_0^y + g_1^y + g_2^y + g_3^y + g_{12}^y + g_{23}^y + g_{13}^y + g_{123}^y)/8 \tag{6}$$

From GACF2 and GACF3, we can find that geometric product and inner product can not only directly model the historical interaction between users and items, but also capture the complex interactions between different parts of multivectors of users and items. Therefore, GACF2 and GACF3 have stronger modeling capability than MF. Compared with GACF2, GACF3 has better representation capability, because 3-grade multivector has more components than 2-grade multivector. Besides, the asymmetry property of geometric product also conforms to the asymmetric relationship between users and items, which makes GACF2 and GACF3 have great potential to capture the asymmetry in the recommender system. We will verify this conclusion in experiments.

2.2   Model Optimization

In this paper, we mainly focus on top-K recommendation. So, we use the common pairwise BPR loss to optimize GACF2 and GACF3. Specifically, the loss function is as follows:

$$\mathcal{L}_{GACF} = \sum_{(u,i,j)\in D} -\ln \sigma \left(\hat{y}_{ui} - \hat{y}_{uj} \right) + \lambda \|\Theta_{GACF}\|_2^2 \tag{7}$$

where $D = \{(u,i,j)|(u,i) \in \mathcal{R}_+, (u,j) \in \mathcal{R}_-\}$ stands for training dataset with observed interactions set $\mathcal{R}_+$ and the unobserved interactions set $\mathcal{R}_-$, the $\sigma(x)$ is sigmoid function, λ is the L_2 regularization coefficient, Θ_{GACF} is the parameter of GACF model.

Table 1. Datasets statistics.

Dataset	#User	#Item	#Interactions	Density
Book-Crossing	6,754	13,670	374,325	0.0040
ML-1M	6,034	3,125	574,376	0.0304

3   Experiments

3.1   Experimental Setup

Datasets. We conduct experiments on two widely used benchmark datasets: Book-Crossing and ML-1M. Information about these datasets is shown in Table 1. **Book-Crossing** is a dataset about user ratings for books [9]. For each user, we take score greater than 0 as positive feedback and use 10-core setting [9]. **ML-1M** is a dataset about user ratings for movies provided by GroupLens research [9]. For each user, we take score greater than 3 as positive feedback. We use 5-core setting [9] to ensure the quality of dataset.

Baselines. We compare proposed model GACF with a number of common strong baselines. For models in real-valued space, we report MF [7], MLP [4], NeuMF [4], JRL [15], DMF [13]. CCF and QCF [14] respectively use complex numbers and quaternions to represent users and items.

Evaluation Setup and Parameter Settings. For each dataset, we randomly choose 80% of interaction records for training, 10% for validation, and 10% for test. Models are evaluated using two common evaluation metrics Recall@K and NDCG@K ($K \in \{10, 20, 30\}$) by the all-ranking protocol. We perform 10 times and report the average result. For fair comparison, we set loss function of all baselines as pairwise BPR loss, the embedding size d as 64 and the optimal settings of baselines are obtained by tunning parameter on each dataset. For GACF2 and GACF3, we use the normal initialization to initialize the model parameter Θ_{GACF}, use Adam as the optimizer, and set batch size as 1024. And we adapt Bayesian HyperOpt [9] to perform hyper-parameter optimization on learning rate and coefficients of L_2 regularization term w.r.t. NDCG@20 on each dataset for 30 trails.

3.2   Performance Study

Performance Comparison. The comparison results of all models are shown in Tables 2 with the best result hightlighted in bold. From the table, we can draw the following conclusions:

- In some cases, the models using simple inner product, including MF, CCF, QCF, GACF2 and GACF3, perform better than those using nonlinear neural networks. The result shows the validity of inner product in modeling interactions, in line with the conclusion of previous studies [3, 8, 9].

Table 2. Overall performance of baselines on Book-Crossing and ML-1M datasets with the best result highlighted in bold.

	Book-Crossing						ML-1M					
	Recall@K			NDCG@K			Recall@K			NDCG@K		
	K = 10	K = 20	K = 30	K = 10	K = 20	K = 30	K = 10	K = 20	K = 30	K = 10	K = 20	K = 30
MF	0.0012	0.0020	0.0024	0.0007	0.0009	0.0010	0.0629	0.1120	0.1515	0.0537	0.0687	0.0810
MLP	0.0011	0.0021	0.0024	0.0007	0.0010	0.0011	0.0520	0.0922	0.1272	0.0444	0.0570	0.0679
NeuMF	0.0009	0.0017	0.0021	0.0006	0.0008	0.0009	0.0565	0.1002	0.1375	0.0476	0.0611	0.0728
JRL	0.0009	0.0018	0.0021	0.0006	0.0009	0.0010	0.0516	0.0937	0.1273	0.0456	0.0584	0.0691
DMF	0.0011	0.0024	0.0030	0.0008	0.0012	0.0013	0.0562	0.1016	0.1379	0.0484	0.0623	0.0734
CCF	0.0016	0.0026	0.0031	0.0008	0.0011	0.0013	0.0642	0.1110	0.1487	0.0543	0.0688	0.0806
QCF	0.0017	0.0028	0.0034	0.0011	0.0014	0.0015	0.0614	0.1052	0.1426	0.0504	0.0640	0.0757
GACF2	0.0015	0.0028	0.0031	0.0009	0.0013	0.0014	0.0628	0.1119	0.1511	0.0529	0.0680	0.0802
GACF3	**0.0019**	**0.0034**	**0.0041**	**0.0011**	**0.0015**	**0.0017**	**0.0647**	**0.1122**	**0.1518**	**0.0549**	**0.0695**	**0.0817**

- The models in the non-real space perform better than the models in real space. CCF, QCF, GACF2 and GACF3 perform better than MF in two datasets. The reason is that representations and operations in a non-real space can better capture complex interactions between users and items.
- In general, GACF3 performs the best among all models. Compared with GACF2, GACF3 uses 3-grade multivector to represent users and items. So, GACF3 has better representation capability and achieves better performance than GACF2 in most case. Similarly, compared with CCF and QCF, GACF2 and GACF3 use multivector as embeddings of users and items and have more components. So, GACF2 and GACF3 have better representation capability and achieve better performance than CCF and QCF in most case.

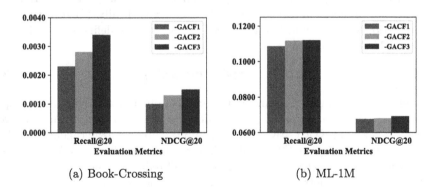

(a) Book-Crossing   (b) ML-1M

Fig. 2. Impact of grade of multivector.

Impact of Grade of Multivector. Multivectors with different grade have different degree of freedom and different expressiveness ability. The proposed model GACF can use multivectors with any grade. Here we explore the impact

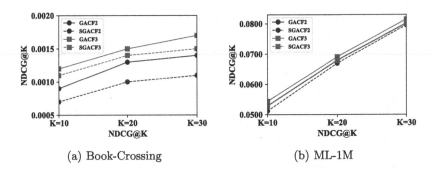

Fig. 3. Impact of asymmetrical geometric product.

of grade of multivectors on GACF model. We call the model using N-grade multivectors as GACFN. The experimental results of two datasets with respect of Recall@20 and NDCG@20 are shown in Fig. 2. We can find GACF2 and GACF3 have better performance than GACF1, because multivectors with big grade have larger degree of freedom and better expressiveness ability. GACF3 has limited improvement than GACF2 on ML-1M, which may be caused by the enough expressiveness ability of GACF2. Therefore, when we use GACF, we should consider the size of dataset to choose a suitable GACFN.

Impact of Asymmetrical Geometric Product. For the geometric product defined in geometric algebra, it has an asymmetry property. The asymmetry property of geometric product conforms to the asymmetry between users and items in recommender system. In order to verify the influence of asymmetry property of geometric product on proposed model GACF, we design a variant of GACF with symmetry property, namely SGACF. Specifically, when calculating the prediction score of model as in formula 2 and 5, we redefine the prediction score in SGACF as follows:

$$G^y = (\mathbf{G^u} \otimes \mathbf{G^v} + \mathbf{G^v} \otimes \mathbf{G^u})/2 \tag{8}$$

Through above formula, we can eliminate the influence of asymmetry property of geometric product on SGACF model. We carry out experiments on two datasets in terms of NDCG@K (K $\in \{10, 20, 30\}$), and the experimental results of GACF2, GACF3, SGACF2 and SGACF3 are shown in the Fig. 3. We can find eliminating the asymmetry property of model will reduce the performance of the model. So the asymmetry property of the geometric product is helpful to the GACF.

4   Conclusion

We proposes a novel collaborative filtering model GACF in geometric algebra space to solve implicit-feedback based top-K recommendation. GACF represents users and items as multivectors in geometric algebra and captures the complex relations between users and items through inner product and asymmetrical geometric product. Extensive experiments on two datasets show that GACF achieves very promising results.

Acknowledgement. The work was supported by National Natural Science Foundation of China (62172086, 61872074, 62106039)

References

1. Bassey, J., Qian, L., Li, X.: A survey of complex-valued neural networks. arXiv preprint arXiv:2101.12249 (2021)
2. Bayro-Corrochano, E.: Geometric Algebra Applications, Vol. I: Computer Vision, Graphics and Neurocomputing (2018)
3. Dacrema, M.F., Cremonesi, P., Jannach, D.: Are we really making much progress? A worrying analysis of recent neural recommendation approaches. In: RecSys, pp. 101–109 (2019)
4. He, X., Liao, L., Zhang, H., Nie, L., Hu, X., Chua, T.: Neural collaborative filtering. In: WWW (2017)
5. Li, Z., Xu, Q., Jiang, Y., Cao, X., Huang, Q.: Quaternion-based knowledge graph network for recommendation. In: MM, pp. 880–888 (2020)
6. Parcollet, T., Morchid, M., Linarès, G.: A survey of quaternion neural networks. Artif. Intell. Rev. **53**(4), 2957–2982 (2019). https://doi.org/10.1007/s10462-019-09752-1
7. Rendle, S., Freudenthaler, C., Gantner, Z., Schmidt-Thieme, L.: BPR: Bayesian personalized ranking from implicit feedback. In: UAI (2009)
8. Rendle, S., Krichene, W., Zhang, L., Anderson, J.R.: Neural collaborative filtering vs. matrix factorization revisited. In: RecSys, pp. 240–248 (2020)
9. Sun, Z., et al.: Are we evaluating rigorously? Benchmarking recommendation for reproducible evaluation and fair comparison. In: RecSys, pp. 23–32 (2020)
10. Tran, T., You, D., Lee, K.: Quaternion-based self-attentive long short-term user preference encoding for recommendation. In: CIKM, pp. 1455–1464 (2020)
11. Wang, R., Wang, K., Cao, W., Wang, X.: Geometric algebra in signal and image processing: a survey. IEEE Access **7**, 156315–156325 (2019)
12. Wang, X., He, X., Wang, M., Feng, F., Chua, T.: Neural graph collaborative filtering. In: SIGIR (2019)
13. Xue, H., Dai, X., Zhang, J., Huang, S., Chen, J.: Deep matrix factorization models for recommender systems. In: IJCAI (2017)
14. Zhang, S., Yao, L., Tran, L.V., Zhang, A., Tay, Y.: Quaternion collaborative filtering for recommendation. In: IJCAI (2019)
15. Zhang, Y., Ai, Q., Chen, X., Croft, W.B.: Joint representation learning for top-N recommendation with heterogeneous information sources. In: CIKM (2017)

Graph Neural Networks with Dynamic and Static Representations for Social Recommendation

Junfa Lin, Siyuan Chen, and Jiahai Wang[✉]

School of Computer Science and Engineering, Sun Yat-sen University,
Guangzhou, China
{linjf26,chensy47}@mail2.sysu.edu.cn, wangjiah@mail.sysu.edu.cn

Abstract. Recommender systems based on graph neural networks receive increasing research interest due to their excellent ability to learn a variety of side information including social networks. However, previous works usually focus on modeling users, not much attention is paid to items. Moreover, the possible changes in the attraction of items over time, which is like the dynamic interest of users are rarely considered, and neither do the correlations among items. To overcome these limitations, this paper proposes graph neural networks with dynamic and static representations for social recommendation (GNN-DSR), which considers both dynamic and static representations of users and items and incorporates their relational influence. GNN-DSR models the short-term dynamic and long-term static interactional representations of the user's interest and the item's attraction, respectively. The attention mechanism is used to aggregate the social influence of users on the target user and the correlative items' influence on a given item. The final latent factors of user and item are combined to make a prediction. Experiments on three real-world recommender system datasets validate the effectiveness of GNN-DSR.

Keywords: Social recommendation · Social network · Item correlative network · Graph neural network

1 Introduction

According to the social homophily hypothesis [9], two closely related users in the social network may have common or similar interests. Social recommender systems are increasing rapidly by considering social influences [13]. In addition to these explicit user social influences, the correlation between two items is also important. It provides extra information that may describe the items since items are likely to be similar or related [10]. Most of the information including social networks essentially has a graph structure, and graph neural networks (GNNs) have a powerful capability in graph representation learning, thus the field of utilizing GNNs in recommender systems is flourishing [16]. Existing methods can be classified into session-based and social recommendation. Session-based methods

© The Author(s), under exclusive license to Springer Nature Switzerland AG 2022
A. Bhattacharya et al. (Eds.): DASFAA 2022, LNCS 13246, pp. 264–271, 2022.
https://doi.org/10.1007/978-3-031-00126-0_18

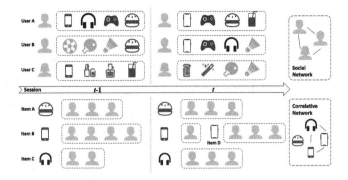

Fig. 1. An example of possible changes in the interest of users and the attraction of items over time. More detailed description is shown on the extended version of this paper [7] (https://arxiv.org/abs/2201.10751).

[6,8,11] mainly focus on the user-item sequential interactions. Social recommendation [1,2,5] attempt to leverage social networks to improve the results.

However, previous works usually focus on modeling users' behaviors with little attention to items. For the modeling of items, they have rarely modeled the possible changes in the attraction of items over time and the correlations among items, which are like the users' interests and their social influence, respectively. Figure 1 presents a motivating example that the interest of users and the attraction of items may change over time and be influenced by their relationships.

To overcome these limitations and to focus on the attraction of items over time, we propose graph neural networks with dynamic and static representations of users and items for social recommendation (GNN-DSR). GNN-DSR consists of two main components: interaction aggregation and relational graph aggregation. For both users and items in the interaction aggregation, the recurrent neural networks (RNNs) are used to model the short-term dynamic representations and the graph attention mechanism is utilized on historical user-item interactions to model the long-term static representations. In the relational graph aggregation, the influences from the user-user graph and the item-item graph that are termed as the relational graphs, are aggregated via the graph attention mechanism over users' or items' representations. The experimental results on three real-world data sets verify the effectiveness of the proposed model.

2 Problem Definition

Let $U = \{u_1, \ldots, u_n\}$ and $V = \{v_1, \ldots, v_m\}$ denote the set of users and items, respectively, where n is the number of users and m is the number of items. $\mathbf{R} \in \mathbb{R}^{n \times m}$ is the user-item interaction graph. r_{ij} represents the rating value given by user u_i on item v_j. $G_U = (U, E_U)$ is defined as the user social graph, where E_U is the set of edges connecting users. The set of u_i's friends is denoted as $\mathcal{N}_i^{uu}$, with the superscript 'uu' representing an undirected edge between two users. $\mathcal{N}_i^{uv}$ is

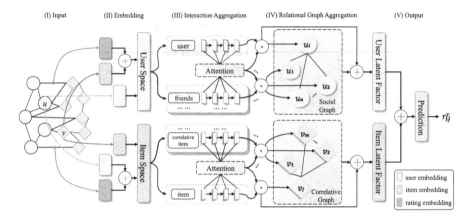

Fig. 2. The overview of GNN-DSR.

defined as the set of items with which u_i has interacted, while the superscript 'uv' indicates a directed edge from a user to an item. Meanwhile, the item correlative graph is denoted as $G_V = (V, E_V)$, where E_V is the set of edges connecting related items. $\mathcal{N}_j^{vv}$ denotes the set of items that are correlative with v_j. $\mathcal{N}_j^{vu}$ is defined as the set of users who have interacted with v_j. Each interaction is recorded as a triple (u_i, v_j, τ), with τ as the time stamp. Let $\mathcal{T}_i^v$ be an ascending sequence of all time stamps of u_i's interactions, then the item sequence that u_i has interacted with can be written as $S_i^v = (v_{j(\tau)})_{\tau \in \mathcal{T}_i^v}$. Similarly, let $\mathcal{T}_j^u$ denote an ascending sequence of all time stamps that users interact with v_j, then the corresponding user sequence can be written as $S_j^u = (u_{i(\tau)})_{\tau \in \mathcal{T}_j^u}$.

The social recommendation problem can be defined as [13]: given the observed user-item interaction $\mathbf{R}$ and the social relationship G_U, the recommender should predict the users' unknown interactions in $\mathbf{R}$, i.e., the rating value of unobserved item v_j user u_i will score (for explicit feedback) or the probability of an unobserved candidate item v_j user u_i will click on (for implicit feedback).

3   The Proposed Framework

The overview of GNN-DSR is shown in Fig. 2. The input layer constructs the item correlative graph which connects two items with high similarities by using cosine similarity. The embeddings of user and item are transformed from the preprocessed data. These embeddings are fed to the interaction aggregation and the relational graph aggregation, which capture the user-item interactions and the relational influences, respectively. The final output layer integrates the latent factor of user and item to predict the ratings.

3.1   User and Item Embedding

The user-item interaction graph $\mathbf{R}$ not only contains the interactions between the user and the item, but it also contains the user's ratings or opinions (denoted by r) of the item. These ratings of items not only reveal the user's preferences and interests in the items but also reflect the attraction of the item to users, thus helping to model the user's and item's latent factors. Typically, the rating r_{ij} takes a discrete value. For example, each $r_{ij} \in \{1, 2, 3, 4, 5\}$ in a 5-star rating system. Inspired by GraphRec [2], our method embeds each rating to a D dimensional vector and uses $\mathbf{e}_{ij}$ to represent the embedding of r_{ij}.

Let $\mathbf{p}_i, \mathbf{q}_j \in \mathbb{R}^D$ be the embeddings of user u_i and item v_j, respectively. For the interaction between u_i and v_j with rating r_{ij}, the interaction embeddings of u_i and v_j are computed as follows:

$$\mathbf{x}_{i \leftarrow j} = g^{uv}([\mathbf{e}_{ij}, \mathbf{q}_j]), \quad \mathbf{y}_{j \leftarrow i} = g^{vu}([\mathbf{e}_{ij}, \mathbf{p}_i]), \tag{1}$$

where $\mathbf{x}_{i \leftarrow j}$ and $\mathbf{y}_{j \leftarrow i}$ denotes the interaction embedding from v_j to u_i, and from u_i to v_j, respectively. $[\cdot, \cdot]$ denotes the concatenation operation of two vectors. Both g^{uv} and g^{vu} are two-layer perceptrons.

3.2   Interaction Aggregation

Short-Term Dynamic Representation. After obtaining the user's interaction embedding $\mathbf{x}_{i \leftarrow j}$, the items that u_i interacts within all sequences can be written as $\mathbf{X}(i) = (\mathbf{x}_{i \leftarrow j(\tau)})_{\tau \in \mathcal{T}_i^v}$, and so does the user interaction sequence of item, $\mathbf{Y}(j) = (\mathbf{y}_{j \leftarrow i(\tau)})_{\tau \in \mathcal{T}_j^u}$. To capture the dynamic representations, RNNs are used to model the sequence, since RNNs have good modeling capabilities for sequential data. The long short-term memory (LSTM) [4] is applied in our method:

$$\mathbf{h}_i^S = \mathbf{LSTM}(\mathbf{X}(i)), \quad \mathbf{h}_j^S = \mathbf{LSTM}(\mathbf{Y}(j)), \tag{2}$$

where $\mathbf{h}_i^S$ and $\mathbf{h}_j^S$ denotes the dynamic representation of the user's interest, and of the item's attraction, respectively. For long item sequences, the lengths are truncated to a fixed value to reduce the computational cost.

Long-Term Static Representation. Regarding each interaction embedding $\mathbf{x}_{i \leftarrow j}$ or $\mathbf{y}_{j \leftarrow i}$ as an edge representation of the user-item graph, our method can aggregate the edge representations via an attention mechanism, as follows:

$$\mathbf{h}_i^L = \sigma(\mathbf{W}_0^{uv} \sum_{j \in \mathcal{N}_i^{uv}} \alpha_{ij} \mathbf{x}_{i \leftarrow j} + \mathbf{b}_0^{uv}), \quad \mathbf{h}_j^L = \sigma(\mathbf{W}_0^{vu} \sum_{i \in \mathcal{N}_j^{vu}} \alpha_{ji} \mathbf{y}_{j \leftarrow i} + \mathbf{b}_0^{vu}), \tag{3}$$

where $\mathbf{h}_i^L$ is the static interest representation of the target user u_i, and $\mathbf{h}_j^L$ is the static attraction of the item. σ is a nonlinear activation function, and $\mathbf{W}_0$ and $\mathbf{b}_0$ are the weight and bias of the network. Different from the self-attention in graph attention networks (GAT) [15] that ignore the edge features, α_{ij} and α_{ji} represent the learned attentive propagation weights over the central node representations $(\mathbf{p}_i, \mathbf{q}_j)$ and the edge representations $(\mathbf{x}_{i \leftarrow j}, \mathbf{y}_{j \leftarrow i})$. Formally, taking α_{ij} as an example, they are calculated as:

$$\alpha_{ij} = \frac{\exp(\mathbf{W}_2^{uv} \cdot \sigma(\mathbf{W}_1^{uv} \cdot [\mathbf{p}_i, \mathbf{x}_{i \leftarrow j}] + \mathbf{b}_1^{uv}) + \mathbf{b}_2^{uv})}{\sum_{j \in \mathcal{N}_i^{uv}} \exp(\mathbf{W}_2^{uv} \cdot \sigma(\mathbf{W}_1^{uv} \cdot [\mathbf{p}_i, \mathbf{x}_{i \leftarrow j}] + \mathbf{b}_1^{uv}) + \mathbf{b}_2^{uv})}, \tag{4}$$

where $(\mathbf{W}_1, \mathbf{b}_1)$ and $(\mathbf{W}_2, \mathbf{b}_2)$ are the weights and biases of the first and second layers of the attention network, respectively.

For simplicity, Eqs. (3)–(4) can be written in a compact form,

$$\mathbf{h}_i^L = f^{uv}\left(\mathbf{p}_i, \{\mathbf{x}_{i \leftarrow j} : j \in \mathcal{N}_i^{uv}\}\right), \quad \mathbf{h}_j^L = f^{vu}\left(\mathbf{q}_j, \{\mathbf{y}_{j \leftarrow i} : i \in \mathcal{N}_j^{vu}\}\right). \tag{5}$$

Interactional Representation. The dynamic representations and static representations are directly combined to get the interest representation $\mathbf{h}_i^I$ of user and the attraction representation $\mathbf{h}_j^A$ of item via Hadamard product, i.e.,

$$\mathbf{h}_i^I = \mathbf{h}_i^S \odot \mathbf{h}_i^L, \quad \mathbf{h}_j^A = \mathbf{h}_j^S \odot \mathbf{h}_j^L. \tag{6}$$

Similarly, for a target user u_i's friends $u_o, o \in \mathcal{N}_i^{uu}$ in the social graph or an item $v_k, k \in \mathcal{N}_j^{vv}$ related to v_j in the item correlative graph, we can use the aforementioned method, summarized in Eq. (6), to obtain their interest representations $\mathbf{h}_o^I$ or attraction representation $\mathbf{h}_k^A$, respectively.

3.3 Relational Graph Aggregation

Social Aggregation for User. Modeling users' latent factors under social network information should take into account the heterogeneity of social relationship strength. Therefore, the social influence $\mathbf{h}_i^N$ of user is aggregated by the attention mechanism from the social graph, which is calculated as follows:

$$\mathbf{h}_i^N = f^{uu}\left(\mathbf{p}_i, \{\mathbf{h}_o^I : o \in \mathcal{N}_i^{uu}\}\right). \tag{7}$$

Correlative Aggregation for Item. Since items are not independent, there are likely other similar or correlative items. To further enrich the item latent factors from the item correlative graph G_V is reasonable. Similar to the user's one, item correlative representation $\mathbf{h}_j^N$ is calculated as follows:

$$\mathbf{h}_j^N = f^{vv}\left(\mathbf{q}_j, \{\mathbf{h}_k^A : k \in \mathcal{N}_j^{vv}\}\right). \tag{8}$$

3.4 Output Layer

To better model the latent factors of user and item, we need to consider the interaction-based representations and the relational-based representations together, since both the user-item interaction and the relational graph provide different perspectives of information about users and items. Therefore, the latent factor $\mathbf{h}_i^u$ of u_i and $\mathbf{h}_j^v$ of v_j are defined as

$$\mathbf{h}_i^u = g^{uu}\left([\mathbf{h}_i^I, \mathbf{h}_i^N]\right), \quad \mathbf{h}_j^v = g^{vv}\left([\mathbf{h}_j^A, \mathbf{h}_j^N]\right). \tag{9}$$

Our method is mainly applied to the recommendation task of rating prediction, which calculate the predicted rating $\hat{r}_{ij}$ as follows:

$$\hat{r}_{ij} = g_{\text{output}}\left([\mathbf{h}_i^u, \mathbf{h}_j^v]\right), \tag{10}$$

where g_{output} is a multilayer perceptron with three layers.

3.5   Training

Let $\mathcal{P} = \{(i,j) : r_{ij} \neq 0\}$ be the set of known ratings in the dataset. The mean squared error is used for training the model:

$$\text{MSELoss} = \frac{1}{2|\mathcal{P}|} \sum_{(i,j)\in\mathcal{P}} (\hat{r}_{ij} - r_{ij})^2, \tag{11}$$

where r_{ij} is the true rating score. GNN-DSR is optimized using gradient descent. To alleviate the overfitting problem, the Dropout [12] is applied to our work.

4   Experiments

4.1   Experimental Settings

Datasets. Three public real-world datasets are used to experimentally evaluate the proposed approach: Ciao, Epinions[1] and Delicious.[2] These datasets all contain user interactions of items and social information.

Baselines. Three classes of recommenders are used for comparison: (A) social recommenders **SocialMF** [5] and **DeepSoR** [1], which take into account the social influences; (B) session-based recommenders **NARM** [6] and **STAMP** [8], which model user sequential interests in sessions; (C) GNN-based recommendation methods **GraphRec** [2], **DGRec** [11] and **GraphRec+** [3], which utilize GNN to capture complex interactions among users and items. **GraphRec** and **GraphRec+** are also social recommenders and **DGRec** is session-based one.

Parameter Settings. The dimensions of users and items embedding are set to 128 and the batch size is 256. The lengths of user sequence and item sequence are both truncated to 30. The sample neighbors sizes of the social graph and correlative graph are both set to 30. The RMSprop [14] optimizer is used to train the models with a 0.001 learning rate. Dropouts with rates of 0.5 for rating prediction and 0.4 for item ranking are used to avoid overfitting.

Evaluation Metrics. For rating prediction, the Mean Absolute Error (MAE) and the Root Mean Square Error (RMSE) are used to evaluate prediction accuracy. For item ranking, Mean Reciprocal Rank at K (MRR@K) and Normalized Discounted Cumulative Gain at K (NDCG@K) are adopted to evaluate the performance, where K is 20.

---

[1] Ciao and Epinions available from http://www.cse.msu.edu/~tangjili/trust.html.
[2] Delicious available from https://grouplens.org/datasets/hetrec-2011/.

Table 1. Rating prediction performance of different methods. ↓ means lower is better. ↑ means higher is better. * indicates statistically significant improvements ($p < 0.01$) over the best baseline.

Models	Ciao		Epinions		Delicious	
	RMSE ↓	MAE ↓	RMSE ↓	MAE ↓	MRR ↑	NDCG ↑
SocialMF	1.0501	0.8270	1.1328	0.8837	–	–
DeepSoR	1.0316	0.7739	1.0972	0.8383	–	–
NARM	1.0540	0.8349	1.1050	0.8648	0.2074	0.2680
STAMP	1.0827	0.9558	1.0829	0.8820	0.2053	0.2626
GraphRec	0.9894	0.7486	1.0673	0.8123	0.1527	0.2243
DGRec	0.9943	0.8029	1.0684	0.8511	0.2080	0.2944
GraphRec+	0.9750	0.7431	1.0627	0.8113	0.1659	0.2392
GNN-DSR	**0.9444***	**0.6978***	**1.0579***	**0.8016***	**0.2254***	**0.3164***

4.2 Quantitative Results

The performance is shown in Table 1. For rating prediction on Ciao and Epinions, social and session-based approaches have similar performance, but all the GNN-based recommenders perform better than the previous ones. GraphRec and GraphRec+ perform slightly better than DGRec due to their better integration of rating information, which is an advantage for rating prediction. GraphRec+ performs the best for rating prediction among all baselines, as it not only exploits the social influence from users but also considers the correlations between items. Our proposed method GNN-DSR outperforms all baselines. Compared to the GNN-based methods, our method provides advanced components for integrating user-item temporal interactions and social/correlative graph information.

For item ranking on Delicious, GraphRec+, the best social recommendation baseline in rating prediction, performs much worse than the session-based methods in the item ranking experiment since it do not take into account the user's temporal information. DGRec performs the best in the baseline models, which demonstrates that dynamically incorporating social information helps to improve the performance of the recommender system. Our method performs the best not only because it captures both dynamic and static information, but also it models the effect of both social and correlative influences. More study and analysis of GNN-DSR are shown on this paper's extended version [7].

5   Conclusion

This paper proposes a GNN-based social recommendation called GNN-DSR. In particular, we consider both dynamic and static representations of users and items effectively. GNN-DSR models the short-term dynamic representations and the long-term static representations in interaction aggregation via RNNs and attention mechanism, respectively. The relational influences from the user social graph or item correlative graph are aggregated via the graph attention mechanism over users' or items' representations in the relational graph aggregation. The experimental results of three real-world datasets demonstrate the effectiveness of our method, verifying that items have dynamic and static attraction, and that the correlations among items benefit recommendation.

Acknowledgements. This work is supported by the National Key R&D Program of China (2018AAA0101203), and the National Natural Science Foundation of China (62072483).

References

1. Fan, W., Li, Q., Cheng, M.: Deep modeling of social relations for recommendation. In: Thirty-Second AAAI Conference on Artificial Intelligence, pp. 8075–8076 (2018)
2. Fan, W., et al.: Graph neural networks for social recommendation. In: The World Wide Web Conference, pp. 417–426 (2019)
3. Fan, W., et al.: A graph neural network framework for social recommendations. IEEE Trans. Knowl. Data Eng. (2020)
4. Hochreiter, S., Schmidhuber, J.: Long short-term memory. Neural Comput. **9**(8), 1735–1780 (1997)
5. Jamali, M., Ester, M.: A matrix factorization technique with trust propagation for recommendation in social networks. In: The 4th ACM Conference on Recommender Systems, pp. 135–142 (2010)
6. Li, J., et al.: Neural attentive session-based recommendation. In: The 2017 ACM on Conference on Information and Knowledge Management, pp. 1419–1428 (2017)
7. Lin, J., Chen, S., Wang, J.: Graph neural networks with dynamic and static representations for social recommendation. arXiv preprint arXiv:2201.10751 (2022)
8. Liu, Q., Zeng, Y., Mokhosi, R., Zhang, H.: Stamp: short-term attention/memory priority model for session-based recommendation. In: The ACM SIGKDD International Conference on Knowledge Discovery & Data Mining, pp. 1831–1839 (2018)
9. McPherson, M., Smith-Lovin, L., Cook, J.M.: Birds of a feather: homophily in social networks. Annu. Rev. Soc. **27**(1), 415–444 (2001)
10. Sarwar, B., Karypis, G., Konstan, J., Riedl, J.: Item-based collaborative filtering recommendation algorithms. In: The 10th International Conference on World Wide Web, pp. 285–295 (2001)
11. Song, W., et al.: Session-based social recommendation via dynamic graph attention networks. In: The 12th ACM International Conference on Web Search and Data Mining, pp. 555–563 (2019)
12. Srivastava, N., Hinton, G., Krizhevsky, A., Sutskever, I., Salakhutdinov, R.: Dropout: a simple way to prevent neural networks from overfitting. J. Mach. Learn. Res. **15**(1), 1929–1958 (2014)
13. Tang, J., Hu, X., Liu, H.: Social recommendation: a review. Soc. Netw. Anal. Min. **3**(4), 1113–1133 (2013). https://doi.org/10.1007/s13278-013-0141-9
14. Tieleman, T., Hinton, G., et al.: Lecture 6.5-rmsprop: divide the gradient by a running average of its recent magnitude. COURSERA Neural Netw. Mach. Learn. **4**(2), 26–31 (2012)
15. Veličković, P., et al.: Graph attention networks. In: International Conference on Learning Representations (2018)
16. Wu, S., Zhang, W., Sun, F., Cui, B.: Graph neural networks in recommender systems: a survey. arXiv preprint arXiv:2011.02260 (2020)

Toward Paper Recommendation by Jointly Exploiting Diversity and Dynamics in Heterogeneous Information Networks

Jie Wang, Jinya Zhou$^{(\boxtimes)}$, Zhen Wu, and Xigang Sun

School of Computer Science and Technology, Soochow University, Suzhou, China
{jwang24,zwu1024,xgsun27}@stu.suda.edu.cn, jy_zhou@suda.edu.cn

Abstract. Current recommendation works mainly rely on the semantic information of meta-paths sampled from the heterogeneous information network (HIN). However, the diversity of meta-path sampling has not been well guaranteed. Moreover, changes in user's reading preferences and paper's audiences in the short term are often overshadowed by long-term fixed trends. In this paper, we propose a paper recommendation model, called **COMRec**, where the diversity and dynamics are jointly exploited in HIN. To enhance the semantic diversity of meta-path, we propose a novel in-out degree sampling method that can comprehensively capture the diverse semantic relationships between different types of entities. To incorporate the dynamic changes into the recommended results, we propose a compensation mechanism based on the Bi-directional Long Short-Term Memory Recurrent Neural Network (Bi-LSTM) to mine the dynamic trend. Extensive experiments results demonstrate that COMRec outperforms the representative baselines.

Keywords: Paper recommender system · Heterogeneous information networks · Graph neural networks

1 Introduction

With the development of science and technology, electronic literature has become extremely rich. However, the rapid growth of the number of papers and the gradual subdivision of research fields have forced researchers to spend a lot of time and efforts looking for papers that they are really interested in. Therefore, the personalized paper recommendation becomes very urgent.

Previous work mainly utilize collaborative filtering [10,15,17] to make recommendations. In today's age of machine learning, various graph learning based methods have been employed in many recommender systems and become a new research paradigm, where the heterogeneous information network (HIN) is a very promising direction [12]. A HIN consists of multiple types of nodes and links that are used to represent different entities including users, items and their relationships. The basic idea of most existing HIN-based recommendation methods is to make use of path-based semantic relevance between users and items for recommendation [7,9]. For example, HERec [11] mines latent structural features of users and items through a meta-path based random walk and leverages the matrix factorization technique to optimize the recommendation results. Though many efforts have been devoted to the recommendation study, they still suffer from two essential challenges:

Diversity of Meta-path Sampling: Existing works [3,11,16] mainly treat directed edges as undirected ones, which is easy to collect invalid information or even noise information, while others [2,18] use backtracking sampling instead, but it cannot comprehensively collect paths. Then the diversity of meta-path sampling is poor.

Dynamically Changing Information: Few current studies take into account the dynamic factors, e.g., changes in both users' reading interests and papers' audiences.

To address the above challenges, we propose a dynamic compensation based paper recommendation model COMRec that jointly exploits diversity and dynamics in HIN. Our main contributions are summarized as follows:

- We present an in-out degree sampling method by leveraging both in-degree and out-degree of nodes to guide the meta-path sampling in the directed HIN, which effectively guarantees the diversity of meta-path sampling.
- We propose a compensation recommendation scheme to enable COMRec to not only grasp the user's overall reading preferences and the paper's overall audience but also capture short-term change trends.
- We conduct extensive experiments on real-world datasets [14]. The experimental results demonstrate the superiority of our model over the representative baselines.

2   Problem Statement

Given a target user, and his/her interaction data in HIN, our goal is to learn the prediction function f to calculate the user's probability $\hat{r}$ of reading a paper, where $\hat{r} \in [0, 1]$, i.e., $f : u, p \longrightarrow \hat{r}_{up}$, where p denotes a paper that user u never read. When $\hat{r}_{up}$ is greater than a certain value, we recommend p to u.

3   Our Proposed Model

To solve the aforementioned paper recommendation problem, we propose a novel paper recommendation model named COMRec. It includes four steps as illustrated in Fig. 1.

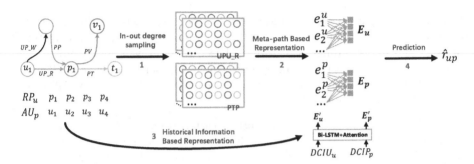

Fig. 1. The overview of COMRec model.

3.1   Information Extraction

In this section, we elaborate on how to build a HIN [13] for information extraction. We model the paper recommender system network as an academic HIN: $G = \{O, E\}$, including an object set O and a link set E. Specifically, the constructed HIN includes four types of object (user (U), paper (P), topic (T), and venue (V)), and a series of link relationships between them (UP_R, UP_W, PP, PT, and PV). To identify the reading relationship, a user is assumed to have read the paper before he/she cites it in his/her own article, and then we connect the user to the cited paper via a reading edge, denoted by UP_R. Accordingly, UP_W represents a user writes a paper. PP represents the citation relationship between papers. PT and PV indicate the subordination relationship.

To exploit the dynamically changing information, we still have to extract necessary information by preprocessing the dataset, including a collection of papers read by each user, a set of audiences for each paper, and the time of reading/being read. For each user u and paper p, we sort the corresponding collection in chronological order, i.e., $RP_u = \{\{p_1\}, \{p_2, p_3\}, \ldots, \{p_n\}\}$ and $AU_p = \{\{u_1\}, \{u_2, u_3\}, \ldots, \{u_n\}\}$, where RP_u represents the user u's reading paper list, and AU_p represents the paper p's audience list. $\{p_2, p_3\}$ indicates that user u read p_2 and p_3 in the same time, $\{u_2, u_3\}$ indicates that paper p is read by u_2 and u_3 in the same time.

3.2   In-out Degree Meta-path Sampling

Based on the academic HIN, the meta-path is denoted as $o_1 \overset{e_1}{\leftrightarrow} o_2 \overset{e_2}{\leftrightarrow} \ldots \overset{e_k}{\leftrightarrow} o_k$, where $\leftrightarrow$ denotes a directed edge with two optional directions.

Different from the classic definition of meta-path, our meta-path is directional. For example, the meta-path $U \overset{UP\_R}{\rightarrow} P \overset{PT}{\rightarrow} T \overset{PT}{\leftarrow} P \overset{UP\_R}{\leftarrow} U$ indicates that two individual users read articles on the same topic, and it can be abbreviated as "UPTPU" for short. There are two relationships between U and P, to avoid confusion, we use "UPU_R" to represent the meta-path $U \overset{UP\_R}{\rightarrow} P \overset{UP\_R}{\leftarrow} U$. We primarily focus on the entities of both U and P and then the meta-paths are divided into two types: the set MP_U of user meta-paths and the set MP_P of paper meta-paths, which start and end with the same node type (U or P).

We propose a novel sampling method called in-out degree sampling. Starting from to the source node, the sampling strictly follows the connection direction specified by the meta-path, when the direction is →, it means the next node must come from the outgoing neighbors of the source node, while the direction is ←, then the next node must come from the incoming neighbors. The next node's type should also be consistent with the type specified by the meta-path. If the sampling path length exceeds the meta-path pattern, it also requires the selection of subsequent node strictly follows the node type and connection type specified by the meta-path. With the help of in-out degree sampling, we are able to not only significantly avoid sampling noise information but also obtain diverse meta-paths with rich semantics.

3.3   Heterogeneous Entity Representation Learning

Meta-path Based Representation. The set of meta-paths designed by starting and ending with user type node is MP_U, i.e., $MP_U = \{mp_1, mp_2, mp_3, ...\}$. After sampling by each meta-path, many existing embedding methods can be employed for the initialization of node embedding: $e_k^u = Emb(mp_k)$, where $Emb(*)$ represents the employed embedding method, $e_k^u \in \mathbb{R}^{d_0 \times 1}$ is the embedded representation of user u learned from the k-th meta-path and d_0 is the initial embedding dimension.

Next, we need to integrate the initial embeddings learned from each meta-path. For each user, there will be different meta-path preferences. Some users pay more attention to a certain venue, while others may be more concerned about a topic. Therefore, we use a fully connected layer to combine the initial embeddings of each user: $E_u = \sum_k^{|MP_U|} W_k^U e_k^u$, where $|MP_U|$ is the number of meta-paths designed by starting and ending with a user, W^U is a set of learnable weight matrices for all users, $W_k^U \in \mathbb{R}^{d_1 \times d_0}$ is the k-th matrix in the set W^U, d_1 is the embedded dimension after the fully connected layer, and $E_u \in \mathbb{R}^{d_1 \times 1}$ is the final representation obtained by learning all the meta-paths of user u. Similarly, the representation of each paper p (E_p) is obtained by the same process.

Historical Information Based Representation. The representation learned by meta-path cannot highlight the changes in user preferences and the changes in the audience of the paper over time. Thereby we still have to capture these changes by mining historical information.

Assume that we have the sorted reading list of user u, and then we use the first $a\%$ of the reading data as the positive samples of the training set (P_{train}), and the last $(1-a)\%$ of the data is the positive samples of the testing set (P_{test}). According to the actual situation, we should grasp the recent reading trend and ensure the scale of the data input into the Bi-LSTM [6] is the same, so we use the last $b\%$ of P_{train} to form the data of user u's information containing dynamic changes:

$$RP_u = \{\{p_1\}, \{p_2, p_3\}, \ldots, \{p_n\}\}, P_{train} = \{p_1, p_2, \ldots, p_{n \times a\%}\},$$
$$P_{test} = \{p_{n \times a\% + 1}, \ldots, p_n\}, DCIU_u = \{p_{n \times a\% \times (1-b\%)}, \ldots, p_{n \times a\%}\}, \quad (1)$$

where $DCIU_u$ represents user u's information containing dynamic changes. If the location where the segmentation occurs falls on a set, we take a random sorting of the set and then perform the segmentation. If we can get the specific reading time, it will greatly avoid the occurrence of this situation. Similarly, the paper $p's$ information containing dynamic changes ($DCIP_p$) is treated in the same way.

We design an attention-aware Bi-LSTM to capture dynamic information of $DCIU_u$. Considering that different papers should not be treated equally, we integrate attentions in Bi-LSTM to fuse the representations of entities that include dynamically changing information. Formally, the $DCIU_u$ based representation of user u is given below:

$$\boldsymbol{E'_u} = \sum_{p_i \in DCIU_u} att(\boldsymbol{E_{p_i}}) \left[\overrightarrow{\text{LSTM}} \{\boldsymbol{E_{p_i}}\} \oplus \overleftarrow{\text{LSTM}} \{\boldsymbol{E_{p_i}}\} \right], \quad (2)$$

where $\boldsymbol{E'_u} \in \mathbb{R}^{1 \times 2d_1}$ is the representation of user u based on the historical information, including recent trends in reading preferences, and the operator $\oplus$ denotes concatenation. The above architecture captures "deep" relation between the papers read by user u from both directions, where the attention $att(*)$ is leveraged to fuse all hidden states to obtain the final representation based on the information containing dynamic changes. Here, we define the attention weights as follows:

$$att(\boldsymbol{E_{p_i}}) = \frac{\exp(\sigma(\boldsymbol{aE_{p_i}}))}{\sum_{p_j \in DCIU_u} \exp(\sigma(\boldsymbol{aE_{p_j}}))}, \quad (3)$$

where $\boldsymbol{a}$ denotes a trainable attention vector, and σ is a nonlinear activation function, we choose Sigmoid function [5]. Similarly, the hidden information representation of the paper p ($\boldsymbol{E'_p}$) can be also obtained in the same way.

3.4   Probability Prediction

After having two types of representation for each user, i.e., $\boldsymbol{E_u}$ and $\boldsymbol{E'_u}$ and each paper, i.e., $\boldsymbol{E_p}$ and $\boldsymbol{E'_p}$, we are able to start predicting the connection probability $\hat{r}$ between user u and paper p. We define the connection probability $\hat{r}$ as follows:

$$\hat{r}_{up} = f(u, p) = \begin{cases} \sigma\left(\boldsymbol{E'_u}^\top \boldsymbol{E'_p}\right) & \text{if } \left(\sigma\left(\boldsymbol{E_u}^\top \boldsymbol{E_p}\right)\right) < \eta, \\ \sigma\left(\boldsymbol{E_u}^\top \boldsymbol{E_p}\right) & \text{otherwise} . \end{cases} \quad (4)$$

The value of $\sigma\left(\boldsymbol{E_u}^\top \boldsymbol{E_p}\right)$ is the basis for prediction, when it is lower than the threshold η, indicating that the user $u's$ overall reading preference does not match the paper p. Then we use the user $u's$ recent reading preferences to match p, i.e., $\sigma\left({E'_u}^{\top}E'_p\right)$. If $\sigma\left({E'_u}^{\top}E'_p\right)$ is still lower than η, it means that the user $u's$ recent reading preference also does not match the paper p, so we will not recommend paper p to user u.

To train COMRec, the training set D consists of the training data extracted from the HIN and is built in the form of (u, p, r), where r has only two possible values: 0 or 1, $r = 0$ indicates that user u has not read the paper p, otherwise, u has read it. We set a cross-entropy loss function Θ as the objective for model training. All parameters are updated by the back-propagation in conjunction with stochastic gradient, attempting to minimize the objective function Θ, which is the average of $\Theta(u, p, r)$ for each training data:

$$\Theta = -\frac{1}{|D|} \sum_{(u,p,r)\in D} [r \log \hat{r} + (1 - r) \log (1 - \hat{r})]. \tag{5}$$

4 Experiments

4.1 Datasets

Our datasets come from Aminer [14], where T1 and T2 are datasets in two different time periods in the DBLP citation network (T1: 2000–2005, T2: 2005–2015). The detailed statistics are shown in Table 1. In addition to positive samples, we generate negative samples for each user by randomly selecting 6 papers such that they have not been read by the user.

Table 1. The statistics of datasets.

Item	User	Paper	Topic	Venue	PU_W	PU_R	PP	PT	PV	Cold start user
T1	16863	9566	3844	752	29266	54803	13978	32320	9566	4957
T2	52685	28930	7126	1983	99816	220815	47872	89921	28929	16566
MP_U	"UPU_R" "UPTPU" "UPVPU" "UPU_W"									
MP_P	"PP" "PTP" "PVP" "PUP_R" "PUP_W"									

4.2 Evaluation Metrics and Settings

We use three classic evaluation metrics, i.e., **F1-score**, **AUC** and **NDCG**, as indicators to evaluate the performance of a recommendation model. For all metrics, a higher value indicates a better performance.

For all experiments, we perform 10 times. Each experiment stops whenever the loss value converges and the final indicators are reported in terms of the average of 10 results. The parameters in our model are set as follows: a is set to 80, b is set to 25, and in Eq. (4), η is set to 0.5. For the learning rate we set to 0.0004 and the batch size is set to 500. Dimensions d_0 takes 64, d_1 takes 45.

4.3   Comparison Study

Baseline Methods. To evaluate the performance of COMRec, we compare with several representative baselines: Deepwalk [8], Node2vec [4], and Metapath2vec [1] are classic methods for node representation learning using the skip-gram model; HERec [11] non-linearly merges the embeddings obtained by the meta-path; We also implement two variants of COMRec, i.e., $COMRec_{mp}$ and $COMRec_{lstm}$. The former only relies on the embeddings learned from meta-paths, and the latter uses LSTM instead of Bi-LSTM to connect dynamic information for compensation recommendation.

Table 2. Performance comparison with baselines (MPS means meta-path sampling method, IO means in-out degree method).

Models	MPS	F1-score (%)		AUC (%)		NDCG	
		T1	T2	T1	T2	T1	T2
Deepwalk	Original	54.76	60.74	50.04	50.02	0.488	0.482
	IO	56.51	63.07	50.16	50.29	0.503	0.502
Node2vec	Original	55.14	63.58	50.28	50.36	0.497	0.505
	IO	59.47	64.28	53.34	52.28	0.506	0.507
Metapath2vec	Original	60.82	67.32	53.92	62.93	0.511	0.510
	IO	64.35	71.95	58.51	69.53	0.501	0.513
HERec	Original	86.65	85.47	88.76	86.25	0.548	0.549
	IO	88.55	86.81	91.81	86.84	0.571	0.556
$COMRec_{mp}$	IO	87.25	87.07	90.49	87.19	0.553	0.561
$COMRec_{lstm}$	IO	89.08	87.28	93.10	89.73	0.572	0.573
COMRec	IO	**89.44**	**88.1**	**95.24**	**90.26**	**0.574**	**0.576**

Comparison Results. The results of comparison are shown in Table 2. All HIN-based methods perform better than homogeneous graph based methods Deepwalk and Node2vec. Among these HIN-based methods, HERec, COMRec, and its variants perform better than Metapath2vec whose recommendation results are directly obtained based on heterogeneous sampling. $COMRec_{mp}$ performs better than HERec, so compared with the non-linear layer the fully connected layer can be a better choice for the embedding fusion. $COMRec_{lstm}$'s superiority over $COMRec_{mp}$ further illustrates that the compensation recommendation regarding dynamic changes can help our model get better recommendation results. $COMRec_{lstm}$ is defeated by COMRec by a narrow margin. Besides attention, our modified Bi-LSTM relies on two-way operation to better capture the dynamic trend of both user's reading preference and paper's audience in short term and contributes more useful embeddings to make accurate recommendations.

5   Conclusion

In this paper, we investigated the problem of personalized paper recommendation and presented a novel recommendation model, called COMRec. We jointly exploited the diversity and dynamics to enable COMRec to not only mine users' long-term reading preferences but also capture their recent interest changes. We conducted extensive experiments on real-world datasets. The results demonstrated the superiority of COMRec over the state-of-the-art baselines. As future work, we intend to incorporate explicit time feature embedding to further improve the recommendation performance.

Acknowledgements. This work was partially supported by the National Natural Science Foundation of China under Grant Nos. 61972272, 62172291, 62072321, and U1905211, the Natural Science Foundation of the Jiangsu Higher Education Institutions of China under Grant No. 21KJA520008, and the Postgraduate Research & Practice Innovation Program of Jiangsu Province under Grant No. SJCX21_1344.

References

1. Dong, Y., Chawla, N.V., Swami, A.: metapath2vec: scalable representation learning for heterogeneous networks. In: SIGKDD, pp. 135–144 (2017)
2. Fu, T.y., Lee, W.C., Lei, Z.: HIN2Vec: explore meta-paths in heterogeneous information networks for representation learning. In: CIKM, pp. 1797–1806 (2017)
3. Gong, J., Wang, S.: Attentional graph convolutional networks for knowledge concept recommendation in MOOCs in a heterogeneous view. In: SIGIR (2020)
4. Grover, A., Leskovec, J.: node2vec: scalable feature learning for networks. In: SIGKDD, pp. 855–864 (2016)
5. Han, J., Moraga, C.: The influence of the sigmoid function parameters on the speed of backpropagation learning. In: IWANN, pp. 195–201 (1995)
6. Hochreiter, S., Schmidhuber, J.: Long short-term memory. Neural Comput. **9**(8), 1735–1780 (1997)
7. Ma, X., Zhang, Y., Zeng, J.: Newly published scientific papers recommendation in heterogeneous information networks. Mob. Netw. Appl. **24**(1), 69–79 (2019)
8. Perozzi, B., Al-Rfou, R., Skiena, S.: DeepWalk: online learning of social representations. In: SIGKDD, pp. 701–710 (2014)
9. Ren, X., Liu, J., Yu, X., Khandelwal, U.: ClusCite: effective citation recommendation by information network-based clustering. In: SIGKDD, pp. 821–830 (2014)
10. Schafer, J.B., Frankowski, D., Herlocker, J., Sen, S.: Collaborative filtering recommender systems, pp. 291–324 (2007)
11. Shi, C., Hu, B., Zhao, W.X., Yu, P.S.: Heterogeneous information network embedding for recommendation. TKDE **31**(2), 357–370 (2019)
12. Shi, C., Li, Y., Zhang, J., Sun, Y., Yu, P.S.: A survey of heterogeneous information network analysis. IEEE TKDE **29**(1), 17–37 (2017)
13. Sun, Y.: Mining heterogeneous information networks: principles and methodologies. Synth. Lect. Data Min. Knowl. Discov., **3** 1–159 (2012)
14. Tang, J., et al.: ArnetMiner: extraction and mining of academic social networks. In: SIGKDD, pp. 990–998 (2008)
15. Wang, C., Blei, D.M.: Collaborative topic modeling for recommending scientific articles. In: SIGKDD, pp. 448–456 (2011)

16. Ma, X., Wang, R.: Personalized scientific paper recommendation based on heterogeneous graph representation. IEEE Access **7**, 79887–79894 (2019)
17. Yang, Z., Yin, D., Davison, B.D.: Recommendation in academia: a joint multi-relational model. In: ASONAM, pp. 566–571 (2014)
18. Zhang, C., Song, D., Huang, C., Swami, A., Chawla, N.V.: Heterogeneous graph neural network. In: SIGKDD, pp. 793–803 (2019)

Enhancing Session-Based Recommendation with Global Context Information and Knowledge Graph

Xiaohui Zhang[1], Huifang Ma[1,2(✉)], Zihao Gao[1], Zhixin Li[2], and Liang Chang[3]

[1] College of Computer Science and Engineering, Northwest Normal University,
Lanzhou 730070, Gansu, China
mahuifang@yeah.net
[2] Guangxi Key Lab of Multi-source Information Mining and Security,
Guangxi Normal University, Guilin 541004, Guangxi, China
[3] Guangxi Key Laboratory of Trusted Software, Guilin University of Electronic
Technology, Guilin 541004, Guangxi, China

Abstract. Predicting a user's next click by utilizing a short anonymous behavior is a challenging problem in the real-life session-based recommendation (SBR). Most existing methods usually learn the users' preference from current session. However, they seldom consider global context information or knowledge graph and failed to distill high-quality item from similar sessions. In this work, we combine Global Context information with Knowledge Graph, and develop a new framework to enhance session-based recommendation (GCKG). Technically, we model a global knowledge graph, exploiting a knowledge aware attention mechanism for better learning item embeddings. Then, we leverage an attention network and a gated recurrent unit to learn session representations. Furthermore, session representations are augmented simultaneously through constructing a similar session referral circle. Comprehensive experiments demonstrate that GCKG significantly outperforms the state-of-the-art methods of existing SBR.

Keywords: Session-based recommendation · Attentive network · Global context information · Knowledge graph · Gated recurrent unit

1 Introduction

With the thriving of online networks in the mobile internet, an emerging session-based recommendation (SBR) scenario has played a vital role in protecting user's privacy. Unlike traditional recommendation methods, SBR aims at predicting the next item, based on an anonymous user clicking sequences within one visit. To perform SBR, a lot of models have been proposed. Early SBR that incorporates Markov-chains-based methods [7] predict the user's next behavior solely based on the previous one. Afterwards, numerous researches apply recurrent neural networks (RNNs) to summarize all previous behaviors via a hidden state and

A. Bhattacharya et al. (Eds.): DASFAA 2022, LNCS 13246, pp. 281–288, 2022.
https://doi.org/10.1007/978-3-031-00126-0_20

obtain promising results [2]. Attention networks are also a powerful tool to capture user interest in each session [4]. Recently, graph neural networks (GNNs) achieved state-of-the-art performance and efficiency for SBR tasks [6,9,10]. Nevertheless, most existing studies mainly concentrate on model user's preference based on current session.

As a matter of fact, it is of great necessity to utilize global context information and item knowledge for alleviating the data sparsity issue and filtering noisy preference signals. In light of this, in this work, we primarily pay attention to tackling the following two challenges: How to incorporate other session's information and item knowledge for fully capturing complex dependency relationships among items? How to construct a unique similar session reference circle (SRC) for target session to extract high quality items.

To overcome these challenges, a novel SBR model which combines the Global Context information with Knowledge Graph (GCKG) is proposed in this paper. Specifically, a global knowledge graph is built based on the global graph and knowledge graph. Then a knowledge-aware attention mechanism is developed to capture the item correlations and item knowledge from the global knowledge graph. Next, the global and local preference signals are separately captured via an attention net and a gated recurrent unit (GRU) model. Finally, a similar session referral circle is built to extract high quality items for recommendation. As such, coupled with the global context information and item knowledge, GCKG effectively obtains item and session representations for accurate SBR. In summary, our contributions are three folds: (1) We innovatively construct a global knowledge graph and leverage a knowledge-aware attention mechanism to capture the global context information and the item knowledge, which can effectively ease data sparsity issues existing in the SBR. (2) We propose to construct a unique session reference circle for the target session, which can effectively filter out the noisy signals and extract high quality items for SBR. (3) Extensive experiments show that the proposed model has superiority over the state-of-the-art baselines and achieves significant improvements on real-world datasets.

2   Notations and Problem Statement

In this section, we give out the definition of the SBR problem.

Session Sequence: let $S = [s_1, s_2, \cdots, s_{|S|}]$ denote a set of sessions over an item set $V = \{v_1, v_2, \cdots, v_{|V|}\}$. An anonymous session $s_t = [v_1^t, v_2^t, \cdots, v_n^t] \in S$ is a sequence of items ordered by timestamps, where $v_j^t \in V$ is the j-th clicked item and n is the length of session s_t, which may contain duplicated items.

Global Graph: let $G_G = \{(v_{i-1}, trainsition, v_i)|v_{i-1}, v_i \in V\}$ is the Global Graph (GG), where the item set V contains all distinct items appearing in S, transition means that a time sequence relation from item v_{i-1} to item v_i in any session of S. When processing the individual session s_t, we need to sample a Session Graph (SG) from the GG, $G_{s_t} = (V_{s_t}, E_{s_t})$, where V_{s_t} contains the unique items in s_t and edge set E_{s_t} contains an edge $(v_{i-1}^t, v_i^t) \in E_{s_t} (2 \leq i \leq n)$

if there is a transition from item v_{i-1}^t to item v_i^t in s_t. Please note that, an item often recurs in a session. As a result, $|V_{s_t}| \leq n$.

Knowledge Graph: The item side information are organized in the form of knowledge graph(KG) $G_k = \{(h, r, t)|h, t \in \mathcal{E}, r \in R\}$, where $\mathcal{E}$ represents all entities node, each triplet describes that there is a relationship r from head entity h to tail entity t. Moreover, we establish a set of item-entity alignments $\mathcal{A} = \{(v, e)|v \in V, e \in \varepsilon\}$, where (v, e) indicates that item v can be aligned with an entity e in KG.

Problem Statement: The goal of our model is to take all session sequences S, and knowledge graph as input, given a target session $s_t = [v_1^t, v_2^t, \cdots, v_n^t]$, returns a list of top-$N$ candidate items to be consumed as the next one v_{n+1}^t.

3 Method

In this section, we present our proposed model that exploits Global Context information and Knowledge Graph for session-based recommendation (GCKG). Our model contains three tasks. First, a global knowledge graph G_{GK} is constructed by a global graph G_G and a knowledge graph G_K, then GCKG learns item correlations from global knowledge graph by a knowledge-aware attention mechanism and encode them into item representations. Next, A GRU and an attention net is utilized to learn the session embedding (Sect. 3.1). Second, we construct a unique session reference circle(SRC) for the target session, and then leverage a novel attention feature aggregator to learn the influence session embedding (Sect. 3.2). Finally, the prediction layer aggregates the learned session, item and the influence session embedding, and outputs the prediction score of the target session-item pair (Sect. 3.3).

3.1 Global Knowledge Graph

Constructing Global Knowledge Graph. To utilize and incorporate Global context information, item knowledge and solve the first challenge, we propose to construct a Global Knowledge Graph (GKG) G_{GK} to alleviate the data sparsity issue. Based on the item-entity alignment set $\mathcal{A}$, the global graph G_G can be seamlessly integrated with knowledge graph G_K as a unified relation graph $G_{GK} = \{(v, r, e)|v \in V, e \in \mathcal{E}', r \in R'\}$, where $\mathcal{E}' = V \cup \mathcal{E}, R' = R \cup \{\text{transition}\}$.

Learning Item Embeddings. In order to capture the global context information and item knowledge, we follow KGAT [8] to leverage a knowledge-aware attention mechanism for learning the item embeddings. After performing L layers knowledge-aware attention mechanism, we obtain multiple representations for item node v, i.e., $\{\mathbf{e}_v^1 \cdots, \mathbf{e}_v^L\}$. We hence adopt the layer-aggregation mechanism to concatenate the representations at each step into the knowledge embeddings of v: $\mathbf{v} = \mathbf{e}_v^0||\cdots||\mathbf{e}_v^L$, where $||$ is the concatenation operation. In this way, we obtain the item knowledge embedding matrix $\mathbf{I} \in \mathbb{R}^{|V| \times Ld}$.

Generating Session Embeddings. Although the item embeddings capture the global context information in all sessions and the item knowledge, it does not capture the session-specific context information. Therefore, It is necessary to preserve the original session s_t' information and capture the user's current preferences by learning session-level embedding. Specifically, given a target session $s_t = [v_1^t, v_2^t, \cdots, v_n^t]$, session embedding learning involves two tasks: (1) Perform an operation called embedding lookup to extract the s_t-specific embedding matrix from the item knowledge embedding matrix $\mathbf{I}$, $\mathbf{I}_s = [\mathbf{v}_1, \mathbf{v}_2, \cdots, \mathbf{v}_n]$, where $\mathbf{I}_s \in \mathbb{R}^{n \times Ld}$, and $\mathbf{v}_k \in \mathbb{R}^{Ld}$ is the knowledge embedding of the k-th item in session s_t. (2) Generat a representation of the session s_t according to the extracted knowledge embeddings. First, as GRU overcomes the vanishing gradients problem of RNN and is faster than LSTM, we adopt GRU to learn short-term and evolving preferences $\mathbf{s}_t^{recent}$:

$$\mathbf{s}_t^{recent} = \mathbf{h}_n = GRU(\mathbf{h}_{n-1}, \mathbf{v}_n, \phi_{GRU}) \tag{1}$$

where $\mathbf{h}_n$ is the hidden state (vector) in the nth step output by GRU, which is calculated based on $\mathbf{v}_n$ and the hidden state in the $(n-1)th$ step $\mathbf{h}_{n-1}$, ϕ_{GRU} denotes all GRU parameters.

Then, we consider the global preference $\mathbf{s}_t^{global}$ of the session graph G_{s_t} by aggregating all node vectors. Consider information in these embedding may have different levels of priority, we further adopt the attention network to better represent the global session preference:

$$\mathbf{s}_t^{global} = \sum_i a_{i,n} \mathbf{v}_i = \sum_i softmax(\mathbf{v}_i{}^{T} \mathbf{h}_n) \mathbf{v}_i, \tag{2}$$

where i indexes node v_i^t in the session graph G_{s_t}, $a_{i,n}$ is defined as the similarity score between the ith item and the last item in session s_t.

Finally, we compute the hybrid embedding $\mathbf{s}_t^{current}$ by taking linear transformation over the concatenation of the local and global embedding vectors:

$$\mathbf{s}_t^{current} = \mathbf{W}_3[\mathbf{s}_t^{recent}||\mathbf{s}_t^{global}], \tag{3}$$

3.2 Similar Session Reference Circle

To cope with the second challenge, we propose to leverage the similar SRC for target session to distill high quality items for filtering out the noisy signals.

Constructing the Similar SRC. To construct the similar SRC for target session, we need to find sessions that are both similar to the target session and have interacted with the target item. Let $Cs_t(v_k) = \{s_j | s_j \in S, sim(s_t, s_j) > \theta, v_k \in s_j\}$ is the similar SRC about target session $s_t \in S$ and the target item $v_k \in V$, where $sim(s_t, s_j)$ is the similarity between s_t and s_j, θ is a threshold that controls similarity. Here, we adopt Radial Basis Function (RBF) kernel [3] as our similarity definition, $sim(s_t, s_j) = e^{-\gamma||\mathbf{s}_t - \mathbf{s}_j||_2^2}$, where $\mathbf{s}_t$ and $\mathbf{s}_j$ are the session embedding of s_t and s_j obtained from section (Sect. 3.1), γ is a positive parameter, in real implementation we set $\gamma = 1/|S|$ as suggested in [3].

Coupling Influence Representation. Given target session s_t, target item v_k, and the related session reference circle $Cs_t(v_k)$, we couple the influence of each session $s_j \in Cs_t(v_k)$ and item v_k as: $c_{<s_j,v_k>} = MLP(\mathbf{s}_j||\mathbf{v}_k)$, where $MLP(\mathbf{s}_j||\mathbf{v}_k)$ is the computation of a one-layer neural network fed with the concatenation of session $\mathbf{s}_j$ and item $\mathbf{v}_k$.

Attentive Influence Degree. To obtain the attentive influence degree of $\mathbf{c}_{<s_j,v_k>}$ on the target session s_t, the coupled influence representation $\mathbf{c}_{<s_j,v_k>}$ and the target session representation $\mathbf{s}_t$ are used to calculate the following attention score: $d_{s_t \leftarrow <s_j,v_k>} = softmax(MLP(\mathbf{c}_{<s_j,v_k>}||\mathbf{s}_t))$, Where $MLP(.)$ denotes a two-layers feed-forward neural network with the *LeakyRELU* as activation function.

Since the influences of similar sessions propagate from the similar SRC, we attentively sum the coupled influences of the influential sessions and item v_k on the target session s_t: $\mathbf{s}_t^{influence} = \sum_{s_j \in Cs_t(v_k)} d_{s_t \leftarrow <s_j,v_k>} \mathbf{c}_{<s_j,v_k>}$. This equation models guarantees that the influence propagating among the similar session and the item can be encoded into the influence representation $\mathbf{s}_t^{influence}$.

3.3 Making Recommendation and Model Training

Intuitively, the closer an item is to the preference of the current session, the more important it is to the recommendation. After obtaining the embedding of each session, we compute the score $\hat{y}_{s_t,v_k}$ for each candidate item $v_k \in V$ by concatenation its embedding $\mathbf{v}_k$ and session representation $s_t^{current}, s_t^{influence}$. Here we utilize a MLP: $\hat{y}_{s_t,v_k} = softmax(MLP(\mathbf{s}_t^{current}||\mathbf{s}_t^{influence}||\mathbf{v}_k))$.

For each session, the loss function is defined as the cross-entropy of the prediction and the ground truth. It can be written as follows:

$$L_S = - \sum_{s_t \in S} \sum_{v_k \in V} \{y_{s_t,v_k} \log(\hat{y}_{s_t,v_k}) + (1 - y_{s_t,v_k}) \log(1 - \hat{y}_{s_t,v_k})\} \quad (4)$$

where y_{s_t,v_k} denotes the one-hot encoding vector of the ground truth item.

The training of TransR considers the relative order between valid triplets and broken ones, and encourages their discrimination through a pairwise ranking loss:

$$L_{KG} = \sum_{(h,r,t,t') \in T} - \ln \sigma \left(g(h,r,t') - g(h,r,t) \right) \quad (5)$$

where $T = \{(h,r,t,t')|(h,r,t) \in G_{ck}, (h,r,t') \in G\}$, and (h,r,t') is a broken triplet constructed by replacing one entity in a valid triplet randomly; $\sigma(\cdot)$ is the sigmoid function. This layer models the entities and relations on the granularity of triples, working as a regularizer and injecting the direct connections into representations, thus increasing the model representation ability.

Finally, we have the objective function to learn Eqs. (4) and (5) jointly, as follows:

$$L_{GCKG} = L_S + L_{KG} + \lambda ||\Theta||_2^2 \quad (6)$$

where Θ is the set of model parameters, and $||\Theta||_2^2$ is the L2-regularization that parameterized by λ to prevent over-fitting. At last, we compute the loss and adopt Adam optimization to optimize our model parameters.

4   Experiments

4.1   Datasets

We evaluate our model on a music dataset and an e-commerce dataset: **KKBOX**[1] and **JDATA**.[2] For both datasets, we follow [5] to filter out all sessions of length 1 and items appearing less than 3 times in datasets. For KKBOX dataset, the item knowledge is music attributes. For JDATA dataset, the item knowledge is the product attributes. We also divided two sparse JDATA datasets: Demo and Demo(N)containing some cold-start items.

4.2   Compared Models

To demonstrate the effectiveness our proposed method, we compare it with the following representative and state-of-the-art methods: (1) RNN-based and Attention-based methods including GRU4REC [2] and STAMP [4]. (2) Graph-based method including SR-GNN [9], FGNN [6], MKM-SR [5] and SERec[1]. We evaluate all models with two widely used ranking-based metrics in previous SBR evaluations: Hit@k and MRR@k. By default, we set $k = 20$.

4.3   Over Performance

Table 1 summarizes the results of all methods on two datasets in terms of Hit@20 and MRR@20. From the results, we have following observations. First, GCKG achieves the best performance on all metrics of all datasets. It demonstrates the effectiveness of the proposed method. Second, SERec achieves the best results among all baseline models, since it constructs a heterogeneous knowledge graph to capture the social information and cross-session item transitions. However, SERec is outperformed by our model in all cases. The primary reason is that our model explicitly utilizes item knowledge and similar session's information to infer user preference while SERec only relies on the social information. This comparison indicates that the effectiveness of our model.

4.4   Ablation Study

Effectiveness of Global Knowledge Graph. To investigate whether GKG construction is necessary, we introduce three variants: (1) GCKG-GKG has no global knowledge graph, the item embeddings are learned only from the session graph; (2) GCKG-GG does not consider the global context information;

---

[1] https://www.kaggle.com/c/kkbox-music-recommendation-challenge/data.
[2] https://jdata.jd.com/html/detail.html?id=8.

(3) GCKG-KG does not consider the item's side information. The results on all datasets in Table 1. From the results, we can find that: especially in Demo datasets, GCKG consistently achieves the best performance against other variants, demonstrating that CKG component are necessary to alleviate the sparsity problem of cold-start items.

Table 1. The performance of all methods on all datasets.

	KKBOX		JDATA		Demo		Demo(N)	
	Hit@20	MRR@20	Hit@20	MRR@20	Hit@20	MRR@20	Hit@20	MRR@20
GRU4REC	11.304	3.985	34.672	11.823	12.007	4.367	9.432	4.308
STAMP	13.662	4.682	35.042	12.248	14.109	5.048	9.786	3.154
SR-GNN	14.093	4.372	39.235	14.770	14.566	7.182	10.318	4.463
FGNN	15.137	5.846	41.558	16.758	16.489	7.556	11.248	4.956
MKM-SR	21.472	7.078	42.101	17.251	_24.592_	_9.629_	_14.010_	_6.028_
SERec	_22.597_	_7.706_	_42.653_	_17.698_	23.294	9.116	13.220	5.825
GCKG-GKG	16.116	6.17	41.606	16.795	18.231	7.691	11.549	5.058
GCKG-GG	18.135	7.09	42.215	17.035	20.326	7.849	12.047	5.353
GCKG-KG	21.621	7.68	43.106	17.847	23.654	8.213	13.623	5.754
GCKG	**23.374**	**7.865**	**43.976**	**18.015**	**25.263**	**9.812**	**14.362**	**6.127**
Improv. (%)	3.43	2.06	3.10	1.79	2.72	1.90	2.51	1.64

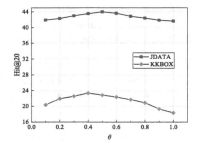 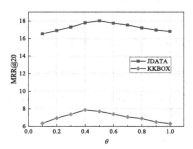

Fig. 1. The GCKG performance by changing the value of SRC controlling threshold θ.

Effectiveness of SRC. To verify the significance of SRC, we start experiments with different controlling threshold θ on two datasets to check on its influence. We evaluate θ in $[0.1, 0.2, \ldots, 0.8, 0.9, 1.0]$ and the result is presented in Fig. 1. From the results, we find that the recommendation performance increases and then decreases as θ increases. If it is too small, the model may introduce noisy data, while if it is too large, the similar sessions in SRC would become so sparse, which leads the role of SRC can not be reflected. When $\theta = 1$, the model does not consider SRC component, so it performs bad.

5   Conclusion

This paper presents exploiting global context information and knowledge graph for session-based recommendation. To complement the drawback of existing models, we construct a global knowledge graph and a SRC to fully capture global context information and the knowledge information from different perspectives. Empirical results on two real-world datasets demonstrate the effectiveness of GCKG compared with state-of-of-the art baselines.

Acknowledgment. This work is supported by Gansu Natural Science Foundation Project (21JR7RA114), the National Natural Science Foundation of China (61762078, 61363058, U1811264, 61966004), Northwest Normal University Young Teachers Research Capacity Promotion Plan (NWNU-LKQN2019-2) and Northwest Normal University Postgraduate Research Funding Project (2021KYZZ02107).

References

1. Chen, T., Wong, R.C.W.: An efficient and effective framework for session-based social recommendation. In: Proceedings of the 14th ACM International Conference on Web Search and Data Mining, pp. 400–408 (2021)
2. Hidasi, B., Karatzoglou, A., Baltrunas, L., Tikk, D.: Session-based recommendations with recurrent neural networks. In: Proceedings of the 4th International Conference on Learning Representations (2016)
3. Hsu, C.C., Lai, Y.A., Chen, W.H., Feng, M.H., Lin, S.D.: Unsupervised ranking using graph structures and node attributes. In: Proceedings of the 10th ACM International Conference on Web Search and Data Mining, pp. 771–779 (2017)
4. Liu, Q., Zeng, Y., Mokhosi, R., Zhang, H.: Stamp: short-term attention/memory priority model for session-based recommendation. In: Proceedings of the 24th ACM SIGKDD International Conference on Knowledge Discovery & Data Mining, pp. 1831–1839 (2018)
5. Meng, W., Yang, D., Xiao, Y.: Incorporating user micro-behaviors and item knowledge into multi-task learning for session-based recommendation. In: Proceedings of the 43rd International ACM SIGIR Conference on Research and Development in Information Retrieval, pp. 1091–1100 (2020)
6. Qiu, R., Huang, Z., Li, J., Yin, H.: Exploiting cross-session information for session-based recommendation with graph neural networks. ACM Trans. Inf. Syst. (TOIS) **38**(3), 1–23 (2020)
7. Rendle, S., Freudenthaler, C., Schmidt-Thieme, L.: Factorizing personalized Markov chains for next-basket recommendation. In: Proceedings of the 19th International Conference on World Wide Web, pp. 811–820 (2010)
8. Wang, X., He, X., Cao, Y., Liu, M., Chua, T.S.: KGAT: knowledge graph attention network for recommendation. In: Proceedings of the 25th ACM SIGKDD International Conference on Knowledge Discovery and Data Mining, pp. 950–958 (2019)
9. Wu, S., Tang, Y., Zhu, Y., Wang, L., Xie, X., Tan, T.: Session-based recommendation with graph neural networks. In: Proceedings of the AAAI Conference on Artificial Intelligence, vol. 33, pp. 346–353 (2019)
10. Zhou, H., Tan, Q., Huang, X., Zhou, K., Wang, X.: Temporal augmented graph neural networks for session-based recommendations. In: Proceedings of the 44th International ACM SIGIR Conference on Research and Development in Information Retrieval, pp. 1798–1802 (2021)

GISDCN: A Graph-Based Interpolation Sequential Recommender with Deformable Convolutional Network

Yalei Zang[1], Yi Liu[1], Weitong Chen[2], Bohan Li[1(✉)], Aoran Li[1], Lin Yue[2], and Weihua Ma[1]

[1] Nanjing University of Aeronautics and Astronautics, Nanjing, China
bhli@nuaa.edu.cn
[2] The University of Queensland, Brisbane, Australia

Abstract. Sequential recommendation systems aim to predict users' next actions based on the preferences learned from their historical behaviors. There are still fundamental challenges for sequential recommender. First, with the popularization of online services, recommender needs to serve both the warm- and cold-start users. However, most existing models depending on user-item interactions lose merits due to the difficulty of learning sequential dependencies with limited interactions. Second, users' behaviors in their historical sequences are often implicit and complex due to the objective variability of reality and the subjective randomness of users' intentions. It is difficult to capture the dynamic transition patterns from these user-item interactions. In this work, we propose a graph-based interpolation enhanced sequential recommender with deformable convolutional network (GISDCN). For cold-start users, we re-construct item sequences into a graph to infer users' possible preferences. To capture the complex sequential dependencies, we employ the deformable convolutional network to generate more robust and flexible filters. Finally, we conduct comprehensive experiments and verify the effectiveness of our model. The experimental results demonstrate that GISDCN outperforms most of the state-of-the-art models at cold-start conditions.

Keywords: Sequential recommendation · Cold-start user · Graph neural network · Deformable convolutions

1 Introduction

Recommendation systems (RS) aim to learn user preferences and choose useful information for users. Different from the most existing RS which represent users general preferences as the static patterns, sequential recommendation systems (SRS) capture the dynamic dependencies from user's sequential interactions and attract increasing attention in recent years.

To capture the sequential dependency, some early methods such as FPMC [1] used Markov Chain. However, Markov Chain-based models cannot extract the

A. Bhattacharya et al. (Eds.): DASFAA 2022, LNCS 13246, pp. 289–297, 2022.
https://doi.org/10.1007/978-3-031-00126-0_21

high-order dependencies from sequences. Recently, neural network based methods have been employed to model the sequential dependencies since the rich representation power, such as recurrent neural network (RNN) [11,12], convolutional neural network (CNN) [2–4], and graph neural network (GNN) [14,15].

Although existing works have achieved good performance, there are still fundamental challenges for sequential recommenders. First of all, since most SRS models mainly employ user-item interactions to learn users' preferences, the performance of these models relies on the sufficiency of historical interactions. It is difficult to extract the sequential dependencies from the short sequences which only have a few items. In addition, the sequential dependencies are complex and include many different patterns [2]. To overcome these challenges, methods with different kinds of algorithms are proposed. For the cold-start users, most existing works [6,7] exploit side information. To get rid of the additional auxiliary knowledge, MetaTL [8] models the transition patterns of users through meta-learning and enables fast learning for cold-start users with only limited interactions. However, these methods hardly consider the long-range dependencies in warm-start scenarios and users' core preferences from global sequences. For the complex sequential dependencies, CNN-based methods employ the convolution operation but are limited to the fixed filters [2]. RNN-based models are constrained by the strong order assumption and cannot adapt to the complex and changeable data flexibly [13]. Therefore, it is necessary to introduce a new sequential recommender that can extract the complex dependencies and has the ability to satisfy the needs of cold-start users.

In this work, we proposed a novel sequential recommender which improves the extraction of complex sequential dependencies and reduces the difficulty of modeling cold-start users' preferences. First, We design a graph neural network-based interpolation method to infer users' possible preferences from global sequences and alleviate the problem of cold-start users. Then, we improve the dynamic filter networks for sequential recommendation and employ the deformable convolutional network to enhance the ability of extracting the complex dependencies from the sequences. Experimental results on three real-world datasets show that our proposed approach significantly outperforms baseline methods.

2   Proposed Method

In this section, we introduce the proposed GISDCN (Fig. 1) which applies graph neural network (GNN) to make up interactions for cold-start users and employs deformable convolutional network to extract the sequential dependencies.

2.1   Graph-Based Interpolation Module

To alleviate the insufficiency of records in sequences, we try to utilize the core interests of all users and make up interactions for these short sequences. We employ GNN to find the core interest node and distinguish the suitable items from the candidate nodes to fit users' possible preferences in short sequences.

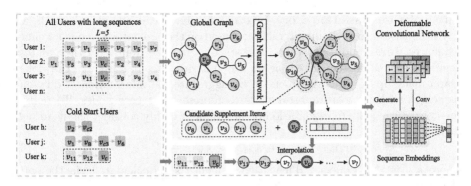

Fig. 1. The framework of our model GISDCN. Users' interactions are represented by rectangular boxes. The global graph is constructed by all long sequences and node embeddings are learned through the graph neural network. The candidate items are selected and interpolated to the cold-start users' sequences. Finally, the deformable convolutional network is used to capture sequential dependencies.

Construction of Global Graph. Firstly, each sequence is divided by its length L. The long sequences are used to construct the global graph and the short sequences need to be interpolated later. Then we construct the directed global graph $G_g = (V_g, \mathcal{E}_g)$. Each node $v_i \in V_g$ corresponds to one item and each edge $(v_{i-1}, v_i) \in \mathcal{E}_g$ means that a user clicks item v_i after v_{i-1}. We mark the importance of edges with the occurrence of the edge divided by the outdegree of the edge' s start node.

Node Embedding Learning. After constructing the graph, we present how to encode the item transition and learn the node embeddings via graph neural networks. In our work, we utilize $GGNNs$ [9] to learn of the nodes' representations.

For details, $\{h_1, h_2, \cdots, h_n\}$ is the embedding vectors which dimension is d, n is the length of sequence. Then for each long sequence, the learning process of all nodes can be illustrated as [9,16]:

$$h_i^t = GGNNs\left([h_1, h_2 \ldots, h_n], A_s\right) \tag{1}$$

where A_s is the matrix that describes the connection among nodes. The calculation function is as follow:

$$A_s = \left(\begin{bmatrix} q_{11} & \cdots & q_{1n} \\ \vdots & \ddots & \vdots \\ q_{1n} & \cdots & q_{nn} \end{bmatrix} \begin{bmatrix} \frac{1}{q_1^{out}} \\ \cdots \\ \frac{1}{q_n^{out}} \end{bmatrix}\right) \Big\| \left(\begin{bmatrix} q_{11} & \cdots & q_{1n} \\ \vdots & \ddots & \vdots \\ q_{1n} & \cdots & q_{nn} \end{bmatrix}^T \begin{bmatrix} \frac{1}{q_1^{in}} \\ \cdots \\ \frac{1}{q_n^{in}} \end{bmatrix}\right), \tag{2}$$

where $\|$ is the concatenation operation, q_{ij} denotes the number of edges from item i_i to i_j, q_i^{out} and q_i^{in} are respectively the outdegree and indegree of the node i_i. After the learning of all nodes, we denote h_i' as the new representation which aggregates information from its neighbors base on the graph.

Interpolation Based on Core Interest Node. Given the global graph, we give our method to make up interactions based on the core interest node for short sequences corresponding to these cold-start users. The item with the greatest degree in each short sequence is as the core interest node of a user. In the global graph made up of all long sequences, we aggregate the information of different sequences with the core interest node to represent the core nodes. To reduce the noise during the representation of core node, we only consider the nearest $\mathcal{E}$ items:

$$s_j = \{v_i \mid v_{i'} = v_{\text{core}} \in s_{\text{ori}}, v_i \in s_{\text{ori}}, i \in [i' - \varepsilon, i' + \varepsilon]\}, \tag{3}$$

where s_{ori} is the sequence. Then each sequence s_j is calculated via the soft-attention mechanism:

$$H_{s_j} = \sum_{v_i \in s_j} \left(W_3^T \sigma \left(W_1 h'_{v_{last}} + W_2 h'_{v_i} + b\right)\right) h'_{v_i}, \tag{4}$$

where W_* is weight matrix, b is the bias, v_{last} is the last item in s_j.

To distinguish the importance of the core interest node in different sequence and obtain the robust representation, we assign a weight to each embedding of sequence s_j. H'_{s_j} is the sequence representation without core node. Then, the representation of core interest node is calculated by adding different sequence representation. Finally, we select items which are adjacent to the core interest node in the global graph as the candidate items and get score by multiplying its embedding with the representation of code interest node:

$$score_i = h_i'^T \left(\sum_{s_j \in S_{\text{agg}}} \left(1 - \cos\left(w \odot H_{s_j}, w \odot H'_{s_j}\right)\right) H_{s_j} \right), \tag{5}$$

We select the top N items with higher scores for the interpolation of the short sequence with the core interest node.

2.2 Deformable Convolutional and Prediction Module

Deformable Convolutional Network. After the interpolation, we get the sequences with more interactions. Next, we need to capture user preferences from the sequence. Inspired by the deformable kernels [17], we propose to use offsets to comprehensively generate the kernels according to more items and improve the flexibility of kernels based on the dynamic convolutional network [4,18]. The structure is like Fig. 2.

For details, the sequence with more interactions is mapped into the embedding vectors: $\{h_1, h_2, \ldots, h_n\}$, where $h_i \in R^d$ and n is the length of sequence. To make convolutional filters more flexible, we need to get offsets based on the feature of the current input $E_l^i = \{h_{i-l+1}, \ldots, h_{i-1}, h_i\}$, $E_l^i \in R^{d \times l}$, where l is the scope size that determines the extent of convolution kernel can shift. The offsets and original kernel space are calculated as follows:

$$\Delta p = \text{linear}_1 \left(E_l^i\right), \tag{6}$$

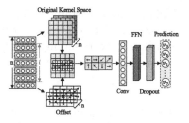

Fig. 2. The structure of deformable convolution module.

$$W_{\text{ori}} = \text{linear}_2\left(E_l^i\right), \tag{7}$$

where $linear_*(*)$ is one linear layer. $\Delta p \in R^{1 \times 2k}$ is a set of kernel offsets that is projected by the input, $W_{\text{ori}} \in R^{l \times l}$ is the original kernel space used to generate the final kernels.

Then we employ the offsets on the discrete kernel positions to make the fixed kernel flexible and get the final adaptive convolutional filter:

$$w_i = \mathcal{G}\left(W_{ori}, (p_0 + \Delta p)\right), \tag{8}$$

where $\mathcal{G}(\alpha, \beta)$ is the process of obtaining the convolutional kernel from original kernel space α according to index β, $w_i \in R^{(n-1) \times k}$ is the final kernel. The bilinear sampling[17] is used when sampling in the kernel space. The convolutional process is illustrated as:

$$\text{DefConv}\left(E, W, i\right) = DepthwiseConv\left(E, w_i, i\right) = w_i * E^{i-k+1:i}, \tag{9}$$

where $W = \{w_1, w_2, \ldots, w_{n-1}\}$ is convolutional filters matrix, $DepthwiseConv$ is depth-wise convolutions and employs the same filter on all channels [4].

Prediction Layer. After getting the output of the convolution layer, we calculate the prediction scores. The point-wise function is used as the loss function[4,18]:

$$L_{\text{loss}} = \sum_{n=0}^{N} \sum_{l=2}^{L} \left[\log\left(\sigma\left(\cos\left(o_l^n, e_l^n\right)\right)\right) + \sum_{\bar{e} \in \bar{E}^u} \log\left(1 - \sigma\left(\cos\left(o_l^n, \bar{e}\right)\right)\right)\right] \tag{10}$$

where N is the number of sequence, L is the length of sequence, e_l^n is the ground-truth at position l, $\sigma(*)$ is the sigmoid function. $\bar{E}^u$ is the negative samples.

3   Experiments

3.1   Datasets and Processing

We evaluate our model on four public datasets. The basic statistics of the these datasets are shown in Table 1. For all datasets, the interaction records are divided by user id and sorted ascendingly by timestamps. For the movie dataset which has no such short sequence, we generate the sequences by randomly removing items from long sequences. Finally, we randomly select 70%, 10% and 20% of sequences as the training, validation and test set.

Table 1. Statistics of RS datasets.

Datasets	ml-1m	ml-20m	Music	Book
#Users	6038	23212	102652	36475
#Items	3706	13070	5735	5999
#Actions	959527	1485966	6717493	442433
#Avg. action/user	159	64	65	12
#Sparsity	0.9571	0.9893	0.9890	0.9979

3.2 Baselines and Experimental Setup

We compare our model with Markov Chains method (FPMC [1]), RNN method (GRU4Rec [11]), CNN methods (Caser [2], NextItNet [3], DynamicRec [4]), GNN method (SR-GNN [9]) and Meta-learning method (MetaTL [8]). We use three widely adopted accuracy metrics: MRR, NDCG and HR. For our model, the batch size is set to 64. We set the item embedding size d = 200 for all datasets. The maximum length for ml-1m and book is set to 10 and for ml-20m and music, we set it to 40. The learning rate is set to 0.001. For the graph-based interpolation module, the hidden layer size is set to 100. The filter size of the convolution module is 5 and the scope size is set to 7.

3.3 Performance Comparison

The overall results of all models are presented in Table 2. The best results on all evaluation metrics are boldfaced. We compare our model with the GNN-based model and the meta-learning based model. Our model has better performance than SR-GNN. The possible reasons we considered are twofold. Firstly, SR-GNN does not make good use of the information in position. Secondly, it does nothing for these sequences of cold-start users. Compared with MetaTL, since we reconstruct item sequences into a global graph to consider the information in long sequences, our model can make better use of the long-range dependencies when dealing with these cold-start users. Meanwhile, from the results we can observe that our model does not achieve good performance in ml-1m. A possible reason is that ml-1m has a small amount of data which may not fully support the training of our model.

Table 2. The Performance of all methods over four datasets.

Datasets	Metrics	FPMC	GRU4 -Rec	Caser	NextIt -Net	Dynamic -Rec	SR- GNN	Meta -TL	GIS -CN	Improv.
Book	MRR@10	0.0501	0.0522	0.0454	0.0573	0.0663	0.0740	0.0769	**0.0800**	4.03%
	NDCG@10	0.0655	0.0523	0.0631	0.0786	0.0931	0.0940	0.0950	**0.1071**	12.74%
	HR@10	0.1163	0.1191	0.1219	0.1490	0.1811	0.1544	0.1800	**0.1956**	8.01%
Music	MRR@10	0.0599	0.0522	0.0634	0.0743	0.0845	0.0803	0.0835	**0.0917**	8.52%
	NDCG@10	0.0874	0.0523	0.0967	0.1066	0.1179	0.1103	0.1150	**0.1270**	7.72%
	HR@10	0.1793	0.1191	0.2080	0.2140	0.2287	0.2098	0.2134	**0.2438**	6.60%
ml-1m	MRR@10	0.0565	0.0570	0.0686	0.0705	0.0691	0.0754	0.0830	**0.0849**	2.29%
	NDCG@10	0.0804	0.0660	0.0958	0.0989	0.0975	0.0926	**0.1070**	0.1058	−1.12%
	HR@10	0.1591	0.1622	0.1858	0.1930	0.1917	0.1569	**0.1968**	0.1937	−1.58
ml-20m	MRR@10	0.0370	0.0457	0.0377	0.0617	0.0511	0.0432	0.0641	**0.0669**	4.37%
	NDCG@10	0.0495	0.0511	0.0523	0.0818	0.0857	0.0837	0.0854	**0.0901**	5.13%
	HR@10	0.0913	0.1233	0.1007	0.1486	0.1665	0.1484	0.1793	**0.1874**	4.5%

3.4 Model Discussions

Ablation Study. In this section, we examine the impact of the graph-based interpolation module and deformable convolutional network based recommender. In Table 3, We can observe that, without the order restriction, the performance of the model is even reduced. This may be due to the noise caused by the unordered interpolation. In addition, we can observe that with the enhancement to the representation of the core node, the model can increase the positive effect of the interpolation and further improve the performance. Table 4 proves the effectiveness of dynamic filters in capturing the complex transition patterns from sequences. The deformable convolutional network achieves the best performance which dominates that comprehensively considering several items can better generate the robust filters and improve the extractability of dynamics from sequences.

Table 3. Effect of interpolation module.

Datasets	Interpolation module	MRR@10	NDCG@10
Book	Baseline[1]	0.0611	0.0863
	No order restriction	0.0582(−4.75%)	0.083(−3.82%)
	No global representation	0.0623(+1.96%)	0.0898(+4.06%)
	Graph-based interpolation	0.0682(+11.62%)	0.0952(+10.31%)
ml-1m	Baseline[1]	0.0747	0.0981
	No order restriction	0.0714(−4.42%)	0.0963(−1.83%)
	No global representation	0.0759(+1.61%)	0.1029(+4.89%)
	Graph-based interpolation	0.078(+4.42%)	0.1042(+6.22%)

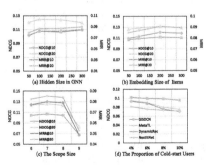

(a) Hidden Size in GNN
(b) Embedding Size of Items
(c) The Scope Size
(d) The Proportion of Cold-start Users

Fig. 3. The Performance of different parameter settings.

Table 4. Effect of convolution methods.

Datasets	Convolutional methods	MRR@10	NDCG@10
Book	Baseline[2]	0.0653	0.0872
	Dynamic Conv	0.0726(+11.18%)	0.0987(+13.19%)
	Deformable Conv	0.0839(+28.48%)	0.1088(+24.77%)
ml-1m	Baseline[2]	0.0713	0.0963
	Dynamic Conv	0.0776(+8.84%)	0.1052(+9.24%)
	Deformable Conv	0.0810(+13.6%)	0.1101(+14.33%)

Hyper-Parameter Sensitivity Analysis. We perform hyper-parameter sensitivity experiments over a number of parameters in order to better understand their impacts. In Fig. 3(a) and Fig. 3(b), with the increase of hidden size and Embedding Size, NDCG and MRR are both improved. Intuitively, increasing the size can enhance items' presentation and get better performance, but higher embedding dimension does not always imply better model performance probably because of the problem of overfitting. In Fig. 3(c), with the increase of scope size in the initial phase, the performance of our model is not improved constantly with the increase of scope size because of the noise. The Fig. 3(d) compares performance of GISDCN and three baselines under different cold-start scenarios. As confirmed by the results, our model constantly outperforms the other baselines in all conditions, indicating the effectiveness of our model for alleviating the cold-start problem.

4 Conclusion

In this paper, we propose a novel framework for the cold-start users and further incorporate the deformable filters to extract the complex sequential dependencies. We make up interactions for the short sequences by re-constructing item sequences into a global graph and selecting the interpolated items that are adjacent to the core interest nodes via GNN. Then to extract the dynamic transition patterns, we utilize the offset to generate more robust filters and improve the flexibility of convolution. Experimental results with four real datasets validate the effectiveness of our approach and our model outperforms most of the state-of-the-art methods.

References

1. Rendle, S., Freudenthaler, C., Schmidt-Thieme, L.: Factorizing personalized Markov chains for next-basket recommendation. In: WWW, pp. 811–820 (2010)
2. Tang, J., Wang, K.: Personalized Top-N sequential recommendation via convolutional sequence embedding. In: WSDM, pp. 565–573 (2018)
3. Yuan, F., Karatzoglou, A., Arapakis, I., et al.: A simple convolutional generative network for next item recommendation. In: WSDM, pp. 582–590 (2019)
4. Tanjim, M., Ayyubi, H., Cottrell, G.: DynamicRec: a dynamic convolutional network for next item recommendation. In: CIKM, pp. 2237–2240 (2020)
5. Hu, G., Zhang, Y., Yang, Q.: CoNet: collaborative cross networks for cross-domain recommendation. In: CIKM, pp. 667–676 (2018)
6. Yao, H., Liu, Y., Wei, W., et al.: Learning from multiple cities: a meta-learning approach for spatial-temporal prediction. In: WWW, pp. 2181–2191 (2019)
7. Du, Z., Wang, X., Yang, H., Zhou, J., Tang, J.: Sequential scenario-specific meta learner for online recommendation. In: KDD, pp. 2895–2904 (2019)
8. Wang, J., Ding, K., Caverlee, J.: Sequential recommendation for cold-start users with meta transitional learning. In: SIGIR, pp. 1783–1787 (2021)
9. Wu, S., et al.: Session-based recommendation with graph neural networks. In: AAAI, pp. 346–353 (2019)

10. Chang, J., et al.: Sequential recommendation with graph neural networks. In: SIGIR, pp. 378–387 (2021)
11. Hidasi, B., Karatzoglou, A.: Recurrent neural networks with top-k gains for session-based recommendations. In: CIKM, pp. 843–852 (2018)
12. Wu, C., Ahmed, A., Beutel, A., Smola, A., Jing, H.: Recurrent recommender networks. In: WSDM, pp. 495–503 (2017)
13. Wang, S., Hu, L., Wang, Y., Cao, L., et al.: Sequential recommender systems: challenges, progress and prospects. In: IJCAI, pp. 6332–6338 (2019)
14. Wu, S., Tang, Y., Zhu, Y., Wang, L., Xie, X., Tan, T.: Session-based recommendation with graph neural networks. In: AAAI, pp. 346–353 (2019)
15. Wang, Z., Wei, W., Cong, G., Li, X., et al.: Global context enhanced graph neural networks for session-based recommendation. In: SIGIR, pp. 169–178 (2020)
16. Li, Y., Tarlow, D., et al.: Gated graph sequence neural networks. In: ICLR (Poster) (2016)
17. Gao, H., Zhu, X., Lin, S., Dai, J.: deformable kernels: adapting effective receptive fields for object deformation. In: ICLR, pp. 1–15 (2020)
18. Liu, Y., Li, B., et al.: A Knowledge-aware recommender with attention-enhanced dynamic convolutional network. In: CIKM, pp. 1079–1088 (2021)

Deep Graph Mutual Learning
for Cross-domain Recommendation

Yifan Wang[1], Yongkang Li[1], Shuai Li[1], Weiping Song[1], Jiangke Fan[2],
Shan Gao[2], Ling Ma[2], Bing Cheng[2], Xunliang Cai[2], Sheng Wang[3],
and Ming Zhang[1(✉)]

[1] School of Computer Science, Peking University, Beijing, China
{yifanwang,liyongkang,songweiping,mzhang_cs}@pku.edu.cn,
lishuai@stu.pku.edu.cn
[2] Meituan, Beijing, China
{jiangke.fan,mount.gao,maling10,bing.cheng,caixunliang}@meituan.com
[3] Paul G. Allen School of Computer Science, University of Washington, Seattle, USA
swang@cs.washington.edu

Abstract. Cross-domain recommender systems have been increasingly important for helping users find satisfying items from different domains. However, existing approaches mostly share/map user features among different domains to transfer the knowledge. In fact, user-item interactions can be formulated as a bipartite graph and knowledge transferring through the graph is a more explicit way. Meanwhile, these approaches mostly focus on capturing users' common interests, overlooking domain-specific preferences. In this paper, we propose a novel Deep Graph Mutual Learning framework (DGML) for cross-domain recommendation. In particular, we first separately construct domain-shared and domain-specific interaction graphs, and develop a parallel graph neural network to extract user preference in corresponding graph. Then the mutual learning procedure uses extracted preferences to form a more comprehensive user preference. Our extensive experiments on two real-world datasets demonstrate significant improvements over state-of-the-art approaches.

Keywords: Cross-domain recommendation · Collaborative filtering · Graph neural networks · Mutual learning

1 Introduction

Collaborative Filtering (CF), which assumes that users who have made similar choices before tend to prefer similar items in the future, always suffer from the cold-start and data sparsity issues since very few interactions of inactive users are insufficient to yield high-quality embeddings for recommendation. Fortunately, in real-world applications, a user often interacts with multiple domains to acquire different information services.

Cross-domain recommendation aims at improving the recommendation performance of the target domain using user preferences from other domains. For example, CodeBook Transfer (CBT) [7] first compresses the dense rating matrix

A. Bhattacharya et al. (Eds.): DASFAA 2022, LNCS 13246, pp. 298–305, 2022.
https://doi.org/10.1007/978-3-031-00126-0_22

of auxiliary domain into an implicit cluster level rating pattern and then transfers the knowledge by sharing this pattern with target domain. Collective Matrix Factorization (CMF) [12] collectively factorizes the rating matrix of different domains with the same user (item) and transfers the knowledge by sharing user's (item's) latent features. However, most of these works only focus on features shared among domains, overlooking domain-specific features.

In this paper, we propose a novel **Deep Graph Mutual Learning** (DGML) framework for cross-domain recommendation. Instead of transferring knowledge across domains by sharing/mapping the embeddings, we build different graph neural networks (GNNs) to extract both domain-shared and domain-specific features and combine these GNNs by the carefully designed mutual regularization strategies so that different models can complement each other with their extracted knowledge. Specifically, we separately construct domain-specific useritem interaction graph for each domain and link them by the overlapping users to construct domain-shared graph. Given a target user-item interaction as input, we associate each user and item with a type-specific embedding and propose a novel parallel GNN to independently aggregate neighborhood embeddings from different hops. Then, we arrange different GNN models to mutually learn in constructed graphs, and apply mutual regularization strategies to enforce user's specific preferences to be distinguished from each other while keeping some common preferences shared among different domains. Finally, we integrate the domainshared and domain-specific preference of a user to calculate the matching score for prediction. Extensive experiments are conducted on two real-world datasets to evaluate the proposed approach and experimental results demonstrate that our approach not only outperforms the existing state-of-the-art models, but also provides good explainability for recommendation.

2 Related Work

Cross-domain Recommendation. Cross-domain recommendation leverages relevant source domains as auxiliary information and transfers corresponding knowledge for the target domain. Shallow cross-domain models CBT [7], CMF [12] and CDFM [10] transfer codebook, latent factors and domain contexts respectively for recommendation. Deep learning based models CoNet [6] and DDTCDR [8] leverage cross-stitch network and autoencoder respectively to enhance knowledge transfer across domains. However, most of these works focus on sharing/mapping embeddings of users, and lack of explicit transferring knowledge across domains.

GNNs-based Recommendation. Inspired by the great success of GNNs, there are some efforts leverage GNNs framework to exploit high-order connectivity in the constructed user-item graph [3,13,15]. Moreover, some works also integrate other context information to construct heterogeneous user-item interaction graph and leverage GNNs to strengthen user's (item's) representations [1,14]. Recently, there are some works leveraging GNNs for knowledge transformation across domains [9,18], but only focus on the overlapping users.

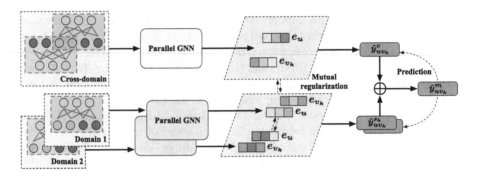

Fig. 1. A schematic view of the DGML, where gray circle represents items in two domains, yellow circle represents the overlapping users between domains. We use different Parallel GNN represented as different colors to extract domain-shared and domain-specific features. (Color figure online)

Knowledge Distillation and Mutual Learning. Knowledge Distillation (KD) is a model compression strategy to improve the learning of a new small student model by transferring knowledge from a previously trained larger teacher model [5,17,19]. As a special case of KD, mutual learning [16] builds an ensemble of student networks to teach each other with distillation losses of Kullback-Leibler (KL) divergence. Our work is partially inspired from KD and serves as the first attempt to apply mutual learning for cross-domain recommendation.

3   The Proposed Model

3.1   Overview

The basic idea of DGML is to link interactions from different domains into a whole graph, and use a cohort of proposed GNNs with mutual learning strategies to extract domain-shared and domain-specific user features respectively. As shown in Fig. 1, there are three components in our DGML framework: 1) an ensemble of parallel GNN models that extracts domain-shared and domain-specific user features by embedding propagation in the constructed user-item bipartite graphs; 2) a mutual learning framework which optimizes parallel GNN models collaboratively to transfer the knowledge across domains and ensure that the specific user features of each domain will be considered; and 3) an ensemble of prediction layers which leverages the refined embeddings from corresponding parallel GNN models and output the matching scores of the user-item pairs.

3.2   Parallel GNN

Influence of neighbors for the target node in the user-item interaction graph always varies depending on their types and distance to the target node. To address the heterogeneity of user-item interaction graph $\mathcal{G}$, we design a Parallel Graph Neural Network mechanism (Parallel GNN). Given a target node t, we sample its neighbors up to L hops away and update target node's embedding by aggregating independently from each of the l-hop neighborhoods $N_t^{(l)}$. Then we take all the aggregated neighborhood embeddings as the target node's features and propose a hop-wise multi-head self-attention layer to propagate information between hops. In this way, all these hops are able to propagate their aggregated embeddings to each other.

3.3   Recommendation

We treat user and item as target nodes and the output of parallel GNN are a set of refined embedding vectors $\{\hat{z}_u^{(0)}, \hat{z}_u^{(1)}, \ldots, \hat{z}_u^{(L)}\}$ and $\{\hat{z}_v^{(0)}, \hat{z}_v^{(1)}, \ldots, \hat{z}_v^{(L)}\}$ (each embedding dimension is d), representing user/item features learned in different hops. We then concatenate them to constitute the final embedding for $\hat{e}_u$ and $\hat{e}_v$ respectively for the user/item. For the top-N recommendation, we simply concatenate the final embeddings of the target user and item, and use an MLP to output the matching score:

$$\hat{y}_{uv} = w_2^{\mathrm{T}} \mathrm{ReLU}\big((\hat{e}_u \parallel \hat{e}_v) W_1 + b_1\big) \tag{1}$$

where $W_1 \in \mathbb{R}^{2Ld \times d}, b_1 \in \mathbb{R}^d, w_2 \in \mathbb{R}^d$ are the parameters of the MLP which maps the final concatenated embedding to a scalar matching score. And the loss function to learn the parameters Θ of the model is as follows:

$$\mathcal{L}_\Theta = \sum_{(u,v)\in\mathcal{Y}\cup\mathcal{Y}^-} \mathcal{L}_{\mathrm{ce}}(y_{uv}, \hat{y}_{uv}), \tag{2}$$

where $\mathcal{Y}$ is positive instances and $\mathcal{Y}^-$ is negative instances uniformly sampled from unobserved interactions. y_{uv} is the label of the matching score. $\mathcal{L}_{ce}$ is the binary cross-entropy defined as:

$$\mathcal{L}_{\mathrm{ce}}(\hat{y}, y) = -y \log \hat{y} - (1 - y) \log(1 - \hat{y}). \tag{3}$$

3.4   Mutual Regularization

To transfer the knowledge across different domains, we arrange the parallel GNN models to mutually learn in domain-shared graph $\mathcal{G}_c$ as well as in domain-specific graphs $\{\mathcal{G}_{s_1}, \mathcal{G}_{s_2}, \ldots, \mathcal{G}_{s_K}\}$ constructed from K different domains and propose two mutual learning regularization strategies for all the user in user set $\mathcal{U}$.

Domain-shared Preference Regularization. The idea of the domain-shared preference regularization is that the common user preference shared across domains should be close to the preference in each domain. Thus, we pull user features extracted in $\mathcal{G}_c$ and $\{\mathcal{G}_{s_1}, \mathcal{G}_{s_2}, \ldots, \mathcal{G}_{s_K}\}$ close. Specially, for $\mathcal{G}_c$ and k-th domain-specific graph $\mathcal{G}_{s_k}$, we apply different parallel GNN models to get refined user embeddings $\hat{e}_u^c$ and $\hat{e}_u^{s_k}$ and minimize the distance of these two embeddings by cosine similarity:

$$\mathcal{D}_{\cos}(c, s_k) = \sum_{u \in \mathcal{U}} 1 - \cos(\hat{e}_u^c, \hat{e}_u^{s_k}). \tag{4}$$

Domain-specific Preference Regularization. The strategy we mentioned above aims to maintain common preference by limiting the distance between domain-shared and domain-specific user features. This will lead to a result that domain-specific preferences are pulled all together. In order to distinguish domain-specific preferences from each other, we encourage the features extracted via parallel GNNs from different domains vary from each other. Thus, we include the following orthogonal constraint as part of the loss.

$$\mathcal{D}_{\text{orth}}(s_k, s_j) = \sum_{u \in \mathcal{U}} |\hat{e}_u^{s_k \text{T}} \hat{e}_u^{s_j}|. \tag{5}$$

where $|\cdot|$ is the L_1 norm. $\hat{e}_u^{s_j}$ is the refined user embedding from other j-th domain.

Shared-specific Preference Mutual Learning. We take both domain-shared and domain-specific user preferences into consideration. For $\mathcal{G}_c$ and $\mathcal{G}_{s_k}$, we denote the domain-shared and domain-specific preference as the matching scores learned in corresponding graph and ensemble the two kinds of user preferences by a gating component g_u to get their complete preference.

$$\hat{y}_{uv_k}^m = g_u \cdot \hat{y}_{uv_k}^c + (1 - g_u) \cdot \hat{y}_{uv_k}^{s_k}. \tag{6}$$

specially,

$$g_u = w_{g_2}^{\text{T}} \text{ReLU}\big((\hat{e}_u^c \parallel \hat{e}_u^{s_k}) W_{g_1} + b_{g_1}\big), \tag{7}$$

where $\hat{y}_{uv_k}^c, \hat{y}_{uv_k}^{s_k}$ and $\hat{y}_{uv_k}^m$ are the domain-shared, domain-specific and complete preference of the user, $W_{g_1} \in \mathbb{R}^{2Ld \times d}, b_{g_1} \in \mathbb{R}^d, w_{g_2} \in \mathbb{R}^d$ are the parameters of the gating component. We consider $\hat{y}_{uv_k}^m$ as an online teacher and distill the preference back into two graphs in a closed-loop form. In contrast to hard labels, soft labels convey the subtle difference between models and therefore can help the student model generalize better than directly learning from hard labels. The alignment between shared (specific) and complete preference can be defined as:

$$\mathcal{D}_{\text{ce}}(c, s_k, m) = \sum_{(u,v) \in \mathcal{Y} \cup \mathcal{Y}^-} \mathcal{L}_{\text{ce}}\big(\sigma(\frac{\hat{y}_{uv_k}^c}{\tau}), \sigma(\frac{\hat{y}_{uv_k}^m}{\tau})\big) + \mathcal{L}_{\text{ce}}\big(\sigma(\frac{\hat{y}_{uv_k}^{s_k}}{\tau}), \sigma(\frac{\hat{y}_{uv_k}^m}{\tau})\big), \tag{8}$$

where σ is the sigmoid function, τ is the temperature parameter to produce a softer probability distribution of labels.

3.5　Model Optimization

By integrating the recommendation loss and mutual regularization together. The overall loss function of parallel GNN model in the domain-shared graph $\mathcal{G}_c$ and k-th specific graph $\mathcal{G}_{s_k}$ are defined as:

$$\mathcal{L}_c = \mathcal{L}_{\Theta_c} + \sum_{j=1}^{K} \mathcal{D}_{\cos}(c, s_j),$$

$$\mathcal{L}_{s_k} = \mathcal{L}_{\Theta_{s_k}} + \sum_{j=1, j \neq k}^{K} \mathcal{D}_{orth}(s_k, s_j), \tag{9}$$

Table 1. Descriptive statistics of two datasets.

	Dianping			Amazon		
	POI	Feeds	Total	Cell	Elec	Total
User	18,636	10,631	21,737	21,182	27,915	27,946
Item	11,372	24,460	35,832	4,846	17,962	22,808
Interactions	133,016	256,244	389,260	69,798	337,511	407,309

and overall loss function of parallel GNN models for the mutual learning procedure is defined as:

$$\mathcal{L} = \mathcal{L}_c + \mathcal{L}_{s_k} + \mathcal{L}_{\Theta_m} + \tau^2 * \mathcal{D}_{ce}(c, s_k, m), \qquad (10)$$

where $\mathcal{L}_{\Theta_c}$, $\mathcal{L}_{\Theta_{s_k}}$ and $\mathcal{L}_{\Theta_m}$ are the recommendation loss based on domain-shared, domain-specific and complete user preferences. We multiply the last loss term by a factor τ^2 to ensure that the relative contributions of ground-truth and teacher probability distributions remain roughly unchanged.

4   Experiment

Experimental Settings. We conduct experiments on two datasets collected from real-world platforms with varying domains. **Dianping**[1]: including POI and Feeds domains. **Amazon**[2]: including Cell Phones (Cell) and Electronics (Elec) domains [2]. The statistics of the datasets are summarized in Table 1. We rank the interactions in chronological order for all the datasets and select the first 80% of historical interactions as the training set with the remaining 10%, 10% as the validation and test set respectively. To demonstrate the effectiveness, we compare DGML with three classes of methods: (A) classical CF methods: including **BPR-MF** [11] and **NMF** [4]; (B) GNNs-based recommendation methods: including **PinSAGE** [15], **NGCF** [13] and **LightGCN** [3]; (C) cross-domain recommendation methods: including **CMF** [12], **CDFM** [10] and **DDTCDR** [8]. We follow the leave-one out evaluation method that randomly samples 99 negative items and use HR@10 and NDCG@10 as evaluation metrics.

Performance Comparison. We summarize the results by comparing the performance of all the methods shown in Table 2 and we can find cross-domain methods outperforms most classifical CF methods and GNNs-based methods. DGML performs consistently better than other baselines on all the datasets. By mutually learn our proposed paralled GNN model in domain-shared and domain-specific graph, DGML is specially designed for cross-domain recommendation and capable to explore both domain-shared and domain-specific user preferences. The significant improvement also indicates the positive effect on achieving better representations for recommendation.

---

[1] https://www.dianping.com.
[2] http://jmcauley.ucsd.edu/data/amazon/.

Table 2. Results on two different datasets. ** and * indicates the statistical significance for $p <= 0.01$ and $p <= 0.05$ compared with the best baseline.

Domain	Metrics	Classical		GNNs-based			Cross-domain			Ours
		bprmf	nmf	pinsage	ngcf	lightgcn	cmf	cdfm	ddtcdr	dgml
POI	HR	0.5516	0.6702	0.4815	0.4918	0.5736	0.7126	0.7067	0.7035	**0.7139**
	NDCG	0.3765	0.4459	0.3214	0.3515	0.3865	0.4845	0.4820	0.4758	**0.4996**∗∗
Feeds	HR	0.5160	0.5502	0.4933	0.5103	0.5291	0.6205	0.5924	0.6023	**0.6533**∗∗
	NDCG	0.3289	0.3385	0.3099	0.3231	0.3330	0.4026	0.3896	0.3868	**0.4307**∗∗
Cell	HR	0.3198	0.3632	0.3573	0.3203	0.3874	0.3924	0.3923	0.3701	**0.3956**∗
	NDCG	0.1936	0.2131	0.2202	0.1947	0.2321	0.2358	0.2355	0.2207	**0.2429**∗∗
Elec	HR	0.3381	0.3471	0.3647	0.3556	0.3848	0.4149	0.4015	0.3873	**0.4235**∗∗
	NDCG	0.2022	0.2041	0.2140	0.2115	0.2295	0.2477	0.2415	0.2326	**0.2557**∗∗

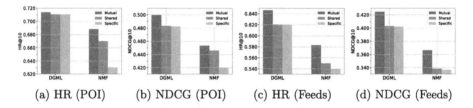

(a) HR (POI)　　　(b) NDCG (POI)　　　(c) HR (Feeds)　　　(d) NDCG (Feeds)

Fig. 2. Effect of mutual regularization on *Dianping*.

Study of DGML. To verify the effect of the mutual regularization, we incorporate NMF as well as our DGML with mutual regularization for the learning process and compare the recommendation performance under three conditions, which learn the model on the data of one domain (denoted as Specific), the data of all domains (denoted as Shared) and the data of all domains with mutual regularization (denoted as Mutual). Figure 2 shows the results and we find that performance learned in domain-shared data is better compared with that learned in one specific domain. Moreover, mutual regularization further improves the performance which demonstrates the effectiveness of mutually learning domain-shared and domain-specific user preferences for cross-domain recommendation.

5   Conclusions

In this work, we propose a Deep Graph Mutual Learning (DGML) framework for cross-domain recommendation. To encode both common and specific user preferences among different domains, we leverage mutual regularization strategies to build the complete preference learned from proposed parallel GNN models for recommendation, which encourage domain-specific features different across domains while remaining close to the domain-shared feature. Extensive experiments on two real-world datasets demonstrate the effectiveness and interpretability of our model.

Acknowledgement. This paper is partially supported by National Key Research and Development Program of China with Grant No. 2018AAA0101902 and the National Natural Science Foundation of China (NSFC Grant No. 62106008 and No. 62006004).

References

1. Fan, S., et al.: Metapath-guided heterogeneous graph neural network for intent recommendation. In: SIGKDD, pp. 2478–2486 (2019)
2. He, R., McAuley, J.J.: Ups and downs: modeling the visual evolution of fashion trends with one-class collaborative filtering. In: WWW, pp. 507–517 (2016)
3. He, X., Deng, K., Wang, X., Li, Y., Zhang, Y., Wang, M.: Lightgcn: simplifying and powering graph convolution network for recommendation. In: SIGIR, pp. 639–648 (2020)
4. He, X., Liao, L., Zhang, H., Nie, L., Hu, X., Chua, T.S.: Neural collaborative filtering. In: WWW, pp. 173–182 (2017)
5. Hinton, G., Vinyals, O., Dean, J.: Distilling the knowledge in a neural network. arXiv preprint arXiv:1503.02531 (2015)
6. Hu, G., Zhang, Y., Yang, Q.: Conet: collaborative cross networks for cross-domain recommendation. In: CIKM, pp. 667–676 (2018)
7. Li, B., Yang, Q., Xue, X.: Can movies and books collaborate? cross-domain collaborative filtering for sparsity reduction. In: IJCAI (2009)
8. Li, P., Tuzhilin, A.: DDTCDR: deep dual transfer cross domain recommendation. In: WSDM, pp. 331–339 (2020)
9. Liu, M., Li, J., Li, G., Pan, P.: Cross domain recommendation via bi-directional transfer graph collaborative filtering networks. In: CIKM, pp. 885–894 (2020)
10. Loni, B., Shi, Y., Larson, M., Hanjalic, A.: Cross-domain collaborative filtering with factorization machines. In: ECIR, pp. 656–661 (2014)
11. Rendle, S., Freudenthaler, C., Gantner, Z., Schmidt-Thieme, L.: BPR: Bayesian personalized ranking from implicit feedback. In: UAI, pp. 452–461 (2009)
12. Singh, A.P., Gordon, G.J.: Relational learning via collective matrix factorization. In: SIGKDD, pp. 650–658 (2008)
13. Wang, X., He, X., Wang, M., Feng, F., Chua, T.S.: Neural graph collaborative filtering. In: SIGIR, pp. 165–174 (2019)
14. Wang, Y., Tang, S., Lei, Y., Song, W., Wang, S., Zhang, M.: Disenhan: disentangled heterogeneous graph attention network for recommendation. In: CIKM, pp. 1605–1614 (2020)
15. Ying, R., He, R., Chen, K., Eksombatchai, P., Hamilton, W.L., Leskovec, J.: Graph convolutional neural networks for web-scale recommender systems. In: SIGKDD, pp. 974–983 (2018)
16. Zhang, Y., Xiang, T., Hospedales, T.M., Lu, H.: Deep mutual learning. In: CVPR, pp. 4320–4328 (2018)
17. Zhang, Y., Xu, X., Zhou, H., Zhang, Y.: Distilling structured knowledge into embeddings for explainable and accurate recommendation. In: WSDM, pp. 735–743 (2020)
18. Zhao, C., Li, C., Fu, C.: Cross-domain recommendation via preference propagation graphnet. In: CIKM, pp. 2165–2168 (2019)
19. Zhu, J., et al.: Ensembled CTR prediction via knowledge distillation. In: CIKM, pp. 2941–2958 (2020)

Core Interests Focused Self-attention
for Sequential Recommendation

Zhengyang Ai[1,2], Shupeng Wang[1(✉)], Siyu Jia[1], and Shu Guo[3(✉)]

[1] Institute of Information Engineering, Chinese Academy of Sciences, Beijing, China
{aizhengyang,wangshupeng,jiasiyu}@iie.ac.cn
[2] School of Cyber Security, University of Chinese Academy of Sciences,
Beijing, China
[3] National Computer Network Emergency Response Technical Team/Coordination
Center of China, Beijing, China
guoshu@cert.org.cn

Abstract. Recent studies identify that sequential recommender systems (SRSs) are improved by self-attention mechanism due to its ability to capture the correlation between interactions. However, two major limitations remain unaddressed by existing works. First, user behaviors in long sequences contain many implicit and noisy preference signals that cannot sufficiently reflect users' actual preferences. Therefore, modeling all the interactive behaviors will worsen the representation of their actual interests. Second, most models only consider the interaction histories as ordered sequences, while ignoring the time interval between interactions, which leads to a loss of effective information. To tackle these issues, we herein propose CIFARec (Core Interests Focused self-Attention based sequential recommendation), which can explicitly extract those interest-relevant interactions from users' implicit feedback information and focus on users' core interests adaptively. Meanwhile, our model takes into account time intervals to retain valid information. Extensive experiments on five benchmark datasets show that our model outperforms various state-of-the-art sequential models consistently.

Keywords: Recommender system · Sequential user behaviors · Attention model

1 Introduction

Sequential recommender systems (SRSs) seek to exploit the order of users' interactions to predict their next action. Markov Chain-based approaches [1,7] are typical examples, which make a simplifying assumption that the next action is conditioned on one or several most recent actions. Another representative work is the use of Recurrent Neural Networks (RNNs) for SRSs [3,4]. Given a sequence of historical interactions of a user, an RNN-based SRS seeks to predict the next interaction by modeling the sequential dependencies. In recent years, inspired by Transformer [11], adopting a self-attention mechanism to SRSs has become a research

A. Bhattacharya et al. (Eds.): DASFAA 2022, LNCS 13246, pp. 306–314, 2022.
https://doi.org/10.1007/978-3-031-00126-0_23

trend. Self-attention-based models [5,6,13] can emphasize those truly relevant and important interactions in a sequence while diminishing the irrelevant ones.

Generally, when modeling the users' interaction sequences, we hope to generate their interest representations, based on which the prediction is yielded. However, in real-life scenarios, not all interactions between users and items reflect the interests of users. Interaction sequences usually contain user interests drift caused by unintended clicks. We collectively refer to these interactions that cannot represent the users' real interests and do not have any impact on the users' subsequent behaviors as *noise interactions*. In contrast, core interests reflect users' deep preferences for items and dominate their selection of candidate items. Therefore, discovering those interactions that represent the users' core interests from their interaction sequences is crucial for generating user interest representations and making next item recommendations. Furthermore, a simplifying assumption made by most of the previous models is to treat the interaction histories as ordered sequences, without considering the time intervals between each interaction. Such an approach leads to a loss of valid information, since the time intervals between interactions are also part of the user behavior patterns and should be included in the user interest representations.

To tackle these above limitations, we propose CIFARec (Core Interests Focused self-Attention based sequential recommendation). Figure 1 gives an overview of the CIFARec model. Specifically, we propose a measurement that adaptively measures the relevancy between interactions and users' core interests. Under this measurement, core interests-related interactions and noise interactions can be explicitly distinguished, and their attention weights are obtained in different ways: for the former, attention weights are calculated as in canonical self-attention, whereas for the latter, a default distribution is directly adopted as a substitute. In this way, our model can focus on those items that are relevant to users' core interests to the maximum extent possible by eliminating the influence of irrelevant ones. As for the time intervals, we calculate the normalized time differences and inject them into the embedding layer as additional information. Compared to models based on canonical self-attention, our model uncovers deeper interests and preferences of users without losing the information contained in the time intervals between interactions. Extensive experiments conducted on both dense and sparse datasets demonstrate the effectiveness of our model over state-of-the-art algorithms.

2 Methodology

2.1 Embedding Layer

The input representation of CIFARec consists of three separate parts, a scalar projection, position embeddings and time interval embeddings.

Scalar Projection: For the training sequence $(S_1^u, S_2^u, \ldots, S_{n_u-1}^u)$, we transform it into a fixed-length sequence $(s_1, s_2, \ldots, s_l)$. For each item s_t in the

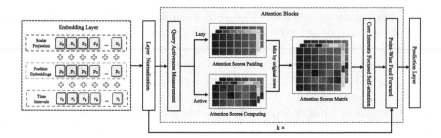

Fig. 1. The overview architecture of the proposed CIFARec.

sequence, to align the dimension, we project the scalar context into d-dim vector $\mathbf{x}_t$ with 1-D convolutional filters where d is the latent dimensionality. After stacking them together, we get the item embedding matrix $\mathbf{X} \in \mathbb{R}^{l \times d}$.

Position Embeddings: Like [5], we inject a learnable position embedding matrix $\mathbf{P} \in \mathbb{R}^{l \times d}$ into the sequence to preserve the local context.

Time Interval Embeddings: Similar to the interaction sequence, the time sequence $(T_1^u, T_2^u, \ldots, T_{n_u-1}^u)$ is transformed into a fixed-length time sequence $(t_1, t_2, \ldots, t_l)$. For each user u in the user set U, we are only concerned about the relative length of the time interval. Therefore, for all time intervals, we divide them by the shortest time interval in the user sequence (except 0) for personalized time intervals as $r_{i(i+1)}^u = \left\lfloor \frac{t_{i+1}-t_i}{r_{min}^u} \right\rfloor$. At last, after projection and stacking, the time interval embedding of user u is denoted as $\mathbf{R} \in \mathbb{R}^{l \times d}$.

The final embedding matrix is the sum of the above three parts:

$$\hat{\mathbf{E}} = \mathbf{X} + \mathbf{P} + \mathbf{R}. \tag{1}$$

2.2   Core Interests Focused Self-attention

The canonical self-attention is defined as $\mathrm{A}(\mathbf{Q}, \mathbf{K}, \mathbf{V}) = \mathrm{softmax}\left(\mathbf{Q}\mathbf{K}^\top/\sqrt{d}\right)\mathbf{V}$ [11], where $\mathbf{Q} \in \mathbb{R}^{L_Q \times d}$, $\mathbf{K} \in \mathbb{R}^{L_K \times d}$, $\mathbf{V} \in \mathbb{R}^{L_V \times d}$ represent queries, keys, and values respectively, and are converted from the embedding $\hat{\mathbf{E}}$ through linear projections. For further discussion, let $\mathbf{q}_i$, $\mathbf{k}_i$, $\mathbf{v}_i$ denote the i-th row in $\mathbf{Q}$, $\mathbf{K}$, $\mathbf{V}$ respectively. Following the formulation in [10], the attention of $\mathbf{q}_i$ is performed as a kernel smoother in a probability form:

$$\mathrm{A}(\mathbf{q}_i, \mathbf{K}, \mathbf{V}) = \sum_j \frac{k(\mathbf{q}_i, \mathbf{k}_j)}{\sum_l k(\mathbf{q}_i, \mathbf{k}_l)}\mathbf{v}_j = \mathbb{E}_{p(\mathbf{k}_j|\mathbf{q}_i)}[\mathbf{v}_j], \tag{2}$$

where $p(\mathbf{k}_j|\mathbf{q}_i) = k(\mathbf{q}_i, \mathbf{k}_j)/\sum_l k(\mathbf{q}_i, \mathbf{k}_l)$ and $k(\mathbf{q}_i, \mathbf{k}_j)$ selects the asymmetric exponential kernel $\exp(\mathbf{q}_i\mathbf{k}_j^\top/\sqrt{d})$. For each $\mathbf{q}_i$ in $\mathbf{Q}$, the self-attention computes the probability $p(\mathbf{k}_j|\mathbf{q}_i)$, based on which the values are combined for the final outputs. As mentioned earlier, the introduction of noise interactions is inevitable with such an approach. Therefore, our motivation is to calculate the probability only for those important $\mathbf{q}_i$, while for the rest, a fixed default distribution is directly used as an alternative. Then, the next question is how to distinguish them?

Query Activeness Measurement: Query activeness is a concept we put forward to denote the relevancy between items and users' core interests. The more active the query, the more its corresponding item can represent the user's core interests. From Eq. (2), the i-th query's attention scores on all keys are defined as a probability $p(\mathbf{k}_j|\mathbf{q}_i)$, and the output is its combination with the value $\mathbf{v}$. There is a wide variety of interests among different users. Hence, if an item reflects a user's core interests, the probability $p(\mathbf{k}_j|\mathbf{q}_i)$ of the query corresponding to it should also be very different from that of other users. In other words, suppose we define a globally fixed distribution $q(\mathbf{k}_j|\mathbf{q}_i)$ as the default distribution, if a query's attention distribution differs little from distribution q, then the item it corresponds to is likely to be a noise interaction. We set this default distribution as a scaled exponential function as $q(\mathbf{k}_j|\mathbf{q}_i) = e^{\mu j}/\sum_{j=1}^{L_k} e^{\mu j}$, where μ is a constant that controls the importance of recent behaviors. Naturally, the "similarity" between distributions p and q can be used to distinguish "active" queries. We measure the "similarity" by Kullback-Leibler divergence:

$$KL(q\,\|\,p) = \log \sum_{j=1}^{L_k} e^{\frac{\mathbf{q}_i \mathbf{k}_j^\top}{\sqrt{d}}} - \sum_{j=1}^{L_k} \log q(\mathbf{k}_j|\mathbf{q}_i)\frac{\mathbf{q}_i \mathbf{k}_j^\top}{\sqrt{d}} + \sum_{j=1}^{L_k} q(\mathbf{k}_j|\mathbf{q}_i) \log q(\mathbf{k}_j|\mathbf{q}_i).$$

$$(3)$$

Dropping the constant term, we define the i-th query's activeness measurement as:

$$M(\mathbf{q}_i, \mathbf{K}) = \log \sum_{j=1}^{L_k} e^{\frac{\mathbf{q}_i \mathbf{k}_j^\top}{\sqrt{d}}} - \sum_{j=1}^{L_k} \frac{e^{\mu j}\frac{\mathbf{q}_i \mathbf{k}_j^\top}{\sqrt{d}}}{\sum_{L_k} e^{\mu j}}. \tag{4}$$

If the i-th query obtains a larger $M(\mathbf{q}_i, \mathbf{K})$, it is more likely to represent the item that the user is really interested in.

Multi-head Core Interests Focused Self-attention: Based on $M(\mathbf{q}_i, \mathbf{K})$, we allow each key to focus on only m active queries and then get the attention:

$$\overline{\mathcal{A}}(\mathbf{Q}, \mathbf{K}, \mathbf{V}) = \mathrm{Softmax}\left(\frac{\overline{\mathbf{Q}}\mathbf{K}^\top}{\sqrt{d}}\right)\mathbf{V}, \tag{5}$$

where $\overline{\mathbf{Q}}$ is a sampling matrix of $\mathbf{Q}$ that only contains the Top-m queries under $M(\mathbf{q}_i, \mathbf{K})$. We set $m = \lceil c \cdot L_Q \rceil$, where $c \in (0, 1]$ is a constant sampling factor that represents the proportion of interests-related interactions. It is worth mentioning that under the multi-head perspective, this attention extracts different active query-key pairs for each head, avoiding severe information loss. As for the remaining $(L_Q - m)$ lazy queries, we use the default distribution $q(\mathbf{k}_j|\mathbf{q}_i)$ as their attention scores.

Point-Wise Feed-Forward Network: After each attention layer, we apply a two-layer feed-forward network with ReLU activation in between:

$$\text{FFN}(\mathbf{S}) = \text{ReLU}(\mathbf{S}\mathbf{W}^{(1)} + \mathbf{b}^{(1)})\mathbf{W}^{(2)} + \mathbf{b}^{(2)}, \tag{6}$$

where $\mathbf{W}^{(1)}, \mathbf{W}^{(2)} \in \mathbb{R}^{d \times d}$ and $\mathbf{b}^{(1)}, \mathbf{b}^{(2)} \in \mathbb{R}^d$. To make the whole network more robust, we utilize layer normalization, residual connections and dropout :

$$\mathbf{S} = \mathbf{S} + \text{Dropout}(\text{FFN}(\text{LayerNorm}(\mathbf{S}))). \tag{7}$$

2.3 Prediction Layer

After k stacked attention blocks, we employ a latent factor model to calculate users' preference score of item i as shown below:

$$R_{i,t} = \mathbf{S}_t \mathbf{X}_i^I, \tag{8}$$

where $\mathbf{X}_i^I \in \mathbb{R}^d$ is the embedding vector of item i and $\mathbf{S}_t$ is the user interests representation generated after the previous t items (i.e., $s_1, s_2, \ldots, s_t$) are given.

3 Experiments

3.1 Settings

Datasets. We study the effectiveness of our proposed approach on two real-world datasets, *i.e.*, *Amazon*[1] and *MovieLens*.[2] For Amazon dataset, we adopt four subsets, 'Movies and TV', 'CDs and Vinyl', 'Video Games' and 'Beauty'. For MovieLens dataset, we adopt the version (MovieLens-1M) that includes 1 million user ratings. To ensure the quality of the dataset, we discard users and items with fewer than 5 related actions. Following [5], for each user, we divide his historical sequence into three parts: the most recent item for testing, the second recent item for validation, and the remaining items for training.

Evaluation Metrics. To evaluate the recommendation performance, we employ two widely adopted Top-N metrics including Hit@10 and NDCG@10 [2]. Hit@10 indicates what percentage of the ground-truth items emerge in the top 10 recommended items. NDCG@10 is the normalized discounted cumulative gain at 10, which assigns larger weights on higher positions. Following [5], for each user u, we rank the 100 randomly sampled negative items together with the ground-truth item. The Hit@10 and NDCG@10 are calculated based on the rankings of these 101 items.

---

[1] http://jmcauley.ucsd.edu/data/amazon/.
[2] https://grouplens.org/datasets/movielens/1m/.

Table 1. Performance comparison of different methods. The best performing method is boldfaced, and the second best method is underlined.

Dataset	Metric	POP	BPR	FPMC	GRU4Rec+	Caser	MARank	SASRec	TiSASRec	CIFARec
ML-1m	NDCG@10	0.2401	0.3249	0.5216	0.5543	0.5488	0.5517	0.5952	0.6091	**0.6289**
	Hit@10	0.4391	0.5813	0.7218	0.7678	0.7986	0.7993	0.8265	0.8283	**0.8439**
Movies & TV	NDCG@10	0.2712	0.3479	0.3371	0.2980	0.3211	0.4598	0.4882	0.4970	**0.5336**
	Hit@10	0.4514	0.5597	0.5098	0.4897	0.5007	0.6531	0.7035	0.7113	**0.7429**
CDs & Vinyl	NDCG@10	0.1912	0.3686	0.3401	0.2130	0.2315	0.4382	0.4983	0.5051	**0.5248**
	Hit@10	0.3345	0.5614	0.5098	0.3516	0.3841	0.6510	0.7151	0.7218	**0.7248**
Video games	NDCG@10	0.2319	0.2831	0.3509	0.2622	0.2801	0.4630	0.5108	0.5139	**0.5243**
	Hit@10	0.4116	0.4434	0.5310	0.4170	0.4714	0.6608	0.7378	0.7421	**0.7501**
Beauty	NDCG@10	0.1818	0.1591	0.1817	0.1843	0.1441	0.2324	0.3219	0.3297	**0.3554**
	Hit@10	0.3195	0.2511	0.2826	0.3181	0.2719	0.3692	0.4854	0.4941	**0.5234**

Baselines. Traditional baselines include (a) frequency-based methods POP, (b) factorization-based methods Bayesian personalized ranking (BPR) [8] and (c) factorizing personalized Markov chain model (FPMC) [7]. We also consider deep learning baselines, including RNN-based recommender model GRU4Rec+ [3], CNN-based recommender model Caser [9], and attention-based models MARank [12], SASRec [5] and TiSASRec [6].

Parameter Setup. We implement CIFARec with *PyTorch* and fine-tune hyperparameters on the validation set. For the architecture in the default version, we set the attention block number $k = 2$ and head number $h = 2$. For all datasets, the dimensionality of latent vectors d is set to 64. The maximum sequence length l is set to 300 for ML-1m and 50 for the other four Amazon datasets. We tune the query's sampling factor c using the validation set, resulting in $c = 0.5$ for Amazon datasets and $c = 0.8$ for ML-1m.

3.2 Recommendation Performance

Table 1 shows the recommendation performance of all the methods on the five datasets, and we have the following observations.

The non-personalized POP method has the worst performance on all datasets because it does not model user preferences based on the user's interaction history. Among all the baseline methods, sequential methods generally outperform non-sequential methods, proving that considering the sequential information of historical interaction helps improve the recommendation performances.

Among sequential recommendation methods, neural methods significantly perform better than traditional methods on dense datasets due to their ability to capture long-term sequential patterns, which is important for dense datasets. Methods based on self-attention mechanism like SASRec and TiSASRec consistently outperform the RNN/CNN-based methods, demonstrating the superiority of self-attention mechanism in modeling sequential information.

Our method improves over the best baseline methods on all dense and sparse datasets with respect to the two metrics. The main reason is that our model is the only one that can explicitly eliminate the negative effects of noise interactions.

3.3 Ablation Study

As described in Sect. 2.2, we define a default distribution $q(\mathbf{k}_j|\mathbf{q}_i)$ that serves as a substitute for lazy queries' attention scores. Given our motivation for conducting a more in-depth model analysis study, we replace the exponential function with a uniform distribution $q_{uni}(\mathbf{k}_j|\mathbf{q}_i) = 1/L_K$ for comparison. Correspondingly, we modify Eq. (4) as:

$$M_{uni}(\mathbf{q}_i, \mathbf{K}) = \log \sum_{j=1}^{L_k} e^{\frac{\mathbf{q}_i \mathbf{k}_j^\top}{\sqrt{d}}} - \frac{1}{L_K} \sum_{j=1}^{L_K} \frac{\mathbf{q}_i \mathbf{k}_j^\top}{\sqrt{d}}. \tag{9}$$

We adopt q_{uni} as the default distribution and M_{uni} as the query activeness measurement to obtain a new model. We denote this model as CIFARec-Uni and add it to the ablation study as a variant of CIFARec.

Table 2 shows the results of our default version ($k = 2, h = 2$) and its seven variants as well as SASRec [5] on four datasets with dimensionality $d = 64$. We introduce the variants and analyze their effects respectively:

Table 2. NDCG@10 on four datasets. Bold score indicates performance better than the default version, while ↓ indicates performance drop more than 5%.

Architecture	Beauty	Games	Movie&TV	ML-1m
CIFARec	0.3554	0.5243	0.5336	0.6289
SASRec	0.3219↓	0.5108	0.4882↓	0.5952↓
w/o TI	0.3542	0.5220	0.5291	0.6152
w/o CIFA	0.3433	0.5179	0.5227	0.6167
CIFARec-Uni	0.3486	0.5152	0.5307	0.5598↓
1 block ($k = 1$)	**0.3570**	**0.5266**	0.5201	0.6068
4 blocks ($k = 4$)	0.3426	0.5103	**0.5352**	**0.6316**
1 head ($h = 1$)	0.1721↓	0.2939↓	0.4571↓	0.6197
4 heads ($h = 4$)	**0.3603**	**0.5255**	0.5317	0.6203

TI. The results show that removing the time interval embedding has little effect on the model on sparse datasets, but causes a noticeable performance degradation on long sequence datasets. This is because on long sequence datasets, the model needs more sequential information to model user's behaviors, and the time interval facilitates this process as additional information.

CIFA. Our model employs the canonical self-attention in [11] after removing the proposed core interests focused self-attention mechanism. The results indicate that the performance degrades more significantly on sparse datasets. Presumably this is because there are fewer interactions related to users' interests in sparse datasets, which makes them more susceptible to noise interactions.

CIFARec-Uni. Not surprisingly, after the default distribution is set to a uniform distribution, the performance is slightly worse than the default model on sparse datasets, but decreases dramatically on the densest dataset (ML-1m). This is because in a long sequence scenario, adopting uniform distribution would greatly weaken the impact of recent items, which does not correspond to reality.

Number of blocks k. We observe that dense datasets (*e.g.*., ML-1m) benefit from a larger k while sparse datasets (*e.g.*., Beauty, Games) prefer a smaller k. The results imply the overfitting problem is less severe on dense datasets.

Head number h. The single-head architecture significantly weakens the performance on sparse datasets. This may be due to the fact that for short sequence data, multi-head attention is required to eliminate the severe information loss caused by the sampling of queries. In contrast, for dense datasets, the sequence length is long enough to still provide sufficient information after sampling.

4 Conclusion

In this paper, we propose a core interests focused self-attention model for sequential recommendation (CIFARec). Compared with the models based on canonical self-attention, our model can tap into users' deeper interests and explicitly eliminate the influence of noise interactions without losing the information contained in the time intervals. Extensive experiments on five real-world datasets verify that our model consistently outperforms the state-of-the-art methods.

References

1. He, R., Kang, W.C., McAuley, J.: Translation-based recommendation. In: RecSys, pp. 161–169 (2017)
2. He, X., Liao, L., Zhang, H., Nie, L., Hu, X., Chua, T.S.: Neural collaborative filtering. In: WWW, pp. 173–182 (2017)
3. Hidasi, B., Karatzoglou, A.: Recurrent neural networks with top-k gains for session-based recommendations. In: CIKM, pp. 843–852 (2018)
4. Hidasi, B., Karatzoglou, A., Baltrunas, L., Tikk, D.: Session-based recommendations with recurrent neural networks. arXiv preprint arXiv:1511.06939 (2015)
5. Kang, W.C., McAuley, J.: Self-attentive sequential recommendation. In: ICDM, pp. 197–206. IEEE (2018)
6. Li, J., Wang, Y., McAuley, J.: Time interval aware self-attention for sequential recommendation. In: WSDM, pp. 322–330 (2020)
7. Rendle, S., Freudenthaler, C., Schmidt-Thieme, L.: Factorizing personalized Markov chains for next-basket recommendation. In: WWW (2010)
8. Rendle, S., Freudenthaler, C., Gantner, Z., Schmidt-Thieme, L.: BPR: Bayesian personalized ranking from implicit feedback. arXiv preprint arXiv:1205.2618 (2012)

9. Tang, J., Wang, K.: Personalized top-n sequential recommendation via convolutional sequence embedding. In: WSDM, pp. 565–573 (2018)
10. Tsai, Y.H.H., Bai, S., Yamada, M., Morency, L.P., Salakhutdinov, R.: Transformer dissection: a unified understanding of transformer's attention via the lens of kernel. arXiv preprint arXiv:1908.11775 (2019)
11. Vaswani, A., et al.: Attention is all you need. In: NIPS, pp. 5998–6008 (2017)
12. Yu, L., Zhang, C., Liang, S., Zhang, X.: Multi-order attentive ranking model for sequential recommendation. In: AAAI, vol. 33, pp. 5709–5716 (2019)
13. Zhou, C., et al.: ATRank: an attention-based user behavior modeling framework for recommendation. In: AAAI, vol. 32 (2018)

SAER: Sentiment-Opinion Alignment Explainable Recommendation

Xiaoning Zong[1], Yong Liu[2], Yonghui Xu[3(✉)], Yixin Zhang[1], Zhiqi Shen[4], Yonghua Yang[5], and Lizhen Cui[1,3(✉)]

[1] School of Software, Shandong University, Jinan, China
{xiaoning.zong,yixinzhang}@mail.sdu.edu.cn, clz@sdu.edu.cn
[2] Alibaba-NTU Singapore Joint Research Institute, Nanyang Technological University, Singapore, Singapore
stephenliu@ntu.edu.sg
[3] Joint SDU-NTU Centre for Artificial Intelligence Research (C-FAIR), Shandong University, Jinan, China
xu.yonghui@hotmail.com
[4] School of Computer Science and Engineering, Nanyang Technological University, Singapore, Singapore
zqshen@ntu.edu.sg
[5] Alibaba Group, Hangzhou, China
huazai.yyh@alibaba-inc.com

Abstract. Explainable recommendation systems not only provide users with recommended results but also explain why they are recommended. Most existing explainable recommendation methods leverage sentiment analysis to help users understand reasons for recommendation results. They either convert particular preferences into sentiment scores or simply introduce the rating as the overall sentiment into the model. However, the simple rating information cannot provide users with more detailed reasons for recommendations in the explanation. To encode more sentiment information, some methods introduce user opinions into the explanations. As the opinion-based explainable recommendation system does not utilize supervision from sentiment, the generated explanations are generally limited to templates. To solve these issues, we propose a model called Sentiment-opinion Alignment Explainable Recommendation (SAER), which combines sentiment and opinion to ensure that the opinion in the explanation is consistent with the user's sentiment to the product. Moreover, SAER provides informative explanations with diverse opinions for recommended items. Experiments on real datasets demonstrate that the proposed SAER model outperforms state-of-the-art explainable recommendation methods.

Keywords: Recommender systems · Explainable recommendation · Sentiment analysis

1 Introduction

Recommender systems have been successfully applied in different scenarios, e.g., e-commerce [5,9], location-based services [2], and online course platforms [8].

A. Bhattacharya et al. (Eds.): DASFAA 2022, LNCS 13246, pp. 315–322, 2022.
https://doi.org/10.1007/978-3-031-00126-0_24

One emerging research topic is the explainable recommendation system, which can not only provide users with the recommendation lists, but also intuitive explanations about why these items are recommended.

To improve the performance of explainable recommendation systems, recent efforts [4] focus on using sentiment analysis of reviews, which can encode integral insights into people's perspectives. Some methods [1,6] leverage the predicted ratings from the recommendation module to adjust the sentiment of the generated explanations. However, even in reviews with a high rating score, users may be dissatisfied with a certain aspect of the product. Recently, some approaches [13] take into account that users may comment on multiple aspects of the product. They present sentiment in polarity. However, these methods cannot learn accurate sentiment representations for providing explanations, because they only use simple rating information for sentiment analysis. To obtain detailed sentiment information, many approaches start to encode opinion into the explainable recommendation. Opinion is the word-level representation of sentiment. Although some methods try to recognize the importance of generating opinionated text, it is still limited with the form of the template. While the rating is simple, if we do not consider both sentiment and opinion, the similarity of opinions in different explanations would be extremely high.

To address these problems, we propose a Sentiment-opinion Alignment Explainable Recommendation (SAER) method. Differing from previous explainable recommendation models [1], which either use sentiment analysis or introduce opinions, We try to capture the user's preference on different attributes by utilizing sentiment analysis and opinion mining simultaneously and keeping them consistent. Specifically, we use the sentiment of the user as the basis for learning opinion representation. Therefore, we introduce the sentiment representation covering the user's specific preferences and encode the sentiment into the opinion representation to achieve alignment. By introducing the sentiment-opinion alignment module, our model improves the performance of the explainable recommendation system.

2   The Proposed Model

Problem Definition. We denote the training set by $\mathcal{T} = \{\mathcal{U}, \mathcal{I}, \mathcal{R}, \mathcal{F}, \mathcal{O}, \mathcal{G}\}$, which includes a set of users $\mathcal{U}$, items $\mathcal{I}$, ratings $\mathcal{R}$, features $\mathcal{F}$, opinions $\mathcal{O}$ and explanation text $\mathcal{G}$. Besides, we define a vocabulary set $\mathcal{V} = \{w_1, w_2, ..., w_{|\mathcal{V}|}\}$ for explanation generation. As we utilize features and opinions from user historical reviews, the feature set $\mathcal{F}$ and opinion set $\mathcal{O}$ are subsets of $\mathcal{V}$. For a given user and item pair (u, i), the objective of SAER is predict the user u's preference rating on the item i and provide an informative explanation $g_{u,i} = \{w_1, w_2, ..., w_n\}$ for prediction.

Rating Prediction. This section predicts the user u's rating $\hat{r}_{u,i}$ on the item i. We fully consider that users may not share the same sentiment for different features. Therefore, we introduce the corresponding feature set as one of the

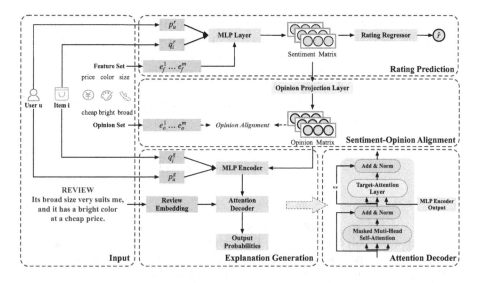

Fig. 1. The overall structure of the proposed SAER model.

input, such as price, color, and size. As follows, we use e_f^m to represent the embedding of the m-th feature in the feature set $\mathcal{F}$. Besides, the ID of user u and item i are given. p_u^r and q_i^r are the respective embeddings.

The final predicted rating value is the average of the users' combined scores for all features and should be close to the users' overall ratings. We adopt MSE loss as the objective function:

$$\mathcal{L}_r = \frac{1}{|\mathcal{T}|} \sum_{u,i \in \mathcal{T}} (r_{u,i} - \hat{r}_{u,i})^2 \tag{1}$$

where $\mathcal{T}$ is the training data set, $r_{u,i}$ and $\hat{r}_{u,i}$ denote the ground truth and the prediction, respectively. Since the simple rating lacks detailed sentiment information, it may lead to the sentiment in the generated explanation failing to cover the user's preference on a particular aspect. We utilize the combination of sentiment and opinion to solve the problem.

Sentiment-Opinion Alignment. To solve the problem that ratings cannot represent users' specific preferences, we will use the sentiment matrix as the premise of the opinion representation. At the same time, we will achieve alignment between both of them.

First, we utilize the opinion projection layer to transfer the sentiment matrix into the opinion matrix,

$$\hat{O}_{u,i} = \sigma(W_o S_{u,i} + b_o) \tag{2}$$

where $W_o \in \mathbb{R}^{d \times d}$, $b_o \in \mathbb{R}^d$. $\hat{O}_{u,i}$ represents the opinion matrix, which represents the user u's opinions on features of the item i. Here $\sigma(\cdot)$ is the nonlinear activation function ReLU [10].

To make the learned opinion matrix close to real opinions, we use mutual information maximization to minimize the distance between the representation of the learned opinion matrix and the real opinion representation during the training stage. Specifically, we perform negative sampling on the opinion set $\mathcal{O}$. In this way, the learned opinions $\hat{O}_{u,i}$ are close to the opinions $O_{u,i}$ and far away from negative samples. Based on real opinions $O_{u,i}$ and negative samples $\tilde{O}_{u,i}$, the loss of alignment training is defined as:

$$\mathcal{L}_o = -log\frac{exp[f_\theta(O_{u,i}, \hat{O}_{u,i})]}{\sum_{\tilde{O}_{u,i} \in \mathcal{O}} exp[f_\theta(\tilde{O}_{u,i}, \hat{O}_{u,i})]} \tag{3}$$

$$f_\theta(x, y) = \sigma(x \cdot W_\theta \cdot y) \tag{4}$$

where $W_\theta \in \mathbb{R}^{d \times d}$, $\sigma(\cdot)$ denotes the sigmoid activation function.

Note that we only use $\mathcal{O}$ during the training process. Because we align real opinions with learned opinions. After that, we need to figure out how to use existing information to generate informative explanations.

Explanation Generation. We have obtained the aligned opinion representation by encoding the sentiment matrix and optimizing mutual information between real opinions and learned opinions. In this section, we will use the learned opinion matrix to generate explanations related to recommended item i for the user u. The explanation generation module is based on the encoder-decoder framework. We will divide this module into MLP encoder and attention decoder.

MLP Encoder. To guarantee the generation of personalized reviews, we introduce user and item again, with its representation p_u^g and q_i^g distinguish from the recommendation task. Besides, the learned opinion matrix is used as the supervised signal for the generation procedure.

$$z^g = \begin{bmatrix} p_u^g || q_i^g \\ p_u^g || q_i^g \\ \cdots \\ p_u^g || q_i^g \end{bmatrix} \hat{O}_{u,i} \tag{5}$$

We employ MLP with one hidden layer as encoder,

$$Q_{enc} = \phi(W_e z^g + b_e) \tag{6}$$

where $W_e \in \mathbb{R}^{d \times 3d}$ and $b_e \in \mathbb{R}^d$ are model parameters, $\phi(\cdot)$ is hyperbolic tangent function. Since the decoder is based on the vanilla-attention mechanism [12], we use Q_{enc} to denote the output of the encoder.

Attention Decoder. As shown in the attention decoder in Fig. 1, we use the encoder output as Q and the output of the masked self-attention in the decoder as K and V. Then we learn more complex transitions using a multi-layer attention mechanism.

Table 1. The statistics of experimental datasets.

Dataset	# users	# items	# reviews	# features	# opinions
Amazon	7,506	7,360	441,783	8,650	8,103
Yelp	27,147	20,266	1,293,247	11,224	11,183

Objective Function. In this module, we optimize the generated results using the cross-entropy loss function:

$$\mathcal{L}_g = -\frac{1}{|\mathcal{T}|} \sum_{u,i \in \mathcal{T}} \frac{1}{|g_{u,i}|} \sum_{n=1}^{|g_{u,i}|} \log p\left(y_n\right) \tag{7}$$

where $g_{u,i} = \left\{w_1, w_2, ..., w_{|g|}\right\}$ is the ground-truth explanation for the user u and the item i, $|g_{u,i}|$ is the length of it, and $p(y_n)$ denotes the output probability of word w_n. In the generation module, we introduce user and item a second time to ensure the diversity of generated explanation. Besides, we expect the feature and related opinion to be present in the generated sentences. That could improve the similarity with original reviews.

Optimization. We optimize three modules together to form end-to-end training. The joint objective function is as follows:

$$\mathcal{J} = \min_{\Theta} \left(\mathcal{L}_r + \lambda_o \mathcal{L}_o + \lambda_g \mathcal{L}_g + \lambda_n \|\Theta\|^2\right) \tag{8}$$

where Θ is the set of model parameters, and λ_o, λ_g are proportions of each module in the total task. λ_n is the regularization weight for all parameters. It is worth mentioning that the attention decoder does not know the input during the test stage. As a result, we make an assumption of the output in advance. Specifically, we take the start symbol as input and use its output as the input sequence for the test stage.

3   Experiments

Experimental Settings. We perform the experiments on two different public datasets: **Amazon** [3] and **Yelp**[1]. As for the data processing, we follow the setting in [6]. We randomly divide each dataset into training, testing, and validation sets following 8:1:1. The statistics about the dataset are summarized in Table 1. We compare the proposed model with the following explainable recommendation methods: Attribute-to-Sequence (Att2Seq) [1], Neural Rating and Tips generation (NRT) [7], and Neural Template Explanations Generation (NETE) [6]. In

---

[1] https://www.yelp.com/dataset.

Table 2. BLEU and ROUGE score on the dataset. The best results are in bold.

Dataset	Model	BLEU(%)		ROUGE-1(%)			ROUGE-2(%)		
		BLEU-1	BLEU-4	Precision	Recall	F1	Precision	Recall	F1
Amazon	Att2Seq	12.21	0.97	18.97	12.02	14.34	2.32	1.79	1.98
	NRT	13.86	0.48	22.2	13.44	15.71	2.38	1.64	1.87
	NETE	18.17	2.34	32.16	20.91	23.99	7.33	4.62	5.27
	SAER-w/o opinion	23.15	3.08	34.88	26.16	28.07	7.51	6.51	6.41
	SAER	**25.03**	**3.97**	**35.41**	**27.22**	**28.93**	**8.42**	**6.85**	**6.52**
Yelp	Att2Seq	10.42	0.58	17.89	11.39	12.85	1.62	1.11	1.21
	NRT	13.91	0.15	18.94	8.01	10.13	1.21	0.45	0.59
	NETE	19.45	2.71	33.35	22.61	25.44	8.74	5.58	6.29
	SAER-w/o opinion	22.88	2.95	33.96	26.11	27.54	7.13	6.49	6.64
	SAER	**24.57**	**3.16**	**34.27**	**26.89**	**28.21**	**9.02**	**7.33**	**7.14**

the text generation task, we apply the commonly used machine translation evaluation metrics BLEU and ROUGE to measure the similarity between the predicted results and the ground truth. SAER is implemented using PyTorch [11] and trained with batch size of 128. For the joint objective learning, we search the parameters λ_o and λ_g from $[0.1, 0.5, 1]$, and λ_n is set to 0.0001. Besides, the learning rate is set to 0.0001.

Experimental Results. In this section, we focus on evaluating the quality of the generated explanations from the perspective of similarity. The corresponding experimental results are shown in Table 2. It can be seen that our method outperforms all baseline methods. SAER and NETE perform better in terms of BLEU and ROUGE. Because both of these methods introduce additional supervision signals for explanation generation. Moreover, we consider sentiment and opinion at the same time and align them. In this way, SAER can ensure that opinions in the explanations are diverse and make them more similar to real opinions, so as to achieve better experimental results. To verify whether the sentiment-opinion alignment module is effective, we first remove it and then use the remaining part as the new model SAER-w/o opinion. We use SAER-w/o opinion to perform the same experiment. The experimental results show that even without the sentiment-opinion alignment module, our model is still better than other methods in generating explanations. However, SAER-w/o opinion is slightly worse than SAER. This is because the introduction of opinions can enrich the content of the generated explanations. The alignment between the sentiment and the opinion increases the diversity of opinions in the generated text. Therefore, the sentiment-opinion alignment module can improve the explainable recommendation performance.

To verify that SAER is an effective explainable recommendation model, we analyze generated results on the Amazon Movies & TV dataset. Table 3 show the set of examples. According to the results of rating prediction, it can be seen that SAER is better than the baseline method NETE in the recommendation task. NETE only considers the user's overall sentiment for the item. However, we believe that users allocate their sentiments in all aspects of the item. As a result, considering user preferences from multiple aspects can improve the effectiveness of the recommendation system. It can be seen from the generated explanations that our model can capture the given features. However, NETE utilizes features as the supervision signal for generating text, there are still cases where the supervision factor cannot be captured. This is because we introduce the opinion matrix to the generation task, and the sentiment matrix learns the feature representation. The generated results of our model variant SAER-w/o opinion can only capture the user's sentiment towards the item. And its results have no informative explanations and no detailed reasons. Therefore, it is necessary to introduce the sentiment-opinion alignment module.

Table 3. Case study of explanation generations with the ratings 2, 3, and 5.

Rating	Feature (F) & Opinion (O)	Ground truth	Model	Generated explanation
2 NETE: 3.54 **SAER: 2.79**	F: performance O: wonderful	Justin bartha's performance is wonderful	NETE	I gave this movie 5 stars because it was a very good
			SAER-w/o opinion	**I was surprised to see the performance**
			SAER	**The performance of the cast is wonderful**
3 NETE: 3.53 **SAER: 3.77**	F: characters O: nice	The characters are nicely fleshed out	NETE	The characters are very good
			SAER-w/o opinion	**The characters are interesting and very good**
			SAER	**The characters is interesting and a bit nice**
5 NETE: 4.89 **SAER: 4.97**	F: movie O: awesome	Fast five is a awesome movie	NETE	It has a great cast
			SAER-w/o opinion	**This movie is awesome**
			SAER	**This movie is awesome action**

4   Conclusion and Future Work

In this paper, we propose a Sentiment-opinion Alignment Explainable Recommendation model (SAER), which can provide recommendations and generate detailed explanations. We consider user preferences from different aspects for the recommendation. Moreover, SAER combines sentiment and opinion to improve the effect of explainable recommendation. In addition, we introduce the sentiment-opinion alignment module to help generate more informative explanations. For future work, we would like to investigate the method of evaluating sentiment expression and further improve the quality of the generated text.

Acknowledgements. This work is supported, in part, by National Key R&D Program of China No. 2021YFF0900804, National Natural Science Foundation of China No. 91846205, the Innovation Method Fund of China No. 2018IM020200, the Fundamental Research Funds of Shandong University. This work is also supported, in part, by Alibaba Group through Alibaba Innovative Research (AIR) Program and Alibaba-NTU Singapore Joint Research Institute (JRI), Nanyang Technological University, Singapore.

References

1. Dong, L., Huang, S., Wei, F., Lapata, M., Zhou, M., Xu, K.: Learning to generate product reviews from attributes. In: Proceedings of the 15th Conference of the European Chapter of the Association for Computational Linguistics, vol. 1, Long Papers, pp. 623–632 (2017)
2. Han, P., et al.: Contextualized point-of-interest recommendation. In: Proceedings of the Twenty-Ninth International Joint Conference on Artificial Intelligence, pp. 2484–2490 (2020)
3. He, R., McAuley, J.: Ups and downs: modeling the visual evolution of fashion trends with one-class collaborative filtering. In: Proceedings of the 25th International Conference on World Wide Web, pp. 507–517 (2016)
4. Ito, T., Tsubouchi, K., Sakaji, H., Yamashita, T., Izumi, K.: Contextual sentiment neural network for document sentiment analysis. Data Sci. Eng. **5**(2), 180–192 (2020)
5. Lei, C., et al.: Semi: a sequential multi-modal information transfer network for e-commerce micro-video recommendations. In: Proceedings of the 27th ACM SIGKDD Conference on Knowledge Discovery & Data Mining, pp. 3161–3171 (2021)
6. Li, L., Zhang, Y., Chen, L.: Generate neural template explanations for recommendation. In: Proceedings of the 29th ACM International Conference on Information & Knowledge Management, pp. 755–764 (2020)
7. Li, P., Wang, Z., Ren, Z., Bing, L., Lam, W.: Neural rating regression with abstractive tips generation for recommendation. In: Proceedings of the 40th International ACM SIGIR conference on Research and Development in Information Retrieval, pp. 345–354 (2017)
8. Lin, Y., Feng, S., Lin, F., Zeng, W., Liu, Y., Wu, P.: Adaptive course recommendation in MOOCs. Knowl.-Based Syst. **224**, 107085 (2021)
9. Liu, Y., et al.: Pre-training graph transformer with multimodal side information for recommendation. In: Proceedings of the 29th ACM International Conference on Multimedia, pp. 2853–2861 (2021)
10. Nair, V., Hinton, G.E.: Rectified linear units improve restricted Boltzmann machines. In: ICML (2010)
11. Paszke, A., et al.: PyTorch: an imperative style, high-performance deep learning library. Adv. Neural. Inf. Process. Syst. **32**, 8026–8037 (2019)
12. Vaswani, A., et al.: Attention is all you need. arXiv preprint arXiv:1706.03762 (2017)
13. Zhang, Y., Lai, G., Zhang, M., Zhang, Y., Liu, Y., Ma, S.: Explicit factor models for explainable recommendation based on phrase-level sentiment analysis. In: Proceedings of the 37th International ACM SIGIR Conference on Research & Development in Information Retrieval, pp. 83–92 (2014)

Toward Auto-Learning Hyperparameters for Deep Learning-Based Recommender Systems

Bo Sun[1,2], Di Wu[1,3(✉)], Mingsheng Shang[1], and Yi He[4]

[1] Chongqing Institute of Green and Intelligent Technology, Chinese Academy of Sciences, Chongqing 400714, China
wudi.cigit@gmail.com

[2] Chongqing School, University of Chinese Academy of Sciences, Chongqing 400714, China

[3] Institute of Artificial Intelligence and Blockchain, Guangzhou University, Guangzhou 510006, China

[4] Department of Computer Science, Old Dominion University, Norfolk, VA 23462, USA

Abstract. Deep learning (DL)-based recommendation system (RS) has drawn extensive attention during the past years. Its performance heavily relies on hyperparameter tuning. However, the most common approach of hyperparameters tuning is still Grid Search—a tedious task that consumes immerse computational resources and human efforts. To aid this issue, this paper proposes a general hyperparameter optimization framework for existing DL-based RSs based on differential evolution (DE), named DE-Opt. Its main idea is to incorporate DE into a DL-based RS model's training process to auto-learn its hyperparameters λ (regularization coefficient) and η (learning rate) simultaneously at layer-granularity. Empirical studies on three benchmark datasets verify that: 1) DE-Opt is compatible with and can automate the training of the most recent DL-based RSs by making their λ and η adaptively learned, and 2) DE-Opt significantly outperforms the state-of-the-art hyperparameter searching competitors in terms of both higher learning performance and lower runtime.

Keywords: Recommender systems · Hyperparameter tuning · Grid search · Deep learning · Differential evolution

1 Introduction

Recommender system (RS) has become an indispensable building block in online service, which boosts business and elevates user experience [1]. Traditional RSs tend to have inferior performance in dealing with data sparsity and cold-start issues and balancing the recommendation qualities against different evaluation metrics [2–4]. To aid the issues, recently advanced RSs manifesting remarkable results mostly advocated deep learning (DL) techniques [3, 5], thanks to their strong capability of modeling user-item interactions that are complex and non-linear in practice [5, 6].

A. Bhattacharya et al. (Eds.): DASFAA 2022, LNCS 13246, pp. 323–331, 2022.
https://doi.org/10.1007/978-3-031-00126-0_25

Despite their successes, existing DL-based RSs mainly suffer from a common draw-back—the tedious process of hyperparameter tuning [7], for two reasons. On the one hand, due to the black-box nature of neural architecture, the hyperparameters in a DL-based RS directly decide the system properties in both training and trained states, e.g., how fast can the RS converge and how well the converged equilibrium optimizes the designed learning objective. On the other hand, the ubiquity of DL-based RSs exacerbates this drawback. In fact, various regularization terms have been engineered to incorporate side information of the items and users (e.g., textual description of items and geograph-ical relations among users) in a wide range of recommendation applications, where each such term added to the learning objective introduces a new hyperparameter, and a well-tuned DL-based RS can usually lead to significantly superior performance [7–9]. However, the most common approach of tuning the RS hyperparameters is still Grid Search [10, 15, 16], which is a tedious process that consumes immerse computational resources and human efforts.

To alleviate the drawback of Grid Search, some efforts are made to tune hyperpa-rameters of RSs automatically. A classic study is SGDA [11] that adaptively tunes λ (regularization coefficient) based on alternating optimization. Inspired by SGDA, λOpt is proposed [8] to tune λ adaptively by enforcing regularization during training. Besides, to tune the learning rate η, an η-adaptive optimizer can be adopted. Such as Adam [17], AMSGrad [12], and AdaMod [13]. However, these approaches can only tune a single hyperparameter λ or η adaptively.

In this paper, we propose a general hyperparameter optimization framework for DL-based RSs based on differential evolution (DE) [18, 19], named DE-Opt, with its key ideas being twofold. First, DE-Opt decomposes and controls the RS training process in a finer level of granularity by allowing the hyperparameters to differ across neural layers. This enlarges the search space for optimal hyperparameters yet alleviates the impact of each hyperparameter so as to better the resultant models. Second, DE-Opt frames the hyperparameter tuning task as an optimization problem and embeds it into the RS training framework, so that the model parameters and the hyperparameters of an RS are jointly trained. As such, DE-Opt allows to train the RSs *only once*, thereby substantially improving the training efficiency of DL-based RSs by attaining their optimal RS models and hyperparameters all at once.

Specific Contributions of this paper are summarized as follows:

- This is the first work to propose the auto-learning of the model parameters and hyperparameters of DL-based RSs at layer-wise granularity.
- The proposed DE-Opt framework is agnostic to the learning objectives and thus is compatible with most existing DL-based RSs, boosting their training efficacy by making hyperparameters auto-learned along with the model parameters jointly.
- Extensive experiments on three real-world datasets are carried out to show that our DE-Opt 1) can automate the training of the state-of-the-art DL-based RS models, and 2) outperforms the state-of-the-art hyperparameter searching competitors.
- To promote reproducible DL research, we open-access our code implementation by a link at https://github.com/Wuziqiao/DE-Opt.

2   The Proposed DE-Opt Optimization Framework

We design the architecture of DE-Opt. It consists of two parts, as shown in Fig. 1.

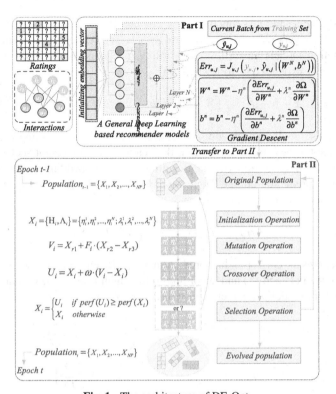

Fig. 1. The architecture of DE-Opt.

2.1   Part I: RS Training with Layer-Wise Granular Hyperparameter Control

To make different layers be finely controlled, we match η and λ with model depth (the number of hidden layers), as shown in Table 1. H and Λ corresponding to the η and λ vectors, respectively, where η^n denotes η at n-th layer and λ^n denotes λ at n-th layer, $n \in \{1, 2, .. , N\}$.

Table 1. The vectorized hyperparameters of matching to model depth.

Hyperparameters	Layer 1	...	Layer n	...	Layer N
Learning rate: H	η^1		η^n		η^N
Regularization coefficient: Λ	λ^1		λ^n		λ^N

When the observed ground-truth $y_{u,j}$ on interaction (u, j) of user u and item j is input from the current batch of the training set, the current objective function is:

$$L_{u,j}(\Theta) = L_{u,j}(W, b) = J_{u,j}(W, b) + \Lambda \otimes \Omega(W, b), \tag{1}$$

where $J_{u,j}(\cdot)$ denotes the loss function on $y_{u,j}$ and $\otimes$ denotes the element-wise product between Λ and layers (λ^n corresponding to n-th layer). Although different DL-based recommendation models have different working mechanisms, their DL architectures can be uniformly expressed as follows [5, 15]:

$$a^1 = W^1 x + b^1, \; a^n = f\left(W^n a^{n-1} + b^n\right), \; a^N = f\left(W^N a^{N-1} + b^N\right), \tag{2}$$

where x denotes initializing embedding vector, a^n denotes the embedding vector of n-th layer, $f(\cdot)$ denotes the activation function, W^n denotes the n-th weight vector or matrix, b^n denotes the n-th bias term. After aggregating the last embedding vector a^N, we can obtain the final prediction output. Then, we can calculate the error $Err_{u,j}$ of the output of the last N-th layer on (u, j) as follows:

$$Err_{u,j} = J_{u,j}\left(y_{u,j}, \hat{y}_{u,j} | \left(W^N, b^N\right)\right). \tag{3}$$

where $\hat{y}_{u,j}$ denotes the corresponding prediction of $y_{u,j}$. By employing stochastic gradient descent (SGD) to (1) and (3) and according to the chain rule [5–7], we can obtain the updating rules of model parameters of n-th layer as follows:

$$W^n \leftarrow W^n - \eta^n \frac{\partial L_{u,j}}{\partial W^n} = W^n - \eta^n\left(\delta^n a^{n-1} + \lambda^n \frac{\partial \Omega}{\partial W^n}\right), \tag{4}$$

$$b^n \leftarrow b^n - \eta^n \frac{\partial L_{u,j}}{\partial b^n} = b^n - \eta^n\left(\delta^n + \lambda^n \frac{\partial \Omega}{\partial b^n}\right), \tag{5}$$

Formulas (4) and (5) show that each layer is finely controlled with different hyperparameters at layer-granularity.

2.2 Part II: DE-Based Hyperparameter Auto-Learning

This part is to employ DE to auto-learn H and Λ simultaneously, including four operations of Initialization, Mutation, Crossover, and Selection [18, 19].

Initialization. At training epoch $t - 1$, part II starts with a pair of initial H and Λ with NP individuals, where each pair is called a 'target vector' as follows:

$$X_i = \{H_i, \Lambda_i\} = \left\{\left[\eta_i^1, \eta_i^2, ..., \eta_i^N\right], \left[\lambda_i^1, \lambda_i^2, ..., \lambda_i^N\right]\right\}, \tag{6}$$

where X_i stands for the i-th individual, $i \in \{1, 2, ... , NP\}$. All the NP individuals group together to form an original population.

Mutation. Mutant vector V_i is generated for each target vector X_i. We choose the *DE/Rand/1* mutation strategy to generate V_i for the sake of robustness [18, 19] as:

$$V_i = X_{r1} + F_i \cdot (X_{r2} - X_{r3}), \tag{7}$$

where $r1, r2, r3 \in \{1, 2, ..., NP\}$ are randomly different from each other and i, and F_i is the scaling factor that positively controls the scaling of various vectors. The setting of scaling factor F_i decides the balance of convergence speed and the optimal performance. We adapt the scale factor local search in DE (SFLSDE) to set F_i self-adaptively according to [18, 19].

Crossover. After mutation, crossover uses V_i to disturb X_i to increase population diversity. X_i and V_i are subject to the crossover operation to generate a new trial vector U_i. We adopt the *DE/CurrentToRand/1* strategy [18, 19] to generate the trial vector U_i, which linearly combines X_i and V_i as follows:

$$U_i = X_i + \omega \cdot (V_i - X_i), \tag{8}$$

where ω is a random number between 0 and 1. Joining (7) into (8), we arrive at:

$$U_i = X_i + \omega \cdot (X_{r1} - X_i) + F_i \cdot (X_{r2} - X_{r3}). \tag{9}$$

Selection. Next, we test a DL-based recommendation model's performance with U_i or X_i on validation dataset as follows:

$$X_i = \begin{cases} U_i \text{ if } perf(U_i) \geq perf(X_i) \\ X_i \text{ otherwise} \end{cases}, \tag{10}$$

where function $perf(\cdot)$ denotes evaluating a DL-based recommendation model's performance on a validation dataset.

After all the NP target vectors (individuals) are evolved at training epoch $t - 1$, DE-Opt moves to the next training epoch t.

3 Experiments

We mainly aim to investigate the following research questions:

RQ.1. Can DE-Opt improve the state-of-the-art (SoTA) models with hyperparameter searched and fixed ad-hoc?
RQ.2. Can DE-Opt outperform the SoTA hyperparameter searching competitors?

3.1 General Settings

Datasets. Three real-world datasets [5], i.e., *FilmTrust*,[1] *MovieLens 1M* (See footnote 1), and *Each Movie* (See footnote 1), are adopted to benchmark our DE-Opt, with their statistics summarized in Table 2. Each dataset is randomly split into the training/validation/testing sets that contain 70%/10%/20% observed entries, respectively.

---
[1] https://grouplens.org/datasets/movielens/.

Table 2. Statistics of the studied datasets

No.	Name	#User	#Item	#Interaction	Density
D1	Film trust	1508	2071	37,919	1.21%
D2	MovieLens 1 M	6040	3706	1,000,209	4.47%
D3	Each movie	72,916	1628	2,811,983	2.36%

Baselines. We compare our DE-Opt with the following state-of-the-art models.

Fixed Hyperparameter Models. Three item ranking models: NeuMF (WWW 2017) [14], LRML (WWW 2018) [16], and NGCF (SIGIR 2019) [20].

Adaptive Hyperparameter Models. Two λ-adaptive models: SGDA (WSDM 2012) [11] and λOpt (KDD 2019) [8] whose η is set by Grid Search. We respectively adopt Adam (ICLR 2015) [17], AMSGrad (ICLR 2018) [12], and AdaMod (2019) [13] to optimize λOpt to implement three λ-η-adaptive models: Adam-λOpt, AMSGrad-λOpt, and AdaMod-λOpt.

Evaluation Metrics. To evaluate item ranking accuracy, we adopt three widely-used evaluation protocols [5], i.e., NDCG@K, Recall@K, and Hit Ratio (HR)@K, where K is a cutoff parameter in the top-ranked list.

Implementation Details. The searching space of Grid Search and DE-Opt are the same. We adopt the ten times amplification principle for setting searching space, i.e., the searching space is ten times the optimal values set by Grid Search. Except for λ and η, all the other hyperparameters of involved models are set according to original papers. The default optimizer is SGD.

4   Comparison with Fixed Hyperparameter Models (RQ.1)

Table 3 records the detailed comparison results of item ranking accuracy. We conduct the Wilcoxon signed-ranks test [21] on the comparison results of Table 3 with a significance level of 0.05. The results demonstrate that DE-Opt can significantly improve NeuMF and LRML in item ranking accuracy. Although the hypothesis is not accepted on NGCF, we can see that DE-Opt outperforms Grid Search in most cases. Besides, Fig. 2 shows the comparison results of CPU running time in item ranking tasks. We find that DE-Opt consumes significantly less time than Grid Search.

4.1   Comparison with Adaptive Hyperparameter Models (RQ.2)

Table 4 presents the comparison results. These comparisons are also checked by the Wilcoxon signed-ranks test [21] with a significance level of 0.05. Table 4 clearly shows that DE-Opt achieves significantly higher item ranking accuracy than all the comparison models. Besides, we also compare their computational efficiencies. The results are shown in Fig. 3, where we see that DE-Opt significantly outperforms the adaptive hyperparameter models in computational efficiency.

Table 3. The comparison results of item ranking accuracy

Dataset	Metric*	NeuMF		LRML		NGCF	
		Grid search	DE-Opt	Grid search	DE-Opt	Grid search	DE-Opt
D1	NDCG@10	0.4513	**0.4641**	**0.3781**	0.3746	0.3734	**0.3773**
	Recall@10	0.5404	**0.5500**	0.4385	**0.4460**	**0.4002**	0.3987
D2	NDCG@10	0.3035	**0.3047**	0.2136	**0.2234**	0.3541	**0.3544**
	Recall@10	0.1241	**0.1262**	0.0806	**0.1296**	0.1366	**0.1392**
D3	NDCG@10	0.1078	**0.1105**	0.0655	**0.0752**	0.0523	**0.0528**
	Recall@10	0.1908	**0.1921**	0.1138	**0.1315**	0.0651	**0.0663**
Statistic	*p-value*	**0.0156**		**0.0313**		0.1094	

* A higher value indicates a higher item ranking accuracy

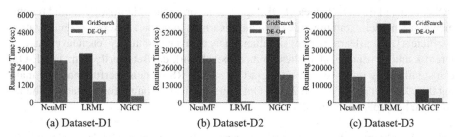

(a) Dataset-D1   (b) Dataset-D2   (c) Dataset-D3

Fig. 2. The comparison results of GPU running time in item ranking task.

Table 4. The comparison results with adaptive hyperparameter models

Dataset	Metric*	SGDA	λOpt	Adam-λOpt	AMSGrad-λOp	AdaMod-λOpt	**DE-Opt**
D1	HR@50	0.8696	0.8724	0.8738	0.8749	0.8765	**0.8791**
	NDCG@50	0.4515	0.4692	0.4916	**0.5003**	0.4981	0.4993
D2	HR@50	0.3877	0.4115	0.4213	0.4302	0.4242	**0.4349**
	NDCG@50	0.1335	0.1471	0.1524	0.1553	0.1529	**0.1556**
D3	HR@50	0.4430	0.4720	0.5123	0.5217	0.5285	**0.5339**
	NDCG@50	0.1278	0.1351	0.1462	0.1478	0.1484	**0.1535**
Statistic	*p-value*	**0.0156**	**0.0156**	**0.0156**	**0.0469**	**0.0156**	—

* A higher value indicates a higher item ranking accuracy

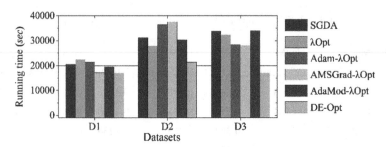

Fig. 3. The comparison results of GPU running time

5   Conclusion

This paper proposes a general hyperparameter optimization framework based on differential evolution (DE) for existing deep learning (DL)-based recommender systems (RSs), named DE-Opt. Its idea of DE-Opt is to incorporate DE into a DL-based recommendation model's training process to auto-learn its hyperparameters λ (regularization coefficient) and η (learning rate) simultaneously at layer-granularity (each layer has different λ and η). Empirical studies on three benchmark datasets verify that: 1) DE-Opt can significantly improve state-of-the-art DL-based RS models' performance by making their λ and η adaptive, and 2) DE-Opt also significantly outperforms state-of-the-art adaptive hyperparameter models.

Acknowledgements. This work is supported in part by the National Natural Science Foundation of China under Grants 62176070, 62072429, 62002337, and 61902370, in part by the Chinese Academy of Sciences "Light of West China" Program, and in part by the Key Cooperation Project of Chongqing Municipal Education Commission (HZ2021008).

References

1. Wu, D., Shang, M., Luo, X., Wang, Z.: An L1-and-L2-norm-oriented latent factor model for recommender systems. IEEE Trans. Neural Netw. Learn. Syst. (2021). https://doi.org/10.1109/TNNLS.2021.3071392
2. Wei, T., et al.: Fast adaptation for cold-start collaborative filtering with meta-learning. In: ICDM, pp. 661–670 (2020)
3. Meng, L., Shi, C., Hao, S., Su, X.: DCAN: deep co-attention network by modeling user preference and news lifecycle for news recommendation. In: DASFAA, pp. 100–114 (2021)
4. Wu, D., Luo, X., Shang, M., He, Y., Wang, G., Wu, X.: A data-characteristic-aware latent factor model for web service QoS prediction. IEEE Trans. Knowl. Data Eng. (2020). https://doi.org/10.1109/TKDE.2020.3014302
5. Zhang, S., Yao, L., Sun, A., Tay, Y.: Deep learning based recommender system: a survey and new perspectives. ACM Comput. Surv. **52**(1), 1–38 (2019)
6. Covington, P., Adams, J., Sargin, E.: Deep neural networks for youtube recommendations. In: RecSys, pp. 191–198 (2016)
7. Maher, M., Sakr, S.: SmartML: a meta learning-based framework for automated selection and hyperparameter tuning for machine learning algorithms. In: EDBT, pp. 554–557 (2019)

8. Chen, Y., et al: λopt: learn to regularize recommender models in finer levels. In: SIGKDD, pp. 978–986 (2019)
9. Dong, X., Shen, J., Wang, W., Shao, L., Ling, H., Porikli, F.: Dynamical hyperparameter optimization via deep reinforcement learning in tracking. IEEE Trans. Pattern Anal. Mach Intell. **43**(5), 1515–1529 (2021)
10. Ndiaye, E., Le, T., Fercoq, O., Salmon, J., Takeuchi, I.: Safe grid search with optimal complexity. In: ICML, pp. 4771–4780 (2019)
11. Rendle, S.: Learning recommender systems with adaptive regularization. In: WSDM, pp. 133–142 (2012)
12. Reddi, S.J., Kale, S., Kumar, S.: On the convergence of adam and beyond. In: ICLR (2018)
13. Ding, J., Ren, X., Luo, R., Sun, X.: An adaptive and momental bound method for stochastic learning. arXiv preprint arXiv:1910.12249 (2019)
14. He, X., Liao, L., Zhang, H., Nie, L., Hu, X., Chua, T.S.: Neural collaborative filtering. In: WWW, pp. 173–182 (2017)
15. Fu, Z., et al.: Deep learning for search and recommender systems in practice. In: SIGKDD, pp. 3515–3516 (2020)
16. Tay, Y., Anh Tuan, L., Hui, S.C.: Latent relational metric learning via memory-based attention for collaborative ranking. In: WWW, pp. 729–739 (2018)
17. Kingma, D.P., Ba, J.: Adam: a method for stochastic optimization. In: ICLR (2015)
18. Wu, D., Luo, X., Wang, G., Shang, M., Yuan, Y., Yan, H.: A highly-accurate framework for self-labeled semi-supervised classification in industrial applications. IEEE Trans. Ind. Informat. **14**(3), 909–920 (2018)
19. Neri, F., Tirronen, V.: Scale factor local search in differential evolution. Memetic Comput. **1**(2), 153–171 (2009)
20. Wang, X., He, X., Wang, M., Feng, F., Chua, T.S.: Neural graph collaborative filtering. In: SIGIR, pp. 165–174 (2019)
21. Demšar, J.: Statistical comparisons of classifiers over multiple data sets. J. Mach. Learn. Res. **7**, 1–30 (2006)

GELibRec: Third-Party Libraries Recommendation Using Graph Neural Network

Chengming Zou[1,3] and Zhenfeng Fan[2(✉)]

[1] Hubei Key Laboratory of Transportation Internet of Things,
Wuhan University of Technology, Wuhan, China
`zoucm@whut.edu.cn`
[2] School of Computer Science and Technology, Wuhan University of Technology,
Wuhan, China
`fanzhenfeng@whut.edu.cn`
[3] Peng Cheng National Laboratory, Shenzhen, China

Abstract. Third-party libraries have become an indispensable part of the software. The function provided by well-tested third-party libraries can be reused through their programming interfaces, significantly increasing developers' software quality and productivity of developers. However, the vast number of third-party libraries and the complex dependencies are the main obstacles to efficiently exploiting the available resources. So, advanced methods are needed to explore the dependencies between projects and third-party libraries to make meaningful recommendations. This paper proposes GELibRec, which combines graph embedding and collaborative filtering to recommend libraries to developers. We extract a dataset from the open-source dataset libraries.io, and the experimental results show that GELibRec outperforms these methods concerning various quality metrics.

Keywords: Third-party libraries · Recommender system · Graph embedding · Collaborative filtering

1 Introduction

To facilitate the development process, developers often search for available resources such as code snippets [13]. In addition, third-party libraries provide developers with well-tested function [6], which can significantly improve the usability, security, and reliability of software.

However, due to the increasing number of reusable libraries, it is a tedious task for developers to search for relevant libraries. Therefore, better tools are needed to support developers to facilitate the search for suitable libraries. Among the existing methods, CrossRec [7], AppLibRec [17] are the most advanced techniques for library recommendation.

In this paper, we propose GELibRec, a novel approach to recommend third-party libraries using graph neural networks [16] and collaborative filtering [11].

A. Bhattacharya et al. (Eds.): DASFAA 2022, LNCS 13246, pp. 332–340, 2022.
https://doi.org/10.1007/978-3-031-00126-0_26

First, we construct a bipartite graph based on the dependencies between projects and third-party libraries. We use Graph Neural Network to extract the features of nodes in the graph. Finally, we use collaborative filtering to recommend available libraries for projects. In addition, an empirical evaluation shows that GELibRec outperforms CrossRec, AppLibRec, and other classic methods of recommender systems concerning different quality indicators. To this end, the significant contributions of this paper include:

- Building a recommender system GELibRec to select suitable libraries for software developers;
- Extracting a dataset and performing extensive experiments on the performance of GELibRec in comparison with four baselines using the dataset, exploiting various quality metrics;

The rest of the paper is structured as follows. Section 2 discusses existing approaches relevant to the work in this paper. Section 3 introduces our proposed method GELibRec. Sections 4 and 5 illustrate the experiment and results. Finally, Sect. 6 summarizes the paper and outlines the future work direction.

2   Related Work

Third-party library prediction is an important research direction in software engineering. Librec [14] and Libcup [10] are based on association rule mining. However, the co-occurrence probability of third-party libraries has a great influence on association rule mining. So, they can not effectively alleviate the problem of the cold start of the project.

Some scholars used the text description of projects as the initial feature to solve the cold start problem in varying degrees. LibFinder [8] and AppLibRec [15] are representative recommendation algorithms. However, the recommendation will be poor when the text is of poor quality.

Other work, such as CrossRec [7] and Libseek [4], uses techniques to calculate the importance of third-party libraries to projects as weights to calculate similar projects. Then use collaborative filtering techniques to make recommendations for projects. Sun et al. [12] proposed a deep learning framework, Req2Lib, which is the first time that researchers use neural network technology to recommend usable third-party libraries for projects. It also provides new ideas for academic researchers.

3   Proposed Approach

3.1   Lib-Based Analysis

3.1.1   Definition and Notation

Definition 1 (Bipartite Graph). Firstly, we abstract the interaction between projects and third-party libraries into a bipartite graph. Let G = (U, V, E) denotes an undirected bipartite graph, where U and V are two sets of different types of nodes and E is the set of links. Suppose there are n nodes of type U $= \{u_1, ..., u_n\}$. and m nodes of type V $= \{v_1, ..., v_m\}$. For a bipartite graph, we denote its adjacency matrix as A $\in R^{n \times m}$. Each edge e_{ij} is associated with a weight a_{ij}, where the value $a_{ij} = 1$ if u_i and v_j are connected and $a_{ij} = 0$ otherwise.

Definition 2 (Local structural similarity). Local structural similarity represents the similarity between directly connected nodes. For any two adjacent nodes u_i and v_j, the similarity can be defined as a_{ij}, where $a_{ij} > 0$.

Definition 3 (Global structural similarity). Global structural similarity represents the similarity of neighborhood structure of node pairs. For any node $u_i \in U$, the global structure is formulated as: $\{a_{i*}|a_{i*} > 0, \forall v_* \in V, u_i \in U\}$. For any node $v_j \in V$, the global structure is defined as: $\{a_{*j}|a_{*j} > 0, \forall u_* \in U, v_j \in V\}$.

3.1.2 Structural Feature Extractor
This step extracts the structural information from adjacency matrix. We extend the adjacency matrix A to A', $A' \in R^{(n+m) \times (n+m)}$ as shown in Formula 1.

$$A' = \begin{bmatrix} 0 & A \\ A^T & 0 \end{bmatrix} \tag{1}$$

Next, we input A' into the neural network. The framework is shown on the left of Fig. 1. We reconstruct the neighborhood structure of each node using a deep autoencoder. We denote the input data and reconstructed data as $X = \{x_i\}_{i=1}^{n+m}$ and $\hat{X} = \{\hat{x}_i\}_{i=1}^{n+m}$, and given an input data $x_i = a_i'$ and the number of layers K of the encoder, the representations of each layer in the encoding process are as:

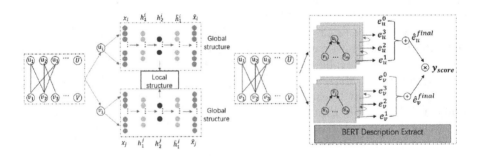

Fig. 1. Feature extractor components of GELibRec

$$h_1^i = f(W_1 x_i + b_1), h_k^i = f(W_k h_{k-1}^i + b_k), k = 2, ...K, \tag{2}$$

where W_k and b_k are the weight matrix and bias of layer k respectively.

Correspondingly, the decoding part corresponding to node x_i is as follows:

$$\hat{h}_{k-1}^i = f(\hat{W}_k \hat{x}_i + \hat{b}_k), \hat{x}_i = f(\hat{W}_1 \hat{h}_1 + \hat{b}_1), k = 2, ..., K, \tag{3}$$

where $\hat{W}_k$, $\hat{b}_k$ denote weight matrix and bias of the $(k\text{-}1)$-th layer respectively.

The goal of the autoencoder is to minimize the reconstruction error of input and output. The global loss function can be expressed as follows:

$$L_{global} = \sum_{i=1}^{n+m} ||(\hat{x}_i - x_i) \odot b_i|| = ||(\hat{X} - X) \odot B||_F^2 \tag{4}$$

where $\odot$ is the Hadamard product, $b_i = \{b_{i,j}\}_{j=1}^{n+m}$. If $a'_{i,j} = 0$, $b_{i,j} = 1$, else $b_{i,j} = \beta > 1$, $a'_{i,j}$ is the j-th element of a'_i and β is hyper-parameter. Minimizing loss function L_{global} can make two nodes with similar neighborhood structures in the original network be similar in the embedded space.

In addition, we use first-order similarity to represent local network structure.

$$L_{local} = \sum_{i=1}^{n} \sum_{j=1}^{m} a_{i,j} ||h_k^i - h_k^j||_2^2 \tag{5}$$

By minimizing L_{local}, two directly connected nodes are also similar in the embedding space. To maintain the global and local structure of the network, we propose the joint objective function as follows:

$$L = L_{global} + \lambda_1 L_{local} + \lambda_2 L_{reg} \tag{6}$$

where λ_1 and λ_2 are hyperparameters. L_{reg} is a regularization term that prevents overfitting, which is formulated as follows:

$$L_{reg} = \sum_{k=1}^{K} (||W_k||_2^2 + ||\hat{W}_k||_2^2 + ||b_k||_2^2 + ||\hat{b}_k||_2^2) \tag{7}$$

3.2   Desc-Based Analysis

3.2.1   Textual Feature Extractor

This step trains a Graph Convolution Neural (GCN) network model and extracts textual features. First, we use Bert [3] to extract the descriptions of projects and libraries as input for the GCN model, and then the GCN model is used for information aggregation. The framework is shown in the right of Fig. 1.

The design of the graph convolution operation is as follows:

$$e_u^{(k+1)} = \sum_{i \in N_u} \frac{1}{\sqrt{|N_u|}\sqrt{|N_i|}} e_i^{(k)}, e_v^{(k+1)} = \sum_{j \in N_v} \frac{1}{\sqrt{|N_v|}\sqrt{|N_j|}} e_j^{(k)}, \tag{8}$$

where $e_u^{(k)}$ and $e_v^{(k)}$ are the K-th layer embedding of project u and library v. N is all neighbors of the node. The normalization term $\frac{1}{\sqrt{|N_u|}\sqrt{|N_i|}}$ follows standard

GCN design. We aggregate only the neighbor feature of the nodes for each layer without considering the nodes themselves. We will take into account the original feature of nodes when we finally aggregate each layer representation.

After the K-layer embedding is obtained, feature aggregation is carried out through the following formula:

$$e_u = \sum_{k=0}^{K} \alpha_k e_u^{(k)}, e_v = \sum_{k=0}^{K} \alpha_k e_v^{(k)}, \tag{9}$$

where α_k represents the importance of the k-th layer embedding in the final embedding. Experimentally, it has been found that setting α_k to $1/(K+1)$ usually results in good performance.

The function of the GCN network is to smooth the features of neighbor nodes. We aggregate the embedding of different layers to get the final embedding. Finally, we take the inner product of the final representations of projects and libraries as prediction results of the model:

$$\hat{y}_{uv} = e_u^T e_v. \tag{10}$$

We use Bayesian personalized ranking (BPR) [9] loss to optimize the model.

$$L_{BPR} = -\sum_{u=1}^{M} \sum_{i \in N_u} \sum_{j \notin N_u} ln\sigma(\hat{y}_{ui} - \hat{y}_{uj}) + \lambda||E^{(0)}||^2, \tag{11}$$

where λ is regularization coefficient. We use Adam optimizer for optimization.

In the setting of hyperparameters, we set the embedding size of the model as 768 dimensions, the learning rate as 0.001, the L_2 regularization coefficient as 1e-4, and the number of convolution layers as $K = 3$.

3.3   Find Similar Projects

Given a project in the test set, we calculate distances between the feature vector and all vectors of training set projects.

$$Cosine(Proj_i, Proj_j) = \frac{\theta_{Proj_i} \cdot \theta_{Proj_j}}{|\theta_{Proj_i}||\theta_{Proj_j}|} \tag{12}$$

where $\cdot$ is the dot product. We choose the top k highest cosine similarity scores as K-nearest neighbors of the project.

Finally, we collect all the libraries used by the K-nearest neighbors and calculate the scores of these libraries as follows:

$$Score_{Lib-based/Desc-based}(L) = \frac{Count(L)}{K} \tag{13}$$

where the Count(L) is the number of nearest neighbors that have used the library L, and K is the number of nearest neighbors. We recommend third-party libraries with the highest scores for the project.

3.4 Aggregator Components

We are to get the final score by combining $Score_{Lib-based}$ and $Score_{Desc-based}$, denoted as $Score_{GELibrec}$. For each library, the final score is defined as follows:

$$Score_{GELibrec} = \alpha \times Score_{Lib-based} + \beta \times Score_{Desc-based} \qquad (14)$$

we set $\alpha + \beta = 1$. The top n libraries which have highest score are recommended to developers.

4 Experiment Setup

4.1 Dataset

We extract the data set from libraries.io [5], we selected the projects according to the following criteria.

- The project should contain at least ten third-party libraries.
- The project should contain at least 50 words a textual description.
- The project should be unique. It cannot be forked from other projects.

We select 5,932 projects, 12,222 libraries and 208,882 dependencies. we randomly select 80% of libraries as training set and the rest as the test set for each project.

4.2 Baselines

- **LDA** [2]: LDA is a topic model that extracts topic vector for document. In the experiment, we take the description of all projects as a document set.
- **UCF** [1]: In this paper, we use user-based collaborative filtering.
- **CrossRec** [7]: CrossRec uses IDF to minimize impact of popular libraries on the recommendation results when calculating the similarity of projects.
- **AppLibRec** [17]: Applibrec contains RM-based and Lib-based component.

5 Research and Evaluation

We are interested in answering the following questions:

(1) How effective is our proposed GELibRec?

Figure 2 shows that GELibRec is superior to other methods. In particular, when n¡20, the performance is most obviously improved in quality metrics. The performance of the content-based LDA is the worst. The reason may be that there are few descriptions in the libraries.io dataset, so LDA cannot adequately extract the feature of projects.

(2) Are the two components necessary for our proposed GELibRec?

We have done experiments on the two components. It can be seen from Table 1 that the combination of Lib-based and Desc-based can improve performance.

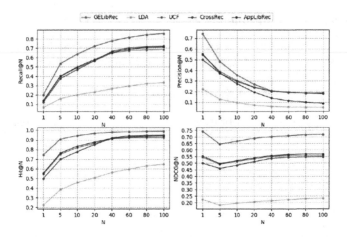

Fig. 2. Comparison of top-N recommendation on the dataset.

Table 1. Performance comparison of GELibRec with two components

Matrics	k	GELibRec	GELibRec-LibBased		GELibRec-DescBased	
			Value	Improved	Value	Improved
Recall@k	1	0.199	0.197	1.02%	0.153	30.07%
	5	0.535	0.510	4.90%	0.467	14.56%
	10	0.633	0.590	7.23%	0.586	8.02%
	20	0.720	0.652	10.42%	0.682	5.57%
Precision@k	1	0.743	0.737	0.81%	0.599	24.04%
	5	0.481	0.464	3.66%	0.432	11.34%
	10	0.354	0.342	3.51%	0.336	5.36%
	20	0.265	0.259	2.32%	0.262	1.15%
Hit@k	1	0.743	0.737	0.81%	0.599	24.04%
	5	0.908	0.888	2.25%	0.847	7.20%
	10	0.943	0.918	2.72%	0.915	3.06%
	20	0.966	0.936	3.21%	0.952	1.47%
NDCG@k	1	0.743	0.736	0.95%	0.597	24.46%
	5	0.645	0.625	3.20%	0.554	16.43%
	10	0.668	0.638	4.70%	0.587	13.80%
	20	0.690	0.652	5.83%	0.614	12.38%

6   Conclusion and Future Work

The primary purpose of our research is to consider the dependency relationships of the third-party libraries and the text description of the applications and the third-party libraries. Experimental results show that our method outperforms the existing methods. In the future, we plan to use readme documents of projects and libraries to improve the performance of GELibRec further.

References

1. Babu, M.S.P., Kumar, B.R.S.: An implementation of the user-based collaborative filtering algorithm. IJCSIT Int. J. Comput. Sci. Inf. Technol. **2**(3), 1283–1286 (2011)
2. Blei, D.M., Ng, A.Y., Jordan, M.I.: Latent Dirichlet allocation. J. Mach. Learn. Res. **3**, 993–1022 (2003)
3. Devlin, J., Chang, M.W., Lee, K., Toutanova, K.: BERT: pre-training of deep bidirectional transformers for language understanding. arXiv preprint arXiv:1810.04805 (2018)
4. He, Q., Li, B., Chen, F., Grundy, J., Xia, X., Yang, Y.: Diversified third-party library prediction for mobile app development. IEEE Trans. Softw. Eng. **48**(1), 150–165 (2020)
5. Katz, J.: Libraries. IO open source repository and dependency metadata (2018)
6. Li, M., et al.: LIBD: scalable and precise third-party library detection in android markets. In: 2017 IEEE/ACM 39th International Conference on Software Engineering (ICSE), pp. 335–346. IEEE (2017)
7. Nguyen, P.T., Di Rocco, J., Di Ruscio, D., Di Penta, M.: CrossRec: supporting software developers by recommending third-party libraries. J. Syst. Softw. **161**, 110460 (2020)
8. Ouni, A., Kula, R.G., Kessentini, M., Ishio, T., German, D.M., Inoue, K.: Search-based software library recommendation using multi-objective optimization. Inf. Softw. Technol. **83**, 55–75 (2017)
9. Rendle, S., Freudenthaler, C., Gantner, Z., Schmidt-Thieme, L.: BPR: Bayesian Personalized Ranking from Implicit Feedback. AUAI Press (2012)
10. Saied, M.A., Ouni, A., Sahraoui, H., Kula, R.G., Inoue, K., Lo, D.: Improving reusability of software libraries through usage pattern mining. J. Syst. Softw. **145**, 164–179 (2018)
11. Schafer, J.B., Frankowski, D., Herlocker, J., Sen, S.: Collaborative filtering recommender systems. In: Brusilovsky, P., Kobsa, A., Nejdl, W. (eds.) The Adaptive Web. LNCS, vol. 4321, pp. 291–324. Springer, Heidelberg (2007). https://doi.org/10.1007/978-3-540-72079-9_9
12. Sun, Z., Liu, Y., Cheng, Z., Yang, C., Che, P.: Req2Lib: a semantic neural model for software library recommendation. In: 2020 IEEE 27th International Conference on Software Analysis, Evolution and Reengineering (SANER), pp. 542–546. IEEE (2020)
13. Thummalapenta, S., Xie, T.: PARSEWeb: a programmer assistant for reusing open source code on the web. In: Proceedings of the Twenty-Second IEEE/ACM International Conference on Automated Software Engineering, pp. 204–213 (2007)
14. Thung, F., Lo, D., Lawall, J.: Automated library recommendation. In: 2013 20th Working Conference on Reverse Engineering (WCRE), pp. 182–191. IEEE (2013)

15. Yu, H., Xia, X., Zhao, X., Qiu, W.: Combining collaborative filtering and topic modeling for more accurate android mobile app library recommendation. In: Proceedings of the 9th Asia-Pacific Symposium on Internetware, pp. 1–6 (2017)
16. Zhang, Z., Cui, P., Zhu, W.: Deep learning on graphs: a survey. IEEE Trans. Knowl. Data Eng. **34**, 249–270 (2020)
17. Zhao, X., Li, S., Yu, H., Wang, Y., Qiu, W.: Accurate library recommendation using combining collaborative filtering and topic model for mobile development. IEICE Trans. Inf. Syst. **102**(3), 522–536 (2019)

Applications of Machine Learning

Hierarchical Attention Factorization Machine for CTR Prediction

Lianjie Long[1], Yunfei Yin[1(✉)], and Faliang Huang[1,2(✉)]

[1] College of Computer Science, Chongqing University, Chongqing 400044, China
{longlianjie,yinyunfei}@cqu.edu.cn
[2] Guangxi Key Lab of Human-Machine Interaction and Intelligent Decision, Nanning Normal University, Nanning 530001, China
faliang.huang@gmail.com

Abstract. Click-through rate (CTR) prediction is a crucial task in recommender systems and online advertising. The most critical step in this task is to perform feature interaction. Factorization machines are proposed to complete the second-order interaction of features to improve the prediction accuracy, but they are not competent for high-order feature interactions. In recent years, many state-of-the-art models employ shallow neural networks to capture high-order feature interactions to improve prediction accuracy. However, some studies have proven that the addictive feature interactions using feedforward neural networks are inefficient in capturing common high-order feature interactions. To solve this problem, we propose a new way, a hierarchical attention network, to capture high-order feature interactions. Through the hierarchical attention network, we can refine the feature representation of the previous layer at each layer, making it become a new feature representation containing the feature interaction information of the previous layer, and this representation will be refined again through the next layer. We also use shallow neural networks to perform higher-order non-linear interactions on feature interaction terms to further improve prediction accuracy. The experiment results on four real-world datasets demonstrate that our proposed HFM model outperforms state-of-the-art models.

Keywords: Recommender systems · Factorization machines · Hierarchical attention · CTR prediction

1 Introduction

Click-through rate (CTR) prediction plays a significant role in recommender systems, such as online advertising and short video recommendation, and it's attracted a lot of attention from researchers over the last decade [8,16,17]. The recommendation task here is to estimate the probability that a user will click on a displayed ad or a recommended item, So the predicted performance will

Supported by the Natural Science Foun dation of China under Grant 61962038, and by the Guangxi Bagui Teams for Innovation and Research under Grant 201979.

directly affect the final revenue of the system provider and the user's experience in the system.

The input of these prediction problems is often a variety of features, and features can be crossed to form new features. For example, we can cross feature $occupation = \{teacher, doctor\}$ with feature $gender = \{male, female\}$ and generate a new feature $gender\_occ = \{m\_teacher, f\_teacher, m\_doctor, f\_doctor\}$. A model with feature interaction can achieve better prediction accuracy, but it takes too many parameters to generate all cross-features [4]. To solve this issue, factorization machines (FMs) were proposed [23] to model the influence of feature interaction via the inner product of the embedding vectors of the features. By learning the embedding vector of each feature, FM can calculate the weight of any cross-feature. Owing to their efficient learning ability, FMs have been successfully applied to various applications, from recommendation systems [18] to natural language processing [21]. However, limited by its polynomial fitting time, FM is only effective for modeling low-order feature interaction and impractical to capture high-order feature interaction, which may make it insufficient for data representation in the real world with complex nonlinear underlying structures. Although high-order FMs have been proposed, they are still based on linear models, and their massive model parameters and inefficient training time are difficult to estimate.

NFM [8] designed a new operation, Bilinear Interaction (Bi-Interaction) pooling, to cancel the final summing operation of FM according to the embedded size k. The model stacks a multi-layer perceptron (MLP) on the Bi-Interaction layer to deepen the shallow linear FM, modeling high-order and nonlinear feature interactions effectively to improve FM's expressiveness. This is the first work to introduce the deep neural network (DNN) to FMs. Although promising, it uses the same weight to model all factorized interactions, just like FM, which may result in the inability to differentiate the importance of feature interaction. In fact, in real-world applications, different users may choose an item for different reasons, so the contribution of different features is different. Not all features include useful signals for estimating the target, and so do cross-features. For example, the feature *male* and the cross-feature *young_blue* are relatively less crucial for click probability in an instance: $\{young, male, teacher, blue, notebook\}$. Therefore many studies have tried directly to use DNN for feature interaction to improve the prediction accuracy of FMs. FNN [31] utilizes DNN to learn high-order feature interactions, and it uses pre-trained FM to learn feature embedding before applying DNN. PNN [22] further introduces a product layer between the embedding layer and DNN layer and does not rely on pre-trained FM. AFM [28] enhances FMs by learning the importance of each feature interaction from data via a neural attention network. FFM [12] was proposed to consider the field information to model the different interaction effects of features from different field pairs. Wide&Deep [5] and DeepFM [7] jointly learn low-order and high-order feature interaction by introducing a hybrid architecture containing shallow components and deep components. Similarly, xDeepFM [14] proposed a compressed interaction network part, which aims to model the low-order and high-order feature interaction at the vector-wise level in an explicit fashion. IFM [30] and DIFM [19] learn a more flexible and informative representation of a given feature according to different input instances, and the latter can adaptively reweight the original feature representation at the bit-wise and vector-wise levels simultaneously. MaskNet [27] is a high-performance ranking system by proposing a MaskBlock, which transforms the feedforward layer of the DNN model into a mixture of addictive and multiplicative feature interactions. Most of these models employ shallow MLP layers to implicitly model high-order interactions, which is an important part of the current most advanced ranking system. However, Alex Beutel et al. [3] have proved that addictive feature interactions, especially feedforward neural networks, are inefficient in capturing common feature interactions. Recently, Rendle et al. [24] redid the NCF [9] experiment and found that the traditional inner product is much better than the MLP layer in simulating interaction in collaborative filtering.

Even in the era of neural networks, the inner product is still one of the most effective ways of feature interaction. It can interact with features from low-order to high-order, but in each interaction, it will enumerate all possible interactions, which will increase the expense of the model exponentially. Therefore, there is an urgent need for a new type of high-order feature interaction that can effectively complete the low-order to high-order feature interaction without causing large expense. In linguistics, linguists have discovered that language has a hierarchical structure [10], from basic components to compound components and then to quite complex components. In other words, the sound units and meaning units of the language can be arranged in an order of increasing complexity. For example, English has only 10 number symbols and 26 letter symbols, but these letters and numbers can construct about 40,000 words we are using, and these words can construct countless sentences. Analogous to the hierarchical structure of linguistics, in our scenario, it seems that we can treat various features as symbols in linguistics, and construct feature interactions from low-order to high-order through a hierarchical structure.

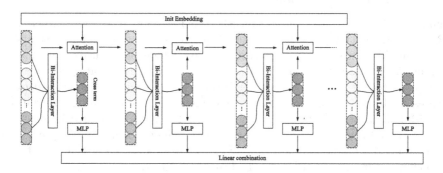

Fig. 1. The network architecture of our proposed Hierarchical Attention Factorization Machine model. The final output $\hat{y}$ of the model comes from the linear summation at the bottom.

To address the above issues, In this work, we propose a novel model for sparse data prediction named Hierarchical Attention Factorization Machine (HFM) as shown in Fig. 1, which enhances FMs by taking advantage of high-order feature interaction to refine feature representation hierarchically. By designing an attention network module, we use the output of Bi-Interaction Layer to learn a weight for each initial feature to refine its representation. After that, the refined feature representations can be refined again through the Bi-Interaction Layer and attention network, and features representations are updated hierarchically and iteratively in this way, which is the core of the hierarchical attention model we proposed. More importantly, after each feature representation is updated, the feature interaction information of the previous layer will be transferred to the next layer, thereby forming higher-order feature interaction. Furthermore, we stack a multi-layer perceptron (MLP) after each Bi-Interaction layer to further model high-order and nonlinear feature interactions effectively. To summarize, we make the following key contributions:

- To the best of our knowledge, this is the first work to directly combine hierarchical attention networks with FMs and refine feature representation through hierarchical attention for the first time.
- We propose a novel network model HFM for CTR prediction. Particularly, HFM forms new feature representations that contain high-order feature interaction information through iterative feature interaction information by hierarchical attention networks and employs MLP to further model high-order nonlinear interactions after each layer of feature interactions.
- Extensive experiments are conducted on four real-world datasets and the experiment results demonstrate that our proposed HFM outperforms state-of-the-art models significantly.

2 Preliminaries

2.1 Problem Formulation

Suppose the dataset for training consists of massive instances (x, y), where $x \in \mathbf{R}^n$ denotes the concatenation of user u's features and item v's features, and n is the number of concatenated features. x may include categorical features (e.g., gender, occupation) and continuous features (e.g., age). Each categorical feature is represented as a vector of one-hot encoding, and each continuous feature is represented as the value itself, or a vector of one-hot encoding after discretization. $y \in \{0, 1\}$ is the label indicating the user's click behavior, where $y = 1$ means the user u clicked the item v and $y = 0$ otherwise. The problem of CTR prediction aims to build a prediction model $\hat{y} = \mathcal{F}(\mathbf{x} \mid \Theta)$ to predict the probability of user u clicking on item v according to the feature vector $\mathbf{x}$.

2.2 Factorization Machines

Factorization machines are proposed to learn feature interactions for sparse data effectively. FM estimates thetarget by modelling all interactions between each pair of features:

$$\hat{y}_{FM}(\mathbf{x}) = w_0 + \sum_{i=1}^{n} w_i x_i + \sum_{i=1}^{n} \sum_{j=i+1}^{n} \langle \mathbf{v}_i, \mathbf{v}_j \rangle x_i x_j \tag{1}$$

where w_0 is the global bias, w_i and $\mathbf{v}_i \in \mathbf{R}^k$ denote the scalar weight and the k-dimensional embedding vector of the i-th feature, respectively. $\langle \mathbf{v}_i, \mathbf{v}_j \rangle$ is the dot product of two vectors of size k, which models the interaction between the i-th and j-th features. Note that the feature interactions can be reformulated [23] as:

$$\sum_{i=1}^{n} \sum_{j=i+1}^{n} \langle \mathbf{v}_i, \mathbf{v}_j \rangle x_i x_j = \sum_{f=1}^{k} \left(\left(\sum_{i=1}^{n} v_{i,f} x_i \right)^2 - \sum_{i=1}^{n} v_{i,f}^2 x_i^2 \right) \tag{2}$$

where $v_{j,f}$ denotes the f-th element in $\mathbf{v}_j$. The time complexity of Eq. 1 is $O(kn^2)$, but with reformulating it drops to linear time complexity $O(kn)$.

3   Our Approach

In this section, we present the details of the proposed HFM model as shown in Fig. 1.

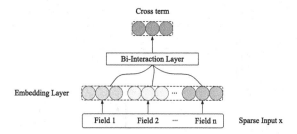

Fig. 2. The network architecture of Bi-Interaction.

3.1   The HFM Model

Sparse Input and Embedding Layer. The sparse input layer represents user's features and item's features as a sparse vector, which is the concatenation of all feature fields. Since the feature representation of category features is very sparse and high-dimensional, a common method is to use the embedding layer to embed sparse features into low-dimensional, dense real-valued representation vectors. The output of the embedding layer is a set of concatenated feature embedding vectors $\mathbf{x}$:

$$\mathbf{x} = \left[\mathbf{v}_1^T, \mathbf{v}_2^T, \cdots, \mathbf{v}_i^T, \cdots, \mathbf{v}_n^T \right] \tag{3}$$

where n denotes the number of feature fields while $\mathbf{v}_i \in \mathbf{R}^k$ denotes the embedding vector of the i-th feature field, and k is the embedding size.

Bi-interaction Layer. In this part, we feed the embedding vector $\mathbf{x}$ into the Bi-interaction layer to complete the second-order feature interaction.

$$f_{BI}(\mathbf{x}) = \sum_{i=1}^{n} \sum_{j=i+1}^{n} \langle \mathbf{v}_i, \mathbf{v}_j \rangle x_i x_j \tag{4}$$

Like Eq. 2, we reformulate Eq. 4 by canceling the final summation by embedding size, that is, the Bi-Interaction pooling in NFM, so that we can get the output *Cross_term* of the Bi-Interaction layer through f_{BI} as shown in Fig. 2.

$$f_{BI}(\mathbf{x}) = \frac{1}{2} \left(\left(\sum_{i=1}^{n} v_{i,k} \right)^2 - \sum_{i=1}^{n} v_{i,k}^2 \right) \tag{5}$$

Where k is the embedding size and $v_{i,k}$ denotes the k-th dimension of the embedding vector of the i-th feature. After the Bi-Interaction layer, our input feature embedding vector set can be pooled into a k-dimensional cross term. Therefore, the cross term contains a wealth of feature interaction information, which can be used in our subsequent hierarchical attention calculations. More importantly, the Bi-Interaction layer does not introduce additional model parameters, and it can be calculated efficiently in linear time.

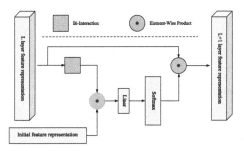

Fig. 3. The network architecture of Hierarchical Attention.

Hierarchical Attention. In recent years, attention mechanisms have become very popular in fields such as computer vision [29] and natural language processing [1,11,26]. The key to attention is that humans selectively focus on a portion of all information while ignoring other visible information. Attention mechanism has good performance and can provide interpretability of deep learning model. Additionally, HANN [6] uses hierarchical attention to automatically distinguish the importance of word-level and sentence-level in item reviews to capture accurate portraits of users and items. Inspired by this, we use hierarchical attention for feature representation learning as shown in Fig. 3. The hierarchical attention layer captures rich information from feature interactions and refines feature representations, thereby obtaining new feature representations containing feature interaction information from low-order interaction to high-order interaction.

As mentioned earlier, in real-world applications, different features contribute differently to the user's choice. Not all features include useful signals for estimating the target, and so does feature interaction. In this part, we try to employ the cross term that is output by Bi-Interaction to find the contribution of each feature in order to refine the feature representation later. Then we use a scoring function $g : \mathbf{R}^k \times \mathbf{R}^k \rightarrow \mathbf{R}$ to compute the score between the feature $\mathbf{v_i}$ and the cross term $Cross\_term$ as follows:

$$\alpha_{\mathbf{v_i}}^{(l)} = W(\mathbf{v_i} \odot Cross\_term^{(l)}) + b \qquad (6)$$

where $\mathbf{v_i} \in \mathbf{R}^k$ and $Cross\_term^{(l)} \in \mathbf{R}^k$ are the initial representation of the i-th feature as in Eq. 3 and cross term in the l-th layer respectively, l denotes the current attention layer, and k denotes the embedding size. W and b are transformation weight and bias. To better characterize feature interactions, we normalize the score:

$$\tilde{\alpha}_{\mathbf{v_i}}^{(l)} = \frac{\exp\left(\alpha_{\mathbf{v_i}}^{(l)}\right)}{\sum_{\mathbf{v_i} \in \mathbf{x}} \exp\left(\alpha_{\mathbf{v_i}}^{(l)}\right)} \qquad (7)$$

Then the feature representation of the current l-th layer can be updated through the attention score to obtain the feature representation of the next layer.

$$\mathbf{x}^{(l+1)} = \tilde{\alpha}^{(l)}\mathbf{x}^{(l)} \qquad (8)$$

It is worth noting that in each attention layer we will get a cross term containing the second-order feature interaction of the current layer. Feed the cross term into MLP to complete more complex and higher-order nonlinear feature interactions and get an output $o^{(l)}$ of this layer.

$$o^{(l)} = h^T(W_L(\ldots(W_1(Cross\_term^{(l)}) + b_1)\ldots) + b_L) + b \qquad (9)$$

Where L denotes the number of hidden layers, W_i, b_i, h and δ are transformation weight, bias, projection weight and activation function for the i-th layer perceptron, respectively. Finally we can get the output O of all layers.

$$O = \left[o^{(0)}, o^{(1)}, o^{(2)}, \cdots, o^{(l)}, \cdots, o^{(s)}\right] \qquad (10)$$

Where s denotes the number of layers of attention. Note that our index starts from 0. When it is 0, it means that the 0th layer is directly the output of the initial embedding feature interaction as shown in Fig. 1. Finally, we compute the linear combination the outputs of all layers and then project them into the prediction score $\hat{y}_{HFM}$ through an activation function.

$$\hat{y}_{HFM}(\mathbf{x}) = \delta(WO + b) \qquad (11)$$

Where δ is the activation function such as sigmoid, and W and b are transformation weight and bias. For binary classifications, the common loss function is the log loss:

$$\min_{\Theta} \mathcal{L} = -\frac{1}{N} \sum_{i=1}^{N} y_i \log \hat{y}_i + (1 - y_i) \log (1 - \hat{y}_i) + \lambda \|\Theta\|_2^2 \qquad (12)$$

where Θ is the total parameter space, including all embeddings and variables of attention networks. N is the total number of training instances and $\lambda \|\Theta\|_2^2$ is the L2-regularizer.

Table 1. A summary of model complexities.

Model name	Number of parameters
LR	m
Poly2	$m + \frac{m(m-1)}{2}$
FM	$m + mk$
NFM	$m + mk + L_{MLP}k$
AFM	$m + mk + \frac{m(m-1)}{2}k$
FFM	$m + m(n-1)k$
HFM	$mk + (nL_{HA} + L_{MLP})k$

3.2 Complexity Analysis

The number of parameters in FM is $m + mK$, where m accounts for the weights for each feature in the linear part $\{w_i | i = 1, \ldots, m\}$ and mk accounts for the embedding vectors for all the features $\{\mathbf{v_i} | i = 1, \ldots, m\}$. HFM learns a weight for each feature domain through the attention layer, and the cross term input into the MLP is a k-dimensional vector. The MLP neuron setting in this paper is consistent with the feature embedding size k. So the number of parameters is $mk + (nL_{HA} + L_{MLP})k$, Where n is the number of feature fields, L_{HA} is the the number of attention layers and L_{MLP} is the number of MLP layers. For FFM, the number of parameters is $m + m(n-1)k$ since each feature has $n-1$ embedding vectors. Given that usually $n \ll m$, the parameter number of HFM is comparable with that of FM and significantly less than that of FFMs. In Table 1, we compare the model complexity of some other models, noting that the bias term b is ignored.

4 Experiments

In this section, we compare our proposed HFM and the other state-of-the-art models empirically on four different datasets.

Table 2. The basic statistics of the four datasets

Dataset	Criteo	Avazu	MovieLens-1M	Frappe
#samples	45,840,617	40,428,967	739,012	288,609
#fields	39	23	7	8
#features	1,327,180	1,544,488	3,529	5,382

4.1 Experiment Setup

Datasets. We use four public real-world datasets to evaluate our model. **Criteo**[1] is a benchmark dataset for CTR prediction. This dataset has 45 million records of users clicking on ads, which contains 26 categorical feature fields and 13 numerical feature fields. **Avazu**[2] was used in the Avazu CTR prediction competition, which consists of several days of ad click through data. For each click data, there are 23 feature fields spanning from user/device features to ad attributes. **MovieLens-1M**[3] is the most popular dataset for recommendation systems, which contains users' ratings on movies. **Frappe**[4] is a context-aware app discovery tool. The dataset is constructed by Baltrunas et al. [2], which contains 96,203 app usage logs of users under diferent contexts. We randomly split instances by 8:1:1 for training, validation and test while the statistics of the datasets are summarized in Table 2.

Evaluation Metrics. We use two evaluation metrics in our experiments: AUC (Area Under ROC [15]) and Logloss (cross entropy [19]).

Baselines. We compare the performance of the following CTR estimation models with our proposed approach: LR, FM, FNN, WD, NFM, AFM, FFM, DeepFM, xDeepFM, IFM, and DIFM, all of which are discussed in Sect. 1.

Parameter Settings. We train our model using Pytorch. To make a fair comparison, all models are learned by optimizing the log loss as in Eq. 11 using the Adam [13] (Learning rate: 0.001) optimizer. For all models, the embedding size is set as 16, except FFM is set as 4, because it is limited by the number of model parameters as shown in Table 1. The other hyperparameter settings of the baseline are the same as those reported in the original paper, or the default settings in the code. For models with DNN part, if there is no default setting in the original implementation, the depth of hidden layers is set to 2, the number of neurons per layer is 16, all activation function is ReLU, which is also the parameter setting of the DNN part of our HFM.

---

[1] https://www.kaggle.com/c/criteo-display-ad-challenge.
[2] https://www.kaggle.com/c/avazu-ctr-prediction.
[3] https://grouplens.org/datasets/movielens.
[4] https://baltrunas.info/research-menu/frappe.

Table 3. The results of AUC and LogLoss in CTR prediction in on four Datasets

Model name	Criteo		Avazu		MovieLens-1M		Frappe	
	AUC	LogLoss	AUC	LogLoss	AUC	LogLoss	AUC	LogLoss
LR	0.7942	0.4561	0.7576	0.3922	0.7900	0.5426	0.9318	0.3371
FM (2010)	0.8035	0.4482	0.7810	0.3794	0.8112	0.5252	0.9628	0.2189
FNN (2016)	0.8010	0.4538	0.7810	0.3793	0.8008	0.5305	0.9767	0.1995
Wide&Deep (2016)	0.8010	0.4500	0.7817	0.3788	0.8001	0.5307	0.9753	0.2400
NFM (2017)	0.8034	0.4477	0.7853	0.3766	0.8037	0.5277	0.9756	0.1939
AFM (2017)	0.7918	0.4582	0.7769	0.3825	0.8014	0.5306	0.9511	0.2555
FFM (2017)	0.8089	0.4431	0.7859	0.3775	0.7928	0.5392	0.9684	0.2035
DeepFM (2017)	0.8069	0.4445	0.7860	0.3764	0.8104	0.5224	0.9757	0.1836
xDeepFM (2018)	0.8087	0.4433	0.7854	0.3778	0.8070	0.5247	0.9744	0.2361
IFM (2019)	0.8077	0.4439	0.7842	0.3782	0.8093	0.5228	0.9745	0.1943
DIFM (2020)	0.8088	0.4428	0.7844	0.3775	0.8074	0.5233	0.9766	0.1947
MaskNet (2021)	0.8109	0.4419	0.7876	0.3751	0.8046	0.5264	0.9771	0.1810
HFM (ours)	**0.8111**	**0.4411**	**0.7896**	**0.3740**	**0.8143**	**0.5157**	**0.9820**	**0.1600**

4.2   Performance Comparison

Effectiveness. The overall performance of different models on the four datasets is shown in Table 3. In particular, in the CTR prediction task, the performance gain of AUC or Logloss reaching 10^{-3} magnitude is considered to be a huge improvement [25, 27]. From the experimental results we can observe that:

a). In general, our proposed model achieves better performance on all four datasets and obtains significant improvements over the state-of-the-art methods. Compared to state-of-the-art FM-based methods, our model obtain 0.02–0.49% (0.29% average) and 0.18–11.61% (3.34% average) relative improvements in AUC and LogLoss metrics, respectively.

b). Compared with LR, our model, like all other FM-based models, is significantly better than this model without feature interactions, which illustrates the necessity of feature interaction in CTR estimation.

c). Compared with FM and AFM, our model, like other high-order feature interaction models, is significantly better than low-order feature interaction models, which shows that high-order feature interaction can greatly improve FM.

d). Compared with other FM-based high-order feature interaction models and attention models, our model is significantly better than them. This shows that our proposed hierarchical attention network, a novel way from low-order feature interaction to high-order feature interaction iteration, has obvious effects on modeling high-order feature interactions.

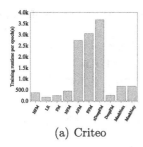
(a) Criteo

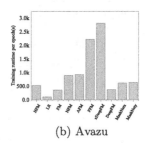
(b) Avazu

Fig. 4. Efficiency comparison between HFM and other baseline and state-of-the-art models on average runtime per epoch (8 GB Nvidia GeForce GTX 1080 GPU). MaskNetx and MaskNety refer to Serial MaskNet and Parallel MaskNet respectively.

Efficiency. Figure 4 shows the running time comparison of HFM and baseline models and state-of-the-art models on Criteo and Avazu. It is worth noting that MovieLens-1M and Frappe are not used for efficiency testing because they are relatively small in size and computational overhead accounts for most of the running time. The vertical axis shows the average running time of each epoch of all models and the hardware settings of all models are consistent. From the figure, we observe that HFM displays an outstanding efficiency advantage by spending relatively short running time for each epoch among models. Like NFM, we use the Bi-Interaction layer to avoid enumerating all possible feature interactions $m(m-1)/2$ times. This greatly reduces the time consumption of each epoch compared with AFM and FFM. For other models that directly use DNN to complete feature interaction, the running time mostly depends on the scale of the neural network used in the model. In this paper, we mainly explore the effectiveness and efficiency of hierarchical attention networks, so the neural network scale used is small. Therefore, the average running time of our HFM (4-layer attention network) and other DNN models is not much different in each epoch, except for xDeepFM due to its complex compressed interaction network.

4.3   Ablation Study of Hierarchical Attention

In order to better investigate the impact of each component in HFM, we perform ablation experiments over key components of HFM by only removing one of them to observe the performance change, including hierarchical attention network (HAN) and MLP after Bi-Interaction layer. In particular, after removing the attention layer for the former, we can regard this model as a mixture of multiple experts, just as MMoE [20] does. And the number of experts is the same as the number of attention layers of our full version model. Table 4 shows the results of our full version model and its variant with only one component removed.

From the results in Table 4 we can observe that removal of the hierarchical attention network will result in a decrease in the performance of the model, indicating that hierarchical attention is a necessary component in HFM for its

Table 4. Overall performance (AUC) of HFM models removing different component

Model name	Full	-HAN	-MLP
Criteo	0.8111	0.8074	0.8112
Avazu	0.7896	0.7861	0.7895
MovieLens-1M	0.8143	0.8090	0.8136
Frappe	0.9820	0.9810	0.9808

effectiveness. Other models loaded with MLP are used to complete high-order feature interaction. When we observe from the results in Table 3, MLP is a necessary component. However, our model has used hierarchical attention networks for high-order feature interaction, and it has proven to be more effective than MLP. Therefore, stacking MLP after the Bi-Interaction layer aims to further complete the nonlinear feature interaction, which is icing on the cake. It can be seen from the results that removing MLP will have different effects on the performance of the model. For smaller datasets such as MovieLens-1M and Frappe, removing MLP will cause the performance of the model to decrease. We guess that these datasets may have fewer feature fields and features, and MLP is needed to further interact with non-linear and high-order features. Conversely, for the Criteo and Avazu datasets with a large number of features and a large number of feature fields, MLP may have minimal gain to them and sometimes may bring a little noise. In order to achieve better generalization of the model in most scenarios, we ignore this effect, and we include MLP as a component of proposed model HFM.

4.4 Hyper-Parameter Study

In the following part of the paper, we study the impacts of hyperparameters on HFM, including *1)* the number of attention layers, *2)* the feature embedding size and *3)* the dropout ratio of neural networks. The experiment was carried out on 4 datasets, by changing a hyperparameter while keeping other parameter settings.

Number of Attention Layers. In order to investigate the impact of the number of layers of the hierarchical attention network on the model's performance, we conducted a hierarchical experiment from 0 to 5 on the HFM model and the experimental results are shown in Fig. 5. The experimental results show that the performance gradually improves with the increase of the number of layers, and can reach the highest performance. On four datasets, the optimal number of attention layers of our proposed model is 4 layers. When the number of layers is set to 0, it is equivalent to removing our hierarchical attention network and adding MLP directly behind the Bi-Interaction layer for feature interaction learning and output. At this time, the performance of the model is very poor.

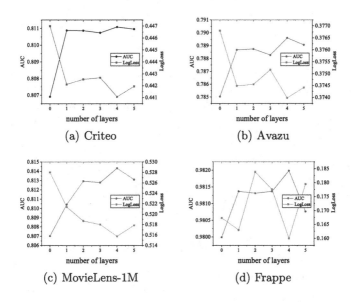

Fig. 5. The performance with different number of attention layers.

Once we add 1 layer of attention network, performance has improved qualitatively. This proves once again that addictive feature interaction is inefficient in capturing feature interaction, and the hierarchical attention we proposed is effective.

Table 5. The performance (AUC) with different embedding size k

k	4	8	16	32	64	128
Criteo	0.8073	0.8100	0.8111	0.8110	0.8112	0.8113
Avazu	0.7834	0.7860	0.7896	0.7914	0.7940	0.7955
MovieLens-1M	0.8080	0.8128	0.8143	0.8110	0.8115	0.8102
Frappe	0.9760	0.9784	0.9820	0.9832	0.9852	0.9858

Feature Embedding Size. The results in Table 5 show the effect of feature embedding size on model performance. It can be observed from the results that the performance of the model mostly increases with the increase of the embedding size. However, when the embedding size is set to be greater than 16 for the MovieLens-1M dataset, the performance of the model will decrease. For other datasets, when the embedding size reaches a certain value and continues to increase, the performance of the model improves slowly. Experimental results show that in most scenarios, a larger feature embedding size is conducive to better performance of the model, but will also bring more model parameters.

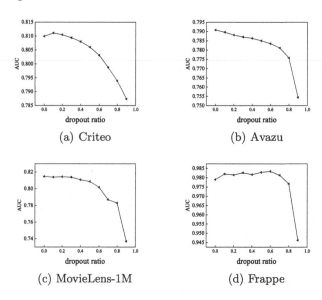

Fig. 6. The performance with different dropout ratio.

Dropout Ratio. Dropout is an effective regularization technique for neural networks to prevent overfitting. The results in Fig. 6 show the impact of the dropout rate of the network layer (including the Bi-Interaction layer, the attention network, and the MLP) on the performance of the model. It can be observed from the results that the performance of the model mostly decreases as the dropout rate increases. However, the optimal dropout rate of our proposed model is different for different datasets. Criteo is set to 0.1, Frappe is set to 0.6, Avazu and MovieLens-1M are both set to 0. It is worth noting that in the model comparison stage, for fairness, we set the same dropout rate as other models, and did not select the optimal dropout rate of our model.

5   Conlusion

In this paper, we propose a new hierarchical attention network for FMs. The network can iteratively refine the feature representation, and then model feature interaction from low-order to high-order. Different from the traditional employ of DNN for high-order feature interaction, the hierarchical feature interaction we proposed can better capture feature interaction information and achieve better performance. Furthermore, in order to increase the generalization performance of the model, we also added MLP components to the model to learn higher-order nonlinear feature interactions. The experiment results on four real-world datasets demonstrate that our proposed models outperform state-of-the-art models.

References

1. Bahdanau, D., Cho, K., Bengio, Y.: Neural machine translation by jointly learning to align and translate. arXiv preprint arXiv:1409.0473 (2014)
2. Baltrunas, L., Church, K., Karatzoglou, A., Oliver, N.: Frappe: understanding the usage and perception of mobile app recommendations in-the-wild. arXiv preprint arXiv:1505.03014 (2015)
3. Beutel, A., et al.: Latent cross: making use of context in recurrent recommender systems. In: Proceedings of the Eleventh ACM International Conference on Web Search and Data Mining, pp. 46–54 (2018)
4. Chapelle, O., Manavoglu, E., Rosales, R.: Simple and scalable response prediction for display advertising. ACM Trans. Intell. Syst. Technol. (TIST) 5(4), 1–34 (2014)
5. Cheng, H.T., et al.: Wide & deep learning for recommender systems. In: Proceedings of the 1st Workshop on Deep Learning for Recommender Systems, pp. 7–10 (2016)
6. Cong, D., et al.: Hierarchical attention based neural network for explainable recommendation. In: Proceedings of the 2019 on International Conference on Multimedia Retrieval, pp. 373–381 (2019)
7. Guo, H., Tang, R., Ye, Y., Li, Z., He, X.: DeepFM: a factorization-machine based neural network for CTR prediction. arXiv preprint arXiv:1703.04247 (2017)
8. He, X., Chua, T.S.: Neural factorization machines for sparse predictive analytics. In: Proceedings of the 40th International ACM SIGIR conference on Research and Development in Information Retrieval, pp. 355–364 (2017)
9. He, X., Liao, L., Zhang, H., Nie, L., Hu, X., Chua, T.S.: Neural collaborative filtering. In: Proceedings of the 26th International Conference on World Wide Web, pp. 173–182 (2017)
10. Hockett, C.F.: Linguistic elements and their relations. Language 37(1), 29–53 (1961)
11. Huang, F., Li, X., Yuan, C., Zhang, S., Zhang, J., Qiao, S.: Attention-emotion-enhanced convolutional LSTM for sentiment analysis. IEEE Trans. Neural Netw. Learn. Syst. (2021, online). https://doi.org/10.1109/TNNLS.2021.3056664
12. Juan, Y., Lefortier, D., Chapelle, O.: Field-aware factorization machines in a real-world online advertising system. In: Proceedings of the 26th International Conference on World Wide Web Companion, pp. 680–688 (2017)
13. Kingma, D.P., Ba, J.: Adam: a method for stochastic optimization. arXiv preprint arXiv:1412.6980 (2014)
14. Lian, J., Zhou, X., Zhang, F., Chen, Z., Xie, X., Sun, G.: xDeepFM: combining explicit and implicit feature interactions for recommender systems. In: Proceedings of the 24th ACM SIGKDD International Conference on Knowledge Discovery & Data Mining, pp. 1754–1763 (2018)
15. Lobo, J.M., Jiménez-Valverde, A., Real, R.: AUC: a misleading measure of the performance of predictive distribution models. Glob. Ecol. Biogeogr. 17(2), 145–151 (2008)
16. Long, L., Huang, F., Yin, Y., Xu, Y.: Multi-task learning for collaborative filtering. Int. J. Mach. Learn. Cybern., 1–14 (2021). https://doi.org/10.1007/s13042-021-01451-0
17. Long, L., Yin, Y., Huang, F.: Graph-aware collaborative filtering for top-N recommendation. In: 2021 International Joint Conference on Neural Networks (IJCNN), pp. 1–8. IEEE (2021)

18. Loni, B., Shi, Y., Larson, M., Hanjalic, A.: Cross-domain collaborative filtering with factorization machines. In: de Rijke, M., et al. (eds.) ECIR 2014. LNCS, vol. 8416, pp. 656–661. Springer, Cham (2014). https://doi.org/10.1007/978-3-319-06028-6_72

19. Lu, W., Yu, Y., Chang, Y., Wang, Z., Li, C., Yuan, B.: A dual input-aware factorization machine for CTR prediction. In: Proceedings of the 29th International Joint Conference on Artificial Intelligence, pp. 3139–3145 (2020)

20. Ma, J., Zhao, Z., Yi, X., Chen, J., Hong, L., Chi, E.H.: Modeling task relationships in multi-task learning with multi-gate mixture-of-experts. In: Proceedings of the 24th ACM SIGKDD International Conference on Knowledge Discovery & Data Mining, pp. 1930–1939 (2018)

21. Petroni, F., Corro, L.D., Gemulla, R.: CORE: context-aware open relation extraction with factorization machines. Association for Computational Linguistics (2015)

22. Qu, Y., et al.: Product-based neural networks for user response prediction. In: 2016 IEEE 16th International Conference on Data Mining (ICDM), pp. 1149–1154. IEEE (2016)

23. Rendle, S.: Factorization machines. In: 2010 IEEE International Conference on Data Mining, pp. 995–1000. IEEE (2010)

24. Rendle, S., Krichene, W., Zhang, L., Anderson, J.: Neural collaborative filtering vs. matrix factorization revisited. In: Fourteenth ACM Conference on Recommender Systems, pp. 240–248 (2020)

25. Sun, Y., Pan, J., Zhang, A., Flores, A.: FM2: field-matrixed factorization machines for recommender systems. In: Proceedings of the Web Conference 2021, pp. 2828–2837 (2021)

26. Vaswani, A., et al.: Attention is all you need. In: Advances in Neural Information Processing Systems, pp. 5998–6008 (2017)

27. Wang, Z., She, Q., Zhang, J.: MaskNet: introducing feature-wise multiplication to CTR ranking models by instance-guided mask. arXiv preprint arXiv:2102.07619 (2021)

28. Xiao, J., Ye, H., He, X., Zhang, H., Wu, F., Chua, T.S.: Attentional factorization machines: learning the weight of feature interactions via attention networks. arXiv preprint arXiv:1708.04617 (2017)

29. Xu, K., et al.: Show, attend and tell: neural image caption generation with visual attention. In: International Conference on Machine Learning, pp. 2048–2057 (2015)

30. Yu, Y., Wang, Z., Yuan, B.: An input-aware factorization machine for sparse prediction. In: IJCAI, pp. 1466–1472 (2019)

31. Zhang, W., Du, T., Wang, J.: Deep learning over multi-field categorical data. In: Ferro, N., et al. (eds.) ECIR 2016. LNCS, vol. 9626, pp. 45–57. Springer, Cham (2016). https://doi.org/10.1007/978-3-319-30671-1_4

MCRF: Enhancing CTR Prediction Models via Multi-channel Feature Refinement Framework

Fangye Wang[1,2], Hansu Gu[3], Dongsheng Li[4], Tun Lu[1,2], Peng Zhang[1,2(✉)], and Ning Gu[1,2]

[1] School of Computer Science, Fudan University, Shanghai, China
{fywang18,lutun,zhangpeng_,ninggu}@fudan.edu.cn
[2] Shanghai Key Laboratory of Data Science, Fudan University, Shanghai, China
[3] Seattle, USA
[4] Microsoft Research Asia, Shanghai, China
dongsli@microsoft.com

Abstract. Deep learning methods have recently achieved huge success in Click-Through Rate (CTR) prediction tasks mainly due to the ability to model arbitrary-order feature interactions. However, most of the existing methods directly perform feature interaction operations on top of raw representation without considering whether raw feature representation only obtained by embedding layer is sufficiently accurate or not, which leads to suboptimal performance. To address this issue, we design a model-agnostic and lightweight structure named Gated Feature Refinement Layer (GFRL), which dynamically generates flexible and informative feature representation by absorbing both intra-field and contextual information based on specific input instances. We further propose a generalized framework Multi-Channel Feature Refinement Framework (MCRF), which utilizes GFRLs to generate multiple groups of embedding for richer interactions without exploding the parameter space. GFRL and MCRF can be well generalized in many existing CTR prediction methods to improve their performance. Extensive experiments on four real-world datasets show that GFRL and MCRF can achieve statistically significant performance improvements when applied to mainstream CTR prediction algorithms. The integration of MCRF and FM can further push the state-of-the-art performance of CTR prediction tasks. We also identify that refining feature representation is another fundamental but vital direction to boost the performance of CTR prediction models.

Keywords: CTR prediction · Feature refinement · Recommendation

1 Introduction

CTR prediction, which aims to estimate the likelihood of an ad or item being clicked, is essential to the success of several applications such as computational

© The Author(s), under exclusive license to Springer Nature Switzerland AG 2022
A. Bhattacharya et al. (Eds.): DASFAA 2022, LNCS 13246, pp. 359–374, 2022.
https://doi.org/10.1007/978-3-031-00126-0_28

advertising [13], and recommender systems [1]. Accurate CTR prediction plays a vital role to improve revenue, and customer engagement of the platform [4,12]. However, in CTR tasks, the input data is typically high-dimensional and sparse multi-field categorical features, which takes great difficulty for making an accurate prediction. Many models have been proposed such as Logistic Regression (LR) [17], and Factorization Machines (FM) based methods [5,10,16,19,23].

With its strong ability to learn high-order feature interaction information, deep learning-based models gain popularity and achieve great success in CTR prediction tasks [1,2,4,5,25]. Meanwhile, various researchers improve performance by modeling sophisticated high-order feature interaction, including DCN [20], DeepFM [4], xDeepFM [11], FiBiNET [9], AutoInt [18], TFNet [21], etc. Most CTR models comprise of three components: 1) embedding layer; 2) feature interaction layer; 3) prediction layer. However, despite the great success of the above-mentioned methods, two key research challenges remain unsolved.

Firstly, obtaining more informative and flexible feature representation before performing feature interaction remains challenging. Precise feature representation is the basis of feature interaction and prediction layer. However, raw feature representation only obtained by embedding layer may not be sufficiently accurate. NON utilizes intra-field information to refine raw features [13]. However, for the same feature in different input instances, its representation is still fixed in NON. In the NLP area, contextual information is leveraged in Bert [3] and other methods to obtain word embedding dynamically based on the sentence context where the word appears. It is also reasonable to refine feature representation according to other concurrent features in each input instance for CTR prediction tasks. For example, we have two input instances: {*female, student, red, lipstick*} and {*female, student, white, computer*}. Apparently, the feature *"female"* is more important for click probability in the former instance, which is affected by the instance it appears [24]. So it remains a big challenge to learn more flexible feature representation for downstream feature interaction and prediction layer.

Secondly, given the huge possible combinations of raw features, interactions are usually numerous and sparse. FM [16] and most deep methods utilize a single embedding for each feature but fail to capture multiple aspects of the same feature. Field-aware factorization machine (FFM) introduces a separate feature embedding for each field [10]. However, it sometimes creates more parameters than necessary. Meanwhile, most parallel-structured models (i.e., DeepFM [4], DCN [20]) perform different sub-networks to capturing feature interaction based on the unique shared embedding. Intuitively, different interaction functions should contain discriminative embedding for optimal performance. Those bring up the question of how to generate multiple groups of embeddings for richer feature interaction information without exploding the parameter space.

To address the above two issues, we firstly propose a model-agnostic and lightweight structure named **G**ated **F**eature **R**efinement **L**ayer (**GFRL**), which dynamically refines raw feature's embedding by absorbing both intra-field and contextual information. Based on GFRL, we further propose a generalized **M**ulti-**C**hannel Feature **R**efinement **F**ramework (**MCRF**), which is comprised of three layers. In multi-channel refinement layer, we employ several independent GFRLs to generate multiple refined embeddings for further feature interactions. In aggregation layer, we adopt mainstream feature interaction operations in each channel and aggregate all generated interactions using attention mechanism. Finally, we utilize a DNN to generate the final output in prediction layer.

In summary, our main contributions are as follows: (1) We design a lightweight and model-agnostic module GFRL to refine feature representation by integrating intra-field and contextual information simultaneously. (2) Based on GFRL, we propose MCRF, which enables richer feature interaction information by generating multiple discriminative refined embedding in different channels. (3) Comprehensive experiments are conducted to demonstrate the effectiveness and compatibility of GFRL and MCRF on four datasets.

2 Related Work

FM-based methods are widely used to address CTR prediction tasks. FM [16] models all pairwise interactions using factorized parameters, which can handle large-scale sparse data via dimension reduction. Because of its efficiency and robustness, there are many subsequent works to improve FM from different angles, such as FFM [10], Field-weighted Factorization Machines (FwFM) [14], Attentional FM (AFM) [22], Neural FM (NFM) [5], Input-aware FM (IFM) [24], etc.

In recent years, leveraging deep learning to model complex feature interaction draws more interest. Models, such as FNN [25], WDL [1], DeepFM [4] utilizes plain DNN module to learn higher-order feature interactions. Meanwhile, DeepFM replaces the wide component of WDL with FM to reduce expensive manual feature engineering. xDeepFM [11] improve DeepFM by a novel Compressed Interaction Network (CIN) which learns high-order feature interactions explicitly. PNN [15] feeds the concatenation of embedding and pairwise interactions into DNN. DCN [20] model feature interactions of bounded-degree by using a cross network. And self-attention is a commonly used network to capture high-order interaction, such as AutoInt [18].

All aforementioned CTR prediction methods feed the raw embeddings to feature interaction layer for modeling arbitrary-order feature interaction. Differ to model feature interactions, researchers start to focus on refining feature representation before the feature interactions. FiBiNET [9] utilize the Squeeze-Excitation Network (SENET) [8] to generate SENET-like embedding for further feature interactions by assigning each feature unique importance. Similarly, IFM [24] also improves FM by learning feature importance with the proposed Factor Estimating Network (FEN). NON [13] leverages field-wise network to extract intra-field information and enrich raw features. In our work, GFRL considers intra-field information and contextual information simultaneously and refines original embedding more flexible and fine-grained.

3 The Structure of MCRF

We aim to refine raw features by encoding both intra-field information and contextual information and enable multiple embedding channels for richer useful

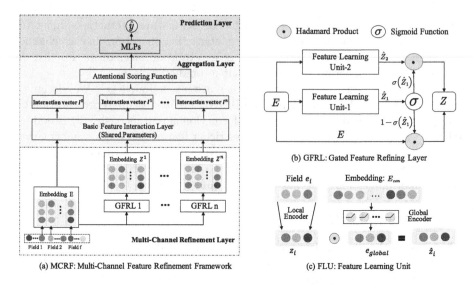

Fig. 1. Architectures of the proposed (a) MCRF, (b) GFRL, and (c) FLU.

feature interactions. To this end, we propose GFRL and MCRF for CTR prediction. As depicted in Fig. 1a, MCRF is composed of three layers:

- **Multi-Channel Refinement Layer.** In multi-channel refinement layer, we propose a model-agnostic structure named GFRL, which is applied on top of Embedding Layer to refine original embedding. GFRL contains Feature Learning Unit (FLU), and Gate Unit. FLU focuses on integrating both intra-field and contextual information and Gate Unit controls the selection probability of the supplementary embedding and original embedding for generating final refined representation.
- **Aggregation Layer.** In aggregation layer, we utilize attention mechanism to select and aggregate useful feature interactions based on multiple refined embeddings enabled by the multi-channel design.
- **Prediction Layer.** In prediction layer, a DNN is used to output the final prediction based on the features from the aggregation layer.

3.1 Embedding Layer

In most CTR prediction tasks, the raw data is typically in a multi-field categorical format, which is usually represented as a high-dimensional sparse vector (binary) by one-hot encoding [4,12]. For example, an input instance (Gender = Female, Item = Lipstick, Color = Red, Name = Amy) can be represented as:

$$\underbrace{(0,1)}_{\text{Gender=Female}} \underbrace{(0,\ldots,1,0,0)}_{\text{Item=Lipstick}} \underbrace{(0,1,\ldots,0,0)}_{\text{Color=Red}} \underbrace{(1,0,0,\ldots,0)}_{\text{Name=Amy}}. \tag{1}$$

In most DNN-based methods, an embedding layer is used to transform high-dimensional sparse feature data into low-dimensional dense vectors. Each input instance has f feature fields, after inputting into the embedding layer, each field i ($1 \leq i \leq f$) is represented by low-dimensional vector $e_i \in \mathbb{R}^d$. Each input instance can be represented by embedding matrix $E = [e_1, ..., e_i, ..., e_f] \in \mathbb{R}^{f \times d}$, where d is the dimension of embedding layer, e_i is the embedding of field i.

3.2 Gated Feature Refinement Layer

GFRL dynamically refines raw feature embedding by absorbing intra-field and contextual information simultaneously. And more informative and expressive representation can boost the performance of other basic models. As shown in Fig. 1b, GFRL consists of **FLU** and **Gate Unit** presented as follows.

Feature Learning Unit. FLU focuses on encoding intra-field and contextual information explicitly. We use two independent FLUs to generate supplementary features and a group of weight matrix. As illustrated in Fig. 1c, FLU is comprised of two steps: encoder step and combination step.

Encoder. This step encodes intra-field and contextual information by a local and global encoder. In local encoder, we assign a separate Fully Connected (FC) layer for each field. The parameters in FCs are used to store the intra-field information, then the output is computed by:

$$z_i = FC_i(e_i), \tag{2}$$

where e_i, z_i and FC_i are the original embedding, output embedding, and FC of the i-th field respectively. In practice, each field has the same structure as the FC, so we calculate them all together by applying matrix multiplication once (stacking inputs and weights of each FC [13]). We denote $B_i \in \mathbb{R}^{b \times d}$ and $W_i \in \mathbb{R}^{d \times d}$ as the batch of inputs and weights of the FC_i, where b is the mini-batch size. Formally, we formulate the local encoder as follows:

$$Z_{ori} = stack[B_1, ..., B_i, ..., B_f] \in \mathbb{R}^{f \times b \times d}, \tag{3}$$

$$W_{local} = stack([W_1, ..., W_i, ..., W_f]) \in \mathbb{R}^{f \times d \times d}, \tag{4}$$

$$Z_{local} = matmul(Z_{ori}, W_{local}) + b_l \in \mathbb{R}^{f \times b \times d}, \tag{5}$$

where d is the input and output dimension of FC_i. $b_l \in \mathbb{R}^{f \times 1 \times d}$ is bias term and *matmul* is batch matrix multiplication which is supported by Pytorch.[1] $Z_{local} = [z_1, ..., z_i, ..., z_f]$ is the output of local encoder.

After encoding intra-field information, we proceed to capture contextual information with a global encoder. Contextual information indicates the environment in which a feature is located. The concatenate all the embedding is represented by $E_{con} \in \mathbb{R}^{fd}$, and feed them to a simple DNN as follows:

$$e_{global} = \sigma_2(W_2\sigma_1(W_1E_{con} + b_1) + b_2), \tag{6}$$

---

[1] https://pytorch.org/docs/1.1.0/torch.html?#torch.matmul.

where $W_1 \in \mathbb{R}^{d_k \times fd}$, $W_2 \in \mathbb{R}^{d \times d_k}$, b_1 and b_2 are learnable parameters; $\sigma(\cdot)$ is the ReLU function. We model e_{global} as the contextual information, which represents the specific context information of each instance.

Combination. In this step, we combine intra-field and contextual information, using the following equation:

$$\hat{z}_i = F(z_i, e_{global}) = z_i \odot e_{global} \in \mathbb{R}^d, \tag{7}$$

where $F(\cdot)$ is a function, we choose the Hadamard product denoted by $\odot$ here, which is parameter-free and widely used in recommendation systems [4,15,16]. Finally, the supplementary features are generated as $\hat{Z} = [\hat{z}_1, ..., \hat{z}_i, ..., \hat{z}_f] \in \mathbb{R}^{f \times d}$, which explicitly integrates both intra-field and contextual information.

Compared to NON, FLU further absorbs contextual information, enabling the same feature to obtain various feature representations dynamically. It is also affected by the specific instance it appears, represented by contextual information here. To further enhance the expressive power of feature learning, we design a Gate Unit to select salient features from the original and supplementary embedding adaptively.

Gate Unit. Inspired by Long Short Term Memory Network (LSTM) [7], we leverage gate mechanism to obtain refined features by controlling the selection probability of supplementary features and raw features. As shown in Fig. 1b, we compute one supplementary embedding matrix $\hat{Z}_2$ and one group of weight matrix $\hat{Z}_1$ by two independent FLUs. Then we generate a group of probabilities σ by adding a sigmoid function after matrix $\hat{Z}_1$, which can control information flow in bit-level. Finally, we obtain the refined feature embedding Z as follows:

$$\hat{Z}_1 = FLU_1(E) \in \mathbb{R}^{f \times d}, \tag{8}$$

$$\hat{Z}_2 = FLU_2(E) \in \mathbb{R}^{f \times d}, \tag{9}$$

$$\sigma = sigmoid(\hat{Z}_1) \in \mathbb{R}^{f \times d}, \tag{10}$$

$$Z = \hat{Z}_2 \odot \sigma + E \odot (1 - \sigma), \tag{11}$$

where $Z \in \mathbb{R}^{f \times d}$ denotes the refined features. σ is the weight matrix, which controls the information flow like a gate in bit-level. $\hat{Z}_2 \odot \sigma$ represents the portion of each element that can be retained in supplementary features. $E \odot (1 - \sigma)$ denotes a selective "forgetness" of the raw features, and it only retains the important portion of the raw features. These two parts are complementary to each other. It should be noted that we do not use activation function in *local encoder*. As we have applied ReLU in Eq. 6. If functions like ReLU were used in *local encoder* again, the outputs of FLU would be greater than 0, which would cause all the probabilities to be greater than 0.5. In that case, the gate function would not help identify useful features. We believe that the use of gate unit helps improve the stability of the GFRL module.

In summary, GFRL utilizes two FLUs to generate supplementary features and weights matrix by integrating intra-field and contextual information and

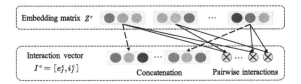

Fig. 2. The interaction layer of IPNN [15].

leverages gate unit to choose important information from both the raw and the supplementary features. It can be used as a building block in a plug-and-play fashion to improve base models' performance, we will discuss this in Sect. 4.3.

3.3 Multi-channel Feature Refinement Framework

To enable multiple embedding channels for richer feature interactions, we design MCRF (Fig. 1a) based on GFRL. It is comprised of three layers: Multi-Channel Refinement Layer, Aggregation Layer, and Prediction Layer.

Multi-channel Refinement Layer. Given the numerous combinations of raw features, feature interactions usually suffer from sparsity issues. FM-based and most deep methods address the issue by proposing feature embeddings, improving the model's generalization performance but performing feature interaction only based on the raw embedding matrix. FFM learns a separate feature embedding for each field, although achieving a significant performance boost, cause an explosion of parameters and compromise memory usage. We utilize multiple GFRLs to output multiple refined feature matrices that only scale the parameter space based on the number of parameters within GFRL and then perform feature interaction independently. In this way, we leverage richer embeddings through multiple channels without exploding the parameter space. We input the raw embedding matrix E into multiple GFRLs as follows:

$$Z^c = GFRL(E) \in \mathbb{R}^{f \times d}, \tag{12}$$

where Z^c denotes refined embedding in c-th ($1 \leq c \leq n$) channel. We denote $\mathbb{Z} = (E, Z^1, ..., Z^c, ..., Z^n)$ as all the embedding matrices, and we also keep the raw embedding E. When channel number is zero, we directly use the base model.

Aggregation Layer. In the previous layer, we obtain $n+1$ embedding matrices, n is the channel numbers. In this layer, we first apply basic feature interaction operation in each channel to get a feature vector. Then all vectors are aggregated by the attention mechanism. To be mentioned, any other advanced interaction operations can be applied to generate feature vector, e.g., bi-interaction in NFM [5], attention in AFM [22], etc. For ease of presentation, we take the interaction operation in IPNN as an example, as shown in Fig. 2.

IPNN concatenates pairwise inner-product and input vectors to model feature interaction. We denote e_f^c and i_f^c as the embedded features and the inner-product features in the c-th channel. We can calculate e_f^c and i_f^c as follow:

$$e_f^c = concat(Z^c) = concat([z_1^c; ...; z_i^c; ...; z_f^c]), \qquad (13)$$

$$i_f^c = [p_{(1,2)}, p_{(2,3)}, ..., p_{(f-1,f)}], \qquad (14)$$

$$p_{(i,j)} = < z_i^c, z_j^c >, \qquad (15)$$

where $Z^c \in \mathbb{R}^{f \times d}$ and z_i^c are the embedding matrix and i-th field embedding in c-th channel respectively. $p_{(i,j)}$ is the value of the "inner-product" operation between the i-th and j-th features. Concatenating e_f^c and i_f^c obtains the interaction vector $I^c = [e_f^c, i_f^c] \in \mathbb{R}^{n_f}$ in c-th channel, where $n_f = fd + \frac{(f-1)f}{2}$. Considering the input $\mathbb{Z}$, we obtain a total of $n + 1$ interaction vectors $\mathbb{I} = [I^0, I^1, ..., I^c, ..., I^n] \in \mathbb{R}^{(n+1) \times n_f}$.

As the embedding matrix Z^c is different in each channel, each interaction vector captures different interaction information. To aggregate all interaction vectors, we adopt attention on vector set $\mathbb{I}$. Formally, the attentional aggregation representation $\mathbb{I}_{agg}$ can be described as follows:

$$\mathbb{I}_{agg} = AttentionAgg(\mathbb{I}) = \sum_{c=0}^{n} \alpha^c I^c, \qquad (16)$$

$$\alpha^c = \frac{\exp\left(h^T \text{ReLU}\left(W^c I^c\right)\right)}{\sum_{c' \in n+1} \exp\left(h^T \text{ReLU}\left(W^c I^{c'}\right)\right)}, \qquad (17)$$

where α^c denotes the attention score in c-th channel; $W^c \in \mathbb{R}^{s \times n_f}$ are learnable parameters; $h \in \mathbb{R}^s$ is the context vector; and s is the attention size. Experiments show attention size does not affect results apparently, we set s equals to n_f.

Briefly, we implement basic interaction operation in each channel for generating the interaction vector, and each vector stores supplementary interaction information. For simplicity, we share the learning parameters in basic feature interaction layer. Actually, most operations like inner product, outer product and bi-interaction are parameter-free. Finally, we utilize an attention mechanism to aggregate all the interaction vectors generated in multiple channels.

Prediction Layer. After the aggregation layer, we collect the aggregated representation $\mathbb{I}_{agg}$. Then, we feed it into a DNN for the final prediction $\hat{y}$ as follows:

$$\hat{y} = \text{sigmoid}\left(\text{DNN}\left(\mathbb{I}_{agg}\right)\right). \qquad (18)$$

The learning process aims to minimize the following objective function:

$$\mathcal{L}(y, \hat{y}) = -\frac{1}{N} \sum_{i=1}^{N} \left(y_i \log\left(\hat{y}_i\right) + (1 - y_i) * \log\left(1 - \hat{y}_i\right)\right) \qquad (19)$$

where $y_i \in \{0, 1\}$ is the true label and $\hat{y}_i \in (0, 1)$ is the predicted CTR, and N is the total size of samples.

4   Experiments

To comprehensively evaluate MCRF and its major component GFRL, we conduct experiments by answering five crucial research questions. During the process, we essentially break down and evaluate different components of MCRF, so the whole experiment serves as a comprehensive ablation study.

- (Q1) How do GFRL and MCRF perform compared to mainstream methods while applying them to FM?
- (Q2) Can GFRL and MCRF improve the performance of other state-of-the-art CTR prediction methods?
- (Q3) What are the important components (FLU and Gate Unit) in GFRL?
- (Q4) How does GFRL perform compared to other structures?
- (Q5) How does the number of channels impact the performance of MCRF?

4.1   Experimental Setup

Datasets. We conduct empirical experiments on four real-world datasets. Table 1 lists the statistics of four real-world datasets.

Criteo[2] is a famous benchmark dataset, which contains 13 continuous feature fields and 26 categorical feature fields. It includes 45 million user click records. We use the last 5 million sequential records for test and the rest for training and validation [6]. We further remove the infrequent feature categories and treat them as a "$\langle unknown \rangle$" category, where the threshold is set to 10. And numerical features are discretized by the function $discrete(x) = \lfloor log^2(x) \rfloor$, where $\lfloor . \rfloor$ is the floor function, which is proposed by the winner of Criteo Competition.[3]

Avazu[4] is provided by Avazu to predict whether a mobile ad will be clicked. We also remove the infrequent feature categories, the threshold is set to five [18].

Frappe[5] is used for context-aware recommendation. Its target value indicates whether the user has used the app [2].

MovieLens[6] is used for personalized tag recommendation. Its target value denotes whether the user has assigned a particular tag to the movie [2]. We strictly follow AFM [22] and NFM [5] to divide the training, validation and test set for MovieLens and Frappe.

Evaluation Metrics. We use **AUC** (Area Under the ROC curve) and **Logloss** (cross-entropy) as the metrics. Higher AUC and lower Logloss indicate better performance [2,12,13]. It should be noted that **0.001**-level improvement in AUC or Logloss is considered significant [2,13,20] and likely to lead to a significant increase in online CTR prediction [4,12]. Considering the huge daily turnover of platforms, even a few lifts in CTR bring extra millions of dollars each year [12].

---

[2] https://www.kaggle.com/c/criteo-display-ad-challenge.
[3] https://www.csie.ntu.edu.tw/~r01922136/kaggle-2014-criteo.pdf.
[4] https://www.kaggle.com/c/avazu-ctr-prediction.
[5] https://www.baltrunas.info/context-aware/frappe.
[6] https://grouplens.org/datasets/movielens/.

Table 1. Datasets statistics (K indicates thousand.)

Datasets	Positive	Training	Validation	Test	Features	Fields
Criteo	26%	35,840K	5,000K	5,000K	1,087K	39
Avazu	17%	32,343K	4,043K	4,043K	1,544K	23
Frappe	33%	202K	58K	29K	5K	10
MovieLens	33%	1,404K	401K	201K	90K	3

Comparison Methods. We select the following existing CTR prediction models for comparison, including FM [16], IFM [24], FFM [10], FwFM [14], WDL [1], NFM [5], DeepFM [4], xDeepFM [11], IPNN [15], OPNN [15], FiBiNET [9], AutoInt+ [18], AFN+ [2], TFNet [21], NON [13]. As xDeepFM, FiBiNET and AFN has outperformed LR, FNN, AFM, DCN, we do not present their results. Meanwhile, our experiments also show the same conclusions.

Reproducibility. All methods are optimized by Adam optimizer with a learning rate of 0.001, and a mini-batch size of 4096. To avoid overfitting, early-stopping is performed according to the AUC on the validation set. We fix the dimension of field embedding for all models to be a 10 for Criteo and Avazu, 20 for Frappe and MovieLens, respectively. Borrowed from previous work [2, 4, 9], for models with the DNN, the depth of layer is set to 3, and the number of neurons per layer is 400. ReLU activation functions are used, and the dropout rate is set to 0.5. For other models, we fetch the best setting from its original literature.

To ensure fair comparison, for each model on each dataset, we run the experiments five times and report the average value. Notably, all the standard deviations are in the order of **1e−4**, showing our results are very stable. Furthermore, the two-tailed pairwise t-test is performed to detect significant differences between our proposed methods and the best baseline methods.

4.2   Overall Performance Comparison (Q1)

To verify the effectiveness of GFRL and MCRF, we choose FM as the basic method, represented by FM+GFRL and FM+MCRF-4.[7] FM is the most simple and effective feature interaction method. The performance of different models on the test set is summarized in Table 2. The key observations are as follow:

Firstly, most of the deep-learning models outperform shallow models. One of the reasons is that many deep models apply well-designed feature interaction operations, such as CIN in xDeepFM, self-attention in AutoInt, etc.

Secondly, by applying GFRL on top of embedding layer, FM+GFRL outperforms FM and other mainstream models, only second to FM+MCRF-4. It

---

[7] As the output of FM is a scalar value, we omit the attention layer when we apply MCRF with FM. Meanwhile, we feed all outputs in multiple channels to a simple LR than DNN in the prediction layer, which makes FM+MCRF more efficient.

Table 2. Overall performance comparison. ($\star : p < 10^{-2}$; $\star\star : p < 10^{-4}$)

Datasets		Criteo		Avazu		Frappe		MovieLens	
Model		AUC	Logloss	AUC	Logloss	AUC	Logloss	AUC	Logloss
Shallow models	FM	0.8028	0.4514	0.7812	0.3796	0.9708	0.1934	0.9391	0.2856
	FFM	0.8066	0.4477	0.7873	0.3758	0.9780	0.1980	0.9485	0.2682
	FwFM	0.8072	0.4470	0.7855	0.3770	0.9740	0.2164	0.9457	0.2840
	IFM	0.8066	0.4470	0.7839	0.3778	0.9765	0.1896	0.9471	0.2853
Deep models	WDL	0.8068	0.4474	0.7880	0.3753	0.9776	0.1895	0.9403	0.3045
	NFM	0.8057	0.4483	0.7828	0.3782	0.9746	0.1915	0.9437	0.2945
	DeepFM	0.8084	0.4458	0.7875	0.3758	0.9789	0.1770	0.9465	0.3079
	xDeepFM	0.8086	0.4456	0.7884	0.3751	0.9792	0.1889	0.9480	0.2889
	IPNN	0.8088	0.4454	0.7882	0.3754	0.9791	<u>0.1759</u>	0.9490	0.2785
	OPNN	<u>0.8096</u>	<u>0.4446</u>	0.7885	0.3752	<u>0.9795</u>	0.1805	0.9497	0.2704
	FiBiNET	0.8089	0.4453	<u>0.7887</u>	<u>0.3746</u>	0.9756	0.2767	0.9435	0.3427
	AutoInt+	0.8088	0.4456	0.7882	0.3751	0.9786	0.1890	0.9501	0.2813
	AFN+	0.8095	0.4447	0.7886	0.3747	0.9791	0.1824	<u>0.9509</u>	<u>0.2583</u>
	TFNet	0.8092	0.4449	0.7885	0.3745	0.9787	0.1942	0.9493	0.2714
	NON	0.8095	0.4446	0.7886	0.3748	0.9792	0.1813	0.9505	0.2625
Our models	FM+GFRL	0.8107**	0.4439**	0.7889**	0.3746**	0.9803*	0.1774*	0.9549*	0.2453*
	FM+MCRF-4	**0.8114****	**0.4433****	**0.7898****	**0.3740****	**0.9821***	**0.1720***	**0.9589****	**0.2409****

demonstrates that the refined feature representation is extremely efficient, which captures more information by absorbing intra-field and contextual information. And compared to modeling feature interaction, refining feature representation is more effective, considering FM only models second-order feature interaction.

Finally, with the help of four groups of refined embeddings, FM+MCRF-4 achieves the best performance. Specifically, it significantly outperforms basic FM 1.07%, 1.10%, 1.16% and 2.11% in terms of AUC(1.79%, 1.48%, 11.07%, and 15.65% in terms of Logloss) on four datasets. Those facts verify that the additional group of embeddings can provide more information. In subsection 4.6, we will discuss the impact of channel numbers in detail.

4.3   Compatibility of GFRL and MCRF with Different Models (Q2)

We implement GFRL and MCRF on five base models: FM, AFM, NFM, IPNN, and OPNN. GFRL can be applied after embedding layer to generate informative and expressive embedding. For a fair comparison, we only use one channel in MCRF to obtain another group feature embedding. The results are presented in Table 3. We focus on BASE, GFRL, and MCRF-1 and draw the following conclusions.

All base models achieve better performance after applying GFRL and MCRF, demonstrating the effectiveness and compatibility of GFRL and MCRF. Meanwhile, MCRF-1 consistently outperforms the building block GFRL, which verifies that MCRF-1 successfully aggregates more useful interactions from raw features and refined features by attention mechanism.

Compared to IPNN and OPNN, FM-based models get better improvements by applying GFRL and MCRF-1. Especially, FM+GFRL and FM+MCRF-1

Table 3. Model performance of different modules based upon various basic mainstream models on Criteo and Avazu datasets.

(a) Criteo Dataset

Models	FM		AFM		NFM		IPNN		OPNN	
Modules	AUC	Logloss	AUC	Logloss	AUC	Logloss	AUC	Logloss	AUC	Logloss
BASE	0.8028	0.4514	0.7999	0.4535	0.8057	0.4483	0.8088	0.4454	0.8096	0.4444
SENET	0.8057	0.4484	0.7990	0.4544	0.8052	0.4486	0.8090	0.4451	0.8093	0.4449
FEN	0.8066	0.4470	0.7954	0.4570	0.8063	0.4475	0.8092	0.4451	0.8088	0.4454
FWN	0.8049	0.4491	0.7884	0.4637	0.8070	0.4476	0.8090	0.4451	0.8098	0.4446
FLU	0.8098	0.4446	0.8082	0.4461	0.8097	0.4445	0.8093	0.4450	0.8100	0.4445
GFRL	0.8107	0.4439	0.8091	0.4453	0.8103	0.4441	0.8097	0.4444	0.8104	0.4440
MCRF-1	**0.8110**	**0.4435**	**0.8092**	**0.4452**	**0.8106**	**0.4437**	**0.8099**	**0.4444**	**0.8107**	**0.4439**

(b) Avazu dataset.

Models	FM		AFM		NFM		IPNN		OPNN	
Modules	AUC	Logloss	AUC	Logloss	AUC	Logloss	AUC	Logloss	AUC	Logloss
BASE	0.7812	0.3796	0.7727	0.3849	0.7828	0.3782	0.7882	0.3754	0.7885	0.3752
SENET	0.7831	0.3785	0.7743	0.3847	0.7847	0.3743	0.7883	0.3751	0.7887	0.3747
FEN	0.7839	0.3778	0.7731	0.3854	0.7830	0.3783	0.7886	0.3746	0.7885	0.3751
FWN	0.7846	0.3776	0.7731	0.3854	0.7853	0.3768	0.7884	0.3750	0.7886	0.3748
FLU	0.7875	0.3755	0.7840	0.3780	0.7881	0.3750	0.7885	0.3750	0.7888	0.3746
GFRL	0.7889	0.3746	0.7854	0.3773	0.7883	0.3750	0.7887	0.3746	0.7890	0.3745
MCRF-1	**0.7891**	**0.3745**	**0.7856**	**0.3768**	**0.7887**	**0.3747**	**0.7889**	**0.3745**	**0.7895**	**0.3742**

achieve the best result on Criteo dataset, although the other four basic models have more complex structures. Those facts show that inaccurate feature representation limits the effectiveness of these models, and learning precise and expressive representation is more basic but practical.

The key structure GFRL can improve the performance of base models and accelerate the convergence. Figure 3 exhibits the test AUC of three models during the training process before and after applying GFRL. Base models with GFRL can speed up the training process significantly. Specifically, within five epochs, base models with GFRL outperform the best results of the base models, indicating that refined representation generated by GFRL can capture better features and optimize the subsequent operations, and lead to more accurate prediction. This justifies the rationality of GFRL's design of integrating intra-field and contextual information before performing feature interaction, which is the core contribution of our work. It is worth mentioning that, MCRF-1 shows better result than GFRL, but for simplicity, we just show GFRL's training process.

4.4 Effectiveness of GFRL Variants (Q3)

We conduct ablation experiments to explore the specific contributions of FLU and Gate Unit in GFRL based on five models. FWN is proposed in NON [13], which only captures intra-field information. FLU adds contextual information

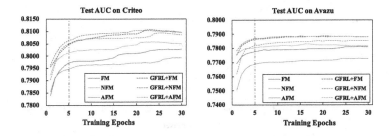

Fig. 3. The convergence curves of training process.

based on FWN. And GFRL is comprised of two FLUs and a Gate Unit. As observed from Table 3, both FLU and Gate Unit are necessary for GFRL.

Firstly, FLU is always better than FWN, which directly certifies our idea that contextual information is necessary. The same feature has different representations in various instances by absorbing contextual information, which makes it more flexible and expressive.

Secondly, GFRL always outperforms FLU and FWN, which shows the importance of Gate Unit. It further confirms the effectiveness of the Gate Unit, which can select useful features from original and supplementary representation with learned weight in bit-level. As a comparison, both SENET and FEN learn the importance in vector-level.

4.5   Superiority of GFRL Compared to Other Structures (Q4)

We select several modules from other models as comparisons to verify the superiority of GFRL. SENET [9] and FEN [24] are proposed to learn the importance of features. With FWN, we have four different modules: 1) SENET; 2) FWN; 3) FEN; 4) GFRL. Those are applied after the embedding layer to generate new feature representation. As shown in Table 3, we reach the following conclusions:

(1) In most cases, all four modules can boost base models' performance. It proves that refining feature representation before performing feature interaction is necessary and deserves further study. Meanwhile, from these unoptimized examples, we can know that randomly adding a module to a specific model sometimes doesn't lead to incremental results.

(2) With the help of GFRL, all the base models are improved significantly. Meanwhile, GFRL consistently outperforms other modules (FWN, SENET, FEN). Furthermore, GFRL is the only structure that improves the performance of all models. All those facts demonstrate the compatibility and robustness of GFRL, which can be well generalized in the majority of mainstream CTR prediction models to boost their performance.

(3) By checking the learning process of AFM, we find it updates slowly. This situation does not get modified either after using SENET, FWN, and FEN, or even worse. However, GFRL solves the problem, we show the training process of AFM, GFRL+AFM in Fig. 3. The possible reason is that Gate

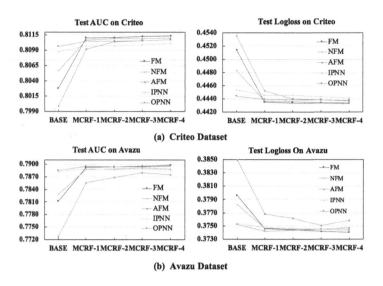

Fig. 4. The performance of different number of channels in MCRF based upon various base models on Criteo and Avazu datasets.

Unit allows the gradient to propagate through three paths in GFRL. This illustrates the soundness and rationality of the GFRL's design again.

4.6 The Impact of Channel Numbers of MCRF (Q5)

The number of channels is the critical hyper-parameter in MCRF. In previous sections, we have shown that MCRF-1 improves the performance of several mainstream models and outperforms GFRL with one additional channel. Here, we change the number of channels from 1 to 4 to show how it affects their performance. Note that we treat the base model as #channel = 0. Fig. 4 depicted the experimental results and have the following observations:

(1) As the number of channels increases, the performance of base models is consistently improved, especially for FM, AFM, and NFM. Unlike several parallel-structured models (e.g., DeepFM, xDeepFM), MCRF performs the same operation in different channels based on discriminative feature representations generated by GFRLs. Our results verify our design that it is effective to absorb more informative feature interactions from multiple groups of embeddings.

(2) Surprisingly, FM+MCRF achieves the best performance on two datasets. Especially on the Criteo dataset, with the help of MCRF, FM and NFM always outperform IPNN and OPNN. It proves that the FM successfully captures more useful feature interactions from multiple refined representations. Furthermore, considering the results of FM+GFRL, those experiments further verify that the idea of refining feature representation is highly efficient

and reasonable. With informative representation, even simple interaction operations (such as the sum of inner product in FM) can get better results than some complicated feature interaction operations. FM+MCRF may be a good candidate for achieving state-of-the-art CTR prediction models with ease of implementation and high efficiency.

5 Conclusion

Unlike modeling feature interaction, we focus on learning flexible and informative feature representation according to the specific instance. We design a model-agnostic module named GFRL to refine raw embedding by simultaneously absorbing intra-field and contextual information. Based on GFRL, we further propose a generalized framework MCRF, which utilizes GFRLs to generate multiple groups of embedding for richer interactions without exploding the parameter space. A detailed ablation study shows that each component of GFRL contributes significantly to the overall performance. Extensive experiments show that while applying GFRL and MCRF in other CTR prediction models, better performance is always achieved, which shows the efficiency and compatibility of GFRL and MCRF. Most importantly, with the help of GFRL and MCRF, FM achieves the best performance compared to other methods for modeling complex feature interaction. Those facts identify a fundamental direction for CTR prediction tasks that it is essential to learn more informative and expressive feature representation on top of the embedding layer.

Acknowledgements. This work was supported by the National Natural Science Foundation of China (NSFC) under Grants 61932007 and 62172106.

References

1. Cheng, H.T., et al.: Wide & deep learning for recommender systems. In: Proceedings of the 1st Workshop on Deep Learning for Recommender Systems, pp. 7–10 (2016)
2. Cheng, W., Shen, Y., Huang, L.: Adaptive factorization network: learning adaptive-order feature interactions. In: Proceedings of the AAAI Conference on Artificial Intelligence, vol. 34, pp. 3609–3616 (2020)
3. Devlin, J., Chang, M.W., Lee, K., Toutanova, K.: BERT: pre-training of deep bidirectional transformers for language understanding. arXiv preprint arXiv:1810.04805 (2018)
4. Guo, H., Tang, R., Ye, Y., Li, Z., He, X.: DeepFM: a factorization-machine based neural network for CTR prediction. arXiv preprint arXiv:1703.04247 (2017)
5. He, X., Chua, T.S.: Neural factorization machines for sparse predictive analytics (2017)
6. He, X., Du, X., Wang, X., Tian, F., Tang, J., Chua, T.S.: Outer product-based neural collaborative filtering. arXiv preprint arXiv:1808.03912 (2018)
7. Hochreiter, S., urgen Schmidhuber, J., Elvezia, C.: Long short-term memory. Neural Comput. 9(8), 1735–1780 (1997)

8. Hu, J., Shen, L., Sun, G.: Squeeze-and-excitation networks. In: Proceedings of the IEEE Conference on Computer Vision and Pattern Recognition, pp. 7132–7141 (2018)

9. Huang, T., Zhang, Z., Zhang, J.: FiBiNET: combining feature importance and bilinear feature interaction for click-through rate prediction. In: Proceedings of the 13th ACM Conference on Recommender Systems, pp. 169–177 (2019)

10. Juan, Y., Zhuang, Y., Chin, W.S., Lin, C.J.: Field-aware factorization machines for CTR prediction. In: Proceedings of the 10th ACM Conference on Recommender Systems, pp. 43–50 (2016)

11. Lian, J., Zhou, X., Zhang, F., Chen, Z., Xie, X., Sun, G.: xDeepFM: combining explicit and implicit feature interactions for recommender systems. In: Proceedings of the 24th ACM SIGKDD International Conference on Knowledge Discovery & Data Mining, pp. 1754–1763 (2018)

12. Liu, B., Tang, R., Chen, Y., Yu, J., Guo, H., Zhang, Y.: Feature generation by convolutional neural network for click-through rate prediction. In: The World Wide Web Conference, pp. 1119–1129 (2019)

13. Luo, Y., Zhou, H., Tu, W., Chen, Y., Dai, W., Yang, Q.: Network on network for tabular data classification in real-world applications. arXiv preprint arXiv:2005.10114 (2020)

14. Pan, J., et al.: Field-weighted factorization machines for click-through rate prediction in display advertising. In: Proceedings of the 2018 World Wide Web Conference, pp. 1349–1357 (2018)

15. Qu, Y., et al.: Product-based neural networks for user response prediction over multi-field categorical data. ACM Trans. Inf. Syst. (TOIS) $37(1)$, 1–35 (2018)

16. Rendle, S.: Factorization machines. In: 2010 IEEE International Conference on Data Mining, pp. 995–1000. IEEE (2010)

17. Richardson, M., Dominowska, E., Ragno, R.: Predicting clicks: estimating the click-through rate for new ads. In: Proceedings of the 16th International Conference on World Wide Web, pp. 521–530 (2007)

18. Song, W., et al.: AutoInt: automatic feature interaction learning via self-attentive neural networks. In: Proceedings of the 28th ACM International Conference on Information and Knowledge Management, pp. 1161–1170 (2019)

19. Tang, W., Lu, T., Li, D., Gu, H., Gu, N.: Hierarchical attentional factorization machines for expert recommendation in community question answering. IEEE Access 8, 35331–35343 (2020)

20. Wang, R., Fu, B., Fu, G., Wang, M.: Deep & cross network for ad click predictions. In: Proceedings of the ADKDD 2017, pp. 1–7 (2017)

21. Wu, S., et al.: TFNet: multi-semantic feature interaction for CTR prediction. In: Proceedings of the 43rd International ACM SIGIR Conference on Research and Development in Information Retrieval, pp. 1885–1888 (2020)

22. Xiao, J., Ye, H., He, X., Zhang, H., Wu, F., Chua, T.S.: Attentional factorization machines: learning the weight of feature interactions via attention networks. arXiv preprint arXiv:1708.04617 (2017)

23. Yang, B., Chen, J., Kang, Z., Li, D.: Memory-aware gated factorization machine for top-N recommendation. Knowl.-Based Syst. 201, 106048 (2020)

24. Yu, Y., Wang, Z., Yuan, B.: An input-aware factorization machine for sparse prediction. In: IJCAI, pp. 1466–1472 (2019)

25. Zhang, W., Du, T., Wang, J.: Deep learning over multi-field categorical data. In: Ferro, N., et al. (eds.) ECIR 2016. LNCS, vol. 9626, pp. 45–57. Springer, Cham (2016). https://doi.org/10.1007/978-3-319-30671-1_4

CaSS: A Channel-Aware Self-supervised Representation Learning Framework for Multivariate Time Series Classification

Yijiang Chen[1(✉)], Xiangdong Zhou[1(✉)], Zhen Xing[1], Zhidan Liu[1], and Minyang Xu[1,2]

[1] School of Computer Science, Fudan University, Shanghai 200433, China
{chenyj20,xdzhou,zxing20,zdliu20,16110240027}@fudan.edu.cn
[2] Arcplus Group PLC, Shanghai 200041, China

Abstract. Self-supervised representation learning of Multivariate Time Series (MTS) is a challenging task and attracts increasing research interests in recent years. Many previous works focus on the pretext task of self-supervised learning and usually neglect the complex problem of MTS encoding, leading to unpromising results. In this paper, we tackle this challenge from two aspects: encoder and pretext task, and propose a unified channel-aware self-supervised learning framework CaSS. Specifically, we first design a new Transformer-based encoder Channel-aware Transformer (CaT) to capture the complex relationships between different time channels of MTS. Second, we combine two novel pretext tasks Next Trend Prediction (NTP) and Contextual Similarity (CS) for the self-supervised representation learning with our proposed encoder. Extensive experiments are conducted on several commonly used benchmark datasets. The experimental results show that our framework achieves new state-of-the-art comparing with previous self-supervised MTS representation learning methods (up to +7.70% improvement on LSST dataset) and can be well applied to the downstream MTS classification.

1 Introduction

With the fast progress of IoT and coming 5Gs, multivariate time series widely exists in medical, financial, industrial and other fields as an increasingly important data form [6,8]. Compared with univariate time series, multivariate time series usually contains more information and brings more potentials for data mining, knowledge discovery and decision making, etc. However, MTS not only contains time-wise patterns, but also has complex relationships between different channels, which makes MTS analysis much more difficult.

In recent years, the self-supervised learning attracts more and more attention from research and industry communities. The self-supervised pre-training demonstrates its success in the fields of Natural Language Processing (NLP) [17] and Computer Visions (CV) [29]. Especially in NLP, adopting a pre-trained language model is de facto the first step of almost all the NLP tasks. Likewise,

A. Bhattacharya et al. (Eds.): DASFAA 2022, LNCS 13246, pp. 375–390, 2022.
https://doi.org/10.1007/978-3-031-00126-0_29

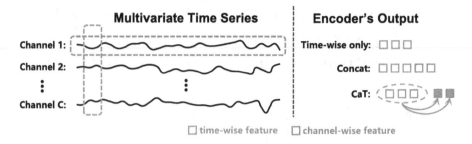

Fig. 1. A sample of encoding of multivariate time series.

the self-supervised representation learning of time series not only brings performance improvements, but also helps to close the gap between the increasing amount of data (more abstraction and complexity) and the expensive cost of manually labeling for supervised learning tasks. Many research efforts have been devoted to the self-supervised representation learning of time series [12, 15, 34] and promising results have been achieved. However, the previous works for MTS are limited and the challenge still exists.

The self-supervised representation learning usually consists of two aspects: encoder and pretext task. As shown in Fig. 1, in the past works, most of them only focus on time-wise features where all channel values of one or several time steps are fused through convolution or fully connected layer directly in the embedding process. There is a lack of deliberate investigation of the relationships between channel-wise features, which affects the encoder's ability to capture the whole characteristics of the MTS. To deal with the problem, the methods combining with Recurrent Neural Network (RNN) are presented to capture the individual feature of each channel [3, 35]. The recent work of [21] employs Transformer [30] to integrate the features of time-wise and channel-wise. Among these solutions, RNN seems not very suitable for self-supervised leaning due to the consumption and is usually employed in prediction task [14]. Transformer is becoming more and more popular and is suitable for time series [34], however the previous Transformer-based MTS embedding provokes the problem of high complexity of computing and space, which prevents it from real applications. It inspires us to design a more effective Transformer-based for MTS to take advantages of the strong encoding ability of Transformer. In the aspect of pretext task, most of the previous works adopt the traditional time series embedding based on time-wise features. How to integrate channel-wise features with pretext task is a challenge.

In this paper, we propose a novel self-supervised learning framework CaSS from the aspects of encoder and pretext task. First, we propose a new Transformer-based encoder Channel-aware Transformer (CaT) for MTS encoding which investigates time-wise and channel-wise features simultaneously. It is noticed that in practice the number of channels of MTS is fixed while the time length can be unlimited and the number of channels is usually much less than the time length. Therefore as shown in Fig. 1, different from previous work [21], we integrate the time-wise features into the channel-wise features and concatenate all these novel channel-wise features as the representation of the sample. Second, we design a new self-supervised pretext task Next Trend Prediction (NTP) from the perspective of channel-wise for the first time in self-supervised MTS representation learning. It is considered that in many cases only the rise and fall of future time rather than the specific value of the time series is necessary. So we cut the multivariate time series from the middle, and using the previous sequences of all rest channels to predict the trend for each channel. Different from fitting the specific value (regression), the prediction of trend (rise and fall) is more suitable for arbitrary data. We also demonstrate through experiments that compared with fitting specific values, prediction of trend is more efficient. In addition, we employ another task called Contextual Similarity (CS) which combines a novel data augmentation strategy to maximize the similarity between similar samples and learn together with NTP task. The CS task focuses on the difference between samples while NTP task focuses on the sample itself and helps to learn the complex internal characteristics.

In summary, the main contributions of our work are as follows:

- We propose a new Transformer-based encoder Channel-aware Transformer. It can efficiently integrate the time-wise features to the channel-wise representation.
- We design two novel pretext tasks, Next Trend Prediction and Contextual Similarity for our CaSS framework. To the best of our knowledge, Next Trend Prediction task is conducted from the perspective of channel-wise for the first time in self-supervised MTS representation learning.
- We conduct extensive experiments on several commonly used benchmark datasets from different fields. Compared with the state-of-the-art self-supervised MTS representation learning methods, our method achieves new state-of-the-art in MTS classification (up to +7.70% improvement on LSST dataset). We also demonstrate its ability in few-shot learning.

2   Related Work

2.1   Encoders for Time Series Classification

A variety of methods have been proposed for time series classification. Early works employ traditional machine learning methods to solve the problem, like combining Dynamic Time Warping (DTW) [32] with Support Vector Machine (SVM) [9]. Time Series Forest [11] introduces an approach based on Random Forest [7]. Bag of Patterns (BOP) [18] and Bag of SFA Symbols (BOSS) [28] construct a dictionary-based classifier. Although these early works can deal with the problem to some extent, they need heavy crafting on data preprocessing and feature engineering. The emergence of deep learning greatly reduces feature engineering and boosts the performance of many machine learning tasks. So far Convolutional Neural Network (CNN) is popular in time series classification due to its balance between its effect and cost, such as Multi-scale Convolutional Neural Network (MCNN) [10] for univariate time series classification, Multi-channels Deep Convolutional Neural Networks (MC-DCNN) [38] for multivariate time series classification and so on [20,36,37]. Hierarchical Attention-based Temporal Convolutional Network (HA-TCN) [19] and WaveATTentionNet (WATTNet) [25] apply dilated causal convolution to improve the encoder's effect. Among these CNN methods, Fully Convolutional Network (FCN) and Residual Network (ResNet) [31] have been proved to be the most powerful encoders in multivariate time series classification task [14]. Due to the high computation complexity, RNN based encoders are rarely applied solely to the time series classification [22,35]. It is often combined with CNN to form a two tower structure [3,16]. In recent years, more and more works have tried to apply Transformer to time series [26,33,39]. However, most of them are designed for prediction task, and few works cover the classification problem [21,24,27,34].

2.2 Pretext Tasks for Time Series

Manually labeling is a long lasting challenge for the supervised learning, and recently self-supervised training (no manually labeling) becomes more and more popular in many research fields including time series analysis. To name a few, [15] employs the idea of word2vec [23] which regards part of the time series as word, the rest as context, and part of other time series as negative samples for training. [12] employs the idea of contrastive learning where two positive samples are generated by weak augmentation and strong augmentation to predict each other while the similarity among different augmentations of the same sample is maximized. [13] is designed for univariate time series. It samples several segments of the time series and labels each segment pairs according to their relative distance in the origin series. It also adds the task of judging whether two segments are generated by the same sample. [4] is based on sampling pairs of time windows and predicting whether time windows are close in time by setting thresholds to learn EEG features. [34] is a Transformer-based method which employs the idea of mask language model [17]. The mask operation is performed for multivariate time series and the encoder is trained by predicting the masked value. However, the previous works usually focus on time-wise features and need to continuously obtain the features of several time steps [12,15,34], which makes them difficult to be applied to the novel MTS representation.

3   The Framework

In this work, we focus on self-supervised multivariate time series representation learning. Given M multivariate time series $X = \{x_0,...,x_M\}$, where $x_i \in \mathbb{R}^{C \times T}$ refers to the i-th time series which has C channels and T time steps. For each multivariate time series x_i, our goal is to generate a proper representation z_i which is applicable to the subsequent task.

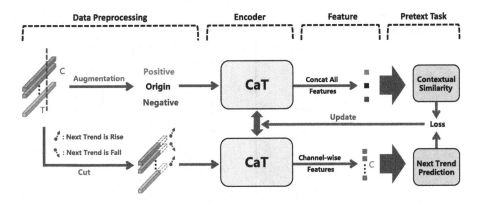

Fig. 2. Overall architecture of the channel-aware self-supervised framework CaSS.

Our proposed novel encoder Channel-aware Transformer (CaT) and pretext tasks constitute our channel-aware self-supervised learning framework CaSS. The overall architecture is shown in Fig. 2. In our framework, CaT is to generate the novel channel-wise features of the MTS sample, and the generated features are served as the inputs of Next Trend Prediction task and Contextual Similarity task. Specifically, we first preprocess the time series samples and apply the two pretext tasks to learn the encoder (representations), then we employ the learnt representations to the MTS classification task by freezing the encoder.

4   Channel-Aware Transformer

This section describes our proposed Transformer-based encoder Channel-aware Transformer which is served as the encoder in our self-supervised learning framework. As shown in Fig. 3, It consists of Embedding Layer, Co-Transformer Layer and Aggregate Layer. The two Transformer structures in Co-Transformer Layer encode the time-wise and channel-wise features respectively, and interact with each other during encoding. Finally, we fuse the time-wise features into channel-wise features through Aggregate Layer to generate the final representation.

4.1   Embedding Layer

Given an input sample $x \in \mathbb{R}^{C \times T}$, we map it to the D-dimension time vector space and channel vector space respectively to obtain the time embedding $e_t \in \mathbb{R}^{T \times D}$ and channel embedding $e_c \in \mathbb{R}^{C \times D}$:

$$e_t = x^T W_t + e_{pos}, \tag{1}$$

$$e_c = x W_c, \tag{2}$$

where $W_t \in \mathbb{R}^{C \times D}$, $W_c \in \mathbb{R}^{T \times D}$ are learnable embedding matrices. $e_{pos} \in \mathbb{R}^{T \times D}$ is the positional embedding applying the design of [30].

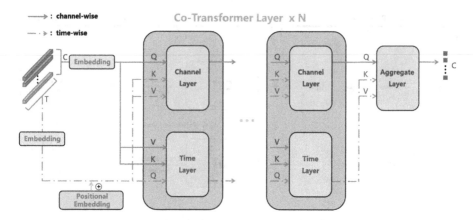

Fig. 3. Overall architecture of Channel-aware Transformer (CaT). Q, K, V represent the query vector, key vector, value vector in attention operation respectively.

4.2   Co-transformer Layer

This part adopts a N-layer two tower structure based on Transformer [21]. Each layer is composed of Time Layer and Channel Layer, focusing on time-wise and channel-wise respectively. Supposing in the i-th layer ($i = 0, ..., N-1$), we obtain the input $a_t^i \in \mathbb{R}^{T \times D}$ for Time Layer and $a_c^i \in \mathbb{R}^{C \times D}$ for Channel Layer. The process in Time Layer is:

$$Q_t^i = a_t^i W_{qt}^i, \quad K_t^i = a_c^i W_{kt}^i, \quad V_t^i = a_c^i W_{vt}^i, \tag{3}$$

$$b_t^i = \text{LayerNorm}(\text{MHA}(Q_t^i, K_t^i, V_t^i) + a_t^i), \tag{4}$$

$$a_t^{i+1} = \text{LayerNorm}(\text{FFN}(b_t^i) + b_t^i), \tag{5}$$

and the process in Channel Layer is:

$$Q_c^i = a_c^i W_{qc}^i, \quad K_c^i = a_t^i W_{kc}^i, \quad V_c^i = a_t^i W_{vc}^i, \tag{6}$$

$$b_c^i = \text{LayerNorm}(\text{MHA}(Q_c^i, K_c^i, V_c^i) + a_c^i), \tag{7}$$

$$a_c^{i+1} = \text{LayerNorm}(\text{FFN}(b_c^i) + b_c^i), \tag{8}$$

where W_{qt}^i, W_{kt}^i, W_{vt}^i, W_{qc}^i, W_{kc}^i, $W_{vc}^i \in \mathbb{R}^{D \times D}$ are learnable matrices. MHA is the abbreviation of multi-head attention and FFN is the abbreviation of feed forward network. Specifically, $a_t^0 = e_t$, $a_c^0 = e_c$.

This interactive way helps reduce the time complexity from $o((T^2 + C^2)D)$ to $o(2TCD)$ compared with the non-interactive two tower Transformer-based encoder which applies self-attention mechanism [30]. In the real world, T is usually much larger than C, so it can further boost the speed of the encoder.

4.3 Aggregate Layer

Through Co-Transformer Layer, we can obtain time-wise features a_t^N and channel-wise features a_c^N. If all features are forcibly concatenated, the final dimension of the representation will be too large to be applied in the subsequent work. In real applications, the channel length is usually much less than the time length, therefore we integrate the time-wise features into the channel-wise features through attention operation. Finally, we concatenate these novel channel-wise features as the final representation $z \in \mathbb{R}^{1 \times (C \cdot D)}$:

$$Q_c^N = a_c^N W_{qc}^N, \quad K_c^N = a_t^N W_{kc}^N, \quad V_c^N = a_t^N W_{vc}^N, \tag{9}$$

$$a_c = \text{MHA}(Q_c^N, K_c^N, V_c^N), \tag{10}$$

$$z = [a_c^1, a_c^2, ..., a_c^C], \tag{11}$$

where $W_{qc}^N, W_{kc}^N, W_{vc}^N \in \mathbb{R}^{D \times D}$ are learnable matrices. $[\cdot, \cdot]$ is the concatenation operation.

5 Pretext Task

In order to enable our encoder to carry out self-supervised learning more efficiently, we design two novel pretext tasks Next Trend Prediction (NTP) and Contextual Similarity (CS) based on our novel channel-wise representation.

5.1 Next Trend Prediction

Given a sample $x_i \in \mathbb{R}^{C \times T}$, we randomly select a time point $t \in [1, T-1]$ for truncation. The sequence before t time step $x_i^{NTP(t)} \in \mathbb{R}^{C \times t}$ is regarded as the input of the NTP task and the data after t is padded to T with zeros. For each channel j, we adopt the trend of the $t+1$ time step as the label $y_{i,j}^{NTP(t)}$ for training:

$$y_{i,j}^{NTP(t)} = \begin{cases} 1, & \text{if } x_i[j, t+1] \geq x_i[j, t] \\ 0, & \text{if } x_i[j, t+1] < x_i[j, t] \end{cases}, \tag{12}$$

where $x_i[j, t]$ represents the value of t time step of channel j in x_i.

After inputting the NTP sample x_i into our encoder, we can obtain the representation $z_i^{NTP(t)} \in \mathbb{R}^{C \times D}$ where $z_{i,j}^{NTP(t)} \in \mathbb{R}^D$ represents the representation of the j-th channel. Finally, a projection head is applied to predict the probability of the rise and fall. Assuming that every sample generated K_{NTP} input samples and the corresponding truncating time point set is $S \in \mathbb{R}^{K_{NTP}}$. For x_i, the loss of the NTP task can be obtained through the following formula:

$$\ell_{NTP} = \sum_{t \in S} \sum_{j}^{C} \text{CE}(\varphi_0(z_{i,j}^{NTP(t)}), y_{i,j}^{NTP(t)}), \tag{13}$$

where φ_0 is the projection head of the NTP task and CE is the Cross Entropy loss function.

5.2   Contextual Similarity

The purpose of NTP task is to enable the sample to learn the relationships between its internal channels. Meanwhile, we need to ensure the independence between different samples, so we further employ the Contextual Similarity task. The main difference between CS task and Contextual Contrasting task in TS-TCC [12] lies in the design of the augmentation method. To help the framework focus more on the dependencies between channels, we further apply asynchronous permutation strategy to generate negative samples. And we also add the original sample to the self-supervised training to enhance the learning ability.

In this task, we generate several positive samples and negative samples for each sample through augmentation strategy. For positive samples, we adopt the Interval Adjustment Strategy which randomly selects a series of intervals, and then adjust all values by jittering. Further, we also adopt a Synchronous Permutation Strategy which segments the whole time series and disrupts the segment order. It helps to maintain the relations between segments and the generated samples are regarded as positive. For negative samples, we adopt an Asynchronous Permutation Strategy which randomly segments and disrupts the segment order for each channel in different ways. In experiments, for each sample, we use the interval adjustment strategy and the synchronous disorder strategy to generate one positive sample respectively, and we use the asynchronous disorder strategy to generate two negative samples. Assuming that the batch size is B, we can generate extra $4B$ augmented samples. Therefore the total number of the sample in a batch is $5B$.

With the i-th sample x_i in a batch our encoder can obtain its representation $z_i^0 \in \mathbb{R}^{1 \times (C \cdot D)}$. The representations of inputs except itself are $z_i^* \in \mathbb{R}^{(5B-1) \times (C \cdot D)}$, where $z_i^{*,m} \in \mathbb{R}^{1 \times (C \cdot D)}$ is the m-th representation of z_i^*. Among z_i^*, the two positive samples are $z_i^{+,1}, z_i^{+,2} \in \mathbb{R}^{1 \times (C \cdot D)}$. Therefore, for i-th sample, the loss of the CS task can be obtained through the following formula:

$$\ell_{CS} = -\sum_{n=1}^{2} \log \frac{\exp(\text{sim}(\varphi_1(z_i^0), \varphi_1(z_i^{+,n}))/\tau)}{\sum_{m=1}^{5B-1} \exp(\text{sim}(\varphi_1(z_i^0), \varphi_1(z_i^{*,m}))/\tau)}, \tag{14}$$

where τ is a hyperparameter, φ_1 is the projection head of the CS task, sim is the cosine similarity.

The final self-supervised loss is the combination of the NTP loss and CS loss as follows:

$$\ell = \alpha_1 \cdot \ell_{NTP} + \alpha_2 \cdot \ell_{CS}, \tag{15}$$

where α_1 and α_2 are hyperparameters.

6   Experiments

6.1   Datasets

To demonstrate the effectiveness of our self-supervised framework, we use the following four datasets from different fields:

- **UCI HAR**[1]: The UCI HAR dataset [1] contains sensor readings of 6 human activity types. They are collected with a sampling rate of 50 Hz.
- **LSST** [2]: The LSST dataset [2] is an open data to classify simulated astronomical time series data in preparation for observations from the Large Synoptic Survey Telescope.
- **ArabicDigits**[3]: The ArabicDigits dataset [5] contains time series of mel-frequency cepstrum coefficients corresponding to spoken Arabic digits. It includes data from 44 male and 44 female native Arabic speakers.
- **JapaneseVowels** (See footnote 3): In the JapaneseVowels dataset [5], several Japanese male speakers are recorded saying the vowels 'a' and 'e'. A '12-degree linear prediction analysis' is applied to the raw recordings to obtain time-series with 12 dimensions.

The detailed information is shown on Table 1.

Table 1. Detailed information of the used datasets.

Dataset	Train	Test	Time	Channel	Class
UCI HAR	7352	2947	128	9	6
LSST	2459	2466	36	6	14
ArabicDigits	6600	2200	93	13	10
JapaneseVowels	270	370	29	12	9

---

[1] https://archive.ics.uci.edu/ml/datasets/Human+Activity+Recognition+Using+Smartphones.

[2] http://www.timeseriesclassification.com/description.php?Dataset=LSST.

[3] http://www.mustafabaydogan.com/.

6.2 Experimental Settings

For supervised learning, we set $D = 512$, $N = 8$, attention head $= 8$, batch size $= 4$, dropout $= 0.2$. We use Adam optimizer with a learning rate of $1e-4$.

For self-supervised learning, we set $K_{NTP} = 10$, $D = 512$, $N = 8$, attention head $= 8$, batch size $= 10$, dropout $= 0.2$, $\tau = 0.2$, $\alpha_1 = 2$, $\alpha_2 = 1$. We use Adam optimizer with a learning rate of $5e-5$.

For fine-tuning after self-supervised learning, we train a single fully connected layer on top of the frozen self-supervised pre-trained encoder to evaluate the effect of our self-supervised framework. We set batch size $= 4$ and use Adam optimizer with a learning rate of $1e-3$.

We evaluate the performance using two metrics: Accuracy (ACC) and Macro-F1 score (MF1). Every result is generated by repeating 5 times with 5 different seeds.

We conduct our experiments using PyTorch 1.7 and train models on a NVIDIA GeForce RTX 2080 Ti GPU.

6.3 Baselines

We compare our framework against the following self-supervised methods. It is noted that we apply the default hyperparameters of each compared method from the original paper or the code. The detailed information and the reason why we choose these methods are as followed:

(1) W2V [15]: This method employs the idea of word2vec. It combines an encoder based on causal dilated convolutions with a triplet loss and time-based negative sampling. Finally, they train a SVM on top of the frozen self-supervised pre-trained encoder. It achieves great results leveraging unsupervised learning for univariate and multivariate classification datasets. We select $K = 10, 20$ from the original experiments.

(2) W2V+: It applies our proposed two tower Transformer-based model as the encoder while training with the pretext task of W2V. For the requirement of the time-wise feature, we replace Channel-aware Transformer with Time-aware Transformer (TaT) which integrates the channel-wise features into the time-wise features in the Aggregate Layer.

(3) TS-TCC [12]: This method employs the idea of contrastive learning using a convolutional architecture as encoder. After self-supervised learning, they train a fully connected layer on top of the frozen self-supervised pre-trained encoder. It is the state-of-the-art contrastive learning method in the field of self-supervised of time series.

(4) TS-TCC+: It applies TaT as the encoder like W2V+ while training with the pretext task of TS-TCC.

(5) TST [34]: This method employs the idea of masked language model using a Transformer-based encoder. In their work, it achieves outstanding results by finetuning the whole encoder after pre-training it. For fair comparison, we freeze the encoder while finetuning.

(6) TST+: It applies TaT as the encoder like W2V+ while training with the pretext task of TST.

(7) NVP+CS: To compare with the regression, in our framework we replace Next Trend Prediction with Next Value Predict (NVP) and regard it as a new strong baseline. We select 15% time steps to predict the values of the next time step.

(8) Supervised: Supervised learning on both encoder and fully connected layer.

Table 2. Comparison between CaSS and other self-supervised methods. ↑ mark indicates that the self-supervised result performs better than the supervised result.

Method	HAR		LSST		ArabicDigits		JapaneseVowels	
	ACC	MF1	ACC	MF1	ACC	MF1	ACC	MF1
W2V K=10	90.37 ± 0.34	90.67 ± 0.97	57.24 ± 0.24	36.97 ± 0.49	90.16 ± 0.57	90.22 ± 0.52	97.98 ± 0.40	97.73 ± 0.48
W2V K=20	90.08 ± 0.18	90.10 ± 0.16	53.47 ± 0.75	32.84 ± 1.25	90.55 ± 0.77	90.60 ± 0.75	97.98 ± 0.40	97.89 ± 0.43
W2V+	84.55 ± 0.59	84.34 ± 0.69	54.90 ± 0.58	33.97 ± 0.79	87.59 ± 0.48	87.56 ± 0.49	95.27 ± 0.38	95.25 ± 0.39
TS-TCC	90.74 ± 0.25	90.23 ± 0.29	40.38 ± 0.35	23.93 ± 1.93	95.64 ± 0.37	95.43 ± 0.37	82.25 ± 1.16	82.04 ± 1.17
TS-TCC+	90.87 ± 0.31	90.86 ± 0.30	50.75 ± 0.23	32.87 ± 0.65	96.87 ± 0.34	96.80 ± 0.28	84.36 ± 0.27	84.09 ± 0.21
TST	77.62 ± 2.48	78.05 ± 2.56	32.89 ± 0.04	7.86 ± 1.63	90.73 ± 0.36	90.90 ± 0.33	97.30 ± 0.27	97.47 ± 0.34
TST+	87.39 ± 0.49	87.77 ± 0.11	34.49 ± 0.38	14.62 ± 0.32	96.82 ± 0.23	96.82 ± 0.22	97.87 ± 0.23[↑]	97.51 ± 0.22
NVP+CS	92.47 ± 0.20	92.38 ± 0.19	32.42 ± 0.14	6.02 ± 0.30	96.50 ± 0.45	96.51 ± 0.55	96.34 ± 0.43	95.49 ± 0.11
CaSS	**92.57 ± 0.24**[↑]	**92.40 ± 0.17**[↑]	**64.94 ± 0.02**	**46.11 ± 0.55**	**97.07 ± 0.20**	**97.07 ± 0.20**	**98.11 ± 0.27**[↑]	**98.14 ± 0.08**[↑]
Supervised	92.35 ± 0.63	92.40 ± 0.58	66.57 ± 0.38	51.60 ± 1.26	98.07 ± 0.38	98.07 ± 0.38	97.71 ± 0.13	97.56 ± 0.07

6.4 Results and Analysis

Comparison with Self-supervised Methods. The experimental results are shown in Table 2. Overall, our self-supervised learning framework can significantly surpass the previous state-of-the-art methods. Especially in LSST and JapaneseVowels whose time lengths are relative short, methods based on fitting specific values or features like TST, TS-TCC and NVP cannot perform well, while our framework can obtain promising and stable performances. It demonstrates that simple trend predicting is more efficient than regression. W2V and our framework are suitable for both short and long time length datasets, and our performances can significantly surpass W2V. This demonstrates the powerful representation learning ability of our framework. Moreover, our self-supervised framework is shown to be superior to the supervised way in two of four datasets while in the other two datasets can achieve similar results. It also proves that our self-supervised learning framework is capable of learning not only complex characteristics between samples but also within samples.

For a more convincing comparison, we conduct experiments by replacing the origin encoder of the previous methods with our proposed encoder. It helps to offer a more detailed view on the aspects of both encoder and pretext task. It is shown in Table 2 that applying our encoder like TS-TCC+ and TST+ helps to improve their effectiveness, which demonstrates the encoding ability of our encoder. W2V is a method which pays more attention on the local information, so the causal dilated convolutions which only focuses on previous information is more suitable than W2V+ which encodes the global information. In the aspect of pretext task, when combining NTP task and CS task with our encoder, the self-supervised learning ability is further improved by a large margin compared to other pretext tasks.

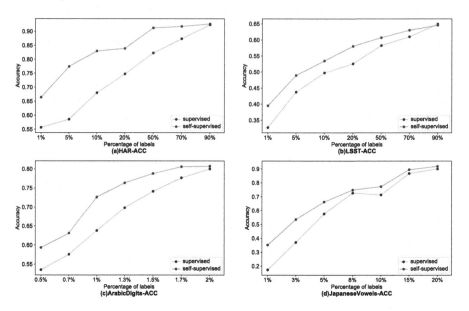

Fig. 4. Few-shot learning results. We report the comparison between our self-supervised framework and the supervised way.

Few-Shot Learning. To further prove the effect of our self-supervised framework, we conduct few-shot learning experiments comparing with the supervised learning. In HAR and LSST datasets, we choose 1%, 5%, 10%, 20%, 50%, 70%, 90% percentage of labeled samples for model training respectively. For ArabicDigits and JapaneseVowels datasets, we adopt a set of smaller percentages in order to compare the performance of few-shot learning more clearly. The results are shown in Fig. 4. Among these datasets, our self-supervised framework can significantly surpass the supervised learning by training a single fully connected layer with limited labeled samples.

6.5 Ablation Study

Ablation Study on Pretext Task. To analyze the role of each of our pretext tasks, we apply the following variants as comparisons:

(1) **-NTP**: It only applies Contextual Similarity task to self-supervised learning.
(2) **-CS**: It only applies Next Trend Prediction task to self-supervised learning.

(3) -neg augment: It removes the negative samples generated by the asynchronous disorder strategy.

(4) reverse neg: It regards the samples generated by the asynchronous disorder strategy as positive samples.

The experimental results are shown in Table 3. It can be seen that in our pretext tasks, NTP task and CS task can well cooperate with each other. Specifically, NTP task occupies the most important position in small datasets while CS task is more important in datasets with large number of samples. On the one hand, it shows the importance of the internal relationships between channels to multivariate time series. On the other hand, it demonstrates that self-supervised learning should not only focus on the characteristics of the sample itself, but also need to maintain the independence from other samples. In addition, negative sample enhancement can bring a more stable effect to the encoder and further enhance the effect of self-supervised learning method. The conversion of the negative samples into positive samples will lead to the decline of the effect, which shows that in the time series, the time relationships between channels can not be disturbed.

Table 3. Ablation study on pretext task.

Method	HAR		LSST		ArabicDigits		JapaneseVowels	
	ACC	MF1	ACC	MF1	ACC	MF1	ACC	MF1
-NTP	90.96 ± 0.59	90.71 ± 0.59	61.05 ± 1.20	41.30 ± 0.40	95.55 ± 0.36	95.50 ± 0.41	94.87 ± 0.27	94.87 ± 0.22
-CS	76.94 ± 0.66	75.96 ± 0.67	53.00 ± 0.81	33.50 ± 1.97	90.71 ± 0.06	90.72 ± 0.08	97.71 ± 0.13	97.69 ± 0.13
-neg augment	87.06 ± 0.25	86.79 ± 0.23	61.88 ± 0.57	45.21 ± 0.10	96.32 ± 0.32	96.32 ± 0.32	97.98 ± 0.13	97.79 ± 0.16
reverse neg	83.02 ± 0.22	82.65 ± 0.15	63.30 ± 0.65	45.40 ± 0.50	92.62 ± 0.43	92.61 ± 0.44	97.42 ± 0.16	97.22 ± 0.09
CaSS	**92.57 ± 0.24**	**92.40 ± 0.17**	**64.94 ± 0.02**	**46.11 ± 0.55**	**97.07 ± 0.20**	**97.07 ± 0.20**	**98.11 ± 0.27**	**98.14 ± 0.08**

Ablation Study on Encoder. To analyze the effect of each component in the encoder, we apply the following variants for comparisons by supervised learning:

(1) Self Aggregate: It contains two Transformers with self-attention mechanism to encode time-wise and channel-wise features independently. Finally the time-wise features are fused into channel-wise features through Aggregate Layer.

(2) Channel Self: A single Transformer is applied with self-attention mechanism to encode channel-wise features.

(3) -Aggregate Layer: The channel-wise features of the last Co-Transformer Layer are applied without fusing the time-wise features.

The experimental results are shown in Table 4. It can be seen that, if time-wise and channel-wise features are only fused in the last aggregate layer without interactions in the previous stage, they cannot be well integrated and bring the loss of information. As contrast, the interactions between the two Transformers can significantly bring the performance improvement. The results of Channel Self illustrate the importance of each channel's independent time pattern. Finally, the existence of Aggregate Layer can also better integrate the features of time-wise and channel-wise while alleviating the redundancy of features.

Table 4. Ablation study on encoder.

Method	HAR		LSST		ArabicDigits		JapaneseVowels	
	ACC	MF1	ACC	MF1	ACC	MF1	ACC	MF1
Self aggregate	91.88 ± 0.35	91.85 ± 0.38	66.10 ± 0.24	42.80 ± 0.83	96.64 ± 0.22	96.63 ± 0.23	97.17 ± 0.13	96.94 ± 0.10
Channel self	**93.15 ± 0.91**	**93.13 ± 0.95**	56.45 ± 0.48	31.31 ± 0.50	97.48 ± 0.11	97.47 ± 0.12	97.57 ± 0.27	97.42 ± 0.22
-Aggregate layer	92.35 ± 0.12	92.36 ± 0.13	63.97 ± 0.59	42.65 ± 0.18	97.68 ± 0.32	97.68 ± 0.32	97.57 ± 0.27	97.38 ± 0.23
CaT	92.35 ± 0.63	92.40 ± 0.58	**66.57 ± 0.38**	**51.60 ± 1.26**	**98.07 ± 0.38**	**98.07 ± 0.38**	**97.71 ± 0.13**	**97.56 ± 0.07**

7   Conclusion

Self-supervised learning is essential for multivariate time series. In this work, we propose a new self-supervised learning framework CaSS to learn the complex representations of MTS. For the encoder, we propose a new Transformer-based encoder Channel-aware Transformer to capture the time-wise and channel-wise features more efficiently. For the pretext task, we propose Next Trend Prediction from the perspective of channel-wise for the first time and combine it with Contextual Similarity task. These novel pretext tasks can well cooperate with our encoder to learn the characteristics. Our self-supervised learning framework demonstrates significant improvement on MTS classification comparing with previous works, and can significantly surpass supervised learning with limited labeled samples.

Acknowledgment. This work was supported by the National Key Research and Development Program of China, No.2018YFB1402600.

References

1. Anguita, D., Ghio, A., Oneto, L., Parra Perez, X., Reyes Ortiz, J.L.: A public domain dataset for human activity recognition using smartphones. In: Proceedings of the 21th International European Symposium on Artificial Neural Networks, Computational Intelligence and Machine Learning, pp. 437–442 (2013)
2. Bagnall, A., et al.: The UEA multivariate time series classification archive. arXiv preprint arXiv:1811.00075 (2018)
3. Bai, Y., Wang, L., Tao, Z., Li, S., Fu, Y.: Correlative channel-aware fusion for multi-view time series classification. In: Proceedings of the AAAI Conference on Artificial Intelligence, vol. 35, pp. 6714–6722 (2021)
4. Banville, H., Albuquerque, I., Hyvärinen, A., Moffat, G., Engemann, D.A., Gramfort, A.: Self-supervised representation learning from electroencephalography signals. In: 2019 IEEE 29th International Workshop on Machine Learning for Signal Processing (MLSP), pp. 1–6. IEEE (2019)
5. Baydogan, M.G.: Multivariate time series classification datasets (2019)
6. Binkowski, M., Marti, G., Donnat, P.: Autoregressive convolutional neural networks for asynchronous time series. In: International Conference on Machine Learning, pp. 580–589. PMLR (2018)
7. Breiman, L.: Random forests. Mach. Learn. **45**(1), 5–32 (2001)
8. Che, Z., Purushotham, S., Cho, K., Sontag, D., Liu, Y.: Recurrent neural networks for multivariate time series with missing values. Sci. Rep. **8**(1), 1–12 (2018)

9. Cortes, C., Vapnik, V.: Support-vector networks. Mach. Learn. **20**(3), 273–297 (1995)
10. Cui, Z., Chen, W., Chen, Y.: Multi-scale convolutional neural networks for time series classification. arXiv preprint arXiv:1603.06995 (2016)
11. Deng, H., Runger, G., Tuv, E., Vladimir, M.: A time series forest for classification and feature extraction. Inf. Sci. **239**, 142–153 (2013)
12. Eldele, E., et al.: Time-series representation learning via temporal and contextual contrasting. In: International Joint Conference on Artificial Intelligence (IJCAI 2021) (2021)
13. Fan, H., Zhang, F., Gao, Y.: Self-supervised time series representation learning by inter-intra relational reasoning. arXiv preprint arXiv:2011.13548 (2020)
14. Ismail Fawaz, H., Forestier, G., Weber, J., Idoumghar, L., Muller, P.-A.: Deep learning for time series classification: a review. Data Min. Knowl. Disc. **33**(4), 917–963 (2019). https://doi.org/10.1007/s10618-019-00619-1
15. Franceschi, J.Y., Dieuleveut, A., Jaggi, M.: Unsupervised scalable representation learning for multivariate time series. In: Thirty-Third Conference on Neural Information Processing Systems, vol. 32. Curran Associates, Inc. (2019)
16. Karim, F., Majumdar, S., Darabi, H., Chen, S.: LSTM fully convolutional networks for time series classification. IEEE Access **6**, 1662–1669 (2017)
17. Kenton, J.D.M.W.C., Toutanova, L.K.: BERT: pre-training of deep bidirectional transformers for language understanding. In: Proceedings of NAACL-HLT, pp. 4171–4186 (2019)
18. Lin, J., Khade, R., Li, Y.: Rotation-invariant similarity in time series using bag-of-patterns representation. J. Intell. Inf. Syst. **39**(2), 287–315 (2012)
19. Lin, L., Xu, B., Wu, W., Richardson, T.W., Bernal, E.A.: Medical time series classification with hierarchical attention-based temporal convolutional networks: a case study of myotonic dystrophy diagnosis. In: CVPR Workshops, pp. 83–86 (2019)
20. Liu, C.L., Hsaio, W.H., Tu, Y.C.: Time series classification with multivariate convolutional neural network. IEEE Trans. Industr. Electron. **66**(6), 4788–4797 (2018)
21. Liu, M., et al.: Gated transformer networks for multivariate time series classification. arXiv preprint arXiv:2103.14438 (2021)
22. Ma, F., Chitta, R., Zhou, J., You, Q., Sun, T., Gao, J.: Dipole: diagnosis prediction in healthcare via attention-based bidirectional recurrent neural networks. In: Proceedings of the 23rd ACM SIGKDD International Conference on Knowledge Discovery and Data Mining, pp. 1903–1911 (2017)
23. Mikolov, T., Chen, K., Corrado, G., Dean, J.: Efficient estimation of word representations in vector space. In: 1st International Conference on Learning Representations, ICLR (2013)
24. Oh, J., Wang, J., Wiens, J.: Learning to exploit invariances in clinical time-series data using sequence transformer networks. In: Machine Learning for Healthcare Conference, pp. 332–347. PMLR (2018)
25. Poli, M., Park, J., Ilievski, I.: WATTNet: learning to trade FX via hierarchical spatio-temporal representation of highly multivariate time series. In: Twenty-Ninth International Joint Conference on Artificial Intelligence and Seventeenth Pacific Rim International Conference on Artificial Intelligence, IJCAI 2020 (2020)
26. Rasul, K., Sheikh, A.S., Schuster, I., Bergmann, U.M., Vollgraf, R.: Multivariate probabilistic time series forecasting via conditioned normalizing flows. In: International Conference on Learning Representations (2020)
27. Rußwurm, M., Körner, M.: Self-attention for raw optical satellite time series classification. ISPRS J. Photogramm. Remote. Sens. **169**, 421–435 (2020)

28. Schäfer, P.: The boss is concerned with time series classification in the presence of noise. Data Min. Knowl. Disc. **29**(6), 1505–1530 (2015)

29. Sun, C., Myers, A., Vondrick, C., Murphy, K., Schmid, C.: VideoBERT: a joint model for video and language representation learning. In: Proceedings of the IEEE/CVF International Conference on Computer Vision, pp. 7464–7473 (2019)

30. Vaswani, A., et al.: Attention is all you need. In: Advances in Neural Information Processing Systems, pp. 5998–6008 (2017)

31. Wang, Z., Yan, W., Oates, T.: Time series classification from scratch with deep neural networks: a strong baseline. In: 2017 International Joint Conference on Neural Networks (IJCNN), pp. 1578–1585. IEEE (2017)

32. Xi, X., Keogh, E., Shelton, C., Wei, L., Ratanamahatana, C.A.: Fast time series classification using numerosity reduction. In: Proceedings of the 23rd International Conference on Machine Learning, pp. 1033–1040 (2006)

33. Xu, J., Wang, J., Long, M., et al.: Autoformer: decomposition transformers with auto-correlation for long-term series forecasting. In: Advances in Neural Information Processing Systems, vol. 34 (2021)

34. Zerveas, G., Jayaraman, S., Patel, D., Bhamidipaty, A., Eickhoff, C.: A transformer-based framework for multivariate time series representation learning. In: Proceedings of the 27th ACM SIGKDD Conference on Knowledge Discovery & Data Mining, pp. 2114–2124 (2021)

35. Zhang, Y., et al.: Memory-gated recurrent networks. In: Thirty-Fifth AAAI Conference on Artificial Intelligence (AAAI 2021) (2020)

36. Zhao, B., Lu, H., Chen, S., Liu, J., Wu, D.: Convolutional neural networks for time series classification. J. Syst. Eng. Electron. **28**(1), 162–169 (2017)

37. Zheng, Y., Liu, Q., Chen, E., Ge, Y., Zhao, J.L.: Time series classification using multi-channels deep convolutional neural networks. In: Li, F., Li, G., Hwang, S., Yao, B., Zhang, Z. (eds.) WAIM 2014. LNCS, vol. 8485, pp. 298–310. Springer, Cham (2014). https://doi.org/10.1007/978-3-319-08010-9_33

38. Zheng, Y., Liu, Q., Chen, E., Ge, Y., Zhao, J.L.: Exploiting multi-channels deep convolutional neural networks for multivariate time series classification. Front. Comp. Sci. **10**(1), 96–112 (2016). https://doi.org/10.1007/s11704-015-4478-2

39. Zhou, H., et al.: Informer: beyond efficient transformer for long sequence time-series forecasting. In: Proceedings of AAAI (2021)

Temporal Knowledge Graph Entity Alignment
via Representation Learning

Xiuting Song, Luyi Bai[(✉)], Rongke Liu, and Han Zhang

School of Computer and Communication Engineering, Northeastern University (Qinhuangdao),
Qinhuangdao 066004, China
baily@neuq.edu.cn

Abstract. Entity alignment aims to construct a complete knowledge graph (KG) by matching the same entities in multi-source KGs. Existing methods mainly focused on the static KG, which assumes that the relationship between entities is permanent. However, almost every KG will evolve over time in practical applications, resulting in the need for entity alignment between such temporal knowledge graphs (TKGs). In this paper, we propose a novel entity alignment framework suitable for TKGs, namely Tem-EA. To incorporate temporal information, we use recurrent neural networks to learn temporal sequence representations. Furthermore, we use graph convolutional network (GCN) and translation-based embedding model to fully learn structural information representation and attribute information representation. Based on these two representations, the entity similarity is calculated separately and combined using linear weighting. To improve the accuracy of entity alignment, we also propose a concept of nearest neighbor matching, which matches the most similar entity pair according to distance matrix. Experiments show that our proposed model has a significant improvement compared to previous methods.

Keywords: Temporal knowledge graph · Entity alignment · Structure embedding · Attribute embedding

1 Introduction

Knowledge graph (KG) describes various entities, concepts and their relationships in the real-world. They often serve intelligent systems such as question answering [28], knowledge-representation system [1] and recommender systems [5] as knowledge base. In recent years, the number of temporal knowledge graphs (TKGs) containing temporal information of each fact has increased rapidly. Some typical TKGs including ICEWS, YAGO3 [14] and Wikidata [8].

At present, researchers have made considerable progress on TKGs embedding. TKG embedding models encode temporal information in their embedding to generate more accurate link predictions. Most existing TKG embedding methods [10, 12, 25] are derived by extending temporal information on TransE [2]. In addition, a few methods [10, 20] are temporal extension based on DistMult [26]. All the above studies have proved that the

© The Author(s), under exclusive license to Springer Nature Switzerland AG 2022
A. Bhattacharya et al. (Eds.): DASFAA 2022, LNCS 13246, pp. 391–406, 2022.
https://doi.org/10.1007/978-3-031-00126-0_30

performance of KG embedding models can be further improved by extending temporal information. In practical applications, the KGs embedding model is also the core part of embedding-based entity alignment techniques. Entity alignment aims to construct more complete KGs by linking the shared entities between complementary KGs, and it has received extensive attention in the integration of multi-source KGs.

However, current entity alignment approaches ignore a large amount of temporal information of facts and assume that KGs are static. Obviously, this is not in line with reality because the relationship between entities is dynamically changing. Existing researches can be divided into translation-based and graph neural network (GNN)-based. Most traditional work (e.g. [3, 17, 31]) relies on translation-based model to encode KG entities and relationships, and calculate the distance between entity vectors to match entities. In addition, some methods (e.g. [18, 27, 32]) also use attributes, entity names, and entity context information to refine the embedding. However, translation-based models usually rely on a large number of pre-aligned entities and are less efficient in capturing complex structural information. In order to make better use of the neighborhood information of entities, some methods [22–24] have been proposed to improve entity alignment by using graph neural networks. GNN-based methods can better capture the neighborhood structure information of entities and can handle multi-relation KGs well.

Although GNN-based models are very effective in learning neighborhood-aware embedding of entities, it has shortcomings in the utilization of attribute information. Most methods (e.g. [23, 24]) ignore the importance of attribute triples and only use the relational structure of KGs for entity alignment. GCN-Align [22] regards attribute triples as relational triples and uses GCN for embedding, without independent representation of attribute information. Furthermore, structural information in KGs is complex graph relational network, while attribute information is just simple sequence information. Current research does not use different methods for independent representation learning according to their respective characteristics. A single use of GNN cannot fully model multiple information of KGs, and may lead to incorrect modeling results of different information. Another problem is that the above models are all designed for static KGs and cannot be used for entity alignment between TKGs.

In response to the above-mentioned problems, in this paper we propose a novel entity alignment framework (Tem-EA) for TKGs. Tem-EA is based on long short-term memory neural network (LSTM) to encode temporal sequence information, and combined with structural information embedding learned by GCN. To improve the utilization of structural information and attribute information, we use graph convolutional networks and translation-based embedding models to represent them independently, and jointly perform entity alignment. Besides, we propose a concept of nearest neighbor matching, which requires the aligned entities to be closest to each other in addition to the embedding distance less than a set threshold, thus improving the accuracy of entity alignment. We summarize the main contributions of this paper as follows:

- We propose a novel entity alignment method Tem-EA between TKGs. Tem-EA uses recurrent neural networks to learn temporal sequence representation, and then combines with structural information representation to form complete entity embedding results.

- According to the different characteristics of structural information and attribute information, our model uses GCNs and translation-based methods to learn them respectively. We also propose a concept of nearest neighbor matching to improve the accuracy of entity alignment.
- We evaluate the proposed approach on two large-scale datasets. The experimental results showed that Tem-EA significantly outperformed five state-of-art models on TKGs entity alignment task in terms of Hits@k (k = 1, 5, 10) and MRR.

The rest of this paper is organized as follows. We first introduce the related work on TKGs embedding and entity alignment in Sect. 2. We describe our approach in detail in Sect. 3, and report experimental results in Sect. 4. Finally, we conclude this paper in Sect. 5.

2 Related Work

2.1 Temporal Knowledge Graph Embedding

In recent years, the representation learning of static KGs has been extensively studied. Some examples of typical embedding models include TransE [2], DistMult [26], RESCAL [15], ComplEx [19] and QuatE [30]. These methods have been proved to be effective for the completion task of static KGs, but cannot model the temporal evolution of KGs due to their neglect of temporal information.

Some recent attempts on TKGs embedding learning has shown that the use of temporal information can further improve the performance of embedding models. TTransE [12] is an extension of TransE by adding temporal information embedding in the translation distance score function. Similar to TTransE, HyTE [7] projects the embedding of entities and relationships to the temporal hyperplane and then applies TransE on the projected embedding representation. Know-Evolve [20] models the development of events as a temporal point process. TA-TransE and TA-DistMult [10] regard relationship and temporal as a temporal predicate sequence, and use recurrent neural networks to learn the embedding of the sequence. However, this model does not consider concurrent facts that occur within the same time interval. DE-SimplE [9] provides a diachronic entity feature embedding function for static KGs embedding model, and the function is model-agnostic and can be combined with any static model. ATiSE [25] incorporates temporal information into entity and relationship representations through additive temporal series decomposition. The above methods are all reasoning about the past time, and cannot predict the facts that will happen in the future. RE-Net [11] uses recurrent event encoder and neighborhood aggregator two modules to infer future events in a sequential manner. DBKGE [13] is a dynamic Bayesian embedding model that can dynamically infer the semantic representation of entities evolving over time and make predictions about the future. TigeCMN [33] learns node representations from a series of time interactions to learn the various attributes of entities.

2.2 Embedding-Based Entity Alignment

Existing entity alignment methods embed entities of different KGs into the same vector space, and find the aligned entities by measuring the distance between these embedding.

These studies can be divided into two kinds: translation-based models and graph neural network (GNN)-based.

Most translation-based methods use TransE or its variants to learn the embedding of entities and relationships. MTransE [3] represents different KGs as independent embedding, and learns transformation between KGs via five alignment models. In order to solve the problem of lack of seed alignment, IPTransE [31] and BootEA [17] proposed an iterative strategy to repeatedly mark new aligned entities as training data. In addition to using relational triples, some methods also introduce other features of entities. JAPE [18] and AttrE [21] use attribute information to improve entity embedding. MultiKE [32] and KDCoE [4] integrate entity name and description information to learn entity embedding. However, translation-based models are affected by the seed entity and structural differences, resulting in poor performance of KGs embedding. Considering that entities with similar neighborhoods are likely to be aligned, some recent work attempts to improve entity alignment by using graph convolutional networks (GCN). GCN-Align [22] is the first to use GCN to model the neighborhood information of entities. HGCN [23] jointly learns entity and relationship representation, which is often ignored in GCN-based entity alignment techniques. RDGCN [24] constructed a dual relation graph to merge relationship information. Considering the heterogeneous situation of KGs, MuGNN [6] proposes a multi-channel graph neural network model, and each channel uses a different relational weighting scheme to encode KG. AliNet [16] aggregates long-distance neighborhood information by using gating mechanism and attention mechanism. Moreover, RNM [29] proposes to use neighborhood matching to enhance entity alignment. However, all the above methods do not use different methods to embed structure information and attribute information respectively, and they are all designed for static KGs and cannot be used for TKGs. Therefore, we propose a novel Tem-EA to study the entity alignment of TKGs.

3 Methodology

3.1 Problem Formulation

A temporal knowledge graph (TKG) can be formalized as $TKG = (E, R, A, T, T_R, T_A)$, where E, R, A are the sets of entities, relations and attributes, respectively. The T is temporal information set. $T_R \subseteq E \times R \times E$ denotes the set of relation triples, $T_A \subseteq E \times A \times V$ is the set of attribute triples, where V is the set of attribute values. We consider the entity alignment task between two TKGs TKG_1 and TKG_2. Let E_1 and E_2 denote their entity sets, respectively. The goal is to find a set of identical entities $L = \{(e_1, e_2) \in E_1 \times E_2 | e_1 \equiv e_2\}$, where $\equiv$ denotes an equivalence relation in which two entities refer to the same thing.

3.2 Overview

The framework of our model Tem-EA is given in Fig. 1. The inputs are two temporal knowledge graphs TKG_1 and TKG_2, and some pre-aligned entity pairs (called seed entity alignment). The outputs are new equivalent entities. Tem-EA uses both relation triples and attribute triples for entity alignment. For relational triples with temporal information, we first use long short-term memory (LSTM) to learn time-aware representations, and use GCNs for structural embedding of entities and relationships. Then we calculate the joint entity representation from the above two embedding results and get the distance matrix based on the structure embedding. For attributional triples, we use TransE model for embedding and calculate attribute-based entity similarity. Finally, we merge the entity similarity calculated based on structure and attribute into linear weighting. We also proposed a concept of nearest neighbor matching, which requires matching entities to be closest to each other in addition to the embedding distance less than a set threshold. According to the distance matrix of the entity pair, we calculate the final aligned entity set.

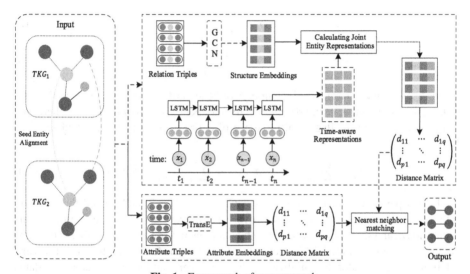

Fig. 1. Framework of our approach.

3.3 GCN-Based TKGs Structure Embedding

With the development of GCNs in recent years, GCN has been proven to effectively capture complex and huge structural information from KGs. Entities in KGs tend to be adjacent to equivalent entities through the same type of relationship. Therefore, the use of GCNs can effectively model the neighborhood information of the entity and improve the embedding of the entity. Here, we also use GCN to project the structural information of TKGs into a low-dimensional vector space, where equivalent entities are close to each other.

In order to aggregate multi-hop neighbors, our model consists of multiple stacked GCN layers. We first weight all the relationship types in the two TKGs, and generate a weighted connected matrix as the input structure feature of the GCNs. The $l + 1$ layer of GCN takes the $X^{(l)}$ of the l layer as the entity feature input, and the output entity represents $X^{(l+1)}$, which is calculated as follows:

$$X^{(l+1)} = \xi\left(\hat{D}^{-\frac{1}{2}}\hat{A}\hat{D}^{-\frac{1}{2}}X^{(l)}W^{(l)}\right) \tag{1}$$

where ξ is the *ReLU* activation function; $\widehat{A} = A + I$, A is the relational matrix that represents the structure information of TKGs, and I is an identity matrix; $\widehat{D}$ is the diagonal node degree matrix of $\widehat{A}$; $W^{(l)} \in \mathbb{R}^{d^{(l)} \times d^{(l+1)}}$ is the weight matrix of a specific layer of GCNs.

Considering that traditional GCN can only deal with undirected single-relationship networks, it is not suitable for KGs with multiple relations. Since the probability of the equivalence of two entities largely depends on the relationship that exists between the aligned entities, we first assign weight to the relationship between entities, and use A_{ij} to measure the impact of entity i on entity j degree. Each relationship is measured by head entity, tail entity, and triples, and is defined as follows:

$$deg(r) = \frac{h\_num_r}{tri\_num_r} \tag{2}$$

$$ideg(r) = \frac{t\_num_r}{tri\_num_r} \tag{3}$$

where h_num_r and t_num_r are the numbers of head entities and tail entities in the relational triples containing r, and tri_num_r is the number of triples of relation r. Therefore, in order to measure the degree of influence of entity i on entity j, we calculate A_{ij} as follows:

$$A_{ij} = \sum_{(e_i,r,e_j)\in G} ideg(r) + \sum_{(e_j,r,e_i)\in G} deg(r) \tag{4}$$

In this paper, we set the number of GCN layers $l = 2$, and embed the structure information of TKG_1 and TKG_2 respectively. Each layer sets the same feature vectors dimension d, and shares two layers of weight matrices W to ensure embedding in the same vector space. When embedding structural information, the feature vectors of entities and relationships are randomly initialized in the 0-th layer, and then continuously updated during training.

3.4 Time-Aware Representation and Joint Entity Representation

The temporal information in the TKGs is sequence data, so we use LSTM to encode it separately, and embed it into a unified vector space with structural information. The equations defining an LSTM are: Recurrent Neural Network (RNN) is a type of neural network used to process and predict sequence data. Long short-term memory (LSTM) is an improved RNN that can solve the problems of gradient disappearance and gradient explosion in the training process of long sequences. It also has a wide range of applications in time series prediction problems. Therefore, we use LSTM to model the temporal data in TKGs. The calculation formula of LSTM is as follows:

$$g_t = tanh(U_g h_{t-1} + W_g x_t) \tag{5}$$

$$f_t = \sigma(U_f h_{t-1} + W_f x_t) \tag{6}$$

$$i_t = \sigma(U_i h_{t-1} + W_i x_t) \tag{7}$$

$$o_t = \sigma\left(U_o h_{t-1} + W_o x_t\right) \tag{8}$$

$$s_t = s_{t-1} \odot f_t + g_t \odot i_t \tag{9}$$

$$h_t = tanh(s_t) \odot o_t \tag{10}$$

where g_t, f_t, i_t and o_t are the input modulation, forget, input and output gates, respectively. h_t is the value of the hidden layer of LSTM. x_t is the feature vector of the t-th element at the current time. σ and $tanh$ are activation functions, and $\odot$ is element-wise multiplication. To combine temporal information embedding with structural embedding, we also set the embedding dimension of LSTM to d.

We denote the result of structure embedding based on GCN as h_s, and the LSTM-based temporal information embedding result as h_t. Formally, for each entity $e \in E$, its complete entity representation can be calculated as:

$$e = [h_s; h_t] \tag{11}$$

where ; represents the connection operation. After obtaining the complete embedding result of the entity carrying temporal information, we use the seed alignment set S as training data to train the model. We use a margin-based scoring function as the training objective, to make the distance between aligned entity pairs to be as close as possible. The loss function L_{SE} is defined as:

$$L_{SE} = \sum_{(e_1,e_2)\in S} \sum_{\left(e_1', e_2'\right)\in S'} max\left\{0, \gamma_1 + f(e_1, e_2) - f\left(e_1', e_2'\right)\right\} \tag{12}$$

where $\gamma_1 > 0$ is a margin hyper-parameter, S' is obtained by randomly replacing a certain entity of entity pairs in the seed alignment S. In our work, entity alignment is performed by simply measuring the distance between two entity nodes on their embedding space. For entities e_i in TKG_1 and e_j in TKG_2, we calculate the distance between them based on structure information:

$$d_r\left(e_i, e_j\right) = \left\|e_i - e_j\right\|_{L_1} \tag{13}$$

where $\|\cdot\|_{L_1}$ denotes the 1-norm measure for vectors.

3.5 Translation-Based TKGs Attribute Embedding

Although GCN is very effective in learning structural information, it is not as good as traditional translation-based embedding methods in learning simple sequence information. The attribute information does not contain complex relational networks, so we use the simple TransE model to embed them. For all attribute triples (e, a, v), we interpret attribute a as a translation from the head entity e to the attribute value v and define the relationship of each element in an attribute triple as $e + a \approx v$. To learn the attribute embedding, we minimize the following margin-based objective function L_{AE}:

$$L_{AE} = \sum_{t_a \in T_a} \sum_{t_a' \in T_a'} max\left\{0, \gamma_2 + f(t_a) - f\left(t_a'\right)\right\} \tag{14}$$

where T_a is the set of valid attribute triples from the training dataset, while T_a' is the set of corrupted attribute triples. The corrupted triples are used as negative samples by randomly replacing an element in the attribute triple (e, a, v). $f(t_a) = || e + a - v ||$ is the plausibility score based on the vector representation of entity e, attribute a and attribute value v. We denote the result of attribute embedding as $\boldsymbol{h_a}$. For entities e_i in TKG_1 and e_j in TKG_2, we calculate the distance between them based on attribute information:

$$d_a(e_i, e_j) = || \boldsymbol{h_a}(e_i) - \boldsymbol{h_a}(e_j) ||_{L_1} \tag{15}$$

In order to further improve the accuracy of the entity alignment results, we propose a concept of nearest neighbor matching. We have added new constraints to the previous calculation of entity similarity, requiring the aligned entities to be closest to each other, and the distance between them must be less than the set threshold. As shown in Fig. 2, u_2 and v_3 are the nearest neighbors to each other in the distance matrices M and M' respectively, the similarity distances d_{23} and d_{32} between them are both smaller than the threshold θ. Then we consider u_2 and v_3 to be equivalent entities. We have defined this concept as follows.

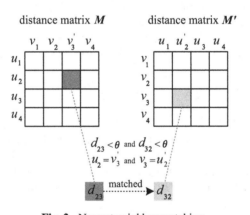

Fig. 2. Nearest neighbor matching.

Definition 1. Given two entities u and v, the nearest neighbor matching is defined as: for each $u \in E_1$ and $\mathrm{argmin}_{v' \in E_2} M(u, v') = v$, if $\mathrm{argmin}_{u' \in E_2} M(v, u') = u$ and $M(u, v) < \theta$, then $(u, v) \to L$, where $M(u, v)$ is the entity in u-th row and v-th column of the similarity matrix M, θ is the given distance threshold and $L = \{(u, v)|u \in E_1, v \in E_2\}$ is the set of entity matching pairs. The process of nearest neighbor matching is shown in Algorithm 1.

Algorithm 1: Nearest neighbor matching.

Input: entity sets E_1 and E_2, given threshold θ, distance matrix M.
Output: entity matching results L.
01. **foreach** $u \in E_1$ **do**
02. $v = \arg\min_{v' \in E_2} M(u, v')$;
03.   **if** $\arg\min_{u' \in E_2} M(v, u') = u$ **and** $M(u, v) < \theta$ **then**
04. $L = L \cup (u, v)$
05.   **end if**
06. **end for**
07. **return** L.

3.6 Entity Alignment of TKGs

As mentioned above, we calculated the similarity distance between entities based on structural information and attribute information respectively. According to the distance d_r and d_a, we use the nearest neighbor matching strategy to calculate the entity distance matrix $M_r \in \mathbb{R}^d$ and $M_a \in \mathbb{R}^d$. Then we incorporate the similarity distances of the corresponding entities in these two matrices into linear weighting to obtain the final entity alignment similarity. For the source entity $e_i \in TKG_1$ and the target entity $e_j \in TKG_2$, their final similarity is calculated as follows:

$$Sim(e_i, e_j) = \alpha \cdot Sim_r(e_i, e_j) + (1 - \alpha) \cdot Sim_a(e_i, e_j) \tag{16}$$

where Sim_r is the entity similarity in M_r based on structure information, and Sim_a represents the entity similarity in M_a based on attribute information. $\alpha \in (0, 1)$ is a hyper-parameter that balances the importance of Sim_r and Sim_a.

4 Experiments

In this section, we report our experiments and results compared with several state-of-art methods on a set of real world TKG datasets. We use TensorFlow to develop our approach. The experiments are conducted on a personal workstation with an Intel(R) Core(TM) i9-9900 2.80 GHz CPU, 32 GB memory and GeForce GTX 2080Ti.

4.1 Experimental Settings

Datasets. We use DWY100K as source datasets, which were built in [17]. DWY100K contains two subsets: DBP-WD and DBP-YG. DBP-WD is extracted from DBpedia and Wikidata datasets, and DBP-YG is extracted from DBpedia and YAGO datasets. Each dataset has 100 thousand pre-aligned entity pairs. We extract the temporal information of each relationship in datasets and process it into a unified format, such as [2009-10-08]. In the experiment, we use two temporal datasets obtained after processing: DBP-WDT and DBP-YGT. The statistics of the two datasets are shown in Table 1.

Table 1. Statistics of datasets.

Datasets		Entities	Relations	Attributes	Temporal	Rel.triples	Attr.triples
DBP-WDT	DBpedia	100,000	330	351	95,142	463,294	381,166
	Wikidata	100,000	220	729	93,161	448,774	789,815
DBP-YGT	DBpedia	100,000	302	334	70,414	428,952	451,646
	YAGO3	100,000	31	23	101,265	502,563	118,376

Baselines. In the experiments, we compared our proposed model Tem-EA with five recent embedding-based entity alignment methods, including MTransE [3], IPTransE [31], JAPE [18], AttrE [21] and GCN-Align [22]. Among them, MTransE, IPTransE, JAPE and AttrE are translation-based methods, and GCN-Align is GCN-based method. Specifically, MTransE has five variants, and the fourth performs best according its authors. Therefore, we chose this variant as a comparative experiment. IPTransE is an iterative method, we chose the best variant with parameter sharing and iterative alignment. JAPE, AttrE and GCN-Align use both structure and attribute embedding for entity alignment, we used their complete models. We run the source code of these models on our two temporal datasets and obtain the results of comparative experiments.

Evaluation Metrics. Following the conventions, we chose Hits@k ($k = 1, 5, 10$) and mean reciprocal rank (MRR) as the evaluation metrics. Hits@k measures the percentage of correct alignment ranked at top k. MRR calculate the mean reciprocal of these ranks. Naturally, a higher Hits@k and MRR scores indicate better performance. Note that, Hits@1 should be more preferable, and it is equivalent to precision widely-used in conventional entity alignment.

Parameter Settings. To make a fair comparison, we try our best to unify the experimental settings. For all comparison methods, we use 30% of the reference entity alignment for training and 70% of them for testing. The division of data used for training and testing is the same for all methods. For the structural information embedding part, we use the Adam optimizer to optimize the model. For parameters settings: the learning rate is 0.01, the margin-based loss function $\gamma_1 = 3$, the embedding dimension $d_r = 200$, and the batch size is 2000. For the attribute information embedding part, we adopt stochastic gradient descent (SGD) as the optimization method. The embedding dimension of attribute information $d_a = 200$ and epochs = 2000. For the hyper-parameters, we set $\gamma_2 = 1.0$ and the learning rate is 0.1.

4.2   Results

Table 2 shows the results of Tem-EA and all compared approaches on DBP-WDT and DBP-YGT datasets. We observed that our method consistently outperforms all baseline methods on the two temporal datasets. For example, the Hits@1 and MRR score of Tem-EA on DBP-YGT is 39.62% and 0.51. Compared with MTransE, it has increased by 18.27% and 0.228, and compared with the best-performing baseline method GCN-Align, it has improved by 3.87% and 0.053. It proves the effectiveness of Tem-EA on the entity alignment task of TKGs.

Among all baseline methods, the GCN-based methods GCN-Align has a small advantage over other translation-based methods (e.g. JAPE and AttrE) for the reason that GCN provides a better ability to capture the inherent graph structure information in TKGs. Tem-EA also uses GCN to fully learn structural information embedding. However, it is different from the GCN-Align model in the use of attribute information. Tem-EA uses the translation-based method TransE to independently embed attribute triples, while GCN-Align treats attribute triples as relational triples and uses GCN for learning. Compared with GCN-Align, our method Tem-EA is better on all metrics of the two datasets. This shows that using different methods according to the characteristics of the information can more fully learn the various information of TKGs, and generate more aligned entities. In addition, among the four translation-based methods, AttrE works better because it performs character-level embedding on attribute value and union relation triples improve the alignment performance. IPTransE is better than JAPE because it adds an iterative process to label new entity pairs as training data.

Table 2. Result comparison on entity alignment.

Methods	DBP-WDT				DBP-YGT			
	Hits@1	Hits@5	Hits@10	MRR	Hits@1	Hits@5	Hits@10	MRR
MTransE	24.87	41.46	51.76	0.334	21.35	35.83	49.41	0.282
IPTransE	29.74	55.18	55.84	0.386	22.54	36.28	50.85	0.285
JAPE	28.57	47.48	58.88	0.320	21.57	36.94	51.38	0.291
AttrE	31.65	55.04	63.84	0.411	28.90	47.18	62.18	0.386
GCN-Align	32.43	55.96	65.47	0.432	35.75	58.56	67.24	0.457
Tem-EA	**36.94**	**61.86**	**70.29**	**0.480**	**39.62**	**64.09**	**72.94**	**0.510**

It is worth noting that the results of these methods on the two datasets are slightly different. As shown in Table 1, the types of relationships and attributes in DBP-WDT are much more than those in DBP-YGT. Therefore, the translation-based methods (MTransE, IPTransE, JAPE and AttrE) that use entity relationship or attribute information perform better on DBP-WDT. On the contrary, Tem-EA and GCN-Align perform better on DBP-YGT, which is caused by the YAGO3 dataset containing fewer relationships. It can be seen from Table 1 that YAGO3 has so few relationship types but the number of relationship triples is huge, which shows that YAGO3 contains a large number of multiple mapping relationships and the entities have strong associations. Tem-EA and GCN-Align both use GCN learning relationship triples embedding, which can capture the neighborhood information of the relationship better than translation-based methods.

In order to explore the impact of training data size on our framework, we further compared Tem-EA with GCN-Align by changing the proportion of pre-aligned entity pairs from 10% to 50% with a step of 10%, and all remaining pre-aligned entities are used for testing. Figures 3, 4, 5, and 6 show that as the proportion of training data increases, the performance of these two models on both datasets gradually improves. Compared with GCN-Align, Tem-EA always achieves better results. It can be seen that the performance is best when half of the pre-aligned entities are used as training data, Hits@k ($k = 1, 5,$ 10) and MRR of Tem-EA model on DBP-YGT are 48.35%, 73.19%, 80.27% and 0.6 respectively. Moreover, even if the proportion of training data is small, such as 10%, Tem-EA still achieves good results. These results further confirm the robustness of our method, especially in the case of limited seed entity alignment.

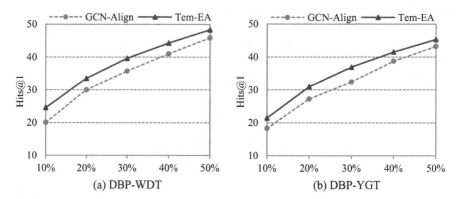

Fig. 3. Hits@1 of entity alignment results with different sizes of training data.

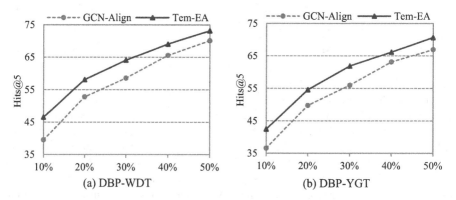

Fig. 4. Hits@5 of entity alignment results with different sizes of training data.

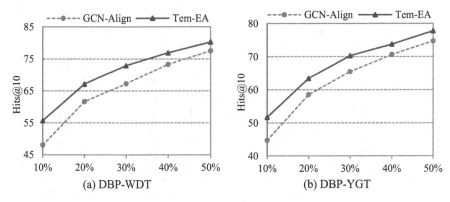

Fig. 5. Hits@10 of entity alignment results with different sizes of training data.

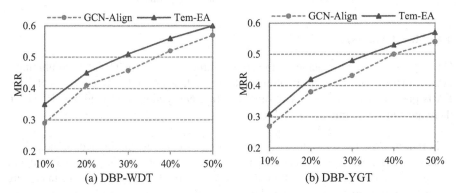

Fig. 6. MRR of entity alignment results with different sizes of training data.

4.3 Ablation Study

In order to further evaluate the effectiveness of proposed modules, we construct several ablation studies on Tem-EA, namely Tem-EA (w/o NNM) and Tem-EA (w/o Attr). Tem-EA (w/o NNM) denotes that Tem-EA does not consider the nearest neighbor matching strategy in the entity similarity calculation module. Tem-EA (w/o Attr) means to delete the translation module that obtains the attribute representation, and only use relation-ship information to get the final entity representation. The result is shown in Fig. 7. Compared with Tem-EA, the performance of Tem-EA (w/o NNM) and Tem-EA (w/o Attr) on two datasets has decreased, which proves the effectiveness of these two mod-ules. Specifically, we observed that the performance of Tem-EA (w/o Attr) degrades the most, which confirms that in addition to structural information, attribute information is also very important for the effect of entity alignment. Moreover, Tem-EA consistently outperforms Tem-EA (w/o NNM), which shows that adding the nearest neighbor match-ing condition to the entity similarity calculation module helps to improve the accuracy of entity alignment.

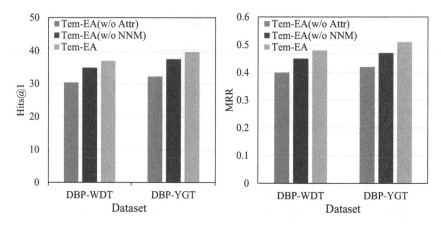

Fig. 7. The results of ablation experiment.

4.4 Model Analysis

Our proposed model has achieved the entity alignment between TKGs by combining a large amount of temporal information contained in facts. And the effectiveness of Tem-EA is proved by comparison with other existing entity alignment methods. However, there are still some limitations of Tem-EA:

– In real-world TKGs, due to the heterogeneity of patterns, the immediate neighborhoods of equivalent entities are usually different. The graph convolutional network used in Tem-EA performs well in identifying isomorphic subgraphs and can capture direct neighborhood information, but some non-isomorphic knowledge graphs may affect the performance of our model.
– As there are a large amount of structural information and attribute information in TKGs, Tem-EA mainly uses these two kinds of information for entity alignment. In fact, there are some other useful entity characteristics, such as entity name and entity context. In the future, we can try to combine these features to further improve Tem-EA.

5   Conclusion

In this paper, we propose a novel framework called Tem-EA for entity alignment between TKGs. We first use recurrent neural network to learn the representation of each temporal sequence and connect it with the corresponding relational triple embedding. Then we adopt more effective methods to independently embed structure information and attribute information according to their different characteristics, and finally jointly perform entity alignment. In addition, we propose a concept of nearest neighbor matching to improve the alignment results. The experimental results on two real-world temporal datasets demonstrate the effectiveness of our proposed model.

Acknowledgment. The work was supported by the National Natural Science Foundation of China (61402087), the Natural Science Foundation of Hebei Province (F2019501030), the Key Project of Scientific Research Funds in Colleges and Universities of Hebei Education Department (ZD2020402), the Fundamental Research Funds for the Central Universities (N2023019), and in part by the Program for 333 Talents in Hebei Province (A202001066).

References

1. Asamoah, C., Tao, L., Gai, K., Jiang, N.: Powering filtration process of cyber security ecosystem using knowledge graph. In: CSCloud, pp. 240–246 (2016)
2. Bordes, A., Usunier, N., García-Durán, A., Weston, J., Yakhnenko, O.: Translating embeddings for modeling multi-relational data. In: NIPS, pp. 2787–2795 (2013)
3. Chen, M., Tian, Y., Yang, M., Zaniolo, C.: Multilingual knowledge graph embeddings for cross-lingual knowledge alignment. In: IJCAI, pp. 1511–1517 (2017)
4. Chen, M., Tian, Y.N., Chang, K., Skiena, S., Zaniolo, C.: Co-training embeddings of knowledge graphs and entity descriptions for cross-lingual entity alignment. In: IJCAI, pp. 3998–4004 (2018)
5. Cao, Y., Wang, X., He, X., Hu, Z., Chua, T. S.: Unifying knowledge graph learning and recommendation: Towards a better understanding of user preferences. In: WWW, pp. 151–161 (2019)
6. Cao, Y., Liu, Z., Li, C., Liu, Z., Li, J., Chua, T.S.: Multi-channel graph neural network for entity alignment. In ACL, pp. 1452–1461 (2019)
7. Dasgupta, S.S., Ray, S.N., Talukdar, P.P.: HyTE: hyperplane-based temporally aware knowledge graph embedding. In: EMNLP, pp. 2001–2011 (2018)
8. Erxleben, F., Günther, M., Krötzsch, M., Mendez, J., Vrandečić, D.: Introducing wikidata to the linked data web. In: Mika, P., et al. (eds.) ISWC 2014. LNCS, vol. 8796, pp. 50–65. Springer, Cham (2014). https://doi.org/10.1007/978-3-319-11964-9_4
9. Goel, R., Kazemi, S.M., Brubaker, M., Poupart, P.: Diachronic embedding for temporal knowledge graph completion. In: AAAI, pp. 3988–3995 (2020)
10. García-Durán, A., Dumančić S., Niepert, M.: Learning sequence encoders for temporal knowledge graph completion. In: EMNLP, pp. 4816–4821 (2018)
11. Jin, W., Qu, M., Jin, X., Ren, X.: Recurrent event network: autoregressive structure inference over temporal knowledge graphs. In: EMNLP (2020)
12. Leblay, J., Chekol, M.W.: Deriving validity time in knowledge graph. In: Champin, pp. 1771–1776 (2018)
13. Liao, S., Liang, S., Meng, Z., Zhang, Q.: Learning dynamic embeddings for temporal knowledge graphs. In: WSDM, pp. 535–543 (2021)
14. Mahdisoltani, F., Biega, J., Suchanek, F.M.: Yago3: A knowledge base from multilingual wikipedias. In: CIDR (2013)
15. Nickel, M., Tresp, V., Kriegel, H.P.: A three-way model for collective learning on multi-relational data. In: ICML (2011)
16. Sun, Z., Wang, C., Hu, W., Chen, M., Dai, J., Zhang, W., Qu, Y.: Knowledge graph alignment network with gated multi-hop neighborhood aggregation. In: AAAI, pp. 222–229 (2020)
17. Sun, Z., Hu, W., Zhang, Q., Qu, Y.: Bootstrapping entity alignment with knowledge graph embedding. In: IJCAI, pp. 4396–4402 (2018)
18. Sun, Z., Hu, W., Li, C.: Cross-lingual entity alignment via joint attribute-preserving embedding. In: d'Amato, C., et al. (eds.) ISWC 2017. LNCS, vol. 10587, pp. 628–644. Springer, Cham (2017). https://doi.org/10.1007/978-3-319-68288-4_37

19. Trouillon, T., Welbl, J., Riedel, S., Gaussier, E., Bouchard, G.: Complex embeddings for simple link prediction. In: ICML, pp. 2071–2080 (2016)

20. Trivedi, R., Dai, H., Wang, Y., Song, L.: Know-evolve: Deep temporal reasoning for dynamic knowledge graphs. In: ICML, pp. 3462–3471 (2017)

21. Trsedya, B.D., Qi, J., Rui, Z.: Entity alignment between knowledge graphs using attribute embeddings. In: AAAI, pp. 297–304 (2019)

22. Wang, Z., Lv, Q., Lan, X., Zhang, Y.: Cross-lingual knowledge graph alignment via graph convolutional networks. In: EMNLP, pp. 349–357 (2018)

23. Wu, Y., Liu, X., Feng, Y., Wang, Z., Zhao, D.: Jointly learning entity and relation representations for entity alignment. In: EMNLP (2019)

24. Wu, Y., Liu, X., Feng, Y., Wang, Z., Yan, R., Zhao, D.: Relation-aware entity alignment for heterogeneous knowledge graphs. In: IJCAI (2019)

25. Xu, C., Nayyeri, M., Alkhoury, F., Yazdi, H. S., Lehmann, J.: Temporal knowledge graph embedding model based on additive time series decomposition. arXiv preprint arXiv:1911.07893 (2019)

26. Yang, B., Yih, W., He, X., Gao, J., Deng, L.: Embedding entities and relations for learning and inference in knowledge bases. In: ICLR (2015)

27. Yan, Z., Peng, R., Wang, Y., Li, W.: CTEA: Context and topic enhanced entity alignment for knowledge graphs. In: Neurocomputing, pp. 419–431 (2020)

28. Yih, W.T., Chang, M.W., He, X., Gao J.: Semantic parsing via staged query graph generation: question answering with knowledge base. In: ACL, pp. 1321–1331 (2015)

29. Zhu, Y., Liu, H., Wu, Z., Du, Y.: Relation-aware neighborhood matching model for entity alignment. In: AAAI, pp. 4749–4756 (2021)

30. Zhang, S., Tay, Y., Yao, L., Liu, Q.: Quaternion knowledge graph embeddings. In: NIPS, pp. 2731–2741 (2019)

31. Zhu, H., Xie, R., Liu, Z., Sun, M.: Iterative entity alignment via joint knowledge embeddings. In: IJCAI, pp. 4258–4264 (2017)

32. Zhang, Q., Sun, Z., Hu, W., Chen, M., Guo, L., Qu, Y.: Multi-view knowledge graph embedding for entity alignment. In: IJCAI, pp. 5429–5435 (2019)

33. Zhang, Z., Bu, J., Li, Z., Yao, C., Wang, C., Wu, J.: TigeCMN: On exploration of temporal interaction graph embedding via coupled memory neural networks. In: Neural Networks, pp. 13–26 (2021)

Similarity-Aware Collaborative Learning for Patient Outcome Prediction

Fuqiang Yu[1,2], Lizhen Cui[1,2(✉)], Yiming Cao[1,2], Ning Liu[1(✉)],
Weiming Huang[1,2], and Yonghui Xu[2]

[1] School of Software, Shandong University, Jinan, China
liun21cs@sdu.edu.cn
[2] Joint SDU-NTU Centre for Artificial Intelligence Research (C-FAIR),
Shandong University, Jinan, China
{yfq,caoyiming}@mail.sdu.edu.cn, {clz,weiming}@sdu.edu.cn

Abstract. The rapid growth in the use of electronic health records (EHR) offers promises for predicting patient outcomes. Previous works on this task focus on exploiting temporal patterns from sequential EHR data. Nevertheless, such approaches model patients independently, missing out on the similarities between patients, which are crucial for patients' health risk assessment. Moreover, they fail to capture the fine-grained progression of patients' status, which assist in inferring the patients' future status. In this work, we propose a similarity-aware collaborative learning model *SiaCo* for patient outcome prediction. In particular, we design two similarity measurers and two global knowledge matrices to separately calculate the similarity of patients with different information levels and support collaborative learning between patients. To capture the more fine-grained progression of patients' status, we design a parallelized LSTM to model the temporal-dependent patterns of patient status. Finally, SiaCo integrates the information learned from two measures and the parallelized LSTM to predict patient outcomes. Extensive experiments are conducted on two real disease datasets. The experimental results demonstrate that SiaCo outperforms the state-of-the-art models for two typical tasks of patient outcome prediction.

Keywords: Medical data mining · Patient outcome prediction · EHR

1 Introduction

Electronic health records (EHR) accumulate considerable amounts of healthcare data, covering laboratory tests and results, diagnoses, prescriptions, and procedures [3, 16, 22–24]. The rapid growth in the use of EHR provides a great opportunity for healthcare providers and researchers to perform diverse healthcare tasks to improve care for individual patients. Patient outcome prediction (e.g., mortality risk prediction, diagnosis prediction) is one of the important healthcare tasks, as physicians can select personalized follow-up treatments to avoid adverse outcomes accordingly [18].

© The Author(s), under exclusive license to Springer Nature Switzerland AG 2022
A. Bhattacharya et al. (Eds.): DASFAA 2022, LNCS 13246, pp. 407–422, 2022.
https://doi.org/10.1007/978-3-031-00126-0_31

Most existing works [1,2,10,17,19] attempt to capture temporal patterns from sequential patient data, since it is crucial for understanding the progression of the patients' status and inferring patient outcomes. However, there are still unsolved critical challenges as follows.

The Importance of Similar Patients to Current Patient Modeling and Patient Outcome is Ignored. Existing patient outcome prediction methods model each patient independently, overlooking the similar patients that are crucial for a physician to make decisions (e.g., infer patient's health status). Figure 1 demonstrates how patient similarities participate in the physician's decision-making process. The experience base EB denotes the set of all the patients known to the physician. With knowledge of a current patient p_i's information, the physician makes the preliminary decisions for the patient using his/her professional knowledge. Furthermore, the physician will fine-tune the preliminary decisions and make final decisions, referring to the information of patient $p_j \in EB$ whose status progression is most similar to that of patient p_i, since the patients with similar status progression may have similar disease conditions in the future. Irrespective of experience (EB) means that it is difficult to consider patient similarities, which may affect the accuracy of the physician's decisions. Similarly, modeling each patient independently limits the model's ability to understand the temporal patterns of patients' status and assess the risk of patient health (i.e., patient outcome prediction). Therefore, we argue that the effective use of patient similarities can improve the accuracy of predicting patient outcomes. However, there are two challenges of incorporating patient similarities: (1) how to measure the similarities between patients; (2) how to inject patient similarities into the process of risk assessment.

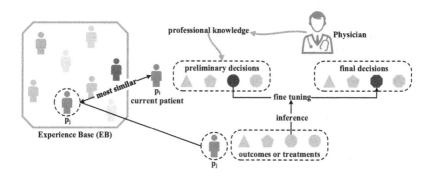

Fig. 1. Physician's decision-making process

Fine-Grained Progression of Patients' Status is Ignored. The fine-grained progression of a patient's status plays an important role in inferring the patients' future status. The progression is memorized in the patient events at the encounter (admission) level and the medical concept level. Specifically, at the encounter level, each patient's events are represented as a temporal sequence of patient encounters, where each encounter is composed of ordered and unordered medical concepts (e.g., diagnoses, medications, procedures). At the medical concept level (i.e., within each encounter), the treatments (e.g., medications) form a temporal sequence of more fine-grained events. The temporal pattern at the encounter level reflects the overall health status of a patient, while the sequences of treatments within an encounter represent the fine-grained progression of the disease. For example, if the patient is mainly taking auxiliary anti-inflammatory medications in the second half of the medication sequence, it generally indicates that the patient's status is evolving in a positive direction. Considering the temporal influence at the encounter level and the medical concept level enables us to capture the more fine-grained progression of the patients' status. However, it is challenging to model the sequences of the encounter level's events and the sequences of the medical concept level's more-grained events simultaneously.

To address these challenges, we propose a similarity-aware collaborative learning model *SiaCo* for patient outcome prediction. SiaCo consists of an encounter-level similarity measurer, a concept-level measurer and a parallelized LSTM module. The two similarity measurers derive the patient similarities at different information levels, i.e., the encounter level and the medical concept level. Furthermore, they use two global knowledge matrices KM and KM' to support collaborative learning between patients. According to the calculated patient similarities, the measurers can search and utilize the knowledge vectors of other patients from KM for the modeling process of each patient. Therefore, by allowing patients with similar health status to participate in the process of each other's modeling and health risk assessment, SiaCo overcomes the limitations of independently modeling patients. The parallelized LSTM is devised to exploit the more fine-grained progression of the patients' health status. Subsequently, SiaCo combines the information learned from the three modules to make patient outcome prediction. This work makes the following main contributions:

- We propose to measure the patient similarities and build global knowledge matrices to support collaborative learning between patients.
- We design a parallelized LSTM module to exploit the more fine-grained progression of the patients' health status.
- The comparative experimental results on MIMIC-III confirmed the indispensability of patient similarities and fine-grained progression of the patients' health status in the patient outcome prediction.

2   Related Work

To capture the temporal influence, some works attempt to employ sequence-based models to sequential patient records for predictive analysis of patient outcomes. For example, Cheng et al. [4] propose a four-layer Convolutional Neural Network (CNN) model to make predictions. They build a temporal matrix with time on one dimension and event on the other dimension for each patient. Choi et al. [9] employ RNN to model the multilevel structure of EHR data.

Moreover, Choi et al. [7] develop an LSTM-based model RETAIN to predict disease (i.e., heart failure). RETAIN builds two-level neural attention to detect influential past encounter.

Recently, rich and useful information is gradually being embedded in medical concept representations for healthcare analytics. To alleviate the high-dimensional and discrete characteristics of one-hot vectors, Choi et al. [5] propose Med2Vec to learn code-level and visit-level representations of medical concepts. Choi et al. [6] propose a graph-based attention model, i.e., GRAM, to learn efficient healthcare representations from limited training data. Moreover, some works [1,10,18,25] have been conducted on the phenotypes representation learning. For instance, Yin et al. [25] propose a Collective Non-negative Tensor Factorization model CNTF to learn a temporal tensor for each patient, where the phenotype definitions are shared across all patients. Ma et al. [18] makes use of the features adaptively in different situations while improving the model's prediction performance. Moreover, Gao et al. [10] design a Stage-aware neural Network model StageNet to make mortality prediction. StageNet integrates inter-visit time information to extract disease stage information from patient EHR data. StageNet achieves high performance in mortality risk prediction.

However, modeling each patient independently limits the performance to assess health risk. Additionally, these works suffer from capture the fine-grained progression of patients' status, as they only consider the temporal influence at the encounter level.

3   Problem Formulation

Let $P = \{p_1, p_2, \ldots, p_{|P|}\}$ be a set of patients in EHR data. For medical concepts, we use $D = \{d_1, d_2, \ldots, d_{|D|}\}$ to denote a set of diagnosis codes, $M = \{m_1, m_2, \ldots, m_{|M|}\}$ a set of medication codes, and $R = \{r_1, r_2, \ldots, r_{|R|}\}$ a set of procedure codes. So $|D|$, $|M|$ and $|R|$ are the number of diagnoses, medications and procedure codes, respectively.

Definition 1 (Encounter). *Each patient encounter is represented as ordered and disordered collection of medical concepts. Each encounter is formalized as*

$$((d_1, d_2, \ldots, d_{pd}), \langle m_1, m_2, \ldots, m_{t'} \rangle, \langle r_1, r_2, \ldots, r_{t''} \rangle),$$

where $d_i \in D$, $m_j \in M$ and $r_k \in R$ for $1 \leq i \leq pd$, $1 \leq j \leq t'$ and $1 \leq k \leq t''$. Note that medication codes form a temporal sequence, while diagnosis codes and procedure codes both form an unordered set.

Definition 2 (Patient Records). *In sequential EHR data, each patient's records is a sequence of encounters. We use $V_p = \langle v_1^p, v_2^p, \ldots, v_t^p \rangle$ to denote a patient p's full sequence of t encounters. The superscript p is omitted whenever the context is clear.*

We define two typical tasks of patient outcome prediction solved by SiaCo.

Task 1 (Diagnosis Prediction). *Given a patient p's encounter sequence $V = \langle v_1, v_2, \ldots, v_t \rangle$, this task returns N diagnosis codes $(d_{rec_1}, d_{rec_2}, \ldots, d_{rec_N})$, i.e., the diagnostic information of the next step $t + 1$.*

Task 2 (Mortality Prediction). *Given a patient p's encounter sequence $V = \langle v_1, v_2, \ldots, v_t \rangle$, this task returns the risk $y_{risk} \in \{0, 1\}$ of patient death in hospital.*

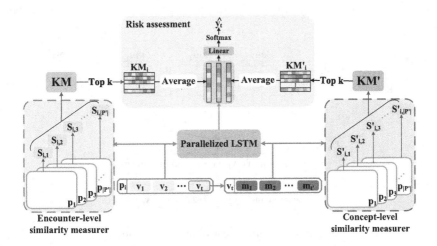

Fig. 2. Similarity-aware collaborative learning model

4   Methodology

Considering all the challenges analyzed in the fourth and fifth paragraphs of Sect. 1, we propose a similarity-aware collaborative learning model SiaCo for patient outcome prediction. Figure 2 shows the high-level overview of the proposed SiaCo. It contains two similarity measurers to calculate the patient similarities at different information levels and a parallelized LSTM module to model patient encounter sequences and medication sequences. Furthermore, SiaCo combines the information learned from these parts to make risk assessments for patient health, i.e., patient outcome prediction.

4.1   Patient Similarity Measurement

Patient similarity is crucial for inferring patients' health status, since the patients with similar status progression may have similar disease conditions in the future. We design two similarity measurers to calculate patient similarities at both the encounter level and the medical concept level. Also, they support collaborative learning between patients.

Encounter-Level Similarity Measurer. Figure 3 shows the process of similarity measurement at the encounter level. We embed all the EHR data into a information vector PE_i for each patient $p_i \in P$. PE_i contains three parts as follows:

$$PE_i = concat(\mathcal{N}^{|D|}, \mathcal{N}^{|R|}, \mathcal{N}^{|M|}), \tag{1}$$

where $\mathcal{N}$ is the set of natural numbers. The fist part $\mathcal{N}^{|D|}$ is generated according to the diagnostic information in p_i's records $V_t = \langle v_1, v_2, \ldots, v_t \rangle$, where the n-th element is the number of occurrences of diagnosis $d_n \in D$ in V_t. Similarly, $\mathcal{N}^{|R|}$ and $\mathcal{N}^{|M|}$ are derived according to the medications and procedures, respectively.

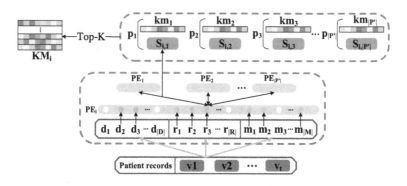

Fig. 3. Encounter-level similarity measurer

Given two information vectors, i.e., PE_i of the current patient $p_i \in P$ and PE_j of the known patient $p_j \in P'$, we employ a efficient indicator **Cosine Similarity** to measure the similarities $S_{i,j}$ between them.

$$S_{i,j} = \frac{PE_i \cdot PE_j}{||PE_i|| \cdot ||PE_j||} = \frac{\sum_{k=1}^{n} i_k j_k}{\sqrt{\sum_{k=1}^{n} i_k^2} \sqrt{\sum_{k=1}^{n} j_k^2}}, \tag{2}$$

where i_k and j_k are the k-th elements of PE_i and PE_j, respectively. P' is a set of known patients and $|P'|$ is the number of known patients. It is noteworthy that we take all the patients in training set as known patients.

Moreover, we aim to learn patient representations collaboratively instead of modeling them independently. To achieve this goal, we build a global knowledge matrix $KM \in \mathbb{R}^{s \times |P'|}$ to allow patients to participate in the process of each other's modeling and health risk assessment. Each column vector $km_j \in \mathbb{R}^s$ of KM represents the knowledge of a known patient p_j. Specifically, given a patient p_i, we obtain top-K known patients most similar to p_i and their knowledge vectors according to similarity S:

$$\{(S_{i,i_1}, km_{i_1}), (S_{i,i_2}, km_{i_2}), \cdots, (S_{i,i_K}, km_{i_K})\}, \tag{3}$$

where $km_{i_k} \in KM$ for $0 \le k \le K$ and $0 \le i_k \le |P'|$. Next, we use these knowledge vectors to form a new matrix $KM_i \in \mathbb{R}^{s \times K}$ for patient p_i as follows:

$$KM_i = [km_{i_1}; km_{i_2}; \cdots; km_{i_K}] \tag{4}$$

KM_i is used as a reference to assess the patient's risk detailed in Sect. 4.3.

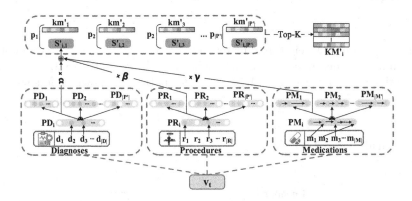

Fig. 4. Concept-level similarity measurer

Concept-Level Similarity Measurer. As shown in Fig. 4, we consider the patient similarities at the medical concept level within the latest patient encounter v_t, since the diagnoses and treatments (i.e., medications and procedures) in the latest encounter reflect more specific disease progression and are likely to have the greatest impact on assessing patients' health risks. Similar to the first and second part of PE_i, for each patient p_i, we embed his/her diagnoses and procedures in v_t into two information vectors, i.e., PD_i and PR_i. p_i's temporal medication sequence PM_i is defined as follows:

$$PM_i = \langle m_{i_1}, m_{i_2}, \cdots, m_{i_t} \rangle \tag{5}$$

Given a current patient $p_i \in P$ and a known patient $p_j \in P'$, we use Cosine Similarity to calculate the diagnosis similarities $S_{i,j}^D$ and procedure similarities $S_{i,j}^R$.

However, it is challenging to measure the medication similarities $S_{i,j}^M$, as it is difficult for Cosine Similarity to measure the similarity between temporal sequences. Moreover, the inconsistency of sequence length also brings challenges. To this end, Dynamic Time Warping (DTW) [13] is adopt. Given two medication sequences, i.e., PM_i of a new patient p_i and PM_j of a known patient p_j, the similarity $D_{dtw}(PM_i, PM_j)$ is calculated as follows:

$$D(PM_i, PM_j) = \begin{cases} 0 & k = g = 0 \\ \infty & k = 0 | 0\, g = 0 \\ E(PM_i^1, PM_j^1) + min \begin{cases} D(R(PM_i) + R(PM_j)) \\ D(R(PM_i) + PM_j)) \\ D(PM_i + R(PM_j)) \end{cases} & otherwise \end{cases}, \tag{6}$$

where PM_i^k and PM_j^g are the k-th and g-th elements of PM_i and PM_j, respectively. $E(PM_i^k, PM_j^g)$ denotes the Euclidean distance between PM_i^k and PM_j^g. $R(X)$ is the remaining sequence after the first point of sequence X is removed.

Furthermore, to distinguish the impact of different information (i.e., diagnoses, procedures, and medications), we calculate the patient similarity $S'_{i,j}$ by assigning different weights to the three types of information:

$$S'_{i,j} = \alpha S_{i,j}^D + \beta S_{i,j}^R + \gamma S_{i,j}^M \tag{7}$$

where α, β, γ are the weight parameters for $\alpha + \beta + \gamma = 1$. Similar to encounter-level similarity measurer, we build a global knowledge matrix $KM' \in \mathbb{R}^{s \times |P'|}$ and derive a new knowledge $KM'_i \in \mathbb{R}^{s \times K}$ for patient p_i shown as follows:

$$KM'_i = [km'_{i_1}; km'_{i_2}; \cdots ; km'_{i_K}] \tag{8}$$

4.2 Parallelized LSTM

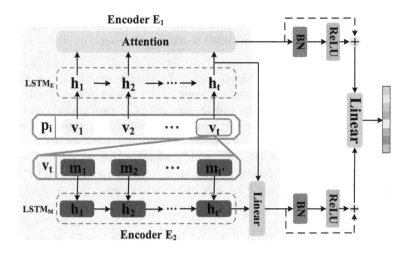

Fig. 5. Parallelized LSTM

Capturing the fine-grained progression of patients' health status is the key to perform health risk assessments. To this end, we devise a parallelized LSTM module (illustrated in Fig. 5) to model patient encounter sequences and medication sequences, as LSTM is efficient at processing sequential data [11,20].

We process each encounter into a information vector $v_j \in \mathcal{N}^{|D|+|R|+|M|}$ in a similar way as Eq. 1. Given a patient p_i's records $V_t = \langle v_1, v_2, \ldots, v_t \rangle$, the process of encoder E_1 is as follows:

$$x_j = \mathcal{V}v_j \tag{9}$$

$$h_j, S_j = LSTM_E(x_j, S_{j-1}); \ 1 \le j \le t , \tag{10}$$

where $\mathcal{V} \in \mathbb{R}^{s \times (|D|+|R|+|M|)}$ is the embedding matrix of medical concepts to learn. $h_j \in \mathbb{R}^s$ and $S_j \in \mathbb{R}^s$ are the LSTM cell's hidden state and cell state, respectively. We obtain a sequence of hidden states $\langle h_1, h_2, \cdots, h_t \rangle$, which contains the information of patient status. To further extract the progression of patient status, we design *Attention* mechanism to investigate the impact of different encounters:

$$H_e = \boldsymbol{W_a} \text{concat}(h_1, h_2, \cdots, h_t) + b_a, \tag{11}$$

where $W_a \in \mathcal{R}^{s \times ts}$ is a attention weight and b_a is a bias vector. Moreover, we build a encoder E_2 to encode the fine-grained progression of patients' status within the latest encounter. Given a medication sequence $\langle m_1, m_2, \cdots, m_{t'} \rangle$ of the latest encounter, the process of encoder E_2 is as follows:

$$x'_j = \boldsymbol{\mathcal{M}} m_j \tag{12}$$

$$h'_j, S'_j = LSTM_M(x_j, S'_{j-1}); \ 1 \leq j \leq t' , \tag{13}$$

where m'_j is the one-hot vector of m_j and $\mathcal{M} \in \mathbb{R}^{s \times |M|}$ is the embedding matrix of medications to learn. We linearly combine the final state H_t of v_t and $H'_{t'}$ to enhance the influence of latest encounter:

$$H_m = Linear(h_t, h'_{t'}), \tag{14}$$

where $Linear(x, y)$ is linearly combining function, which defined as $w_1 x + w_2 y + b$.

After parallelized encoding, we fuse the encoded information shown in follows:

$$\hat{H}_m = H_m + ReLU(BN(H_m)) \tag{15}$$

$$\hat{H}_e = H_e + ReLU(BN(H_e)) \tag{16}$$

$$H = Linear(\hat{H}_m, \hat{H}_e), \tag{17}$$

where BN is batch normalization and $ReLU(x) = max(x, 0)$ is a rectified linear unit that maintains the sparsity of the network by setting some neurons to 0. The temporal-dependent patterns of patients' status are embedded in the final output $H \in \mathbb{R}^s$.

4.3   Methods for Patient Outcome Prediction

Patient Representations. We generate patient representations Y by integrating the information learned from two measures and the parallelized LSTM:

$$A_{average}(KM_i) = \frac{\sum_{n=1}^{N} km_{i_n}}{N} \tag{18}$$

$$A_{average}(KM'_i) = \frac{\sum_{n=1}^{N} km'_{i_n}}{N} \tag{19}$$

$$Y = Linear(A_{average}(KM_i), A_{average}(KM'_i), H), \tag{20}$$

where KM_i and KM'_i are the knowledge of patient p_i at the encounter level and the medical concept level, respectively. H is the information embedding learned by the parallelized LSTM. km_{i_n} and km'_{i_n} are the column vectors of KM_i and KM'_i, respectively. Y is used to solve two typical tasks of patient outcome prediction.

Diagnosis Prediction. Given a sequential records $V = \langle v_1, v_2, \ldots, v_t \rangle$ of a patient p, this task predicts the $(t+1)$-th encounter's diagnosis code $\hat{y}_{d,p}$:

$$\hat{y}_{d,p} = Softmax(W_{y_d} Y + b_{y_d}), \tag{21}$$

where $W_{y_d} \in \mathbb{R}^{|D| \times s}$ and $b_{y_d} \in \mathbb{R}^{|D|}$ are parameters to learn. Cross-entropy is used to calculate the loss between $\hat{y}_{d,p}$ and the ground truth $y_{d,p}$ for all patients:

$$\mathcal{L}_1 = -\frac{1}{|P|} \sum_{p=1}^{P} (y_{d,p}^{\mathrm{T}} log(\hat{y}_{d,p}) + (1 - y_{d,p})^{\mathrm{T}} log(1 - \hat{y}_{d,p})) \tag{22}$$

Mortality Prediction. Given a sequential records $V = \langle v_1, v_2, \ldots, v_t \rangle$ of a patient p, this binary classification task predicts the risk $\hat{y}_p$ of patient death during the current encounter shown as follows:

$$\hat{y}_p = Sigmoid(W_{y_p}^{\mathrm{T}} Y + b_{y_p}), \tag{23}$$

where $W_{y_p} \in \mathbb{R}^s$ and $b_{y_p} \in \mathbb{R}$ are parameters to be learned. We calculate the loss between $\hat{y}_p$ and the ground truth y_p for all patients:

$$\mathcal{L}_2 = -\frac{1}{|P|} \sum_{p=1}^{P} (y log(\hat{y}) + (1 - y) log(1 - \hat{y})), \tag{24}$$

5   Experiments

5.1   Experimental Setup

Datasets. We conduct experiments to evaluate models on EHR data of circulatory-system diseases and heart failures from an open-source dataset MIMIC [12]. We select the patients with circulatory disease codes in their diagnosis to form dataset CS and the patients with heart failure codes in their diagnosis to form dataset HF. We remove drugs and procedures that appear less than 5 times, as they are outliers in the data. After pre-processing, CS contains 25,169 patients with 35,474 hospital admissions and HF contains 10,347 patients with 13,878 admissions, which are illustrated in Table 1.

Baselines. To validate the performance of SiaCo for patient outcome prediction, we compare it with several state-of-the-art models. We choose representative works as baselines. Table 2 demonstrates the assigned tasks of these approaches according to their designed purpose, and the character whether they employ RNN to capture temporal influence. Moreover, we evaluate the quality of the medical concept representations learned by SiaCo and other representative methods (i.e., MCE, CBOW and Skip-Gram) on diagnosis prediction.

Table 1. Statistics of CS and HF

Dataset	Circulatory (CS)	Heart failure (HF)
# of patients	25,169	10,347
# of encounters	34,474	13,878
Max # of encounter per patient	42	19
# of diagnosis codes	5,529	3,974
# of medication codes	336	321
# of procedure codes	1,667	1,198

Table 2. Baselines and their assigned tasks

	Approach	Diagnosis prediction	Mortality prediction	Temporal influence	RNN-based
Predictive Model	LSTM [11]	✓	✓	✓	✓
	Dipole [17]	✓	✗	✓	✓
	T-LSTM [1]	✓	✓	✓	✓
	Med2Vec [5]	✓	✗	✓	✗
	RETAIN [7]	✓	✓	✓	✓
	StageNet [10]	✗	✓	✓	✓
Representation learning for medical concepts	MCE [3]	✓	✗	✓	✗
	CBOW [21]	✓	✗	✗	✗
	Skip-Gram [21]	✓	✗	✗	✗

Performance Metrics. We adopt $Pre@N$ and $Rec@N$ to evaluate diagnosis prediction tasks, where N is the number of predicted diagnoses. Given the set of predicted result D_p^R and the groundtruth D_p^T, i.e., the diagnoses of the next encounter of patient p, $Pre@N$ and $Rec@N$ are defined as follows:

$$Pre@N = \frac{1}{|P|} \sum_{p \in P} \frac{|D_p^T \cap D_p^R|}{N}; \quad Rec@N = \frac{1}{|P|} \sum_{p \in P} \frac{|D_p^T \cap D_p^R|}{|D_p^T|} \tag{25}$$

To evaluate the binary classification task (i.e., mortality prediction), we employ $AUROC$ [15], $Accuracy$ and $F1$-score.

Implementation Details. Each dataset is randomly divided into train, validation, and test set in 7:1:2 ratio. We employ adaptive moment estimation (Adam) [14] to compute adaptive learning rates for each parameter. The batch size is set 500 for diagnosis prediction and 20 for mortality prediction. The decision threshold is set 0.55 for $Accuracy$ and $F1$-score. For Eq. 7, we set $\alpha = 0.8$, $\beta = 0.1$ and $\gamma = 0.1$.

5.2 Results of Diagnosis Prediction

Table 3 shows the performance results of SiaCo and baselines on CS and HF for diagnosis prediction with different predicting numbers of diagnosis codes, i.e.,

$N \in \{5, 10, 15\}$. We observe that LSTM and T-LSTM perform worst. The reason might be that LSTM and T-LSTM only utilize the final output of LSTM to make a prediction, ignoring a large amount of information contained in EHR data. Although Dipoles employ bidirectional recurrent neural networks to model patient encounter sequence, it overlooks the healthcare information (e.g., temporal influence) within encounters, i.e., the medical concept level. Therefore, Dipole fails to achieve the ideal performance. Compared with these methods, Med2Vec and RETAIN achieve considerably better performance. The reason is that Med2Vec and RETAIN consider the visit-level and code-level influence. SiaCo outperforms all the baselines because it focuses on the collaborative learning between patients, which is overlooked by these approaches.

Table 3. Comparison of diagnosis prediction performance on CS and HF

Dataset	Approach	Pre@5	Rec@5	Pre@10	Rec@10	Pre@15	Rec@15
Circulatory (CS)	Med2Vec	0.4057	0.1287	0.3145	0.1949	0.2674	0.2457
	LSTM	0.3298	0.1047	0.2577	0.1601	0.2219	0.2035
	T-LSTM	0.3379	0.1068	0.2717	0.1687	0.2360	0.2162
	Dipole	0.3881	0.1230	0.3096	0.1917	0.2647	0.2433
	RETAIN	0.4786	0.1551	0.3717	0.2346	0.3122	0.2915
	SiaCo	**0.4931**	**0.1599**	**0.3818**	**0.2416**	**0.3190**	**0.2989**
Heart Failure (HF)	Med2Vec	0.5410	0.1447	0.4173	0.2216	0.3554	0.2815
	LSTM	0.5033	0.1352	0.3848	0.2039	0.3340	0.2633
	T-LSTM	0.5151	0.1378	0.3753	0.1979	0.3225	0.2520
	Dipole	0.4729	0.1274	0.3447	0.1812	0.2759	0.2151
	RETAIN	0.6467	0.1756	0.5089	0.2717	0.4263	0.3379
	SiaCo	**0.6600**	**0.1787**	**0.5122**	**0.2719**	**0.4273**	**0.3380**

5.3   Evaluation of Representation Learning

To verify that the medical concept representations learned by SiaCo contain rich information. We conduct experiments to evaluate the quality of the medical concept representations learned by SiaCo with that learned by the representation learning methods. We select three representative baselines, namely CBOW, Skip-Gram and MCE. CBOW and Skip-Gram based methods can learn medical concept representations for diagnosis prediction [6,8]. Moreover, MCE focus on the time windows of medical concept sequences in the learning process. The experimental results are presented in Table 4. Although MCE considers the temporal influence, it is not optimistic to employ the representation learned by MCE to predict diagnosis. The reason is that MCE misses out on the temporal influence at different information levels. In addition, both Skip-Gram and CBOW perform worse than SiaCo, which demonstrates the importance of exploiting the progression of patients' status for patient outcome prediction.

Table 4. Evaluation of the medical concept representations

Dataset	Approach	Pre@5	Rec@5	Pre@10	Rec@10	Pre@15	Rec@15
Circulatory (CS)	MCE	0.3255	0.1042	0.2444	0.1524	0.2091	0.1921
	Skip-Gram	0.3967	0.1291	0.3017	0.1914	0.2536	0.2383
	CBOW	0.4118	0.1338	0.3224	0.2062	0.2701	0.2562
	SiaCo	**0.4931**	**0.1599**	**0.3818**	**0.2416**	**0.3190**	**0.2989**
Heart Failure (HF)	MCE	0.5158	0.1393	0.3843	0.2048	0.3114	0.2476
	Skip-Gram	0.5702	0.1539	0.4230	0.2248	0.3586	0.2830
	CBOW	0.5556	0.1501	0.4318	0.2295	0.3592	0.2840
	SiaCo	**0.6600**	**0.1787**	**0.5122**	**0.2719**	**0.4273**	**0.3380**

5.4 Results of Mortality Prediction

Table 5 compares the model performance on datasets CS and HF. Our SiaCo achieves high performance on both datasets. Both RETAIN and T-LSTM perform worse than SiaCo, as only considering the temporal influence results in poor performance improvement on this task. StageNet get a high accuracy but low AUROC score, indicating that it generalizes poorly for mortality prediction. The experimental results demonstrate that patient similarities play an important role in mortality prediction.

Table 5. Performance of mortality prediction on CS and HF

	CS			HF		
	AUROC	ACC	F1-score	AUROC	ACC	F1-score
LSTM	0.8543	0.8803	0.5063	0.7750	0.9085	0.4414
RETAIN	0.8408	0.8950	0.5209	0.8390	0.8975	0.5108
T-LSTM	0.8844	0.8917	0.5338	0.8748	0.9155	0.5255
StageNet	0.8960	0.9032	**0.5997**	0.9005	0.9117	0.5877
SiaCo	**0.9147**	**0.9220**	0.5942	**0.9221**	**0.9210**	**0.5934**

5.5 Ablation Study

To verify the effectiveness of several key modules of the proposed SiaCo, we evaluate the variants of SiaCo, which exclude particular designs. (1) **SiaCo$_{E_1-}$** discards the design Encoder E_1 in the parallelized LSTM, which captures the temporal-dependent patterns of patient status at the encounter level. (2) **SiaCo$_{E_2-}$** discards the design Encoder E_2 in the parallelized LSTM, which captures the temporal-dependent patterns of patient status at the medical concept level. (3) **SiaCo$_{KM_i-}$** excludes the patient similarities at the encounter level. (4) **SiaCo$_{KM_i'-}$** excludes the patient similarities at the medical concept level.

Table 6 reports the experimental results on diagnosis prediction. SiaCo outperforms SiaCo$_{E_1-}$ and SiaCo$_{E_2-}$, showing the importance of modeling the more fine-grained progression of patients' status. Moreover, both SiaCo_{KM_i-} and $\text{SiaCo}_{KM_i'-}$ perform worse than SiaCo, demonstrating effectively measuring patient similarities and collaborative learning between patients can improve the model's ability for health risk assessments.

Table 6. Ablation study on diagnosis prediction

Dataset	Approach	Pre@5	Rec@5	Pre@10	Rec@10	Pre@15	Rec@15
Circulatory (CS)	SiaCo_{E_1-}	0.4573	0.1475	0.3563	0.2257	0.2998	0.2814
	SiaCo_{E_2-}	0.4863	0.1577	0.3786	0.2378	0.3134	0.2929
	SiaCo_{KM_i-}	0.4748	0.1529	0.3716	0.2358	0.3102	0.2880
	$\text{SiaCo}_{KM_i'-}$	0.4819	0.1529	0.3745	0.2347	0.3115	0.2883
Heart Failure (HF)	SiaCo_{E_1-}	0.6294	0.1710	0.4994	0.2672	0.4231	0.3365
	SiaCo_{E_2-}	0.6487	0.1759	0.5084	0.2708	0.4241	0.3359
	SiaCo_{KM_i-}	0.6387	0.1734	0.5031	0.2686	0.4260	0.3380
	$\text{SiaCo}_{KM_i'-}$	0.6514	0.1759	0.5084	0.2708	0.4241	0.3360

5.6 Study of Weights α, β and γ

To distinguish the impact of diagnoses, procedures and medications on patient outcome prediction, we assign different weights to them, which are α, β and γ, respectively. To find a better strategy of weight assignment, we select several representative strategies and report the corresponding experimental results of diagnosis prediction in Fig. 6. SiaCo performs better than other cases when α is assigned a larger weight. We attribute this to that the diagnostic information

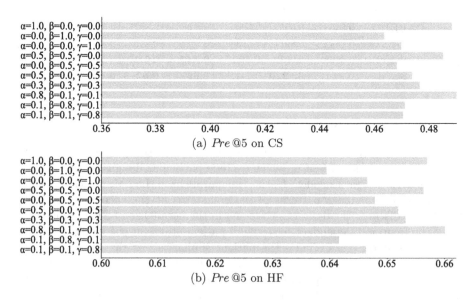

Fig. 6. Evaluation of the weights α, β and γ

has a much greater impact on health risk assessments than procedures and medications. However, the inclusion of procedures and medications also improves performance, shown by the comparison of the results of $\alpha = 0.8, \beta = 0.1, \gamma = 0.1$ and $\alpha = 1.0, \beta = 0, \gamma = 0$. Thus, we set $\alpha = 0.8$, $\beta = 0.1$ and $\gamma = 0.1$.

6   Conclusions

In this paper, we propose a similarity-aware collaborative learning model SiaCo for patient outcome prediction. SiaCo includes an encounter-level similarity measurer, a concept-level measurer and a parallelized LSTM module. The two measurers separately calculate patient similarities at different information levels and build two global knowledge matrices to support collaborative learning between patients. Moreover, the parallelized LSTM module is designed to exploit the more fine-grained progression of the patients' status. The comparative results on two real disease datasets demonstrate that SiaCo outperforms the state-of-the-art models for patient outcome prediction. For future work, it is interesting to design effective restraints to bring similar patient representations closer.

Acknowledgements. This work is supported by the National Key R&D Program of China (No. 2021YFF0900800), the major Science and Technology Innovation of Shandong Province grant (No. 2021CXGC010108), the Fundamental Research Funds of Shandong University, and partially supported by the NSFC (No. 91846205).

References

1. Baytas, I.M., Xiao, C., Zhang, X., Wang, F., Jain, A.K., Zhou, J.: Patient subtyping via time-aware LSTM networks. In: SIGKDD, pp. 65–74 (2017)
2. Bhattacharya, S., Rajan, V., Shrivastava, H.: ICU mortality prediction: a classification algorithm for imbalanced datasets. In: AAAI, pp. 1288–1294 (2017)
3. Cai, X., Gao, J., Ngiam, K.Y., Ooi, B.C., Zhang, Y., Yuan, X.: Medical concept embedding with time-aware attention. In: IJCAI, pp. 3984–3990 (2018)
4. Cheng, Y., Wang, F., Zhang, P., Hu, J.: Risk prediction with electronic health records: a deep learning approach. In: SIAM, pp. 432–440 (2016)
5. Choi, E., et al.: Multi-layer representation learning for medical concepts. In: SIGKDD, pp. 1495–1504 (2016)
6. Choi, E., Bahadori, M.T., Song, L., Stewart, W.F., Sun, J.: GRAM: graph-based attention model for healthcare representation learning. In: SIGKDD, pp. 787–795 (2017)
7. Choi, E., Bahadori, M.T., Sun, J., Kulas, J., Schuetz, A., Stewart, W.F.: RETAIN: an interpretable predictive model for healthcare using reverse time attention mechanism. In: Advances in Neural Information Processing Systems, vol. 29, pp. 3504–3512 (2016)
8. Choi, E., Schuetz, A., Stewart, W.F., Sun, J.: Medical concept representation learning from electronic health records and its application on heart failure prediction. arXiv (2016)
9. Choi, E., Xiao, C., Stewart, W.F., Sun, J.: MiME: multilevel medical embedding of electronic health records for predictive healthcare. In: Advances in Neural Information Processing Systems, vol. 31, pp. 4552–4562 (2018)

10. Gao, J., Xiao, C., Wang, Y., Tang, W., Glass, L.M., Sun, J.: StageNet: stage-aware neural networks for health risk prediction. In: Proceedings of the Web Conference (WWW), pp. 530–540 (2020)
11. Hochreiter, S., Schmidhuber, J.: Long short-term memory. Neural Comput. **9**, 1735–1780 (1997)
12. Johnson, A.E., et al.: MIMIC-III, a freely accessible critical care database. Sci. Data **3**, 1–9 (2016)
13. Keogh, E.J., Pazzani, M.J.: Derivative dynamic time warping. In: Proceedings of the 2001 SIAM International Conference on Data Mining, pp. 1–11. SIAM (2001)
14. Kingma, D.P., Ba, J.: Adam: a method for stochastic optimization. In: Bengio, Y., LeCun, Y. (eds.) ICLR (2015)
15. Liu, L., Shen, J., Zhang, M., Wang, Z., Tang, J.: Learning the joint representation of heterogeneous temporal events for clinical endpoint prediction. In: AAAI, pp. 109–116 (2018)
16. Liu, N., Lu, P., Zhang, W., Wang, J.: Knowledge-aware deep dual networks for text-based mortality prediction. In: ICDE, pp. 1406–1417 (2019)
17. Ma, F., Chitta, R., Zhou, J., You, Q., Sun, T., Gao, J.: Dipole: diagnosis prediction in healthcare via attention-based bidirectional recurrent neural networks. In: SIGKDD, pp. 1903–1911 (2017)
18. Ma, L., et al.: AdaCare: explainable clinical health status representation learning via scale-adaptive feature extraction and recalibration. In: AAAI, pp. 825–832 (2020)
19. Ma, L., et al.: ConCare: personalized clinical feature embedding via capturing the healthcare context. In: AAAI, pp. 833–840 (2020)
20. Mahajan, R., Mansotra, V.: Predicting geolocation of tweets: using combination of CNN and BiLSTM. Data Sci. Eng. **6**(4), 402–410 (2021)
21. Mikolov, T., Chen, K., Corrado, G., Dean, J.: Efficient estimation of word representations in vector space. In: ICLR (2013)
22. Umemoto, K., Goda, K., Mitsutake, N., Kitsuregawa, M.: A prescription trend analysis using medical insurance claim big data. In: ICDE, pp. 1928–1939 (2019)
23. Wang, F., Lee, N., Hu, J., Sun, J., Ebadollahi, S., Laine, A.F.: A framework for mining signatures from event sequences and its applications in healthcare data. IEEE Trans. Pattern Anal. Mach. Intell. **35**, 272–285 (2012)
24. Wawrzinek, J., Pinto, J.M.G., Wiehr, O., Balke, W.: Exploiting latent semantic subspaces to derive associations for specific pharmaceutical semantics. Data Sci. Eng. **5**(4), 333–345 (2020)
25. Yin, K., Qian, D., Cheung, W.K., Fung, B.C., Poon, J.: Learning phenotypes and dynamic patient representations via RNN regularized collective non-negative tensor factorization. In: AAAI, pp. 1246–1253 (2019)

Semi-supervised Graph Learning
with Few Labeled Nodes

Cong Zhang[1], Ting Bai[1,2(✉)], and Bin Wu[1,2]

[1] Beijing University of Posts and Telecommunications, Beijing, China
{zhangc137,baiting,wubin}@bupt.edu.cn
[2] Beijing Key Laboratory of Intelligent Telecommunications Software and
Multimedia, Beijing, China

Abstract. Graph-based semi-supervised learning, utilizing both a few labeled nodes and massive unlabeled nodes, has aroused extensive attention in the research community. However, for the graph with few labeled nodes, the performance of Graph Convolutional Networks (GCNs) will suffer from a catastrophic decline due to its intrinsic shallow architecture limitation and insufficient supervision signals. To accommodate this issue, we propose a novel Self-Training model (ST-LPGCN) which reinforces the pseudo label generation on the GCNs with Label Propagation algorithm (LPA). By making full use of the advantages of GCNs in aggregating the local node features and LPA in propagating the global label information, our ST-LPGCN improves the generalization performance of GCNs with few labeled nodes. Specifically, we design a pseudo label generator to pick out the nodes assigned with the same pseudo labels by GCN and LPA, and add them to the labeled data for the next self-training process. To reduce the error propagation of labels, we optimize the transition probability between nodes in LPA under the supervision of the pseudo labels. The extensive experimental results on four real-world datasets validate the superiority of ST-LPGCN for the node classification task with few labeled nodes.

Keywords: Graph neural networks · Label propagation · Few labels · Semi-supervised learning

1 Introduction

Recently, graph representation learning had been successfully applied into many research areas, such as social networks [8], physical process [18], knowledge graph [7], and biological networks [17]. However, due to the data privacy policy and the high cost of data annotation, the node labels on graphs are usually sparse in many real-world scenarios. To alleviate the lack of labeled data

This work is supported by the National Natural Science Foundation of China under Grant No. 62102038; the National Natural Science Foundation of China under Grant No. 61972047, the NSFC-General Technology Basic Research Joint Funds under Grant U1936220.

problem, graph-based semi-supervised learning (GSSL), which trains numerous unlabeled data along with a small amount of labeled data, has attracted increasing interests in the research community. And it had been successfully developed in various approaches, such as label propagation [36], graph regularization [35], graph autoencoder [13] and graph neural networks [8,14,25]. The success of GSSL depends on the simple assumption that the nearby nodes on a graph tend to have the same labels [21]. It spreads the labels of a few labeled nodes to the remaining massive unlabeled nodes according to the adjacent relationship within the graph to accurately classify the unlabeled nodes. The mainstream method Graph Convolutional Networks (GCNs) learning representations of each node by aggregating the information of its neighbors to facilitate the label assignments, have been demonstrated to significantly outperform the other classic GSSL methods [5,36].

Despite the noticeable progress of GCNs and the variants [9,14,31] in graph-based semi-supervised learning tasks, it still faces adverse conditions when the labeled data is extremely limited. The performance of GCNs would suffer from a catastrophic decline when the labeled data in GSSL is minimal. GCNs iteratively update each node's representation via aggregating the representations of its neighbors, and assign pseudo labels to the unlabeled nodes in the training process. GCNs with shallow architecture will restrict the efficient propagation of label signals, but when GCNs is equipped with many convolutional layers, it will suffer from the over-smoothing problem [15,30], resulting in the restriction of the performance of GCNs in the case of few labels. Some previous studies [12,15,22] expand the labeled data by assigning unlabeled nodes with pseudo labels or further use the unsupervised cluster information. Although they reduce the error information contained in pseudo labels, they can not effectively involve the global structure information, and the limited label signals can not be efficiently propagated to the entire graph.

To address this problem, we propose a novel Self-Training model (ST-LPGCN) which reinforces the pseudo label generation on the GCNs with Label Propagation algorithm (LPA), then generates pseudo labels to expand the training set (i.e., labeled nodes). In ST-LPGCN, we utilize GCNs to extract the node features and local structure information and assign the unlabeled nodes with pseudo labels. However, as described above, the shallow GCNs cannot propagate the label information through the entire graph. We further use LPA to spread the labels to the entire graph to reinforces the pseudo labels generation. By doing that, we not only solve the over smoothing problem of GCNs in propagating labels from high-order neighbors, but also incorporate both local and global semantic information, which is helpful to predict the pseudo labels of nodes. Specifically, to reduce the error propagation of labels, we learn the transition probability matrix in LPA under the supervision of the pseudo labels to optimize the transition probability between nodes. Then we carefully design a pseudo label generator to pick out the nodes assigned with the same pseudo labels by GCN and LPA, and add them to the labeled data for the next self-training process. Our contributions are summarized as follows:

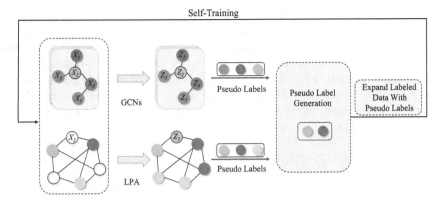

Fig. 1. An overall framework of our proposed ST-LPGCN. ST-LPGCN consists of three components: a GCN generates pseudo labels based on the embeddings learned from node features and local structure; a LPA component assigns unlabeled nodes with pseudo labels according to the global structure and label information; and the pseudo label generation component picks nodes which assigned with the same pseudo labels by GCN and LPA. The selected nodes are added to the labeled data for the next self-training process.

- We propose a novel self-training algorithm for the semi-supervised node classi-fication tasks called ST-LPGCN, which reinforces the pseudo label generation on the GCNs with Label Propagation algorithm (LPA).
- To reduce the error propagation of labels, we optimize the transition proba-bility between nodes in LPA under the supervision of the pseudo labels. The nodes assigned with the same pseudo labels by GCN and LPA are added to the labeled data for the next self-training process.
- We conduct extensive experiments on four benchmark graph datasets, and the experimental results demonstrate that our proposed ST-LPGCN outper-forms the state-of-the-art baselines, especially in the cases with extremely few labeled data.

2   Related Works

Graph-Based Semi-supervised Learning. Graph-based semi-supervised learning (GSSL) has attracted much attention in recent years due to the wide range of applications, scalability to large-scale data, and promising performance. The early GSSL methods are mainly based on the cluster assumption that the nearby nodes on a graph tend to possess the same label. Researches along this line mainly consist of label propagation [36] and its variants such as Gaussian random fields [37], local and global consistence [34] and modified adsorption [23]. Another line is to predict the labels based on the embeddings learned from input graphs [32], such as factorization-based methods [1,3] and random-walk-based methods [5,24] which learn node embeddings based on the graph structure.

However, these methods fail to leverage the rich node features, which are significantly critical for GSSL. To jointly model the data features and graph structure, autoencoder-based methods have been proposed including SDNE [27], GAE and VGAE [13]. Later, with the introduction of GCN [14], the GNN-based methods become the dominant solution. GNNs take advantage of message passing based on the Laplacian smoothing assumption, in which messages are exchanged and updated between each pair of the nodes and capture more facts to make a more robust prediction on the unlabeled nodes. To improve the performance of GNNs in the case of few labels, Li [15] combined GCNs and self-training to expand supervision signals, M3S [22] proposed multi-stage self-training and utilized clustering method to eliminate the pseudo labels that may be incorrect. CGCN [12] further utilized an optimized graph clustering approach to strengthen the performance.

Graph Convolutional Networks. Inspired by the success of convolutional networks on grid-structure data, GCNs have been proposed to generalize the CNNs to graph-structure data and achieved state-of-the-art performance in many graph mining tasks. Generally, GCNs can be divided into spectral-based and spatial-based methods. Spectral-based approaches define the graph convolutions in the Fourier domain based on the graph signal processing. The general graph convolution framework is first proposed by Bruna [2] which needs to compute the Laplacian eigenvectors and then ChebyNet [4] using Chebyshev polynomials to optimize this method. Afterwards, Kipf [14] utilized the localized first-order approximation of spectral graph convolutions further simplified this model. Furthermore, to overcome time-consuming computation on approximated Laplace eigenvalues, spatial approaches define convolutions directly on the graph based on nodes' spatial relations. The graph convolution is defined as the weighted average function on the target node's neighbors such as GraphSAGE [8] and GAT [25]. Although GCNs and their variants have shown promising results, the performance of most existing GCNs will suffer from a catastrophic decline when the labeled data is limited. To address this problem, BGCN [33] incorporates a Bayesian method into GCN to get a more robust and generalized model for node classification. LCGNN [29] directly uses label information to reconstruct the aggregation matrix in GNNs to optimize the aggregation process. CG^3 [26] leverages contrastive learning and a hierarchical GCN model to capture nodes' information from local to global perspectives to learn better representations for the classification task.

Label Propagation Algorithm. Label propagation algorithm is a classic graph-based semi-supervised learning method that regards the labeled nodes as guides that lead the label information to propagate through the edges within the graph to assign the unlabeled nodes with predicted labels. Gaussian random fields [37] and local and global consistency [6] are the early typical work using label propagation. With the proposal of GNNs, several researches try to incorporate GNNs and LPA because they are both based on the message passing model. TPN [16] uses GNNs generating soft labels and propagates them by label propagation, GCN-LPA [28] uses the label propagation as the regularization for parameter matrix in GCN. UniMP [20] incorporates label information to the feature propagation process in GCN. Recent work Correct and Smooth [11] notes that the key to improve the performance is using labels directly, utilizes label propagation twice for correcting and smoothing and achieves performance comparable to GCNs.

3   The Proposed Model

In this section, we introduce our Self-Training framework (ST-LPGCN) which reinforces the pseudo label generation on the GCNs with Label Propagation algorithm. The overall architecture is shown in Fig. 1. Before introducing the ST-LPGCN, we provide the preliminary notations in this paper.

3.1   Preliminary Notations

Given a graph $\mathcal{G} = (\mathcal{V}, \mathcal{E}, X)$, where $\mathcal{V} = \{v_1, v_2, ..., v_n\}$ is the set of nodes, $\mathcal{E}$ represents the set of edges. The feature vector of node v_i is denoted as $\mathbf{x}_i \in \mathbb{R}^{1 \times d}$ and $\mathbf{X} = [\mathbf{x}_1; \mathbf{x}_2; ...; \mathbf{x}_n] \in \mathbb{R}^{n \times d}$ denotes the feature matrix of all nodes. $\mathbf{A} \in \{0, 1\}^{n \times n}$ denotes the adjacency matrix of the $\mathcal{G}$, in which each entry A_{ij} represents the state of connection between node v_i and v_j. $A_{ij} = 1$ indicates that there exists an edge between v_i and v_j; otherwise, $A_{ij} = 0$. The label of a node v_i is represented as a one-hot vector $\mathbf{y}_i \in \mathbb{R}^{1 \times C}$, where C is the number of classes. Given a node v_i, our task is to predict the label of v_i. If $y_{ij} = 1$, the label of node v_i is $j \in C$.

3.2   Generating Pseudo Labels with GCN

GCN is a widely used message passing model for semi-supervised node classification. A GCN model usually contains multiple layers, and each layer aggregates the first-order neighbors' information and generate a low dimensional vector for each node. The simple message passing process can be formulated as follows [33]:

$$\begin{aligned}
\mathbf{H}^{(1)} &= \sigma(\hat{\mathbf{A}} \mathbf{X} \mathbf{W}^{(0)}), \\
\mathbf{H}^{(l+1)} &= \sigma(\hat{\mathbf{A}} \mathbf{H}^{(l)} \mathbf{W}^{(l)}),
\end{aligned} \tag{1}$$

where $\hat{\mathbf{A}}$ is the normalized adjacency matrix, defined as $\widetilde{\mathbf{D}}^{-1} \widetilde{\mathbf{A}}$. $\widetilde{\mathbf{A}}$ is the adjacent matrix with self-loops and $\widetilde{\mathbf{D}}$ is the degree matrix of $\widetilde{\mathbf{A}}$. $\mathbf{W}^{(l)}$ is the parameter matrix at layer l and $\mathbf{H}^{(l)}$ is the output features matrix from layer $l - 1$. σ is a non-linear activation function.

We adopted the vanilla GCN [14] with two convolutional layers in our paper, formulated as:

$$\mathbf{Z} = softmax(\hat{\mathbf{A}}(ReLU(\hat{\mathbf{A}}\mathbf{X}\mathbf{W}^{(0)}))\mathbf{W}^{(1)}), \tag{2}$$

where $\mathbf{W}^{(0)} \in \mathbb{R}^{d \times h}$ and $\mathbf{W}^{(1)} \in \mathbb{R}^{h \times c}$ are the parameter matrices, and h is the dimension of the hidden layer. The non-linear activation function in first layer is rectified linear unit (ReLU). $\mathbf{Z} \in \mathbb{R}^{n \times c}$ is a probability distribution, where each row represents the probabilities that the node belongs to the corresponding labels.

The GCN model is optimized by minimizing the cross-entropy loss function:

$$\mathcal{L}_{gcn} = -\sum_{i \in L} \sum_{c=1}^{C} Y_{ic} \ln Z_{ic}, \tag{3}$$

where L is the set of the labeled nodes and C is the number of the labels.

The node is finally classified into the class with the maximal probability, and the pseudo labels assigned by GCN is $\hat{\mathbf{Z}}$. Following [15], we select top-t confident nodes in each class and assign the pseudo labels to them. Instead of adding them to the labeled data directly, we feed the pseudo labels generated by GCN into pseudo label generation component for further selection.

3.3 Generating Pseudo Labels with LPA

Traditional LPA. Label propagation algorithm (LPA) is a classic GSSL approach, which propagates the node labels according to the similarity between nodes. Given an initial label matrix $\mathbf{Y}^{(0)}$, which consists of one-hot label indicator vectors $y_i^{(0)}$ for the few labeled nodes and zero vectors for the unlabeled ones. The process of traditional LPA can be formulated as following:

$$\hat{\mathbf{Y}}^{(k+1)} = \mathbf{D}^{-1}\mathbf{A}\hat{\mathbf{Y}}^{(k)}, \tag{4}$$

where $\mathbf{A}$ and $\mathbf{D}$ are the adjacent matrix and its degree matrix respectively. The labeled nodes propagate labels to their neighbors according to the normalized adjacent matrix $\mathbf{D}^{-1}\mathbf{A}$, namely transition probability matrix. The greater the similarity between each node and its neighbors, the greater the probability propagated from the label of its neighbors.

However, in the above equation, the fixed transition probability of labels in LPA will lead to the avalanche effect that the initial errors will be magnified with the iterative process, which would have a catastrophic impact on the predictions. In addition, the connection of nodes from different classes in the graph may be a noise for LPA, which would further degrade the performance of LPA. To reduce the error propagation of labels, we optimize the traditional LPA by making the transition probability matrix learnable under the supervision of the pseudo labels.

Trainable LPA. To propagate the label information more correctly, we propose a trainable LPA method to learn a optimized transition probability matrix where the nodes in the same class connect more strongly, while the nodes from different classes connect weakly. The trainable LPA can be formulated as:

$$\hat{\mathbf{Y}}^{(k+1)} = \mathbf{D}^{-1}\mathbf{A}\mathbf{W}^{(k)}\hat{\mathbf{Y}}^{(k)}, \tag{5}$$

where $\mathbf{W}^{(k)} \in \mathbb{R}^{n \times n}$ is the parameter matrix at the k-th iteration. The trainable LPA are optimized by cross-entropy loss function:

$$\mathcal{L}_{lpa} = -\sum_{i \in L}\sum_{c=1}^{C} Y_{ic}\ln\hat{Y}_{ic}, \tag{6}$$

where the $\hat{Y}_{ic}$ is the predicted label by LPA.

In this way, the transition probability matrix can be corrected by the pseudo labels, which alleviates the errors magnification problem in the iterative process. We select top-t confident nodes in each class and assign the pseudo labels to them. The same as the pseudo labels in GCN, we feed the pseudo labels generated by trainable LPA into the pseudo label generation component for further selection.

Algorithm 1: ST-LPGCN Algorithm

Input: Adjacent matrix $\mathbf{A}$, feature matrix $\mathbf{X}$, label matrix $\mathbf{Y}$, initial labeled and unlabeled set Y_0, U_0.

Output: Predicted labels for each unlabeled node

1   Random initialize the parameter matrices of GCN and trainable LPA;
2   **for** $k = 1, 2, ..., K$ **do**
3   Initial the pseudo-label sets $\hat{Z}_{gcn} = \emptyset$ and $\hat{Y}_{lpa} = \emptyset$;
4   Train GCN and LPA on the labeled set Y_{k-1} to obtain the predictions $\hat{Z}$ and $\hat{Y}$;
5   Sort nodes according to the confidence scores in the unlabeled set U_{k-1};
6   **for** *each class c* **do**
7   Select the top t nodes from $\hat{Z}$ and $\hat{Y}$, and add them to $\hat{Y}_{gcn}$ and $\hat{Y}_{lpa}$ respectively;
8   **for** *each node* **do**
9   **if** *the label of node v are same in both $\hat{Y}_{gcn}$ and $\hat{Y}_{lpa}$* **then**
10   Add it to the labeled data Y_{k-1} with pseudo label c;
11   Delete it from the unlabeled set U_{k-1};
12   Train the GCN and LPA on the new labeled data Y_k;
13 Conduct label prediction based on the final trained GCN;

3.4 Pseudo Label Generation

The core of self-training lies in the accuracy of pseudo labels assigned on the nodes. It is important to generate reliable labels to avoid error propagation. In this component, we use the pseudo labels from GCN and LPA components for mutual checking. The GCN pseudo labels are generated based on the nodes' feature similarities in the feature space. While the LPA pseudo labels are generated by using the label information and the global structure information, which makes up for the limitations of the shallow GCN.

Specifically, given a node, we check their pseudo-labels generated in the GCN and trainable LPA. If the labels are the same, we assign the label of the node to the labeled data for the next self-training process; otherwise, we remove them to diminish potential error information. Given a node v which had been generated pseudo labels in GCN and LPA, the reliable pseudo label can be generated by:

$$\hat{\mathbf{Y}}_{final}(v) = \left\{ v | \hat{\mathbf{Z}}_{gcn}(v) = \hat{\mathbf{Y}}_{lpa}(v) \right\}, \tag{7}$$

where $\hat{\mathbf{Z}}_{gcn}(v)$, $\hat{\mathbf{Y}}_{lpa}(v)$ denote the pseudo labels of node v generated by GCN and LPA respectively. $\hat{\mathbf{Y}}_{final}(v)$ is the final pseudo label of v that will be added to the labeled data for next self-training process.

3.5 Self-training Process

Following M3S [22], we repeat K times of the self-training process to provides more supervision information from the learnable pseudo labels. The final loss function to optimized is:

$$\mathcal{L} = \mathcal{L}_{gcn} + \lambda \mathcal{L}_{lpa}, \tag{8}$$

where λ is the trade-off hyper-parameter. The GCN and LPA component are trained simultaneously (see in Algorithm 1).

4 Experiments

4.1 Experiment Setup

Datasets. The four widely used benchmark datasets are Cora, CiteSeer, Pubmed [19] and ogbn-arxiv [10]. In these datasets, each node represents a document and has a feature vector. The label of each node represents the topic that it belongs to. The statistics of the datasets are shown in Table 1.

Baseline Methods. In the experiments, we compare our method with the state-of-the-art semi-supervised node classification methods, including:

- **LPA** [36]: Label propagation is a classical semi-supervised learning algorithm. It iteratively assigns labels to the unlabeled nodes by propagating the labels through the graph.

Table 1. Statistics of the datasets.

Dataset	Nodes	Edges	Classes	Features
CiteSeer	3,327	4,732	6	3,703
Cora	2,708	5,429	7	1,433
PubMed	19,717	44,338	3	500
ogbn-arxiv	169,343	1,166,243	40	128

- **GCN** [14]: GCN is a widely used graph neural network, and it learns node representations based on the first-order approximation of spectral graph convolutions.
- **GAT** [25]: GAT leverages attention mechanism to improve the model performance for node classification task.
- **Union and Intersection** [15]: Two methods **Co-training** and **Self-training** are proposed in [15] for semi-supervised node classification. Co-training uses ParWalk to select the most confident nodes and add their pseudo labels to the labeled data to train GCN. While **Self-Training** selects the pseudo node labels from GCN to boost the model performance. **Union and Intersection** takes the union or intersection of the pseudo labels generated by Co-training and Self-training as the additional supervision information.
- **M3S** [22]: It leverages the deep cluster to check the pseudo labels assigned by GCN and add to the labeled data for self-training on GCN.
- **GCN-LPA** [28]: It learns the optimal edge weights by LPA, which are leveraged as the regularization in GCN to separate different node classes.
- **LCGNN** [29]: The LCGCN and LCGAT use label information to reconstruct the aggregation matrix based on the label-consistency for GCN and GAT, which alleviate noise from connected nodes with different labels.
- **BGCN** [33]: It combines Bayesian approach with the GCN model, and views the observed graph as a realization from a parametric family of random graphs.
- **CGCN** [12]: It generates pseudo labels by combining variational graph autoencoder with Gaussian mixture models, which can be used to boost the performance of semi-supervised learning.
- **CG^3** [26]: It leverages the contrastive learning on different views of graph and captures the similarity and graph topology information to facilitate the node classification task.

The above methods cover different kinds of approaches in graph-based semi-supervised node classification task. LPA is a classic GSSL method that only propagates the label signals to the entire graph directly. GCN and GAT are the widely used GNN-based model which aggregates node features and local structure. GCN-LPA, LCGNN and CG^3 further leverage the label information or global structure information to improve the performance of GCNs. However, they cannot be fully trained with few labeled nodes, which leads to the underperformance. Self-Training model, M3S and CGCN all utilize the self-training process based on the GCNs to expand the labeled data with pseudo labels.

The most similar work to our method is M3S. It uses the unsupervised method cluster to diminish the potential errors in pseudo label generation. The differences between our model and M3S lie in: (1) M3S does not leverage the global structure and label information of graph; (2) M3S needs more layers to incorporate the useful features on GCN which is time-consuming and would lead to the over-smoothing problem of GCN. Our ST-LPGCN leverages a LPA to make full use of the global structure information and propagate the label information to enrich the pseudo label signals. The trainable LPA enables ST-LPGCN to alleviate the potential errors in label propagation simultaneously, which would help to improve the accuracy of pseudo labels assigned on the nodes in the self-training process.

Parameters Settings. We follow the experimental settings in M3S [22] for fair comparation. We conducted experiments with different label rates, i.e., 0.5%, 1%, 2%, 3%, 4% on Cora and CiteSeer, 0.03%, 0.05%, 0.1% on PubMed and 1%, 2%, 5%, 10% on ogbn-arxiv dataset respectively. The layers of GCN is 3, 3, 2, 2 and the number of propagation times in LPA is 3 in Cora, CiteSeer, PubMed and ogbn-arxiv datasets respectively in each self-training stage. For the top-t most confident pseudo labels, t is set to 60 in GCN and LPA in all datasets. The times K repeated in self-training process are 3, 4, 5 in CiteSeer, PubMed and ogbn-arxiv datasets. While for Cora dataset, it varies depending on the label rates, which is set to 5, 4, 4, 2, 2. For each baseline method, grid search is applied to find the optimal settings, and all the results are the mean accuracy of 10 runs on all datasets. The hyperparameters of GCN are set as follows: the learning rate is 0.002, dropout rate is 0.6, L2 regularization weight is 5×10^{-4}, and the dimension of hidden layers in GCNs is 16. For other baseline methods, we adopt their public code and tune hyperparameters for the best performance. The trade-off hyperparameter λ in Eq. 8 is set to 1.

4.2 Result Analysis

We compute the accuracy of node classification task on different methods. The results on four datasets with the different label rates are shown in the Table 2 and 3. For the baseline methods which are restricted by the memory size to deal with the large-scale dataset ogbn-arxiv, comparisons are only carried out on Cora, CiteSeer and Pubmed datasets. We have the following observations:

(1) LPA performs worst on CiteSeer and ogbn-arxiv datasets. Although it can propagate the global label information through the entire graph to obtain relatively strong supervision information, without the node feature information, it can not achieve good performance due to the lack of using node semantic information.
(2) GCN and GAT perform worst on Cora and PubMed datasets. They aggregate the local node features and structure information, but they can not provide sufficient supervision information for model training with few labels, especially when the labeled data is extremely scarce.

Table 2. Node classification accuracy on Cora.

Label rate	Cora					CiteSeer					PubMed		
	0.5%	1%	2%	3%	4%	0.5%	1%	2%	3%	4%	0.03%	0.05%	0.1%
LPA	56.8	62.0	64.2	66.3	69.6	39.6	42.2	44.0	44.6	45.1	57.6	62.0	64.5
GCN	50.1	60.3	69.8	75.5	76.7	44.7	54.2	62.3	68.0	69.5	50.9	58.2	68.1
GAT	50.3	61.2	70.6	76.4	77.1	45.3	56.4	61.9	69.6	71.2	52.1	59.8	70.4
Co-training	55.1	61.3	70.1	74.9	76.8	45.6	54.0	59.6	63.5	64.8	58.2	65.7	69.8
Self-training	56.6	62.4	71.7	76.8	77.1	43.6	57.5	64.2	67.2	68.5	56.7	65.5	70.1
Union	57.0	67.6	74.5	77.1	78.7	47.0	59.3	63.1	65.9	66.2	57.1	65.2	68.8
Intersection	51.3	63.2	71.1	76.2	77.3	43.3	60.4	65.8	69.8	70.2	56.0	59.9	68.0
GCN-LPA	52.5	66.9	72.0	72.9	74.3	42.1	50.3	59.4	67.8	69.0	63.6	64.4	69.5
BGCN	56.7	69.8	74.8	76.9	78.1	52.0	58.6	68.6	70.8	72.3	63.4	66.2	70.2
M3S	61.5	67.2	75.6	77.8	78.0	56.1	62.1	66.4	70.3	70.5	59.2	64.4	70.6
LCGCN	69.7	73.4	75.6	79.1	80.0	61.1	69.1	70.3	71.1	_72.9_	_68.8_	69.4	75.9
LCGAT	_70.9_	_75.8_	_77.1_	79.6	81.3	62.3	68.9	70.4	71.3	72.2	68.6	69.0	75.7
CGCN	64.3	72.4	76.8	79.8	81.3	59.3	63.1	69.5	_72.4_	72.7	64.7	69.2	_77.8_
CG^3	69.3	74.1	76.6	_79.9_	_81.4_	_62.7_	_70.6_	_70.9_	71.3	72.5	68.3	_70.1_	73.2
ST-LPGCN	**75.9**	**78.0**	**80.3**	**81.0**	**81.6**	**65.2**	**70.7**	**71.3**	**72.6**	**73.4**	**75.8**	**77.6**	**80.1**

(3) The performance of Self-training, Co-Training, Union and Intersection is not consistent through different datasets and label rates. The variant models with self-training process achieve better performance, showing the usefulness of leveraging the pseudo labels in model training.

(4) M3S and CGCN with self-training outperform the algorithm without self-training such as GCN-LPA and BGCN on Cora, CiteSeer and PubMed datasets, which also indicates that the self-training can provide more supervision information to improve the model performance. Besides, M3S and CGCN utilize the clustering method to reduce the incorrect label information, which leads to better performance. It indicates that reducing the propagation error of pseudo labels can improve the model performance in self-training process.

Table 3. Node classification accuracy on ogbn-arxiv.

Label rate	1%	2%	5%	10%
LPA	51.0	56.2	62.0	65.6
GCN	60.7	63.3	65.0	65.8
Self-training	62.2	63.9	66.1	66.5
M3S	_63.1_	_64.3_	_66.9_	_68.6_
ST-LPGCN	**64.2**	**65.8**	**67.9**	**69.2**

(5) LCGCN, LCGAT and CG^3 outperforms M3S and CGCN on Cora, CiteSeer and PubMed datasets. LCGCN (LCGAT) directly uses the label information

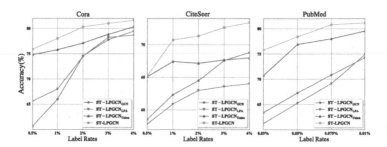

Fig. 2. Accuracy of ST-LPGCN on different datasets using different pseudo label sets.

to reconstruct the aggregation matrix and then train the GCN (GAT) by aggregating the node feature information. These approaches further use label information to supervise the aggregation process of GCN (GAT), which achieves better performance compared with the model learning from pseudo labels. CG^3 designs a hierarchical GCN model to capture nodes' information from local to global perspectives. It incorporates the global structure information to learn a powerful node representation for the classification task.

(6) ST-LPGCN outperforms the state-of-the-art baselines on all datasets with different label rates, verifying the effectiveness of our proposed method. The trainable LPA in ST-LPGCN makes it possible to capture the global structure and label information, and meanwhile alleviate the error propagation of the pseudo labels in self-training process.

(7) As the number of labels decreasing, the performance of all methods becomes worse on all datasets. However, the improvements of accuracy of our ST-LPGCN becomes increasing compared with the state-of-the-art baseline, showing the effectiveness of our model to deal with the data with extremely few labels.

4.3 Ablation Study

In this section, we first make ablation studies to demonstrate the effectiveness of our pseudo label generation strategy and the trainable LPA in our paper. Then we analyze the impact of the amount of pseudo labels added in the self-training.

Pseudo Label Generation Strategy. In the pseudo label generation, we select the nodes with the same pseudo labels generated by the GCN and LPA, and add them to the training set for the next training process. To verify the effectiveness of the pseudo label generation strategy, we adopt the following strategies to make comparing.

- $ST\text{-}LPGCN_{GCN}$: the pseudo labels generated by the GCN.
- $ST\text{-}LPGCN_{LPA}$: the pseudo labels generated by the LPA.
- $ST\text{-}LPGCN_{Union}$: the pseudo labels are generated by adding all pseudo labels in GCN or LPA

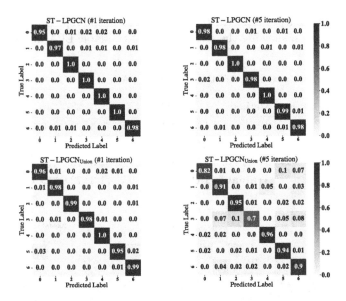

Fig. 3. Confusion matrix (the reliability of pseudo label prediction compared with true label) for ST-LPGCN and ST-LPGCN$_{Union}$ in different iterations. The larger diagonal value is, the more reliable of the predicted pseudo label. The left column represents the first iteration and the right column represents the fifth iteration of ST-LPGCN and ST-LPGCN$_{Union}$ respectively.

As shown in Fig. 2, ST-LPGCN$_{Union}$ obtains better performance than ST-LPGCN$_{GCN}$ and ST-LPGCN$_{LPA}$ due to that it uses additional label (node features) information. Our ST-LPGCN achieves the best performance, showing the effectiveness of our generation strategy to select more reliable pseudo labels of nodes, so as to alleviate the error propagation in self-training process.

As shown in Fig. 3, we further analyze the error propagation of pseudo labels in the training process, and visualize the confidence scores of pseudo labels compared with the true labels at the first iteration and the fifth iteration. We can see that the decreasing of confidence scores is more in the first iteration in ST-LPGCN$_{Union}$. And the error of pseudo labels is magnified as the iteration, leading to the large gap in the fifth iteration compared with the pseudo label generation in ST-LPGCN.

Effectiveness of Trainable LPA. The trainable LPA designed in ST-LPGCN can assign higher transition probability to the node labels belonging to the same class and reduce the probability of node labels in different classes, which may reduce the propagation of error information from the possible noise in the adjacency matrix itself. To verify this, we conducted experiments to analyze the effectiveness of the trainable transition probability matrix in LPA. We keep the same settings and change the pseudo labels generated by traditional LPA for a fair comparison. The results are reported in Table 4. We find that the trainable

LPA consistently outperforms the traditional LPA on two representative datasets with all label rates, showing the effectiveness of trainable LPA in ST-LPGCN.

Table 4. Effect of the trainable LPA in ST-LPGCN.

Dataset	Method	0.5%	1%	2%	3%	4%
Cora	Traditional LPA	72.6	74.4	77.6	79.3	80.1
	Trainable LPA	**75.9**	**78.0**	**80.3**	**81.0**	**81.6**
CiteSeer	Traditional LPA	61.4	67.8	69.5	70.1	71.2
	Trainable LPA	**65.2**	**70.7**	**71.3**	**72.6**	**73.4**

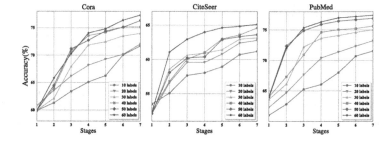

Fig. 4. Sensitivity analysis of the model to the number of pseudo labels in the training process.

Sensitivity of the Number of Pseudo Labels. In this subsection, we explore the sensitivity of the number of pseudo labels added to the labeled data. We conducted experiments taking different numbers of pseudo labels in each self-training process with a label rate of 0.5% on all datasets. As shown in Fig. 4, at the beginning of the training process, the more pseudo-labels generated during each training process, the better the performance of our model. As the training iteration increasing, the performance will approach stabilization with different numbers of pseudo labels. The performance of our model gains slight improvements when a certain number of pseudo labels are added to training data. Based on this observation, we select 60 pseudo labels to expand our training data in experiments.

5   Conclusion

Graph-based semi-supervised learning is a hot topic, but there is relatively little work focusing on semi-supervised learning tasks when the labeled data is quite few, leading to a significant decline in the performance of many existing approaches. In this paper, we propose a novel Self-Training model (ST-LPGCN) which reinforces the pseudo label generation on the GCNs with Label Propagation algorithm. Our ST-LPGCN improves the effect of GCNs in propagating labels from high-order neighbors with shallow architecture, and incorporates both local and global semantic information, which is helpful to predict the pseudo labels of nodes. In the future work, we will further verify the effectiveness of ST-LPGCN in other graph learning tasks, for example, link prediction and so on.

References

1. Ahmed, A., Shervashidze, N., Narayanamurthy, S.M., Josifovski, V., Smola, A.J.: Distributed large-scale natural graph factorization. In: WWW, pp. 37–48 (2013)
2. Bruna, J., Zaremba, W., Szlam, A., LeCun, Y.: Spectral networks and locally connected networks on graphs. In: ICLR (2014)
3. Cao, S., Lu, W., Xu, Q.: GraRep: learning graph representations with global structural information. In: CIKM, pp. 891–900 (2015)
4. Defferrard, M., Bresson, X., Vandergheynst, P.: Convolutional neural networks on graphs with fast localized spectral filtering. In: NIPS, pp. 3837–3845 (2016)
5. Grover, A., Leskovec, J.: node2vec: scalable feature learning for networks. In: KDD, pp. 855–864 (2016)
6. Gui, J., Hu, R., Zhao, Z., Jia, W.: Semi-supervised learning with local and global consistency. Int. J. Comput. Math. **91**(11), 2389–2402 (2014)
7. Hamaguchi, T., Oiwa, H., Shimbo, M., Matsumoto, Y.: Knowledge transfer for out-of-knowledge-base entities: a graph neural network approach. In: IJCAI, pp. 1802–1808 (2017)
8. Hamilton, W.L., Ying, R., Leskovec, J.: Inductive representation learning on large graphs. In: NIPS, pp. 1025–1035 (2017)
9. Hong, H., Guo, H., Lin, Y., Yang, X., Li, Z., Ye, J.: An attention-based graph neural network for heterogeneous structural learning. In: AAAI 2020, pp. 4132–4139 (2020)
10. Hu, W., et al.: Open graph benchmark: datasets for machine learning on graphs. arXiv preprint arXiv:2005.00687 (2020)
11. Huang, Q., He, H., Singh, A., Lim, S.N., Benson, A.R.: Combining label propagation and simple models out-performs graph neural networks. arXiv preprint arXiv:2010.13993 (2020)
12. Hui, B., Zhu, P., Hu, Q.: Collaborative graph convolutional networks: unsupervised learning meets semi-supervised learning. In: AAAI 2020, pp. 4215–4222 (2020)
13. Kipf, T.N., Welling, M.: Variational graph auto-encoders. arXiv preprint arXiv:1611.07308 (2016)
14. Kipf, T.N., Welling, M.: Semi-supervised classification with graph convolutional networks. In: ICLR (2017)
15. Li, Q., Han, Z., Wu, X.: Deeper insights into graph convolutional networks for semi-supervised learning. In: AAAI, pp. 3538–3545 (2018)
16. Liu, Y., et al.: Learning to propagate labels: transductive propagation network for few-shot learning. In: ICLR (2019)
17. Pavlopoulos, G.A.: Using graph theory to analyze biological networks. BioData Min. **4**, 10 (2011)
18. Sanchez-Gonzalez, A., et al.: Graph networks as learnable physics engines for inference and control. In: ICML, pp. 4467–4476 (2018)

19. Sen, P., Namata, G., Bilgic, M., Getoor, L., Gallagher, B., Eliassi-Rad, T.: Collective classification in network data. AI Mag. **29**(3), 93–106 (2008)
20. Shi, Y., Huang, Z., Wang, W., Zhong, H., Feng, S., Sun, Y.: Masked label prediction: unified message passing model for semi-supervised classification. arXiv preprint arXiv:2009.03509 (2020)
21. Song, Z., Yang, X., Xu, Z., King, I.: Graph-based semi-supervised learning: a comprehensive review. arXiv preprint arXiv:2102.13303 (2021)
22. Sun, K., Lin, Z., Zhu, Z.: Multi-stage self-supervised learning for graph convolutional networks on graphs with few labeled nodes. In: AAAI, pp. 5892–5899 (2020)
23. Talukdar, P.P., Crammer, K.: New regularized algorithms for transductive learning. In: Buntine, W., Grobelnik, M., Mladenić, D., Shawe-Taylor, J. (eds.) ECML PKDD 2009. LNCS (LNAI), vol. 5782, pp. 442–457. Springer, Heidelberg (2009). https://doi.org/10.1007/978-3-642-04174-7_29
24. Tang, J., Qu, M., Wang, M., Zhang, M., Yan, J., Mei, Q.: LINE: large-scale information network embedding. In: WWW, pp. 1067–1077 (2015)
25. Velickovic, P., Cucurull, G., Casanova, A., Romero, A., Liò, P., Bengio, Y.: Graph attention networks. In: ICLR (2018)
26. Wan, S., Pan, S., Yang, J., Gong, C.: Contrastive and generative graph convolutional networks for graph-based semi-supervised learning. In: AAAI, vol. 35, pp. 10049–10057 (2021)
27. Wang, D., Cui, P., Zhu, W.: Structural deep network embedding. In: KDD, pp. 1225–1234 (2016)
28. Wang, H., Leskovec, J.: Unifying graph convolutional neural networks and label propagation. arXiv preprint arXiv:2002.06755 (2020)
29. Xu, B., Huang, J., Hou, L., Shen, H., Gao, J., Cheng, X.: Label-consistency based graph neural networks for semi-supervised node classification. In: SIGIR, pp. 1897–1900 (2020)
30. Xu, K., Li, C., Tian, Y., Sonobe, T., Kawarabayashi, K., Jegelka, S.: Representation learning on graphs with jumping knowledge networks. In: ICML, pp. 5449–5458 (2018)
31. Xu, N., Wang, P., Chen, L., Tao, J., Zhao, J.: MR-GNN: multi-resolution and dual graph neural network for predicting structured entity interactions. In: IJCAI, pp. 3968–3974 (2019)
32. Yang, Z., Cohen, W., Salakhudinov, R.: Revisiting semi-supervised learning with graph embeddings. In: ICML, pp. 40–48. PMLR (2016)
33. Zhang, Y., Pal, S., Coates, M., Üstebay, D.: Bayesian graph convolutional neural networks for semi-supervised classification. In: AAAI 2019, pp. 5829–5836 (2019)
34. Zhou, D., Bousquet, O., Lal, T.N., Weston, J., Schölkopf, B.: Learning with local and global consistency. In: NIPS, pp. 321–328 (2004)
35. Zhou, D., Huang, J., Schölkopf, B.: Learning from labeled and unlabeled data on a directed graph. In: ICML, vol. 119, pp. 1036–1043 (2005)
36. Zhu, X., Ghahramani, Z.: Learning from labeled and unlabeled data with label propagation (2003)
37. Zhu, X., Ghahramani, Z., Lafferty, J.D.: Semi-supervised learning using Gaussian fields and harmonic functions. In: ICML, pp. 912–919 (2003)

Human Mobility Identification by Deep Behavior Relevant Location Representation

Tao Sun[1,2], Fei Wang[1], Zhao Zhang[1], Lin Wu[1], and Yongjun Xu[1(✉)]

[1] Institute of Computing Technology, Chinese Academy of Sciences, Beijing, China
{suntao,wangfei,zhangzhao2021,wulinshuxue,xyj}@ict.ac.cn
[2] School of Computing Science and Technology,
University of Chinese Academy of Sciences, Beijing, China

Abstract. This paper focuses on Trajectory User Link (TUL), which aims at identifying user identities through exploiting their mobility patterns. Existing TUL approaches are based on location representation, a way to learn location associations by embedding vectors that can indicate the level of semantic similarity between the locations. However, existing methods for location representation don't consider the semantic diversity of locations, which will lead to a misunderstanding of the semantic information of trajectory when linking anonymous trajectories to candidate users. To solve this problem, in this paper, we propose Deep Behavior Relevant Location representation (DBRLr) to map the polysemous locations into distinct vectors, from the perspective of users' behavior to reflect the semantic polysemy of locations. To learn this representation, we build a Location Prediction-based Movement Model (LP-based MM), which learns user behavior representation at each visited location from a large history trajectory corpora. LP-based MM considers both Continuity and Cyclicity characteristics of user's movement. We employ the combination of the intermediate layer representation in LP-based MM as DBRLr. An effective recurrent neural network is used to link anonymous trajectories with candidate users. Experiments are conducted on two real-world datasets, and the result shows that our method performs beyond existing methods.

Keywords: Human mobility identification · Trajectory-user link · Location representation · Polysemous location

1 Introduction

The plentiful location-based applications make it possible to accumulate lots of users' movement data. Massive anonymous trajectories, which we do not know who created them, are collected, bringing in many problems to trajectory-based analysis. Trajectory User Link (TUL) [8] aims to solve this problem, to identify anonymous trajectories and associate them with candidate users. Because the user's behavior information is difficult to fully analyze, TUL is still a challenging problem. In recent years, there have been many works focused on TUL [7,8,13, 20,22,28,29].

Existing TUL approaches are based on location representation. Similar to word embedding [14] in natural language processing and node embedding in graph learning [9], the location representation learns embedding vectors that can indicate semantic similarity between the locations from large historical trajectory corpora. By location representation, the original anonymous trajectory, represented by longitude and latitude, can be converted into a sequence composed of semantic locations, which benefits a better understanding of the whole trajectory information. Existing location representation methods consider sequential information [1,11] and spatial information [5,31]. All those approaches are based on a single semantic location representation, establishing a one-to-one correspondence between locations and their semantic vector representations.

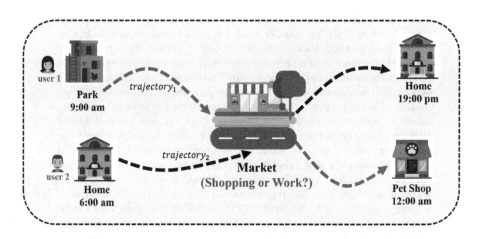

Fig. 1. A demonstration of polysemous location.

Nevertheless, we hold that the semantic information of locations is polysemous. Precisely, the same location unfolds varying effects for different users at different time. To illustrate our motivation more precisely, we set up a simple example. As shown in Fig. 1, there are two trajectories. Suppose the two trajectories belong to two users (In fact, we do not know this message because those trajectories are anonymous). Both user 1 and user 2 have reached a market. We can conjecture from their complete trajectories that they have completely different behaviors on the market. In this example, the exact location (market) generates specific semantic information in different trajectories, which is a universal phenomenon. However, existing TUL methods ignore the polysemy of locations, which would lead to a biased understanding of trajectory patterns and misjudge the candidate user corresponding to this trajectory.

To solve this problem, we propose **DBRLr**, **D**eep **B**ehavior **R**elevant **L**ocation representation, which learns the semantic information of user's behavior on location to represent the polysemous location embedding. To learn this representation, we build a **L**ocation **P**rediction-based **M**ovement **M**odel (LP-based MM) to learn user behavior from a large number of historical trajectories. LP-based MM considers **2C** characteristics of users' behavior: continuity and cyclicity, and employs the combination of the intermediate layer representation as DBRLr. After that, we establish the connection between anonymous trajectory and candidate users based on a deep recurrent neural network called Linker. The Linker takes the representation of the anonymous trajectory as input and outputs the probability of each candidate user. We conduct experiments on three real-world datasets, and the result shows that our method performs beyond existing approaches. Our contributions are as follow:

- We propose **DBRLr**, a polysemous location representation method. DBRLr utilizes the behavior characteristics of the user's visiting location to represent the location. As a result, the same location can reflect different semantic information.
- We establish LP-based **MM** to depict the user movement pattern on trajectories. LP-based MM learns from numerous historical trajectory corpora and embeds locations dynamically based on complete trajectory information.
- We employ DBRLr for **TUL** problem and conduct experiments on three real-world datasets. Extra trajectories of history is used for training MM. The results show that employing MM for location representation improves TUL performance. Our source codes are publicly available at https://github. com/taos123/TUL_by_DBRLr.

2   Related Works

2.1   Trajectory Classification

TUL is one kind of trajectory classification problem if we regard one user as one category of trajectory classification. Trajectory classification has been widely studied and applied [23]. We mainly introduce two kinds of trajectory classification methods. One is based on trajectory similarity measure metrics such as *Euclidean Distance*, *Hausdorff Distance*, *Dynamic Time Warping Distance (DTW)*, *Longest Common Subsequence (LCSS) Distance*, and *Fréchet Distance*, or trajectory feature extraction method of trajectory [6]. Another is based on deep learning, which employs deep neural networks to learn a trajectory representation [3,8,10,12]. These methods mostly use recurrent neural networks to extract trajectory characteristics and classify trajectories, which can effectively deal with the problem of the uncertain length of the trajectory. In recent years, there are some pertinent trajectory classification approaches for TUL problems, such as TULER [8], TULVAE [29], TULAR [22] and AdattTUL [7]. Reference [8] is the first work to put forward the TUL problem formally and employs location embedding, in which locations represented by longitude and latitude are learned into semantic information representation. Variational autoencoder is employed into TUL in [29] to improve TUL performance. Adversarial neural networks are employed [7] for generating more training trajectories. Both [13] and [22] consider the influence of different locations by attention mechanism.

2.2 Location Representation

A graph-based embedding model is proposed in [24], jointly capturing the sequential effect, geographical influence, temporal cyclic effect, and semantic effect in a unified way. Reference [27] proposes a model to capture the semantics information of place types. The model is based on Word2Vec, which augments the spatial contexts of POI types by using distance and information-theoretic approaches to generate embedding. Reference [5] proposes POI2Vec, a latent representation model, which uses the geographical influence of POIs to learn latent representations. Reference [16] introduces a model called Deepcity, which is based on deep learning. It is used to learn features for user and location profiling. Reference [4] employs SkipGram [15] algorithm to predict a location's context given the location itself, which contains the representation of the location. Reference [30] learns latent representations of places by directly model movements between places with large-scale movement data. Reference [26] proposes an unsupervised machine language translation method to translate location representations across different cities. Reference [21] proposes fine-grained location embedding by leveraging hierarchical spatial information according to the local density of observed data points to overcome the data sparsity problem. Existing work considers location embedding from different perspectives of structural information but does not consider the diversity of location semantics. To the best of our knowledge, it is the first time to propose polysemous location representation in trajectory data mining domains.

3   Methodology

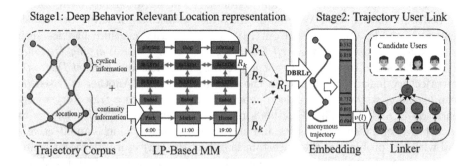

Fig. 2. An overview structure of trajectory user link with deep behavior relevant location representation.

We introduce the technical details of DBRLr for TUL in this section. Our method consists of two main stages: 1) Deep behavior relevant location representation. In this part, we build a LP-based MM based on continuity and cyclicity of user' movement and train it, taking advantage of a lot of history trajectory corpora. We extract the intermediate layer representation of MM as DBRLr. 2) Trajectory-User Link. In this stage, the anonymous trajectory is embedded into semantic vectors by DBRLr. A two-layer biLSTM and a simple classifier are used to capture characteristics and link the anonymous trajectories to candidate users. Figure 2 presents an overview of our method.

3.1   Preliminary

We will present the mathematical notations and problem statement first. Let $U = \{u_1, u_2, \cdots, u_M\}$ be user set. Let $L = \{l_1, l_2, \cdots, l_K\}$ be all of locations that all users visited. For each user, the user's movement through space produces a sequence of locations, which is denoted by $T = \{l_1, l_2, \cdots, l_i, \cdots, l_N\}$, where $l_i \in L$ and N is the length of trajectory. Non-anonymous trajectories will contain user's identity information. For an anonymous trajectory, we need to infer the corresponding user identify.

Given an anonymous trajectory T, the trajectory-user link aims to indicate the most likely user from U that is most likely to produce the anonymous trajectory. TUL learns a mapping function that links trajectories to users: $T \rightarrow U$ [29].

3.2   Empirical Analysis

We first intuitively investigate two problems: Is users' behavior learnable? How to learn a user's behavior in a given location? Due to the restriction of trajectory acquisition technology and privacy protection, users' behaviors in locations are not captured in almost all public trajectory datasets. Though, the user's behaviors are potentially indicated in the user's trajectories. As it is shown in Fig. 3, we summarize users' movement adhering to **2C** principles: **C**ontinuity and **C**yclicity. Continuity means that where a user stays is affected by his previous locations and where he plans to go [18,31]. Cyclicity means that users will be influenced by regular habits when they decide where to go [19,24]. Through the above two characteristics, we can obtain the similarity of users' behavior in locations. Therefore, if we can establish a model to comprehensively consider the two characteristics based on an extensive trajectory data corpora, we can learn the similar relationship between users' behaviors and get the behavior representation of users on a location.

3.3   Location Prediction Based Movement Model

Inspired by the language model [17], we propose a location prediction-based movement model to learn the users' movement. We first describe the parts of the

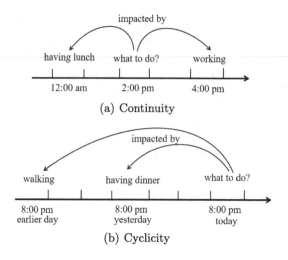

Fig. 3. 2C users' movement principles.

continuity. We can intuitively discover that a user's current location is affected by his historical location. For a trajectory T in historical trajectory corpora, T contains N locations, $(l_1, l_2, \cdots, l_N)$. A forward movement model computes the probability of the trajectory by modeling the probability of location l_k given the preceding locations (l_1, l_2, l_{k-1}). The probability is shown as Eq. 1.

$$p(l_1, l_2, \cdots, l_N) = \prod_{k=1}^{N} p(l_k | l_1, l_2, \cdots, l_{k-1}) \tag{1}$$

At the same time, we also consider the users' visiting backward model, which means the user's current location is affected by the location he will visit later. A backward movement model is similar to a forward movement model, except it runs over the sequence in reverse, predicting the current locations given the future locations. The backward model is shown in Eq. 2.

$$p(l_1, l_2, \cdots, l_N) = \prod_{k=1}^{N} p(l_k | l_{k+1}, l_{k+2}, \cdots, l_N) \tag{2}$$

We employ the location embedding by reference [8] then pass it through L layers of forward LSTMs. At each location l_k, each LSTM layer outputs a context-dependent representation $\overrightarrow{h}_{k,j}^{MM}$ where j denotes the LSTM layer and $j = 1, 2, \cdots, L$. Let $\overrightarrow{h}_{k,L}^{MM}$ denote the top layer LSTM output, which is used to predict p_{k+1} with a softmax layer which is shown in Eq. 3.

$$p(l_k) = \frac{\overrightarrow{h}_{k,L}^{MM}}{\sum_{i=1}^{N} \overrightarrow{h}_{i,L}^{MM}} \tag{3}$$

It can be implemented in an simple way to a forward model, with each backward LSTM layer j in a L layers deep model producing representation $\overleftarrow{h}_{k,j}^{MM}$ of l_k given $(l_{k+1}, l_{k+2}, \cdots, l_N)$.

We employ a biLSTM combining both forward and backward models as continuity parts. We maximize the log likelihood of the forward and backward model by Eq. 4.

$$\sum_{k=1}^{N}(\log p(l_k, \cdots, l_{k+1}; \Theta_x, \overrightarrow{\Theta}_{LSTM}, \Theta_s)$$

$$+ \log p(l_{k+1}, \cdots, l_n; \Theta_x, \overleftarrow{\Theta}_{LSTM}, \Theta_s)) \tag{4}$$

We consider the cyclicity of movement from global perspective. For all trajectory $\{T_1^k, T_2^k, \ldots, T_n^k\}$ which contain location l_k, we extract locations $\{l_n^{T_j^k}\}$ in the same period with l_k, where $l_j^{T_i^k} \in T_i^k$ and $l_j^{T_i^k}$ is visited in same periods with l_k. Similarly, we need to establish the optimal location probability under this set of periodic correlation sets. The probability is shown as Eq. 5.

$$p(l_1, l_2, \cdots, l_n) = \prod_{k=1}^{N} \prod_{j=1}^{n_k} p(l_k | l_j^{T_n^k}) \tag{5}$$

The cyclicity can also implemented by the same biLSTM. The maximizes the log likelihood of the cyclicity of movement model is shown as formulation 6:

$$\sum_{k=1}^{N}(\log p(l_k; \Theta_x, \widetilde{\Theta}_{LSTM}, \Theta_s) \tag{6}$$

The LP-based MM, considering both continuity and cyclicity, can be trained under jointly maximized log likelihood with formulation 4 and formulation 6.

3.4 Deep Behavior Relevant Location Representations

After training LP-based MM, we use the combination of the intermediate layer representation in biLSTM. For each location l_k, a L-layer LP-based MM computes a set of 2L+1 representation.

$$R_k = \left\{ X_k^{MM}, \overrightarrow{h}_{k,j}^{MM}, \overleftarrow{h}_{k,j}^{MM}, \widetilde{h}_{k,j}^{MM} | j = 1, \ldots, L \right\}$$

$$= \{ h_{k,j}^{MM} | j = 0, \ldots, L \} \tag{7}$$

Where $h_{k,0}^{MM}$ is the first location layer and $h_{k,j}^{MM} = \left[\overrightarrow{h}_{k,j}^{MM}; \overleftarrow{h}_{k,j}^{MM}; \widetilde{h}_{k,j}^{MM} \right]$, for each biLSTM layer.

DBRLr collapses all layers in R into a single vector, $DBRLr_k = E(R_k; \Theta_\varepsilon)$. In the simplest case, DBRLr just selects the top layer.

Given a trajectory $T = (l_1, l_2, \cdots, l_N)$, the location l_k in T represented by DBRLr is $v(l_k^T)$.

3.5   Trajectory-User Linker

For the anonymous trajectory datasets to be identified, we first input each anonymous trajectory into the behavior representation layer to obtain the trajectory representation with DBRLr, which are represented as $\left(v(l_1^T), v(l_2^T), \cdots, v(l_N^T)\right)$.

To process the long-term variable-length location sequence, we employ a biLSTM to control input and output of location embedding. For the input trajectory $Tra_j = \{l_1, l_2, \cdots, l_N\}$, let $\{h_1, h_2, \cdots, h_N\}$ denote the output status of biLSTM as $h_t = biLSTM(v(l_t))$. In this way, we can get the every time of the biLSTM outputs. We use a weighted average formula to fuse every time information by Eq. 8.

$$v\left(Tra_j\right) = \sum_{t=1}^{N} a_t h_t \tag{8}$$

where a_t is the weight of location on t, reflecting the influence of this l_t to the whole trajectory. We calculate a_t by the following Eq. 9.

$$a_t = \tanh\left(W_1 h_t + W_2 \overline{h}_s\right) \tag{9}$$

where W_1 and W_2 are parameters to learn and $\overline{h}_s$ is the mean value of $\{h_t | t = 1, 2, \cdots, N\}$. Then we employ fully connection layer with parameters $W_3 \in \mathbb{R}^{N \times M}$ and $b \in \mathbb{R}^{1 \times M}$ to mapping trajectory information to the dimensions of the candidate user set.

$$v\left(Tra_j\right)^{user} = v\left(Tra_j\right) * W_3 + b \tag{10}$$

Let $p\left(u_i | Tra_j\right)$ denote the probability that trajectory Tra_j belongs to user u_i, which is calculated as Eq. 11. The softmax function converts logits into probabilities.

$$p\left(u_i | Tra_j\right) = \frac{v\left(Tra_j\right)^{user}}{\sum_{k=1}^{M} v\left(Tra_j\right)^{user}} \tag{11}$$

To measure the distance from the truth values, we compute softmax cross entropy between logits and labels as Eq. 12. In the training process, our objective is to minimize the loss function.

$$\mathcal{L} = \frac{1}{N} \sum_{j=1}^{N} \left(v\left(u_i\right) - \sum_{i=1}^{M} \log\left(p\left(u_i | Tra_j\right)\right)\right) \tag{12}$$

Finally, given an anonymous trajectory Tra, the corresponding predicted user u_i is calculated.

$$\underset{1 < i < M}{\arg\max}\, p\left(u_i | Tra_j\right) = \{u_i \in U : p\left(u_i | Tra_j\right)\} \tag{13}$$

4   Experiments

In this section, we conduct experiments to evaluate the accuracy of the proposed DBRLr by answering the following three key research questions.

- **Q1**: Does DBRLr outperform the existing TUL baselines in real-world datasets?
- **Q2**: Does the LP-based MM improve the performance of DBRLr? How does DBRLr perform with single-direction movement models?
- **Q3**: Does DBRLr distinguish between polysemous locations? How does the polysemous location representation improve the performance of TUL?

4.1   Datasets

We conduct our experiments on three benchmark datasets [2]: **Gowalla**[1], **Brightkite**[2] and **Foursquare**[3], which are publicly available. Both of them are collected from location-based social networking websites where users share their locations by checking in. The data recorded information such as [**user id, check-in time, longitude, latitude, location id**]. We randomly select a set of users in Gowalla, Brightkite and Foursquare, which are the same number as [29]. The statistics of datasets are summarized in Table 1.

Table 1. Datasets description and statistics

Datasets	$|U|$	$|T|$	$|C|$	$|Ave|$
Gowalla	201	19968	1958	99.34
Brightkite	92	19904	471	216.34
Foursquare	300	13281	162	44.27

$|U|$ is the number of users in the datasets. $|T|$ denotes the number of trajectories sets. As we can see $|T| \gg |U|$. $|C|$ is the number of check-in locations. $|Ave|$ represents the average number of check-in locations per trajectory, which is calculated by dividing $|T|$ by $|U|$.

4.2   Baseline Algorithms

We compare our method with both classical trajectory classification methods and TUL approaches. The following baseline models are evaluated.

- **TULER** [8]. TULER uses the check-in location embedding method to reinforce the check-in location information. TULER employs LSTM, GRU, and their variants as the RNN model, which is called: TULER-L, TULER-G, and Bi-TULER. We employ open-source TULER in github[4].

---

[1] Gowalla: http://snap.stanford.edu/data/loc-Gowalla.html.
[2] Brightkite: http://snap.stanford.edu/data/loc-Brightkite.html.
[3] https://sites.google.com/site/yangdingqi/home.
[4] TULER: https://github.com/gcooq/TUL.

- **TULVAE** [29]. TULVAE learns the human mobility in a neural generative architecture with stochastic latent variables than span hidden states in RNN. We employ open-source TULVAE in github[5].
- **AdattTUL** [7]. AdattTUL is a semi-supervised method, which makes adversarial mobility learning for human trajectory classification, which is an end-to-end framework modeling human moving patterns.
- **TULAR** [13,22]. TULAR considers the influence of different locations and introduces trajectory attention mechanism. TULAR is the state-of-the-art method for TUL. We employ open-source TULAR in github[6].

4.3   Evaluation Metrics

We employ Acc1, Acc5, and macro-F1 as the evaluation metrics, which are the standard metrics of TUL problem [29]. The definitions of those metrics are shown in the following equations.

$$AccK = \frac{\#correctly\ linked\ trajectories\ @K}{\#trajectories} \tag{14}$$

where the *#correctly linked trajectories @K* is the correct users at top K candidates, and *#trajectories* is the total number of anonymous trajectories. In addition, because TUL is a multi-classification task, we also need to consider macro-R, macro-P, and macro-f1. The macro-R is the mean of recall value of every classification, and macro-P is the mean of the precision value of every classification. The macro-f1 is defined as follows.

$$macro - F1 = 2 \times \frac{macro - P \times macro - R}{macro - P + macro - R} \tag{15}$$

4.4   Parameter Setup

The training process of our model includes two stages: DBRLr training and Linker training. In the DBRLr training process, we employ a 2-lay bi-LSTM for realizing LP-based MM with input dimension 256. We slice the original trajectory data at 6-hour intervals. We set at least ten training epochs for movement model training. We introduce the regularization method dropout with a dropout rate of 0.1. In the Linker training process, we set the input dimension of Linker to 256. We set the initial learning rate as 0.001, and after 20 to 30 iterations, we reduce the learning rate by half. We use Adam as the optimizer. What's more, we shuffle the training trajectory data set before the initialization of the model. We'll expose the code later in GitHub for reproduction.

---

[5] TULVAE: https://github.com/AI-World/IJCAI-TULVAE.
[6] TULAR: https://github.com/taos123/TULAR.

4.5   Overall Performance (Q1)

Table 2 exhibits the overall results compared our method with baselines. It can be seen that our method has a significant improvement for TUL. More specifically, on Gowalla and Foursquare, our method is higher than baselines at one percentage point, with over 2% performance improvement in Acc@1. On Brightkite, our method is higher than baselines at nine percentage points, with over 18% performance improvement in Acc@1. This means that we have achieved a recognition accuracy of more than 50%, which will significantly improve TUL availability in practical application scenarios. It can also be seen that our method improves the effect more on Brightkite than on Gowalla and Foursquare.

In has been verified in [29] that the classical trajectory classification method has poor performance for TUL because those classical trajectory classification methods measure the geospatial similarity between trajectories, while TUL needs to find the behavioral similarity between trajectories. There is a slight discrepancy between the results of TULER, TULVAE, AdattTUL, and TULAR. This is because these approaches focus on Linker model improvements to capture trajectory patterns. However, as their input layer, single semantic location representation will cause errors in the semantic information of some locations, resulting in biased recognitions for some trajectories. The improvement of our method benefits from a more accurate understanding of the semantic information of trajectory. DBRLr infers the specific semantic information of each location according to the knowledge learned from the historical trajectory corpora and the context information of the whole trajectory.

4.6   Ablation Experiments (Q2)

In order to explore how LP-based MM impacts the performance of DBRLr, we design five contrast movement models to retrain DBRLr and compare those performances on TUL. The model variants and descriptions are shown in Table 3. We train these above four MMs in the same history trajectory corpora and the same experimental environment. Then we extract location representations from the trained movement model. Applying the above different location representation in TUL with Linker in Sect. 3.4, the performance comparison is shown in Fig. 4.

Figure 4 shows the comparison between different contrast models. We can see that DBRLr with a 2-layer bi-direction movement model achieves the best performance. The location embedding effect without the MM model is generally lower than that with the MM model. Compared with the single-direction movement model, and the bi-direction model can better capture global information of trajectory, which significantly improves the accuracy of TUL. It can also be observed that DBRLr with two-layer performance is better than one layer, indicating that high-level exploring might obtain more practical information from human movement. Nevertheless, this gap is not very obvious. We think the higher layer can capture the information for trajectory. Furthermore, a two-direction of movement model can describe movement patterns with advantages.

Table 2. Performance comparison on Gowalla and Brightkite

Method	Metric								
	Acc@1	Acc@5	macro-F1	Acc@1	Acc@5	Macro-F1	Acc@1	Acc@5	Macro-F1
	Gowalla			Brightkite			Foursquare		
TULER-L	0.4179	0.5789	0.3243	0.4124	0.5688	0.3007	0.5122	0.5911	0.4566
TULER-G	0.4261	0.5795	0.3391	0.4085	0.5731	0.2864	0.5091	0.5887	0.4560
Bi-TULER	0.4267	0.5954	0.3215	0.4195	0.5758	0.3190	0.5388	0.6141	0.4873
TULVAE	0.4435	0.6446	0.3621	0.4540	0.6239	0.3541	0.5428	0.6169	49.22
AdattTUL-G	0.4692	0.6364	0.3726	0.4838	0.6496	0.4269	0.5783	0.6463	0.5364
AdattTUL-L	0.4761	0.6464	0.3774	0.4891	0.6544	0.4335	0.5812	0.6470	0.5385
TULAR-L	0.3265	0.4613	0.2702	0.3012	0.3913	0.2302	0.5881	0.6474	0.5293
TULAR-G	0.3786	0.4928	0.3408	0.4050	0.5338	0.3998	0.5853	0.6521	0.5301
TULAR-B	0.4125	0.5550	0.3432	0.4207	0.6146	0.3659	0.5807	0.6533	0.5358
Our Method	**0.4875**	**0.7227**	**0.4055**	**0.5798**	**0.7750**	**0.5511**	**0.6072**	**0.6933**	**0.5541**

On Gowalla, our method higher than existing methods than 2.39%, 11.80%, 7.45% in terms of Acc@1, Acc@5, F1. On Brightkite, our method higher than existing methods than 18.54%, 18.42%, 27.12% in terms of Acc@1, Acc@5, F1. On Foursquare, our method higher than existing methods than 4.56%, 6.12%, 3.41% in terms of Acc@1, Acc@5, F1.

Table 3. Ablation experiment model settings

Model variants	Description
F-MM	Location prediction-based movement model with forward prediction only
B-MM	Location prediction-based movement model with back prediction only
Bi-MM	Location prediction-based movement model with bi-direction prediction
Bi-MM-2L	Location prediction-based movement model with 2-layer bi-direction prediction
Non-MM	Location embedding [8] without MM

4.7 Case Study (Q3)

We select some specific trajectories from real-world datasets for analysis and visualization. First, we verify whether DBRLr can distinguish polysemous locations. We select a location[7] that is embedded to multiple vectors in different directions under DBRLr, and two trajectories contained the location mentioned above. In the single semantic location representation, these two trajectories are connected to unmatched users. While under DBRLr, T_1 and T_2 are respectively linked to the correct corresponding candidate users. We visualize the two trajectories and the above location l_3 shown in Fig. 5. By looking up in Google maps[8], we find that l_3 may be a bank, and T_1 passes through l_3 between two restaurants.

---

[7] Location ID is efa6e44dfa0145249be273ecd84a97f534b04920 in Brightkite.
[8] https://www.google.com/maps.

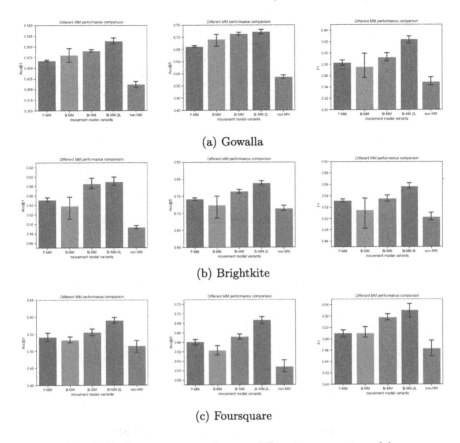

(a) Gowalla

(b) Brightkite

(c) Foursquare

Fig. 4. Performance comparison on different movement models.

We speculate that the user is handling business at l_3. On the contrary, T_2 passes another band and stays in l_3 for a long time, and we speculate that the user is working at l_3.

Secondly, we verify whether DBRLr contributes to the improvement of TUL recognition accuracy. We input four users' trajectories into the Linker under different location representations and exact the output layer for visualization, which is shown in Fig. 6. We notice that under DBRLr, the trajectories of the same users are more compact and clustered. However, in a single representation, the clusters are more scattered. This indicates that outliers decreases, while the clustered points increases under DBRLr, which is a practical demonstration that DBRLr can overcome misunderstandings of trajectory semantic information when linking anonymous trajectories to candidate users.

Fig. 5. Visualization of polysemous location in Brightkite.

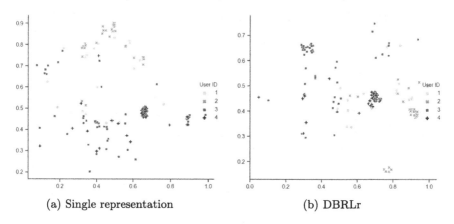

(a) Single representation   (b) DBRLr

Fig. 6. Trajectory embedding visualization.

5 Conclusions

In this paper, we improved TUL performance by using a polysemous location representation model called DBRLr. To learn this representation, we build a LP-based MM and train it historical trajectory corpora. Compared with the previous work, the trajectory with DBRLr can better describe the behavior and movement characteristics of a user. The experiment results confirm this view. We sincerely believe that DBRLr can be used not only in TUL but also in other trajectory data mining fields [25], such as next visited location prediction or location-based recommendation. Our follow-up work will examine our ideas on more datasets and more trajectory analysis tasks.

Acknowledgements. This work is partially supported by NSFC No. 61902376.

References

1. Chang, B., Park, Y., Park, D., Kim, S., Kang, J.: Content-aware hierarchical point-of-interest embedding model for successive poi recommendation. In: 27th IJCAI, vol. 2018 (2018)
2. Cho, E., Myers, S.A., Leskovec, J.: Friendship and mobility: user movement in location-based social networks. In: Proceedings of the 17th ACM SIGKDD International Conference on Knowledge Discovery and Data Mining, pp. 1082–1090 (2011)
3. Endo, Y., Toda, H., Nishida, K., Ikedo, J.: Classifying spatial trajectories using representation learning. Int. J. Data Sci. Anal. **2**(3), 107–117 (2016). https://doi.org/10.1007/s41060-016-0014-1
4. Feng, J., et al.: DeepMove: predicting human mobility with attentional recurrent networks. In: Proceedings of the 2018 World Wide Web Conference, pp. 1459–1468 (2018)
5. Feng, S., Cong, G., An, B., Chee, Y.M.: POI2Vec: geographical latent representation for predicting future visitors. In: Proceedings of the AAAI Conference on Artificial Intelligence, vol. 31 (2017)
6. Ferrero, C.A., Alvares, L.O., Zalewski, W., Bogorny, V.: MOVELETS: exploring relevant subtrajectories for robust trajectory classification. In: Proceedings of the 33rd Annual ACM Symposium on Applied Computing, pp. 849–856 (2018)
7. Gao, Q., Zhang, F., Yao, F., Li, A., Mei, L., Zhou, F.: Adversarial mobility learning for human trajectory classification. IEEE Access **8**, 20563–20576 (2020)
8. Gao, Q., Zhou, F., Zhang, K., Trajcevski, G., Luo, X., Zhang, F.: Identifying human mobility via trajectory embeddings. In: IJCAI, vol. 17, pp. 1689–1695 (2017)
9. Grover, A., Leskovec, J.: node2vec: scalable feature learning for networks. In: Proceedings of the 22nd ACM SIGKDD International Conference on Knowledge Discovery and Data Mining, pp. 855–864 (2016)
10. Jiang, X., de Souza, E.N., Pesaranghader, A., Hu, B., Silver, D.L., Matwin, S.: TrajectoryNet: an embedded GPS trajectory representation for point-based classification using recurrent neural networks. arXiv preprint arXiv:1705.02636 (2017)
11. Liu, X., Liu, Y., Li, X.: Exploring the context of locations for personalized location recommendations. In: IJCAI, pp. 1188–1194 (2016)
12. Petry, L.M., Da Silva, C.L., Esuli, A., Renso, C., Bogorny, V.: MARC: a robust method for multiple-aspect trajectory classification via space, time, and semantic embeddings. Int. J. Geog. Inf. Sci. **34**(7), 1428–1450 (2020)
13. Miao, C., Wang, J., Yu, H., Zhang, W., Qi, Y.: Trajectory-user linking with attentive recurrent network. In: Proceedings of the 19th International Conference on Autonomous Agents and MultiAgent Systems, pp. 878–886 (2020)
14. Mikolov, T., Sutskever, I., Chen, K., Corrado, G.S., Dean, J.: Distributed representations of words and phrases and their compositionality. In: Advances in Neural Information Processing Systems, pp. 3111–3119 (2013)
15. Mikolov, T., Yih, W., Zweig, G.: Linguistic regularities in continuous space word representations. In: Proceedings of the 2013 Conference of the North American Chapter of the Association for Computational Linguistics: Human Language Technologies, pp. 746–751 (2013)
16. Pang, J., Zhang, Y.: DeepCity: a feature learning framework for mining location check-ins. In: Proceedings of the International AAAI Conference on Web and Social Media, vol. 11 (2017)

17. Peters, M.E., et al.: Deep contextualized word representations. arXiv preprint arXiv:1802.05365 (2018)
18. Qian, T., Wang, F., Xu, Y., Jiang, Yu., Sun, T., Yu, Y.: CABIN: a novel cooperative attention based location prediction network using internal-external trajectory dependencies. In: Farkaš, I., Masulli, P., Wermter, S. (eds.) ICANN 2020. LNCS, vol. 12397, pp. 521–532. Springer, Cham (2020). https://doi.org/10.1007/978-3-030-61616-8_42
19. Qian, T., Liu, B., Nguyen, Q.V.H., Yin, H.: Spatiotemporal representation learning for translation-based POI recommendation. ACM Trans. Inf. Syst. **37**(2), 1–24 (2019). https://doi.org/10.1145/3295499
20. Seglem, E., Züfle, A., Stutzki, J., Borutta, F., Faerman, E., Schubert, M.: On privacy in spatio-temporal data: user identification using microblog data. In: Gertz, M., et al. (eds.) SSTD 2017. LNCS, vol. 10411, pp. 43–61. Springer, Cham (2017). https://doi.org/10.1007/978-3-319-64367-0_3
21. Shimizu, T., Yabe, T., Tsubouchi, K.: Improving land use classification using human mobility-based hierarchical place embeddings. In: 2021 IEEE International Conference on Pervasive Computing and Communications Workshops and other Affiliated Events (PerCom Workshops), pp. 305–311. IEEE (2021)
22. Sun, T., Xu, Y., Wang, F., Wu, L., Qian, T., Shao, Z.: Trajectory-user link with attention recurrent networks. In: 2020 25th International Conference on Pattern Recognition (ICPR), pp. 4589–4596. IEEE (2021)
23. Wang, S., Cao, J., Yu, P.: Deep learning for spatio-temporal data mining: a survey. IEEE Trans. Knowl. Data Eng. (2020)
24. Xie, M., Yin, H., Wang, H., Xu, F., Chen, W., Wang, S.: Learning graph-based POI embedding for location-based recommendation. In: Proceedings of the 25th ACM International on Conference on Information and Knowledge Management, pp. 15–24 (2016)
25. Xu, Y., Liu, X., Cao, X., et al.: Artificial intelligence: a powerful paradigm for scientific research. Innovation **2**(4), 100179 (2021). https://doi.org/10.1016/j.xinn.2021.100179. https://www.sciencedirect.com/science/article/pii/S2666675821001041
26. Yabe, T., Tsubouchi, K., Shimizu, T., Sekimoto, Y., Ukkusuri, S.V.: City2City: translating place representations across cities. In: Proceedings of the 27th ACM SIGSPATIAL International Conference on Advances in Geographic Information Systems, pp. 412–415 (2019)
27. Yan, B., Janowicz, K., Mai, G., Gao, S.: From ITDL to Place2Vec: reasoning about place type similarity and relatedness by learning embeddings from augmented spatial contexts. In: Proceedings of the 25th ACM SIGSPATIAL International Conference on Advances in Geographic Information Systems, pp. 1–10 (2017)
28. Yu, Y., et al.: TULSN: siamese network for trajectory-user linking. In: 2020 International Joint Conference on Neural Networks (IJCNN), pp. 1–8. IEEE (2020)
29. Zhou, F., Gao, Q., Trajcevski, G., Zhang, K., Zhong, T., Zhang, F.: Trajectory-user linking via variational autoencoder. In: IJCAI, pp. 3212–3218 (2018)
30. Zhou, Y., Huang, Y.: DeepMove: learning place representations through large scale movement data. In: 2018 IEEE International Conference on Big Data (Big Data), pp. 2403–2412. IEEE (2018)
31. Zhu, M., et al.: Location2vec: a situation-aware representation for visual exploration of urban locations. IEEE Trans. Intell. Transp. Syst. **20**(10), 3981–3990 (2019). https://doi.org/10.1109/tits.2019.2901117

Heterogeneous Federated Learning via Grouped Sequential-to-Parallel Training

Shenglai Zeng[1], Zonghang Li[1,3], Hongfang Yu[1(✉)], Yihong He[1], Zenglin Xu[2], Dusit Niyato[3], and Han Yu[3]

[1] School of Information and Communication Engineering, University of Electronic Science and Technology of China, Chengdu, China
heyh.uestc@gmail.com, yuhf@uestc.edu.cn
[2] School of Computer Science and Technology, Harbin Institute of Technology, Shenzhen, China
xuzenglin@hit.edu.cn
[3] School of Computer Science and Engineering, Nanyang Technological University, Singapore, Singapore
{dniyato,han.yu}@ntu.edu.sg

Abstract. Federated learning (FL) is a rapidly growing privacy preserving collaborative machine learning paradigm. In practical FL applications, local data from each data silo reflect local usage patterns. Therefore, there exists heterogeneity of data distributions among data owners (a.k.a. FL clients). If not handled properly, this can lead to model performance degradation. This challenge has inspired the research field of heterogeneous federated learning, which currently remains open. In this paper, we propose a data heterogeneity-robust FL approach, FEDGSP, to address this challenge by leveraging on a novel concept of dynamic Sequential-to-Parallel (STP) collaborative training. FEDGSP assigns FL clients to homogeneous groups to minimize the overall distribution divergence among groups, and increases the degree of parallelism by reassigning more groups in each round. It is also incorporated with a novel Inter-Cluster Grouping (ICG) algorithm to assist in group assignment, which uses the centroid equivalence theorem to simplify the NP-hard grouping problem to make it solvable. Extensive experiments have been conducted on the non-i.i.d. FEMNIST dataset. The results show that FEDGSP improves the accuracy by 3.7% on average compared with seven state-of-the-art approaches, and reduces the training time and communication overhead by more than 90%.

Keywords: Federated learning · Distributed data mining · Heterogeneous data · Clustering-based learning

---

S. Zeng and Z. Li—Equal contributions.

A. Bhattacharya et al. (Eds.): DASFAA 2022, LNCS 13246, pp. 455–471, 2022.
https://doi.org/10.1007/978-3-031-00126-0_34

1   Introduction

Federated learning (FL) [1], as a privacy-preserving collaborative paradigm for training machine learning (ML) models with data scattered across a large number of data owners, has attracted increasing attention from both academia and industry. Under FL, data owners (a.k.a. FL clients) submit their local ML models to the FL server for aggregation, while local data remain private. FL has been applied in fields which are highly sensitive to data privacy, including healthcare [2], manufacturing [3] and next generation communication networks [4]. In practical applications, FL clients' local data distributions can be highly heterogeneous due to diverse usage patterns. This problem is referred to as the non-independent and identically distributed (non-i.i.d.) data challenge, which negatively affects training convergence and the performance of the resulting FL model [5].

Recently, heterogeneous federated learning approaches have been proposed in an attempt to address this challenge. These works try to make class distributions of different FL clients similar to improve the performance of the resulting FL model. In [5–7], FL clients share a small portion of local data to build a common meta dataset to help correct deviations caused by non-i.i.d. data. In [8,9], data augmentation is performed for categories with fewer samples to reduce the skew of local datasets. These methods are vulnerable to privacy attacks as misbehaving FL servers or clients can easily compromise the shared private data and the augmentation process. To align client data distributions without exposing the FL process to privacy risks, we group together heterogeneous FL clients so that each group can be perceived as a homogeneous "client" to participate in FL. This process does not involve any manipulation of private data itself and is therefore more secure.

An intuitive approach to achieve this goal is to assign FL clients to groups with similar overall class distribution, and use collaborative training to coordinate model training within and among groups. However, designing such an approach is not trivial due to the following two challenges. Firstly, assigning FL clients to a specified number of groups of equal group sizes to minimize the data divergence among groups (which can be reduced from the well-known bin packing problem [10]) is an NP-hard problem. Moreover, such group assignment process needs to be performed periodically in a dynamic FL environment, which introduces higher requirements for its effectiveness and execution efficiency. Secondly, even if the data distributions among groups are forced to be homogeneous, the data within each group can still be skewed. Due to the robustness of sequential training mode (STM) to data heterogeneity, some collaborative training approaches (e.g., [9]) adopt STM within a group to train on skewed client data. Then, the typical parallel training mode (PTM) can be applied among homogeneous groups. These methods are promising, but are still limited due to their static properties, which prevents them from adapting to the changing needs of FL at different stages. In FL, STM should be emphasized in the early stage to achieve a rapid increase in accuracy in the presence of non-i.i.d. data, while PTM should be emphasized in the later stage to promote convergence. In the static mode, the above parallelism degree must be carefully designed to realize a proper trade-off between sensitivity to heterogeneous data of PTM and overfitting of STM. Otherwise, the FL model performance may suffer.

To address these challenges, this paper proposes a new concept of dynamic collaborative Sequential-to-Parallel (STP) training to improve FL model performance in the presence of non-i.i.d. data. The core idea of STP is to force STM to be gradually transformed into PTM as FL model training progresses. In this way, STP can better refine unbiased model knowledge in the early stage, and promote convergence while avoiding overfitting in the later stage. To support the proposed STP, we propose a Federated Grouped Sequential-to-Parallel (FEDGSP) training framework. FEDGSP allows reassignment of FL clients into more groups in each training round, and introduces group managers to manage the dynamically growing number of groups. It also coordinates model training and transmission within and among groups. In addition, we propose a novel Inter-Cluster Grouping (ICG) method to assign FL clients to a pre-specified number of groups, which uses the centroid equivalence theorem to simplify the original NP-hard grouping problem into a solvable constrained clustering problem with equal group size constraint. ICG can find an effective solution with high efficiency (with a time complexity of $\mathcal{O}(\frac{K^6 \mathcal{F}_T}{M^2} \log Kd)$). We evaluate FEDGSP on the most widely adopted non-i.i.d. benchmark dataset FEMNIST [11] and compare it with seven state-of-the-art approaches including FedProx [12], FedMMD [13], FedFusion [14], IDA [15], FedAdam, FedAdagrad and FedYogi [16]. The results show that FEDGSP improves model accuracy by 3.7% on average, and reduces training time and communication overhead by more than 90%. To the best of our knowledge, FEDGSP is the first dynamic collaborative training approach for FL.

2   Related Work

Existing heterogeneous FL solutions can be divided into three main categories: 1) data augmentation, 2) clustering-based learning, and 3) adaptive optimization.

Data Augmentation: Zhao et al. [5] proved that the FL model accuracy degradation due to heterogeneous local data can be quantified by the earth move distance (EMD) between the client and global data distributions. This result motivates some research works to balance the sample size of each class through data augmentation. Zhao et al. [5] proposed to build a globally shared dataset to expand client data. Jeong et al. [8] used the conditional generative network to generate new samples for categories with fewer samples. Similarly, Duan et al. [9] used augmentation techniques such as random cropping and rotation to expand client data. These methods are effective in improving the FL model accuracy by reducing data skew. However, they involve modifying clients' local data, which can lead to serious privacy risks.

Clustering-Based Learning: Another promising way to reduce data heterogeneity is through clustering-based FL. Sattler et al. [17] groups FL clients with similar class distributions into one cluster, so that FL clients with dissimilar data distributions do not interfere with each other. This method works well in personalized FL [18] where FL is perform within each cluster and an FL model is produced for each cluster. However, it is not the same as our goal which is to train one shared FL model that can be generalized to all FL clients. Duan et al. [9] makes the KullbackLeibler divergence of class distributions similar among clusters, and proposed a greedy best-fit strategy to assign FL clients.

Adaptive Optimization: Other research explores adaptive methods to better merge and optimize client- and server-side models. On the client side, Li et al. [12] added a proximal penalty term to the local loss function to constrain the local model to be closer to the global model. Yao et al. [13] adopted a two-stream framework and used transfer learning to transfer knowledge from the global model to the local model. A feature fusion method has been further proposed to better merge the features of local and global models [14]. On the server side, Yeganeh et al. [15] weighed less out-of-distribution models based on inverse distance coefficients during aggregation. Instead, Reddi et al. [16] focused on server-side optimization and introduced three advanced adaptive optimizers (Adagrad, Adam and Yogi) to obtain FedAdagrad, FedAdam and FedYogi, respectively. These methods perform well in improving FL model convergence.

Solutions based on data augmentation are at risky due to potential data leakage, while solutions based on adaptive optimization do not solve the problem of class distribution divergence causing FL model performance to degrade. FEDGSP focuses on clustering-based learning. Different from existing research, it takes a novel approach of dynamic collaborative training, which allows dynamic scheduling and reassignment of clients into groups according to the changing needs of FL.

3   Federated Grouped Sequential-to-Parallel Learning

In this section, we first describe the concept and design of the STP approach. Then, we present the FEDGSP framework which is used to support STP. Finally, we mathematically formulate the group assignment problem in STP, and present our practical solution ICG.

3.1   STP: The Sequential-to-Parallel Training Mode

Under our grouped FL setting, FL clients are grouped such that clients in the same group have heterogeneous data but the overall data distributions among the groups are homogeneous. Due to the difference in data heterogeneity, the training modes within and among groups are designed separately. We refer to this jointly designed FL training mode as the "collaborative training mode".

Intuitively, the homogeneous groups can be trained in a simple parallel mode PTM because the heterogeneity of their data has been eliminated by client grouping. Instead, for FL clients in the same group whose local data are still skewed, the sequential mode STM can be useful. In STM, FL clients train the model in a sequential manner. A FL client receives the model from its predecessor client and delivers the local trained model to its successor client to continue training. In the special case of training with only one local epoch (i.e., $e = 1$), STM is equivalent to centralized SGD, which gives it robustness against data heterogeneity.

This naive collaborative training mode is static and has limitations. Therefore, we extend it to propose a more dynamic approach STP. As shown in Fig. 1, STP reassigns FL clients into $f(r)$ groups and shuffles their order in each round r, where f is a pre-specified group number growth function, with the goal to dynamically adjust the degree of parallelism. Then, STP can be smoothly transformed from (full) sequential mode to (full) parallel mode. This design can prevent catastrophic forgetting caused by the long "chain of clients" that causes the FL model to forget the data of previous clients and overfit the data of subsequent clients, and can also prevent the FL model from learning interfering information such as the order of clients. Moreover, the growing number of groups improves the parallelism efficiency, which promotes convergence and speeds up training when the global FL model is close to convergence.

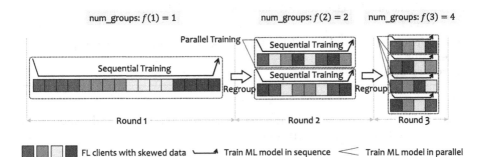

Fig. 1. An example of STP. The ML model in each group is trained in sequence, while ML models among groups are trained in parallel. In each round r, the group number grows according to function f, and FL clients are regrouped and shuffled.

The pseudo code of STP is given in Algorithm 1. In round r, STP divides all FL clients into $f(r)$ groups using the ICG grouping algorithm (Line 3), which will be described in Sect. 3.3. Due to the similarity of data among groups, each group can independently represent the global distribution, so only a small proportion of κ groups are required to participate in each round of training (Line 5). The first FL client in each group pulls the global model from the FL server (Line 7), and trains its local model using mini-batch SGD for one epoch (Line 9). The trained local model is then delivered to the next FL client to continue training (Line 10), until the last FL client is reached. The last FL client in each group sends the trained model to the FL server (Line 12). Models from all groups are aggregated to update the global FL model (Line 14). The above steps repeat until the maximum training round R is reached. Finally, the well-trained global FL model is obtained (Line 16).

The choice of the growth function for the number of groups, f, is critical for the performance of STP. We give three representative growth functions, including

---

Algorithm 1. ⬛ **Sequential-To-Parallel** (main)

Input: All FL clients $\mathcal{C}$, the total number of FL clients K, the maximum training rounds R, the group number growth function f, the group sampling rate κ.
Output: The well-trained global FL model ω^R_{global}.

1: Initialize the global FL model ω^0_{global};
2: **for** each round $r = 1, \cdots, R$ **do**
3:     Reassign all FL clients $\mathcal{C}$ to $f(r)$ groups to obtain $\mathcal{G}$,
4:         $\mathcal{G} \leftarrow$ ⬛ **Inter-Cluster-Grouping** $(\mathcal{C}, K, f, r)$;
5:     Randomly sample a subset of groups $\tilde{\mathcal{G}} \subset \mathcal{G}$ with proportion κ;
6:     **for** each group $\mathcal{G}_m$ in $\tilde{\mathcal{G}}$ in parallel **do**
7:         The first FL client $\mathcal{C}^1_m$ in $\mathcal{G}_m$ initializes $\omega^1_m \leftarrow \omega^{r-1}_{\text{global}}$;
8:         **for** each FL client $\mathcal{C}^k_m$ in $\mathcal{G}_m$ in sequence **do**
9:             Train ω^k_m on local data $\mathcal{D}^k_m$ using mini-batch SGD for one epoch;
10:             Send the trained $\omega^{k+1}_m \leftarrow \omega^k_m$ to the next FL client $\mathcal{C}^{k+1}_m$;
11:         **end for**
12:         The last FL client $\mathcal{C}^{K/f(r)}_m$ in $\mathcal{G}_m$ uploads $\omega^{K/f(r)}_m$;
13:     **end for**
14:     Update the global FL model using the aggregation $\omega^r_{\text{global}} \leftarrow \frac{\sum_{\forall \mathcal{G}_m \in \tilde{\mathcal{G}}} (\omega^{K/f(r)}_m)}{f(r)}$;
15: **end for**
16: **return** ω^R_{global};

---

linear (smooth grow), logarithmic (fast first and slow later), and exponential (slow first and fast later) growth functions:

$$\text{Linear Growth Function}: \quad f(r) = \beta \lfloor \alpha(r-1)+1 \rfloor, \tag{1}$$

$$\text{Log Growth Function}: \quad f(r) = \beta \lfloor \alpha \ln r + 1 \rfloor, \tag{2}$$

$$\text{Exp Growth Function}: \quad f(r) = \beta \lfloor (1+\alpha)^{r-1} \rfloor, \tag{3}$$

where the real number coefficient α controls the growth rate, and the integer coefficient β controls the initial number of groups and the growth span. We recommend to initialize α, β to a moderate value and explore the best setting in an empirical manner.

3.2   FEDGSP: The Grouped FL Framework to Enable STP

In this section, we describe the FEDGSP framework that enables dynamic STP. FEDGSP is generally a grouped FL framework that supports dynamic group management, as shown in Fig. 2. The basic components include a top server (which acts as an FL server and performs functions related to group assignment) and a large number of FL clients. FL clients can be smart devices with certain available computing and communication capabilities, such as smart phones, laptops, mobile robots and drones. They collect data from the surrounding environment and use the data to train local ML models.

In addition, FEDGSP creates group managers to facilitate the management of the growing number of groups in STP. The group managers can be virtual

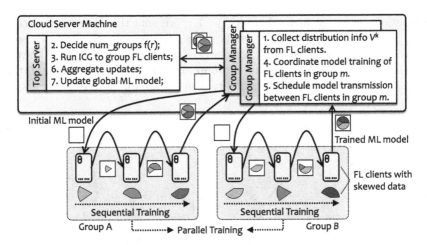

Fig. 2. An overview of the FEDGSP framework.

function nodes deployed in the same machine as the top server. Whenever a new group is built, a new group manager is created to assist the top server to manage this group by performing the following tasks:

1. Collect distribution information. The group manager needs to collect class distributions of FL clients and report them to the top server. These meta information will be used to assign FL clients to $f(r)$ groups via ICG.

2. Coordinate model training. The group manager needs to coordinate the sequential training of FL clients in its group, as well as the parallel training with other groups, according to the rules of STP. Specifically, it needs to shuffle the order of clients and report resulting model to the top server for aggregation.

3. Schedule model transmission. In applications such as Industrial IoT systems, wireless devices can directly communicate with each other through wireless sensor networks (WSNs). However, this cannot be realized in most scenarios. Therefore, the group manager needs to act as a communication relay to schedule the transmission of ML models from one client to another.

3.3   ICG: The Inter-cluster Grouping Algorithm

As required by STP, the equally sized groups containing heterogeneous FL clients should have similar overall class distributions. To achieve this goal, in this section, we first formalize the FL client grouping problem which is NP-hard, and then explain how to simplify to propose the ICG approach.

(A) Problem Modeling

Considering an $\mathcal{F}$-class classification task involving K FL clients, STP needs to assign these clients to M groups, where M is determined by the group number growth function f and the current round r. Our goal is to find a grouping strategy $\mathbf{x} \in \mathbb{I}^{M \times K}$ in the 0–1 space $\mathbb{I} = \{0, 1\}$ to minimize the difference in class distributions of all groups, where $\mathbf{x}_m^k = 1$ represents the device k is assigned to the group m, $\mathcal{V} \in (\mathbb{Z}^+)^{\mathcal{F} \times K}$ is the class distribution matrix composed of $\mathcal{F}$-dimensional class distribution vectors of K FL clients, $\mathcal{V}_m \in (\mathbb{Z}^+)^{\mathcal{F} \times 1}$ represents the overall class distribution of group m, and $\langle \cdot, \cdot \rangle$ represents the distance between two class distributions. The problem can be formalized as follows:

$$\underset{\mathbf{x}}{\text{minimize}} \quad z = \sum_{m_1=1}^{M-1} \sum_{m_2=m_1+1}^{M} < \mathcal{V}_{m_1}, \mathcal{V}_{m_2} >, \tag{4}$$

$$\text{s.t.} \quad M = f(r), \tag{5}$$

$$\sum_{k=1}^{K} \mathbf{x}_m^k \le \left\lceil \frac{K}{M} \right\rceil \quad \forall m = 1, \cdots, M, \tag{6}$$

$$\sum_{m=1}^{M} \mathbf{x}_m^k = 1 \quad \forall k = 1, \cdots, K, \tag{7}$$

$$\mathcal{V}_m = \sum_{k=1}^{K} \mathbf{x}_m^k \mathcal{V}^k \quad \forall m = 1, \cdots, M, \tag{8}$$

$$\mathbf{x}_m^k \in \{0, 1\}, \; k \in [1, K], \; m \in [1, M]. \tag{9}$$

Constraint (5) ensures that the number of groups M meets $f(r)$ required by STP. Constraint (6) ensures that the groups have similar or equal size $\lceil \frac{K}{M} \rceil$. Constraint (7) ensures that each client can only be assigned to one group at a time. The overall class distribution $\mathcal{V}_m$ of the group m is defined by Eq. (8), where $\mathcal{V}^k \in \mathcal{V}$ is the class distribution vector of client k. Constraint (9) restricts the decision variable $\mathbf{x}$ to only take up a value of 0 or 1.

Proposition 1. *The NP-hard bin packing problem (BPP) can be reduced to the grouping problem in Eq. (4) to Eq. (9), making it also an NP-hard problem.*

Proof. The problem stated by Eq. (4) to Eq. (9) is actually a BPP with additional constraints, where K items with integer weight $\mathcal{V}^k$ and unit volume should be packed into the minimum number of bins of integer capacity $\lceil \frac{K}{M} \rceil$. The difference is that Eq. (4) to Eq. (9) restricts the number of available bins to M instead of unlimited, and the difference in the bin weights not to exceed ξ. The input and output of BPP and Eq. (4) to Eq. (9) are matched, with only additional $\mathcal{O}(1)$ transformation complexity to set M and ξ to infinity. Therefore, BPP can call the solution of Eq. (4) to Eq. (9) in $\mathcal{O}(1)$ time to obtain its solution, which proves that the NP-hard BPP [10] can be reduced to the problem stated by Eq. (4) to Eq. (9). Therefore, Eq. (4) to Eq. (9) is also an NP-hard problem.

Therefore, it is almost impossible to find the optimal solution within a polynomial time. To address this issue, we adopt the centroid equivalence theorem to simplify the original problem to a constrained clustering problem.

(B) Inter-cluster Grouping (ICG)

Consider a constrained clustering problem with K points and L clusters, where the size of all clusters is strictly the same K/L.

Assumption 1. *We make the following assumptions:*

1. *K is divisible by L;*
2. *Take any point $\mathcal{V}_l^m$ from cluster l, the squared l_2-norm distance $\|\mathcal{V}_l^m - C_l\|_2^2$ between the point $\mathcal{V}_l^m$ and its cluster centroid C_l is bounded by σ_l^2.*
3. *Take one point $\mathcal{V}_l^m$ from each of L clusters at random, the sum of deviations of each point from its cluster centroid $\epsilon^m = \sum_{l=1}^{L}(\mathcal{V}_l^m - C_l)$ meets $\mathbf{E}[\epsilon^m] = 0$.*

Definition 1 (Group Centroid). *Given L clusters of equal size, let group m be constructed from one point randomly sampled from each cluster $\{\mathcal{V}_1^m, \cdots, \mathcal{V}_L^m\}$. Then, the centroid of group m is defined as $C^m = \frac{1}{L}\sum_{l=1}^{L}\mathcal{V}_l^m$.*

Proposition 2. *If Assumption 1 holds, suppose the centroid of cluster l is $C_l = \frac{L}{K}\sum_{i=1}^{K/L}\mathcal{V}_l^i$ and the global centroid is $C_{\text{global}} = \frac{1}{L}\sum_{l=1}^{L}C_l$. We have:*

1. *The group and global centroids are expected to coincide, $\mathbf{E}[C^m] = C_{\text{global}}$.*
2. *The error $\|C^m - C_{\text{global}}\|_2^2$ between the group and global centroids is bounded by $\frac{1}{L^2}\sum_{l=1}^{L}\sigma_l^2$.*

Proof.

$$\mathbf{E}[C^m] = \mathbf{E}[\frac{1}{L}\sum_{l=1}^{L}\mathcal{V}_l^m] = \mathbf{E}[\frac{1}{L}\sum_{l=1}^{L}(\mathcal{V}_l^m - C_l + C_l)]$$

$$= \mathbf{E}[\frac{1}{L}\sum_{l=1}^{L}(\mathcal{V}_l^m - C_l) + \frac{1}{L}\sum_{l=1}^{L}C_l] = \frac{1}{L}\mathbf{E}[\epsilon^m] + C_{\text{global}} = C_{\text{global}},$$

$$\|C^m - C_{\text{global}}\|_2^2 = \|\frac{1}{L}\sum_{l=1}^{L}\mathcal{V}_l^m - \frac{1}{L}\sum_{l=1}^{L}C_l\|_2^2 = \frac{1}{L^2}\|\sum_{l=1}^{L}(\mathcal{V}_l^m - C_l)\|_2^2$$

$$\leq \frac{1}{L^2}\sum_{l=1}^{L}\|\mathcal{V}_l^m - C_l\|_2^2 = \frac{1}{L^2}\sum_{l=1}^{L}\sigma_l^2.$$

Proposition 2 indicates that there exists a grouping strategy $\tilde{\mathbf{x}}$ and $\mathcal{V}_{m_1} = \sum_{k=1}^{K}\tilde{\mathbf{x}}_{m_1}^k\mathcal{V}^k = LC^{m_1}$, $\mathcal{V}_{m_2} = \sum_{k=1}^{K}\tilde{\mathbf{x}}_{m_2}^k\mathcal{V}^k = LC^{m_2}$ $(\forall m_1 \neq m_2)$, so that the objective in Eq. (4) turns to $z = \sum_{m_1 \neq m_2} L < C^{m_1}, C^{m_2} >$ and the expectation value reaches 0. This motivates us to use the constrained clustering model to solve $\tilde{\mathbf{x}}$ in the objective Eq. (4). Therefore, we consider the constrained clustering problem below,

$$\underset{\mathbf{y}}{\text{minimize}} \quad \sum_{k=1}^{K}\sum_{l=1}^{L}\mathbf{y}_l^k \cdot \left(\frac{1}{2}\|\mathcal{V}^k - C_l\|_2^2\right), \tag{10}$$

Algorithm 2. **Inter-Cluster-Grouping**

Input: All FL clients $\mathcal{C}$ (with attribute $\mathcal{V}^k$), the total number of FL clients K, the group number growth function f, the current training round r.

Output: The grouping strategy $\mathcal{G}$.

1: Randomly sample $L \cdot \lfloor \frac{K}{L} \rfloor$ clients from $\mathcal{C}$ to meet Assumption 1, where $L = \lfloor \frac{K}{f(r)} \rfloor$;

2: **repeat**

3:    CLUSTER ASSIGNMENT: Fix the cluster centroid C_l and optimize $\mathbf{y}$ in Eq. (10) to Eq. (13);

4:    CLUSTER UPDATE: Fix $\mathbf{y}$ and update the cluster centroid C_l as follows,

$$C_l \leftarrow \frac{\sum_{k=1}^{K} \mathbf{y}_l^k \mathcal{V}^k}{\sum_{k=1}^{K} \mathbf{y}_l^k} \quad \forall l = 1, \cdots, L;$$

5: **until** C_l converges;

6: GROUP ASSIGNMENT: Randomly sample one client from each cluster without replacement to construct group $\mathcal{G}_m (\forall m = 1, \cdots, f(r))$;

7: **return** $\mathcal{G} = \{\mathcal{G}_1, \cdots, \mathcal{G}_{f(r)}\}$;

$$\text{s.t.} \qquad \sum_{k=1}^{K} \mathbf{y}_l^k = \frac{K}{L} \qquad \forall l = 1, \cdots, L, \tag{11}$$

$$\sum_{l=1}^{L} \mathbf{y}_l^k = 1 \qquad \forall k = 1, \cdots, K, \tag{12}$$

$$\mathbf{y}_l^k \in \{0, 1\}, \ k \in [1, K], \ l \in [1, L], \tag{13}$$

where $\mathbf{y} \in \mathbb{I}^{L \times K}$ is a selector variable, $\mathbf{y}_l^k = 1$ means that client k is assigned to cluster l while 0 means not, C_l represents the centroid of cluster l. Equation (10) is the standard clustering objective, which aims to assign K clients to L clusters so that the sum of the squared l_2-norm distance between the class distribution vector $\mathcal{V}^k$ and its nearest cluster centroid C_l is minimized. Constraint (11) ensures that each cluster has the same size $\frac{K}{L}$. Constraint (12) ensures that each client can only be assigned to one cluster at a time. In this simplified problem, Constraint (7) is relaxed to $\sum_{m=1}^{M} \mathbf{x}_m^k \leq 1$ to satisfy the assumption that K/L is divisible.

The above constrained clustering problem can be modeled as a minimum cost flow (MCF) problem and solved by network simplex algorithms [19], such as SIMPLEMINCOSTFLOW in Google OR-Tools. Then, we can alternately perform cluster assignment and cluster update to optimize $\mathbf{y}_l^k$ and $C_l (\forall k, l)$, respectively. Finally, we construct M groups, each group consists of one client randomly sampled from each cluster without replacement, so that their group centroids are expected to coincide with the global centroid. The pseudo code is given in Algorithm 2. ICG has a complexity of $\mathcal{O}(\frac{K^6 \mathcal{F} \tau}{M^2} \log Kd)$, where $d = \max\{\sigma_l^2 | \forall l \in [1, L]\}$, and $K, M, \mathcal{F}, \tau$ are the number of clients, groups, categories, and iterations, respectively. In our experiment, ICG is quite fast, and it can complete group assignment within only 0.1 s, with $K = 364, M = 52, \mathcal{F} = 62$ and $\tau = 10$.

4 Experimental Evaluation

4.1 Experiment Setup and Evaluation Metrics

Environment and Hyperparameter Setup. The experiment platform contains $K = 368$ FL clients. The most commonly used FEMNIST [11] is selected as the benchmark dataset, which is specially designed for non-i.i.d. FL environment and is constructed by dividing 805,263 digit and character samples into 3,550 FL clients in a non-uniform class distribution, with an average of $n = 226$ samples per client. For the resource-limited mobile devices, a lightweight neural network composed of 2 convolutional layers and 2 fully connected layers with a total of 6.3 million parameters is adopted as the training model. The standard mini-batch SGD is used by FL clients to train their local models, with the learning rate $\eta = 0.01$, the batch size $b = 5$ and the local epoch $e = 1$. We test FEDGSP for $R = 500$ rounds. By default, we set the group sampling rate $\kappa = 0.3$, the group number growth function $f = \text{LOG}$ and the corresponding coefficients $\alpha = 2$, $\beta = 10$. The values of κ, f, α, β will be further tuned in the experiment to observe their performance influence.

Benchmark Algorithms. In order to highlight the effect of the proposed STP and ICG separately, we remove them from FEDGSP to obtain the naive version, NaiveGSP. Then, we compare the performance of the following versions of FEDGSP through ablation studies:

1. *NaiveGSP*: FL clients are randomly assigned to a fixed number of groups, the clients in the group are trained in sequence and the groups are trained in parallel (e.g., Astraea [9]).
2. *NaiveGSP+ICG*: The ICG grouping algorithm is adopted in NaiveGSP to assign FL clients to a fixed number of groups strategically.
3. *NaiveGSP+ICG+STP* (FEDGSP): On the basis of NaiveGSP+ICG, FL clients are reassigned to a growing number of groups in each round as required by STP.

In addition, seven state-of-the-art baselines are experimentally compared with FEDGSP. They are FedProx [12], FedMMD [13], FedFusion [14], IDA [15], and FedAdagrad, FedAdam, FedYogi from [16].

Evaluation Metrics. In addition to the fundamental test accuracy and test loss, we also define the following metrics to assist in performance evaluation.

Class Probability Distance (CPD). The maximum mean discrepancy (MMD) distance is a probability measure in the reproducing kernel Hilbert space. We define CPD as the kernel two-sample estimation with Gaussian radial basis kernel $\mathcal{K}$ [20] to measure the difference in class probability (i.e., normalized class distribution) $\mathcal{P} = \text{norm}(\mathcal{V}_{m_1})$, $\mathcal{Q} = \text{norm}(\mathcal{V}_{m_2})$ between two groups m_1, m_2. Generally, the smaller the CPD, the smaller the data heterogeneity between two groups, and therefore the better the grouping strategy.

$$\text{CPD}(m_1, m_2) = \text{MMD}^2(\mathcal{P}, \mathcal{Q}) \tag{14}$$
$$= \mathbf{E}_{x,x' \sim \mathcal{P}}\left[\mathcal{K}(x, x')\right] - 2\mathbf{E}_{x \sim \mathcal{P}, y \sim \mathcal{Q}}\left[\mathcal{K}(x, y)\right] + \mathbf{E}_{y,y' \sim \mathcal{Q}}\left[\mathcal{K}(y, y')\right].$$

Computational Time. We define T_{comp} in Eq. (15) to estimate the computational time cost, where the number of floating point operations (FLOPs) is $\mathcal{N}_{\text{calc}} = 96M$ FLOPs per sample and $\mathcal{N}_{\text{aggr}} = 6.3M$ FLOPs for global aggregation, and $\mathcal{T}_{\text{FLOPS}} = 567G$ FLOPs per second is the computing throughput of the Qualcomm Snapdragon 835 smartphone chip equipped with Adreno 540 GPU.

$$T_{\text{comp}}(R) = \sum_{r=1}^{R} \left(\underbrace{\frac{\mathcal{N}_{\text{calc}}}{\mathcal{T}_{\text{FLOPS}}} \cdot \frac{neK}{\min\{K, f(r)\}}}_{\text{Local Training}} + \underbrace{\frac{\mathcal{N}_{\text{aggr}}}{\mathcal{T}_{\text{FLOPS}}} \cdot [\kappa f(r) - 1]}_{\text{Global Aggregation}} \right) \text{ (s).} \quad (15)$$

Communication Time and Traffic. We define T_{comm} in Eq. (16) to estimate the communication time cost and D_{comm} in Eq. (17) to estimate the total traffic, where the FL model size is $\mathcal{M} = 25.2\,\text{MB}$, the inbound and outbound transmission rates are $\mathcal{R}_{\text{in}} = \mathcal{R}_{\text{out}} = 567\,\text{Mbps}$ (tested in the Internet by AWS EC2 r4.large 2 vCPUs with disabled enhanced networking). Equation (16) to Eq. (17) consider only the cross-WAN traffic between FL clients and group managers, but the traffic between the top server and group managers is ignored because they are deployed in the same physical machine.

$$T_{\text{comm}}(R) = 8\kappa K \mathcal{M} R \left(\frac{1}{\mathcal{R}_{\text{in}}} + \frac{1}{\mathcal{R}_{\text{out}}} \right) \text{ (s),} \quad (16)$$

$$D_{\text{comm}}(R) = 2\kappa K \mathcal{M} R \text{ (Bytes).} \quad (17)$$

Please note that Eq. (15) to Eq. (16) are theoretical metrics, which do not consider memory I/O cost, network congestion, and platform configurations such as different versions of CUDNN/MKLDNN libraries.

4.2 Results and Discussion

The Effect of ICG and STP. We first compare the CPD of FedAvg [21], NaiveGSP and NaiveGSP+ICG in Fig. 3a. These CPDs are calculated between every pair of FL clients. The results show that NaiveGSP+ICG reduces the median CPD of FedAvg by 82% and NaiveGSP by 41%. We also show their accuracy performance in Fig. 3b. The baseline NaiveGSP quickly converges but only achieves the accuracy similar to FedAvg. Instead, NaiveGSP+ICG improves the accuracy by 6%. This shows that reducing the data heterogeneity among groups can indeed effectively improve FL performance in the presence of non-i.i.d. data. Although NaiveGSP+ICG is already very effective, it still has defects. Figure 3c shows a rise in the loss value of NaiveGSP+ICG, which indicates that it has been overfitted. That is because the training mode of NaiveGSP+ICG is static, it may learn the client order and forget the previous data. Instead, the dynamic FEDGSP overcomes overfitting and eventually converges to a higher accuracy 85.4%, which proves the effectiveness of combining STP and ICG.

The Effect of the Growth Function f **and Its Coefficients** α, β . To explore the performance influence of different group number growth functions f, we conduct a grid search on $f = \{\text{LINEAR}, \text{LOG}, \text{EXP}\}$ and α, β. The test loss heatmap is shown in Fig. 4. The results show that the logarithmic growth function achieves smaller loss 0.453 with $\alpha = 2, \beta = 10$ among 3 candidate functions. Besides, we found that both lower and higher α, β lead to higher loss values. The reasons may be that a slow increase in the number of groups leads to more STM and results in overfitting, while a rapid increase in the number of groups makes FEDGSP degenerate into FedAvg prematurely and suffers the damage of data heterogeneity. Therefore, we recommend $\alpha \cdot \beta$ to be a moderate value, as shown in the green area.

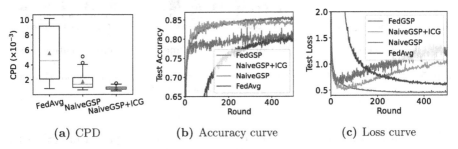

(a) CPD   (b) Accuracy curve   (c) Loss curve

Fig. 3. Comparison among FedAvg, NaiveGSP, NaiveGSP+ICG and FEDGSP in (a) CPD, (b) accuracy curve and (c) loss curve. In subfigure (a), the orange line represents the median value and the green triangle represents the mean value.

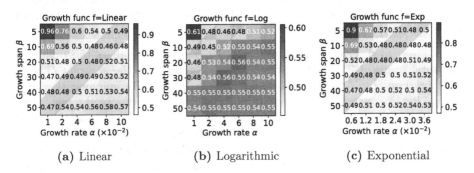

(a) Linear   (b) Logarithmic   (c) Exponential

Fig. 4. Test loss heatmap of (a) linear, (b) logarithmic and (c) exponential growth functions over different α and β settings in FEDGSP.

The Effect of the Group Sampling Rate κ. κ controls the participation rate of groups (also the participation rate of FL clients) in each round. We set $\kappa = \{0.1, 0.2, 0.3, 0.5, 1.0\}$ to observe its effect on accuracy and time cost. Figure 5a shows the robustness of accuracy to different values of κ. This is expected because ICG forces the data of each group to become homogeneous, which enables each group to individually represent the global data. In addition, Figs. 5b and 5c show that κ has a negligible effect on computational time T_{comp}, but a proportional effect on communication time T_{comm} because a larger κ means more model data are involved in data transmission. Therefore, we recommend that only $\kappa \in [0.1, 0.3]$ of groups are sampled to participate in FL in each round to reduce the overall time cost. In our experiments, we set $\kappa = 0.3$ by default.

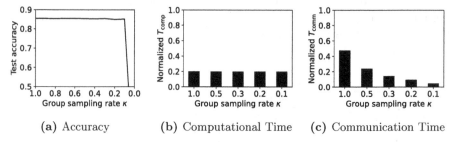

(a) Accuracy   (b) Computational Time   (c) Communication Time

Fig. 5. Comparison of (a) accuracy and the normalized (b) computational time and (c) communication time over different κ settings in FEDGSP.

The Performance Comparison of FedGSP. We compare FEDGSP with seven state-of-the-art approaches and summarize their test accuracy, test loss and training rounds (required to reach the accuracy of 80%) in Table 1. The results show that FEDGSP achieves 5.3% higher accuracy than FedAvg and reaches the accuracy of 80% within only 34 rounds. Moreover, FEDGSP outperforms all the comparison approaches, with an average of 3.7% higher accuracy, 0.123 lower loss and 84% less rounds, which shows its effectiveness to improve FL performance in the presence of non-i.i.d. data.

The Time and Traffic Cost of FedGSP. Figure 6 visualizes the time cost and total traffic of FEDGSP and FedAvg when they reach the accuracy of 80%. The time cost consists of computational time $T_{\mathrm{comp}}(R)$ and communication time $T_{\mathrm{comm}}(R)$, of which $T_{\mathrm{comm}}(R)$ accounts for the majority due to the huge data traffic from hundreds of FL clients has exacerbated the bandwidth bottleneck of the cloud server. Figure 6 also shows that FEDGSP spends 93% less time and traffic than FedAvg, which benefits from a cliff-like reduction in the number of training rounds R (only 34 rounds to reach the accuracy of 80%). Therefore, FEDGSP is not only accurate, but also training- and communication-efficient.

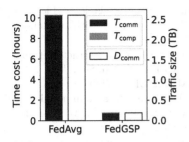

Fig. 6. Comparison of time and traffic cost to reach 80% accuracy.

Table 1. Comparison of accuracy, loss, and rounds required to reach 80% accuracy.

Algorithm	Accuracy	Loss	Rounds
FedAvg	80.1%	0.602	470
FedProx	78.7%	0.633	×
FedMMD	81.7%	0.587	336
FedFusion	82.4%	0.554	230
IDA	82.0%	0.567	256
FedAdagrad	81.9%	0.582	297
FedAdam	82.1%	0.566	87
FedYogi	83.2%	0.543	93
FedGSP	**85.4%**	**0.453**	**34**

5   Conclusions

In this paper, we addressed the problem of FL model performance degradation in the presence of non-i.i.d. data. We proposed a new concept of dynamic STP collaborative training that is robust against data heterogeneity, and a grouped framework FEDGSP to support dynamic management of the continuously growing client groups. In addition, we proposed ICG to support efficient group assignment in STP by solving a constrained clustering problem with equal group size constraint, aiming to minimize the data distribution divergence among groups. We experimentally evaluated FEDGSP on LEAF, a widely adopted FL benchmark platform, with the non-i.i.d. FEMNIST dataset. The results showed that FEDGSP significantly outperforms seven state-of-the-art approaches in terms of model accuracy and convergence speed. In addition, FEDGSP is both training- and communication-efficient, making it suitable for practical applications.

Acknowledgments. This work is supported, in part, by the National Key Research and Development Program of China (2019YFB1802800); PCL Future Greater-Bay Area Network Facilities for Large-Scale Experiments and Applications (LZC0019), China; National Research Foundation, Singapore under its AI Singapore Programme (AISG Award No: AISG2-RP-2020-019); the RIE 2020 Advanced Manufacturing and Engineering (AME) Programmatic Fund (No. A20G8b0102), Singapore; and Nanyang Assistant Professorship (NAP). Any opinions, findings and conclusions or recommendations expressed in this material are those of the authors and do not reflect the views of the funding agencies.

References

1. Kairouz, P., McMahan, H.B., Avent, B., et al.: Advances and open problems in federated learning. Found. Trends Mach. Learn. **14**(1–2), 1–210 (2021). https://doi.org/10.1561/2200000083

2. Xu, J., Glicksberg, B.S., Su, C., Walker, P., Bian, J., Wang, F.: Federated learning for healthcare informatics. J. Healthc. Inf. Res. **5**(1), 1–19 (2021). https://doi.org/10.1007/s41666-020-00082-4

3. Khan, L.U., Saad, W., Han, Z., et al.: Federated learning for internet of things: recent advances, taxonomy, and open challenges. IEEE Commun. Surv. Tut. **23**(3), 1759–1799 (2021). https://doi.org/10.1109/COMST.2021.3090430

4. Lim, W.Y.B., Luong, N.C., Hoang, D.T., et al.: Federated learning in mobile edge networks: a comprehensive survey. IEEE Commun. Surv. Tut. **22**(3), 2031–2063 (2020). https://doi.org/10.1109/COMST.2020.2986024

5. Zhao, Y., Li, M., Lai, L., et al.: Federated learning with non-IID data. arXiv preprint arXiv:1806.00582 (2018)

6. Yao, X., Huang, T., Zhang, R.X., et al.: Federated learning with unbiased gradient aggregation and controllable meta updating. In: Workshop on Federated Learning for Data Privacy and Confidentiality (2019)

7. Yoshida, N., Nishio, T., Morikura, M., et al.: Hybrid-FL for wireless networks: cooperative learning mechanism using non-IID data. In: 2020 IEEE International Conference on Communications (ICC), ICC 2020, pp. 1–7 (2020). https://doi.org/10.1109/ICC40277.2020.9149323

8. Jeong, E., Oh, S., Kim, H., et al.: Communication-efficient on-device machine learning: federated distillation and augmentation under non-IID private data. In: Workshop on Machine Learning on the Phone and other Consumer Devices (2018)

9. Duan, M., Liu, D., Chen, X., et al.: Astraea: self-balancing federated learning for improving classification accuracy of mobile deep learning applications. In: 2019 IEEE 37th International Conference on Computer Design (ICCD), pp. 246–254 (2019). https://doi.org/10.1109/ICCD46524.2019.00038

10. Garey, M.R., Johnson, D.S.: "Strong" NP-completeness results: motivation, examples, and implications. J. Assoc. Comput. Mach. **25**(3), 499–508 (1978). https://doi.org/10.1145/322077.322090

11. Caldas, S., Duddu, S.M.K., Wu, P., et al.: LEAF: a benchmark for federated settings. In: 33rd Conference on Neural Information Processing Systems (NeurIPS) (2019)

12. Li, T., Sahu, A.K., Zaheer, M., et al.: Federated optimization in heterogeneous networks. In: Proceedings of Machine Learning and Systems, vol. 2, pp. 429–450 (2020)

13. Yao, X., Huang, C., Sun, L.: Two-stream federated learning: reduce the communication costs. In: 2018 IEEE Visual Communications and Image Processing (VCIP), pp. 1–4 (2018). https://doi.org/10.1109/VCIP.2018.8698609

14. Yao, X., Huang, T., Wu, C., et al.: Towards faster and better federated learning: a feature fusion approach. In: 2019 IEEE International Conference on Image Processing (ICIP), pp. 175–179 (2019). https://doi.org/10.1109/ICIP.2019.8803001

15. Yeganeh, Y., Farshad, A., Navab, N., et al.: Inverse distance aggregation for federated learning with non-IID data. In: Domain Adaptation and Representation Transfer, and Distributed and Collaborative Learning, pp. 150–159 (2020). https://doi.org/10.1007/978-3-030-60548-3_15

16. Reddi, S., Charles, Z., Zaheer, M., et al.: Adaptive federated optimization. In: International Conference on Learning Representations (2021)

17. Sattler, F., Müller, K.R., Samek, W.: Clustered federated learning: Model-agnostic distributed multitask optimization under privacy constraints. IEEE Trans. Neural Netw. Learn. Syst. **32**(8), 3710–3722 (2021). https://doi.org/10.1109/TNNLS.2020.3015958

18. Fallah, A., Mokhtari, A., Ozdaglar, A.: Personalized federated learning: a meta-learning approach. In: 34th Conference on Neural Information Processing Systems (NeurIPS) (2020)
19. Bradley, P.S., Bennett, K.P., Demiriz, A.: Constrained k-means clustering. Microsoft Research, Redmond, vol. 20 (2000)
20. Gretton, A., Borgwardt, K.M., Rasch, M.J.: A kernel two-sample test. J. Mach. Learn. Res. **13**(1), 723–773 (2012)
21. McMahan, B., Moore, E., Ramage, D., et al.: Communication-efficient learning of deep networks from decentralized data. In: Proceedings of the 20th International Conference on Artificial Intelligence and Statistics, pp. 1273–1282 (2017)

Transportation-Mode Aware Travel Time Estimation via Meta-learning

Yu Fan[1], Jiajie Xu[1(✉)], Rui Zhou[2], and Chengfei Liu[2]

[1] School of Computer Science and Technology, Soochow University, Suzhou, China
20205227013@stu.suda.edu.cn, xujj@suda.edu.cn
[2] Swinburne University of Technology, Melbourne, Australia
{rzhou,cliu}@swin.edu.au

Abstract. Transportation-mode aware travel time estimation (TA-TTE) aims to estimate the travel time of a path in a specific transportation mode (e.g., walking, driving). Different from traditional travel time estimation, TA-TTE requires to consider the heterogeneity of transportation modes due to different moving characteristics in different modes. As a result, when applying classical travel time estimation models, sufficient data is needed for each mode to capture mode-dependent characteristics separately. While in reality, it is hard to obtain enough data in some modes, resulting in a severe data sparsity problem. A practical method to solve this problem is to leverage the mode-independent knowledge (e.g., time for waiting for traffic lights) learned from other modes. To this end, we propose a meta-optimized method called MetaMG, which learns well-generalized initial parameters to support effective knowledge transfer across different modes. Particularly, to avoid negative transfer, we integrate a spatial-temporal memory in meta-learning to cluster trajectories according to spatial-temporal distribution similarity for enhanced knowledge transfer. Besides, a multi-granularity trajectory representation is adopted in our base model to explore more useful features in different spatial granularities while improving the robustness. Finally, comprehensive experiments on real-world datasets demonstrate the superior performance of our proposed method over existing approaches.

Keywords: Travel time estimation · Meta-learning · Trajectory mining

1 Introduction

Travel time estimation (TTE) is essential for many location-based applications such as route planning [19,20], navigation [8] and vehicle dispatching [27]. In real-world applications, users tend to estimate travel time in a specific transportation mode, which raises the problem of transportation-mode aware travel time estimation (TA-TTE). Since trajectories of different modes have different characteristics (e.g., moving speed), TA-TTE needs to consider the heterogeneity of different modes, and accordingly, sufficient data in each mode are needed

A. Bhattacharya et al. (Eds.): DASFAA 2022, LNCS 13246, pp. 472–488, 2022.
https://doi.org/10.1007/978-3-031-00126-0_35

for training. Unfortunately, some modes (e.g., walking, riding) may face a data sparsity problem due to data privacy and difficulty in data collection. This may cause severe over-fitting in existing methods, and thus calls for a novel method that can well support TA-TTE in data-sparse modes.

Meta-learning is a representative few-shot learning method and has shown promising results in various domains such as computer vision [25] and natural language processing [11,13] due to its good generalization ability. Therefore, it provides great opportunities for TA-TTE in data-sparse target modes as well, with the basic idea of transferring knowledge learned from source modes. Fortunately, this is feasible because trajectories of different modes may have some common mode-independent characteristics influenced by several factors, such as road network structure, rush hours and built environment. For example, no matter what mode the trajectory is, more time will be spent on waiting for traffic lights in areas with many intersections. Therefore, we can utilize meta-learning to learn a generalized model from source modes first, and adapt it to estimate travel time in target modes via fine-tuning. However, directly using traditional meta-learning in TA-TTE may face the following challenges.

First, negative transfer may occur in traditional meta-learning [4], because transferred knowledge is learned from trajectories of different regions or time periods in source modes. However, trajectories with different spatial and temporal distributions tend to follow different moving characteristics, which have substantial influence on travel time. For example, a three kilometer trip in a downtown area may have longer travel time than one in a rural area, because there are more intersections in the downtown area. Similarly, the travel time of a trip in peak hour is probably longer than one in off-peak hour due to heavy traffic. Transferring knowledge between trajectories with significantly different spatial-temporal distributions will make no contribution and even hurt the performance. Thus, in order to avoid the risk of negative transfer, it would be beneficial to cluster trajectories according to spatial-temporal distribution similarity, and learn shared mode-independent knowledge in each cluster respectively.

In addition, although the existing TTE methods [3,5,17,28] can be directly applied as base models in meta-learning, they cannot support better knowledge transfer in TA-TTE. These TTE methods for specific routing paths are mainly divided into fine-granularity methods (e.g., road segment based methods or coordinate point based methods) [3,17] and coarse-granularity methods (e.g., grid based methods) [28]. The former can represent trajectories accurately, but they may face over-fitting problem and fail to capture sufficient trajectory features when data is sparse. The later is robust to insufficient data, however, they may fail to capture real movements of trajectories with small sampling intervals because coordinate points may be mapped to the same grid. Therefore, it is necessary to represent trajectories in multiple granularities, so as to not only capture more trajectory features in different spatial granularities, but also improve the robustness of the model.

To address the above challenges, we propose a novel meta-learning framework for TA-TTE named MetaMG, which takes advantage of meta-learning paradigm to solve the few-shot problem in data-sparse transportation mode by transferring knowledge learned from other modes. It first learns a well-generalized initial model from source modes, so that the mode-independent knowledge can be captured and effectively adapted to target modes. Specifically, considering trajectories in different regions or time periods tend to have different moving characteristics, we design a spatial-temporal memory in meta-learning to cluster trajectories according to spatial-temporal distribution similarity, so as to leverage knowledge learned from trajectories with similar spatial-temporal distributions in source modes. The globally shared memory is jointly trained with the base model in an end-to-end manner. Besides, to capture more trajectory features as well as to improve the robustness of our model, we adopt a multi-granularity trajectory representation in our base model. Finally, we evaluate the proposed framework on real-world datasets, the results demonstrate the advantages of our approach compared with several baselines. The main contributions of this paper can be summarized as follows:

- We propose a novel meta-optimized method for TA-TTE, which deals with the cold-start problem in data-sparse modes by transferring knowledge learned from other modes.
- To avoid negative transfer, we design a spatial-temporal memory and properly integrate it in meta-learning to cluster trajectories according to spatial-temporal distribution similarity for enhanced knowledge transfer.
- Moreover, we adopt a multi-granularity trajectory representation in our base model, so as to obtain useful trajectory features in different granularities and improve the robustness of the model simultaneously.
- We conduct extensive experiments on two transportation modes, and the results demonstrate the effectiveness of our proposed framework.

2   Related Work

2.1   Travel Time Estimation

With recent improvements in satellites and GPS-enabled devices, a great amount of trajectory data can be accumulated which provide us much information for trajectory detection [10], destination prediction [21,29] and travel time estimation. As one of the key topics in transportation systems, the existing TTE solutions can be classified as path-aware methods and path-blind methods.

Path-aware methods estimate the travel time for a specific routing path, they can be further classified as fine-granularity methods and coarse-granularity methods. (1) Fine-granularity models estimate the travel time based on the road segments or coordinate points. DeepTTE [17] utilizes a geo-convolution to split the whole path with intermediate GPS points into several sub-paths, then predicts the travel time of each sub-path and the whole path by a multi-task loss function. CompactETA [5] considers road network constraints and applies graph attention network to learn the spatio-temporal dependencies for travel time estimation. (2) Coarse-granularity models first map the coordinate points into grids and learn characteristics of each grid to estimate the travel time.

DeepTravel [3] considers spatial-temporal embeddings, driving state features and traffic features in each grid, and finally uses Bi-LSTMs to predict the travel time. TADNM [22] aims to provide accurate TTE for mixed-mode paths. MTLM [23] first recommends the appropriate transportation mode for users, and then estimates the related travel time of the path.

The path-blind methods deal with the scenario where path information is unavailable, the query only provides the origin and destination locations and external information in TTE task. In contrast with path-aware methods, path-blind query faces great challenges due to uncertain route and travel distance. DeepOD [26] designs novel representations for the OD pair and its corresponding trajectory and encodes such representations to estimate the travel time. [18] simply uses a large number of trajectories to estimate the travel time between source and destination. Although existing methods can support TTE task based on large-scale training trajectories, they fail to achieve satisfactory results when trajectories of a specific transportation mode are insufficient.

2.2   Meta-learning

Meta-learning is a learning paradigm which is also called learning to learn. It aims to learn the general knowledge across a variety of different tasks, and rapidly adapts such knowledge to new tasks with little training data. The state-of-the-art meta-learning algorithms can be divided into three categories: memory-based meta-learning [12,14], metric-based meta-learning [15,16] and optimization-based meta-learning [1,4].

Recently, some studies [2,9,24] have adopted meta-learning in spatial-temporal prediction and travel time estimation tasks for alleviating cold-start problems. SSML [2] attempts to utilize the meta-learning paradigm in en route travel time estimation. By considering each trajectory as a learning task, it aims to capture the potential characteristics from the traveled partial trajectories to estimate the travel time of the remaining routes. MetaST [24] proposes a meta-learning framework for spatial-temporal prediction. It learns a well-generalized initialization of the ST-net from a large number of prediction tasks sampled from multiple source cities, and then adopts the model to a target city. MetaStore [9] is a task-adaptive framework for optimal store placement in new cities with insufficient data by transferring prior knowledge learned from multiple data-rich cities.

Inspired by the above works, this paper proposes a meta-learning approach for transportation-mode aware travel time estimation, aiming to leverage the knowledge from data-rich modes to alleviate the cold-start problem in data-sparse modes. In addition, we design a spatial-temporal memory to better transfer the mode-independent knowledge among trajectories with similar spatial-temporal distributions.

3   Problem Formulation

Definition 1 (Path). A path P is a sequence of intersections in a road network, $P = \{c_1, c_2, ..., c_n\}$, where c_i is the coordinate point of the intersection, represented as $(longitude_i, latitude_i)$. We assume that the path is specified by the user or generated by the route planing applications.

Definition 2 (Trajectory). A trajectory is defined as a tuple of four components, $X = \{P, m, T, \delta\}$, where P is the path that the trajectory passed by, m denotes the transportation mode of the trajectory, $T = \{\tau_1, \tau_2, ..., \tau_n\}$ represents the corresponding timestamps of the coordinate points, and δ is the travel time of the trajectory. For each trajectory, we record its transportation mode and contextual information such as the departure time, the day of the week and the user information.

Problem Statement. Given a query path P of a target data-sparse mode and corresponding contextual information, our goal is to estimate the travel time δ of P by fine-tuning on a generalized model learned from trajectories in other modes. Specifically, we have a source mode set $\mathcal{M}_s$ and a target mode set $\mathcal{M}_t$. For each mode, we select a set of data as support set $\mathcal{D}_m^{spt}$, and the remaining data as the query set $\mathcal{D}_m^{qry}$. Our goal is to estimate the travel time in query set of target modes with the help of knowledge learned from source modes.

4   Model Architecture

In this section, we detail the proposed method MetaMG. First, we introduce our base model MGTA, which adopts a multi-granularity trajectory representation to capture spatial-temporal features in different granularities. Then we present the meta-optimized model MetaMG, which applies a spatial-temporal memory in meta-learning for better knowledge transfer between different modes.

4.1   Base Model

Our base model MGTA is composed of three parts, i.e., a contextual information component, a spatial-temporal network and a multi-task learning component, the framework is shown in Fig. 1.

Contextual Information Component. The travel time of a path is affected by many contextual information which are categorical values and cannot feed to neural network directly, such as departure time, day of the week. Thus this component aims to incorporate the embedding of such factors into our model with the embedding method proposed in [6]. Specifically, the embedding method maps each categorical value to a real space $\mathbf{R}^{E \times 1}$ by multiplying a learnable parameter matrix $W \in \mathbf{R}^{V \times E}$. V represents the size of the original categorical value and E represents the dimension of embedding space. Besides, we further concatenate the obtained embedded vectors together with the distance of the path, and denote the concatenation as e^{CI}.

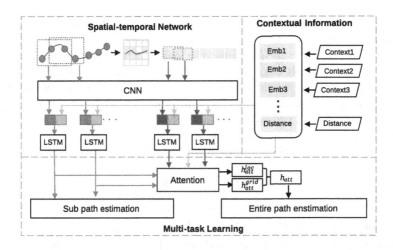

Fig. 1. Base model MGTA

Spatial-Temporal Network. Motivated by the fact that trajectories in different regions and time periods tend to have different characteristics, inspired by DeepTTE [17], we adopt Convolutional Neural Network (CNN) and Long Short-Term Memory (LSTM) to capture the spatial correlations and temporal dependencies for each path. Different from DeepTTE, we obtain the spatial-temporal features of trajectories in different granularities, so as to avoid over-fitting problem caused by sparse data and improve the robustness of the model.

First, we learn spatial-temporal features of paths in fine granularity. In particular, we use a non-liner mapping to map each c_i in $P_{loc} = \{c_1, c_2, ..., c_{n_{loc}}\}$ into vector loc_i as,

$$loc_i = tanh(W_{loc} \cdot [c_i.lat, c_i.lng_i]) \tag{1}$$

W_{loc} is the learnable parameter. Then the loc sequence will be fed into CNN to capture the spatial correlations, i.e.,

$$loc_i^{conv} = \sigma_{cnn}(W_{conv} * loc_{i:i+q-1} + b) \tag{2}$$

$*$ denotes the convolutional operation, W_{conv} and b are the learnable parameters in CNN, $loc_{i:i+q-1}$ is a subsequence in loc sequence from index i to index $i+q-1$, coordinate points in $loc_{i:i+q-1}$ make up a sub-path $sp_{i:i+q-1}$ of P_{loc}, q is the kernel size of the filter and σ_{cnn} is the activation function. Next we take the loc_i^{conv} as input of LSTM, like the way in DeepTTE [17], we incorporate the contextual information to enhance the estimating ability of the recurrent layers, the updating rule can be simply expressed as:

$$h_i^{loc} = \sigma_{lstm}(W_x \cdot loc_i + W_h \cdot h_{i-1}^{loc} + W_a \cdot e^{CI}) \tag{3}$$

where W_x, W_h and W_a are parameter matrices used in the recurrent layer, h_i^{loc} is the hidden state after we processed the i-th sub-path, σ_{lstm} is the non-linear activation function and e^{CI} represents the contextual information. Now we get the sequence $H^{loc} = \{h_1^{loc}, h_2^{loc}, ..., h_{|n_{loc}|-q+1}^{loc}\}$, where h_i^{loc} is the spatial-temporal feature of the i-th sub-path in P_{loc}.

Second, to capture the spatial correlations and temporal dependencies in a coarse granularity simultaneously, we map the coordinate points in P_{loc} to several grids. We partition the whole road network into $N \times N$ disjoint but equal-sized grids. Accordingly, a query path $P_{loc} = \{c_1, c_2, ..., c_{n_{loc}}\}$ can be represented as a sequence of grids it passed by, $P_{grid} = \{g_1, g_2, ..., g_{n_{grid}}\}$, where $n_{(\cdot)}$ is the number of the coordinate points or grids. Note that each grid has its gridID, we embed each g_i as $grid_i$ like the way in contextual information component.

$$grid_i = Emb(g_i) \tag{4}$$

Following the same step in Eq. (2) and Eq. (3), we can get the spatial-temporal features of the P_{grid} as $H^{grid} = \{h_1^{grid}, h_2^{grid}, ..., h_{|n_{grid}|-q+1}^{grid}\}$.

In the next component, the spatial-temporal features in different granularities will be combined to estimate the travel time of the path.

Multi-task Learning Component. This component combines the previous components to estimate the travel time of the input path. We enforce the multi-task learning to estimate the travel time of both entire path and each sub-path.

Since the points in P_{loc} have more accurate timestamp labels than the grids in P_{grid}, we only consider sub-paths in P_{loc} for travel time estimation on sub-paths. Recall that we have obtained the sequence H^{loc}, each $h_i^{loc} \in H^{loc}$ corresponds to the spatial-temporal features of the sub-path $sp_{i:i+q-1}$ in P_{loc}. We use two stacked fully-connected layers to map each h_i^{loc} to a scalar $\hat{r}_i^{loc}$ that represents the travel time estimation of the i-th sub-path in P_{loc}.

To make full use of spatial-temporal features learned in different spatial granularities, the H_{loc} and H_{grid} will be combined to estimate the travel time of the entire path. Above all, the variable length H_{loc} and H_{grid} are needed to be transformed into fixed length vectors. Considering the sub-paths which have more road intersections or traffic lights are more important for the travel time, we adopt the attention mechanism in this step. Formally, we have that

$$h_{att}^{(\cdot)} = \sum_{i=1}^{n_{(\cdot)}-q+1} \alpha_i^{(\cdot)} \cdot h_i^{(\cdot)} \tag{5}$$

and $(\cdot)$ can be either loc or $grid$, $\alpha_i^{(\cdot)}$ is the weight for the i-th sub-path in $P_{(\cdot)}$. We consider the spatial-temporal information as well as the contextual information to learn the parameter $\alpha_i^{(\cdot)}$ like DeepTTE [17]. Thus we devise the attention mechanism as:

$$z_i^{(\cdot)} = \langle \sigma_{att}(attr), h_i^{(\cdot)} \rangle \tag{6}$$

$$\alpha_i^{(\cdot)} = \frac{e^{z_i}}{\Sigma_j e^{z_j^{(\cdot)}}} \tag{7}$$

where $\langle \cdot \rangle$ is the inner product operator and σ_{att} is a non-liner mapping function with $tanh$ activation. In order to use multi-granularity spatial-temporal features of trajectories, we combine the h_{att}^{loc} and h_{att}^{grid} as:

$$h_{att} = \beta \cdot h_{att}^{loc} + (1 - \beta) \cdot h_{att}^{grid} \tag{8}$$

and β is the combination coefficient that balances the tradeoff between h_{att}^{loc} and h_{att}^{grid}. Finally, we pass h_{att} to several fully-connect layers with residual connections FC_{en} to obtain the estimation of the entire path $\hat{r}_{en}$.

$$\hat{r}_{en} = FC_{en}(h_{att}) \tag{9}$$

We estimate the travel time of all the sub-paths and the entire path simultaneously during the training process. The loss for sub-paths and entire path are defined as follows:

$$L_{sub} = \frac{1}{|n_{loc}| - q + 1} \sum_{i=1}^{|n_{loc}| - q + 1} \frac{|r_{sp_i} - \hat{r}_{sp_i}|}{r_{sp_i}} \tag{10}$$

$$L_{en} = \frac{|r_{en} - \hat{r}_{en}|}{r_{en}}. \tag{11}$$

Finally, our model is trained to minimize the weighted combination of two loss terms with the coefficient γ,

$$L = \gamma \cdot L_{sub} + (1 - \gamma) \cdot L_{en} \tag{12}$$

4.2 Meta Optimization

In this part, we elaborate our meta-optimized method MetaMG. MetaMG consists of three parts: initialization adaptation, spatial-temporal memory and knowledge transfer. The first part aims to learn well-generalized initial parameters from source modes. The second part clusters trajectories according to spatial-temporal distribution similarity for better knowledge transfer. The last one transfers the learned initial parameters and memory to estimate the travel time of given path in target mode. The whole framework is represented in Fig. 2.

Initialization Adaptation. Traditionally, the parameters of a neural network are randomly initialized, and then converged to a good local optimum by minimizing the loss based on the training set. However, the training process will take a long time, and the randomly initialized model with limited training data will lead to severe over-fitting. Thanks to the generalization ability of meta-learning, we can solve the cold-start problem in data-sparse target modes in TA-TTE, by learning well-generalized initial parameters Θ from source modes and adapt the parameters to the target modes.

The goal is to obtain global initial parameters Θ that can generalize well on different source modes. As suggested in MAML [4], the base model initialized

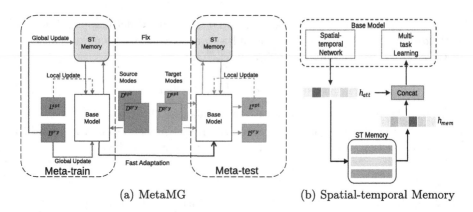

(a) MetaMG   (b) Spatial-temporal Memory

Fig. 2. (a) the framework of MetaMG. Base model and ST memory mean MGTA and spatial-temporal memory. In knowledge transfer, parameters of the base model are updated by the support set in target modes, and the memory is fixed. (b) the composed spatial-temporal memory. The spatial-temporal representation h_{att} captured by the spatial-temporal component of the base model is a query vector, then the attention mechanism is used to get the memory-enhanced representation h_{mem}. h_{att} and h_{mem} are concatenated together for the travel time estimation.

by Θ needs to achieve the minimum of the average generalization losses over all source modes. For each source mode m_s, we have support set $\mathcal{D}_{m_s}^{spt}$ and query set $\mathcal{D}_{m_s}^{qry}$. The global parameters Θ are randomly initialized before meta-training process, and will be updated locally on support set and globally updated on query set respectively.

Specifically, during the local learning phase (i.e., learning on the support set), we initialize the parameters of m_s with Θ, the optimization goal in local training is to minimize the loss of travel time estimation for a single mode. Thus, the local parameters for m_s will be updated by:

$$\theta_{m_s} \leftarrow \theta_{m_s} - \rho \cdot \nabla_{\theta_{m_s}} L_{m_s}^{\mathcal{D}^{spt}} \tag{13}$$

ρ is the learning rate for updating the local parameters. Next, we get the well-generalized initial parameters Θ by minimizing the average generalization losses over all source modes, i.e.,

$$\Theta = min_{\Theta} \sum_{m_{s_i} \in M_s} L_{m_{s_i}}^{\mathcal{D}^{qry}} \tag{14}$$

so the global parameters are updated by

$$\Theta \leftarrow \Theta - \lambda \cdot \sum_{m_{s_i} \in M_s} \nabla L_{m_{s_i}}^{\mathcal{D}^{qry}} \tag{15}$$

λ is the learning rate of global update, and $L_{m_s}^{\mathcal{D}^{spt}}$ is the loss on the query set of source mode m_s.

The global initial parameters can well generalize on source modes by global update in Eq. (15). Therefore, we transfer Θ to the target mode m_t, expecting to achieve superior generalization performance.

Spatial-Temporal Memory. Although traditional meta-learning can provide the well-generalized initial parameters Θ to estimate the travel time in target modes, it may cause negative transfer, because there exists impact from the divergence between trajectories in different regions and time periods. As we mentioned in Sect. 1, trajectories with different spatial-temporal distributions tend to have different moving characteristics. Therefore, the performance can be further improved by transferring the knowledge between trajectories with similar spatial-temporal distributions. For this purpose, we design a spatial-temporal memory in meta-learning, which clusters trajectories according to spatial-temporal distribution similarity and learns the cluster representations in source modes, so as to transfer these cluster representations to target modes for improved estimation accuracy.

First, we construct a parameterized spatial-temporal memory $\mathcal{MEM}_{ST} \in \mathbf{R}^K \times d$, where K denotes the number of spatial-temporal feature clusters, which is predefined before training the model, each row of the memory stores a cluster representation with the dimension of d. Before the training, the spatial-temporal memory is initialized randomly, and is updated during training process.

Next, in order to leverage the knowledge from trajectories with similar spatial-temporal distributions, we distill the cluster representations of spatial-temporal features stored in memory $\mathcal{MEM}_{ST}$ for travel time estimation via attention mechanism. In detail, given a trajectory $X^{(i)} \in \mathcal{D}_{m_s}^{spt}$ in source modes, we can get the spatial-temporal representation $h_{att}^{(i)}$ of it in Eq. (8), and we use the representation $h_{att}^{(i)}$ to query the memory. The attention value $a_i \in \mathbf{R}^K$ is

$$a_i = attention(h_{att}^{(i)}, \mathcal{MEM}_{ST}). \tag{16}$$

Then we calculate the memory-enhanced spatial-temporal representation $h_{mem}^{(i)}$,

$$h_{mem}^{(i)} = \sum_{i=1}^{K} a_i \cdot \mathcal{MEM}_{ST}. \tag{17}$$

To preserve the personalization of each trajectory and include the generalization among trajectories with similar spatial-temporal distributions, we propose to combine the trajectory specific representation $h_{att}^{(i)}$ and the memory-enhanced representation $h_{mem}^{(i)}$. So the input in Eq. (9) is replaced by the combination spatial-temporal representation, i.e.,

$$\hat{r}_{en} = FC_{en}(h_{att}^{(i)} \oplus h_{mem}^{(i)}). \tag{18}$$

Following the idea of Neural Turing Machine (NTM) [7], the memory $\mathcal{MEM}_{ST}$ has a read head to retrieve the memory and a write head to update the memory. The write head will write the updated spatial-temporal features to the memory $\mathcal{MEM}_{ST}$ during meta-training process, i.e.,

$$\mathcal{MEM}_{ST} = \omega(a_i(h_{att}^{(i)})^T) + (1 - \omega)\mathcal{MEM}_{ST} \tag{19}$$

where $(a_i(h_{att}^{(i)})^T)$ is the cross product of a_i and $h_{att}^{(i)}$, ω is a hyper-parameter to control how much new spatial-temporal feature is added. Here we use an attention mask a_i when adding the new information so that the new spatial-temporal information will be attentively added to the memory matrix.

The learned memory will be transferred to target modes, in order to make full use of mode-independent knowledge learned from trajectories with similar spatial-temporal distributions.

Knowledge Transfer. In meta-testing phase on target modes, we transfer the global initialized parameters Θ and the spatial-temporal memory $\mathcal{MEM}_{ST}$. The memory $\mathcal{MEM}_{ST}$ is fixed and the Θ is utilized to initialize the parameters for the target mode θ_{m_t} as:

$$\theta_{m_t} \leftarrow \Theta \tag{20}$$

Then we train the parameters θ_{m_t} on the support set of the target mode $\mathcal{D}_{m_t}^{spt}$:

$$\theta_{m_t} = \theta_{m_t} - \rho\nabla_{\theta_{m_t}} L_{m_t}^{\mathcal{D}^{spt}}. \tag{21}$$

For path P_i in query set $\mathcal{D}_{m_t}^{qry}$, we obtain the spatial-temporal representations by Eq. (8) and the memory-enhanced representation by Eq. (17). Accordingly, the travel time estimation of path P_i is calculated via Eq. (18).

5   Experiments

5.1   Dataset

We use GeoLife dataset [30–32] published by Microsoft Research. The GPS trajectories were collected by 182 users in a period of over five years (From April 2007 to August 2012) and positioned every 1–3 s. 73 users have labeled their trajectories with transportation modes, such as driving, taking bus or taxi, riding bike and walking, each annotation contains a transportation mode with start and end times. Although there are 11 types of annotations, we used only 5 (walking, bus, car, riding and taxi) because trajectories in other modes (e.g., train, airplane) have their specific road network structures, the labels of both taxi and car are regarded as driving. We remove trajectories that users' annotation labels cannot be mapped into, and split the very long trajectories (more than 200 records) into several segments. Detailed statistics of trajectories in different modes are listed in Table 1.

For each source mode, we select 80% data for meta-training and validation, and 20% for meta-testing. For each target mode, we select 20% and 80% as support set and query set respectively. We choose hyper-parameters by conducting meta-train using different hyper-parameters and select the parameters with the best performance, thus we do not set validation sets.

Table 1. Data Statistics of different modes

Transportation mode	Walking	Riding	Bus	Car&Taxi
# of trajectories	5238	2349	5800	6060
# of points	350031	311122	268150	273023
Avg length (m)	1932.61	2765.01	3496.72	5423.43
Avg travel time (s)	3752.58	3693.49	935.38	1416.24

5.2   Evaluation Metrics and Configuration

Three metrics are used to evaluate the performance of our methods, including mean absolute percentage error (MAPE), mean average error (MAE), and root mean square error (RMSE), which are widely used in regression problems.

For our base model MGTA, we partition the whole road network into 600 × 600 grids. In contextual information component, we divide one day into 144 timeslots and embed the departure timeslots to $\mathbf{R}^{16}$, other factors (e.g., day of the week, userID) are embedded to $\mathbf{R}^4$. In spatial-temporal network, the convolutional filter size q is set to 3 and the number of filters is 32, we use ELU function as the activation σ_{cnn} in Eq. 2. The hidden vector of the recurrent is fixed as 128, and the activation is $ReLU$. Finally in the multi-task learning component, we fix the number of residual fully-connected layers as 4 and the size of each layer is 128. The hyper-parameter γ in Eq. (12) and ω in Eq. (19) are set to 0.1 and 0.5 respectively. The performance of MGTA with different combination coefficients β in Eq. (8) will be further evaluated.

For the meta-learning setting, we evaluate our model with different cluster numbers K. The learning rates for local update and global update are 0.0001 and 0.001 respectively.

5.3   Baselines

- **DeepTravel:** [28] maps the coordinate points to grids and feeds the manual characteristics of each grid to Bi-LSTMs to estimate the travel time.
- **TADNM:** [22] aims to provide accurate travel time estimation for mixed-mode paths, it divides the path into several segments according to their modes, and maps each segment into multiple grids to capture the spatial characteristics and temporal correlations in grid sequence respectively.
- **CompactETA:** [5] aims to provide real-time inference, by learning the high level link representations on road network graph by a graph attention network equipped with positional encoding.
- **DeepTTE:** [17] transforms the coordinate sequence to a series of feature maps, then it captures the spatial and temporal features with CNN and LSTM respectively, and estimates the travel time based on such features.
- **MGTA:** is our base model, which is a grid-enhanced method on the basis of DeepTTE, learns a multi-granularity trajectory representation to capture spatial-temporal features in different granularities to improve the travel time estimation accuracy.

- **MGTA-MIX:** is trained with data of all modes directly, without considering the heterogeneity of different modes, and is used to estimate the travel time for trajectories in target mode.
- **MAML:** [4] is a classic meta-learning method, it learns a general initialization from multiple tasks and adapts it to new tasks.

5.4 Performance Comparison

The comparison results of all the evaluated approaches are shown in Table 2.

Table 2. Performance comparison of MetaMG and its competitors

Metrics	Modes					
	Walking			Riding		
	MAPE(%)	MAE(s)	RMSE(s)	MAPE(%)	MAE(s)	RMSE(s)
DeepTravel	46.98	43.92	94.88	47.76	45.25	92.68
CompactETA	28.53	38.84	79.03	31.86	43.18	88.72
DeepTTE	20.92	32.79	68.70	23.17	35.37	85.96
TADNM	19.25	32.65	68.27	22.78	34.24	79.73
MGTA	<u>18.57</u>	<u>30.38</u>	<u>67.60</u>	<u>21.29</u>	<u>32.61</u>	<u>77.31</u>
MGTA-MIX	33.54	35.01	75.59	35.34	39.38	90.75
MAML	17.48	29.08	63.18	19.83	30.41	75.38
MetaMG	**15.55**	**26.67**	**61.56**	**17.37**	**26.97**	**73.26**
Improvement	11.04%	8.29%	2.56%	12.41%	11.31%	2.81%

Base Model Comparison. To evaluate the performance of our base model MGTA, we compare it with state-of-the-art TTE methods, all these methods except MGTA-MIX are trained on support set in each target mode. As shown in Table 2, TADNM and DeepTTE are grid based and coordinate point based model respectively, they achieve suboptimal results, indicating the effectiveness of our base model MGTA, which captures the spatial-temporal features in different granularities. That can be explained as the multi-granularity trajectory representation adopted in base model can not only represent the trajectory well, but also improve the robustness of MGTA on sparse data. Therefore, we consider MGTA as the most effective base model for TA-TTE.

Besides, in order to illustrate the varying patterns belonging to different transportation modes, we further train MGTA with data of all modes directly (i.e., MGTA-MIX), without considering the heterogeneity of different modes. The performance is much worse than MGTA that is trained with data in target mode only, which coincides to our assumption that separating models are needed for different modes for improved estimation.

Meta-learning Strategy Comparison. Then we evaluate the effectiveness of the meta-learning strategies, for fair comparison, the same base model MGTA is used in MAML and MetaMG. When the target mode is walking or riding, the other three modes are set as source modes. From Table 2, we can observe that the traditional meta-learning method MAML can achieve better performance than the base model by capturing the generalization among different modes, indicating that it is feasible to transfer the mode-independent knowledge from other modes to target modes. Furthermore, by adopting a spatial-temporal memory in meta-learning to better transfer knowledge across trajectories with similar spatial-temporal distributions, our method MetaMG successfully achieves 11.04%, 8.29%, 2.56% and 12.41%, 11.31%, 2.81% improvements of MAPE, MAE, RMSE on two modes over MAML.

Influence of Cluster Numbers. In this work, we predefined the dimension K of the spatial-temporal memory, which can be regarded as the number of trajectory clusters according to spatial-temporal distribution similarity. From Fig. 3, we can observe that MetaMG achieves the best result when the cluster number is 4. While it performs worst with the cluster number of 1, which equals to MAML. The obvious improvements against 1 cluster case indicate that incorporating clusters plays an important role in our methods.

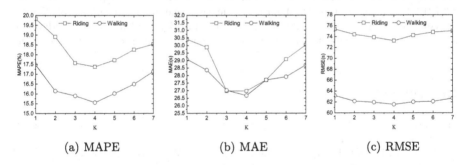

| (a) MAPE | (b) MAE | (c) RMSE |

Fig. 3. Performance of different cluster numbers

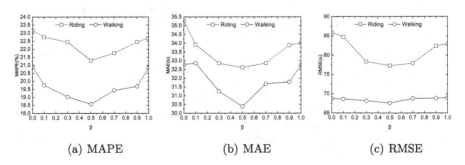

| (a) MAPE | (b) MAE | (c) RMSE |

Fig. 4. Performance of different combination coefficients

Influence of Combination Coefficients. In Fig. 4, we conduct several experiments to investigate the performance of varying combination coefficients from 0 to 1 in MAPE, MAE, RMSE. As we can see, when increasing β, the performance of MetaMG also improves. This is because more spatial-temporal features can be captured in different granularities. However, when β gets to a proper value, increasing β results in worse performance. We set β to the proper value of 0.5.

6   Conclusion

In this paper, we propose a novel meta-optimized model that leverages the learned mode-independent knowledge from other modes. Specifically, we integrate a spatial-temporal memory in meta-learning to cluster trajectories according to spatial-temporal distribution similarity, so that the knowledge can be better transferred across trajectories with similar spatial-temporal distributions to avoid negative transfer. Besides, we apply a multi-granularity trajectory representation in our base mode, to not only capture more spatial-temporal features, but also avoid the over-fitting problem. Finally, we conduct extensive experiments on real-world datasets to verify the superior performance of our proposed method over existing approaches.

Acknowledgements. This work was supported by the National Natural Science Foundation of China projects under grant numbers (No. 61872258, No. 61772356, No. 62072125), the major project of natural science research in universities of Jiangsu Province under grant number 20KJA520005, the priority academic program development of Jiangsu higher education institutions, young scholar program of Cyrus Tang Foundation.

References

1. Andrychowicz, M., et al.: Learning to learn by gradient descent by gradient descent. In: NIPS, pp. 3981–3989 (2016)
2. Fang, X., Huang, J., Wang, F., Liu, L., Sun, Y., Wang, H.: SSML: self-supervised meta-learner for en route travel time estimation at Baidu maps. In: KDD, pp. 2840–2848 (2021)
3. Fang, X., Huang, J., Wang, F., Zeng, L., Liang, H., Wang, H.: ConSTGAT: contextual spatial-temporal graph attention network for travel time estimation at Baidu maps. In: KDD, pp. 2697–2705 (2020)
4. Finn, C., Abbeel, P., Levine, S.: Model-agnostic meta-learning for fast adaptation of deep networks. In: ICML, vol. 70, pp. 1126–1135 (2017)
5. Fu, K., Meng, F., Ye, J., Wang, Z.: CompactETA: a fast inference system for travel time prediction. In: KDD, pp. 3337–3345 (2020)
6. Gal, Y., Ghahramani, Z.: A theoretically grounded application of dropout in recurrent neural networks. In: NIPS, pp. 1019–1027 (2016)
7. Graves, A., Wayne, G., Danihelka, I.: Neural Turing Machines. CoRR abs/1410.5401 (2014)

8. Kisialiou, Y., Gribkovskaia, I., Laporte, G.: The periodic supply vessel planning problem with flexible departure times and coupled vessels. COR **94**, 52–64 (2018)
9. Liu, Y., et al.: MetaStore: a task-adaptative meta-learning model for optimal store placement with multi-city knowledge transfer. TIST **12**(3), 28:1–28:23 (2021)
10. Lv, Z., Xu, J., Zhao, P., Liu, G., Zhao, L., Zhou, X.: Outlier trajectory detection: a trajectory analytics based approach. In: DASFAA, vol. 10177, pp. 231–246 (2017)
11. Madotto, A., Lin, Z., Wu, C., Fung, P.: Personalizing dialogue agents via meta-learning. In: ACL, pp. 5454–5459 (2019)
12. Munkhdalai, T., Yuan, X., Mehri, S., Trischler, A.: Rapid adaptation with conditionally shifted neurons. In: ICML, pp. 3664–3673 (2018)
13. Qian, K., Yu, Z.: Domain adaptive dialog generation via meta learning. arXiv preprint arXiv:1906.03520 (2019)
14. Santoro, A., Bartunov, S., Botvinick, M., Wierstra, D., Lillicrap, T.P.: Meta-learning with memory-augmented neural networks. In: ICML, vol. 48, pp. 1842–1850 (2016)
15. Snell, J., Swersky, K., Zemel, R.S.: Prototypical networks for few-shot learning. arXiv preprint arXiv:1703.05175 (2017)
16. Sung, F., Yang, Y., Zhang, L., Xiang, T., Torr, P.H.S., Hospedales, T.M.: Learning to compare: relation network for few-shot learning. In: CVPR, pp. 1199–1208 (2018)
17. Wang, D., Zhang, J., Cao, W., Li, J., Zheng, Y.: When will you arrive? Estimating travel time based on deep neural networks. In: AAAI, pp. 2500–2507 (2018)
18. Wang, H., Tang, X., Kuo, Y., Kifer, D., Li, Z.: A simple baseline for travel time estimation using large-scale trip data. TIST **10**(2), 19:1–19:22 (2019)
19. Xu, J., Chen, J., Zhou, R., Fang, J., Liu, C.: On workflow aware location-based service composition for personal trip planning. FGCS **98**, 274–285 (2019)
20. Xu, J., Gao, Y., Liu, C., Zhao, L., Ding, Z.: Efficient route search on hierarchical dynamic road networks. DPD **33**(2), 227–252 (2015)
21. Xu, J., Zhao, J., Zhou, R., Liu, C., Zhao, P., Zhao, L.: Predicting destinations by a deep learning based approach. TKDE **33**, 651–666 (2021)
22. Xu, S., Xu, J., Zhou, R., Liu, C., Li, Z., Liu, A.: TADNM: a transportation-mode aware deep neural model for travel time estimation. In: DSFAA, pp. 468–484 (2020)
23. Xu, S., Zhang, R., Cheng, W., Xu, J.: MTLM: a multi-task learning model for travel time estimation. GeoInformatica (2020). https://doi.org/10.1007/s10707-020-00422-x
24. Yao, H., Liu, Y., Wei, Y., Tang, X., Li, Z.: Learning from multiple cities: a meta-learning approach for spatial-temporal prediction. In: WWW (2019)
25. Ye, H.J., Sheng, X.R., Zhan, D.C.: Few-shot learning with adaptively initialized task optimizer: a practical meta-learning approach. ML **109**(3), 643–664 (2020)
26. Yuan, H., Li, G., Bao, Z., Feng, L.: Effective travel time estimation: when historical trajectories over road networks matter. In: SIGMOD, pp. 2135–2149 (2020)
27. Yuan, N.J., Zheng, Y., Zhang, L., Xie, X.: T-Finder: a recommender system for finding passengers and vacant taxis. TKDE **25**(10), 2390–2403 (2013)
28. Zhang, H., Wu, H., Sun, W., Zheng, B.: DeepTravel: a neural network based travel time estimation model with auxiliary supervision. In: IJCAI, pp. 3655–3661 (2018)
29. Zhao, J., Xu, J., Zhou, R., Zhao, P., Liu, C., Zhu, F.: On prediction of user destination by sub-trajectory understanding: a deep learning based approach. In: CIKM, pp. 1413–1422 (2018)
30. Zheng, Y., Li, Q., Chen, Y., Xie, X., Ma, W.: Understanding mobility based on GPS data. In: UbiComp, vol. 344, pp. 312–321 (2008)

31. Zheng, Y., Xie, X., Ma, W.: GeoLife: a collaborative social networking service among user, location and trajectory. DEB **33**(2), 32–39 (2010)
32. Zheng, Y., Zhang, L., Xie, X., Ma, W.: Mining interesting locations and travel sequences from GPS trajectories. In: WWW, pp. 791–800 (2009)

A Deep Reinforcement Learning Based Dynamic Pricing Algorithm in Ride-Hailing

Bing Shi[1,2(✉)], Zhi Cao[1], and Yikai Luo[1]

[1] School of Computer Science and Artificial Intelligence,
Wuhan University of Technology, Wuhan 430070, China
{bingshi,caozhi}@whut.edu.cn
[2] Shenzhen Research Institute of Wuhan University of Technology,
Shenzhen 518000, China

Abstract. Online ride-hailing has become one of the most important transportation ways in the modern city. In the ride-hailing system, the vehicle supply and riding demand is different in different regions, and thus the passengers' willingness to take a riding service will change dynamically. Traditional pricing strategies cannot make reasonable decisions to set the riding prices with respect to the dynamical supply and demand in different regions, and they cannot make adaptive responses to the real-time unbalanced supply and demand. In addition, the ride-hailing platform usually intends to maximize the long-term profit. In this paper, we use deep reinforcement learning to design a multi-region dynamic pricing algorithm to set the differentiate unit price for different regions in order to maximize the long-term profit of the platform. Specifically, we divide the ride-hailing area into several non-overlapping regions, and then propose a model to characterize the passenger's price acceptance probability. We further model the pricing issue as a Markov decision-making process, and then use deep reinforcement learning to design a multi-region dynamic pricing algorithm (**MRDP**) to maximize the platform's long-term profit. We further run extensive experiments based on realistic data to evaluate the effectiveness of the proposed algorithm against some typical benchmark approaches. The experimental results show that **MRDP** can set the price effectively based on supply and demand to make more profit and can balance the supply and demand to some extent.

Keywords: Ride-hailing · Dynamic pricing · Supply and demand · Long-term profit · Reinforcement learning

1 Introduction

Nowadays, various online ride-hailing platforms have emerged, such as DiDi and Uber. In the ride-hailing system, how to set the riding service price is one of the most important issues. It can affect the choice of passengers, thereby affecting the platform's profit and the supply and demand in the future. Currently, the traditional ride-hailing pricing algorithm is based on the combination of basic price and unit price. The online ride-hailing platform set the price by taking into account the order completion time and other factors. These ride-hailing platforms use dynamic pricing to control supply and demand at different locations and times. However, these pricing strategies usually use a uniform way to set the service price for the whole area, and do not consider the supply and demand diversity in different sub-regions. Furthermore, the supply and demand dynamically changes over the time. The current pricing strategy may not be able to adapt the changed supply and demand. Therefore, the ride-hailing system needs to design an effective pricing strategy to set the price differently according to the real-time supply and demand in different sub-regions in order to maximize the long-term profit.

In the ride-hailing market, the supply and demand in different regions are different. Therefore, we first divide the entire area into several non-overlapping sub-regions. The multi-region status reflect the difference between supply and demand more precisely. Secondly, the supply and demand in the same region are dynamically changing. Passengers may have dynamic acceptance probability of the service price. For example, during the peak time, passengers may have a stronger riding demand and are more likely to accept higher price. In contrast, during the off peak time, higher price may cause passengers to give up their ride plans, which will damage the platform's profit. Therefore, the platform should adopt a reasonable dynamic pricing strategy according to different supply and demand. Finally, the platform needs to maximize the long-term profit, rather than the short-term profit.

Although there exist some related works about pricing in the ride-hailing [8,9,13], they did not consider all the above factors. In this paper, we design a dynamic pricing algorithm based on the real-time supply and demand states of different regions to maximize the long-term profit of the platform. The main contributions of this paper are as follows. Firstly, we divide the whole ride-hailing area into several non-overlapping sub-regions, and then collect the supply and demand states in different regions for pricing. Secondly, considering the dynamic change of supply and demand status and passengers' willingness to take a riding service, we propose a multi-region dynamic pricing problem. Since this problem is a sequential decision problem, we model it as a Markov decision process and then propose a multi-region dynamic pricing algorithm (**MRDP**) based on deep reinforcement learning. The algorithm maximizes the platform's long-term profit by taking into account passengers' acceptance probability of prices based on the real-time spatial and temporal distribution information. Finally, we run experiments based on Chengdu ride-hailing order data to evaluate the proposed algorithm. The result shows that **MRDP** can balance the supply and demand, and can make reasonable pricing based on the real-time supply and demand status, which will bring higher profit to the platform.

The rest of this paper is structured as follows. In Sect. 2 we will introduce the related work, in Sect. 3 we will introduce the basic settings of this paper, in

Sect. 4 we will introduce the dynamic pricing algorithm designed in this paper. Finally we will give experimental analysis and summary in Sects. 5 and 6.

2 Related Work

There exist a number of works about pricing in the ride-hailing. In the dynamic pricing of online ride-hailing, Gan et al. [8] proposed a pricing method to incentivize drivers in order to solve the issue that most taxi drivers deliberately avoid providing riding service in the peak time. Chen et al. [5] proposed a dynamic pricing strategy in the intelligent transportation platforms to minimize traffic congestion. Asghari et al. [3] adjusted the prices by considering the future demand of the road network to increase the platform's profit while lowering the prices. However, they did not consider the impact of pricing on the future, and could not maximize the platform's long-term profit. Chen et al. [6] proposed a pricing method for the ride-sharing system, which provides passengers with more choices. Chen et al. [4] proposed a joint framework to optimize both pricing and matching strategies at the same time. However, they only considered discrete pricing action.

There also exist some works on combining mechanism design with pricing strategies. Most of these works focused on preventing strategic behaviors of participants that lead to loss of platform profit. Asghari et al. [1] studied the real-time ride sharing problem and designed a framework based on distributed auctions to maximize the platform's profit without affecting the quality of service. Then Asghari et al. [2] proposed a pricing model based on Vickery auctions to ensure drivers bidding truthfully without sacrificing any profit. Zhang et al. [15] designed a discounted trade reduction mechanism for dynamic pricing in the ride-sharing system. Zheng et al. [16] considered the order price and proposed a constrained optimization problem, which takes the profit of the platform as the optimization goal and controls the detour distance and waiting time of the driver. In order to realize self-motivated bonus bidding of users, Zheng et al. [17] designed an auction mechanism that requires users to submit the information truthfully.

To the best of our knowledge, existing works did not consider the difference of supply and demand in different regions, and did not consider the dynamic changes of supply and demand over the time and the dynamic acceptance probability of the riding price. In this paper, we consider the above factors to design a multi-region dynamic pricing algorithm to maximize the profit of the platform.

3 Basic Settings

In this section we describe basic settings of the online ride-hailing system. First, we introduce how the online ride-hailing system works, and then describe the settings about regions, passengers, vehicles and orders.

Figure 1 show how the online ride-hailing system works. First, the passenger rises the riding demand, and then the platform sets the price for the riding service. The passenger decides whether accepting the service price or not. If so, the riding demand becomes an order, and then the platform matches the order with the idle vehicles.

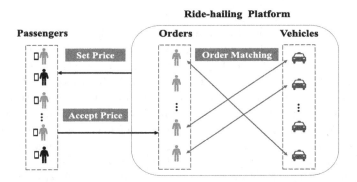

Fig. 1. A ride-hailing system

3.1 Symbols and Definitions

In this paper, we divide the entire online ride-hailing time period into several time steps $\mathcal{T} = \{1, 2, \cdots, T\}$ with the length of each time step denoted as Δt. The relevant definitions are given below.

The entire online ride-hailing area is divided into non-overlapping and inter-connected rectangular regions, numbered as $1, ..., N$. A riding demand $r_i \in R$ is expressed as a tuple $r_i = \left(l_{r_i}^s, l_{r_i}^e, g_{r_i}, t_{r_i}, val_{r_i}\right)$, $l_{r_i}^s$ and $l_{r_i}^e$ represents the pick-up and drop-off location of r_i respectively, g_{r_i} represents the pick-up region, t_{r_i} is the time when r_i is submitted, val_{r_i} is the maximum unit price that the passenger can accept for this trip. In the following, passenger r_i and riding demand r_i have the same meaning.

In addition, we set f_{r_i} as a status identifier, where $f_{r_i} = 0$ means r_i does not accept the riding price provided by the platform, $f_{r_i} = 1$ means that r_i accepts the riding price, $f_{r_i} = 2$ means that the platform has arranged a vehicle to serve the riding demand. val_{r_i} is private to the passenger and cannot be obtained directly by the platform. We assume that val_{r_i} is uniformly distributed within $[p_{\min}, p_{\max}]$ [11]. The unit price in region g is p_g, and the probability that the passenger r_i in region g accepts the platform price p_g is $G_{r_i}(p_g) = 1 - F_{r_i}(p_g)$, where $F_{r_i}(p_g) = P(val_{r_i} \leq p_g)$ is the cumulative distribution function [3,11,13].

Definition 1 (Order). *When the passenger r_i accepts the platform pricing, the riding demand is converted into an order. An order $o_i \in O$ is expressed as $o_i = \left(r_i, t_{r_i}^w, pay_{r_i}\right)$, where r_i is the riding demand information $\left(l_{r_i}^s, l_{r_i}^e, g_{r_i}, t_{r_i}, val_{r_i}\right)$, $t_{r_i}^w$ indicates the maximum waiting time for r_i, pay_{r_i} is the price of the order.*

The price for each order is the unit price multiplied by the shortest travelling distance, that is, $pay_{r_i} = p_{g_{r_i}} \times \text{dis}(l^s_{r_i}, l^e_{r_i})$, $\text{dis}(l^s_{r_i}, l^e_{r_i})$ is the shortest distance between the pick-up and drop-off location of the order.

Definition 2 (Vehicle). *The vehicle $v_i \in V$ is represented as a four-tuple $v_i = (g_{v_i}, l_{v_i}, d_{v_i}, s_{v_i})$, g_{v_i} represents the current region of the vehicle, l_{v_i} indicates the current location of the vehicle, d_{v_i} is the driver's fuel consumption cost.*

We also set an identifier s_{v_i} for vehicle status, where $s_{v_i} = 1$ means that the vehicle is serving other orders, $s_{v_i} = 0$ means that the vehicle is idle. In this paper, the vehicle belongs to the platform, that is, the vehicle will not have spontaneous behavior, which makes it easier for the platform to manage and dispatch, thus improving the efficiency of the platform [14].

Definition 3 (Platform profit). *As we assume that the vehicles are managed by the platform, the profit of the platform is the amount paid by all passengers who have been served minus the cost of the vehicle during the entire time period.*

$$EP = \sum_{T=1}^{T} \sum_{i=1}^{N} \sum_{j=1}^{M^i_T} G\left(p^T_{g_i}\right) \mathbb{I}\left(f_{r_j} = 2\right) \left(p^T_{g_i} \times \text{dis}\left(l^s_{r_j}, l^e_{r_j}\right) - C_{r_j}\right) \quad (1)$$

Where M^i_T is the total number of riding demand in the i-th region at time T, $p^T_{g_i}$ is the unit price in region g_i at T-th time step, and C_{r_j} is the cost of vehicle serving for the riding demand r_j.

3.2   Problem Formulation

Similar to the existing work [16,17], we consider the ride-hailing process in the multiple time steps.

In reality, the platform does not know the acceptance probability of passengers in advance. The acceptance probability should be estimated. Secondly, only the passengers who accept the price will contribute to the total profit of the platform. The decisions of passengers are unknown before the platform determines the unit price, and therefore the platform needs to estimate the expectation of passengers on the unit price. Then the pricing will affect the future vehicle supply, and the supply and demand in turn determines the unit price in each region. Therefore the affections between regions should also be considered. Finally the platform always wishes to find a set of optimal unit price to maximize its total profit. The goal stated above can be formalized to the following problem.

Definition 4 (Multi-region dynamic pricing problem). *In a period of time, given the real-time riding demand set R_t and the idle vehicle set V_t in a continuous time slot, the platform will set the unit price $P_t = (p^1_t, p^2_t, \ldots, p^N_t)$ for each region at each time step according to the real-time supply and demand status to maximize the platform's profit EP over the entire time period.*

4   The Algorithm

We first divide the ride-hailing market into N region, and the platform performs pricing and matching in each region. In this section, we introduce the pricing and matching algorithms.

4.1   Multi-region Dynamic Pricing Algorithm

At each time step, the platform sets unit price for each region based on the dynamic supply and demand information. Pricing will affect the matching results of the platform and thus affect the future supply and demand in different regions. Therefore, this multi-region dynamic pricing problem is a sequential decision-making problem. We model it as a Markov decision process, and then use reinforcement learning to address it.

This Markov decision process can be described as a tuple (S, A, P, r, γ), where S is a set of states, A is a set of actions, P is a transition probability function, r is the immediate reward and γ is a discount factor that decreases the impact of the past reward. We now describe them in details.

State: $s_t = (v_t, c_t, a_{t-1}, m_{t-1}, e_{t-1}) \in S$, where $v_t = (v_t^1, v_t^2, \ldots, v_t^N)$ is the number of idle vehicles in each region at time step t, $c_t = (c_t^1, c_t^2, \ldots, c_t^N)$ is the number of riding demands in each region, $a_{t-1} = (p_{t-1}^1, p_{t-1}^2, \ldots, p_{t-1}^N)$ is the unit price of each region in the last time step, $m_{t-1} = (m_{t-1}^1, m_{t-1}^2, \ldots, m_{t-1}^N)$ is the number of orders successfully matched in each region in the last time step, $e_{t-1} = (e_{t-1}^1, e_{t-1}^2, \ldots, e_{t-1}^N)$ is the profit made in each region in the last time step.

Action: $a_t = (p_t^1, p_t^2, \ldots p_t^N) \in A$ where p_t^i is the unit price set in region i at time step t.

Reward: $r_t(s_t, a_t) = \mu_1 P_t + \mu_2 ratio$, where μ_1 and μ_2 are weights, $P_t = \sum_{i=1}^n pay_t^i$ is the platform's profit and $ratio$ is the ratio of the actual number of orders being served to the maximum possible number of orders being served, which is:

$$ratio = \begin{cases} 0 & \text{if } v_t^i = 0 \text{ or } c_t^i = 0 \\ \frac{1}{N} \sum_{i=0}^N \frac{m_t^i}{\min(v_t^i, c_t^i)} & \text{if } v_t^i \neq 0 \text{ and } c_t^i \neq 0 \end{cases} \tag{2}$$

So $r_t(s_t, a_t) = \mu_1 P_t + \mu_2 ratio$, where μ_1 and μ_2 are weighting factors.

The reason of considering $ratio$ is as follows. When only considering to maximize the platform's profit, it may happen that the platform increases the price (which causes less served orders) to make more profit. Therefore, we take the $ratio$ into account in the reward, to make more the full utilization of idle vehicles.

The pricing action can be regarded as a continuous action. Therefore, we use DDPG (deep deterministic policy gradient) to design a multi-region dynamic pricing algorithm (**MRDP**), which is shown in Algorithm 1:

Algorithm 1 takes the riding demand spatio-temporal information and initial vehicle distribution information as input. A total of κ rounds of training (line 4) will be experienced. In each training, there will be T time steps (line 6). When the

---

Algorithm 1. Multi-region Dynamic Pricing Algorithm (**MRDP**)

Input:
  Riding demand spatio-temporal distribution, initial vehicle distribution
Output:
  Platform's pricing strategy π
  1: Initialize the experience pool $\mathcal{D}$ of the ride-hailing platform;
  2: Initialize the Critic network $\mathcal{Q}\left(s, a \mid \theta^{\mathcal{Q}}\right)$ and Actor network $\mu\left(s \mid \theta^{\mu}\right)$;
  3: Initialize the target network $\mathcal{Q}', \mu'$, and set the weight $\theta^{\mathcal{Q}'} \leftarrow \theta^{\mathcal{Q}}, \theta^{\mu'} \leftarrow \theta^{\mu}$;
  4: **for** $\kappa = 1$ to K **do**
  5:    Initialize the initial state s_1 and the random noise parameter $\mathcal{N}$;
  6:    **for** $t = 1$ to T **do**
  7:       The platform observes the state s_t and selects the action $a_t = \mu\left(s_t \mid \theta^{\mu}\right) + \mathcal{N}_t$
          and executes it to get rewards r_t and transfer to the next state s_{t+1};
  8:       Store the state transition tuple (s_t, a_t, r_t, s_{t+1}) into $\mathcal{D}$;
  9:       Randomly select a set of samples $(s_\chi, a_\chi, r_\chi, s_{\chi+1})$ from $\mathcal{D}$ for training;
  10:       Set $y_\chi = r_\chi + \gamma \mathcal{Q}'(s_{\chi+1}, \mu'(s_{\chi+1} \mid \theta^{\mu'}) \mid \theta^{\mathcal{Q}'})$;
  11:       Update Critic by minimizing the loss $\mathcal{L} = \frac{1}{x}\sum_x (\mathcal{Y}_\chi - \mathcal{Q}(s_\chi, a_\chi \mid \theta^{\mathcal{Q}}))^2$;
  12:       Use the sample policy gradient to update the Actor policy:
          $\nabla_{\theta^\mu} J \approx \frac{1}{X}\sum_\chi \nabla_a Q(s, a \mid \theta^2)|_{s=s_\chi, a=\mu(s_\chi)} \nabla_{\theta^\mu} \mu(s \mid \theta^\mu)|_{s=s_\chi}$;
  13:       Update the target network: $\theta^{\mathcal{Q}'} \leftarrow v\theta^{\mathcal{Q}} + (1-v)\theta^{\mathcal{Q}'}, \theta^{\mu'} \leftarrow v\theta^{\mu} + (1-v)\theta^{\mu'}$;
  14:       Set $s_t = s_{t+1}$;
  15:    **end for**
  16: **end for**

---

platform observes the state s_t, it will add noise to the currently learned strategy as the unit price of each region for the next time step. After the matching is completed, we can calculate the real-time reward r_t of the action, and move to the next state s_{t+1} (line 7), and then store this information in the memory bank $\mathcal{D}$ as training data. We will take a certain number of samples from the memory bank for training each time. After continuous repeated learning and training, a convergent action strategy can be obtained. In the experiment, the experience pool stores the interaction information between the platform and passengers at each time step. **MRDP** can continuously train through the data information collected in the experience pool to learn the passengers acceptance probability under different supply and demand conditions. We treat the platform as an agent. Since the platform can obtain all the information at each time step, the agent will also consider the affections between regions and the impact of current pricing on the future supply and demand. Therefore, we can solve the difficulties mentioned above by using **MRDP**.

4.2 Order Matching Algorithm

In this paper, the matching between passengers and vehicles can be modeled as a bipartite graph maximum weighted matching problem, in order to maximize the platform's profit, i.e.

Algorithm 2. Order Matching Algorithm

Input:
    The order set O_t and the idle vehicle set V_t of each region in each time step
Output:
    Matching result M_t^r, platform profit P_t^{max}
1: Initialize $M_t^r \leftarrow \emptyset$;
2: **for** $g = 1$ to N **do**;
3:     Initialize the bipartite graph G_g;
4:     **for** $<o_i, v_j> \in O_t^g \times V_t^g$ **do**
5:         **if** $current\_time - t_{o_i} \leq t_i^w$ **then**
6:             $w_{ij} = p_{g_{o_i}} \times \text{dis}\left(l_{o_i}^s, l_{o_i}^e\right) - d_{v_j} \times \left(\text{dis}\left(l_{o_i}^s, l_{o_i}^e\right) + \text{dis}\left(l_{v_j}, l_{o_i}^s\right)\right)$;
7:             **if** $w_{ij} \geq 0$ **then**
8:                 Set the weight of edge $<o_i, v_j>$ to w_{ij} and add this edge to G_g;
9:             **end if**
10:        **end if**
11:    **end for**
12:    $M_t^g, P_t^g \leftarrow KM\left(G_g\right), M_t^r \leftarrow M_t^r \cup M_t^g$;
13:    **for** $<o_i, v_j> \in M_t^g$ **do**
14:        Update $f_{o_i} = 2$ and $s_{v_j} = 1$
15:    **end for**
16: **end for**
17: **return** $M_t^r, \sum_{g=1}^N P_t^g$;

$$\underset{x_{ij}}{\text{argmax}} \sum_{i=1}^P \sum_{j=1}^Q x_{ij} w_{ij}, \quad \text{s.t. } \forall i, \sum_{i=0}^P x_{ij} \leq 1; \forall j, \sum_{j=0}^Q x_{ij} \leq 1$$

$$x_{ij} = \begin{cases} 1 & \text{if order } r_i \text{ is matched with vehicle } v_j \\ 0 & \text{if order } r_i \text{ is not matched with vehicle } v_j \end{cases}$$

where P is the number of orders and Q is the number of idle vehicles, w_{ij} is the profit made in the matching of order r_i with vehicle v_j, and thus $w_{ij} = pay_{r_i} - d_{v_j} \times (\text{dis}(l_{r_i}^s, l_{r_i}^e) + \text{dis}(l_{v_j}, l_{r_i}^s))$. We can use Kuhn-Munkre algorithm [10] to solve this bipartite graph maximum weighted matching problem, which is shown in Algorithm 2.

In Algorithm 2, the platform will collect all the riding demand information of each region in the time slot t. For each riding demand r in the region g, if $f_r = 1$, the riding demand will be added to the order set O_t^g, and if $s_v = 0$, the vehicle will be added to the idle vehicle set V_t^g. At the beginning, the bipartite graph is initialized for each region, and then for each matching pair $<o_i, v_j>$ in the region, we will determine whether the current time exceeds the maximum waiting time of order o_i (line 5), and then calculate the platform profit when matching o_i with v_j (line 6). If the profit is non-negative, set the weight of $<o_i, v_j>$ and add to the bipartite graph. Finally, the Kuhn-Munkres algorithm [10] is used to obtain the optimal solution of the bipartite graph to maximize the platform profit in the current time slot (line 12), where M_t^g and P_t^g is the matching result and the profit in region g. The platform will update the status information of vehicles and orders (lines 13–15).

5 Experimental Analysis

In this section, we run experiments to evaluate the proposed algorithm based on the Chengdu Didi ride-hailing data[1]. We select the map near the downtown of Chengdu for the experiment, and collect the order data from 13:00 to 17:00 within this area, including the demand initiation time, starting location and travel distance. After excluding abnormal order data, a total of 31283 demand data were obtained. The heat map of riding demand during this period is shown in Fig. 2. It can be seen that the number of riding demands in the central area is large.

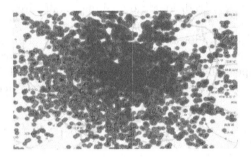

Fig. 2. User demand heat map

Firstly, a rectangular area is selected on the map with the longitude range from 104.03° E to 104.12° E and the latitude range from 30.62° N to 30.71° N. Then the area is divided into 4×4 rectangular regions. We set the length of each time step as 60 s, and vehicles are randomly distributed over 16 regions in the beginning. For each riding demand, the maximum accepted price is uniformly distributed in $[p_{min}, p_{max}]$, where p_{min} is set to $10 + 2 \times \mathrm{dis}(l_r^s, l_r^e)$, and p_{max} is set to 1.5 times p_{min}. That is, for each demand r, its maximum accepted unit price is $val_r \sim U\left(\frac{10+2\times dis(l_r^s,l_r^e)}{dis(l_r^s,l_r^e)}, \frac{15+3\times dis(l_r^s,l_r^e)}{dis(l_r^s,l_r^e)}\right)$. The maximum waiting time is randomly selected within $[\Delta t, 2\Delta t]$. The vehicle fuel consumption v_r cost is randomly selected from $\{1.4, 1.5, 1.6, 1.7\}$CNY/km. In the experiment, the unit pricing range is $[4, 7]$ (CNY/km). The experimental parameter settings are shown in Table 1.

---

[1] https://outreach.didichuxing.com/research/opendata/.

Table 1. Experimental parameters

Parameter	Value
Length of each time step $\Delta t(s)$	60
Map longitude range	$[104.03° E, 104.12° E]$
Map latitude range	$[30.62° N, 30.71° N]$
Number of vehicles $|V|$	$600, 800, 1000, 1200, 1400$
Maximum waiting time	$[\Delta t, 2\Delta t]$
The unit cost of the vehicle c_v (CNY/km)	$1.4, 1.5, 1.6, 1.7$

5.1 Benchmark Approaches and Metrics

We evaluate the proposed multi-region dynamic pricing algorithm against the following benchmark approaches.

FIX (Fixed Pricing). Traditional ride-hailing usually uses a fixed pricing algorithm, where the unit price does not change over time [7]. This is a kind of uniform pricing algorithm for all regions. In this paper, we consider a fixed pricing algorithm, which sets a fixed unit price for all regions over the whole time. Specifically, we repeat the experiment to explore in the whole pricing domain to find the unit price that maximizes the platform's profit, which is:

$$p_{fix} = \underset{p}{\text{argmax}} G(p) \sum_{\tau=1}^{T} \sum_{i=1}^{N} \sum_{j=1}^{m_i} \mathbb{I}(f_{r_j} = 2)(p \times \text{dis}(l_{r_j}^s, l_{r_j}^e) - C_{r_j}) \qquad (3)$$

SDE. This is also multi-region pricing algorithm that dynamically sets unit price based on the different supply and demand. Specifically, according to [13], when the number of vehicles is greater than the number of riding demands, we set $p_b = \underset{p}{\text{argmax}} pG(p)$, i.e. the platform can make the largest profit. The following pricing rules are designed:

$$p = \begin{cases} p_b & if\ |V_t| \geq |M_t| \\ p_b\left(1 + 2e^{|V_t| - |M_t|}\right) & if\ |V_t| < |M_t| \end{cases}$$

where at each time step, the platform determines the unit price according to the supply and demand of each region. The passengers acceptance probability of unit price per kilometer is obtained by studying the historical order data, as shown in Fig. 3, and then we calculate the value of p_b is about 4.8. From Fig. 4, when p is about equal to 4.8, the value of $pG(p)$ reaches the maximum.

GREEDY (Greedy Pricing). Existing works show that greedy algorithms can perform well in the crowdsourcing problems [12]. Therefore, in this paper we also design a greedy based multi-region pricing algorithm, which tries to maximize the platform's profit at the current time step. Similar to the **SDE** algorithm, when the number of vehicles is greater than the number of demands,

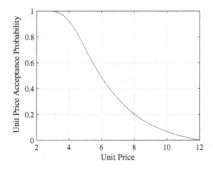

Fig. 3. Passenger acceptance probability of unit price

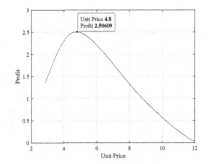

Fig. 4. Total profit when vehicles are sufficient

p_b is the optimal pricing. When the number of vehicles is less than the number of riding demands, we can combine the passenger's price acceptance probability model learned from historical data, and calculate the optimal unit price under the current state based on this model.

In order to evaluate the performance of **MRDP**, we consider the following metrics.

- **Platform profit.** The platform's profit refers to the sum of the actual profit of the platform over all time steps.
- **The number of served orders.** The number of served orders refers to the sum of the number of orders successfully served over all time steps, which can evaluate the effectiveness of the pricing.
- **Average order profit.** The average order profit is the ratio of total profit of the platform to the number of served orders over all time steps. This metrics can evaluate the ability of the platform making profit on each individual order.
- **Market supply and demand changes.** We use the number of idle vehicles in the market minus the number of riding demands to represent the difference between the supply and demand, that is, $\sum_{i=1}^{N} (|V_t| - |R_t|)$, from which we can investigate the impact of pricing strategies on the market supply and demand.

5.2   Experimental Results

The experiments are run on a machine with AMD Ryzen7 4800H processor. In the experiment, we increase the number of vehicles from 600 to 1400 with step 200. For each number of vehicles, the experiment is repeated for 10 times, and then we compute the average results for analysis.

Firstly, the platform's profit of four different algorithms are shown in Fig. 5. We find that as the number of vehicles increases, the platform's profit is also rising. This is because more orders will be served when more vehicles participate, and thus the platform's profit is increased. The supply of vehicles shows a trend from shortage of supply to saturation. We can find that **MRDP** can outperform all other three algorithms. **GREEDY** can perform better than another two algorithms. In more details, we find that multi-region based pricing algorithms (**MRDP**, **SDE** and **GREEDY**) generally can bring more profit to the platform than the **FIX** pricing algorithm, which sets a uniform and fixed price for all regions. Note that when the supply is insufficient, the profit of **SDE** is a little lower than **FIX**. While the supply is gradually saturated, the profit of **SDE** is greatly improved compared with **FIX**. After the gradual saturation of vehicle supply, the performance of **SDE** algorithm is better than **FIX** since it can maximize the profit in the saturation situation [13]. In summary, **MRDP** can achieve the maximum profit with respect to the dynamic supply and demand compared to other algorithms.

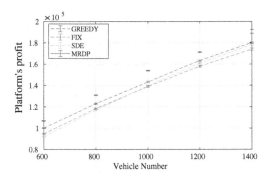

Fig. 5. Platform's profit

From Fig. 6, we find that the number of served orders using the **MRDP** algorithm and the **GREEDY** algorithm are similar, and both are significantly higher than another two algorithms. In combination with Fig. 5, we can see that the **MRDP** algorithm can make more profit than **GREEDY**. The multi-region based pricing algorithms can serve more passengers than the **FIX** pricing algorithm, and brings more profit. Then we can find that differentiated pricing in different regions has greatly improved the efficiency of platform, that is, each vehicle can serve more passengers on average.

From Fig. 7, with the increased number of vehicles, the average profit of a vehicle to complete an order decreases. This is because as the number of vehicles increases, the supply of vehicles in some regions is becoming too sufficient. Therefore the platform decreases the unit price so that more passengers are willing to accept the price. Thus the profit of each single order may decrease. However, the overall profit of the platform increases (see Fig. 5). Furthermore, from Figs. 5 and 6, we find that the **FIX** algorithm has the highest average order profit, which can be concluded that the **FIX** algorithm directly increases the platform profit by increasing the unit price, while the **SDE** algorithm will lose many orders due to high price when there are fewer vehicles. The **GREEDY**

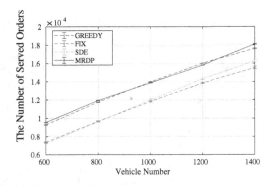

Fig. 6. The number of served orders

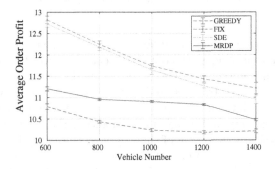

Fig. 7. Average order profit

algorithm and **MRDP** algorithm can set unit price according to the current supply and demand state, and thus can make profit while serving more orders. The **MRDP** algorithm has learned the passenger's acceptance of the unit price, and within the acceptable range of the passenger, it can make higher profit for each order than **GREEDY**, and it will not lose the order due to high pricing.

The market supply and demand changes and the thermal diagram of the prices set by **MRDP** algorithm at a certain time are shown in Fig. 8. The value in each block in Fig. 8(a) is the unit price in each region, and the value in each block in Fig. 8(b) is the difference between the number of riding demands and the number of idle vehicles. We can find that the unit price of each region is consistent with the supply and demand status of each region. For example, we can see that the difference between the number of riding demands and idle vehicles in region 10 is −13. At this time, the supply of vehicles is very sufficient, and the vehicles do not have enough riding demands to serve. Therefore, **MRDP** decreases the unit price to avoid losing orders. The difference between the number of riding demands and the number of idle vehicles in region 7 is 8. At this time, idle vehicles are not enough to meet all the riding demands. Therefore, **MRDP** increases the unit price.

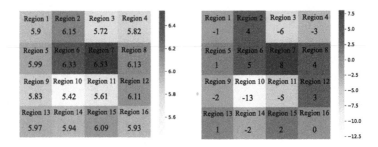

(a) Unit pricing heat map   (b) Heat map of supply and demand in ride-hailing market

Fig. 8. The unit pricing heat map and the supply and demand status

Furthermore, by observing the difference between the number of vehicles and demands at each time step, we can obtain the change curve of supply and demand in the ride-hailing market at 13:40–17:00 when the platform adopts different pricing algorithms with 600, 1000 and 1400 vehicles respectively. This is shown in Fig. 9. When the difference between supply and demand is close to 0, it means that the market supply and demand are balanced. From Fig. 9(a), we can find the market is basically in a state of short supply. However, the **MRDP** supply-demand difference curve is closer to 0, showing a better performance. In Fig. 9(b), the market is relatively stable. The **GREEDY** supply and demand difference curve seems to be more balanced than **MRDP** from 15:00 to 16:30. According to Fig. 6, it can be found that the number of served orders for **GREEDY** is slightly less than that **MRDP**, and most of the supply and demand difference curve of **MRDP** is above **GREEDY**, indicating that **MRDP** use fewer vehicles to achieve better platform profits than **GREEDY**. In Fig. 9(c), the vehicle supply in the entire market is relatively sufficient, and the supply is greater than demand at most of the time. **MRDP** can reach more balanced supply and demand, and and can serve more orders when the supply of vehicles is sufficient (see Fig. 6). In all these cases, we can find that the overall supply and demand of **MRDP** are relatively balanced during the entire time steps, which may imply that the **MRDP** algorithm can help to balance market supply and demand to a certain extent.

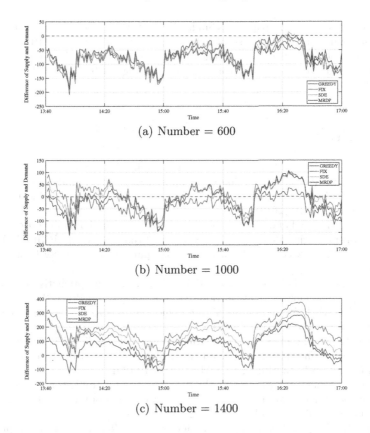

(a) Number = 600

(b) Number = 1000

(c) Number = 1400

Fig. 9. Market supply and demand curve under different number of vehicles

6   Conclusion

In this paper, we propose a multi-region dynamic pricing algorithm (**MRDP**) based on deep reinforcement learning by considering the real-time supply and demand status of different regions in order to maximize the ride-hailing platform's profit. In order to evaluate the effectiveness of **MRDP**, we run experimental analysis based on real Chengdu ride-hailing order data. We evaluate **MRDP** with three typical benchmark algorithms. The experimental results show that **MRDP** can outperform other algorithms. We can find that **MRDP** can set the unit price dynamically for each region according to the real-time change of market supply and demand status to maximize the profit, and it can also balance the market supply and demand to some extent.

Acknowledgment. This paper was funded by the Humanity and Social Science Youth Research Foundation of Ministry of Education (Grant No. 19YJC790111), the Philosophy and Social Science Post-Foundation of Ministry of Education (Grant No. 18JHQ060) and Shenzhen Fundamental Research Program (Grant No. JCYJ20190809 175613332).

References

1. Asghari, M., Deng, D., Shahabi, C., Demiryurek, U., Li, Y.: Price-aware real-time ride-sharing at scale: an auction-based approach. In: Proceedings of the 24th ACM SIGSPATIAL International Conference on Advances in Geographic Information Systems, pp. 1–10 (2016)
2. Asghari, M., Shahabi, C.: An on-line truthful and individually rational pricing mechanism for ride-sharing. In: Proceedings of the 25th ACM SIGSPATIAL International Conference on Advances in Geographic Information Systems, pp. 1–10 (2017)
3. Asghari, M., Shahabi, C.: Adapt-pricing: a dynamic and predictive technique for pricing to maximize revenue in ridesharing platforms. In: Proceedings of the 26th ACM SIGSPATIAL International Conference on Advances in Geographic Information Systems, pp. 189–198 (2018)
4. Chen, H., et al.: InBEDE: integrating contextual bandit with TD learning for joint pricing and dispatch of ride-hailing platforms. In: 2019 IEEE International Conference on Data Mining (ICDM), pp. 61–70 (2019)
5. Chen, L., Shang, S., Yao, B., Li, J.: Pay your trip for traffic congestion: dynamic pricing in traffic-aware road networks. In: Proceedings of the AAAI Conference on Artificial Intelligence, vol. 34, pp. 582–589 (2020)
6. Chen, L., Gao, Y., Liu, Z., Xiao, X., Jensen, C.S., Zhu, Y.: PTRider: a price-and-time-aware ridesharing system. Proc. VLDB Endow. **11**(12), 1938–1941 (2018)
7. Chen, M., Shen, W., Tang, P., Zuo, S.: Dispatching through pricing: modeling ride-sharing and designing dynamic prices. In: Proceedings of the 28th International Joint Conference on Artificial Intelligence, IJCAI 2019, Macao, China, 10–16 August 2019, pp. 165–171 (2019)
8. Gan, J., An, B., Wang, H., Sun, X., Shi, Z.: Optimal pricing for improving efficiency of taxi systems. In: 23rd International Joint Conference on Artificial Intelligence, pp. 2811–2818 (2013)
9. Liu, J.X., Ji, Y.D., Lv, W.F., Xu, K.: Budget-aware dynamic incentive mechanism in spatial crowdsourcing. J. Comput. Sci. Technol. **32**(5), 890–904 (2017)
10. Munkres, J.: Algorithms for the assignment and transportation problems. J. Soc. Ind. Appl. Math. **5**(1), 32–38 (1957)
11. Schröder, M., Storch, D.M., Marszal, P., Timme, M.: Anomalous supply shortages from dynamic pricing in on-demand mobility. Nat. Commun. **11**(1), 1–8 (2020)
12. Tong, Y., She, J., Ding, B., Chen, L., Wo, T., Xu, K.: Online minimum matching in real-time spatial data: experiments and analysis. Proc. VLDB Endow. **9**(12), 1053–1064 (2016)
13. Tong, Y., Wang, L., Zhou, Z., Chen, L., Du, B., Ye, J.: Dynamic pricing in spatial crowdsourcing: a matching-based approach. In: Proceedings of the 2018 International Conference on Management of Data, pp. 773–788 (2018)

14. Xu, Z., et al.: Large-scale order dispatch in on-demand ride-hailing platforms: a learning and planning approach. In: Proceedings of the 24th ACM SIGKDD International Conference on Knowledge Discovery & Data Mining, pp. 905–913 (2018)
15. Zhang, J., Wen, D., Zeng, S.: A discounted trade reduction mechanism for dynamic ridesharing pricing. IEEE Trans. Intell. Transp. Syst. $17(6)$, 1586–1595 (2015)
16. Zheng, L., Chen, L., Ye, J.: Order dispatch in price-aware ridesharing. Proc. VLDB Endow. $11(8)$, 853–865 (2018)
17. Zheng, L., Cheng, P., Chen, L.: Auction-based order dispatch and pricing in ridesharing. In: 2019 IEEE 35th International Conference on Data Engineering (ICDE), pp. 1034–1045 (2019)

Peripheral Instance Augmentation for End-to-End Anomaly Detection Using Weighted Adversarial Learning

Weixian Zong[1], Fang Zhou[1(✉)], Martin Pavlovski[2], and Weining Qian[1]

[1] School of Data Science and Engineering,
East China Normal University, Shanghai, China
wxzong@stu.ecnu.edu.cn, {fzhou,wnqian}@dase.ecnu.edu.cn
[2] Temple University, Philadelphia, USA
martin.pavlovski@temple.edu

Abstract. Anomaly detection has been a lasting yet active research area for decades. However, the existing methods are generally biased towards capturing the regularities of high-density normal instances with insufficient learning of peripheral instances. This may cause a failure in finding a representative description of the normal class, leading to high false positives. Thus, we introduce a novel anomaly detection model that utilizes a small number of labelled anomalies to guide the adversarial training. In particular, a weighted generative model is applied to generate peripheral normal instances as supplements to better learn the characteristics of the normal class, while reducing false positives. Additionally, with the help of generated peripheral instances and labelled anomalies, an anomaly score learner simultaneously learns (1) latent representations of instances and (2) anomaly scores, in an end-to-end manner. The experimental results show that our model outperforms the state-of-the-art anomaly detection methods on four publicly available datasets, achieving improvements of 6.15%–44.35% in AUPRC and 2.27%–22.3% in AUROC, on average. Furthermore, we applied the proposed model to a real merchant fraud detection application, which further demonstrates its effectiveness in a real-world setting.

1 Introduction

Anomaly detection, referred to as the process of identifying unexpected patterns from the normal behavior in a dataset, can provide critical help in various applications, such as fraud detection in finance [9], network attack detection in cybersecurity [36] and disease detection in healthcare [27]. A plethora of anomaly detection methods has been introduced over the years. However, such methods mainly focus on improving the accuracy of detecting anomalies with insufficient attention to high false positives. In real applications, when a great number of normal instances are incorrectly reported as anomalies, this may lead to a waste of labor and material resources.

© The Author(s), under exclusive license to Springer Nature Switzerland AG 2022
A. Bhattacharya et al. (Eds.): DASFAA 2022, LNCS 13246, pp. 506–522, 2022.
https://doi.org/10.1007/978-3-031-00126-0_37

Let us take a merchant fraud detection problem as an example. When certain merchants are considered to be involved in fraudulent activities, a subsequent on-the-spot investigation needs to be conducted to verify whether the merchants were indeed involved in fraudulent events. Suppose that a model outputs a huge amount of potential fraudulent merchants (e.g., $>15,000$ normal merchants were predicted as anomalies (see Table 3)), enormous resources are required for investigation and verification, in order to maintain a fair and secure financial environment. Consequently, reducing the number of false positives while enhancing anomaly detection accuracy is a rather challenging task.

In the last decade, research efforts have been focused on unsupervised anomaly detection methods [15,19]. However, such methods typically produce a large amount of false positives due to lack of labelled data. Considering that sufficient amounts of labelled normal instances can be easily collected, many deep learning approaches were proposed to capture the regularities of normal instances [6]. For instance, GAN-based methods [23,31] aim to learn a latent feature space to capture the normality underlying the given data. On the other hand, a few labelled anomalous data with valuable prior knowledge are available in many real-world applications. Approaches such as DevNet [22] utilize labelled anomalies to guarantee a margin between anomalies and normal instances so as to improve detection accuracy. However, these methods may learn suboptimal representations of normal instances as they fail to capture the characteristics of peripheral normal instances, thus leading to high false positives.

To illustrate the aforementioned issue, we present an example in Fig. 1. Figure 1(a) displays a 2-dimensional synthetic dataset, where the orange dots represent the normal instances and the blue dots represent anomalies. The green dots in Fig. 1(b) represent the instances generated by a conventional generative model (WGAN-GP [13]). Evidently, most generated instances are mainly concentrated in the center, which implies that the generative model is biased towards learning features of high-density instances, while overlooking peripheral instances that account for a small portion of the normal class. A mass of methods [20,22,24] suffer from the same problem, which leads to a high number of false positives. An illustrated explanation is presented in Fig. 1(d) and 1(e), which show the prediction results of two recent anomaly detection methods, GANomaly [1] and DevNet [22], respectively. It is obvious that a large amount of peripheral normal instances were incorrectly reported as anomalies (colored in red) by both methods. In addition, compared to Fig. 1(e), more anomalies were incorrectly predicted as normal instances (colored in black) in Fig. 1(d) as GANomaly does not take labelled anomalies into account.

In this work, we develop a weighted generative model by leveraging a few labelled anomalies for anomaly detection, named PIA-WAL (Peripheral Instance Augmentation with Weighted Adversarial Learning). The goal of our model is to learn representative descriptions of normal instances in order to reduce the amount of false positives while maintaining accurate anomaly detection. The model consists of two modules, an anomaly score learner and a weighted generative model. The anomaly score learner utilizes a small number of labelled

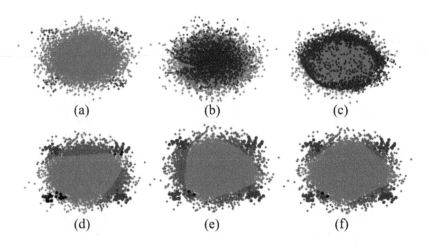

Fig. 1. (a) Synthetic data composed of normal data (colored in orange) as well as anomalies (colored in blue). (b) and (c) show the instances (colored in green) generated by a conventional generative model (WGAN-GP) and our model, respectively. (d), (e) and (f) show the performance of anomaly detection by two recent anomaly detection models, GANomaly and DevNet, and our model PIA-WAL. The false positives and false negatives are colored in red and black, respectively. (Color figure online)

anomalies to distinguish anomalies from normal instances. The scores outputted by the anomaly score learner are expected to reflect the difficulty of instances to be correctly classified. Since the peripheral normal instances constitute a small portion of the data, they are generally challenging to capture and are hard to be correctly predicted. To address this issue, we introduce a weighted generative model guided by the outputs of the anomaly score learner to generate additional *peripheral normal instances* as supplements (see the green dots in Fig. 1(c)), in order to assist the anomaly score learner to better find a representative description of the normal class. Figure 1(f) clearly shows that our model is able to produce a high detection accuracy as well as low false positives.

The advantages of the proposed model are threefold. First, by generating peripheral normal instances as supplements, PIA-WAL is able to capture the complex feature space of normal instances while reducing false positives. Second, PIA-WAL couples the feature representation learning and the anomaly score optimization in an end-to-end schema. Third, PIA-WAL is consistently more robust than the other anomaly detection methods under various anomaly contamination levels.

To verify the effectiveness and robustness of PIA-WAL, we conducted experiments on four publicly available datasets. The results show that PIA-WAL outperforms seven state-of-the-art methods, achieving statistically significant improvements of 2.27%–22.3% in AUROC and 6.15%–44.35% in AUPRC, on average. Furthermore, we applied PIA-WAL to real-world merchant fraud detection, the results of which further demonstrate its effectiveness.

This work makes the following major contributions:

(1) To the best of our knowledge, this work is the first to integrate a small amount of labelled anomalies into an adversarial framework to achieve end-to-end anomaly score learning.
(2) A novel anomaly detection model namely PIA-WAL, is introduced. With the help of generating peripheral normal instances, the detector can find a more representative description of the normal class while accurately detecting anomalies.
(3) The results obtained on publicly available datasets and a real-world merchant fraud dataset demonstrate that PIA-WAL achieves substantial improvements over state-of-the-art methods.

2   Related Work

Anomaly Detection. Anomaly detection is the task of identifying anomalies that deviate significantly from the majority of data objects [21]. Most conventional approaches are unsupervised including distance-based [4], density-based [7], isolation-based methods [15], and so forth. However, such methods are often ineffective in handling high-dimensional or irrelevant features. Recently, deep learning has been explored to improve anomaly detection [3,5]. Autoencoder-based methods [8,35] utilize a bottleneck network to learn a low-dimensional representation space and then use the learned representations to define reconstruction errors as anomaly scores. Anomaly measure-dependent learning methods [19,29] aim at learning feature representations that are specifically optimized for a particular anomaly measure. These deep learning methods can capture more complex feature interactions compared to the traditional shallow methods, however, they learn the feature representations separately from the subsequent anomaly detection, leading to suboptimal detection performance.

The recent advances show that deep anomaly detection can be substantially improved when labelled anomalies are utilized to guarantee a margin between the labelled anomalies and normal instances [25]. For example, ADOA [33] tries to cluster the observed anomalies into k clusters, and detect potential anomalies and reliable normal instances from unlabelled instances. DevNet [22] leverages a few labelled anomalies by defining a deviation loss to perform end-to-end anomaly detection. Deep SAD [24] enforces a margin between the one-class center and the labelled anomalies while minimizing the center-oriented hypersphere. With the help of prior knowledge of anomalies, these models can achieve better detection performance. However, due to the peripheral normal instances constituting a small portion of the training set, these models may fail to capture the characteristics of such instances. This makes it difficult to find a representative description of the normal class, leading to a large number of false positives.

GAN-Based Anomaly Detection. Generative Adversarial Networks (GANs) [12] and the adversarial training framework have been successfully applied to model complex and high-dimensional distributions of real-world data. This

GANs' characteristic suggests they can be used for anomaly detection, although their applications have been only recently explored [10,23]. Anomaly detection using GANs is the task of modeling the normal behavior using the adversarial training process and detecting anomalies by defining some form of a residual between a real instance and a generated instance as an anomaly score. One of the earliest methods is AnoGAN [26] which involves training a standard GAN on normal instances only to enforce the generator to learn the manifold of normal instances. Since the latent space can capture the underlying distribution of normal data, anomalies are expected to be less likely to have highly similar generated counterparts compared to normal instances. One major issue with AnoGAN is the computational inefficiency in the iterative search for the most similar generated instance to the input instances. The model EGBAD [31] and its variant [32] were designed to address this problem by learning the inverse mapping of raw data from the latent space. GANomaly [1] further improves the generator over the previous work by replacing the generator network with an encoder-decoder-encoder network. Different from the aforementioned methods, OCAN [34] and Fence GAN [18] generate fringe instances that are complementary, rather than matching, to the normal data which are used as reference anomalies. However, it cannot guarantee that the generated reference instances well resemble the unknown anomalies. Unlike OCAN, our model generates instances that are close to hard-to-classify normal instances and thus act as normal class supplements.

3   Preliminary: Wasserstein GAN with Gradient Penalty (WGAN-GP)

Generative adversarial networks (GANs) constitute a powerful class of generative models involving an adversarial process in which two modules, a generator G and a discriminator D, are trained simultaneously. The generator G models high dimensional data from a prior noise distribution P_z to learn the real data distribution P_r, while the discriminator D is a binary classifier that estimates the probability that a instance is from the real data x rather than the generated fake data $G(z)$. The objective function of a GAN is: $\min_G \max_D \mathbb{E}_{x \sim P_r} [logD(x)] + \mathbb{E}_{z \sim P_z} [log(1 - D(G(z)))]$.

However, the Jensen-Shannon divergence that GANs aim to minimize is potentially not continuous with respect to the generator's parameters, leading to training difficulties. To solve this problem, [2] propose using the Earth-Mover distance instead in the loss function of the original GAN, thus resulting in the so-called Wasserstein GAN (WGAN) that improves training stability. Under mild assumptions, the Earth-Mover distance is continuous everywhere and differentiable almost everywhere. Thus, WGAN's objective function is constructed using the Kantorovich-Rubinstein duality [28]:

$$\min_G \max_{D \in \mathcal{D}} \mathbb{E}_{x \sim P_r} [D(x)] - \mathbb{E}_{z \sim P_z} [D(G(z))], \qquad (1)$$

where $\mathcal{D}$ is the set of 1-Lipschitz functions and minimizing Eq. (1) with respect to the generator parameters essentially minimizes the Earth-Mover distance between the two distributions. To enforce the Lipschitz constraint on the discriminator, WGAN simply clips the weights of the critic (that is, a discriminator variant) to lie within a compact space $[-c, c]$.

However, WGAN's weight clipping always results in either vanishing or exploding gradients without careful tuning of the clipping threshold c. Considering these issues, WGAN-GP [13] is proposed by introducing gradient penalty to enforce the Lipschitz constraint. The objective function is defined with an additional term forcing the gradient to be smaller than 1:

$$L^{Disc} = \mathop{\mathbb{E}}_{z \sim P_z} [D(G(z))] - \mathop{\mathbb{E}}_{x \sim P_r} [D(x)] + \lambda \mathop{\mathbb{E}}_{\hat{x} \sim P_{\hat{x}}} [(\|\nabla_{\hat{x}} D(\hat{x})\|_2 - 1)^2] \quad (2)$$

$$L^{Gen} = - \mathop{\mathbb{E}}_{z \sim P_z} [D(G(z))], \quad (3)$$

where λ is the penalty coefficient and $P_{\hat{x}}$ is a uniform sampling distribution along straight lines between pairs of points sampled from the original data distribution and the generated data distribution.

4   Method

4.1   Problem Statement

Assume a dataset of m training instances $X = \{x_1, x_2, ..., x_l, x_{l+1}, ..., x_m\}$ with $x_i \in R^D$, where $X^A = \{x_1, x_2, ..., x_l\}$ is a very small set of labelled anomalies and $X^N = \{x_{l+1}, x_{l+2}, ..., x_m\}$ is a set of normal instances, $|X^A| \ll |X^N|$. The goal is to identify whether an instance is anomalous or not.

4.2   Proposed Framework

In this section, we present the details of the proposed model, PIA-WAL[1]. As shown in Fig. 2, the model consists of two major modules, *an anomaly score learner* and *a weighted generative model*. The anomaly score learner, trained on real (and generated) normal instances and a small amount of labelled anomalies, aims to distinguish anomalies from normal instances. A weighted generative model is introduced to assist the anomaly score learner by generating supplemental instances that are close to the peripheral normal instances. With the inclusion of the generated instances, the anomaly score learner is capable to better differentiate between peripheral normal instances and anomalies.

Anomaly Score Learner. Inspired by the DevNet model [22] suggesting that a small number of labelled anomalies can be leveraged into the mechanism of anomaly scoring, the anomaly score learner in our model draws on a similar idea. Specifically, the learner yields an anomaly score for every given input, and defines a reference score for guiding the subsequent anomaly score learning. The scores of anomalies are enforced to significantly deviate from those of normal

---

[1] The code: https://github.com/ZhouF-ECNU/PIA-WAL.

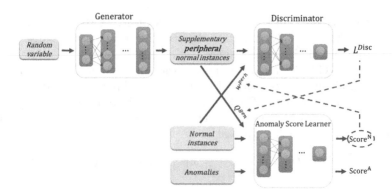

Fig. 2. The framework of PIA-WAL.

instances and lie in the upper tail of the score distribution. Considering that the Gaussian distribution fits the anomaly scores quite well in a variety of datasets [14], we simply use the standard Gaussian distribution as a prior $S \sim N(0,1)$ and set the reference score to 0. The loss function of the anomaly score learner is designed as follows:

$$\ell = \operatorname*{\mathbb{E}}_{x \in X^N} [|\varphi(x)|] + \operatorname*{\mathbb{E}}_{x \in X^A} [max(0, \alpha - \varphi(x))], \tag{4}$$

where $\varphi(\cdot)$ is the output of the anomaly scoring network. The loss in Eq. (4) aims to push the scores of normal instances as close as possible to 0, while enforcing the scores of anomalies to deviate from 0 (that is the mean of the normal instances' scores) at least by α. Note that if an anomaly has a negative anomaly score, the loss will be particularly large. Minimizing the loss encourages large positive deviations for all anomalies. Therefore, the learner can enforce the scores of anomalies to significantly deviate from those of normal instances and to lie in the upper tail of the score distribution. To some degree, the scores of normal instances implicitly indicate the difficulty of the instances to be correctly classified. A normal instance with a high anomaly score indicates that it may be a peripheral normal instance that is hard to be detected by the learner. In addition, there is another reasonable possibility that it may be a noisy instance. Thus, we utilize the anomaly scores for guiding the learning process of the subsequent generative model.

Weighted Generative Model. As previously discussed, most anomaly detection methods do not focus on learning peripheral normal instances due to their complexity and small proportion, which leads to an increase in false positives. Thus, we propose a weighted generative model to generate instances that are close to peripheral normal instances, and thus serve as normal class supplements, to improve the anomaly detection accuracy and reduce false positives. In order to achieve guided instance generation, we utilize the outputs of the anomaly score learner to calculate the degree to which a normal instance is *peripheral*, denoted by w^{perh}. The weight w^{perh} of an observed normal instance $x \in X^N$ is defined as:

$$w^{perh}(x) = |\varphi(x)| * \mathbb{I}[|\varphi(x)| < \alpha], \tag{5}$$

where $\mathbb{I}$ is an indicator function and $|\varphi(x)|$ is the absolute value of an observed normal instance's anomaly score. The more difficult it is for the anomaly score learner to distinguish a normal instance x from the anomalies, the higher the value of $\varphi(x)$ is. Considering that the normal class may contain some anomaly contamination in real-world applications, we use an indicator function to guide the model to not be affected by noise during the learning process. When $|\varphi(x)|$ of an observed normal instance exceeds α (the threshold, same used in Eq. (4)), it indicates that the instance is probably a noisy instance in the proximity of anomalies. The value of $w^{perh}(x)$ is then set to 0, thus not affecting the learning of the generative model.

As mentioned in Sect. 3, the objective of conventional generative models is to minimize the distance between the fake and real data distributions. To make the peripheral normal instances play a more important role in guiding the generator learning, we incorporate the weight w^{perh} into the discriminator loss. We utilize the WGAN-GP model as our basic generative model, due to its excellent data generation capability and stable training. The discriminator loss of WGAN-GP, weighted by w^{perh}, is:

$$\begin{aligned}
L^{Disc} &= \mathop{\mathbb{E}}_{z \sim P_z, x \sim P_r} \left[w^{perh}(x) * (D(G(z)) - D(x)) \right] \\
&\quad + \lambda \mathop{\mathbb{E}}_{\hat{x} \sim P_{\hat{x}}} \left[w^{perh}(x) * (\|\nabla_{\hat{x}} D(\hat{x})\|_2 - 1)^2 \right] \\
&= \mathop{\mathbb{E}}_{z \sim P_z} \left[w^{perh}(x) * D(G(z)) \right] - \mathop{\mathbb{E}}_{x \sim P_r} \left[w^{perh}(x) * D(x) \right] \\
&\quad + \lambda \mathop{\mathbb{E}}_{\hat{x} \sim P_{\hat{x}}} \left[w^{perh}(x) * (\|\nabla_{\hat{x}} D(\hat{x})\|_2 - 1)^2 \right].
\end{aligned} \tag{6}$$

Balanced Sampling. The real peripheral normal instances may account for an extremely small portion of a mini-batch, thus the role of w^{perh} will be weakened in Eq. (6). To this end, we ensure that the distribution of w^{perh} within a batch is uniform by employing a sampling procedure. We first calculate the w^{perh} of a real normal instance x and scale it within the $[0, 1]$ range by $tanh(\cdot)$ for convenient comparison. We then sample ξ uniformly from $[0, 1]$. If $tanh(w^{perh}(x))$ is higher than ξ, the normal instance x is selected and included into the mini-batch. This step is repeated until the mini-batch is full. The instance selection for the next batch continues from the position of the last selected normal instance. Since ξ is uniformly sampled from $[0, 1]$, the distribution of the w^{perh} weights of the selected normal instances is correspondingly uniform. The normal instances with high w^{perh} weights can indeed play an important role in the generator learning.

Generator's Quality. The generated instances are used as supplementary training instances to assist the anomaly score learner to better capture the feature information of the peripheral instances. However, the generated low-quality instances may mislead the anomaly score learner. Therefore, when updating the anomaly score learner's parameters using the generated instances, the quality of the generated instances should be taken into account. In fact, the generated instance's quality is closely related to the generator's quality. Fortunately, because the WGAN model updates the discriminator multiple times before each generator update, it has been shown that the loss function at this point correlates well with the instances' quality [2]. A lower discriminator loss value implies that a generator is capable of generating instances of better quality. The generator's quality is computed as:

$$Q^{gen} = e^{-|L^{Disc}|}.$$ (7)

The weights of the generated instances are set to Q^{gen}. With the inclusion of generated instances as normal class supplements in the training set, the learner's loss function is modified as follows:

$$L^{score} = \underset{x \in X^N}{\mathbb{E}} \left[\|\varphi(x)\| \right] + \underset{x \in X^A}{\mathbb{E}} \left[max(0, \alpha - \varphi(x)) \right] + Q^{gen} * \underset{x \in G(z)}{\mathbb{E}} \left[\|\varphi(x)\| \right].$$ (8)

4.3   Outline of PIA-WAL

In summary, the overall objective function of PIA-WAL can be written as follows:

PIA-WAL's anomaly score learner : $\min\ L^{score}$.

PIA-WAL's weighted generative model : $\underset{G}{\min}\ L^{Gen}, \quad \underset{D}{\max}\ -L^{Disc}$. (9)

The training procedure of PIA-WAL is outlined in Algorithm 1. Given a training dataset, in the first iteration, mini-batches are used to update the parameters of the anomaly score learner so as to minimize ℓ in Eq. (4) (Lines 4–8). Note that each mini-batch is constructed by sampling the same number of abnormal and normal instances. Then in the subsequent iterations, the weighted generative model (Lines 10–14) and anomaly score learner (Lines 15–20) are trained in turns. The outputs of the anomaly score learner trained in the previous iteration are used to calculate the degree to which the instances are peripheral (Line 12), thus assisting the weighted generative model learning. The instances generated by the latest updated generative model act as supplementary training instances to further optimize the learner with the corresponding generator's quality (Line 15). By jointly updating the two modules alternately, a trained anomaly score learner is obtained.

During the testing stage, the optimized anomaly score learner is used to produce an anomaly score for every test instance. Here, a reasonable threshold should be selected to determine whether an instance is normal or not. Due to the standard Gaussian prior assumption of the anomaly scores, we use the scores' quantiles to determine a threshold with a desired confidence level [22]. By applying a confidence level of 99.9%, an instance with an anomaly score of over 3.09 ($z_{0.999} = 3.09$) is considered an anomaly, meaning that the probability of an instance being normal is 0.001. Thus, we can set an appropriate and interpretable threshold to identify anomalies with a high confidence level.

Algorithm 1. PIA-WAL's Training Procedure

Input: Dataset $X = X^A \cup X^N$, margin parameter α, penalty coefficient λ, batch size t, number of training epochs for anomaly score learner $Epoch_{learner}$ and generative model $Epoch_{gen}$

Output: A trained anomaly score learner and a weighted generative model

1: Randomly initialize the parameters of the learner and weighted generative model
2: $n\_batches = (int)(m/t)$
3: **for** i=1 to $Epoch_{learner}$ **do**
4:    **if** i==1 **then**
5:       **for** j=1 to n_batches **do**
6:          Construct a mini-batch by randomly sampling t instances from X^A and X^N, separately
7:          Update the parameters of the anomaly score learner using Eq. (4)
8:       **end for**
9:    **else**
10:      **for** j=1 to $Epoch_{gen}$ **do**
11:         Sample a balanced mini-batch from X^N
12:         Compute w^{perh} for each instance in mini-batch using Eq. (5)
13:         Optimize the discriminator D and generator G using Eq. (6) and Eq. (3), respectively
14:      **end for**
15:      Calculate generator's quality Q^{gen} using Eq. (7)
16:      **for** j=1 to n_batches **do**
17:         Use the latest updated generator G to generate t supplementary instances
18:         Construct a batch with t generated instances, t observed normal instances and t anomalies
19:         Optimize the anomaly score learner using Eq. (8)
20:      **end for**
21:   **end if**
22: **end for**

---

5 Experiments

5.1 Datasets and Baselines

In order to verify the effectiveness of PIA-WAL, we conducted experiments on four widely-used real-word datasets. The NB-15 dataset [17] is a network intrusion dataset containing 107,687 data instances, each being 196-dimensional, in which network attacks are treated as anomalies (21.5%). The Census data [11] extracted from the US Census Bureau database contains 299,285 instances in a 500-dimensional space. The task is to detect the rare high-income individuals, which constitute about 6% of the data. The Celeba dataset [16] contains 201,690 celebrity images in a 39-dimensional space in which the scarce bald celebrities (less than 3%) are treated as anomalies. The Fraud data [9] contains 284,807 credit card transactions in a 29-dimensional space. The task is to detect fraudulent transactions, accounting for 0.17%. We split each dataset into training and test sets with a ratio of 8:2. The number of labelled anomalies used for training was fixed to 70 across all datasets in accordance with the setting in [20].

PIA-WAL is compared against 7 state-of-the-art methods. The first three are GAN-based methods, the fourth is a classical unsupervised approach, while the other three leverage labelled anomalies:

- **ALAD** [32] builds upon bi-directional GANs and uses reconstruction errors to determine if instances are anomalous.
- **GANomaly** [1] employs an adversarial autoencoder within an encoder-decoder-encoder pipeline to capture the distribution of training instances in the original and latent spaces.
- **OCAN** [34] leverages the idea of complementary GANs to generate fringe instances used as reference anomalies and train a one-class discriminator when only normal instances are observed.
- **iForest** [15] is a widely-used unsupervised method that detects anomalies based on the number of steps required to isolate instances by isolation trees.
- **ADOA** [33] follows a two-stage procedure. First, the observed anomalies are clustered while the unlabelled instances are categorized into potential anomalies and reliable normal instances. Then, a weighted classifier is trained for further detection.
- **DevNet** [22] is a deep supervised method that leverages a few labelled anomalies with a Gaussian prior to perform an end-to-end differentiable learning of anomaly scores.
- **Deep SAD** [24] minimizes the distance of normal instances to the one-class center while maximizing the distance of known anomalies to the center.

Experimental Setting. The hyperparameters of the baselines were set to the same values as the ones used in their respective papers. The anomaly score learner used in DevNet was also leveraged in our model and run with one hidden layer of 20 units and a ReLU activation. We optimized the model parameters using RMSprop with a learning rate of 0.0001, batch size $t = 128$, $Epoch^{learner} = 30$, $Epoch^{gen} = 1000$ and $\lambda = 10$. The generator G was set with three layers of 512 units and the discriminator D was set with two hidden layers of 128 and 64 units respectively. We set the margin $\alpha = 5$ to achieve a very high significance level for all labelled anomalies. The PIA-WAL's parameters were initialized randomly. All models were implemented in Keras 2.2.4 and ran on a machine with 32 GB of memory, 6 CPU cores and 1 quadro p400 GPU.

Two widely-used complementary performance metrics, Area Under the Receiver Operating Characteristic Curve (AUROC) and Area Under the Precision-Recall Curve (AUPRC), were used. The paired Wilcoxon signed rank test [30] was used to examine the significance of the performance of PIA-WAL against the seven baselines. All reported results were averaged over 10 independent runs.

5.2   Results and Discussion

Effectiveness on Real-world Datasets. Table 1 shows the performances of PIA-WAL and all baselines on the four datasets in terms of AUROC and AUPRC.

Table 1. AUROC and AUPRC performance (with ± standard deviation) of PIA-WAL and the baselines.

Models	AUPRC						AUROC					
	NB-15	Census	Celeba	Fraud	Average	P-value	NB-15	Census	Celeba	Fraud	Average	P-value
ALAD	70.3±4.0	10.3±1.0	6.8±0.8	44.1±2.7	32.88	<0.0001	87.5±0.7	70.1±2.8	76.9±3.6	95.9±0.5	82.60	<0.0001
GANomaly	73.7±1.6	10.3±2.0	8.3±2.4	42.7±7.8	33.75	<0.0001	86.2±0.9	72.2±3.2	79.0±5.6	94.4±2.1	82.95	<0.0001
OCAN	43.3±3.8	8.3±5.1	4.0±1.2	39.4±8.8	23.75	<0.0001	77.5±1.8	62.3±3.0	62.3±8.2	94.7±1.0	74.20	<0.0001
IForest	24.4±1.7	7.2±0.3	6.4±1.6	16.9±4.8	13.73	<0.0001	57.8±2.3	60.1±2.1	70.7±0.9	95.4±0.3	71.00	<0.0001
ADOA	40.2±3.1	23.9±3.5	29.8±1.0	38.0±0.5	32.98	<0.0001	77.8±2.7	84.5±1.4	90.4±0.4	95.9±0.2	87.15	<0.0001
DevNet	81.0±1.6	36.4±0.9	23.6±1.4	62.8±1.0	50.95	0.0001	92.6±0.3	81.2±2.8	94.5±0.2	95.8±1.3	91.03	<0.0001
Deep SAD	**88.8±0.2**	24.2±6.4	22.6±2.4	**72.1±1.5**	51.93	0.006	**95.7±0.1**	73.2±9.2	92.5±1.4	95.4±1.3	89.20	0.0001
PIA-WAL	**88.8±0.7**	**41.5±1.2**	**30.9±1.6**	71.1±0.6	**58.08**	-	95.1±0.2	**85.3±1.1**	**95.5±0.8**	**97.3±0.7**	**93.30**	-

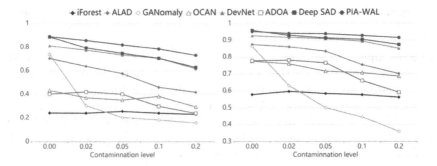

iForest ALAD GANomaly OCAN DevNet ADOA Deep SAD PIA-WAL

Fig. 3. AUPRC (left) and AUROC (right) under different contamination levels on the NB-15 dataset.

It is evident that PIA-WAL obtains substantial improvements over the alternative methods. Particularly, across all datasets, PIA-WAl obtains 27.95% higher average AUPRC than GAN-based methods (ALAD, GANomaly and OCAN), 44.35% higher average AUPRC than the unsupervised method (iForest) and 12.79% higher average AUPRC than supervised methods (ADOA, DevNet and Deep SAD). In terms of AUROC, PIA-WAL produces 13.38% and 22.3% higher average AUROC compared to classic GAN-based methods and iForest, respectively. In addition, PIA-WAL also outperforms the supervised methods, namely, DevNet by 2.27% and ADOA by 6.15%. Compared with Deep SAD, PIA-WAL obtains comparable AUPRC on the NB-15, yet achieves obvious improvements on Census and Celeba and produces a higher AUROC on the Fraud dataset. The improvements are all statistically significant at a 99% confidence level. The reason lies in PIA-WAL's ability to efficiently leverage a limited amount of available anomalies as prior knowledge to enhance the ability of distinguishing anomalies from normal instances. Meanwhile, with the help of generated peripheral normal instances, the model can further reduce the number of false positives.

Robustness Under Anomaly Contamination. To investigate the robustness of PIA-WAL, we polluted the normal training instances from the NB-15 dataset with contamination levels ranging from 0% up to 20%. The results in terms of AUPRC and AUROC are presented in Fig. 3. The detection performance of all methods decreases with increased contamination levels, particularly for the

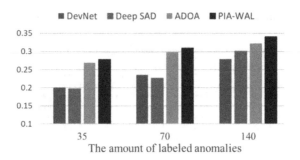

Fig. 4. AUPRC w.r.t the amount of labelled anomalies on the Celeba dataset.

methods trained using only normal instances (e.g., GANomaly and ALAD). Nevertheless, it is obvious that PIA-WAL consistently outperforms its alternatives across different contamination levels. Moreover, as the contamination level increases, the decline rate of PIA-WAL's performance is much lower than that of Deep SAD; even though they perform comparably when the data is pure. This suggests that PIA-WAL has a substantial capability of distinguishing anomalies from normal instances in challenging noisy environments.

Effectiveness w.r.t. the Amount of labelled Anomalies. To account for the difficulty of obtaining labelled anomalies in most applications, we vary the number of labelled anomalies from 35 to 140 and examine the effectiveness of PIA-WAL. Considering that classical GAN-based anomaly detection methods are trained only on normal instances and iForest is an unsupervised method, their performance is invariant to the amount of labelled anomalies. Figure 4 shows the AUPRCs obtained by the other four methods w.r.t. different numbers of labelled anomalies on the Celeba dataset. The performance of these methods generally increases with the increased number of labelled anomalies. As the amount of labelled anomalies increases, it's obvious that PIA-WAL consistently outperforms the alternative models. It should be emphasized that, even when using only 35 labelled anomalies, PIA-WAL still achieves the best performance.

Ablation Study. To study the impact of Q^{gen} (the weights of the generated instances in Eq. (8)) and w^{perh} (the weights of normal instances in Eq. (6)), we further conducted an ablation study of PIA-WAL on the NB-15 dataset, and reported the results in Table 2. When we disable Q^{gen} in PIA-WAL, there is a drop in performance, as expected, since the anomaly score learner is most probably misled by the generated low-quality instances. When we remove w^{perh}, the generator works similarly to a conventional generative model, rather than generating mainly peripheral instances. The performance decreases by 7.7% in AUPRC. This suggests that the use of generated peripheral normal instances can effectively improve the detection accuracy.

6   Application to Merchant Fraud Detection

We applied PIA-WAL to real-world merchant fraud detection. The goal of the task is to predict whether a merchant is involved in fraudulent activities based on their daily transactions. We collected 33,744,737 merchants' transactions made from April to June of 2020 from a mobile payment platform and extracted 182 daily transaction features for each merchant (e.g., payment volume, transaction number). Finally, 272,821 normal instances and 205 fraud instances from April and May were selected for training, while 160,695 normal instances and 106 fraud instances from June were used for testing. Since a subsequent on-the-spot investigation required for verification is time-consuming and labor-intensive, this application would require a fraud detection model that is both accurate and aimed at reducing the number of false positives as much as possible.

Table 2. Model ablation study results on the NB-15 dataset.

Model	AUPRC	AUROC
PIA-WAL	88.8	95.1
PIA-WAL$_{-Q^{gen}}$	87.7	94.4
PIA-WAL$_{-w^{perh}}$	81.1	94.1

Table 3. Merchant fraud detection performance obtained by the baselines and PIA-WAL.

Model	TN	FP	FN	TP	AUPRC
OCAN	$127,151 \pm 13,249$	$33,544 \pm 13,249$	6 ± 5	100 ± 5	1.2 ± 0.3
ALAD	$128,639 \pm 1$	$32,056 \pm 1$	1 ± 0	$\mathbf{105 \pm 0}$	4.7 ± 0.8
GANomaly	$146,171 \pm 3,351$	$14,524 \pm 3,351$	9 ± 2	97 ± 2	1.1 ± 0.1
ADOA	$153,913 \pm 2,051$	$6,782 \pm 2,051$	17 ± 11	89 ± 11	2.7 ± 1.4
IForest	$154,279 \pm 343$	$6,416 \pm 343$	23 ± 2	83 ± 2	2.0 ± 0.1
DevNet	$159,774 \pm 152$	921 ± 152	18 ± 3	88 ± 3	41.6 ± 2.3
Deep SAD	–	–	–	–	54.3 ± 3.2
PIA-WAL	$160,215 \pm 123$	$\mathbf{480 \pm 116}$	20 ± 3	86 ± 3	$\mathbf{56.6 \pm 0.8}$

Table 3 summarizes the results obtained by the seven baselines and PIA-WAL. Note that since the proportion of anomalies and normal instances is extremely unbalanced (1:1516), AUPRC is considered to be more appropriate for evaluating the performance of the models [22]. Evidently, compared with the baselines, PIA-WAL achieves a substantial improvement of 2.3%–50% (with a small standard deviation) in AUPRC, which demonstrates the effectiveness of PIA-WAL in merchant fraud detection. Moreover, we also calculated threshold-specific metrics including TN (True Negatives), FN (False Negatives), FP (False Positives) and TP (True Positives). As Table 3 shows, GAN-based methods obtain high TPs, but tend to produce massive FPs (from 14,524 to 33,544). PIA-WAL obtains comparable TP values when compared to iForest, ADOA and DevNet. However, iForest and ADOA incur high FPs, 6,416 and 6,782, respectively. DevNet achieves the lowest FPs among the baselines, but the number of FPs is two times larger than the one obtained by PIA-WAL. Note that Deep SAD does not provide a comprehensive threshold selection strategy and thus its threshold-specific metric values are omitted from Table 3. Finally, the proposed PIA-WAL is able to significantly reduce the number of false positives while maintaining the anomaly detection accuracy in a real-world application.

7   Conclusion

We introduce a novel end-to-end anomaly detection model, PIA-WAL, which utilizes a few labelled anomalies to guide an adversarial training process. The main contribution is the generation of peripheral normal instances as supplements, which allows for PIA-WAL to learn a more representative description of the normal class. PIA-WAL achieves significant lifts over seven state-of-the-art methods on four public datasets. Even when the anomaly contamination is high or the number of labelled anomalies is low, PIA-WAL still obtains satisfactory performance. When applied to a real merchant fraud detection application, PIA-WAL can indeed reduce the number of false positives and maintain the anomaly detection accuracy, thus avoiding unnecessary labor and material resources.

Acknowledgements. This research was supported in part by NSFC grant 61902127 and Natural Science Foundation of Shanghai 19ZR1415700.

References

1. Akcay, S., Atapour-Abarghouei, A., Breckon, T.P.: GANomaly: semi-supervised anomaly detection via adversarial training. In: Jawahar, C.V., Li, H., Mori, G., Schindler, K. (eds.) ACCV 2018. LNCS, vol. 11363, pp. 622–637. Springer, Cham (2019). https://doi.org/10.1007/978-3-030-20893-6_39
2. Arjovsky, M., Chintala, S., Bottou, L.: Wasserstein generative adversarial networks. In: ICML, pp. 214–223 (2017)
3. Audibert, J., Michiardi, P., Guyard, F., et al.: USAD: unsupervised anomaly detection on multivariate time series. In: SIGKDD, pp. 3395–3404 (2020)
4. Bay, S.D., Schwabacher, M.: Mining distance-based outliers in near linear time with randomization and a simple pruning rule. In: SIGKDD, pp. 29–38 (2003)
5. Bergmann, P., Fauser, M., Sattlegger, D., et al.: MCTec AD - a comprehensive real-world dataset for unsupervised anomaly detection. In: CVPR, pp. 9592–9600 (2019)
6. Boukerche, A., Zheng, L., Alfandi, O.: Outlier detection: methods, models, and classification. ACM Comput. Surv. **53**, 1–37 (2020)
7. Breunig, M.M., Kriegel, H.P., Ng, R.T., Sander, J.: LOF: identifying density-based local outliers. In: SIGMOD, pp. 93–104 (2000)
8. Chen, J., Sathe, S., Aggarwal, C., Turaga, D.: Outlier detection with autoencoder ensembles. In: SDM, pp. 90–98 (2017)

9. Dal Pozzolo, A., Boracchi, G., Caelen, O., Alippi, C., Bontempi, G.: Credit card fraud detection: a realistic modeling and a novel learning strategy. IEEE Trans. Neural Netw. Learn. Syst. **29**, 3784–3797 (2017)
10. Di Mattia, F., Galeone, P., De Simoni, M., Ghelfi, E.: A survey on GANs for anomaly detection. arXiv preprint arXiv:1906.11632 (2019)
11. Dua, D., Graff, C.: UCI machine learning repository (2017). http://archive.ics.uci.edu/ml
12. Goodfellow, I.J., Pouget-Abadie, J., Mirza, M., Xu, B., Warde-Farley, D., Ozair, S., et al.: Generative adversarial networks. In: NIPS, pp. 2672–2680 (2014)
13. Gulrajani, I., Ahmed, F., Arjovsky, M., Dumoulin, V., Courville, A.C.: Improved training of Wasserstein GANs. In: NIPS, pp. 5767–5777 (2017)
14. Kriegel, H.P., Kroger, P., Schubert, E., Zimek, A.: Interpreting and unifying outlier scores. In: SDM, pp. 13–24. SIAM (2011)
15. Liu, F.T., Ting, K.M., Zhou, Z.H.: Isolation-based anomaly detection. TKDD **6**, 1–39 (2012)
16. Liu, Z., Luo, P., Wang, X., Tang, X.: Deep learning face attributes in the wild. In: ICCV, December 2015 (2015)
17. Moustafa, N., Slay, J.: UNSW-NB15: a comprehensive data set for network intrusion detection systems. In: MilCIS, pp. 1–6 (2015)
18. Ngo, P.C., Winarto, A.A., Kou, C.K.L., Park, S., Akram, F., Lee, H.K.: Fence GAN: towards better anomaly detection. In: ICTAI, pp. 141–148 (2019)
19. Pang, G., Cao, L., Chen, L., Liu, H.: Learning representations of ultrahighdimensional data for random distance-based outlier detection. In: SIGKDD, pp. 2041–2050 (2018)
20. Pang, G., van den Hengel, A., Shen, C., Cao, L.: Toward deep supervised anomaly detection: reinforcement learning from partially labeled anomaly data. In: SIGKDD, pp. 1298–1308 (2021)
21. Pang, G., Shen, C., Cao, L., Hengel, A.V.D.: Deep learning for anomaly detection: a review. ACM Comput. Surv. **54**, 1–38 (2021)
22. Pang, G., Shen, C., van den Hengel, A.: Deep anomaly detection with deviation networks. In: SIGKDD, pp. 353–362 (2019)
23. Perera, P., Nallapati, R., Xiang, B.: OCGAN: one-class novelty detection using GANs with constrained latent representations. In: CVPR, pp. 2898–2906 (2019)
24. Ruff, L., et al.: Deep semi-supervised anomaly detection. In: ICLR (2020)
25. Ruff, L., et al.: Deep one-class classification. In: ICML, pp. 4393–4402 (2018)
26. Schlegl, T., Seeböck, P., Waldstein, S.M., Schmidt-Erfurth, U., Langs, G.: Unsupervised anomaly detection with generative adversarial networks to guide marker discovery. In: Niethammer, M., et al. (eds.) IPMI 2017. LNCS, vol. 10265, pp. 146–157. Springer, Cham (2017). https://doi.org/10.1007/978-3-319-59050-9_12
27. Seeböck, P., et al.: Unsupervised identification of disease marker candidates in retinal OCT imaging data. IEEE Trans. Med. Imaging **38**, 1037–1047 (2018)
28. Villani, C.: Optimal Transport: Old and New. Grundlehren der Mathematischen Wissenschaften [Fundamental Principles of Mathematical Sciences], vol. 338. Springer, Heidelberg (2009). https://doi.org/10.1007/978-3-540-71050-9
29. Wang, H., Pang, G., Shen, C., Ma, C.: Unsupervised representation learning by predicting random distances. In: IJCAI, pp. 2950–2956 (2020)
30. Woolson, R.: Wilcoxon signed-rank test. In: Wiley Encyclopedia of Clinical Trials, pp. 1–3 (2007)
31. Zenati, H., Foo, C.S., Lecouat, B., Manek, G., Chandrasekhar, V.R.: Efficient GAN-based anomaly detection. In: ICLR (2018)

32. Zenati, H., Romain, M., Foo, C.S., Lecouat, B., Chandrasekhar, V.: Adversarially learned anomaly detection. In: ICDM. pp. 727–736 (2018)

33. Zhang, Y.L., Li, L., Zhou, J., Li, X., Zhou, Z.H.: Anomaly detection with partially observed anomalies. In: WWW, pp. 639–646 (2018)

34. Zheng, P., Yuan, S., Wu, X., Li, J., Lu, A.: One-class adversarial nets for fraud detection. In: AAAI, pp. 1286–1293 (2019)

35. Zhou, C., Paffenroth, R.C.: Anomaly detection with robust deep autoencoders. In: SIGKDD, pp. 665–674 (2017)

36. Zong, B., Song, Q., Min, M.R., Cheng, W., Lumezanu, C., et al.: Deep autoencoding gaussian mixture model for unsupervised anomaly detection. In: ICLR (2018)

HieNet: Bidirectional Hierarchy Framework for Automated ICD Coding

Shi Wang[1], Daniel Tang[2(✉)], Luchen Zhang[3], Huilin Li[4], and Ding Han[5]

[1] Key Laboratory of Intelligent Information Processing, Institute of Computing Technology, Chinese Academy of Sciences, Beijing, China
wangshi@ict.ac.cn

[2] University of Luxembourg, Interdisciplinary Centre for Security, Reliability and Trust (SNT), TruX, Luxembourg City, Luxembourg
xunzhu.tang@uni.lu

[3] National Computer Network Emergency Response Technical Team/Coordination Center of China, Beijing, China
zlc@cert.org.cn

[4] Department of Civil Engineering, Technical University of Denmark, 2800 Lyngby, Denmark

[5] Huazhong University of Science and Technology, Wuhan, China

Abstract. International Classification of Diseases (ICD) is a set of classification codes for medical records. Automated ICD coding, which assigns unique International Classification of Diseases codes with each medical record, is widely used recently for its efficiency and error-prone avoidance. However, there are challenges that remain such as heterogeneity, label unbalance, and complex relationships between ICD codes. In this work, we proposed a novel Bidirectional Hierarchy Framework(HieNet) to address the challenges. Specifically, a personalized PageRank routine is developed to capture the co-relation of codes, a bidirectional hierarchy passage encoder to capture the codes' hierarchical representations, and a progressive predicting method is then proposed to narrow down the semantic searching space of prediction. We validate our method on two widely used datasets. Experimental results on two authoritative public datasets demonstrate that our proposed method boosts the state-of-the-art performance by a large margin.

Keywords: Structural encoder · ICD coding · Bidirectional passage retriever · Hierarchical embedding · Healthcare · Co-occurrence encoder

1 Introduction

The International Classification of Diseases (ICD) is widely considered as a healthcare multi-label classification system, supported by the World Health Organization (WHO). ICD codes have widely been used for reimbursement, taxonomy of diagnoses and procedures, and monitoring health issues [3, 20]. ICD coding needs coder to assign proper codes to a patient visit, which is composed of multiple long and heterogeneous textual narratives (e.g., discharge diagnosis, procedure notes, event notes), authored by different healthcare professionals, which means it's time-assuming, error-prone, and expensive in manual way. As a result, automated ICD encoding has attracted much attention

---

S. Wang and D. Tang—Represents equal contribution to this work.

© The Author(s), under exclusive license to Springer Nature Switzerland AG 2022
A. Bhattacharya et al. (Eds.): DASFAA 2022, LNCS 13246, pp. 523–539, 2022.
https://doi.org/10.1007/978-3-031-00126-0_38

since it can save time and labor for billing. A number of neural network methods handling automated coding as multi-label task, have been proposed by [5, 19, 33], which converts the ICD coding into a set of binary classification for each code.

ICD codes can be organized in a tree-like structure. If a node N represents a kind of disease, the children of N are the sub-types of this disease. And in many cases, the differences among the siblings from one parent disease are very subtle. There are several challenges to link the discharge summaries with ICD codes: (1) Each admission record has a long and complex discharge summary. And the summaries are usually more than one thousand words long, containing medical history, diagnosis texts, surgical procedures, etc. Thus, it is difficult to assign proper codes to a given clinical note. (2) The label space to predict of ICD codes is very large (e.g., over 18,000 for ICD-9-CM) and the label distribution is extremely unbalanced. However, the average length of training-label codes in MIMIC-III-full is only 15.89 and the length of codes in 80% clinical notes is less than 22. Therefore, we only need to ensure the accuracy of top predicted codes. (3) Mutual exclusivity (*ME*) in ICD codes: As shown in Fig. 1), we use deep and light orange colors to represent *ME* codes. Deeper orange color represents 'parent-child' codes, and lighter orange color represents sibling codes. For example, given a clinical text, if a finer code '521.00' is predicted, the parent of it '521.0' and grand-parent '521' should not occur in the predicted results. Furthermore, some sibling ICD codes should not appear in the same predicted result, such as '464.00' (Acute laryngitis without mention of obstruction) and '464.01' (Acute laryngitis with obstruction), because they are anti-sense. (4) Co-occurrence (*CC*) in ICD codes: We leverage light green and deep green colors for *CC* codes. Light green color represents reasoning co-occurrence codes. For example, '997.91' (hypertension) usually leads to the occurrence of '429.9' (heart disease, unspecified); Deep green color represents *CC* codes caused by common pre-conditions. For example, 'staying up too late' usually causes '997.91' (hypertension) and '784.0' (headache).

Some methods based on CNN [17, 19, 33] were proposed to address the issues from the characteristic (1) above, and they were proved efficient in extracting features from long texts. For characteristic (2), to our knowledge, no previous scheme was proposed to solve it. To address problem with characteristic (3), [32] leverage a sequential tree-lstm architecture to extract structural features of code tree. However, as we know, there is no contextual relationships among siblings in the code tree and it's hard for tree-lstm process long sequential issue. Another approach [5] leverages hyperbolic ball to encode the *ME* feature, but it's hard to measure the real performance of this non-euclidean method on *ME* problem. To address problem with characteristic (4), previous methods (e.g., [5]) employ GCN [14] to encode co-occurrence features of codes. However, GCN is unsuitable for describing the root node's neighborhood and not designed the causal co-occurrence cases. In summary, issues from characteristics (2), (3), (4) remain to be solved.

In this paper, to address issues above, we present a novel method HieNet, which is short for Bidirectional Hierarchy Framework for Automated ICD Coding. HieNet contains three main modules, including a progressive mechanism (*PM* module), a bidirectional hierarchy passage encoder (*BHPE* module), and a personalized PageRank (*PP* module). These modules are designed as the solution of (2), (3), and (4), respectively.

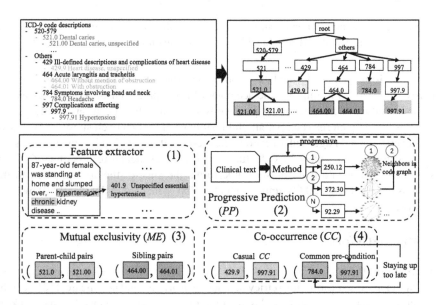

Fig. 1. A schematic diagram of the ICD code tree, matching process between clinical notes and codes (1), progressive mechanism (2), and relationships among ICD codes ((3) and (4)).

Progressive Prediction for (2): Given a clinical text or diagnose description, the previous methods predict all its ICD codes at the same time. However, some codes achieving high scores in their binary predictions can help predict other codes. For example, the neighbors of predicted code '401.9' (unspecified essential hypertension) in code tree usually contains some other potential gold labels, such as '348.4' (compression of brain). To make full use of first predicted, we introduce an approach named progressive mechanism to reduce the difficulty of improving the accuracy of predicted codes (e.g., average 14.3% improvement on Jaccard metric over the average scores of the best baselines).

Hierarchy Features Encoding for (3): There are two patterns in mutual exclusivity (*ME*): parent-child and sibling relationships. As stated above, ICD codes with *ME* relationships should not occur in one clinical note. To address the issue, we propose a bidirectional hierarchy passage encoder (*BHPE*) that contains two sub-modules: bidirectional passage retriever (*BPR*) and tree position encoder (*TPE*). The experimental results indicate that the *BHPE* module improves HieNet by 12.0% on macro-F1 on MIMIC-III full.

Code Co-Occurrence Encoding for (4): Some codes have causal or pre-condition co-occurrence relationship, which is called code co-occurrence (*CC*). Pre-condition *CC* codes are usually caused by common bad habits or hurts. We propose personalized PageRank (*PP*) to encode pre-condition *CC* features. Furthermore, the combination of *PM* and *PP* enable *PP* has the ability to encode causal *CC* relationships among ICD codes. The experimental results show that *PP* module makes the 10.7% improvement on macro-F1 on MIMIC-III full and 13.1% improvement on macro-F1 on MIMIC-II.

Our Contributions: 1) To the best of our knowledge, we are first to propose the progressive mechanism to improve the accuracy of top_K predicted ICD codes. 2) We are first to introduce the bidirectional passage retriever and tree position encoder as the solutions of two patterns (i.e., parent-child, sibling) of mutual exclusivity. 3) We introduce a personalized PageRank to encode the pre-condition *CC* and leverage progressive mechanism to capture casual *CC* features. 4) The experimental results on two widely used datasets illustrate that HieNet outperforms the state-of-the-art compared to the previous methods (e.g., 20.6% improvement on top_{30}-Jaccard (MIMIC-III 50) over the best baseline CAML).

2 Related Work

Automatic ICD Coding: Automatic ICD coding has been studied in a large coverage of areas, including information retrieval, machine learning, and healthcare. [12] treat the ICD coding as a multi-label text classification problem and introduced a label ranking approach based on the features extracted from the clinical notes. [19] proposed the landmark work CAML with attention algorithm and leveraged CNN to capture the key information for each code, and DR-CAML is a updated version of CAML with code description proposed in the same publication. Inspired by CAML, more CNN-based methods are proposed, including [2,26,36]. [28] explored character based on LSTM with attention and [32] applied tree LSTM with hierarchy information for ICD coding. [35] applied multi-modal machine learning to predict ICD codes. [31] transferred the ICD coding into a path generation task and developed an adversarial reinforcement path generation framework for ICD coding.

Hierarchical Encoder: Tree position encoder was proposed by [7] and applied in tree-based transformers. Recent works demonstrate that tree position encoder can handle source code summarization [1] and semantic prediction [8]. To our knowledge, we are the first to take the tree position encoder to capture the position embedding of code tree.

Co-occurrence Encoder: PageRank was proposed by [23] and widely employed in website page ranking [16], latent topics mining [22], key phrase extraction [7], and mutillingual word sense disambiguation [27]. [15] proposed personalized PageRank (PPNP) to address the limit distribution of GCN. We are the first to use the personalized PageRank to encode the code co-occurrence representations for the automated ICD coding task.

3 Proposed Model

3.1 Problem Definition

Following the previous works [5,32], We formulate ICD coding as a multi-label text prediction problem. We make some definitions here to help state the proposed work well.

- **Definition 1: *CC*.** *CC* is short for Code co-occurrence and it contains three cases: (1) pre-condition *CC*. Intuitively, some diseases are usually caused by common bad habits; (2) causal *CC*. (3) other co-occurrence codes without obvious reasons.
- **Definition 2: *ME*.** *ME* is short for Mutual exclusivity and it contains two cases: (1) Parent-child pairs. (2) Complementary sibling pairs, such as '464.00' (without mention of obstruction) and '464.01' (with obstruction).
- **Definition 3: *DP*.** *DP* represents improving the accuracy of top_K predicted codes. Actually, 80% clinical notes have less than 22 codes, so we need to focus on improving the accuracy of top_K predicted codes.

3.2 Model Architecture

Figure 2 shows an overview of the bidirectional hierarchy passage framework. Firstly, we encode the code hierarchy semantic as hierarchical code representations via bidirectional hierarchy passage retriever. Furthermore, we employ a multi-channel CNN to obtain clinical document embeddings and conduct a code-wise operation between document embeddings and hierarchical code representations. Secondly, we introduce a progressive mechanism to improve the accuracy of top_K predicted ICD codes (***DP***). Thirdly, we leverage a personalized PageRank algorithm to calculate the co-occurrence relationships among ICD codes. Finally, we aggregate the results of above modules and conduct a full connected layer with multiple sigmoid functions to generate 0–1 probability distributions for each code.

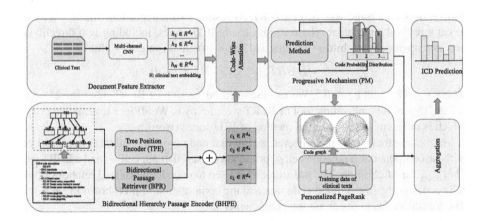

Fig. 2. The architecture of our model.

3.3 Document Feature Extractor (*DFE*)

Given a electronic health record $W = \{w_1, w_2, ..., w_N\}$ (N donates the length of W), we map W into a vector representation $X \in \mathbb{R}^{d_e \times N}$ where d_e indicates the dimension of word embedding. We leverage a multi-channel one-dimensional convolution neural network to encode clinical texts. A convolution operation contains a filter W_f with a window of k words. For example, a feature f_i is generated with a window of words $x_{i:i+k-1}$ by

$$f_i = relu(W_f * x_{n:n+k-1} + b_c) \tag{1}$$

where $relu(\cdot)$ is a non-linear transformation function, $b \in \mathbb{R}^{d_e}$ is a bias, $*$ is the concatenation operator. This filter is applied with multiple filter size to produce a final feature map f:

$$f = f_{1,2,\ldots,n-l+1} = f_1 \oplus f_2 \oplus \cdots \oplus f_{n-l+1}. \tag{2}$$

Then we employ a pooling operation over feature map f and take the maximum value as the final value of f by $\hat{f} = \max(f)$. Regarding l channels (l different window sizes), we concatenate generated l feature maps as a representation H of an clinical text as follows:

$$H = \hat{f}_1 \oplus \hat{f}_2 \oplus \cdots \oplus \hat{f_{n-l+1}}. \tag{3}$$

3.4   Bidirectional Hierarchy Passage Encoder (BHPE)

This section will introduce the hierarchy encoder that includes two main modules: bidirectional passage retriever (**BPR**) and tree position encoder (**TPE**). First, we construct a dimensional vector by averaging the vectors of words of code's description to represent the code. Then, we propose **BPR** to capture parent-child relationships. Moreover, we introduce **TPE** to encode tree positions of codes. Next, we add representations from **BPR** and **TPE** as the final vectors of codes that contain both hierarchical and parent-child contextual features. Finally, we obtain code-wise document representations by conducting code-wise attention between codes vectors and document representations.

Bidirectional Passage Retriever (BPR). BPR uses two independent BERT encoders to capture the hierarchical relationships among ICD codes, including two directional process: up-stream u (child $\rightarrow$ parent) and down-stream d (parent $\rightarrow$ child):

$$e_u = BERT_u(u), e_d = BERT_d(d), \tag{4}$$

where $e_u \in \mathbb{R}^d$ and $e_d \in \mathbb{R}^d$. We use the uncased version of BERT-base; therefore, $d = 768$. The initial embedding tool of BERT we employ is WordPiece tokenization (wp) which is different from the initial method of ICD codes (word2vec)[1].

In the up-stream passage retriever, an internal node (with M children) is comprised of these components: a position cell p for each node, a self-input cell $i_\uparrow$, and a BERT cell $\{b_\uparrow\}_{m=1}^M$ for M children. The position cell is used to encode the related relationship of ICD codes in the hierarchical code tree and the computation of p is shown in Sect. 3.4. The transition equations of among components are:

$$u = Set(p_k + wp(w_k))_{m=1}^M,$$
$$i_\uparrow = \{b_\uparrow\}_{m=1}^M = BERT_u(u), \tag{5}$$

where w_k is the k-th token of M children of node C.

---

[1] Why initial methods of ICD codes and clinical codes are diffident? Answer: ICD codes are usually beyond the vocabulary of BERT because they are professional and technical terms while words in clinical are original that can be covered by the vocabulary of BERT. Therefore, for better representing ICD codes and clinical notes, we leverage word2vec tool and WordPiece tokenization function in BERT to init them, respectively.

In the down-stream passage retriever, for a not-root node, it has such one component: an input cell $i_\downarrow$. The transition equation is:

$$i_\downarrow = BERT_d(p + wp(w)) \tag{6}$$

where p is the position embedding of the parent node and w is the average embedding of the node's description. Since root node has no parent, $i_\downarrow$ cannot be computed using the above equations. Instead, we set $i_\uparrow$ to $i_\downarrow$ here.

Loss function of **BPR**: We call a pair of one up-stream *PR* and one down-stream *PR* an interaction. In one interaction, one node n_j has two representations: $n_j^{i_\uparrow}$ and $n_j^{i_\downarrow}$. The goal of training **BPR** is to reduce the difference between $n_j^{i_\uparrow}$ and $n_j^{i_\downarrow}$ as much as possible. Both $n_j^{i_\uparrow}(X = \{0, 1, 2,..., k,...,d_e\})$ and $n_j^{i_\downarrow}(X)$ can be recognized as two distributions. So we construct a Kullback-Leibler divergence [25] as the loss function of *BPR*:

$$\mathcal{L}_{bpr1} = \frac{1}{L}\Sigma_1^L \Sigma_1^{d_e}[n_j^{i_\uparrow}(x)log(n_j^{i_\uparrow}(x)) - n_j^{i_\uparrow}(x)log(n_j^{i_\downarrow}(x))]$$
$$\mathcal{L}_{bpr2} = \frac{1}{L}\Sigma_1^L \Sigma_1^{d_e}[n_j^{i_\downarrow}(x)log(n_j^{i_\downarrow}(x)) - n_j^{i_\downarrow}(x)log(n_j^{i_\uparrow}(x))] \tag{7}$$
$$\mathcal{L}_{bpr} = \mathcal{L}_{bpr1} + \mathcal{L}_{bpr2} + \|\mathcal{L}_{bpr1} - \mathcal{L}_{bpr2}\|_2^2$$

where L is the number of all ICD codes, d_e represents the dimension of ICD codes, n_j is the *j-th* node of ICD codes, x represents x-th item of n_j, $\mathcal{L}_{bpr1}$ represents child-to-parent KL value between $n_j^{i_\uparrow}$ and $n_j^{i_\downarrow}$, $\mathcal{L}_{bpr2}$ represents parent-to-child KL value between $n_j^{i_\uparrow}$ and $n_j^{i_\downarrow}$.

When the loss value is miner than 0.01, the training stops. Finally, we obtain parent-child code representations $Vt \in \mathbb{R}^{d_e \times N}$.

Tree Position Encoder (*TPE*). Inspired by [29], we propose a tree position encoder to capture the positional features of ICD codes.

$$PE_\beta = A_\phi PE_\alpha \tag{8}$$

Given a position α, we can get the target position β with affine transform operation A_ϕ (Eq. (8)). The path between α and β can be considered as the set of n length-1 paths. Directions of these paths contains "to parent" and "to children". We take "to parent" as function U and "to children" function D. Thus, for any path ϕ, we can obtain the transform A_ϕ by some combinations of U and D. For example, the position encoding of the second child of node x's grandpa can described path $\phi = \langle parent, parent, child\text{-}2\rangle$, which can also be represented as $D_2U^2PE_x$.

We take the root node as zero vector ($\mathbf{0} \in \mathbb{R}^{d_e}$). Then, since all the paths from the root to other positions are downward, we can get embedding of any position x via the Eq. (9).

$$x = D_{b_L}D_{b_{L-1}}...D_{b_1} \tag{9}$$

where L here means the L-th layer the node is located, and b_i represents the chosen path in the i-th layer. We treat tree position encoding as a stack of length-1 component parts.

Every D operation pushes a length-1 path onto the stack, while U pops a length-1 path, which is described as Eq. (10):

$$D_i x = e_i^n \oplus (x \ominus x[n+1:])$$
$$U x = (x \ominus x[:(n-1)]) \oplus \mathbf{0}_n \tag{10}$$

where $\ominus$ means pop operation or truncation, and $\oplus$ indicates push or concatenate operation. In addition, e_i^n is an one-hot encoding with n elements and n is the total number of children of a parent. An example of function D is shown in Fig. 3.

After tree-position encoding, we obtain code tree-position embedding ($Vp \in \mathbb{R}^{d_e*N}$). Then add Vp to Vt to get final code representation Vpt that contains both parent-child relationships and hierarchical features.

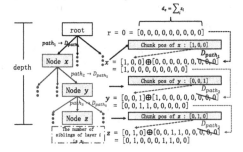

Fig. 3. An example of function D

Code-Wise Attention. Inspired by [5], we use code-wise attention to generate code-aware document representations by using representations outputted from document feature extractor and bidirectional hierarchy passage encoder. The code-wise attention feature a_l for code l is calculated by:

$$s_l = Softmax(\tanh(H \cdot W_a^T + b_a) \cdot v_{pt_l})$$
$$a_l = s_l^T \cdot H \tag{11}$$

where $Softmax$ is the normalized exponential function, s_l donates the attention scores for all elements in document representation H, a_l represents the most relevant information in H about the code l by code-wise attention. Then we get $d_e \times L$ dimensional code-wise adjacent document representations.

3.5   Progressive Mechanism (*PM*) for *DP*

This section introduces a simple progressive mechanism to address the **DP** problem in ICD coding. A former predicted code could help to predict next codes. For example, the neighbours of 'Diabetes mellitus' contain 'heart disease', neighbours of 'heart disease' contain 'cardiovascular disease', then there exists a prediction path <Diabetes mellitus, heart disease, cardiovascular disease>. Given a clinical note, if 'Diabetes mellitus' is true, then we could use this label to predict 'heart disease'. Similarly, we can use 'heart disease' to predict 'cardiovascular disease'. But if 'heart disease' is wrong, the process of progressive prediction will end. The problem is how to use the prior predicted code to predict other codes. We address the problem above by constructing an average-operation between the prior code's hidden output and other codes' hidden outputs with a hyper-parameter λ. Assuming the hidden outputs are described as $f_i \in \mathbb{R}^{1 \times 1}$ and it is proved true, then later hidden output f_j should be calculated as follows:

$$f_j = \lambda f_i + (1 - \lambda) * f_j \tag{12}$$

where λ is a trade-off factor to balance f_j and f_i. Here, f_i decides the influence from a former code to a later code. f_j reflects the current value of node j itself.

Figure 4 shows the dynamic progressive prediction in detail. After *PM*, we obtain the output hidden embedding of ICD coding methods $P = \{p_1, p_2,..., p_L\} \in \mathbb{R}^{d_e * L}$.

3.6   Personalized PageRank (*PP*) for *CC*

Inspired by *PPNP* [15], we build a personalized PageRank [23] to capture the code co-occurrence. Given a clinical text and its golden labels, we can build a strong connected sub-graph, and connect all sub-graphs into a big graph $G = (V, E)$, where V and E are sets of nodes (codes) and edges respectively. Let L denote the number of nodes and m the number of edges. The graph G is described by the matrix $A \in \mathbb{R}^{L*L}$, and $\widetilde{A} = A + I_n$ denotes the matrix A with added self-loops.

In the graph G, let the initial representation of node x as i_x. In addition, We can update node embedding by the recursion equation:

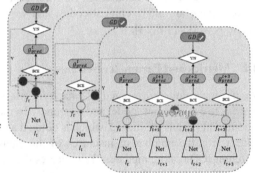

$$PPR(i_x) = (1 - d)\widehat{\widetilde{A}} + d i_x$$

$$= d(I_n - (1 - d)\widehat{\widetilde{A}})^{-1} i_x \tag{13}$$

where $\widehat{\widetilde{A}} = \widetilde{D}^{-\frac{1}{2}} \widetilde{A} \widetilde{D}^{-\frac{1}{2}}$. D is a degree matrix of nodes and $\widetilde{D}$ denotes the degree matrix of graph with added self-loops, where $\widetilde{D}_{i,i} = \Sigma_j \widetilde{A}_{i,j}$

We generate predictions based on each node's own characteristics, and then propagate them

Fig. 4. Overview of progressive mechanism, *Net* represents the network of method for automated ICD coding, *BCE* is binary cross-entropy, *GD* is short for Ground Truth.

through a fully personalized PageRank scheme to generate the pagerank-aware prediction. The pagerank-aware embedding is denoted as $PPR = \{ppr_1, ppr_2,..., ppr_L\} \in \mathbb{R}^{d_e * L}$.

3.7   Aggregation and Training

Aggregation: After exploiting the structural information of code via **BHPE** (i.e., **BPR** and **TPE**) and code co-occurrence via personalized PageRank, we obtain a normal representation (*P*) (from **PM** in Sect. 3.5) and a pagerank-aware representation (*PPR*) (Sect. 3.6), respectively. We concatenate P and PPR as $P3R \in \mathbb{R}^{d_e * 2L}$ and conduct a fully connected layer to reshape the matrix in to $\mathbb{R}^{d_e * L}$ embedding space.

Training: Since the automated icd coding is a multi-label prediction task, we employ a multi-label binary cross-entropy (*BCE*) as the loss of our model:

$$\mathcal{L}_{BCE}(y, \hat{y}) = -\Sigma_{l=1}^{L}[y_l \log(\hat{y}_l) + (1 - y_l) \log(1 - \hat{y}_l)] \tag{14}$$

where y_l is the ground truth and $\hat{y}_l$ is the predicted value, $\hat{y}_l = \sigma(x_l)$. Here, we use Adam optimizer [13] to propagate the parameters of our model.

4   Experimental Setup

4.1   Datasets

MIMIC-II [11] and MIMIC-III [10] are the most widely open-access datasets for evaluating automated ICD encoding methods. In MIMIC-III, there exists two versions. One is MIMIC-III full and the other is MIMIC-III 50. For MIMIC-III full, there are 8,921 unique codes, 47,723 discharge summaries for training, 3,372 summaries for test, and 1,631 for validation. For MIMIC-III 50, we use a set of 8,066 for training, with 1,729 summaries and 1,729 summaries for validation and test, respectively. In MIMIC-II dataset, there are 5,031 clinical codes, and we use the same experimental setting as previous works [4,5,24].

4.2   Metrics and Parameter Settings

We use macro/micro-averaged F1, macro/micro-averaged AUC, and P@N as the main metrics to evaluate our model and baselines. To measure *DP*, we use Jaccard Similarity Coefficient [21] as the metric, which is defined as $Jaccard = \frac{1}{m}\Sigma_i^m |Y_i \cap \hat{Y}_i|/|Y_i \cup \hat{Y}_i|$, where m is the number of instances of the dataset, Y_i is the predicted results from different ICD coding methods, $\hat{Y}_i$ indicates the ground truth ICD set (**note:** Assuming that len(Y) = l_y, len($\hat{Y}$ = K, len(Y$\cap\hat{Y}$) = $l_\cap$, if $K \geq l_y$, let $\hat{Y} = \hat{Y}[: m]$, else let Y = (Y $\cap\hat{Y}$) $\cup$ **0***(K-$l_\cap$).

We set the word embedding size d_e as 100. The size of hidden layer is 128. We set 5 channels in CNN and the filer-sizes of them are 1, 3, 5, 7, 10, respectively. The dropout rate is 0.2. The learning rate is $1e^{-4}$. The batch size is set as 32. The dump rate d in Eq. (13) is 5. The maximize of personalized PageRank loop is 50. The criterion for early stopping is 10. The initial embedding tool for clinical codes is word2vec [18]. The initial embedding tool for clinical text is WordPiece tokenizer in BERT. The reason of why we use different initial embedding methods for clinical codes and nodes are shown in footnote *1* in Sect. 3.3.

4.3   Baselines

In order to demonstrate the effectiveness of HieNet, we compare it with several previous methods, including state-of-the-art models with knowledge graph and GCN.

- **CNN**. CNN is a widely applied method in language modeling [6].

- **CAML & DRCAML** [19]. CAML leverages convolutional attention for automated ICD prediction. DR-CAML is a updated version of CAML with code description.
- **HyperCore** [5]. This method employed hyperbolic representation to capture the code hierarchy and used GCN [14] to encode the semantic of the code co-occurrence.
- **JointLATT** [30]. This method proposed a new label attention method. And the label attention model achieve SOTA results compared with other previous works.
- **MASTATT-KG** [34]. This method utilizes a multi-scale feature attention to select multi-scale features adaptively.
- **MultiResCNN** [17]. MultiResCNN contains a multi-filter convolutional layer to capture various text patterns and a residual convolutional layer to enlarge the receptive field.
- **DCAN** [9]. This method proposes a dilated convolutional attention network, integrating dilated convolutions, residual connections, and label attention, for medical code assignment.

5   Result and Analysis

We focus on answering the following **researching questions (RQs)**:

- **RQ1:** What is the performance of HieNet on ICD encoding task?
- **RQ2:** Why progressive mechanism and how it performs?
- **RQ3:** What are the contributions of the different components?
- **RQ4:** How does the trade-off coefficient (λ in Sect. 3.5) influence the performance?

5.1   Overall Performance (RQ1)

The comparisons between our model and other state-of-the-art on MIMIC-II and MIMIC-III are given in Tables 1 and 2, respectively. Our HieNet model outperforms every single baseline on most of metrics. The CAML architecture is comparable to the DR-CAML, and the CNN baseline essentially performs the worst than all other neural architectures. We recognized P@N as the most intuitive measure to indicate the effectiveness of methods, since it examines the ability of the method to return a high-confidence subset of codes. Moreover, Jaccard is used to measure the accuracy of top_K predicted codes.

For **MIMIC-III Full:** Compared with baselines, HieNet achieves the best performance on macro-F1, micro-F1, and macro-AUC. Since clinical codes are in uneven distribution and macro-F1 emphasizes the performance of rare label, it is difficult to obtain high macro-F1 score. Even in this case, HieNet performs perfectly and achieves 3% improvement compared to the latest state-of-the-art HyperCore method. This demonstrates the effectiveness of HieNet. Furthermore, on Jaccard metric, HieNet improves the performance by a big margin with 7.7% improvement on top_{20} (from 33.7% to 36.3%) and 19.0% improvement on top_{30} (from 23.1% to 27.5%).

For **MIMIC-III 50:** Following the previous work [5,19], we also evaluate our model and baselines on the most common 50 codes set of MIMIC-III. Different from

Table 1. Results (%) of the comparison of our model and other baselines on the MIMIC-III full and MIMIC-III 50. In all tables, the bold number with * indicates the best result compared to other methods.

Model	MIMIC-III full							
	Jaccard		AUC		F1		P@N	
	top_20	top_30	Macro	Micro	Macro	Micro	8	15
CNN	30.2	20.9	80.6	96.9	4.2	41.9	40.2	49.1
CAML	32.4	22.5	88.8	98.4	7.1	51.9	69.7	54.9
DR-CAML	33.7	23.1	89.7	98.5	8.6	52.9	69.0	54.8
HyperCore	-	-	93.0	98.9	9.0	55.1	72.2	57.9
JointLAAT	-	-	92.1	98.8	8.9	55.3	73.5	59.0
MSATT-KG	32.1	22.0	91.0	99.2	9.0	55.3	72.8	58.1
MultiResCNN	-	-	91.0	98.6	8.5	55.2	73.4	58.4
HieNet	$36.3^* \pm 0.2$	$27.5^* \pm 0.3$	$93.3^* \pm 0.4$	$99.2^* \pm 0.2$	$9.3^* \pm 0.1$	$56.6^* \pm 0.7$	$78.3^* \pm 0.5$	$65.0 * 0.3$

Model	MIMIC-III 50							
	Jaccard		AUC		F1		P@N	
	top_20	top_30	Macro	Micro	Macro	Micro	5	-
CNN	31.7	24.3	87.6	90.7	57.6	62.5	62.0	-
CAML	32.5	25.2	87.5	90.9	53.2	61.4	60.9	-
DR-CAML	32.5	24.6	88.4	91.6	57.6	63.3	61.8	-
HyperCore	-	-	89.5	92.9	60.9	66.3	63.2	-
JointLAAT	-	-	92.5	94.6	66.1	71.6	67.1	-
MSATT-KG	33.9	24.7	91.4	93.6	63.8	68.4	64.4	-
MultiResCNN	-	-	89.9	92.8	60.6	67.0	64.1	-
DCAN	-	-	90.2	93.1	61.5	67.1	64.2	-
HieNet	$37.7^* \pm 0.1$	$30.4^* \pm 0.2$	$93.4^* \pm 0.8$	$95.0^* \pm 0.1$	$67.1^* \pm 0.2$	$72.4^* \pm 0.2$	$69.5^* \pm 0.3$	-

MIMIC-III full, MIMIC-III 50 has a relatively smooth distribution, which leads to the possibility of achieving higher macro-F1 scores.

Our method obtains the highest score on macro-AUC, micro-AUC, and P@5 metrics. On MIMIC-III 50, HieNet achieves the best performance on most of the evaluation except micro-F1 and macro-F1. The top_{30}-Jaccard value gets the most significant improvement (i.e., 20.6% over the best baseline CAML). The reason is that codes on datasets on MIMIC-III 50 are more closely related, and progressive mechanism (CC) is just designed for this.

For MIMIC-II: As shown in Table 2, MIMIC-II contains 5,031 labels, and our method HieNet also performs the best on most metrics compared with baselines except macro-F1 value. In addition, latest work HyperCore's macro-AUC, micro-AUC, micro-F1, and P@8 are much lower than HieNet (1.0%, 1.2%, 3.1%, and 5.4% lower). top_{20} and top_{30} Jaccard values indicate that CNN, CAML, and DR-CAML are poor in leveraging relationships among codes while HieNet can predicts a higher coverage of correct ICD codes.

Table 2. Experimental results of our model and other baselines on MIMIC-II.

Model	Jaccard		AUC		F1		P@N
	top_{20}	top_{30}	Macro	Micro	Macro	Micro	8
CNN	13.7	11.3	74.2	94.1	3.0	33.2	38.8
CAML	14.4	12.3	82.0	96.6	4.8	44.2	52.3
DR-CAML	13.8	11.7	82.6	96.6	4.9	45.7	51.1
HyperCore	-	-	88.5	97.1	7.0	47.7	53.7
JointLAAT	-	-	87.1	97.2	6.8	49.1	55.1
MultiResCNN	-	-	85.1	96.8	5.2	46.4	54.4
HieNet	15.7* ± 0.1	13.2* ± 0.3	89.4* ± 0.2	98.3* ± 0.1	7.1* ± 0.5	49.2* ± 0.2	56.6* ± 0.3

5.2   Effectiveness of Progressive Mechanism (RQ2)

We take an example in Fig. 5. For a given clinical text (patient 118299 here), we verify the impact of the first predicted code on predicting the second one, and the effect of the front two predicted codes on predicting the third one.

The gold label of patient 118299 is gl = ['198.3', '348.5', '162.9', '401.9', '272.4', '15.9', '22.0', '20.5'], and the front predicted codes are '401.9' and '15.9'. The node '401.9' has 1,655 neighbors and Fig. 5a just shows the 20 of them. The node '15.9' has 56 neighbors and Fig. 5b only shows 20 of them. The shared neighbors (i/t-1&2 in Fig. 5c) are i/t-1&2 = ['041.19', '22.1', '433.31', '209.79', '013.25', '237.6', '239.6', '43.11', '372.30', '33.22', '23.4', '803.62', '20.5']. As observed, '20.5' occurred in i/t-1&2 is exactly one element in gl. Thus, we only need 13 labels (i/t-1&2) to predict the third code.

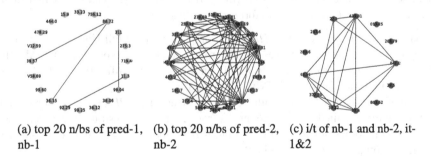

(a) top 20 n/bs of pred-1, nb-1   (b) top 20 n/bs of pred-2, nb-2   (c) i/t of nb-1 and nb-2, it-1&2

Fig. 5. Example of progressive mechanism in HieNet on patient 118299 summaries. n/bs means neighbors, i/t represents interaction, pred-1 indicates the top 1 of predicted codes, nb-i denotes the set of neighbors of i-th predicted codes, and it-1&2 represents the interacted result of nb-1 and nb-2.

Obviously, the neighbors of the former predicted codes can be used help predict the one. In conclusion, the process of above demonstrates the effectiveness of progressive mechanism.

5.3   Ablation Study (RQ3)

We conduct ablation investigation to examine the effectiveness of each part in our model. To evaluate a model, we remove it (denoted as without, w/o) and perform the remaining part on the datasets. The experimental results of ablation study are shown in Table 3.

Table 3. Ablation study by removing the main components.

Models	MIMIC-III full		MIMIC-III 50		MIMIC-II	
	Macro-F1	Micro-F1	Macro-F1	Micro-F1	Macro-F1	Micro-F1
HieNet	**9.3***	**56.6***	**67.1***	**72.4***	**6.9***	**49.2***
w/o *PM*	8.9	56.3	59.8	64.1	6.5	47.7
w/o *BHPE*	8.3	55.2	57.7	62.6	6.1	47.5
w/o *PP*	8.4	56.1	58.6	65.3	6.5	47.3
w GCN for *CC* (w/o *PP*)	9.1	55.3	60.4	65.7	6.7	49.0
w tree-lstm for *DP* (w/o *BHPE*)	8.8	55.2	60.0	64.6	6.6	47.9

Impact of *PM*. We remove the *PM* part from the full model. As shown in Table 3, HieNet without progressive module achieves lower scores of macro-F1, micro-F1 on both MIMIC-II and MIMIC-III.

Impact of *PP*. Compared with the w/o personalized PageRank module, the full HieNet improves the score on Macro-F1 (MIMIC-III full) from 0.084 to 0.093, 0.586 to 0.611 (MIMIC-III 50), and 0.065 to 0.069 (MIMIC-II), respectively. As shown in Table 3, the line 'w GCN for *CC* (w/o *PP*)' performs poorer than HieNet.

Impact of *BHPE*. We remove *BHPE* and compare with the full HieNet. The result is given in Table 3, HieNet improves 10.7% (from 0.084 to 0.093) on macro-F1 (MIMIC-III full), and 14.5% (from 0.586 to 0.671) on macro-F1 (MIMIC-III 50), 6.2% (from 0.065 to 0.069) on macro-F1 (MIMIC-II), etc., respectively.

Table 3 shows that HieNet with tree-lstm instead of *BHPE* performs worse on all metrics on three main datasets. The results from Table 3 demonstrates the effectiveness of different modules in HieNet. In addition, *BHPE* module plays a more important role in HieNet compared with *PM* and *PP*.

5.4 The Impact of λ (RQ4)

The trade-off parameter λ is used to balance the influence factor from former hidden outputs and current hidden output. When λ is larger, HieNet relies more on the former predicted codes to influence the next ones to predict, which

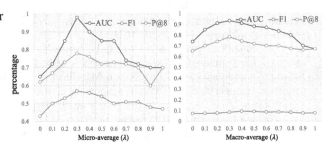

Fig. 6. Performance with different values of λ.

means there exists a strong reasoning relationship in the predicted codes corresponding to the given clinical note. When λ is smaller, HieNet tends to prioritize the current hidden output to predict a proper ICD code during the learning. The value of λ is set from (0, 0.1, 0.2, ..., 1.0) to measure the performance. The results measured on MMIMIC-III-full are shown in Fig. 6.

First, as λ increases, all metrics increase gradually at the beginning, but most of them decrease when $\lambda \geq 0.3$. In addition, F1 of macro-average decreases after $\lambda \geq 0.4$. The best performance is not achieved when $\lambda = 0$ or 1. This demonstrates that progressive mechanism improve the performance of ICD coding methods. Second, Fig. 6 shows that the beginning increases of macro-F1, macro-AUC, micro-F1, and micro-AUC are faster than the afterward decreases. Specially, micro-AUC grows from 0.65 to 0.98 only by the 0.3 added λ-value while micro-AUC drops from 0.98 to 0.7 needs another 0.7 added λ-value. This indicates the information of prior predicted codes can significantly affect the performance of ICD coding.

6 Conclusion and Future Works

In this paper, we propose the HieNet, which employs multi-channel CNN to encode the document representation, bidirectional hierarchy encoder to capture the hierarchy features, progressive mechanism improve the accuracy of top_K predicted codes, and personalized PageRank to obtain code co-occurrence. HieNet yields strong improvements over previous methods, while providing the new state-of-the-art performance on both MIMIC-II and MIMIC-III.

However, for input, the future works need to pay more attention to multi-type input. The input can be medical image (i.e., chest radio graph), structured information (i.e., prescriptions), unstructured data (i.e., clinical texts), etc. Inspired by the effectiveness of the progressive mechanism, we can build a multi-model prediction model in future.

References

1. Ahmad, W.U., Chakraborty, S., Ray, B., Chang, K.: A transformer-based approach for source code summarization. In: Jurafsky, D., Chai, J., Schluter, N., Tetreault, J.R. (eds.) ACL 2020, pp. 4998–5007. Association for Computational Linguistics (2020)

2. Allamanis, M., Peng, H., Sutton, C.: A convolutional attention network for extreme summarization of source code. In: Balcan, M., Weinberger, K.Q. (eds.) ICML 2016. JMLR Workshop and Conference Proceedings, vol. 48, pp. 2091–2100. JMLR.org (2016). http://proceedings.mlr.press/v48/allamanis16.html

3. Avati, A., Jung, K., Harman, S., Downing, L., Ng, A.Y., Shah, N.H.: Improving palliative care with deep learning. BMC Med. Inform. Decis. Mak. **18**(S-4), 55–64 (2018)

4. Baumel, T., Nassour-Kassis, J., Elhadad, M., Elhadad, N.: Multi-label classification of patient notes a case study on ICD code assignment. CoRR abs/1709.09587 (2017)

5. Cao, P., Chen, Y., Liu, K., Zhao, J., Liu, S., Chong, W.: HyperCore: hyperbolic and co-graph representation for automatic ICD coding, pp. 3105–3114. ACL, Online, July 2020. https://doi.org/10.18653/v1/2020.acl-main.282, https://www.aclweb.org/anthology/2020.acl-main.282

6. Dauphin, Y.N., Fan, A., Auli, M., Grangier, D.: Language modeling with gated convolutional networks. In: Precup, D., Teh, Y.W. (eds.) ICML 2017. Proceedings of Machine Learning Research, vol. 70, pp. 933–941. PMLR (2017)

7. Florescu, C., Caragea, C.: A position-biased PageRank algorithm for keyphrase extraction. In: Singh, S.P., Markovitch, S. (eds.) AAAI 2017, pp. 4923–4924. AAAI Press (2017)

8. Huber, P., Carenini, G.: From sentiment annotations to sentiment prediction through discourse augmentation. In: Scott, D., Bel, N., Zong, C. (eds.) COLING 2020, pp. 185–197. International Committee on Computational Linguistics (2020)

9. Ji, S., Cambria, E., Marttinen, P.: Dilated convolutional attention network for medical code assignment from clinical text. arXiv preprint arXiv:2009.14578 (2020)

10. Johnson, A.E., et al.: MIMIC-III, a freely accessible critical care database. Sci. Data **3**(1), 1–9 (2016)

11. Jouhet, V., et al.: Automated classification of free-text pathology reports for registration of incident cases of cancer. Methods Inf. Med. **51**(3), 242 (2012)

12. Kavuluru, R., Rios, A., Lu, Y.: An empirical evaluation of supervised learning approaches in assigning diagnosis codes to electronic medical records. Artif. Intell. Med. **65**(2), 155–166 (2015)

13. Kingma, D.P., Ba, J.: Adam: a method for stochastic optimization. In: Bengio, Y., LeCun, Y. (eds.) ICLR 2015, San Diego, CA, USA, 7–9 May 2015, Conference Track Proceedings (2015)

14. Kipf, T.N., Welling, M.: Semi-supervised classification with graph convolutional networks. In: ICLR 2017. OpenReview.net (2017)

15. Klicpera, J., Bojchevski, A., Günnemann, S.: Predict then propagate: graph neural networks meet personalized PageRank. In: ICLR 2019. OpenReview.net (2019)

16. Langville, A.N., Meyer, C.D.: Survey: deeper inside PageRank. Internet Math. **1**(3), 335–380 (2003)

17. Li, F., Yu, H.: ICD coding from clinical text using multi-filter residual convolutional neural network. In: AAAI, vol. 34, pp. 8180–8187 (2020)

18. Mikolov, T., Chen, K., Corrado, G., Dean, J.: Efficient estimation of word representations in vector space. In: Bengio, Y., LeCun, Y. (eds.) ICLR 2013 (2013)

19. Mullenbach, J., Wiegreffe, S., Duke, J., Sun, J., Eisenstein, J.: Explainable prediction of medical codes from clinical text. In: ACL, June 2018

20. Nadathur, S.G.: Maximising the value of hospital administrative datasets. Aust. Health Rev. **34**(2), 216–223 (2010)

21. Niwattanakul, S., Singthongchai, J., Naenudorn, E., Wanapu, S.: Using of Jaccard coefficient for keywords similarity. In: Proceedings of the International Multiconference of Engineers and Computer Scientists, vol. 1, pp. 380–384 (2013)

22. Ogura, Y., Kobayashi, I.: Text classification based on the latent topics of important sentences extracted by the PageRank algorithm. In: ACL 2013, pp. 46–51. ACL (2013)

23. Page, L., Brin, S., Motwani, R., Winograd, T.: The PageRank citation ranking: bringing order to the web. Technical report, Stanford InfoLab (1999)
24. Perotte, A.J., Pivovarov, R., Natarajan, K., Weiskopf, N.G., Wood, F.D., Elhadad, N.: Diagnosis code assignment: models and evaluation metrics. J. Am. Med. Inform. Assoc. 21(2), 231–237 (2014)
25. Peyré, G.: Entropic approximation of Wasserstein gradient flows. SIAM J. Imaging Sci. 8(4), 2323–2351 (2015)
26. dos Santos, C.N., Tan, M., Xiang, B., Zhou, B.: Attentive pooling networks. CoRR abs/1602.03609 (2016)
27. Scozzafava, F., Maru, M., Brignone, F., Torrisi, G., Navigli, R.: Personalized PageRank with syntagmatic information for multilingual word sense disambiguation. In: Çelikyilmaz, A., Wen, T. (eds.) ACL 2020, pp. 37–46. ACL (2020)
28. Shi, H., Xie, P., Hu, Z., Zhang, M., Xing, E.P.: Towards automated ICD coding using deep learning. arXiv preprint arXiv:1711.04075 (2017)
29. Shiv, V., Quirk, C.: Novel positional encodings to enable tree-based transformers. Adv. Neural Inf. Process. Syst. 32, 12081–12091 (2019)
30. Vu, T., Nguyen, D.Q., Nguyen, A.: A label attention model for ICD coding from clinical text. CoRR abs/2007.06351 (2020). https://arxiv.org/abs/2007.06351
31. Wang, S., et al.: Coding electronic health records with adversarial reinforcement path generation. In: Huang, J., Chang, Y., et al. (eds.) SIGIR 2020, pp. 801–810. ACM (2020)
32. Xie, P., Xing, E.: A neural architecture for automated ICD coding. In: ACL, July 2018
33. Xie, X., Xiong, Y., Yu, P.S., Zhu, Y.: EHR coding with multi-scale feature attention and structured knowledge graph propagation. In: Zhu, W., et al. (eds.) Proceedings of the 28th ACM International Conference on Information and Knowledge Management, CIKM 2019, Beijing, China, 3–7 November 2019
34. Xie, X., Xiong, Y., Yu, P.S., Zhu, Y.: EHR coding with multi-scale feature attention and structured knowledge graph propagation. In: Proceedings of the 28th ACM International Conference on Information and Knowledge Management, pp. 649–658 (2019)
35. Xu, K., et al.: Multimodal machine learning for automated ICD coding. In: Doshi-Velez, F., et al. (eds.) Proceedings of the Machine Learning for Healthcare Conference, MLHC 2019, 9–10 August 2019, Ann Arbor, Michigan, USA. Proceedings of Machine Learning Research, vol. 106, pp. 197–215. PMLR (2019)
36. Yin, W., Schütze, H.: Attentive convolution. CoRR abs/1710.00519 (2017)

Efficient Consensus Motif Discovery of All Lengths in Multiple Time Series

Mingming Zhang[1], Peng Wang[2(✉)], and Wei Wang[2]

[1] School of Software, Fudan University, Shanghai, China
zhangmm19@fudan.edu.cn
[2] School of Computer Science, Fudan University, Shanghai, China
{pengwang5,weiwang1}@fudan.edu.cn

Abstract. Time series motif discovery is an important primitive for the time series data mining. With the explosion of new sensing technology, there is a continuously increasing amount of time series data in every aspect of our lives, from seismology, entomology, human activity monitoring, medicine and so on. Considering the rich information included in time series, motif discovery has become an essential part of many data mining tasks. In recent years, the problem of consensus motif discovery in multiple time series begins to appear in our vision. For this task, the existing approaches can only search the consensus motif of a fixed length. However, variable-length motif mining is more common in real applications. To address this problem, the brute force version of the existing fixed-length approach is prohibitively expensive. In this paper, we propose an efficient, scalable and exact algorithm VACOMI to search the consensus motif of all lengths in a given motif length range. We evaluate the performance of VACOMI on four real datasets. The results show that VACOMI can reduce up to 96% of the running time compared with the state-of-the-art approach.

Keywords: Motif discovery · Variable-length consensus motif · Multiple time series

1 Introduction

Time series motif discovery has become a hot research domain in the data mining for more than two decades [1–5,8,9]. Time series motif is the repeated pattern in one or two time series, and there is always some semantic and meaningful information in it. Time series motif can be applied to many fields, such as seismology, entomology, human activity monitoring, and medicine to help researchers analyse problems. In addition, it can also be used as input in several higher-level data mining tasks, including classification [10], clustering [11], shapelets [12], etc.

In the past decade, there are many works focusing on finding motifs in one or two time series, such as STAMP [13] and STOMP [15] which can find exact

The work is supported by the Ministry of Science and Technology of China, National Key Research and Development Program (No. 2020YFB1710001).

motif of a fixed length. SCRIMP++ [14] is an anytime algorithm which improves the convergence speed of motif search. In recent years, the research of consensus motif discovery in multiple time series is increasing. Both Ostinato [6] and Anytime Ostinato [4] can address this problem. Note that the difference between the consensus motif and the traditional motif [13] is that traditional motif is the most similar subsequence pair from one or two time series. In contrast, consensus motif is the common pattern among multiple time series. While, before using these methods, we all need to set a parameter of motif length. However, it is not a trivial task to find the optimal length to search the motif. To solve this problem, VALMOD [8], HIME [5], MOEN [9] and other approaches are proposed to find motifs of all lengths for a given motif length range. However, they can only search the variable-length motif in one or two time series. For the variable-length consensus motif discovery in multiple time series, there is still no effective solution.

We illustrate the variable-length consensus motif discovery by considering an example shown in Fig. 1 on a real dataset EPG [7]. There are four time series recording the behavior data of an insect Asian citrus psyllid. These two consensus motifs of length 800 and 1724 are both discovered by our algorithm. And we are surprised that they represent different insect behaviors. Thus, it demonstrates that in many circumstances, searching with a fixed length is not enough, because other motifs of variable lengths can also be included in the data.

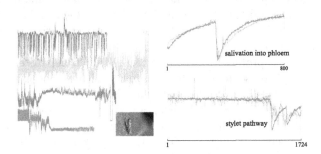

Fig. 1. left) Four time series of insect telemetry. right) Two consensus motifs of different lengths are well conserved. The top one corresponds to the "salivation into phloem" behavior. The bottom one corresponds to the "stylet pathway" behavior.

The biggest challenge of this task is the massive time cost in a larger motif length searching range. Thus, we propose an efficient algorithm, VACOMI, to search the exact consensus motif of all lengths in multiple time series for a given motif length range.

The contributions of our work are the following:

- We define a new problem of the variable-length consensus motif discovery in multiple time series, which extends the application of motif discovery.
- We propose an efficient, scalable and exact algorithm VACOMI to search the consensus motif of all lengths in a given motif length range.

- We propose a novel tight lower bound for the distance of multiple subsequences taken from different time series. It can be calculated in linear time.
- We propose a time auto-tuning method which can maintain the performance of help us adjust the running time to optimal status.
- The experiment results of VACOMI on the four real datasets show that VACOMI can reduce up to 96% of the running time compared with the state-of-the-art. And the pruning rate of our lower bound is 70× to 2000× more than the existing method.

The rest of this paper is organized as follows. Section 2 introduces the related work of motif discovery. Section 3 elaborates some definitions and preliminary knowledge. Section 4 describes the detail of the proposed algorithm VACOMI. Section 5 gives experimental setting and analysis the results. Section 6 concludes this paper.

2   Related Work

Motif discovery for time series was introduced in [2]. Since then, it has created a flurry of research activities. The research type of motif discovery is mainly divided into fixed-length and variable-length motif discovery.

For the fixed-length research, STAMP [13] is an anytime algorithm which can transfer the calculation of distance profile from time domain to frequency domain by FFT method. The time complexity is $O(n^2 logn)$. STOMP [15] is the state of the art algorithm to find motif in one or two time series. It evaluates the distance profiles in-order by exploiting the computation dependency between consecutive distance profiles, which only cost $O(n^2)$ time. Recently, the algorithm Ostinato [6] is presented to find repeated structures among more than two time series. It proposes a method to calculate the distance among multiple subsequences from different time series using a fast pruning strategy, and the seed subsequence with the minimum distance is the exact consensus motif. Based on this, An anytime version of Ostinato algorithm is presented [4]. It calculates the radius profile by exploiting the fact that the distance matrix is symmetric, which can avoid half of all distance calculations and effectively obtain the consensus motif by calculating the distances on diagonals same as the SCRIMP++ [14].

For the variable-length research, the goal is to find motifs of different lengths in one or two time series. VALMOD proposes a lower bound which can solve this problem efficiently. But except the motif length range, it's not parameter-free. And if the parameter is not suitable, it can affect the running time a lot. In the work of Mueen [9], the authors introduce an algorithm called MOEN which uses the upper and lower bounds on the similarity function to prune most of computations to improve the efficiency. In order to achieve faster speed, HIME [5] presents an approximate method to detect variable-length motifs based on grammar induction. But it can't get the exact results.

Thus, to the best of our knowledge, there is no work on finding exact variable-length repeated structures in a set of more than two time series.

3   Preliminary Knowledge

In this section, we first introduce some necessary definitions, and then outline the approach to find the fixed-length consensus motif.

3.1   Definitions and Problem Statement

Definition 1 (Time series). *A time series $T = (t_1, ..., t_n)$ of length n is a real-valued sequence composed of the value of a certain observation on each times-tamp.*

Definition 2 (Subsequence). *A subsequence is the sequence intercepted by a sliding window of length l in time series T^a. We denote a subsequence which starts from the position i and ends at the position $i + l - 1$ as $T^a_{i,l}$.*

Definition 3 (Distance profile). *A distance profile (DP for short) of a subsequence $T^a_{i,l}$ is a vector that stores the z-normalized Euclidean distances [8] between $T^a_{i,l}$ and each subsequence of length l in a time series T^b. The z-normalized Euclidean distance can be calculated as:*

$$d(T^a_{i,l}, T^b_{j,l}) = \sqrt{2l\left(1 - \frac{Q_{i,j} - l\mu_i\mu_j}{l\sigma_i\sigma_j}\right)} \qquad (1)$$

Here, l is the subsequence length, $Q_{i,j}$ is the dot product between $T^a_{i,l}$ and $T^b_{j,l}$, μ_i and μ_j are the mean of $T^a_{i,l}$ and $T^b_{j,l}$, σ_i and σ_j are the standard deviation of $T^a_{i,l}$ and $T^b_{j,l}$. Formally, $DP(T^a_{i,l}, T^b) = \left\{d(T^a_{i,l}, T^b_{1,l}), d(T^a_{i,l}, T^b_{2,l}), \cdots, d(T^a_{i,l}, T^b_{n-l+1,l})\right\}$. Besides, if using Eq.(1), there is no need to normalize each subsequence in advance [13].

Definition 4 (Matrix profile). *A matrix profile (MP for short) of all subsequences is a vector that stores the minimum of each DP. The minimum of MP is the distance between two most similar subsequences, and the indices of these two subsequences are the locations of motif.*

Definition 5 (Mindist). *Given two time series T^a and T^b of length n, for a subsequence $T^a_{i,l}$ within T^a, $mindist(T^a_{i,l}, T^b)$ is the minimum distance between $T^a_{i,l}$ and all subsequences of T^b. Also, It is the minimum of $DP(T^a_{i,l}, T^b)$.*

$$mindist(T^a_{i,l}, T^b) = \min_{j \in [1, n-l+1]}(d(T^a_{i,l}, T^b_{j,l})) \qquad (2)$$

Definition 6 (Kdist). *The kdist of a subsequence $T^a_{i,l}$ within time series T^a is the maximum of $k - 1$ mindist between $T^a_{i,l}$ and other time series except T^a, that is,*

$$kdist(T^a_{i,l}) = \max_{1 \le b \le k, b \ne a}(mindist(T^a_{i,l}, T^b)) \qquad (3)$$

The visual example is shown in Fig. 2.

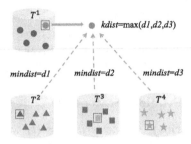

Fig. 2. Assume there are four time series, all the subsequences of each time series are divided in to a bucket and represented by different shapes separately. For a subsequence in T^1, there are $k-1$ *mindist* from other buckets to it, so the *kdist* of this subsequence is max$(d1, d2, d3)$.

Definition 7 (Consensus motif). *Given a set of time series* $\{T^1, T^2, \cdots, T^k\}$, *consensus motif of length l, denoted as CM_l, is the subsequence with the smallest kdist among all the subsequences. And a consensus motif group of length l, denoted as CMG_l, is an array which stores the indices of the consensus motif and its best matching subsequences from other $k-1$ time series.*

Problem 1 (**Variable-length consensus motif discovery**). Given a set of long time series $\{T^1, T^2, \cdots, T^k\}$ of different lengths, a subsequence length range $[l_{min}, l_{max}]$, and a step size of length growth *step*, our goal is to find the CM_l for each length l $(l_{min} \leq l \leq l_{max})$.

3.2 Fixed-Length Consensus Motif Discovery

In this section, we first briefly outline the approach to search the fixed-length consensus motif [6] of length l.

Firstly, we begin by connecting all k time series into a single time series S with "null" markers to discriminate different time series, and the length of S is N. Then we store every pairwise distance of all subsequences into a distance matrix D. As shown in Fig. 3, both row and column of D present a subsequence and each $D_{i,j}$ is the z-normalized Euclidean distance between i-th and j-th subsequence. Moreover, the relation between D and CM_l is observable. For example, as to subsequence T^2_{10}, $kdist(T^2_{10})$ is the max *mindist* from it to its best matching subsequences in other time series (*i.e.* $max\{d(T^2_{10,l}, T^1_{53,l}), d(T^2_{10,l}, T^3_{27,l}), d(T^2_{10,l}, T^4_{81,l})\}$. If $kdist(T^2_{10})$ is smallest among all the subsequences, then $T^2_{10,l}$ is the CM_l we want to search, and [53,10,27,81] is the CMG_l.

Specifically, we need to calculate the matrix profile of every time series as the lower bound of *kdist* for each subsequence, and keep the subsequence whose lower bound is smaller than the current smallest *kdist* (denoted as *bsfkdist*) as our candidate. Then we calculate $k-1$ *minidsts* from each candidate to other

$k - 1$ time series. The maximum of these $minidists$ is the $kdist$ of each candidate subsequence. If the $kdist$ is smaller than the current $bsfkdist$, we need to the update the current $bsfkdist$ to this $kdist$. After calculating all the candidates, the candidate with the $bsfkdist$ is the final consensus motif we want to search.

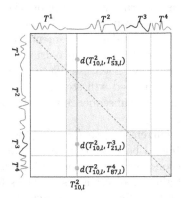

Fig. 3. An example of the consensus motif discovery under the single length l in four time series As the purple line sliding across all the columns of D, we calculate the minimum distances of three regions except T^2, and the max value of them is $kdist$ of $T^2_{10,l}$. The distances on the blue dashed line are always zero and the distances in the grey regions are no need to be calculated. (Color figure online)

Analysis. For Problem 1, if we use this method to search consensus motif of each length in $[l_{min}, l_{max}]$, it's very time-consuming. It is the best case that we calculate k full matrix profiles and one single column of D with $O(\frac{rN^2}{k} + rN log \frac{N}{k})$ time complexity, where $r = l_{max} - l_{min} + 1$. For the worst case, it requires $O(\frac{rN^2}{k} + rN^2 log \frac{N}{k})$ to calculate all the columns of the D. So, how to reduce the time cost is a big challenge for this problem.

4   Proposed Method

In this section, we first introduce an overview of our approach. Then we discuss the details of it.

4.1   Approach Overview

We propose an algorithm VACOMI (**va**riable-length **co**nsensus **m**otif d**i**scovery) to solve the Problem 1. In our algorithm, we define a novel lower bound of $kdist$, denoted as klb. With the subsequence length increasing, klb can help us prune more subsequences in linear time. Our algorithm contains four parts: lower bound base information collection, lower bound filtering, candidate refinement and time auto-tuning.

Given a subsequence length range $[l_{min}, l_{max}]$, we loop all lengths to calculate the consensus motifs we want to search. Here, for each length l, VACOMI has two ways to calculate CM_l. The first way is computing CM_l by running the process of the lower bound base information collection, which uses the optimized fixed-length consensus motif discovery algorithm to obtain CM_l and collect some lower bound base information to support the calculation of klb in the subsequent lengths. The second way is computing CM_l by running the process of lower bound filtering and candidate refinement. In the part of lower bound filtering, we calculate klb for each subsequence using the lower bound base information collected in the first way. Then use klb to prune the subsequences. The subsequences which are not pruned will step into the candidate refinement to compute the $kdist$ of them. The subsequence with the smallest $kdist$ is the CM_l we want to search.

In general, when the current length $l = l_{min}$, we choose the first way to get CM_l and the lower bound base information of klb. When $l_{min} < l \leq l_{max}$, we usually choose the second way to compute CM_l since it is faster than the first way. However, the tightness of klb is less tight as length increases and the pruning performance of klb decline. If we always choose the second way, the efficiency of VACOMI will reduce. In this case, we should choose the first way to calculate CM_l and update the lower bound base information to reinforce our lower bound. Thus, before calculating CM_l, we need to decide which is the best way to compute CM_l with the help of the time auto-tuning strategy, so that we can minimize the overall running time.

4.2   Lower Bound Base Information Collection

In this paper, we propose a novel lower bound to avoid $kdist$ computation and improve the efficiency of VACOMI. The rationale of the lower bound is as follows. Given a subsequence $T_{i,l}^a$, if we can obtain the lower bound of $mindist$ from $T_{i,l}^a$ to other $k-1$ time series except T^a, the maximum of these lower bounds can be the lower bound of $kdist(T_{i,l}^a)$, denoted as $klb(T_{i,l}^a)$. If $klb(T_{i,l}^a)$ is larger than the current $bsfkdist$, the true $kdist(T_{i,l}^a)$ must be larger than the current $bsfkdist$. Thus, we can prune the subsequence whose klb is larger than the current $bsfkdist$ to reduce the $kdist$ computation.

Before calculating klb, we need to collect some base information about the klb under length l to support the calculation of klb under length $l+s$. The lower bound calculation of $mindist$ is based on the lower bound of variable-length z-normalized Euclidean distance. VALMOD [8] proposed a lower bound of z-normalized Euclidean distance between two variable-length subsequences. Assume there are two subsequences $T_{i,l}^a$ and $T_{j,l}^b$, the z-normalized Euclidean distance between these two subsequences is $d_{i,j}^l$ under length l. When the subsequence length is $l+s(s \in [0, l_{max}-l])$, the lower bound of z-normalized Euclidean $e_{i,j}^{l+s}$ between $T_{i,l}^a$ and $T_{j,l}^b$ is:

$$e_{i,j}^{l+s} = \begin{cases} \sqrt{l} \frac{\sigma_{i,l}}{\sigma_{i,l+s}} & \text{, if } Q_{i,j} \leq 0 \\ \sqrt{l\left(1 - Q_{i,j}^2\right)} \frac{\sigma_{i,l}}{\sigma_{i,l+s}} & \text{, otherwise} \end{cases} \tag{4}$$

where $Q_{i,j} = \frac{\sum_{p=1}^{l} \frac{(t_{i+p-1}t_{j+p-1})}{l} - \mu_{i,l}\mu_{j,l}}{\sigma_{i,l}\sigma_{j,l}}$, $\sigma_{i,l}$ and $\sigma_{i,l+s}$ are the standard deviation of $T_{i,l}^{a}$ and $T_{i,l+s}^{a}$. When $s = 0$, for length l, $\sigma_{i,l}/\sigma_{i,l+s} = 1$, $e_{i,j}^{l+s}$ is equal to $\sqrt{l}$ or $\sqrt{l\left(1 - Q_{i,j}^{2}\right)}$. Thus, if we can obtain $e_{i,j}^{l}$, $e_{i,j}^{l}$ just needs to be multiplied by $\sigma_{i,l}/\sigma_{i,l+s}$ to get $e_{i,j}^{l+s}$ in linear time. And this lower bound is less than or equal to the true z-normalized Euclidean distance.

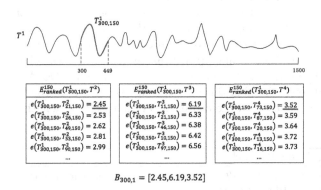

$$B_{300,1} = [2.45, 6.19, 3.52]$$

Fig. 4. An example of the lower bound base information collection. There is a subsequence $T_{300,150}^{1}$ which calculates three lower bound vectors with time series T^{2}, T^{3}, T^{4}. After sorting them, we can find the minimum of them, namely $\{2.45, 6.19, 3.52\}$ and store them into B.

However, if we want to calculate klb of the subsequence under length $l + s$ in linear time, we need to use a special character of this z-normalized Euclidean lower bound. That is, it can keep the same order of lower bound with the subsequence length increasing. Specifically, if a subsequence $T_{i,l}^{a}$ calculates DP with a time series T^{b} under length l, we can calculate z-normalized Euclidean lower bound e of each distance in DP according to Eq. (4) at the same time. After sorting these lower bounds in ascending order, we can obtain a ranked lower bound vector $E_{\text{ranked}}^{l}(T_{i,l}^{a}, T^{b})$. If the z-normalized Euclidean lower bound $e(d_{i,j}^{l})$ between $T_{i,l}^{a}$ and $T_{j,l}^{b}$ is the minimum in $E_{\text{ranked}}^{l}(T_{i,l}^{a}, T^{b})$, $e(d_{i,j}^{l})$ will also be smaller than the minimum of DP. So $e(d_{i,j}^{l})$ can be the lower bound of $mindist(T_{i,l}^{a}, T^{b})$. When the length increases to $l + s$, the z-normalized Euclidean lower bound $e(d_{i,j}^{l+s})$ between $T_{i,l}^{a}$ and $T_{j,l}^{b}$ still is the minimum of $E_{\text{ranked}}^{l+s}(T_{i,l}^{a}, T^{b})$. Thus, due to this special character of lower bound e, we can directly compute $e(d_{i,j}^{l+s})$ based on $e(d_{i,j}^{l})$ as the lower bound of $mindist(T_{i,l+s}^{a}, T^{b})$ without computing $DP(T_{i,l}^{a}, T^{b})$ and $E_{\text{ranked}}^{l+s}(T_{i,l}^{a}, T^{b})$. It can reduce a lot of time.

To compute the $mindist$ lower bound quickly under length $l + s$, we store the $mindist$ lower bound under length l as the lower bound base information. Specifically, if a subsequence $T_{i,l}^{a}$ calculates $m(1 \leq m \leq k - 1)$ DP with other time series under length l, we need to compute m lower bound vectors at the same time. Then store the minimums of these lower bound vectors to a matrix B. The lower bound base information of subsequence $T_{i,l}^a$ is $B_{i,a} = \left\{ \min(E_{ranked}^l(T_{i,l}^a, T^{b_1})), \min(E_{ranked}^l(T_{i,l}^a, T^{b_2})), \cdots, \min(E_{ranked}^l(T_{i,l}^a, T^{b_m})) \right\}$. An Example is shown in Fig. 4.

In terms of algorithm, when execute the process of fixed-length consensus motif discovery, we need to collect B and search the CM_l under length l. As shown in Algorithm 1, in line 1, we set an initial $bsfkdist$ firstly. In line 2–6, we can calculate the matrix profile of T^i and T^w using once STOMP [15] algorithm. And calculate B for each subsequence. Then we connect all MP of all time series into vector $allMP$. In line 7–10, we sort $allMP$ in the ascending order and use it as lower bound of $kdist$ for each subsequence to prune subsequences. For the subsequence which is not pruned, in line 12–18, we calculate $kdist$ of this subsequence and update the current $bsfkdist$, $tsIdx$ and $ssIdx$. At the same time, we also need to collect B of this subsequence. In the end of the loop, we can get B of all subsequences and the consensus motif $T_{ssIdx,l}^{tsIdx}$ of length l.

Algorithm 1. singleSearch

Input: time series set $T^1 \cdots T^k$, total length N, subsequence length l
Output: $kdist, tsIdx, ssIdx, B$

1: $bsfkdist \leftarrow inf$
2: **for** $i \leftarrow 1 : 2 : k$ **do**
3:     $w \leftarrow i + 1 \leq k?i + 1 : 1$
4:     $[MP_i, MP_w] \leftarrow$ stomp(T^i, T^w)
5:     $allMP$.append(MP_i, MP_w)
6:     $B \leftarrow$ computeAndUpdateB(T^i, T^w)
7: $[sortedMP[], ssIdxArr[], tsIdxArr[]] \leftarrow$ sortAndIndex$(allMP)$
8: **for** $i \leftarrow 1$:length$(sortedMP)$ **do**
9:     $kdist \leftarrow sortedMP[i]$
10:     **if** $kdist > bsfkdist$ **then break**
11:     **else**
12:         $offset \leftarrow ssIdxArr[i], t \leftarrow tsIdxArr[i]$
13:         **for** $j \leftarrow 1 : k$ except t and $w \leftarrow t + 1 \leq k?t + 1 : 1$ **do**
14:             $B \leftarrow$ computeAndUpdateB$(T_{offset,l}^t, T^j)$
15:             $kdist \leftarrow$ max$(kdist,$min$($computeDP$(T_{offset,l}^t, T^j)))$
16:             **if** $kdist > bsfkdist$ **then**
17:                 $[bsfkdist, tsIdx, ssIdx] \leftarrow [kdist, t, offset]$
18:             **else break**
19: **return** $[bsfkdist, tsIdx, ssIdx, B]$

4.3 Lower Bound Filtering

When the length increases to $l + s(s \in [1, l_{max} - l])$, we can directly calculate klb for each subsequence using its lower bound base information B collected before in linear time.

As mentioned above, for a subsequence $T^a_{i,l+s}$, if there is $m(1 \leq m \leq k-1)$ time series calculating DP with it, we can directly multiply each value of $B_{i,a}$ by $\sigma_{i,l}/\sigma_{i,l+s}$ using Eq. (4) to get m minimum lower bounds of $mindist$. The maximum of these m lower bounds is the klb of $T^a_{i,l+s}$. It can be calculated as:

$$klb(T^a_{i,l+s}) = \max \left(B_{i,a} * \frac{\sigma_{i,l}}{\sigma_{i,l+s}} \right) \tag{5}$$

where, $\sigma_{i,l}$ and $\sigma_{i,l+s}$ are the standard deviation of $T^a_{i,l}$ and $T^a_{i,l+s}$. In this way, we can obtain the klb of all the subsequences. The time series complexity also can be reduced from $O(\frac{N^2}{k})$ to $O(Nk)$, which is faster than the lower bound calculation (MP calculation) in the fixed-length consensus motif discovery.

Before calculating klb for each subsequence, we also need to set a $bsfkdist$ firstly. We found that CMG between two neighbour subsequence lengths are always similar. Thus, we can calculate $kdist$ of CMG_{l+s-1} found in the previous length under the current length $l + s$ as our initial $bsfkdist$ to prune more subsequences. After calculating klb, we sort klb of all subsequences in ascending order, and prune the subsequence whose klb is greater than the current $bsfkdist$. The retained subsequences are used as our candidates which can enter the part of candidate refinement.

4.4 Candidate Refinement

After screening out the candidates, in this part, we calculate the exact $kdist$ of each candidate subsequence, denoted as $candi$, using an efficient method based on the frequent items. Then find the CM_{l+s} with the smallest $kdist$.

For a $candi$ of T^i, we need to calculate $k-1$ $mindist$ with other time series except T^i. The maximum of these $mindist$ is the exact $kdist$ of this $candi$. However, many $candis$ are pruned in advance if there is a $mindist$ greater than the current $bsfkdist$ before calculating the last $mindist$. Thus, there is an intuition that if we can prune these $candis$ earlier, the number of $mindist$ calculation will be reduced, so that the time efficiency of VACOMI also can be improved. Through a lot of experiments, we found that, when the $candis$ of time series T^i calculate $mindist$ with several other fixed time series, theses $candis$ are more easily pruned. Therefore, we make some optimizations based on the frequent items. We establish a $k \times (k-1)$ matrix FI. Each $FI_{i,j}$ stores the number of the pruned subsequences in T^i when these pruned subsequences calculate $mindist$ with T^j. Then, before a $candi$ of T^i calculating $mindist$ with other time series, we sort FI_i in descending order. This $candi$ can calculate $mindist$ with other $k-1$ time series following the order of sorted FI_i, so that this $candi$ will firstly calculate $mindist$ with the time series which has the highest pruning probability with it. In this way, we can prune more subsequences earlier.

4.5 Time Auto-tuning

As noted above, klb is calculated based on the B. However, the tightness of klb will reduce with the subsequence length increasing, which will lead to an increase in the time cost of VACOMI. The performance of *klb* will be improved only when B is updated. Therefore, we propose a time auto-tuning method based on multi-rounds time prediction to decide when we should update B to improve the overall time efficiency of VACOMI.

First, there is a point we should know. If we update B under length $l + s$ $(1 \leq s \leq l_{max} - l)$, we need to execute the process of the lower bound base information to recalculate B. Nevertheless, it's very time-consuming since the time of matrix profile calculation is non-linear. But after that, the tightness of *klb* will be improved and the time cost of the computation in the subsequent lengths also can be reduced a lot. Thus, we need to calculate the time increment if update B, denoted as *Cost*. And the time decrease brought by *Cost* in the subsequent lengths is denoted as *Gain*. If *Gain* is greater than *Cost*, it's worthy to update B in this round. Otherwise, it means there is no need to update it.

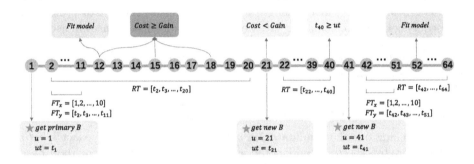

Fig. 5. An example of time auto-tuning

How to calculate the *Cost* and *Gain*? We propose a method of multi-rounds time prediction to solve it. As the example shown in Fig. 5, assume there is 64 lengths in the motif length range [1024,1087]. First, we need to fit a model to predict the time required for the subsequent lengths. At the first round, we need to collect B and calculate CM_{1024} using Algorithm 1. At the same time, we need to record some variables. u represents the round we recently update B. ut is the running time of this round. Then, since 10 fitting data are required by default, we calculate CM without updating B for the next 10 rounds (2–11). And store the running time of them into an array FT_y as the prediction target of our fitting data. $[1, 2, \cdots, 10]$ are stored in FT_x which is the independent variable of our fitting data, each value in FT_x is the difference between round id and u. At the same time, we also need to store the running time of each round into an array RT. It will be used when we calculate *Gain*.

After that, in the 12th round, we can fit the model between FT_x and FT_y using the least square method. For the model $\hat{y}_\theta(x) = \theta_0 + \theta_1 x + \theta_2 x^2, \cdots, \theta_z x^z$, least square method can find a set of parameters $\theta(\theta_0, \theta_1, \cdots, \theta_z)$ to minimize the fitting error of the model. Here, we try 4 different models to fit it using this method, namely, $\hat{y}_\theta(x) = \theta_0 + \theta_1 x$, $\hat{y}_\theta(x) = \theta_0 + \theta_1 x + \theta_2 x^2$, $\hat{y}_\theta(x) = \theta_0 + \theta_1 x +$

$\theta_2 x^2 + \theta_3 x^3$ and $\hat{y}_\theta(x) = \theta_0 + \theta_1 x + \theta_2 x^2 + \theta_3 x^3 + \theta_4 x^4$. Then we choose the model with the minimum RMSE as our current model. The smaller the RMSE, the better the fitting. RMSE can be calculated according to:

$$RMSE = \sqrt{\frac{1}{n}\sum_{i=1}^{n}(\hat{y}_i - y_i)^2} \tag{6}$$

where, n is the length of fitting data ($n = 10$), $\hat{y}_i$ is the predicted value of x_i, y_i is the observed value corresponding to x_i.

After fitting the model, we can calculate *Cost* and *Gain* to decide whether to update B in the 12th round. If we don't update B, the running time of 12th round we predict is $\hat{y}_\theta(12-u)$. Otherwise, we use ut as the reference running time of it. Thus, $ut - \hat{y}_\theta(12 - u)$ is the *Cost* which means the time increment we need to pay if we update B. While, how much *Gain* will *Cost* bring? As mentioned above, we store an array RT to record the running time of each round. If we update B in the 12th round, the current $RT = [t_2, \cdots, t_{11}]$ can be used as the reference running time for the next n rounds. Therefore, the *Gain* is:

$$Gain = \sum_{i=1}^{n}(\hat{y}_\theta(RID_i - u) - RT_i) \tag{7}$$

where, RID is the round id of these n rounds, n is the length of RT.

If $Gain \leq Cost$, there is no need to update B at 12th round. We directly run the process of lower bound filtering and candidate refinement to get CM_{1035}. Then we calculate *Cost* and *Gain* every two rounds until *Gain* is greater than *Cost*. If $Gain > Cost$ at 21th round, we can execute Algorithm 1 to obtain a new B. And u and ut also should be updated at the same time. Besides, since the current length of RT is 19, the *Gain* we predict is the total time decrease of 19 rounds. So, to realize this *Gain*, we need to calculate 19 rounds after 21th round without updating B. However, as round increases, the prediction accuracy of model will reduce, which will lead to the case at 40th round. That is, when the running time t_{40} is greater than ut, we think the prediction performance of the model has become inaccurate and we need to refit it in the next round. Thus, beginning from 41th round, we will repeat the same process as beginning from the first round to refit our model, and continue to compare *Gain* and *Cost* until the end of the algorithm.

5 Experiments

5.1 Setup

Configuration. We implement our algorithm in Matlab (R2019b), and execute the experiments in a machine running a Windows 10 64bit operating system and equipped with the following hardware: Intel(R) Core(TM) i7-7700K CPU @ 4.20 GHz (16 GB of RAM).

Datasets. We conduct our experiments over 4 real datasets including ECG-FiveDays, EthanolLevel, HandOutlines and Shield. The first 3 of them are from the UCR Archive [3] which is a set of time series dataset for time series data mining. Shield [12] is the data converted from the shapes of the shields to time series. The detail information of them is shown in Table 1, where k is the number of time series, $minLen$ is the length of the shortest time series, $maxLen$ is the length of the longest time series and N is the total length of all time series.

Table 1. Detail information of datasets

Dataset	k	$minLen$	$maxLen$	N
ECGFiveDays	442	136	136	60112
Shield	65	994	5996	73994
EthanolLevel	252	1751	1751	441252
HandOutlines	370	2709	2709	1002330

Baseline Approach. Since we propose a new Problem 1 in the domain of motif discovery, there is little work on it. Thus, we use a modified method BFO based on the existing algorithm Ostinato [6] as our baseline approach. That is, for a given subsequence length range, we loop each length using Ostinato to find the consensus motif to solve the Problem 1.

5.2   Results and Analysis

For the dataset HandOutlines and EthanolLevel, since the running time of them are too long when using BFO, we extract two data sets of 13w and 20w length from HandOutlines and EthanolLevel, namely, HandOutlines_13w and EthanolLevel_20w. Besides, the *step* in the experiments is set to 1.

Scalability over Motif Length Range. In Fig. 6, we depict the efficiency of two algorithms when varying the subsequence length. It can be seen that the running time of VACOMI is always better than BFO on four datasets. Especially on the three motif length ranges of Shield, VACOMI reduces the running time by 94%, 93% and 96% respectively compared with BFO. Besides, we observe that the performance of VACOMI remain stable over the four datasets. While, BFO is sensitive to the different datasets. On the contrary, VACOMI is more stable than BFO due to the better pruning performance of lower bound.

Scalability over Time Series Length. In Fig. 7, we run our experiments with variable time series lengths on the EthanolLevel and HandOutlines dataset. The subsequence length range is [1024,1124]. On these two datasets, we can see that the running time of both VACOMI and BFO increases with N growing.

However, the time gap between VACOMI and BFO become bigger and bigger. The running time of BFO on 60w, 80w, 100w dataset of HandOutlines is even more than one day, which is too slow. While, the time increase is more stable using our algorithm VACOMI and it is less than 1 h on the 100w dataset of HandOutlines.

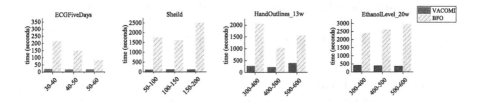

Fig. 6. Scalability over different motif length ranges

Performance Comparison of Lower Bound. In this Experiment, we compare the prunning rate α and the lower bound tightness β of VACOMI and BFO. α is the ratio of the prunned subsequences number to the all subsequences number under each motif length. β is the mean of the ratio between the lower bound and the true $kdist$ of all subsequences under each motif length. As shown in Fig. 8(a), we note that α of VACOMI is much larger than BFO on both two datasets, α of VACOMI can stabilize above 70%, while α of BFO is smaller than 2%, even approaches 0% on the HandOutlines dataset, which is far worse than VACOMI. The reason is shown in Fig. 8(b). We can see that β of VACOMI is always tighter than BFO on these two datasets. Thus, the pruning rate of VACOMI is higher than BFO. In addition, α and β of two algorithms are same under the minimum motif length on both datasets. It is because VACOMI collects B at the minimum subsequence length using Algorithm 1, which uses the same lower bound as BFO to prune the subsequences. Thus, the performance of VACOMI and BFO are the same at minimum subsequence length.

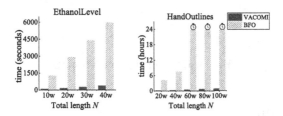

Fig. 7. Scalability over different time series lengths

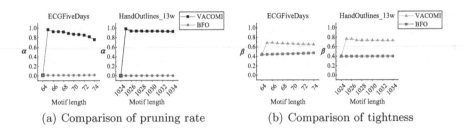

(a) Comparison of pruning rate   (b) Comparison of tightness

Fig. 8. Performance comparison of lower bound

Effect of Time Auto-tuning. To confirm the effectiveness of time auto-tuning, we propose a baseline method VACOMI-δ to compare with VACOMI. There is a difference between them when we decide which round to update B. In VACOMI, we compare *Gain* and *Cost* based on the multi-rounds time prediction to decide when to update B. While, in VACOMI-δ, we decide it according to the number of candidates. That is, before the algorithm begins, we need to set a threshold δ. It is the maximum ratio of the number of *candis* which is obtained after lower bound filtering to the total subsequences number. As the subsequence length increases, if the ratio between the *candis* number and the total subsequences number exceeds δ in a certain round, VACOMI-δ think the lower bound become less tight and update B in this round. As shown in Fig. 9, the red square point on the curve represents the minimum running time of VACOMI-δ under a threshold δ. And the histogram represents the running time of VACOMI using time auto-tuning. It can be seen that on these four datasets, the running time of VACOMI is basically same as the minimum running time of VACOMI-δ, which strongly proves the effectiveness of our time auto-tuning method and eliminates the complexity of threshold tuning in VACOMI-δ.

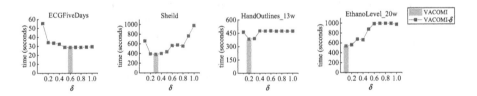

Fig. 9. Effect of time auto-tuning

6   Conclusion

In this paper, we focus on the problem of variable-length consensus motif discovery in multiple time series. To reduce the time cost of the algorithm, we propose an efficient, exact and scalable algorithm VACOMI to improve the efficiency of our method. Due to the novel lower bound we propose, more subsequences can be pruned, so that the time complexity is reduced to linear time. The optimization based on the frequent items and the multi-rounds time prediction in the parts of candidate refinement calculation and time auto-tuning also speed up the overall efficiency of our method. The experimental results show that VACOMI achieves better performance than the baselines.

References

1. Agrawal, R., Faloutsos, C., Swami, A.: Efficient similarity search in sequence databases. In: Lomet, D.B. (ed.) FODO 1993. LNCS, vol. 730, pp. 69–84. Springer, Heidelberg (1993). https://doi.org/10.1007/3-540-57301-1_5
2. Chiu, B., Keogh, E., Lonardi, S.: Probabilistic discovery of time series motifs. In: SIGKDD, pp. 493–498 (2003)
3. Dau, H.A., et al.: The UCR time series archive. IEEE/CAA J. Automatica Sinica 6(6), 1293–1305 (2019)
4. De Paepe, D., Van Hoecke, S.: Mining recurring patterns in real-valued time series using the radius profile. In: ICDM, pp. 984–989. IEEE (2020)
5. Gao, Y., Lin, J.: HIME: discovering variable-length motifs in large-scale time series. KAIS 61(1), 513–542 (2018). https://doi.org/10.1007/s10115-018-1279-6
6. Kamgar, K., Gharghabi, S., Keogh, E.: Matrix profile XV: exploiting time series consensus motifs to find structure in time series sets. In: ICDM, pp. 1156–1161. IEEE (2019)
7. Lei, W., Li, P., Han, Y., Gong, S., Yang, L., Hou, M.: EPG recordings reveal differential feeding behaviors in Sogatella Furcifera in response to plant virus infection and transmission success. Sci. Rep. 6(1), 1–9 (2016)
8. Linardi, M., Zhu, Y., Palpanas, T., Keogh, E.: Matrix profile X: VALMOD-scalable discovery of variable-length motifs in data series. In: SIGMOD, pp. 1053–1066 (2018)
9. Mueen, A., Chavoshi, N.: Enumeration of time series motifs of all lengths. KAIS 45(1), 105–132 (2014). https://doi.org/10.1007/s10115-014-0793-4
10. Paparrizos, J., Gravano, L.: k-Shape: efficient and accurate clustering of time series. In: SIGMOD, pp. 1855–1870 (2015)
11. Wang, X., et al.: RPM: representative pattern mining for efficient time series classification. In: EDBT, pp. 185–196 (2016)
12. Ye, L., Keogh, E.: Time series shapelets: a new primitive for data mining. In: SIGKDD, pp. 947–956 (2009)
13. Yeh, C.C.M., et al.: Matrix profile I: all pairs similarity joins for time series: a unifying view that includes motifs, discords and shapelets. In: ICDM, pp. 1317–1322. IEEE (2016)
14. Zhu, Y., Yeh, C.C.M., Zimmerman, Z., Kamgar, K., Keogh, E.: Matrix profile XI: SCRIMP++: time series motif discovery at interactive speeds. In: ICDM, pp. 837–846. IEEE (2018)
15. Zhu, Y., et al.: Matrix profile II: exploiting a novel algorithm and GPUs to break the one hundred million barrier for time series motifs and joins. In: ICDM, pp. 739–748. IEEE (2016)

LiteWSC: A Lightweight Framework for Web-Scale Spectral Clustering

Geping Yang[1], Sucheng Deng[2], Yiyang Yang[1(✉)], Zhiguo Gong[2(✉)], Xiang Chen[3], and Zhifeng Hao[4]

[1] Faculty of Computer, Guangdong University of Technology, Guangzhou, China
yyygou@gmail.com
[2] State Key Laboratory of Internet of Things for Smart City and Department of Computer and Information Science, University of Macau, Macau, China
fstzgg@um.edu.mo
[3] School of Electronics and Information Technology, Sun Yat-Sen University, Guangzhou, China
[4] College of Engineering, Shantou University, Shantou, China

Abstract. Spectral clustering is an effective clustering method for its excellent performance in partitioning non-linearly distributed data. Nevertheless, it suffers from scalability due to its high computational complexity in constructing the Laplacian graph and computing the corresponding eigendecomposition. In the past decades, many efforts have been made to face this problem. However, the computational bottleneck is still a problem in processing extensive data, especially in web-scale scenarios. We present LiteWSC, a simple yet efficient lightweight spectral clustering framework, to cluster web-scale data with limited resource requirements. Our framework has minimal space overhead and does not require explicit overall embeddings computation. We also analyze the theoretical guarantee and performance boundary of LiteWSC.

LiteWSC is highly flexible with $O(sp)$ memory requirement, where s and p are the number of samples and the number of prototypes, which are adaptive to the available resource. Therefore, LiteWSC can partition web-scale data (e.g., $n = 8,000k$) in an resource-limited host (e.g., memory is restricted to 1 GB). Experiments on real-world, large-scale and web-scale datasets demonstrate both the efficiency and effectiveness of LiteWSC over state-of-the-art methods.

Keywords: Spectral clustering · Data quantization · Machine learning · Scalability

1 Introduction

In the past decades, many clustering methods have been proposed [14,36,38]. Spectral Clustering (SC) [4,8,20,26] has shown great promise in its good performance on adapting to a broader range of geometries and detecting non-convex

---

G. Yang and S. Deng—Equal Contribution.

patterns. However, SC is not scalable well since it exhibits considerable challenges in both computation and memory consumption.

SC generally takes $O(n^2 d)$ time complexity to conduct and $O(n^2)$ memory to store the Laplacian graph, where n is the number of data objects and d denotes the number of feature dimensions. Besides, it requires $O(n^3)$ to perform the eigendecomposition [6]. To alleviate the computational burden, many methods have been presented to speed up SC and reduce the memory requirements in the past decades.

The former eliminates the bottleneck by reducing the scale of eigendecomposition. Variational Nyström [29] (VN) method reduces the scale of eigendecomposition by Nyström approximation. The Power method [19] is used to compute the eigenvectors without explicit eigendecomposition. Weight Kernel k-means (WKKM) [9,21] is used to approximate the SC. Although those methods have dramatically reduced the time complexity of eigendecomposition, they are still resource-consuming in terms of the computational time and memory requirement to construct the Laplacian graph (e.g., $O(n^2)$). Furthermore, sparse representation [1,34] or Nyström [12,31] are applied as well: p ($p \ll n$) landmarks are selected and then used to construct $n \times p$ low-rank affinity matrix. Based on this, an approximated Laplacian graph is built. **Both the sparse representation based methods and the Nyström based methods require $O(np)$ time complexity and $O(np)$ memory.**

Another roadmap of large-scale SC methods aims to find p representatives to represent the entire dataset. k-means-based Spectral Clustering (KASP) [33] uses k-means to generate p representatives, and performs conventional SC on p representatives rather than all data objects. Then, the clustering results are determined by the ones of SC on representatives. In this method, most of the time and memory is spent on finding representatives. It is efficient with $O(npt)$ time complexity and $O(np)$ memory requirement, where t is the iterations of k-means. However, representative selection alone is an expensive process [25]. To summarize, existing SC methods, including large-scale versions, are still resource-consuming, therefore, cannot be adopted to web-scale data, especially for resource-limited hosts.

We propose LiteWSC, a simple and yet efficient Lightweight framework for Web-scale Spectral Clustering, to address these problems with the following properties:

1. Scalable and robust. Our framework can cluster web-scale datasets with empirical cost around $O(spt)$, where s and p are highly flexible according to the available resources, and t denotes the number of iterations in k-means. We provide theoretical analysis to LiteWSC and show its performance is compromised.
2. Memory efficient. In principle, our framework only needs to maintain the information of p prototypes. It can process extensive datasets on a minimal resource machine, e.g., clustering dataset with the size of $8,000k$ on a server with 1 GB memory only.
3. Fast and accurate. We demonstrate with real-world datasets that, LiteWSC typically generates better clustering results by a much faster converge process compared with the state-of-the-art methods.

4. Easy to intercept and implement. Our framework consists of four stages. Each stage is conducted with high interpretation. Our implementation[1] with all four stages described in this work takes less than 100 lines of Matlab code (excluding I/O and evaluation parts).

We compare our framework against existing large-scale SC methods, i.e., k-means Approximated Spectral Clustering (KASP) [33], Variational Nyström (VN) [29], Landmark-based Spectral Clustering (LSC) [1,5], Large-Scale Hypergraph Spectral Clustering (LSHC) [34], Ultra-Scalable Spectral Clustering (U-SPEC) [16]. Results demonstrate that our framework consistently outperforms these methods in three perspectives: clustering accuracy, memory efficiency, and scalability.

2   Related Work

SC applies eigendecomposition on graph Laplacian $L \in \mathbb{R}^{n \times n}$ and obtains its top-k eigenvectors $U \in \mathbb{R}^{n \times k}$ associated with top-k eigenvalues. The top-k eigenvectors U are referred to as embeddings. Usually, k-means [15] is applied to embeddings U and generates the final clustering results. Eigendecomposition on L directly is extremely expensive with $O(n^3)$ time complexity and $O(n^2)$ memory requirement. Many efforts have been designed to resolve this bottleneck.

A natural way is to reduce the scale of eigendecomposition. Variational Nyström method [29] firstly constructs a $n \times n$ sparse Laplacian and selects p column via a random selection or k-means as a sketching matrix. Then, it applies eigendecomposition on a $p \times p$ matrix, that is computed by multiplying sketching matrix and sparse Laplacian, and obtains more informative embeddings. Besides, Power method [19] is a useful tool: it iteratively multiplies the Laplacian matrix by a random matrix until convergence, the resulting matrix can be viewed as the eigenvectors. What's more, Dhillon et al. [9] use the Weighted kernel k-means (WKKM) to approximate SC and provide related theoretical analysis. To further accelerate, Mahesh et al. [21] apply WKKM on partial data points and assign the remaining data points to the resulting centers.

The Nyström method [12] randomly selects p representatives from the dataset and constructs a $n \times p$ affinity matrix to accelerate the SC and reduce its memory requirement further. The eigenvectors of original data objects are approximated by the conducted affinity matrix and the eigenvectors of representatives. Landmarks-based Spectral Clustering (LSC) [1,5] constructs a sparse $n \times p$ affinity matrix to enhance the information embedded by graph Laplacian. Based on it, LSC performs the Laplacian matrix construction and eigenvectors approximation. The clustering quality is enhanced since LSC adopts simple spare coding [10,23]. Following a similar idea, Large Scale Hypergraph Spectral Clustering (LSHC) [34] provides a hypergraph-based solution; a general weighted-incidence matrix is used to describe the affinities. Both LSC and LSHC adopt two ways

---

[1] Implementation of LiteWSC is released on https://github.com/gepingyang/LiteWSC.

Algorithm 1. Procedure of LiteWSC

Input: dataset O; the number of clusters k; the number of samples s; the number of prototypes p; the number of nearest prototypes r

Output: cluster assignment Y

1: $S \leftarrow$ UnifomSample(O, s);
2: $P \leftarrow k$-means(S, p);
3: $Z \leftarrow$ Construct the prototype-sample affinity matrix $Z \in \mathbb{R}^{p \times s}$ (as Equation 3);
4: $U \leftarrow$ Compute the first k eigenvectors of L_p (as Equation 4);
5: k-means(U, k);
6: **for** $B \subset O$; **do**
7:    $Y_B \leftarrow$ ClusterAssign(P, B);
8:    $Y \leftarrow Y \cup Y_B$;
9: **end for**
10: **return** Y

---

for landmarks (hyperedges) selection, random selection and k-means selection, to meet the diverse resource requirements: The former needs less time and has poor results, while the latter leads to more excellent results but with extra cost.

Furthermore, Ultra-Scalable Spectral Clustering (U-SPEC) [16] applies k-means on partial data objects to generate representatives without theoretical guarantee. It uses approximated k-nearest representatives to generate sparse representation fleetly. In contrast, a faster method is to reduce the size of the dataset. k-means-based Spectral Clustering (KASP) [33] uses k-means to decrease the data size from n to s s.t. ($s \ll n$). The resulting p centers are regarded as representatives, and then a conventional SC is executed on representatives. Finally, the clustering results are decided by finding the nearest representative. Besides, Duan et al. [11] apply the deep autoencoder to generate embeddings without constructing the affinity matrix.

3 The Proposed Method

3.1 Preliminaries

Given a set of n data objects $O_1, O_2, ..., O_n \in \mathbb{R}^d$, conventional Spectral Clustering firstly constructs an affinity matrix $W \in \mathbb{R}^{n \times n}$, whose entries w_{ij} denotes the affinity between O_i and O_j. The normalized Laplacian graph L is defined as [20]:

$$L = D^{1/2}(D - W)D^{1/2} = I - D^{1/2}WD^{1/2} \tag{1}$$

where diagonal matrix $D \in \mathbb{R}^{n \times n}$ denotes the degree matrix of W, whose entries are column or row sums of W, $D_{ii} = \sum_{j=1}^{n} w_{ij}$. Moreover, Ng et al. [22] defined an alternative of L as $D^{1/2}WD^{1/2}$.

As introduced in Sect. 2, conducting, storing, and performing eigendecomposition on $L \in \mathbb{R}^{n \times n}$ are resource demanding, especially when n is large. Moreover, concluded by [34], the eigendecomposition is no longer the bottleneck of large-scale SC. However, these methods are still too resource exhausted to be applied on web-scale data: most large-scale SC methods explicitly compute the top-k eigenvectors of Laplacian graph (e.g., L_A) as embeddings $U \in \mathbb{R}^{n \times k}$, according to our analysis to be shown later, that is non-essential.

3.2 Framework Overview

To solve the new barriers of SC, we propose LiteWSC, a simple yet effective lightweight spectral clustering framework for web-scale data. As illustrated in Algorithm 1, LiteWSC consists of four stages:

1. Approximated k-means based Vector Quantization
2. Prototype Laplacian Graph Construction and Eigendecomposition
3. Prototype Clustering
4. Batch Cluster Assignment

In first stage, approximated k-means is executed, and results in p prototypes (Lines 1–2). In the second stage, as other large-scale SC methods, the prototype Laplacian graph is conducted (Line 3), decomposited (Line 4), and leads to the prototype embeddings U. Based on U, the cluster labels of prototypes are detected in the third stage (Line 5). Regarding the final stage, the data objects are processed in terms of the batch, they are assigned to the nearest prototypes.

3.3 Approximated k-means Based Vector Quantization

The primary task of this stage is reducing the demanded resources. Vector quantization is a classical method that aims to find a set of prototypes to represent the entire dataset with minimum data distortion [24], while k-means is widely used if distortion measurement is squared error based [33]. However, applying k-means on web-scale data is expensive [25]. On the other hand, the target of k-means is to detect p cluster centers as prototypes instead of acquiring cluster membership directly. Therefore, LiteWSC randomly selects s ($s \ll n$) data points from the overall dataset without replacement as sample set S. Then, k-means is executed on S to obtain p clusters, while the cluster centers are regarded as prototypes. In other words, the prototypes $P \in \mathbb{R}^{p \times d}$ are excellent representatives of the dataset O.

Clearly, the proposed approximated k-means based Vector Quantization is much more efficient than existing SC methods in terms of computational complexity and memory requirement. We now provide its performance analysis. k-means is a local optimal seeking algorithm. That is, each execution leads to a locally optimal solution. There exists an error to the ideal optimal solution. Thus, k-means can be viewed as an α-approximation algorithm ($\alpha \geq 1$) to the ideal k-means.

Definition 1. *Let $\alpha \geq 1$. $\mathcal{P}$ is an α-approximation algorithm for the k-means problem if the resulting centers, C, return by $\mathcal{P}$ satisfies,*

$$\mathbb{W}(O, C) \leq \alpha \mathbb{W}(O, C_{opt})$$

where $\mathbb{W}(O, C)$ denotes the sum of squared error of every data point $\{O_1, O_2, ..., O_n\}$ to its nearest center in C, and C_{opt} is the ideal optimal centers. The corresponding error bound is given as follow,

Theorem 1. *Let* $\alpha \geq 1$, $0 < \beta < 1$, $0 < \gamma < 1$ *and* $\epsilon > 0$ *be approximation parameters. Let* $\mathcal{P}$ *be an* α*-approximated algorithm of the k-means problem. Given the data object set* $O = \{O_1, O_2, ...O_n\}$, *suppose a subset* $S \subset O$ *of size* s *is uniformly sampled at random without replacement. s.t.*

$$s \geq \ln(\frac{1}{\gamma})(1 + \frac{1}{n})/(\frac{2\beta^2\epsilon^2}{\Delta^2\alpha^2} + \frac{1}{n}\ln(\frac{1}{\gamma}))$$

where $\Delta = \max_{ij} ||O_i - O_j||^2$. *If k-means is run on* S *and obtain the prototypes* P^S, *with probability at least* $1 - \gamma$,

$$\mathbb{W}(S, P^S) \leq 4(\alpha + \beta)\mathbb{W}(O, P_{opt}) + \epsilon \tag{2}$$

where P_{opt} *is the set of optimal prototypes.*

Remark 1. Originally, Theorem 1 is used to address the large-scale weighted kernel k-means [21]. We use similar notations as [21]. If we set the weights of all data points being one and related kernel function $\varphi(O) = O$, where O is the set of data points, the weighted kernel k-means is the same as k-means, thus, Theorem 1 follows.

According to Theorem 1, if the number of samples s is sufficient, the obtained prototypes P are as excellent as the ones that obtained by applying k-means on overall dataset O. In other words, s is a significant parameter that can be used to balance the trade-off between prototype quality and demanded resources. The value setting of s will be discussed in Sect. 5.

3.4 Prototype Laplacian Graph Construction and Eigendecomposition

Based on the obtained prototypes $P \in \mathbb{R}^{p \times d}$, it is possible to conduct a $p \times p$ prototype Laplacian graph L_p which includes the pairwise affinities between prototypes only (as Nyström based methods [30] or KASP [33]). However, the information contained in non-prototypes (i.e. $S - P$) is completely ignored [29].

Thus, inspired by sparse representation-based methods [1,16,29,34], LiteWSC utilizes the affinities between prototypes and samples:

$$Z(i,j) = \exp\left(-\frac{-||P_i - S_j||^2}{2\sigma^2}\right), \tag{3}$$

where σ is a bandwidth parameter, $Z \in \mathbb{R}^{p \times s}$. The entry $Z(i,j)$ denotes the affinity[2] between the i-th prototype and the j-th sample. More significantly, for

---

[2] Besides the Gaussian kernel mentioned above, we can use other kernel functions to formulate the pairwise affinities.

each sample, only the r ($r \ll s$) largest non-zero entries are retained to guarantee the graph sparsity.

Based on Z, a more informative prototype Laplacian graph L_p is defined as:

$$L_p = D_p^{-1/2} W_p D_p^{-1/2} \quad \text{s.t.} \quad W_p = ZZ^T \tag{4}$$

where D_p is a diagonal degree matrix with $D_p(i,i) = \sum_j^p W_p(i,j)$ denotes the affinity sum of the i-th row/column of W_p.

Further, eigendecomposition is performed on L_p and obtain its first k normalized eigenvectors $U \in \mathbb{R}^{p \times k}$. The conducted prototype Laplacian graph L_p is different from that used by existing large-scale SC methods [1,5,16,29,33,34]. Our L_p has a vital advantage: it is highly memory efficient, only the affinities between prototypes P and samples S are considered. This perspective will be discussed in the later section.

3.5   Prototype Clustering and Batch Cluster Assignment

The generated U in the second stage is regarded as the prototype embeddings. And then, a k-means process is applied on U and derive the cluster structure in terms of the prototypes. Finally, LiteWSC loads the web-scale data (probably are from the secondary storage) in batches, and determines the clustering membership of data batches by their nearest prototypes. Different from existing approaches, LiteWSC only records the prototypes rather than that of all data points. This mechanism heavily releases the resources occupied by the entire dataset.

4   Framework Analysis

4.1   Perturbation Analysis

Another critical advantage of LiteWSC is the Batch Cluster Assignment. Existing large-scale methods compute the affinities between prototypes and all data objects (e.g., n). In contrast, our framework only compute partial of them: the affinities between prototypes and samples (e.g., s). We now provide the theoretical analysis to such a perturbation. It is an extension from the study given by Yan et al. [33]. We start from the perturbation of prototypes to the original dataset. Suppose data objects $O \in \mathbb{R}^{n \times d} = \{O_1, O_2, ..., O_n\}$ are independently and identically distributed (i.i.d.), and $\hat{O} \in \mathbb{R}^{n \times d} = \{\hat{O}_1, \hat{O}_2, ..., \hat{O}_n\}$, where $\hat{O}_i$ denotes the nearest prototypes of O_i. Then, we regard ϵ_i as the perturbation between data object O_i and its nearest prototype $\hat{O}_i$ (In terms of the nearest prototype in P), and is defined as:

$$\hat{O}_i = O_i + \epsilon_i \tag{5}$$

To assist in our analysis, we assume ϵ_i has following characters: (1) ϵ_i is independent of O_i; (2) ϵ_i is mutually independent with each others and in a symmetric distribution with mean zero and bound support; (3) the variance of ϵ_i is much small than that of O_i.

Based on the Eq. 3 and Eq. 4, we define four matrices: W, $\hat{W}$, L, and $\hat{L}$ as follows. Let $W \in \mathbb{R}^{n \times n}$ and $L \in \mathbb{R}^{n \times n}$ denote the affinity matrix and Laplacian graph constructed by computing affinities between data objects O and samples S. Besides, $\hat{W} \in \mathbb{R}^{n \times n}$ and $\hat{L} \in \mathbb{R}^{n \times n}$ denote the ones that constructed by prototypes $\hat{O}$ and samples S. We use Frobenius norm to measure the perturbation between $U = \{U_1, U_2, ..., U_k\}$ and $\hat{U} = \{\hat{U}_1, \hat{U}_2, ..., \hat{U}_k\}$, where U_i and $\hat{U}_i$ denote the i-th eigenvector of L, and that of $\hat{L}$ respectively. Especially, it has been proven [32] that $\hat{U}$ is a scalar-multiplication of the top-k eigenvectors of the prototype graph $L_p \in \mathbb{R}^{p \times p}$. The perturbation between U and $\hat{U}$ can be written as:

$$||U - \hat{U}||_F^2 \tag{6}$$

Therefore, a lemma is defined as follows:

Lemma 1. *([27]) Suppose that g_i is the eigengap between the i-th and $(i+1)$-th eigenvalues of L, then the following holds:*

$$||U_i - \hat{U}_i|| \leq \frac{1}{g_i}||L - \hat{L}|| + O(||L - \hat{L})||^2).$$

Based on the lemma above, it is easy to infer the following theorem:

Theorem 2. *Assume that the eigengap g_i ($i = 1, 2, ..., k$) is bound away from zero and the maximum of $\frac{1}{g_i^2}$ is an constant C. If we ignore the high-order terms on the right-hand of Lemma 1, the following holds:*

$$||U - \hat{U}||_F^2 \leq kC||L - \hat{L}||_F^2$$

Proof.

$$||U - \hat{U}||_F^2 \leq \sum_{i=1}^{k} \frac{1}{g_i^2}||L - \hat{L}||^2 \leq kC||L - \hat{L}||_F^2$$

Thus, we have turned our perturbation from eigenvectors form (e.g., $||U - \hat{U}||_F^2$) to the Laplacian form (e.g., $kC||L - \hat{L}||_F^2$). For easy demonstration, let $W = \hat{W} + \Gamma$ and $D = \hat{D} + \Theta$, where D and $\hat{D}$ denote the degree matrix of L and $\hat{L}$ respectively. Then, we have the lemma as follows:

Lemma 2. *([17]) Suppose that $||\Theta D^{-1}|| = o(1)$, the perturbation of L and $\hat{L}$ in the Frobenius norm can be approximated as:*

$$||L - \hat{L}||_F \leq ||D^{-\frac{1}{2}}\Gamma D^{-\frac{1}{2}}||_F + (1 + o(1))||\Theta D^{-\frac{3}{2}}WD^{-\frac{1}{2}}||_F$$

Then, we use Taylor series expansions to work out the bound for $||D^{-\frac{1}{2}}\Gamma D^{-\frac{1}{2}}||_F$ and $||\Theta D^{-\frac{3}{2}}WD^{-\frac{1}{2}}||_F$. Now, we introduce a statistical model for the original data, which is a model for data that fall into two clusters. In the purpose of making our analysis tractable, we assume that the distribution F of the original data objects O can be modeled as two-component mixture model:

$$F = (1 - \eta) \cdot F_1 + \eta \cdot F_2, \tag{7}$$

where $\eta \in 0, 1$ with probabilities $P(\eta = 1) = \theta$ and $P(\eta = 0) = 1 - \theta$. Meanwhile, the approximated prototypes $\hat{O}$ was modeled as $\hat{F} = (1 - \eta) \cdot \hat{F}_1 + \eta \cdot \hat{F}_2$, where $\hat{F}_1$ and $\hat{F}_2$ are generated from Eq. 5. We need the following two Lemmas to further calculate Lemma 2.

Lemma 3. *Suppose that: (1)* $\inf_{1 \le i \le n} \frac{D_i}{n} \ge C_0$ *holds in probability where D_i denotes the i-th non-zero entry of diagonal matrix D and C_0 is a constant for O is generated i.i.d. from Eq. 7, (2)$\|\Theta D^{-1}\| = o(1)$. Then, we have*

$$\|D^{-\frac{1}{2}} \Gamma D^{-\frac{1}{2}}\|_F^2 \le_p C_1 \sigma_\epsilon^2 + C_2 \sigma_\epsilon^4, \tag{8}$$

where "$\le_p$" denotes that inequality holds in probability, C_1 and C_2 are constants as $n \to \infty$, σ_ϵ^2 and σ_ϵ^4 denote the second and the fourth moments of ϵ.

Proof.

$$\|D^{-\frac{1}{2}} \Gamma D^{-\frac{1}{2}}\|_F^2 = \sum_{i=1}^n \sum_{j=1}^n \frac{\Gamma_{ij}^2}{D_i D_j} \le_p \frac{1}{C_0^2 n^2} \sum_{i=1}^n \sum_{j=1}^n \Gamma_{ij}^2$$

$$= \frac{1}{C_0^2 n^2} \sum_{i=1}^n \sum_{j=1}^n \left(\sum_{l=1}^s \left(\exp\left(-\frac{\|O_i + \epsilon_i - S_l\|^2 + \|O_j + \epsilon_j - S_l\|^2}{2\sigma^2} \right) \right.\right.$$

$$\left.\left. - \exp\left(-\frac{\|O_i - S_l\|^2 + \|O_j - S_l\|^2}{2\sigma^2} \right) \right) \right)^2$$

$$\le \frac{s}{C_0^2 n^2} \sum_{i=1}^n \sum_{j=1}^n \sum_{l=1}^s \left(\frac{(O_i - S_l)^T \epsilon_i + (O_j - S_l)^T \epsilon_j}{\sigma^2} + \frac{(O_i - S_l)\epsilon_i^2 + (O_j - S_l)\epsilon_j^2}{2\sigma^2} \right)^2$$

$$\cdot \exp\left(-\frac{\|O_i - S_l\|^2 + \|O_j - S_l\|^2}{\sigma^2} \right)$$

$$\le \frac{4s}{C_0^2 n^2} \sum_{i=1}^n \sum_{j=1}^n \sum_{l=1}^s \left(\frac{(\epsilon_i^2 + \epsilon_j^2)(\|O_i - S_l\|^2 + \|O_j - S_l\|^2)}{\sigma^4} + \frac{(\epsilon_i^4 + \epsilon_j^4)(\|O_i - S_l\|^2 + \|O_j - S_l\|^2)}{4\sigma^4} \right)$$

$$\cdot \exp\left(-\frac{\|O_i - S_l\|^2 + \|O_j - S_l\|^2}{\sigma^2} \right) \le C_1 \sigma_\epsilon^2 + C_2 \sigma_\epsilon^4$$

The forth step holds because we use Taylor expansion of the function $f(x, y) = \exp\left(-\frac{\|a+x\|^2 + \|b+y\|^2}{\sigma^2} \right)$ around $x = 0$ and $y = 0$ on Γ_{ij}, where $x = \epsilon_i$ and $y = \epsilon_j$.

Lemma 4. *Suppose that: (1)* $\inf_{1 \le i \le n} \frac{D_i}{n} \ge C_0$ *holds in probability where D_i denotes the i-th non-zero entry of diagonal matrix D and C_0 is a constant for O is generated i.i.d. from Eq. 7, (2)$\|\Theta D^{-1}\| = o(1)$. Then, we have*

$$\|\Theta D^{-\frac{3}{2}} W D^{-\frac{1}{2}}\|_F^2 \le_p C_1 \sigma_\epsilon^2 + C_2 \sigma_\epsilon^4$$

where C_1 and C_2 are constants as $n \to \infty$, σ_ϵ^2 and σ_ϵ^4 denote the second and the fourth moments of ϵ.

Proof.

$$||\Theta D^{-\frac{3}{2}} W D^{-\frac{1}{2}}||_F^2 = \sum_{i=1}^n \sum_{j=1}^n (\sum_{l=1}^n \frac{W_{ij}^2}{D_i^3 D_j} \Gamma_{il})^2 \leq_P \frac{s^2}{C_0^4 N^2} \sum_{i=1}^N \sum_{l=1}^N \Gamma_{il}^2 \leq C_1 \sigma_\epsilon^2 + C_2 \sigma_\epsilon^4$$

the first step holds for $\Theta_{ii} = \sum_j^n \Gamma_{ij}$, and the third step holds for we use the analysis of Γ_{il} the entry of Γ in Lemma 3.

Together with Lemma 3 and Lemma 4, we can easily get the theorem as follows:

Theorem 3. *Suppose that: (1) $inf_{1 \leq i \leq n} \frac{D_i}{n} \geq C_0$ holds in probability for O is generated i.i.d. from Eq. 7, (2)$||\Theta D^{-1}|| = o(1)$, then*

$$||L - \hat{L}||_F^2 \leq C_1 \sigma_\epsilon^2 + C_2 \sigma_\epsilon^4$$

where C_1 and C_2 are constants as $n \to \infty$, and σ_ϵ^2 and σ_ϵ^4 denote the second and fourth moments of ϵ.

Then, we can use the property of vector quantization [13,35] to further compute the property of $||U - \hat{U}||_F^2$. According to vector quantization, we can characterize precisely the amount of distortion when the prototypes are obtained by k-means if the probability distribution of the original data objects is given. Let a quantizer T be defined as $T = P_1, P_2, ..., P_p$. If original data objects are generated from a random source, the distortion of T can be defined as $\mathbb{D}(T) = E(O - T(O))^u$, which is the mean square error for $u = 2$. The rate of quantizer is defined as $R(T) = log_2 p$. Define the distortion-rate function $\delta(R)$ as

$$\delta(R) = \inf_{T:R(T) \leq R} \mathbb{D}(T).$$

$\delta(R)$ can be characterized in terms of the source density of F and constants d, u by the following theorem.

Theorem 4. *([35]) Suppose that f is the density function for F (defined in Eq. 7). Then, for large rates R, the distortion-rate function of fixed-rate quantization has the following form:*

$$\delta_d(R) \cong y_{u,d} \cdot ||f||_{d/(d+u)} \cdot p^{(-u/d)}$$

where $\cong$ denotes that both sides of the equation are approximately equal, $y_{u,d}$ is a constant depending on u and d, and

$$||f||_{d/(d+u)} = \left(\int f^{d/(d+u)}(O) d(O) \right)^{(d+u)/d}$$

As mentioned above, by Theorem 2, Theorem 3, and Theorem 4, the property of $||U - \hat{U}||_F^2$ can be written as follow.

Theorem 5. *Suppose that the original data objects are generated from a distribution with density f, and assumptions of Theorem 2 and Theorem 3 hold, we have following perturbation of eigenvectors,*

$$||U - \hat{U}||_F^2 = C \cdot y_{2,d} \cdot ||f||_{d/(d+2)} \cdot p^{-2/d} + O(p^{-4/d}) \tag{9}$$

Table 1. Datasets in evaluation.

Name	# instances	# features	# classes
USPS	7,291	256	10
letter	15,000	16	26
MNIST	70,000	784	10
Balanced	131,600	784	47
Covtype	581,012	54	7
Byclass	814,255	784	62
Yahoo	1,400,000	100	10
MNIST8M	8,100,000	784	10

where C is a constant depending on the number of prototypes, the variance of the original data objects, the bandwidth of the Gaussian kernel, the number of samples, and the eigengap of the Laplacian graph of the original data objects.

4.2 Complexity Analysis

In this section, we analyze the time and space consumption of our LiteWSC in the matrix form. Assume that dataset O includes n d-dimensional data objects, and s samples are selected to find p prototypes, and the number of clusters is k, s.t. $k \ll p \ll s \ll n$.

The vector quantization stage requires $O(sp)$ memory storage and $O(spdt)$ runtime, where t denotes the number of iterations of k-means. Regarding the second stage, LiteWSC needs $O(sp)$ memory storage and $O(spd)$ runtime to conduct the prototype Laplacian graph. It requires $O(p^2)$ memory and $O(p^3)$ for the eigendecomposition step. For prototype cluster stage, the costs are $O((p + k)k)$ memory and $O(pk^2t)$ runtime. The last stage requires $O(npd)$ runtime and least $O(pd)$ space. Our overall time complexity is dominated by $O(spdt+p^3+npd)$ with $O(sp)$ maximum memory consumption.

5   Evaluation

To evaluate the performance of LiteWSC, several experiments are conducted on real datasets with different sizes.

Datasets. Datasets are selected from numerous benchmarks, including from LibSVM [3], EMNIST [7] and Yahoo [18]. Details of the employed datasets are described in Table 1.

Baselines. LiteWSC is compared with several state-of-the-art methods:

k-**means** mini batch k-means [25], a fast Web-Scale k-means, we chose the implementation provided by Python scikit-learn.[3]

---
[3] https://scikit-learn.org/.

SC conventional Spectral Clustering [22].

Nyström Nyström approximated Spectral Clustering. Its Matlab version [37][4] is chosen.

VN Variational Nyström approximated Spectral Clustering select prototypes (landmarks) randomly [29].[5]

KASP k-means Approximate Spectral clustering [33], it has a R version online,[6] we implement a Matlab version.

LSC-K and LSC-R Landmark-based Spectral Clustering using k-means(LSC-K) or random sampling approach(LSC-R) for prototypes (landmarks) selection [1].[7]

LSHC-R and LSHC-K Large Scale Hypergraph Spectral Clustering using k-means or random sampling approach(LSHC-R) for prototypes (hyperedges) selection [34].[8]

U-SPEC Ultra-Scalable Spectral Clustering (U-SPEC) [16][9]

Evaluation Metric. The clustering result is evaluated by Accuracy (AC) [2] and Normalized Mutual Information (NMI) [28], which are typical evaluation metrics for clustering. For both metrics, the higher value indicates the better performance. Accuracy (AC) is defined as:

$$AC = \frac{\sum_{i=1}^{n} \delta\left(s_i, map\left(r_i\right)\right)}{n},$$

where n is the number of data objects, s_i and r_i respectively denote the cluster label and the label provided by dataset (ground truth) for vertex i, $map(r_v)$ is a mapping function that maps r_i to its equivalent label in dataset.

Normalized Mutual Information (NMI) is given by

$$NMI(C, C') = \frac{I(C : C')}{[H(C) + H(C')]/2},$$

where C is the set of clusters provided by dataset, C' is the clustering result. $I(C; C')$ is the mutual information between C and C'. $H(C)$ and $H(C')$ are the entropies for C and C' respectively. For both metrics, a higher value indicates the better quality.

Experiment Setup. All experiments are executed on a Win-10 64 bits machine with Intel(R) Core $i5 - 9400F$ CPU (2.90 GHz), 16 GB main memory. k-means is executed on Python 3.7 and scikit-learn 0.20.3. The other methods are executed on MATLAB R2018b.

For a fair comparison, we set the number of iterations $t = 3$ for all k-means sampling applicable approaches, e.g., LSC-K, LSHC-K, LiteWSC, U-SPEC. Regarding the random sampling based approaches, e.g., LSC-R, LSHC-R,

---

[4] https://sites.cs.ucsb.edu/~wychen/sc.

[5] https://eng.ucmerced.edu/people/vladymyrov.

[6] http://www.math.umassd.edu/~dyan/fasp.html.

[7] http://www.cad.zju.edu.cn/home/dengcai/Data/ReproduceExp.html#LSC.

[8] https://github.com/SubaiDeng/LSSHC_matlab.

[9] https://www.researchgate.net/publication/330760669.

VN, and Nyström, we use the same random seeds. Moreover, 10 tests are executed for all methods, and the average result is reported.

5.1 Evaluation on Real Datasets

Evaluation on real datasets is reported in Table 2, Table 3, and Table 4. The bold represents the best result, and the second-best result is highlighted by underline, "N/A" denotes we cannot obtain the clustering result on the experimental host (e.g., Out-of-Memory or Runtime exceeds 12 h). For LiteWSC, the number of samples s is set as a reasonable value adaptive to the number of data objects n: 4,000 (Letter, USPS), 15,000 (Yahoo), 25,000 (MNIST), 3,0000 (Covtype), 50,000 (Byclass, MINST8M, Balanced) except otherwise noted.

Table 2. NMI on real world datasets. The best and second best results are highlighted by **bold** and underline respectively.

Method	USPS	Letter	MNIST	Balanced	Covtype	Byclass	Yahoo	MNIST8M
k-means	60.02	34.69	44.54	39.14	14.66	46.02	55.83	<u>36.99</u>
SC	**81.16**	**46.00**	**78.53**	**62.93**	N/A	N/A	N/A	N/A
Nyström	63.12	38.18	49.32	39.76	11.15	45.32	55.27	N/A
VN	71.70	35.08	56.47	36.43	N/A	N/A	N/A	N/A
KASP	66.32	39.60	53.43	39.63	10.93	46.13	54.88	N/A
LSC-R	73.09	39.41	58.82	38.11	15.29	41.70	57.64	N/A
LSC-K	77.65	42.19	74.07	47.65	17.01	50.92	58.69	N/A
LSHC-R	73.86	38.88	59.44	38.42	15.95	41.94	57.61	N/A
LSHC-K	78.29	42.22	70.81	48.00	16.50	**51.26**	**59.03**	N/A
U-SPEC	77.15	42.82	66.93	42.75	**19.15**	46.94	<u>58.80</u>	N/A
LiteWSC	<u>79.21</u>	<u>43.67</u>	<u>72.40</u>	<u>48.24</u>	<u>17.97</u>	<u>51.21</u>	56.94	**64.11**

Table 3. AC on real world datasets. The best and second best results are highlighted by **bold** and underline respectively.

Method	USPS	Letter	MNIST	Balanced	Covtype	Byclass	Yahoo	MNIST8M
k-means	63.01	25.43	48.47	28.68	28.60	26.22	67.36	<u>45.24</u>
SC	70.10	31.91	<u>74.96</u>	**49.93**	N/A	N/A	N/A	N/A
Nyström	63.80	27.78	54.93	35.93	26.24	25.60	64.77	N/A
VN	67.40	27.28	59.23	28.78	N/A	N/A	N/A	N/A
KASP	67.06	29.11	58.06	27.44	26.47	26.72	69.90	N/A
LSC-R	70.68	29.46	62.45	30.14	33.97	23.91	72.08	N/A
LSC-K	73.48	31.49	70.56	38.89	33.35	29.01	72.59	N/A
LSHC-R	68.71	28.88	63.22	30.44	33.73	23.84	71.47	N/A
LSHC-K	<u>73.53</u>	31.59	72.02	39.09	33.18	<u>29.20</u>	<u>73.35</u>	N/A
U-SPEC	67.38	<u>31.98</u>	66.93	34.19	**36.31**	27.70	73.19	N/A
LiteWSC	**76.75**	**32.08**	**75.02**	<u>39.98</u>	<u>36.05</u>	**30.52**	**73.97**	**66.50**

Table 4. Overall runtime (secs) on real world datasets. The best and second best results are highlighted by **bold** and <u>underline</u> respectively.

Method	USPS	Letter	MNIST	Balanced	Covtype	Byclass	yahoo	MNIST8M
k-means	<u>0.30</u>	0.20	3.63	21.66	<u>4.18</u>	176.07	<u>20.37</u>	<u>1485.08</u>
SC	2.47	7.93	233.88	818.07	N/A	N/A	N/A	N/A
Nyström	1.34	0.71	3.21	8.33	10.89	80.07	98.48	N/A
VN	1.67	8.96	285.35	1220.95	N/A	N/A	N/A	N/A
KASP	0.32	<u>0.19</u>	<u>2.95</u>	<u>5.87</u>	7.73	48.35	126.45	N/A
LSC-R	0.38	1.67	3.86	54.75	11.92	443.51	147.74	N/A
LSC-K	0.46	1.69	5.44	61.22	17.31	480.21	248.85	N/A
LSHC-R	0.39	1.73	4.56	55.28	12.09	439.72	146.54	N/A
LSHC-K	0.45	1.77	5.44	61.01	16.04	482.31	250.12	N/A
U-SPEC	0.60	0.58	3.87	24.26	4.98	229.42	42.57	N/A
LiteWSC	**0.28**	**0.14**	**2.61**	**4.63**	**2.69**	**21.29**	**11.25**	**105.24**

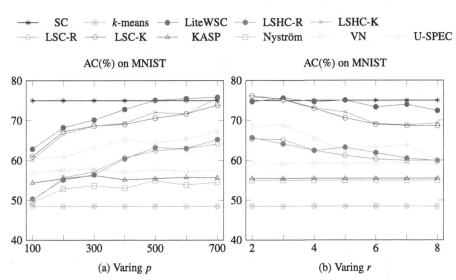

Fig. 1. Performance on MNIST dataset by the number of prototypes p, and the number of retained nearest prototypes r.

These results reveal several exciting findings as follows: (1) As shown in Table 2 and Table 3, LiteWSC achieves the best or the second-best (worse than SC) clustering results in almost all datasets. It demonstrates the robustness and the effectiveness of LiteWSC; (2) Concerning the runtime comparison (as Table 4), there is no doubt that LiteWSC is the fastest algorithm. More significant, the advantage of LiteWSC is more evident as the number of data objects n becomes larger; (3) Only LiteWSC and k-means can process the web-scale dataset MNIST8M in our experimental host with a maximum of 16 GB memory. The other methods failed because of the out-of-memory issue. LiteWSC needs much less memory than k-means. They are further evaluated in the later experiments.

5.2 Effects of the Parameters

In this subsection, we test the method performance by varying the number of prototypes p and the retained nearest prototypes r. Regarding the number of samples s, we use the default setting. As shown in Fig. 1(a), the AC trends are apparent for prototype-based algorithms, including LSC, LSHC, U-SPEC, and our method. Among them, LiteWSC archives the best performance. If $p \geq 500$, LiteWSC outperforms SC. In a word, p is flexible in our method. Thus it can easily be tuned to balance the clustering quality and consumed resource.

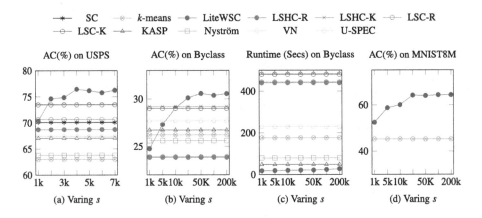

Fig. 2. Performance on USPS, MNIST, Byclss by Varing s

On the other hand, we focus on the effect of r. In Fig. 1 (b), the parameter r varies from 2 to 8 to test the performance of LiteWSC, LSC, LSHC, and U-SPEC only because the remaining do not invoke this parameter (they are plotted as horizontal curves). LiteWSC, LSC and LSHC are significantly better than other methods except for SC. Although, for the mentioned four methods, the AC decreases as r increases. But still, LiteWSC obtains the best results and is more robust than the others.

Effect of the Parameter s.
Now, we concentrate on the effectiveness of the parameter s. Because of the space limitation, we only demonstrate the experiments on three datasets USPS, Byclass, and MNIST8M. According to Fig. 2(a), if $s = 1,000$, LiteWSC outperforms most methods except LSC-K and LSHC-K. As $s = 2,000$, LiteWSC archives the best result, and then AC keeps improving until $s = 4,000$ and becomes stable. A similar trend can be found on the test on dataset Byclass (as Fig. 2(b)), the AC of LiteWSC is terrible when the value of s is small

(e.g., $s = 1,000$). The performance of LiteWSC keeps enhanced as s increasing until converge (e.g., $s = 50k$).

Regarding the runtime comparison (Fig. 2(c)), a larger s costs almost little time. Again, we can infer that LiteWSC is the fastest algorithm. On the other hand, as shown in Fig. 2(d), only two methods can process the dataset MNIST8M. Starts from $s = 1k$, LiteWSC beats the k-means. Furthermore, the AC of LiteWSC continuously increases till stable (e.g., $s = 50k$).

To conclude, LiteWSC is slightly sensitive to the parameter s, especially if the value of s is small (i.e., $s \leq 1k$), but still, the results are acceptable. The results are satisfactory if the samples are sufficient (we suggest s is no more than $50k$ to guarantee the efficiency and minimize the memory consumption). In other words, LiteWSC is quite robust on the parameter s. **Empirically, we might simply set s as a value between $(10k, 50k)$ for large-scale datasets and efficiently obtain a good enough clustering result.**

On the other hand, LiteWSC is extremely memory efficient: constructing a $50k \times 500$ affinity matrix (i.e., $s = 50k$) obviously consumes memory no more than 1 GB. Moreover, even s is set as $10k$ or $5k$, the clustering results of LiteWSC are also good enough against the other methods.

6   Conclusion

In this work, we propose the LiteWSC, a simple and efficient framework of spectral clustering for web-scale data on a resource-limited host. Both the theoretical analysis and evaluation demonstrate the robustness and efficiency of LiteWSC. In the future, a distributed design of LiteWSC is our interesting.

Acknowledgment. We thank the anonymous reviewers for their constructive comments and thoughtful suggestions. This work was supported in part by: National Key D&R Program of China (019YFB1600704, 2021ZD0111501), NSFC (61603101, 61876043, 61976052), NSF of Guangdong Province (2021A1515011941), State's Key Project of Research and Development Plan (2019YFE0196400), NSF for Excellent Young Scholars (62122022), Guangzhou STIC (EF005/FST-GZG/2019/GSTIC), NSFC-Guangdong Joint Fund (U1501254), the Science and Technology Development Fund, Macau SAR (0068/2020/AGJ, 0045/2019/A1, SKL-IOTSC(UM)-2021-2023, GDST (2020B1212030003).

References

1. Cai, D., Chen, X.: Large scale spectral clustering via landmark-based sparse representation. TCYB **45**(8), 1669–1680 (2015)
2. Cai, D., He, X., Han, J.: Document clustering using locality preserving indexing. TKDE **17**(12), 1624–1637 (2005)
3. Chang, C., Lin, C.: LIBSVM: a library for support vector machines. ACM Trans. Intell. Syst. Technol. 2(3), 27:1–27:27 (2011)
4. Chen, P., Wu, L.: Revisiting spectral graph clustering with generative community models. In: Raghavan, V., Aluru, S., Karypis, G., Miele, L., Wu, X. (eds.) ICDM, pp. 51–60 (2017)

5. Chen, X., Cai, D.: Large scale spectral clustering with landmark-based representation. In: AAAI (2011)
6. Chung, F.R., Graham, F.C.: Spectral Graph Theory, no. 92. American Mathematical Society (1997)
7. Cohen, G., Afshar, S., Tapson, J., Van Schaik, A.: EMNIST: extending MNIST to handwritten letters. In: IJCNN, pp. 2921–2926. IEEE (2017)
8. Couillet, R., Chatelain, F., Bihan, N.L.: Two-way kernel matrix puncturing: towards resource-efficient PCA and spectral clustering. In: ICML, vol. 139, pp. 2156–2165 (2021)
9. Dhillon, I.S., Guan, Y., Kulis, B.: Weighted graph cuts without eigenvectors a multilevel approach. TPAMI 29(11), 1944–1957 (2007)
10. Donoho, D.L., Elad, M.: Optimally sparse representation in general (nonorthogonal) dictionaries via l1 minimization. PNAS 100(5), 2197–2202 (2003)
11. Duan, L., Aggarwal, C.C., Ma, S., Sathe, S.: Improving spectral clustering with deep embedding and cluster estimation. In: ICDM, pp. 170–179. IEEE (2019)
12. Fowlkes, C.C., Belongie, S.J., Chung, F.R.K., Malik, J.: Spectral grouping using the Nyström method. TPAMI 26(2), 214–225 (2004)
13. Gray, R.M., Neuhoff, D.L.: Quantization. TIT 44(6), 2325–2383 (1998)
14. Haeffele, B.D., You, C., Vidal, R.: A critique of self-expressive deep subspace clustering. In: ICLR (2021)
15. Hartigan, J.A., Wong, M.A.: A k-means clustering algorithm. J. R. Stat. Soc. Ser. C (Appl. Stat.) 28(1), 100–108 (1979)
16. Huang, D., Wang, C., Wu, J., Lai, J., Kwoh, C.: Ultra-scalable spectral clustering and ensemble clustering. TKDE 32(6), 1212–1226 (2020)
17. Huang, L., Yan, D., Jordan, M.I., Taft, N.: Spectral clustering with perturbed data. In: NIPS. pp. 705–712. Curran Associates, Inc. (2008)
18. Joulin, A., Grave, E., Bojanowski, P., Mikolov, T.: Bag of tricks for efficient text classification. In: Lapata, M., Blunsom, P., Koller, A. (eds.) EACL, pp. 427–431. ACL (2017)
19. Lin, F., Cohen, W.W.: Power iteration clustering. In: Fürnkranz, J., Joachims, T. (eds.) ICML, pp. 655–662. Omnipress (2010)
20. von Luxburg, U.: A tutorial on spectral clustering. Stat. Comput. 17(4), 395–416 (2007)
21. Mohan, M., Monteleoni, C.: Beyond the Nyström approximation: speeding up spectral clustering using uniform sampling and weighted kernel k-means. In: IJCAI (2017)
22. Ng, A.Y., Jordan, M.I., Weiss, Y., et al.: On spectral clustering: analysis and an algorithm. NIPS 2, 849–856 (2002)
23. Olshausen, B.A., Field, D.J.: Emergence of simple-cell receptive field properties by learning a sparse code for natural images. Nature 381(6583), 607–609 (1996)
24. Pan, Z., Fan, H., Zhang, L.: Texture classification using local pattern based on vector quantization. TIP 24(12), 5379–5388 (2015)
25. Sculley, D.: Web-scale k-means clustering. In: Rappa, M., Jones, P., Freire, J., Chakrabarti, S. (eds.) WWW, pp. 1177–1178. ACM (2010)
26. Shi, J., Malik, J.: Normalized cuts and image segmentation. TPAMI 22(8), 888–905 (2000)
27. Stewart, G.W.: Introduction to Matrix Computations. Academic Press, Cambridge (1973)
28. Strehl, A., Ghosh, J.: Cluster ensembles – a knowledge reuse framework for combining multiple partitions. JMLR 3, 583–617 (2002)

29. Vladymyrov, M., Carreira-Perpinan, M.: The variational Nystrom method for large-scale spectral problems. In: ICML, pp. 211–220. PMLR (2016)
30. Wang, S., Zhang, Z.: Improving CUR matrix decomposition and the Nyström approximation via adaptive sampling. JMLR **14**(1), 2729–2769 (2013)
31. Williams, C.K.I., Seeger, M.W.: Using the Nyström method to speed up kernel machines. In: Leen, T.K., Dietterich, T.G., Tresp, V. (eds.) NIPS, pp. 682–688. MIT Press (2000)
32. Yan, D., Huang, L., Jordan, M.: Fast approximate spectral clustering. Technical report UCB/EECS-2009-45, EECS Department, University of California, Berkeley, March 2009
33. Yan, D., Huang, L., Jordan, M.I.: Fast approximate spectral clustering. In: IV, J.F.E., Fogelman-Soulié, F., Flach, P.A., Zaki, M.J. (eds.) KDD, pp. 907–916. ACM (2009)
34. Yang, Y., et al.: GraphLSHC: towards large scale spectral hypergraph clustering. Inf. Sci. **544**, 117–134 (2021)
35. Zador, P.L.: Asymptotic quantization error of continuous signals and the quantization dimension. IEEE Trans. Inf. Theory **28**(2), 139–148 (1982)
36. Zhang, G.-Y., Chen, X.-W., Zhou, Y.-R., Wang, C.-D., Huang, D.: Consistency- and Inconsistency-Aware Multi-view Subspace Clustering. In: Jensen, C.S., et al. (eds.) DASFAA 2021. LNCS, vol. 12682, pp. 291–306. Springer, Cham (2021). https://doi.org/10.1007/978-3-030-73197-7_20
37. Zhang, K., Kwok, J.T.: Clustered Nyström method for large scale manifold learning and dimension reduction. TNN **21**(10), 1576–1587 (2010)
38. Zhang, Z., Lange, K., Xu, J.: Simple and scalable sparse k-means clustering via feature ranking. In: NeurIPS (2020)

Dual Confidence Learning Network for Open-World Time Series Classification

Junwei Lv[1,2], Ying He[1,2], Xuegang Hu[1,2(✉)], Desheng Cai[2], Yuqi Chu[2], and Jun Hu[3]

[1] Key Laboratory of Knowledge Engineering with Big Data, Ministry of Education, Hefei University of Technology, Hefei 230009, China
{junweilv,jsjxhuxg}@hfut.edu.cn
[2] School of Computer Science and Information Engineering, Hefei University of Technology, Hefei 230009, China
ying_he@mail.hfut.edu.cn
[3] National Laboratory of Pattern Recognition, Institute of Automation, Chinese Academy of Sciences, Beijing 100190, China

Abstract. Traditional time series classification methods hold a basic closed-world assumption that the classes of test examples must have been seen by the classifier. However, in the real world, new examples that do not belong to any training class are constantly generated. The existing open-world classification methods cannot be well applied to time series data because they mainly solve problems in the field of computer vision without mining temporal features. In addition, these methods are mostly based on overconfident prediction results, and an overconfident classifier may produce predictive probabilities with high confidence even for incorrect predictions. In this paper, we propose a novel dual confidence learning network for open-world time series classification, which leverages temporal deep neural model to capture temporal features, and designs a dual confidence mechanism to identify which known class or unknown class an example belongs to. Specifically, temporal confidence is learned from the likelihood sequence to reflect the true correctness risk and Weibull distribution confidence is learned to reflect the open space risk. Experimental results evaluated on real-world datasets demonstrate that the proposed model can accurately identify examples of unknown classes without sacrificing the classification performance of known class examples.

Keywords: Dual confidence learning · Open-world classification · Time series

1 Introduction

Nowadays, amount of temporal data from different sensors are available, such as weather readings, stock prices, health monitoring data, etc. resulting in the great demand for time series classification (TSC) methods to help users mine useful information. For example, in the field of healthcare, TSC can help doctors

A. Bhattacharya et al. (Eds.): DASFAA 2022, LNCS 13246, pp. 574–589, 2022.
https://doi.org/10.1007/978-3-031-00126-0_41

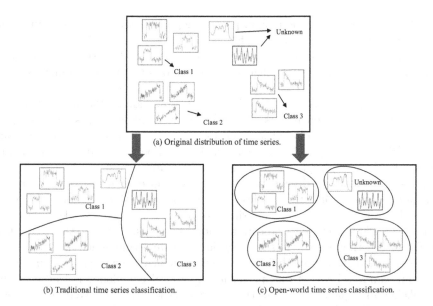

(a) Original distribution of time series.

(b) Traditional time series classification.   (c) Open-world time series classification.

Fig. 1. Comparison of traditional time series classification and open-world time series classification. Open-world classification can predict which known/seen class a test time series belongs to or identify it as an unknown/unseen class.

diagnose diseases by analyzing the heart-beating signals of patients. Thus, TSC has been an important and challenging task both in academia and industry.

Existing methods for TSC hold a basic assumption that classes appearing in testing must have been seen during training processes which refers to a closed-world setting. However, in the real world, new time series belonging to an unseen class are constantly emerging with the growth of information. Thus, traditional classifiers fail to identify these examples that were not trained on. Intuitively, a robust classifier should have a natural characteristic that can not only recognize examples of seen classes but also detect time series that belong to none of the trained classes. For example, when a new disease appears, the classifier can accurately identify it belonging to the unknown disease type, rather than misjudge it as one of the existing diseases by giving high confidence. This problem is called open-world classification [11,40], which demonstrates in Fig. 1.

Recently, there are some open-world research works on computer vision. For example, Bendale et al. [4] argue that the Softmax layer has a closed nature since the probability distribution of trained classes is summed up to one which limits the ability to predict unknown classes, and propose to replace the Softmax layer with an OpenMax layer by applying Nearest Class Mean (NCM) to estimate the open space risk. In the natural language processing area, Shu et al. [32] present an open text classification model called DOC, which uses CNN as an encoder and replaces the Softmax layer with a 1-vs-rest final sigmoid layer to identify unseen classes.

Although existing methods show promising performance, they are insufficient to deal with the two limitations as follows:

Limitation 1: existing open-world classification methods are designed for data without temporal properties such as images and texts, which cannot be well applied to time series data because they are not able to mine temporal features.

Limitation 2: most existing open-world classification methods are based on overconfident prediction results. However, an overconfident classifier may produce high predictive probabilities even for incorrect predictions. Therefore, an open-world model should know when it is likely to be wrong. In other words, it should provide a high quality of confidence estimates.

Based on the above discussions, we propose a novel dual confidence learning network for open-world time series classification, called DCLN for short. Our model leverages a time series classifier based on a temporal stacked convolution encoder architecture to capture temporal patterns. Based on the prediction results of the time series classifier, we design a dual confidence mechanism to revise the prediction probabilities to identify which known class or unknown class an example belongs to. Specifically, temporal confidence is learned from the likelihood sequence to reflect the true correctness risk and Weibull distribution confidence is learned to reflect the open space risk. Finally, an integrator is designed to revise the probability of each class by integrating the probability result of the time series classifier, the temporal confidence and the Weibull distribution confidence.

Our main contributions are summarized below:

- We propose an effectively and efficiently unified framework to address open-world time series classification.
- A dual confidence learning mechanism, including temporal confidence and Weibull distribution confidence, is designed to revise the probability of each class to improve the performance of open-world time series classification.
- Experimental results evaluated on real-world datasets demonstrate that the proposed model can accurately identify examples of unknown classes without sacrificing the classification performance of known class examples.

2   Related Work

2.1   Time Series Classification

Time series classification methods can roughly be categorized into three groups: distance-based methods, feature-based methods, and deep learning methods. Early works mainly measure the distance between two time series and then adopt classifiers such as the k-nearest neighbor (k-NN) or Support Vector Machines (SVM) to do classification. Since the key point of these methods is a distance function, many researchers are devoted to designing a powerful distance function and then exploiting them to classifiers [1,20,29]. Among these methods, dynamic time warping combing with k-NN classifiers are strong baselines [22,38].

Feature-based methods aim to extract discriminative features of time series and then apply classification models to them. Baydogan et al. [14] extract a bag of features based on class probability estimates. Bai et al. [2] utilize both the mean feature and trend feature of time series for ensemble learning. Ben et al. [10] extract thousands of interpretable features from time series such as correlation structure, distribution, entropy, scaling properties, and so on. However, selecting appropriate features is challenging because of the temporal nature of time series. Therefore, it requires heavy domain expert efforts.

Recently, thanks to the development of deep neural networks, the performance of TSC has been significantly improved. Among those deep learning methods, convolution is regarded as the most powerful technique for time series classification. Fully convolutional neural networks (FCNs) are first proposed for time series classification by [39]. Their architecture contains three convolutional networks with global average pooling layers (GAP) which reduce the number of parameters in the network. Inspired by FCN, encoder is proposed to employ an attention layer to replace the GAP layer which has encouraged transfer learning [31]. Multi-scale convolution network (MCNN) [8] employs a traditional CNN model. This method depends on a data preprocessing procedure that contains down-sampling, skip sampling, and sliding slicing. In turn, the performance of the MCNN is heavily dependent on the data preprocessing. Since RNN-based methods show promising successes in natural language processing [23,26], there are also some attempts to employ them on time series classification [25,28]. More recently, researchers attempt to combine LSTM with CNN [9,18,19], where they learn time series representations both by LSTM and FCN, and achieve state-of-art performance. Karim et al. [18] proposed to use the LSTM sub-modules and attention LSTM to enhance FCN for the UTS classification. In [19], the authors extend their previous work to the multivariate time series classification, and gain significant improvements.

2.2 Open-World Classification

In recent years, open-world classification is widely explored by researchers which aims to identify whether a test example belongs to a class that is not been trained on [33,41,43]. There are some similar concepts to open-world classification, such as zero-shot learning [7,37,42], one/few-shot learning [16,34,36] and anomaly detection [5,6]. The main difference between these tasks and open-world classification is that all classes are known in advance. In the field of computer vision, open-world classification is referred to as open set recognition [3,21,35]. Most deep learning-based methods argue that Softmax loss using in neural networks has natural drawbacks of close property for open-world learning [13,33]. Macêdo et al. [24] propose IsoMax loss to replace the general Softmax loss where the loss function is based on a Euclidean distance between input embedding and class prototype. Joseph et al. [17] introduce a contractive clustering loss to the standard loss of object detection. This loss forces the examples of the same class close in latent space and the examples of the different classes away. OpenMax [4] builds a compact abating probability model where first uses EVT to fit the tail examples of each class and then uses a Softmax layer to calculate the class probability of test examples. Recently, there are also some attempts of open-world learning in text classification. DOC [32] builds a multi-class classifier for open-world text classification. This method replaces the final Softmax layer of OpenMax with a 1-vs-rest sigmoid layer. Prakhya et al. [27] utilize OpenMax directly as their final loss for open set text classification.

To the best of our knowledge, we are the first to solve open-world classification on time series. The proposed method can not only classify examples of known classes but identify unknown examples very well.

3   Preliminaries

Time Series Classification: A time series $X = \{x_t\}_{t=1}^{T}$ is an ordered set of observed values, where T is the length of X, $x_t \in \mathbb{R}^d$ denotes the observed value at time step t, and d is the dimension of X. For a time series X, at time step t, only first t values are observed. We use $X_t = [x_1, x_2, \cdots x_t]$ to denote the observed time series from time step 1 to t for the time series X, and X_T to denote the complete time series X. Given a dataset $\mathcal{D} = \{(X_i, y_i)\}_{i=1}^{N}$, where N is the number of time series, $y_i \in \{c_1, c_2, \ldots, c_n\}$ is the label of X_i, and n is the number of classes in the training data, the goal of time series classification is to learn a classifier f from the training data and predict a class label of an test time series, i.e., $f : X \rightarrow y$.

Open-World Classification: Let $\mathcal{K} = \{c_1, c_2 \ldots c_k\}$ denote the set of all known classes and $\mathcal{U} = \{c_{k+1}, \ldots\}$ denote the set of unknown classes, where k is the number of known classes. Let $\mathcal{D}_{train}$ denote the training data containing examples of k known classes and $\mathcal{D}_{test}$ denote the open-world test data that may contain examples belonging to $\mathcal{K} \cup \mathcal{U}$. Open-world classification aims to train a classifier g on $\mathcal{D}_{train}$, and predict which known class a test example belongs to or identify it as an unknown class.

4   Method

The framework of our DCLN is as shown in Fig. 2, which consists of four components: Time Series Classifier, Temporal Confidence, Weibull Distribution Confidence and Integrator. In this section, we will describe these components in detail.

Since two temporal stacked convolution encoders are used both in Time Series Classifier and Temporal Confidence to mine temporal features, this section first introduces the temporal stacked convolution encoder, and then introduces these four components respectively.

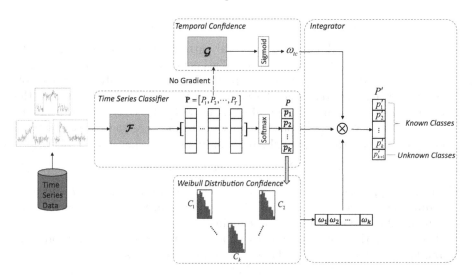

Fig. 2. The overall framework of the DCLN model consists of four components: Time Series Classifier captures temporal patterns from raw time series data and outputs the prediction probabilities, Temporal Confidence estimates the true correctness confidence by mining likelihood sequence, Weibull Distribution Confidence reflects the open space risk and Integrator revises the probability of each class.

4.1   Temporal Stacked Convolution Encoder

In order to mine temporal features, we first design a temporal stacked convolution encoder in close-world to encode time series into feature vectors. In this paper, we use FCN [39] to extract the features of time series since convolution has been proved a powerful operation for feature extraction. The architecture consists of multiple stacked temporal convolution blocks where each block is a convolution layer followed by a ReLU activation layer. The first 2 convolution blocks contain 64 filters with a filter length equal to 30, and the last 2 convolution blocks contain 128 filters with a filter length equal to 30. To alleviate the problem of over-fitting, we apply dropout regularization techniques after each convolution layer. This technique is first proposed by THinton [15], where the neurons in networks are randomly turn off with probability P during the training process. Furthermore, to alleviate the network degradation as the neural networks going deeper, we also apply residual connections [12] across blocks. The key idea of residual connections is to propagate low-layer features to higher layers by adding the last layer input to the output. This operation enables gradients to flow directly to lower layers, which makes training much easier. By stacking multiple convolution layers, the representation of time series preserves more useful features. Then, the results are feed into a GAP layer which performs average pooling over the time dimension and get the m-dimensional feature vector.

4.2   Time Series Classifier

Time Series Classifier $\mathcal{F}$ captures temporal patterns from raw time series data and outputs the prediction probabilities. This component is a temporal stacked convolution encoder mentioned in the previous section. A fully connected followed by a Softmax layer is employed to transform the feature vector to the probability outputs. Since Temporal Confidence in the next section needs to use the probability sequence of all time steps to measure the true correctness confidence, we need to compute the probability of each time step as follows:

$$P_t = \mathcal{F}(X_t) = Softmax(W_f X_t + b_f) \tag{1}$$

Where $P_t = [p_t^1, p_t^2, \cdots, p_t^j, \cdots p_t^k]$ is the probability results at time t and p_t^j is the probability of the jth class, W_f is the weight matrix and b_f is the bias vector, k is the number of known classes on training.

The goal of this component is to fit all the sub-sequences. Therefore, the overall loss $\mathcal{L}_f$ of Time Series Classifier is defined as follows:

$$\mathcal{L}_f = -\frac{1}{NT} \sum_{i=1}^{N} \sum_{t=1}^{T} \sum_{j=1}^{k} y_i^j \log p_{i,t}^j + \lambda_f \|\theta_f\|^2 \tag{2}$$

where $\|\theta_f\|^2$ is the regularization of the learnable parameters θ_f to avoid overfitting and λ_f is the corresponding weight.

4.3   Temporal Confidence

This component estimates the true correctness confidence by mining complex temporal relationship. Since the probability sequence $\mathbf{P} = [P_1, P_2, \cdots, P_T]$ instead of a single probability distribution P_T is taken as the input, this component is called Temporal Confidence. Temporal Confidence $\mathcal{G}$ is also a temporal stacked convolution encoder. Different from the classifier module $\mathcal{F}$, a fully connected followed by a sigmoid layer is employed to transform the feature vector to a confidence value.

The temporal confidence value is computed as follows:

$$\omega_{tc} = \mathcal{G}(\mathbf{P}) = sigmoid(W_g \mathbf{P} + b_g) \tag{3}$$

where W_g is the weight matrix and b_g is the bias vector.

The optimization goal is that when the prediction result of Time Series Classifier is correct, the greater the confidence, the better. Therefore, the overall loss $\mathcal{L}_g$ is defined as follows:

$$\mathcal{L}_g = \frac{1}{N} \sum_{i=1}^{N} (-\delta_i \log \omega_{tc} - (1 - \delta_i) \log(1 - \omega_{tc})) + \lambda_g \|\theta_g\|^2 \tag{4}$$

$$\delta_i = \begin{cases} 1, & if \ \arg\max(P_{i,T}) = y_i \\ 0, & otherwise \end{cases} \tag{5}$$

where $\|\theta_g\|^2$ is the regularization of the learnable parameters θ_g and λ_g is the corresponding weight.

4.4   Weibull Distribution Confidence

Weibull Distribution Confidence reflects the open space risk that is used to estimate whether an example is from an unknown class. In this component, the probability values predicted by Time Series Classifier, instead of the activation vector (the penultimate layer) [4], are used to learn the Weibull distribution confidence, because the probability values are positive and the activation vector may be negative. Negative values will increase their final probabilities after confidence revises.

Based on the correct probability values, let μ_j be the center of class j computed by averaging the correct probability values of class j, and η be the largest distances from μ_j, then the Weibull distribution ρ_j of the η extreme values is computed as follows:

$$\rho_j = (\tau_j, k_j, \lambda_j) = FitHigh(\| \hat{S}_j - \mu_j \|, \eta) \qquad (6)$$

where τ_j, k_j, and λ_j are parameters of ρ_j, $\hat{S}_j$ is the mean vector of η extreme values, $\|\|$ represents distance between $\hat{S}_j$ and μ_j, e.g. Euclidean distance, $FitHigh()$ is a LibMR function [30] used to calculate the Weibull distribution.

Given the Weibull distribution model ρ_j, for an input test example, we first get its corresponding probability vector P through $\mathcal{F}$ (Softmax output), and then compute the distance between P and each class (k number of known classes) respectively. For each distance obtained above, the fitting model ρ_j corresponding to class j is used to predict the fit scores, which refers to the probability that the test example belongs to class j. There are k scores in total, and the probability scores are sorted. Finally, the Weibull distribution confidence $\omega_{wd} = [\omega_{wd}^1, \omega_{wd}^2, \cdots, \omega_{wd}^k]$ is computed as follows:

$$\omega_{wd}^j = 1 - \frac{k - s(j)}{k} e^{-\left(\frac{\| P - \tau_{s(j)} \|}{\lambda_{s(j)}} \right)^{\kappa_{s(j)}}} \qquad (7)$$

where $s(j)$ is the sorted index of the class j.

4.5   Integrator

Integrator integrates the probability results of the time series classifier, the temporal confidence and the Weibull distribution confidence, to revise the probability of each class, so as to identify which known class or unknown class an example belongs to.

The integrated probability p'_j for class j is computed as follows:

$$p'_j = \begin{cases} \omega_{tc} \omega_{wd}^j p_j, & if \ j \leq k \\ 1 - \sum_{c=1}^{k} p'_c, & if \ j = k + 1 \end{cases} \qquad (8)$$

Where class $j + 1$ is the unknown class.

Finally, the prediction result is:

$$\hat{y} = argmax_j \{p'_j\} \qquad (9)$$

5 Experiments and Results

In this section, we evaluate the performance of the proposed method and compare it with other open-world methods.

5.1 Datasets

To evaluate the performance of our proposed DCLN, we do comprehensive experiments on three standard datasets [19]: CharacterTrajectories, PenDigits, and UWave, which have enough classes and examples for open-world classification. Since these datasets are proposed for closed-world classification originally, we extend them by class separation following [44]. We randomly select some classes as known classes and group the remaining classes as the unknown class. In the training phase, only examples from known classes are used for training, while in the testing phase, examples from both known and unknown classes are used for testing. The descriptions of these three datasets are shown as follows.

CharacterTrajectories: this dataset includes a total of 2858 examples from 20 classes, where each class has 143 examples with 3 variables. To recast it for open-world classification, 14 distinct classes are randomly chosen as known classes for training, while the remaining ones are as the unknown class.

PenDigits: PenDigits has a total of 10992 examples from 10 classes, where each class has around 1099 examples with 2 variables. Similarly, 6 distinct classes are randomly chosen as known classes and the remaining ones as unknown.

UWave: UWave has a total of 4478 examples from 8 classes, where each class has around 560 examples with 3 variables. Similarly, 6 distinct classes are randomly chosen as known classes and the remaining ones as unknown.

5.2 Evaluation Metrics and Implementation Details

An open-world classifier should not only remain the capability of performing standard closed-world classification, but also effectively detect the unknown classes, thus we use the overall accuracy and F1-measure for measuring the classification performance. In addition, since it is not known in the training phase how rare or common examples from the unknown classes will be, the area under ROC curve (AUROC) can be used to overcome the effects on the sensitivity of model parameters and thresholds. Therefore, we choose Accuracy, F1-measure and AUROC as our evaluation metrics.

For all datasets, we use 50% of the data for training and the remaining 50% for testing. We implement our method with TensorFlow framework on NVIDIA Tesla M60 GPU and use Adam optimizer with an initial learning rate of $1e-3$. The size of the mini-batch is set to 100, and the dropout rate is set to 0.3. The weights of regularization items λ_g and λ_f are all set to $1e-4$.

5.3   Compared Methods

To show the effectiveness of our method, we compare the performance of the proposed method with the following baselines.

- *Softmax*: Softmax is a commonly used layer of the neural network for getting the probability distribution of k classes in closed-world classification. For open-world classification, we adapt the final output to a $k + 1$ probability distribution, where the $(k + 1)$-th output can be regarded as the probability of the unknown class.
- *Softmax_C*: This method also uses Softmax as the final layer of the neural network, and confidence is constructed based on the network's output distribution to redistribute the probabilities of all classes.
- *OpenMax* [4]: OpenMax is a state-of-the-art method for open-world classification in the field of computer vision. This method replaces the Softmax layer with a well-designed OpenMax layer.

We also design the following variants to show the effects of different components:

- *DCLN_NSTC*: DCLN_NSTC is a variant of DCLN without the component of Temporal Confidence. This variant is designed to demonstrate the effectiveness of the confidence obtained by mining likelihood sequences.
- *DCLN_LTC*: DCLN_LTC is a variant of DCLN, which obtains the temporal confidence by only mining the likelihood distribution of the last time step instead of the likelihood sequences.
- *DCLN_NWDC*: DCLN_NWDC is a variant of DCLN without using probability to fit the Weibull distribution. This variant is designed to demonstrate the effectiveness of replacing the activation vector with the probability distribution.

5.4   Experimental Analysis

Comparative Results. The results of our evaluation experiments are presented in Table 1, 2 and 3. From these results, we have the following observations: (1) Softmax method has the worst performance in accuracy, F1-measure and AUROC, which indicates that only adding an extra class output can not identify unknown classes well, because there is no extra information of unknown classes to be used to optimize the training process. (2) From the comparison results of Softmax and Softmax_C, as well as the comparison results of DCLN_NSTC and DCLN_LTC, we can see that the performance of Softmax and Softmax_C, as well as DCLN_NSTC and DCLN_LTC, are similar. This shows that it is useless to use only the final probability distribution to measure the confidence of the prediction results. (3) OpenMax performing better than Softmax shows that it is effective to use the distance from the sample to the center of each class to predict unknown classes. (4) The proposed DCLN consistently achieves the best performance on all evaluation metrics and all

Table 1. Accuracy results on three real-world datasets

Datasets	CharacterTrajectories	PenDigits	UWave
Softmax	0.69 ± 0.012	0.597 ± 0.004	0.732 ± 0.007
Softmax_C	0.689 ± 0.011	0.597 ± 0.004	0.732 ± 0.008
OpenMax	0.867 ± 0.029	0.782 ± 0.050	0.766 ± 0.044
DCLN_NSTC	0.883 ± 0.031	0.798 ± 0.057	0.793 ± 0.039
DCLN_LTC	0.883 ± 0.031	0.799 ± 0.057	0.793 ± 0.039
DCLN_NWDC	0.908 ± 0.024	0.766 ± 0.059	0.846 ± 0.032
DCLN	**0.917 ± 0.037**	**0.847 ± 0.050**	**0.851 ± 0.036**

Table 2. F1-measure results on three real-world datasets

Datasets	CharacterTrajectories	PenDigits	UWave
Softmax	0.786 ± 0.012	0.667 ± 0.013	0.738 ± 0.011
Softmax_C	0.783 ± 0.010	0.666 ± 0.010	0.740 ± 0.011
OpenMax	0.907 ± 0.016	0.832 ± 0.036	0.802 ± 0.032
DCLN_NSTC	0.917 ± 0.018	0.845 ± 0.037	0.829 ± 0.028
DCLN_LTC	0.917 ± 0.018	0.845 ± 0.037	0.829 ± 0.028
DCLN_NWDC	0.931 ± 0.016	0.809 ± 0.048	0.871 ± 0.023
DCLN	**0.941 ± 0.022**	**0.873 ± 0.037**	**0.877 ± 0.026**

Table 3. AUROC results on three real-world datasets

Datasets	CharacterTrajectories	PenDigits	UWave
Softmax	0.904 ± 0.015	0.789 ± 0.042	0.887 ± 0.017
Softmax_C	0.918 ± 0.014	0.902 ± 0.026	0.932 ± 0.013
OpenMax	0.965 ± 0.012	0.862 ± 0.051	0.894 ± 0.018
DCLN_NSTC	0.973 ± 0.003	0.971 ± 0.007	0.943 ± 0.009
DCLN_LTC	0.973 ± 0.003	0.972 ± 0.007	0.943 ± 0.008
DCLN_NWDC	0.982 ± 0.003	0.917 ± 0.020	0.957 ± 0.017
DCLN	**0.985 ± 0.004**	**0.977 ± 0.008**	**0.963 ± 0.009**

datasets. Compared with OpenMax, DCLN improves the accuracy, F1-measure and AUROC by 5.8–11.1%, 3.7–9.4% and 2.1–13.3%, respectively. The results show that our method can effectively handle the open-world time series classification. (5) The results also show that DCLN consistently outperforms all variants (DCLN_NSTC, DCLN_NWDC and DCLN_LTC), verifying that the dual confidence mechanism can boost the classification. Specifically, with the introduction of temporal confidence, DCLN performs better than DCLN_NSTC, demonstrating that the temporal confidence mechanism can identify unknown classes better. (6) We can also see that DCLN_NSTC outperforms DCLN_NWDC and DCLN_NWDC outperforms OpenMax, which shows that the dual confidence mechanism plays a more important role in improving performance.

Detailed Results. Figure 3 illustrates the obtained confusion matrix of DCLN on the UWav dataset, in which classes 1–6 are known classes and all other unknown classes are taken as class 7. It can be observed that DCLN can not only maintain a high accuracy of known classes, but also recognize unknown classes well. Specifically, most of the wrong samples from known classes are classified into unknown classes instead of other known classes, and about 10% of the examples from unknown classes are wrongly classified into known classes.

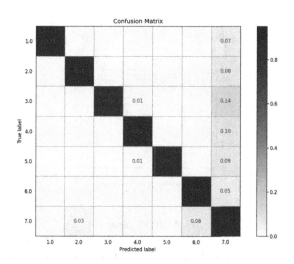

Fig. 3. Confusion matrix of DCLN on the UWav dataset

5.5 Parameter Sensitivity

Another important factor affecting open-world performance is the openness of the problem. We vary the number of unknown classes from 1 to 6 and report the results on the UWav dataset in Fig. 4. The maximum number of unknown classes is 6 because at least 2 known classes are used in model training.

It can be seen from Fig. 4(a) that when there is only one unknown class, the accuracy of all methods is not significantly different. With the increase of unknown classes, the accuracy of DCLN and DCLN_NWDC decrease slowly, while that of other methods decrease obviously, especially that of Softmax and Softmax_C decrease linearly. The results show that DCLN with the dual confidence mechanism is more stable with various numbers of unknown classes in terms of accuracy. Note that the lines of Softmax and Softmax_C overlap.

Figure 4(b) shows that with the increase of unknown classes, the performance trends of all methods in terms of F1-measure are similar to those in terms of accuracy. In terms of AUROC, DCLN is also the most stable with various numbers of unknown classes.

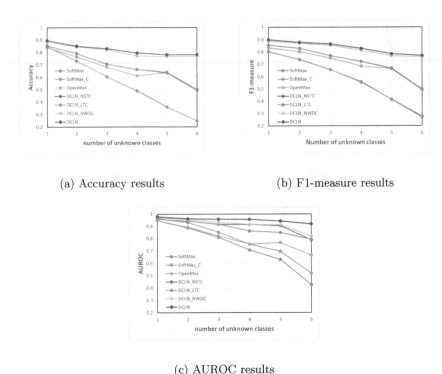

(a) Accuracy results   (b) F1-measure results

(c) AUROC results

Fig. 4. Results under different numbers of unknown classes

6   Conclusions

In this paper, a unified framework DCLN is proposed to address open-world time series classification. Experimental results show that DCLN can accurately identify examples of unknown classes without sacrificing the classification performance of known classes. The results also verify the effectiveness of each component of our proposed framework. In our future work, we plan to take known unknown classes into account which can serve as side information for learning a better representation of time series.

Acknowledgements. This work is supported in part by the Natural Science Foundation of China under grants (62120106008, 61976077, 62076085, 91746209) and the Program for Chang Jiang Scholars and Innovative Research Team in University (PCSIRT) of the Ministry of Education under grant IRT17R32.

References

1. Abanda, A., Mori, U., Lozano, J.A.: A review on distance based time series classification. Data Min. Knowl. Discov. **33**(2), 378–412 (2018). https://doi.org/10.1007/s10618-018-0596-4

2. Bai, B., Li, G., Wang, S., Wu, Z., Yan, W.: Time series classification based on multi-feature dictionary representation and ensemble learning. Expert Syst. Appl. **169**, 114162 (2021)
3. Bendale, A., Boult, T.E.: Towards open world recognition. In: IEEE Conference on Computer Vision and Pattern Recognition, CVPR 2015, Boston, MA, USA, 7–12 June 2015, pp. 1893–1902. IEEE Computer Society (2015)
4. Bendale, A., Boult, T.E.: Towards open set deep networks. In: 2016 IEEE Conference on Computer Vision and Pattern Recognition, CVPR 2016, Las Vegas, NV, USA, 27–30 June 2016, pp. 1563–1572. IEEE Computer Society (2016)
5. Bhuyan, M.H., Bhattacharyya, D.K., Kalita, J.K.: Network anomaly detection: methods, systems and tools. IEEE Commun. Surv. Tutor. **16**(1), 303–336 (2014)
6. Chandola, V., Banerjee, A., Kumar, V.: Anomaly detection: a survey. ACM Comput. Surv. **41**(3), 15:1–15:58 (2009)
7. Changpinyo, S., Chao, W., Gong, B., Sha, F.: Synthesized classifiers for zero-shot learning. In: 2016 IEEE Conference on Computer Vision and Pattern Recognition, CVPR 2016, Las Vegas, NV, USA, 27–30 June 2016, pp. 5327–5336. IEEE Computer Society (2016)
8. Cui, Z., Chen, W., Chen, Y.: Multi-scale convolutional neural networks for time series classification. CoRR arXiv:1603.06995 (2016)
9. Du, Q., Gu, W., Zhang, L., Huang, S.: Attention-based LSTM-CNNs for time-series classification. In: Ramachandran, G.S., Krishnamachari, B. (eds.) Proceedings of the 16th ACM Conference on Embedded Networked Sensor Systems, SenSys 2018, Shenzhen, China, 4–7 November 2018, pp. 410–411. ACM (2018)
10. Fulcher, B.D., Jones, N.S.: Highly comparative feature-based time-series classification. IEEE Trans. Knowl. Data Eng. **26**(12), 3026–3037 (2014)
11. Geng, C., Huang, S., Chen, S.: Recent advances in open set recognition: a survey. CoRR arXiv:1811.08581 (2018)
12. He, K., Zhang, X., Ren, S., Sun, J.: Deep residual learning for image recognition. In: 2016 IEEE Conference on Computer Vision and Pattern Recognition, CVPR 2016, Las Vegas, NV, USA, 27–30 June 2016, pp. 770–778. IEEE Computer Society (2016)
13. Hendrycks, D., Gimpel, K.: A baseline for detecting misclassified and out-of-distribution examples in neural networks. In: 5th International Conference on Learning Representations, ICLR 2017, Toulon, France, 24–26 April 2017, Conference Track Proceedings. OpenReview.net (2017)
14. Hills, J., Lines, J., Baranauskas, E., Mapp, J., Bagnall, A.: Classification of time series by shapelet transformation. Data Min. Knowl. Discov. **28**(4), 851–881 (2013). https://doi.org/10.1007/s10618-013-0322-1
15. Hinton, G.E., Srivastava, N., Krizhevsky, A., Sutskever, I., Salakhutdinov, R.: Improving neural networks by preventing co-adaptation of feature detectors. CoRR arXiv:1207.0580 (2012)
16. Jiang, W., Huang, K., Geng, J., Deng, X.: Multi-scale metric learning for few-shot learning. IEEE Trans. Circuits Syst. Video Technol. **31**(3), 1091–1102 (2021)
17. Joseph, K.J., Khan, S., Khan, F.S., Balasubramanian, V.N.: Towards open world object detection. CoRR arXiv:2103.02603 (2021)
18. Karim, F., Majumdar, S., Darabi, H., Chen, S.: LSTM fully convolutional networks for time series classification. IEEE Access **6**, 1662–1669 (2018)
19. Karim, F., Majumdar, S., Darabi, H., Harford, S.: Multivariate LSTM-FCNs for time series classification. Neural Netw. **116**, 237–245 (2019)

20. Kate, R.J.: Using dynamic time warping distances as features for improved time series classification. Data Min. Knowl. Discov. **30**(2), 283–312 (2015). https://doi.org/10.1007/s10618-015-0418-x

21. Li, F., Wechsler, H.: Open set face recognition using transduction. IEEE Trans. Pattern Anal. Mach. Intell. **27**(11), 1686–1697 (2005)

22. Lines, J., Bagnall, A.: Time series classification with ensembles of elastic distance measures. Data Min. Knowl. Discov. **29**(3), 565–592 (2014). https://doi.org/10.1007/s10618-014-0361-2

23. Ma, Q., Lin, Z., Yan, J., Chen, Z., Yu, L.: MODE-LSTM: a parameter-efficient recurrent network with multi-scale for sentence classification. In: Webber, B., Cohn, T., He, Y., Liu, Y. (eds.) Proceedings of the 2020 Conference on Empirical Methods in Natural Language Processing, EMNLP 2020, 16–20 November 2020, pp. 6705–6715. Association for Computational Linguistics (2020)

24. Macêdo, D., Ludermir, T.B.: Improving entropic out-of-distribution detection using isometric distances and the minimum distance score. CoRR arXiv:2105.14399 (2021)

25. Malhotra, P., TV, V., Vig, L., Agarwal, P., Shroff, G.: TimeNet: pre-trained deep recurrent neural network for time series classification. In: 25th European Symposium on Artificial Neural Networks, ESANN 2017, Bruges, Belgium, 26–28 April 2017 (2017)

26. Merity, S., Keskar, N.S., Socher, R.: Regularizing and optimizing LSTM language models. In: 6th International Conference on Learning Representations, ICLR 2018, Vancouver, BC, Canada, 30 April–3 May 2018, Conference Track Proceedings. OpenReview.net (2018)

27. Prakhya, S., Venkataram, V., Kalita, J.: Open set text classification using CNNs. In: Bandyopadhyay, S. (ed.) Proceedings of the 14th International Conference on Natural Language Processing, ICON 2017, Kolkata, India, 18–21 December 2017, pp. 466–475. NLP Association of India (2017)

28. Rajan, D., Thiagarajan, J.J.: A generative modeling approach to limited channel ECG classification. In: 40th Annual International Conference of the IEEE Engineering in Medicine and Biology Society, EMBC 2018, Honolulu, HI, USA, 18–21 July 2018, pp. 2571–2574. IEEE (2018)

29. Ruiz, A.P., Flynn, M., Large, J., Middlehurst, M., Bagnall, A.: The great multivariate time series classification bake off: a review and experimental evaluation of recent algorithmic advances. Data Min. Knowl. Discov. **35**(2), 401–449 (2020). https://doi.org/10.1007/s10618-020-00727-3

30. Scheirer, W.J., Rocha, A., Micheals, R.J., Boult, T.E.: Meta-recognition: the theory and practice of recognition score analysis. IEEE Trans. Pattern Anal. Mach. Intell. **33**(8), 1689–1695 (2011). https://doi.org/10.1109/TPAMI.2011.54

31. Serrà, J., Pascual, S., Karatzoglou, A.: Towards a universal neural network encoder for time series. In: Falomir, Z., Gibert, K., Plaza, E. (eds.) Artificial Intelligence Research and Development - Current Challenges, New Trends and Applications, CCIA 2018, 21st International Conference of the Catalan Association for Artificial Intelligence, Alt Empordà, Catalonia, Spain, 8–10 October 2018. Frontiers in Artificial Intelligence and Applications, vol. 308, pp. 120–129. IOS Press (2018)

32. Shu, L., Xu, H., Liu, B.: DOC: deep open classification of text documents. In: Palmer, M., Hwa, R., Riedel, S. (eds.) Proceedings of the 2017 Conference on Empirical Methods in Natural Language Processing, EMNLP 2017, Copenhagen, Denmark, 9–11 September 2017, pp. 2911–2916. Association for Computational Linguistics (2017)

33. Shu, L., Xu, H., Liu, B.: Unseen class discovery in open-world classification. CoRR arXiv:1801.05609 (2018)

34. Sun, Q., Liu, Y., Chua, T., Schiele, B.: Meta-transfer learning for few-shot learning. In: IEEE Conference on Computer Vision and Pattern Recognition, CVPR 2019, Long Beach, CA, USA, 16–20 June 2019, pp. 403–412. Computer Vision Foundation/IEEE (2019)

35. Vareto, R.H., Silva, S., de Oliveira Costa, F., Schwartz, W.R.: Towards open-set face recognition using hashing functions. In: 2017 IEEE International Joint Conference on Biometrics, IJCB 2017, Denver, CO, USA, 1–4 October 2017, pp. 634–641. IEEE (2017)

36. Wang, P., Liu, L., Shen, C., Huang, Z., van den Hengel, A., Shen, H.T.: Multi-attention network for one shot learning. In: 2017 IEEE Conference on Computer Vision and Pattern Recognition, CVPR 2017, Honolulu, HI, USA, 21–26 July 2017, pp. 6212–6220. IEEE Computer Society (2017)

37. Wang, W., Zheng, V.W., Yu, H., Miao, C.: A survey of zero-shot learning: settings, methods, and applications. ACM Trans. Intell. Syst. Technol. **10**(2), 13:1–13:37 (2019)

38. Wang, X., Mueen, A., Ding, H., Trajcevski, G., Scheuermann, P., Keogh, E.J.: Experimental comparison of representation methods and distance measures for time series data. Data Min. Knowl. Discov. **26**(2), 275–309 (2013)

39. Wang, Z., Yan, W., Oates, T.: Time series classification from scratch with deep neural networks: a strong baseline. In: 2017 International Joint Conference on Neural Networks, IJCNN 2017, Anchorage, AK, USA, 14–19 May 2017, pp. 1578–1585. IEEE (2017)

40. Wang, Z., Salehi, B., Gritsenko, A., Chowdhury, K.R., Ioannidis, S., Dy, J.G.: Open-world class discovery with kernel networks. In: Plant, C., Wang, H., Cuzzocrea, A., Zaniolo, C., Wu, X. (eds.) 20th IEEE International Conference on Data Mining, ICDM 2020, Sorrento, Italy, 17–20 November 2020, pp. 631–640. IEEE (2020)

41. Wu, M., Pan, S., Zhu, X.: OpenWGL: open-world graph learning. In: Plant, C., Wang, H., Cuzzocrea, A., Zaniolo, C., Wu, X. (eds.) 20th IEEE International Conference on Data Mining, ICDM 2020, Sorrento, Italy, 17–20 November 2020, pp. 681–690. IEEE (2020)

42. Xian, Y., Lampert, C.H., Schiele, B., Akata, Z.: Zero-shot learning - a comprehensive evaluation of the good, the bad and the ugly. IEEE Trans. Pattern Anal. Mach. Intell. **41**(9), 2251–2265 (2019)

43. Xu, H., Liu, B., Shu, L., Yu, P.S.: Open-world learning and application to product classification. In: Liu, L., et al. (eds.) The World Wide Web Conference, WWW 2019, San Francisco, CA, USA, 13–17 May 2019, pp. 3413–3419. ACM (2019)

44. Yoshihashi, R., et al.: Classification-reconstruction learning for open-set recognition. In: IEEE Conference on Computer Vision and Pattern Recognition, CVPR 2019, Long Beach, CA, USA, 16–20 June 2019, pp. 4016–4025. Computer Vision Foundation/IEEE (2019). https://doi.org/10.1109/CVPR.2019.00414

Port Container Throughput Prediction Based on Variational AutoEncoder

Jingze Li[1,2,3], Shengmin Shi[1,2,3], Tongbing Chen[1,2,3], Yu Tian[4], Yihua Ding[5],
Yiyong Xiao[6], and Weiwei Sun[1,2,3(✉)]

[1] School of Computer Science, Fudan University, Shanghai, China
[2] Shanghai Key Laboratory of Data Science, Fudan University, Shanghai, China
[3] Shanghai Institute of Intelligent Electronics and Systems, Fudan University,
Shanghai, China
{jingzeli20,cmshi20,tbchen,wwsun}@fudan.edu.cn
[4] Shanghai International Port (Group) Co., Ltd., Shanghai, China
[5] Shanghai Harbor e-logistics Software Co., Ltd., Shanghai, China
dingyh@hb56.com
[6] NeZha Smart Port and and Shipping Technology (Shanghai) Co., Ltd.,
Shanghai, China
xiaoyiyong@nuzarsurf.com

Abstract. The prediction of port container throughput has a significant impact on many of the port's operations. However, accurate prediction of throughput is a difficult problem due to the complexity of the port environment and the uncertainty of port operations. In this paper, we proposed an approach combining self-attention mechanism and variational autoencoder to forecast the operating time of each container. First, we used self-attention mechanism to capture the features between adjacent containers. Then to reduce the influence of missing data, we designed a variational autoencoder (VAE) module to model the latent variables in the port. Finally, the output layer combined the results of these two parts to obtain the final forecast of the loading and discharging time of containers. The throughput of the entire port can be inferred from the forecasted container operation time. Furthermore, we also proposed dynamic programming algorithms to estimate the distribution of the throughput with the help of variational autoencoder module. Experiment results on port throughput prediction in the real-world datasets show that our approach has superior performance at prediction accuracy. Moreover, experiments conducted at different time intervals demonstrate the effectiveness of our approach on various time scales. And the effectiveness of the dynamic programming algorithms is demonstrated through our case study.

Keywords: Port container throughput prediction · Neural network · Self-attention · Variational autoencoder · Dynamic programming

1 Introduction

With the deepening of economic globalization and the continuous development of international trade, higher requirements have been put forward for the

A. Bhattacharya et al. (Eds.): DASFAA 2022, LNCS 13246, pp. 590–605, 2022.
https://doi.org/10.1007/978-3-031-00126-0_42

international transportation of goods. As an advanced modern transportation means, the container has become the main form of modern international liner transportation with its characteristics of safety, reliability, and efficiency. Statistics from World Bank show that global container throughput reached 795 million twenty-foot equivalent units (TEU) in 2018, an increase of 68% compared to 2015. With the rapid development of container shipping, the importance of reasonable planning of ports is highlighted. And the accurate prediction of port container throughput is the premise of reasonable planning for the port.

Like other prediction problems in complex systems, this task is challenging due to the high complexity and uncertainty. In large container ports, there are usually dozens of vessels operating at the same time, the containers loading and discharging time are closely related to the scheduling strategy of the bridge cranes and transport automated guided vehicles (AGVs), while delays can usually occur for various reasons, such as breakdowns or the congestion of AGVs or operators' mishandling. Under different time scales, the difficulties of throughput prediction also vary. Generally speaking, on longer time scales, as the number of exceptions tends to stabilize, the value of throughput also tends to stabilize, which is easier to obtain more accurate results. In some ways, short time forecasts are a better measure of the strengths and weaknesses of a method. Existing approaches [12,13] typically view throughput as a time series and use deep learning models to make predictions. There are also some works [2] that predict port throughput using macroeconomic variables as inputs. However, these approaches do not analyze the port system in-depth and do not use fine-grained container features, hardly yielding satisfactory results.

In this paper, we attempted to predict the port throughput by forecasting the loading and discharging time of each container first. Our model mainly includes self-attention mechanism and variational autoencoder, which are used to capture the interrelationships between containers and the latent variables in the port, respectively. In particular, the port scheduling system first determines the order of container operations based on the vessels' arrival schedule. Then, we combined the features of the sequentially adjacent containers as input to make a forecast about their operating time. For each container, we fed its sequence features into the self-attention module to establish the relationship with other containers and obtained the self-attention feature vector. The vector is connected directly to the origin input for retaining the attributes of the original containers. In the variational autoencoder module, we used another self-attention layer to extract the mean and variance corresponding to the input containers and sampled them to obtain their respective latent variables. With the latent variables, we decoded them to get a feature vector containing information about both input and latent variables. Two fully-connected layers are added behind these two modules to obtain two forecasts of the operating time and thus the final result.

Based on the operating time, we simulated the completion of loading and discharging tasks within the specified time interval to obtain the predicted value of the throughput. Further, we considered the different results obtained from variational autoencoder as the sampling of the distribution of the predicted time and then used a dynamic programming algorithm to build out a distribution of the port throughput from the discrete samples.

The main contributions of our approach are:

- We proposed a model which combines self-attention and variational autoencoder to forecast the operating time of containers.
- We designed an algorithm to infer the port throughput and proposed a dynamic programming algorithm to obtain the distribution of port throughput based on a discrete sampling of the forecasted operating time distribution.
- Extensive experiments including performance evaluation and case study validate the effectiveness of our approach.

The remainder of this paper is organized as follows. Section 2 presents a brief review of previous works related to port container throughput prediction. Section 3 presents the detailed problem description. We also introduced our model in Sect. 3. Section 4 describes the experimental setup. The experiments and results are presented in Sect. 5. And Sect. 6 gives the conclusion.

2   Related Works

Since the research on container throughput prediction began in the 1980s, numerous studies have proposed approaches to improve the accuracy. Existing methods can be divided into three types.

Some studies use machine learning methods to solve the problem. Traditional time series prediction models include autoregressive integrated moving average (ARIMA) [9], seasonal autoregressive integrated moving average (SARIMA) [6], vector autoregressive (VAR) and decomposition approach, etc. Rashed [9] applies ARIMAX and ARIMA models on container throughput prediction in the port of Antwerp, where the former performed better than the latter. [2] attempts to use bayesian estimation to solve this problem. The approaches above can well fit the container throughput in months. However, short-term throughput changes can hardly be explored easily by traditional time series prediction models.

Deep learning methods have been used to solve this problem in recent years. Compared to the general neural network, recurrent neural network (RNN) [1] generates the memory state of past data when learning time series, which can better handle changes in the series. Then, to overcome the vanishing/exploding gradients problem caused by RNN in long sequences, long short-term memory (LSTM) [4] methods are proposed. Moreover, [12] attempts to combine convolutional neural network (CNN) and LSTM to predict monthly port throughput, and experiments are conducted on five real port datasets.

An increasing number of studies have constructed hybrid prediction methods to improve the accuracy of prediction. Niu et al. [8] present a hybrid decomposition-ensemble model, which decomposes the original container throughput series into low-frequency and high-frequency components. This method includes ARIMA and SVR model to predict the two components correspondingly and generate the final results. Mo et al. [7] propose a combined model composed of SARIMA model and group method of data handling (GMDH) model to predict port container throughput and improved the accuracy of simply SARIMA model. Some scholars have also tried to combine machine learning methods with the neural network. [9] combines random forest and multi-layer perceptron (MLP), and [13] combines ARIMA and MLP, to predict the port throughput.

3   Proposed Approach

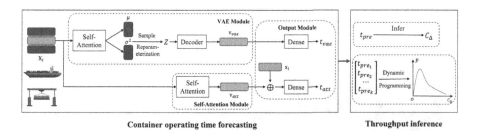

Fig. 1. The framework of our model. The left part is the container operating time forecasting which contains self-attention module, VAE module, and output module. The right part is the throughput inference which provides the prediction for individual throughput values or the distribution of throughput.

In this section, we will first present the problem formulation, and then introduce our proposed model framework for port throughput prediction, which mainly consists of two modules: container operating time forecasting and throughput inference. Specifically, in the first module, we use a self-attention structure to learn features from adjacent containers, and a variational autoencoder structure to learn the latent variables in the entire port system. In the second module, we infer the port throughput based on the previously predicted container operating time. Further, taking advantage of the variational autoencoder, we propose a dynamic programming algorithm to fit the distribution of throughput. The overall architecture of our approach is shown in Fig. 1.

3.1   Problem Formulation

As shown in Fig. 2, the port's transport system consists of three main components: the bridge crane system, the AGVs transport system, and the trucks transport system. For the discharging operations, each container needs to be lifted from the ship to the transit platform by bridge crane, i.e. from the ship side to the land side and waiting for the AGV to transfer the container to the yard, which is used to stack containers. And finally, the containers are delivered from the yard to the rest of the country by trucks. The loading operation is exactly the opposite. In the whole process, the bridge crane system is responsible for dispatching the crane to deliver containers to the AGVs. Specifically, by operating the main trolley, the bridge crane lifts containers to the transit platform, and then sends AGV requirement command to AGVs transport system, the gantry trolley will be used to put containers from the transit platform to AGVs. The AGVs transport system is responsible for dispatching the AGVs to transport containers from bridge crane to yard. Upon receiving the AGV requirement command, the dispatching algorithm selects one of the idle AGVs and schedules a round-trip route. The trucks transport system is responsible for dispatching trucks to deliver containers out of the port. The scheduling algorithm plans the pickup order of the containers and uses appropriate stacking strategies for the yard to ensure that designated containers can be loaded onto trucks faster.

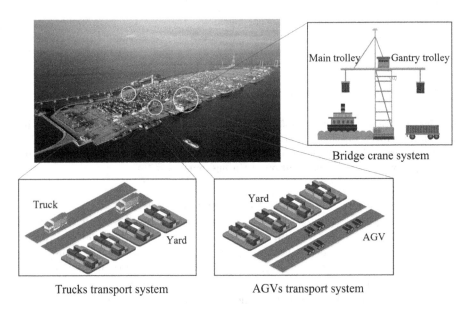

Fig. 2. Port's transport system overview. The diagram on the top left shows the top view of the automated terminal at Yangshan Port. The top right diagram shows the bridge crane system. The bottom right diagram shows the AGVs transport system. Multiple bridge cranes and AGVs are working simultaneously. The diagram on the bottom left shows the trucks transport system.

Once a ship arrives in port, we need to load and discharge the containers on board as soon as possible. Throughout the port, we have dozens of bridge cranes working on multiple ships at the same time, and there may be multiple bridge cranes working on a single ship at the same time. Likewise, we have hundreds of AGVs moving between the bridge cranes and the yard. Since the yard is large enough, with enough trucks and frequent traffic, experience has proved that the main factor affecting the port throughput is the scheduling strategy of the bridge crane system and the AGVs transport system.

However, the behavior of AGVs is more complex. One reason is that AGVs move on complex roads and interfere with each other, sometimes a long wait is necessary to avoid a collision. Another reason is that in an open environment, natural factors such as weather can also affect the speed of the AGVs. Moreover, when an AGV breaks down, an impassable area would be created, which also affects the movement of the AGVs. Unfortunately, we do not have any data related to AGVs, which means it is very difficult to accurately model the behavior of AGVs. In contrast, for the bridge crane system, which follows relatively strict rules, the containers on board must be loaded or discharged layer by layer, and the cranes do not interfere with each other. In practice, we usually determine the loading and discharging sequence of containers before ships arrive. And the bridge crane will follow this sequence to operate basically.

From this, we divided the problem into two parts, container operating time forecast and throughput inference.

Container Operating Time Forecast Problem. Given an order sequence of containers $\mathcal{X} = \{\mathbf{x}_1, \mathbf{x}_2, \cdots\}$, where $\mathbf{x}_i \in \mathbf{R}^d$ is the input feature vector of container i, and d is the dimension of features. As analyzed above, we assume that each container in order sequence would be operated in turn. Let $\mathbf{x}_i^L$ denote the concatenation of some containers' features centered on container i, i.e. $\mathbf{x}_i^L = [\mathbf{x}_{i-\frac{L}{2}}, \mathbf{x}_{i-\frac{L}{2}+1}, \cdots, \mathbf{x}_{i-1}, \mathbf{x}_i, \mathbf{x}_{i+1}, \cdots, \mathbf{x}_{i+\frac{L}{2}-1}, \mathbf{x}_{i+\frac{L}{2}}]$, where L is a hyperparameter, usually taken as an odd number for symmetry reasons. Our model aims to learn a function $f : \mathbf{R}^{L \times d} \mapsto \mathbf{R}$, where $f(\mathbf{x}_i^L)$ is the forecasted time to load or discharge container i. The features include the container type, location, size, etc., which are determined when the ship arrives on shore. The input contains only the features of the containers and no information about AGVs.

Throughput Infer Problem. Given a start time t_s, a time interval Δt and an order sequence of containers $\mathcal{X} = \{\mathbf{x}_1, \mathbf{x}_2, \cdots\}$, our algorithms aim to figure out how many containers could be loaded or discharged within time interval Δt after t_s. Specifically, the throughput $C_{\Delta t}$ during Δt could be written as follows,

$$C_{\Delta t} = |\{\mathbf{x}_i \mid t_{s_i} + f(\mathbf{x}_i^L) \in [t_s, t_s + \Delta t], \mathbf{x}_i \in \mathcal{X}\}|, \tag{1}$$

where t_{s_i} is the start time of $\mathbf{x}_i$, which can be deduced from the end time of the last operation. In the subsequent experiments, we take Δt to be a constant value, i.e., $1\,\mathrm{h}$, $2\,\mathrm{h}$, $1\,\mathrm{day}$, etc. Note that all times are in seconds.

3.2 Container Operating Time Forecasting

In this section, we will introduce our neural network model to forecast container operating time, which consists of a self-attention module to learn the features of multiple adjacent containers, a variational autoencoder module to learn the latent variables of the port, and an output module to combine the above features.

Self-attention Module. The operating time of each container is influenced by the adjacent containers in order sequence. For example, when there are two containers with little difference in size and shape, the crane can adopt a double container crane operation process, i.e., operate two containers at the same time. Similarly, there are triple and quad container crane operations. In this way, the operating time of each container can be cut in half. To capture the interrelationships between the containers, we have designed a self-attention module.

The attention mechanism [11] is formulated as

$$\text{Attention}(Q, K, V) = \text{softmax}(\frac{QK}{\sqrt{d_k}})V, \tag{2}$$

where Q, K, V denote the query, key, and value matrix, d_k is the dimension of queries and keys. For self-attention, we let $Q = K = V$, which is used for learning the internal dependencies of queries and capturing its internal structure.

In our model, the query matrix is set as $\mathbf{x}_i^L \in \mathbf{R}^{L \times d}$. We let L equal to five, for overlong dependency rarely exists. In this way, we obtain the $\mathbf{x}_i$'s feature vector $\mathbf{v}_i^{att}$ that aggregates the information of adjacent containers.

Variational Autoencoder Module. In port transportation, the bridge crane system and the AGVs transport system are strongly coupled. When the AGVs transport system is heavily congested or down, it will significantly reduce the efficiency of the bridge crane system. We lack detailed data on the AGVs transport system and do not have the specific AGV scheduling algorithm. It is difficult to model the AGVs transport system directly. There are more uncertainties in the port, such as operators' errors, which are also difficult to be collected.

Variational autoencoder is a combination of the idea of probabilistic graphical model and autoencoder, which can get the latent variables from the source input and generate the source input again from the latent variables. In this module, we applied variational autoencoder to learn the latent variables in the port, which could be considered as the effects of the AGVs transport system and other uncertainties on the container operating time.

Specifically, we first obtain the mean μ_i and the variance σ_i^2 corresponding to $\mathbf{x}_i$ through an independent self-attention layer. According to the value of mean and variance, we sample the latent variable $\mathbf{Z}_i$ by reparameterization trick which is easy to apply gradients backpropagation algorithm on

$$\mathbf{Z}_i = \mu_i + \epsilon \times \sigma_i, \tag{3}$$

where ϵ is sampled from $\mathcal{N}(0, \mathbf{I})$. After that, a decoder layer is accessed, containing several dense layers, and decodes $\mathbf{Z}_i$ to get feature vector $\mathbf{v}_i^{vae}$. The vector $\mathbf{v}_i^{vae}$ contains the latent knowledge introduced by the whole port and is expected to be as close as possible to the input $\mathbf{x}_i$.

Output Module. In the above modules, vector $\mathbf{v}^{att}$ represents the mutual relationship features between containers, while vector $\mathbf{v}^{vae}$ includes the latent variable features in the whole port system. In this module, we use the feature vectors extracted by self-attention and variational autoencoder to predict the operating time separately and finally add the two together proportionally as the resulting predict time t_i^{pre}. Specifically, we append two non-linear dense layers to generate the time,

$$t_i^{att} = f^{att}(\mathbf{v}_i^{att} \oplus \mathbf{x}_i), \tag{4}$$

$$t_i^{vae} = f^{vae}(\mathbf{v}_i^{vae}), \tag{5}$$

$$t_i^{pre} = \alpha \cdot t_i^{att} + (1 - \alpha) \cdot t_i^{vae}, \tag{6}$$

where $\oplus$ is the concatenate operation of vector, α is a hyperparameter. Here, $\mathbf{x}_i$ is linked to $\mathbf{v}_i^{att}$ to provide the original information about the container i. The resulting prediction function f is composed of the three components, self-attention module, variational autoencoder module, and output module.

Loss Function. The loss consists of two main components. One of them is generated in variational autoencoder module. For $\mathbf{x}_i$ and $\mathbf{v}^{vae}$ are supposed to be similar, we add $D(\mathbf{x}_i, \mathbf{v}_i^{vae})$ as a term of the loss function. Where $D(\cdot)$ is a distance function, we use mean square error in our model. Moreover, in order to restrict the distribution of $\mathbf{Z}_i$, we use a kind of regularization that drives the latent variables to match the prior $\mathcal{N}(0, \mathbf{I})$. And use Kullback-Leibler (KL) divergence to measure the loss. In summary, the overall VAE loss function is formulated as

$$\mathcal{L}_i^{vae} = D(\mathbf{x}_i, \mathbf{v}_i^{vae}) + KL(\mathcal{N}(\mu_i, \sigma_i^2) \| \mathcal{N}(0, \mathbf{I})), \tag{7}$$

where the second components of RSH can be calculated by

$$KL(\mathcal{N}(\mu_i, \sigma_i^2) \| \mathcal{N}(0, \mathbf{I})) = \frac{1}{2}(-\log \sigma_i^2 + \mu_i^2 + \sigma_i^2 - 1), \tag{8}$$

The other part of the loss function is between the predicted value t_i^{pre} and the true value t_i^{true}, which be formulated as MSE loss,

$$\mathcal{L}_i^{mse} = (t_i^{pre} - t_i^{true})^2, \tag{9}$$

The total loss function is

$$\mathcal{L} = \sum_i (\beta \cdot \mathcal{L}_i^{vae} + (1 - \beta) \cdot \mathcal{L}_i^{mse}), \tag{10}$$

where β is hyperparameter denotes the weight of the two loss components.

3.3   Throughput Inference

In the previous section, we predict the operating time of each container. In this section, we propose an algorithm to make an inference about port throughput based on the predicted operating time. Also, with the help of the latent variables modeled by the variational autoencoder, we propose a dynamic programming algorithm to speculate on the distribution of throughput. These algorithms will be described in turn in the following.

Throughput Value Infer Algorithm. We enumerate each ship in the port and the bridge cranes working on each ship, figure out the throughput of each crane separately, and then sum them up. Specifically, we use binary search to find the first container whose arrival time is just after start time t_s, and then the predicted operating time for these containers are cumulated sequentially until the total time is greater than the time interval Δt. The pseudo-code is shown as Algorithm 1. Moreover, we can pre-process the prediction time of each container and the order sequence on each crane in $O(T_f \cdot |\mathcal{X}|)$ to reduce the time complexity. The T_f denotes the time cost for the prediction function f to execute once and can be considered as a constant. The total time complexity after pre-processing is $O(\log |\mathcal{X}| + |\mathcal{X}'|)$, where $\mathcal{X}'$ denotes the set of possible containers.

Algorithm 1: Throughput Value Infer Algorithm.

Input: The predict function $f : \mathbf{R}^{L \times d} \to \mathbf{R}$; Time interval Δt; The order sequence of containers $\mathcal{X} = \{\mathbf{x}_1, \mathbf{x}_2, \cdots\}$; The set of ships S; The dictionary C indicating the set of bridge cranes working on a ship, i.e. C_{ship} denotes the set of cranes working on $ship$; The hyperparameter L;

Output: The prediction of container throughput in time interval Δt: $C_{\Delta t}$;

1   $C_{\Delta t} \leftarrow 0$;
2   **for** $ship \in S$ **do**
3   **for** $crane \in C_{ship}$ **do**
4   Binary search in $\mathcal{X}$ to find the first container waiting for operating on $crane$ after time t_s, called $\mathbf{x}_i$;
5         $t \leftarrow 0$;
6   **while** $t < \Delta t$ **do**
7             $\mathbf{x}_i^L \leftarrow [\mathbf{x}_{i-\frac{L}{2}}, \mathbf{x}_{i-\frac{L}{2}+1}, \cdots, \mathbf{x}_{i-1}, \mathbf{x}_i, \mathbf{x}_{i+1}, \cdots, \mathbf{x}_{i+\frac{L}{2}-1}, \mathbf{x}_{i+\frac{L}{2}}]$;
8             $t \leftarrow t + f(\mathbf{x}_i^L)$;
9             $i \leftarrow i + 1$;
10   **if** $t < \Delta t$ **then**
11               $C_{\Delta t} \leftarrow C_{\Delta t} + 1$;
12   **end**
13   **end**
14   **end**
15 **end**
16 **return** $C_{\Delta t}$

Crane Throughput Distribution Infer Algorithm. We can get the predicted value t_i^{pre} of the operating time for container i by function f. Since we have modeled the latent variables using variational autoencoder, we will get k different results if function f is executed k times for container i. Note them as $T_i = \{(t_i^{pre})_1, (t_i^{pre})_2, \cdots, (t_i^{pre})_k\}$. T_i could be regarded as a discrete sampling of the predicted time distribution of container i. The larger the number of samples k is, the more precisely we can portray the original distribution by the obtained discrete values. With $T_1 \sim T_i$, we can use dynamic programming approach to calculate $f_{i,j}$, which indicates how many ways to complete all operation of first i containers in exactly j seconds. The transfer equation is

$$f_{i,j} = \sum_{p=1}^{k} f_{i-1,j-(t_i^{pre})_p} \tag{11}$$

We can use the value of $f_{i,j}$ to express the probability of spending j seconds to operate the first i containers. Given the minimum value in T_i, which is taken as the predicted operating time of container i, then the number of containers that can be operated in time interval Δt reaches the theoretical maximum, denotes $C_{\Delta t}^{max}$. Similarly, if the maximum in T_i is taken, then the theoretical minimum $C_{\Delta t}^{min}$ can be reached. Summing some of $f_{i,j}$ gives $n_i (i \in [C_{\Delta t}^{min}, C_{\Delta t}^{max}])$, which represents the possible combinations of predicted time fetches for operating exactly i containers in Δt, i.e. the probability of operating the first i containers in Δt. Obviously, we have $\sum_{i=C_{\Delta t}^{min}}^{C_{\Delta t}^{max}} n_i = k^{C_{\Delta t}^{max}}$. The pseudo code is shown as Algorithm 2. The time complexity of this algorithm is $O(C_{\Delta t}^{max} \cdot \Delta t \cdot k)$.

Algorithm 2: Crane Throughput Distribution Infer Algorithm.

Input: The theoretical minimum $C_{\Delta t}^{min}$; The theoretical maximum $C_{\Delta t}^{max}$; The predicted time set $T_1 \sim T_{C_{\Delta t}^{max}}$; The time interval Δt; The number of samples k;

Output: The possible combinations of operating the first i containers in Δt: $n_{C_{\Delta t}^{min}} \sim n_{C_{\Delta t}^{max}}$, which can be simply converted into probabilities;

1   $f_{0,0} \leftarrow 1$;
2   Initialize other $f_{i,j}$ to 0 ;
3   Initialize all n_i to 0 ;
4   **for** $i = 1 \rightarrow C_{\Delta t}^{max}$ **do**
5   **for** $j = 0 \rightarrow \Delta t$ **do**
6   **for** $t_i^{pre} \in T_i$ **do**
7   **if** $j + t_i^{pre} > \Delta t$ **then**
8              $n_{i-1} \leftarrow n_{i-1} + f_{i-1,j} \times k^{C_{\Delta t}^{max}-i}$;
9   **else**
10             $f_{i,j+t_i^{pre}} \leftarrow f_{i,j+t_i^{pre}} + f_{i-1,j}$;
11   **end**
12   **end**
13   **end**
14 **end**
15 **return** $n_{C_{\Delta t}^{min}} \sim n_{C_{\Delta t}^{max}}$

Port Throughput Distribution Infer Algorithm. Using Algorithm 2 for each bridge crane, we can obtain the probability distribution of the throughput of each of them. Let $P_i = \{(j_1, n_{j_1}), \cdots, (j_{q_i}, n_{j_{q_i}})\}$ is a set of pair, where pair (j_k, n_{j_k}) denotes the probability of operating first j_k containers on crane i is n_{j_k}. And there are m cranes in the port, independent from each other. We can borrow the solution of the knapsack problem to obtain the probability distribution of the throughput of the whole port. Let $g_{i,j}$ denotes the probability that the throughput of the first i cranes is j. The transfer equation is

$$g_{i,j} = \sum_{k=1}^{q_i} g_{i-1,j-j_k} \cdot n_{j_k} \tag{12}$$

The value of $g_{m,1\sim C_{\Delta t}}$ is exactly the result throughput distribution. If all the cranes take the maximum throughput, we can get the upper bound of the port's throughput $C_{\Delta t}$. The pseudo-code is shown as Algorithm 3. And the time complexity of this algorithm is $O(m \cdot C_{\Delta t} \cdot \max q_i)$.

Algorithm 3: Port Throughput Distribution Infer Algorithm.

Input: Upper bound of the port's throughput $C_{\Delta t}$; The number of cranes m;
    Pair set $P_1, \cdots, P_m$;
Output: The port throughput distribution: $g_{m,1\sim C_{\Delta t}}$;
1 $g_{i,0} \leftarrow 1$;
2 Initialize other $g_{i,j}$ to 0 ;
3 **for** $i = 1 \rightarrow m$ **do**
4   **for** $j = 1 \rightarrow C_{\Delta t}$ **do**
5   **for** $(v, n_v) \in P_i$ **do**
6   **if** $j - v \geq 0$ **then**
7                  $g_{i,j} \leftarrow g_{i,j} + n_v \times g_{i-1,j-v}$;
8   **end**
9   **end**
10   **end**
11 **end**
12 **return** $g_{m,1\sim C_{\Delta t}}$

---

4 Experimental Setup

To validate the effectiveness of our method, we conduct extensive experiments, we will first introduce experiment setups, including dataset description, baseline methods, and experimental settings.

4.1 Datasets

We extensively perform our experiments on real-world dataset, which is collected at Yangshan Port in Shanghai. Established in 2017, Yangshan Port is one of the world's largest intelligent container terminals with the highest container throughput in the world. The dataset is described below.

YST (**Y**ang**S**han **T**hroughput Dataset): The YST dataset contains most information of the ships in port, including the arrival time of the ship, the type of ship, the properties and layout of the containers on board, etc. Meanwhile, it records the actual loading and discharging time of each container. This dataset contains 2,405,278 pieces of data, which is collected from January 1st to December 31th, 2020 by the sensors on cranes and AGVs. To explore the different granularities of the problem, we created separate datasets as YST1h, YST2h, and YST4h for 1-hour-level, 2-hour-level, and 4-hour-level, which are supposed to evaluate the model performance on short time intervals. Similarly, we have built the datasets YST8h, YST16h, and YST24h to evaluate the model on long time intervals. These six different time scales provide a comprehensive picture of the model's effects. These six datasets are divided into a training set (60%), a validation set (20%), and a test set (20%).

4.2 Baseline Approaches

We used six models to verify our approaches:

- **LR**: a general machine learning method to predict time.
- **SVR** [10]: support vector regression with rbf kernel.
- **GBDT** [3,5]: a gradient boosting decision tree to predict time.
- **CNN-LSTM** [12]: a fusion of CNN model and LSTM model to extract features from time series, and then predicts the port throughput.
- **CNN-LSTM***: the improvement of CNN-LSTM. We applied the CNN-LSTM model to the container operating time forecast problem and use Algorithm 1 for port throughput inference.
- **ARIMA-MLP** [9,13]: throughput is predicted by using ARIMA and MLP respectively, and then combine them.

The fourth and sixth methods predict the port's throughput directly based on time series, while the other methods use features of each container and the order sequence of the containers.

4.3 Experimental Settings

We selected containers' properties, containers' location, AGVs and bridge crane interaction position, and some other features as input. For example, one of the containers' properties is denoted as 45G1, where the first number 4 means its size is 40 feet, 5 means it is a tall container, and G means the container type is normal. We use Adam as the optimizer and fix the learning rate to 0.0001. Our hyperparameters in the experiment are $L = 5, \alpha = 0.5, \beta = 0.5$. The evaluation metrics are root mean squared error (RMSE), mean absolute error (MAE), and mean absolute percentage error (MAPE). For throughput prediction, we trained the model several times and used the average value as the prediction result. For throughput distribution fitting, we performed a case study using Algorithm 2 and Algorithm 3.

5   Result and Discussion

5.1   Performance Evaluation

Table 1 and Table 2 summarizes the main results of all the methods. For each model, we fixed the hyperparameters when it reaches the optimal performance on the validation set. The bolded numbers indicate the best result of each column, the signal - indicates that the value is too large. We have the following observations. First, the method CNN-LSTM and ARIMA-MLP do not use fine-grained container features. Thus they don't work well, especially in the MAE and MAPE indicators for long time intervals. Second, as the time interval becomes larger, the effect of short-term abnormalities on the results becomes weaker and the indicator MAPE becomes smaller. Third, our model outperforms other methods on the MAE and MAPE metrics on most of the datasets and is slightly worse on the RMSE, probably because large outliers affect the final results.

Table 3 shows the experimental results of the hyperparameter L at different values. For all short time intervals, $L = 5$ has the best results for all indicators. While on the long time interval, $L = 5$ performs weaker on RMSE.

Table 1. Results for throughput forecasting of short time intervals.

Methods	YST1h			YST2h			YST4h		
	RMSE	MAE	MAPE	RMSE	MAE	MAPE	RMSE	MAE	MAPE
LR	20.08	15.57	6.786	35.60	27.37	5.690	59.81	45.35	4.580
SVR	29.34	24.08	10.11	54.50	44.87	9.047	97.09	80.05	7.887
GBDT	18.07	13.48	5.919	**32.48**	23.85	4.955	**55.23**	39.24	3.914
CNN-LSTM*	19.21	14.32	6.539	34.92	26.03	5.561	59.94	44.62	4.531
Our Method	**17.73**	**12.66**	**5.547**	32.69	**22.59**	**4.626**	56.15	**37.01**	**3.640**
CNN-LSTM	63.81	48.64	79.44	194.6	162.1	69.58	–	–	75.33
ARIMA-MLP	112.7	97.14	48.03	203.7	173.1	45.42	–	–	39.42

Table 2. Results for throughput forecasting of long time intervals.

Methods	YST8h			YST16h			YST24h		
	RMSE	MAE	MAPE	RMSE	MAE	MAPE	RMSE	MAE	MAPE
LR	92.41	68.23	3.420	123.2	85.64	2.212	135.5	92.51	1.569
SVR	143.9	115.2	5.700	196.1	150.9	3.857	220.5	165.8	2.819
GBDT	**85.91**	59.02	2.932	**119.0**	76.32	1.911	**132.9**	82.24	1.352
CNN-LSTM*	94.58	68.58	3.456	126.0	87.50	2.315	141.14	95.34	1.630
Our Method	88.67	**55.57**	**2.694**	126.8	**74.32**	**1.800**	142.5	**81.69**	**1.284**
CNN-LSTM	–	–	–	–	–	–	–	–	–
ARIMA-MLP	–	–	34.77	–	–	49.35	–	–	30.42

Table 3. Results for different hyperparameters L.

	YST1h			YST2h			YST4h		
	RMSE	MAE	MAPE	RMSE	MAE	MAPE	RMSE	MAE	MAPE
$L = 3$	18.60	13.97	6.137	33.38	24.87	5.181	56.50	41.31	4.154
$L = 5$	**17.73**	**12.66**	**5.547**	**32.69**	**22.59**	**4.626**	**56.15**	**37.01**	**3.640**
$L = 7$	18.40	13.09	5.742	33.94	23.50	4.833	58.91	39.11	3.837
	YST8h			YST16h			YST24h		
	RMSE	MAE	MAPE	RMSE	MAE	MAPE	RMSE	MAE	MAPE
$L = 3$	**88.21**	62.40	3.129	**119.3**	79.56	2.046	**130.9**	86.71	1.454
$L = 5$	88.67	**55.57**	**2.694**	126.8	**74.32**	**1.800**	142.5	**81.69**	**1.284**
$L = 7$	94.95	60.18	2.903	138.9	81.20	1.960	157.3	92.14	1.433

Figure 3 shows the absolute error curves for four different time intervals. For each graph, the horizontal coordinate indicates the start time t_s of the time interval and the vertical coordinate indicates the absolute error. We can observe that our method maintains the position of the minimum error in most cases, and the larger the time interval, the more obvious is the advantage of our method.

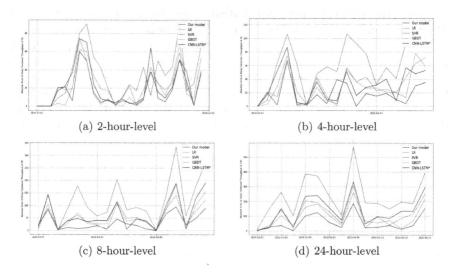

(a) 2-hour-level   (b) 4-hour-level

(c) 8-hour-level   (d) 24-hour-level

Fig. 3. The absolute error curves.

5.2 Ablation Study

We conducted an ablation study to analyze how different components of our method contribute to the final performance. The model ATT is part of our approach, which only uses the self-attention module for training and prediction.

The output t_i^{att} is treated as the final result. Similarly, the VAE model uses only the variational autoencoder module, with t_i^{vae} as the final result. To this end, We evaluate the importance of both the variational autoencoder module and the self-attention module by comparing with the model ATT and VAE. The Results of the ablation experiment are summarized in Table 4. It can be observed that the self-attention module provides a more significant enhancement to the effect, while the variational autoencoder module adds randomness to the model so that the effect of outliers on model performance is reduced. Summarily, using both the variational autoencoder module and the self-attention module achieves the best performance.

Table 4. The ablation results of our method.

	YST1h		YST2h		YST4h		YST8h		YST16h		YST24h	
	MAE	MAPE	MAE	MAPE	MAE	MAPE	MAE	MAPE	MAE	MAPE	MAE	MAPE
ATT	13.48	5.958	25.07	4.999	39.83	3.964	59.30	2.906	75.96	1.914	82.29	1.325
VAE	15.37	6.930	28.22	5.991	49.91	5.065	79.15	3.992	102.3	2.705	112.2	1.940
Ours	**12.66**	**5.547**	**22.59**	**4.626**	**37.09**	**3.639**	**55.57**	**2.694**	**74.32**	**1.800**	**81.69**	**1.284**

5.3   Case Study

We can obtain the probability distribution of the throughput distribution using Algorithm 2 and Algorithm 3. As shown in Fig. 4, the horizontal coordinates indicate the values of throughput, the vertical coordinates indicate the probability values, and the red bars indicate the true values of container throughput of a certain time period. These two cases are taken from the 1-hour-level dataset. We can observe that in these two cases, the true value corresponds to the larger probability value. In fact, every possible value of the throughput is an integer. However, to better show the experimental results, we smoothed the probability distribution and plotted it as the blue curve below.

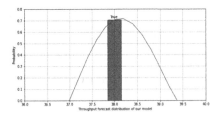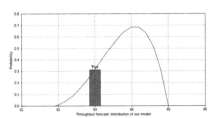

Fig. 4. Throughput forecast distribution of our model. In the left case, the true value is at the position with the highest probability. While in the right case, the probability of the true value is smaller.

6   Conclusion

In this paper, we proposed an approach combining self-attention mechanism and variational autoencoder. The method captures the interrelationship of adjacent containers and models the latent variables in the port. Furthermore, we proposed a dynamic programming algorithm to describe the distribution of port throughput. Experimental results on the real-world dataset YST at different time intervals show the superior performance of our model. Through the case study, we also demonstrated the effectiveness of the dynamic programming algorithm.

Acknowledgements. This work was supported by National Key Research and Development Project (No. 2019YFB1704400); National Natural Science Foundation of China under Grant No. 62172107 and No. 61772138.

References

1. Connor, J.T., Martin, R.D., Atlas, L.E.: Recurrent neural networks and robust time series prediction. IEEE Trans. Neural Netw. **5**(2), 240–254 (1994)
2. Eskafi, M., Kowsari, M., Dastgheib, A., Ulfarsson, G.F., Thorarinsdottir, R.I.: A model for port throughput forecasting using Bayesian estimation. Marit. Econ. Logist. **3**, 1–21 (2021)
3. Friedman, J.H.: Greedy function approximation: a gradient boosting machine. Ann. Stat., 1189–1232 (2001)
4. Huang, Z., Xu, W., Yu, K.: Bidirectional LSTM-CRF models for sequence tagging. arXiv preprint arXiv:1508.01991 (2015)
5. Li, L., et al.: Using improved gradient-boosted decision tree algorithm based on Kalman filter (GBDT-KF) in time series prediction. J. Supercomput., 1–14 (2020)
6. Min, K.C., Ha, H.K.: Forecasting the Korea's port container volumes with Sarima model. J. Korean Soc. Transp. **32**(6), 600–614 (2014)
7. Mo, L., Xie, L., Jiang, X., Teng, G., Xu, L., Xiao, J.: GMDH-based hybrid model for container throughput forecasting: selective combination forecasting in nonlinear subseries. Appl. Soft Comput. **62**, 478–490 (2018)
8. Niu, M., Hu, Y., Sun, S., Liu, Y.: A novel hybrid decomposition-ensemble model based on VMD and HGWO for container throughput forecasting. Appl. Math. Modelling **57**, 163–178 (2018)
9. Rashed, Y., Meersman, H., Van de Voorde, E., Vanelslander, T.: Short-term forecast of container throughout: an ARIMA-intervention model for the port of Antwerp. Marit. Econ. Logist. **19**(4), 749–764 (2017)
10. Su, H., Zhang, L., Yu, S.: Short-term traffic flow prediction based on incremental support vector regression. In: Third International Conference on Natural Computation (ICNC 2007), vol. 1, pp. 640–645. IEEE (2007)
11. Vaswani, A., et al.: Attention is all you need. arXiv preprint arXiv:1706.03762 (2017)
12. Yang, C.H., Chang, P.Y.: Forecasting the demand for container throughput using a mixed-precision neural architecture based on CNN-LSTM. Mathematics **8**(10), 1784 (2020)
13. Zhang, Y., Fu, Y., Li, G.: Research on container throughput forecast based on ARIMA-BP neural network. In: Journal of Physics: Conference Series (2020)

Data Source Selection in Federated Learning: A Submodular Optimization Approach

Ruisheng Zhang[1,2], Yansheng Wang[1,2], Zimu Zhou[3], Ziyao Ren[1,2], Yongxin Tong[1,2(✉)], and Ke Xu[1,2]

[1] State Key Laboratory of Software Development Environment, Beihang University, Beijing, China
{rszhang,arthur_wang,ziyaoren,yxtong,kexu}@buaa.edu.cn
[2] Beijing Advanced Innovation Center for Future Blockchain and Privacy Computing, Beihang University, Beijing, China
[3] Singapore Management University, Singapore, Singapore
zimuzhou@smu.edu.sg

Abstract. Federated learning is a new learning paradigm that jointly trains a model from multiple data sources without sharing raw data. For the practical deployment of federated learning, data source selection is compulsory due to the limited communication cost and budget in real-world applications. The necessity of data source selection is further amplified in presence of data heterogeneity among clients. Prior solutions are either low in efficiency with exponential time cost or lack theoretical guarantees. Inspired by the diminishing marginal accuracy phenomenon in federated learning, we study the problem from the perspective of submodular optimization. In this paper, we aim at efficient data source selection with theoretical guarantees. We prove that data source selection in federated learning is a monotone submodular maximization problem and propose FDSS, an efficient algorithm with a constant approximate ratio. Furthermore, we extend FDSS to FDSS-d for dynamic data source selection. Extensive experiments on CIFAR10 and CIFAR100 validate the efficiency and effectiveness of our algorithms.

Keywords: Federated learning · Data source selection · Submodularity

1 Introduction

Federated learning (FL) [3,13] is an emerging distributed learning paradigm among multiple data sources, where a global model is trained collaboratively without sharing their raw local data. It has been applied in various applications such as cross-hospital medical image classification [13], next-word prediction on smartphones [3,7], information retrieval [10,11], etc. In practice, federated learning often relies on a selective subset of data sources rather than the entire federation. Data source selection, also known as client selection [1], is compulsory due to the massive communication overhead between the data owners (*i.e.* clients) and the server, or simply the budget limit to cover all data sources [2,3,13]. The necessity of data source selection also arises from the heterogeneity of data, whose partition may significantly vary in label distribution and data quality [5].

Prior data source selection methods in federated learning fall into two categories. The first category focuses on evaluating every data source from a theoretical perspective such as the Shapley value [5,8]. These schemes ensure optimal selection, yet at the cost of exponential time complexity, which is prohibitive for practical deployment. The second category exploits heuristics or back-box optimization for selection, which tend to be more efficient, but are prone to low accuracy in case of data heterogeneity. For example, the naive FedAvg [3] adopts a simple random sampling strategy. Others utilize local gradient information [1,2] to approximate the contributions of participants. These solutions lack theoretical guarantees and incur severe performance degradation on heterogeneous data.

In this paper, we aim at efficient data source selection with theoretical guarantees. Our solution is motivated by the empirical observation that the accuracy of deep learning models tends to increase logarithmically with the amount of training samples [9]. Such phenomena inspire us to analyze the data source selection problem in the lens of *monotone submodular maximization* [6]. Our main contributions and results are summarized as follows.

- We theoretically prove that data source selection in federated learning aiming at generalization error minimization can be converted to monotone submodular maximization. To the best of our knowledge, this is the first submodularity analysis directly on the generalization error in federated learning.
- We design an efficient data source selection algorithm called FDSS with an approximate ratio of $1 - \frac{1}{e}$, which can make a better trade-off between accuracy and efficiency. We further propose an extension FDSS-d for data source selection with dynamic participants availability.
- Extensive evaluations on real datasets show that our proposed algorithms outperform the state-of-the-arts [2,5] in terms of test accuracy and communication rounds on heterogeneous data.

2   Problem Statement

2.1   Data Source Selection in Federated Learning

We consider federated learning of model ω over a federation $\mathcal{F} = \{P_1, P_2, \cdots, P_N\}$ of N data sources, where P_i denotes the i-th data source. P_i holds a set of n_i data samples $X_i = \{x_{i,1}, x_{i,2}, \cdots, x_{i,n_i}\}$. Each X_i is independently and identically drawn from a prior distribution π_i. A common objective of federated learning is to minimize the *expected generalization error* $L_g(\pi, \omega) = \mathbb{E}[L_\pi(\omega) - L_P(\omega)]$ on the joint distribution $\pi = \prod_{i=1}^{N} \pi_i$. $L_P(\omega) = \sum_{i=1}^{N} \frac{n_i}{n} L_{P_i}(\omega)$ represents the overall empirical error and $L_{P_i}(\omega)$ is the local empirical error of data source P_i.

We are interested in selecting a subset $F \subset \mathcal{F}$ for federated learning due to the limited budget to recruit all data sources in real-world applications [1–3,8]. We quantify the contribution of a subset F by the evaluation function below.

$$g(F) = L_g(\pi, \omega_0) - L_g(\pi, \omega_F) \tag{1}$$

where $\omega_F = \sum_{P_i \in F} \frac{n_i}{n} \omega_i$ is the aggregated model parameter from F, and ω_0 is the initial model parameter *i.e.* when $F = \emptyset$. Given the evaluation function, we now define the data source selection problem in federated learning.

Definition 1. *Given a federation $\mathcal{F}$, an evaluation function g, and a cardinality constraint c, we define the static data source selection problem as:*

$$\max_{F \subset \mathcal{F}} g(F) \quad s.t. |F| \le c \tag{2}$$

Further assume a time sequence $t = 1, 2, \cdots, T$. Let the available data sources at time t be the subset $\mathcal{F}_t \subset \mathcal{F}$. We can define the dynamic data source selection problem as

$$\max \sum_{F_t \subset \mathcal{F}_t} g(F_t) \quad s.t. |F_t| \le c, \forall t = t_0, t_1, \cdots, T \tag{3}$$

where F_t represents the selected subset at t.

2.2 Submodularity Analysis of Data Source Selection

To analyze the submodularity of the function g, the key idea is to harness the information-theoretic bound [12] for the expected generalization error in federated learning. Our main claim is the following.

Theorem 1. *If for each data source P_i, $n_i = n$, $\pi_i = \mathcal{N}(\nu, \sigma_i^2 I_d)$ and $\sigma_i^2 - \sigma_j^2 \le \frac{1}{2}\sigma_j^2$ for all $i \ne j$, then the evaluation function $g(F)$ in Eq. (1) is both monotone and submodular.*

Proof. We first estimate the generalization error using the bounds in [12]:

$$\frac{1}{n} \sum_{P_i \in F} \sum_{j=1}^{n_i} \psi_{i+}^{*-1}(I(x_{i,j}; \omega_F)) \le L_g(\pi, \omega_F) \le \frac{1}{n} \sum_{P_i \in F} \sum_{j=1}^{n_i} \psi_{i-}^{*-1}(I(x_{i,j}; \omega_F)) \tag{4}$$

where $I(x_{i,j}; \omega_F)$ refers to the mutual information of model parameter w_F and local data $x_{i,j}$, $\psi_+ : [0, b_+) \to \mathbb{R}$ and $\psi_- : [0, b_-) \to \mathbb{R}$ are convex functions. Based on the assumptions of $n_i = n$ and $\pi_i = \mathcal{N}(\nu, \sigma_i^2 I_d)$, we can get $L_g(\pi, \omega_F) = \sum_{P_i \in F} \frac{2d\sigma_i^2}{k^2 n}$. Next, we prove $g(F)$ is monotone. Let F be a subset of $\mathcal{F}$ and $|F| = k \, (k > 1)$, for any P_j such that $P_j \notin F$:

$$g(F \cup P_j) - g(F) \ge \frac{2d(2k+1)\sigma_{min}^2 - 2dk\sigma_{max}^2}{k(k+1)^2 n} \ge \frac{2d(k+2)\sigma_{max}^2}{3k(k+1)^2 n} \ge 0 \tag{5}$$

where $\sigma_{max}(\sigma_{min})$ denotes the maximum(minimum) across all variances. The third inequality results from the bounded data variance of different data sources.

Finally, we prove the submodularity of $g(F)$. Let $\Delta_F^j = g(F \cup P_j) - g(F)$ and $F' = F - P_k(P_k \in F)$,

$$\Delta_{F'}^j - \Delta_F^j = \left(\sum_{P_i \in F'} \frac{2d(2k-1)\sigma_i^2}{k^2(k-1)^2 n} - \frac{2d\sigma_j^2}{k^2 n} \right) - \left(\sum_{P_i \in F} \frac{2d(2k+1)\sigma_i^2}{k^2(k+1)^2 n} - \frac{2d\sigma_j^2}{(k+1)^2 n} \right)$$

$$\geq \frac{2d(6k^2-2)\sigma_{min}^2}{k^2(k-1)(k+1)^2 n} - \frac{2d(4k+2)\sigma_{max}^2}{k^2(k+1)^2 n} \geq \frac{2k+\frac{2}{3}}{k^2(k-1)(k+1)^2 n} 2d\sigma_{max}^2 \geq 0$$

By the arbitrariness of P_k, we can derive the submodularity of $g(F)$.

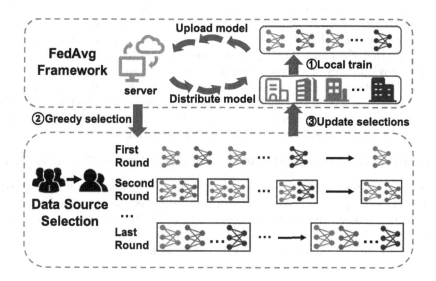

Fig. 1. Illustration of federated learning with data source selection.

3   Data Source Selection Algorithms

Inspired by the submodularity analysis in Sect. 2.2, we devise two greedy data source selection algorithms with constant approximation ratio. The data source selection algorithms can be seamlessly integrated into mainstream federated learning algorithms (see Fig. 1). For ease of presentation, we explain our data source selection algorithms on top of FedAvg [3], but they also function with more advanced federated learning algorithms.

3.1   Static Data Source Selection

As previously mentioned, the monotone submodular maximization nature of the problem ensures a constant approximation ratio by *greedy* selection. To further accelerate the selection process, we also exploit lazy evaluation and approximate the aggregated model. We would like to highlight three aspects of our proposed static algorithm FDSS.

- **Greedy Selection.** The server first initializes ρ as a descending list, and sets the global model and selected federation F to ω_0 and $\emptyset$. Then the algorithm iteratively adds data sources to the federation until c data sources are selected. Let the marginal benefit of g be $\Delta(P_j|F) = g(F \cup P_j) - g(F)$. In each round, the server computes $\Delta(P_j|F)$ and selects the data source that maximizes it, i.e., $F = F \cup \{\arg\max_{P_i \notin F} \Delta(P_i|F)\}$. Afterwards, the server adds the data source to the federation and aggregates global model.

- **Lazy Evaluation.** We use lazy evaluation [4] to accelerate computing the marginal benefit $\Delta(P_j|F)$. Note that the marginal benefit of any data source P_j is monotonically non-increasing during iteration. Hence, we maintain an upper bound list ρ of $\Delta(P_j|F)$ sorted in descending order. In each iteration, we extract the maximal element P_l from the list ρ and update its benefit. If the updated value is the largest in the current list ρ, submodularity will ensure that the marginal benefits of other data sources are lower than P_j.

- **Approximation of Aggregated Model.** Note that computing $\Delta(P_j|F)$ in each iteration requires calculating $g(F_k \cup P_j)$ for every $P_j \notin F$. However, training a new federated model is time-consuming. For further acceleration, we approximate the federated model by aggregating the trained local model and calculating the accuracy on the global validation set V.

Approximation Ratio and Time Complexity. Assume $F^* = \arg\max\limits_{|F| \leq c} g(F)$ and F_c is the final set in our selection algorithm. According to [6], the greedy selection incurs $g(F_c) \leq (1 - \frac{1}{e})g(F^*)$ theoretically. Let the size of global validation set is m. Then the total time complexity of FDSS without lazy evaluation is $m(N + N - 1 + \cdots + N - c + 1) = c\frac{2N-c+1}{2} = O(N^2)$ if $c = O(N)$. Therefore, the worst-case time complexity of FDSS with the accelerations is $O(N^2)$.

3.2 Dynamic Data Source Selection

Now we extend our FDSS algorithm to the dynamic setting. That is, data source selection is performed in a time sequence $t = 1, 2, \cdots, T$. A naive solution is to repeated perform the static data source selection algorithm *i.e.* FDSS in each round. However, this solution can be inefficient because multiple selections would bring more time cost, especially in scenarios with a large T. And the selected data sources in adjacent rounds are often identical, therefore reselecting data sources is not always necessary. In response, we propose a more efficient dynamic data source selection algorithm FDSS-d. Compared with the naive extension, our FDSS-d algorithm makes the following improvements.

- Each time before calling FDSS, a fast verification is conducted to check whether the global model's accuracy is improving. If the model accuracy is still increasing, there is no need to re-select data sources. In this case, the next selection is postponed till the model converges.

- We divide the entire training into $\frac{T}{s}$ stages and identify the data sources who will participate in the next stage at round ks, where k is a positive integer and s is the selection interval. Note that in the dynamic setting, data sources may be unavailable in each round. Thus we only execute the FDSS algorithm for current online data sources to save bandwidth.

Approximation Ratio and Time Complexity. The FDSS-d algorithm selects data sources dynamically. For each selection, the algorithm guarantees a constant approximation ratio. In the worst case, s selections are executed in total. Hence, the time complexity of FDSS-d is $O(\frac{T}{s}N^2)$. Note that the selection is only performed when the model has converged. Thus the actual running time of FDSS-d is much less than the worst case.

Table 1. Accuracy (%) on CIFAR10. The best performance is marked in bold.

Methods	Settings				
	Noise 20%	Noise 30%	Noise 40%	Noise 50%	Noise 60%
SFedAvg	26.23 ± 0.96	29.98 ± 0.33	25.86 ± 0.23	33.44 ± 0.29	32.97 ± 0.75
SS-Fed	46.57 ± 0.55	34.94 ± 0.21	28.98 ± 0.82	24.00 ± 0.48	15.60 ± 0.37
FDSS	$\mathbf{51.01 \pm 0.41}$	$\mathbf{53.54 \pm 0.83}$	$\mathbf{52.85 \pm 0.63}$	44.97 ± 0.20	$\mathbf{50.78 \pm 0.90}$
FedAvg	46.10 ± 1.05	44.46 ± 1.37	41.62 ± 0.82	24.37 ± 0.54	23.38 ± 0.59
Oort	39.32 ± 4.13	34.00 ± 1.66	36.09 ± 2.64	15.53 ± 1.00	34.10 ± 1.86
FDSS-d	$\mathbf{51.38 \pm 0.38}$	$\mathbf{47.98 \pm 0.32}$	$\mathbf{52.52 \pm 0.45}$	$\mathbf{48.60 \pm 0.45}$	$\mathbf{45.70 \pm 0.43}$

Table 2. Accuracy (%) on CIFAR100. The best performance is marked in bold.

Methods	Settings				
	Noise 20%	Noise 30%	Noise 40%	Noise 50%	Noise 60%
SFedAvg	35.91 ± 0.05	34.82 ± 0.03	35.25 ± 0.05	29.27 ± 0.08	30.66 ± 0.04
SS-Fed	36.67 ± 0.01	35.44 ± 0.06	34.87 ± 0.05	29.45 ± 0.04	31.15 ± 0.06
FDSS	$\mathbf{37.32 \pm 0.01}$	$\mathbf{36.32 \pm 0.05}$	$\mathbf{37.38 \pm 0.05}$	$\mathbf{38.44 \pm 0.06}$	$\mathbf{36.35 \pm 0.08}$
FedAvg	$\mathbf{40.56 \pm 0.06}$	37.69 ± 0.21	35.52 ± 0.23	35.16 ± 0.12	31.95 ± 0.73
Oort	39.29 ± 0.23	36.97 ± 0.24	36.06 ± 0.50	33.14 ± 0.90	32.40 ± 1.43
FDSS-d	40.27 ± 0.07	$\mathbf{39.99 \pm 0.07}$	$\mathbf{38.22 \pm 0.11}$	$\mathbf{39.18 \pm 0.10}$	$\mathbf{37.47 \pm 0.31}$

4 Evaluation

4.1 Experiment Settings

Datasets and Models. We compare the performance of different methods on CIFAR10 and CIFAR100. The total number of data sources N is set to 20 and constrained cardinality c is 10. For CIFAR10, We simulate label heterogeneity by allocating data sources with images of different label distributions. The whole dataset has 60,000 images for ten classes. Every data source owns data from two classes randomly. Furthermore, we randomly choose 10 data sources as low data quality enterprises and remap their labels for some training samples [8]. The percentage of noisy data can be used to measure the data quality heterogeneity. For CIFAR100, we make different data sources owning data from the same superclass but different subclasses. We use a two-layer CNN model to recognize images with learning rates 0.02 and 0.01 for CIFAR10 and CIFAR100.

Experimental Environment and Evaluation Metrics. The experiments are conducted on five Intel(R)Xeon(R) Platinum 8269CY 3.10 GHz CPUs each with 4 cores. We use test accuracy and training time as evaluation metrics to evaluate model effectiveness and efficiency.

Baselines. We compare our proposed algorithms with the following methods: (1) FedAvg [3], the vanilla Federated Averaging algorithm; (2) SFedAvg, the static version of FedAvg. It randomly selects data sources to participant in FL at the first round; (3) SS-Fed [5,8], the static version of Shapley-based method. It selects data sources based on Shapley value at the first round; (4) Oort [2], the adaptive selection method based on multi-arm bandit.

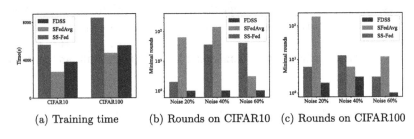

(a) Training time   (b) Rounds on CIFAR10   (c) Rounds on CIFAR100

Fig. 2. Efficiency evaluation

4.2 Experimental Results

Results on CIFAR10. The results on CIFAR10 are shown in Table 1. We evaluate our proposed FDSS and FDSS-d with baselines in five settings, which 20%, 30%, 40%, 50%, 60% of training samples are remapped respectively. The static setting results are shown on the first three rows in Table 1. The final accuracy of SFedAvg is rather unstable since it randomly selects data sources at the beginning. We can see that FDSS outperforms baselines in all scenarios. The fourth row shows the accuracy of FedAvg decreases with the increase of noisy data. Compared with FedAvg and Oort, FDSS-d outperforms in all settings and has a minimum accuracy of 45%, which is much larger than baselines.

Results on CIFAR100. The results on CIFAR100 are shown on Table 2. For the static version, FDSS performs best as on CIFAR10. The difference is that SFedavg and SS-Fed have a smaller gap with FDSS. The reason may be that each data source has data from all superclasses. Thus the impact of data quality heterogeneity dominates model accuracy. For the dynamical version, FDSS-d outperforms the others in most cases. The only exception is when applying 20% noise, but the difference between FDSS-d and the optimal result is only 0.3%. The results match the previous results and indicate the utility of the proposed algorithm in various heterogeneous scenarios.

Efficiency Evaluation. Figure 2 shows our efficiency experiments on two datasets. The dynamical algorithms can be seen as calling on static versions repeatedly, so we only show the results of the static algorithms. From Fig. 2a, SS-Fed takes the most time because $c!$ permutations need to compute. Note that the selection is conducted once in the static setting, the efficiency of SS-Fed can be lower for multiple selections. The time cost of FDSS is much less than SS-Fed and close to SFedAvg. We observe from Fig. 2b and Fig. 2c that FDSS has the smallest minimal rounds under all settings.

5 Conclusion

In this paper, we explore data source selection in FL from a submodular optimization perspective. We formalize the data source selection problem in both the static and dynamic settings and prove that the problem can be converted into a monotone submodular maximization problem. Our theoretical analysis inspires us to devise two greedy-based data source selection algorithms with a constant approximate ratio. Extensive experiments on two real datasets validate the efficiency and effectiveness of our methods.

Acknowledgments. We are grateful to anonymous reviewers for their constructive comments. This work are partially supported by the National Key Research and Development Program of China under Grant No. 2018AAA0101100, the National Science Foundation of China (NSFC) under Grant Nos. U21A20516, 61822201, U1811463 and 62076017, the State Key Laboratory of Software Development Environment Open Funding No. SKLSDE-2020ZX-07, and WeBank Scholars Program.

References

1. Huang, T., et al.: An efficiency-boosting client selection scheme for federated learning with fairness guarantee. IEEE Trans. Parallel Distrib. Syst. **32**, 1552–1564(2020)
2. Lai, F., Zhu, X., Madhyastha, H.V., Chowdhury, M.: Oort: efficient federated learning via guided participant selection. In: OSDI, pp. 19–35 (2021)
3. McMahan, B., Moore, E., Ramage, D., Hampson, S., y Arcas, B.A.: Communication-efficient learning of deep networks from decentralized data. In: AISTATS, pp. 1273–1282 (2017)
4. Minoux, M.: Accelerated greedy algorithms for maximizing submodular set functions. In: Stoer, J. (eds.) Optimization Techniques. LNCIS, vol. 7, pp. 234–243. Springer, Heidelberg (1978). https://doi.org/10.1007/BFb0006528
5. Nagalapatti, L., Narayanam, R.: Game of gradients: Mitigating irrelevant clients in federated learning. In: AAAI, pp. 9046–9054 (2021)

6. Nemhauser, G.L., Wolsey, L.A., Fisher, M.L.: An analysis of approximations for maximizing submodular set functions-I. Math. Program. **14**(1), 265–294 (1978)
7. Shi, Y., et al.: Federated topic discovery: a semantic consistent approach. IEEE Intell. Syst. **36**(5), 96–103 (2021)
8. Song, T., Tong, Y., Wei, S.: Profit allocation for federated learning. In: Big Data, pp. 2577–2586 (2019)
9. Sun, C., Shrivastava, A., Singh, S., Gupta, A.: Revisiting unreasonable effectiveness of data in deep learning era. In: ICCV, October 2017
10. Wang, Y., Tong, Y., Shi, D.: Federated latent Dirichlet allocation: a local differential privacy based framework. In: AAAI, pp. 6283–6290 (2020)
11. Wang, Y., Tong, Y., Shi, D., Xu, K.: An efficient approach for cross-silo federated learning to rank. In: ICDE, pp. 1128–1139 (2021)
12. Yagli, S., Dytso, A., Poor, H.V.: Information-theoretic bounds on the generalization error and privacy leakage in federated learning. In: SPAWC Workshop, pp. 1–5 (2020)
13. Yang, Q., Liu, Y., Chen, T., Tong, Y.: Federated machine learning: concept and applications. ACM Trans. Intell. Syst. Technol. **10**(2), 12:1–12:19 (2019)

MetisRL: A Reinforcement Learning Approach for Dynamic Routing in Data Center Networks

Yuanning Gao, Xiaofeng Gao[✉], and Guihai Chen

MoE Key Lab of Artificial Intelligence, Department of Computer Science and Engineering, Shanghai Jiao Tong University, Shanghai, China
gyuanning@sjtu.edu.cn, {gao-xf,gchen}@cs.sjtu.edu.cn

Abstract. As the cornerstone of cloud services, data centers most commonly adopt multi-rooted tree-like topologies, which provide multiple equal-cost paths but suffer from enormous bandwidth loss due to collisions and congestions. In this paper, we present **MetisRL**, a dynamic flow scheduling system combining the centralized software-defined networking (SDN) controller and reinforcement learning to balance the network traffic. From the global view of the entire network, the SDN controller in MetisRL gathers historical traffic matrices and monitors the link utilization. Then the reinforcement learning component in the SDN controller uses such information to perform dynamic prediction and output flow scheduling decisions accordingly. We evaluate MetisRL with benchmark tests and compare it with classical flow scheduling schemes, which shows that MetisRL performs well in balancing the flow in multiple paths and can efficiently avoid traffic congestions.

Keywords: Data center flow scheduling · Software defined network · Reinforcement learning · Actor-critic

1 Introduction

The rapid outgrowth of cloud service has stimulated the development of data centers, which combine tens of thousands of computing and storage machines and bring tremendous bandwidth to these clusters. Companies like Google, Amazon, and Microsoft have data centers all across the world to support their businesses. The growing demand for cloud-based applications has posed severe challenges to data center's network management and workload balancing ability [5]. Data center designers should not only strive to make full use of limited bandwidth to avoid traffic bottlenecks, but also try to reduce network collisions and congestions to the greatest extent.

This work was supported by the National Key R&D Program of China [2020 YFB1707903]; the National Natural Science Foundation of China [61872238, 61972254], Shanghai Municipal Science and Technology Major Project [2021SHZDZX0102], the Huawei Cloud [TC20201127009] and the China Scholarship Council [202006230179].

A. Bhattacharya et al. (Eds.): DASFAA 2022, LNCS 13246, pp. 615–622, 2022.
https://doi.org/10.1007/978-3-031-00126-0_44

In commercial data centers, flow scheduling is one of the most critical tasks. Since most of the data centers adopt multi-rooted tree topologies like Fat-tree [1], there are a set of paths between each source-destination pair in the network. As a result, the flow scheduling for data center networks is a multipath scheduling problem. Unfortunately, the traditional single path scheduling mechanism, which sets the secondary paths primarily for fault tolerance, performs badly when there are a few elephant flows [3] in the network and is not a good solution for multipath scheduling.

Recently, several methods have been proposed to solve the multipath scheduling problem, which can be divided into mainly two categories: *static methods and dynamic methods*. Static methods [8], which use static flow hashing to assign flows to accessible data links, are simple and easy to implement. But they are load-oblivious and will cause enormous network latency. Dynamic algorithms adjust the scheduling scheme according to the current workload. Therefore, a lot of attempts of dynamic flow scheduling are proposed to improve the overall network performance.

In this paper, we present MetisRL, a reinforcement learning-based flow scheduling system for multi-rooted tree-like data centers. MetisRL firstly collects historical traffic flow information and monitors network utilization rate by a software-defined networking (SDN) centralized controller. In order to compute the optimal routing scheme for current time t, we input the previous m historical traffic matrices into a reinforcement learning framework. A routing decisive agent as part of the SDN controller interacts with the environment to adjust the routing scheme according to the current network resources utilization rate. We conduct the comprehensive experiments based on the Openflow environment, which demonstrates the effectiveness and efficiency of MetisRL compared with other classical network routing schemes.

The remainder of this paper is organized as follows: Some related researches are presented in Sect. 2. We formulate our problem formally in Sect. 3. In Sect. 4, we display our system's overview and elaborate the implementation of reinforcement learning in routing prediction. We present the experiments and performance of our scheme in Sect. 5. Finally, we draw a conclusion in Sect. 6.

2 Related Work

In this section, we first introduce routing in data centers in Sect. 2.1. We then present reinforcement learning algorithms that used in our design in Sect. 2.2.

2.1 Routing in Data Centers

Routing is one of the fundamental tasks of designing data centers. Lots of efforts are invested in settling a myriad of problems like limited bandwidth, low utilization, and high energy cost. In order to reduce network collisions and congestions brought by static routing, many dynamic flow scheduling schemes are proposed. Hedera [2] firstly detected elephant flows, predicted future traffic matrix and then used global first fit and simulated annealing algorithm to schedule the flows dynamically. DLB [9] utilized a greedy algorithm and a deep-first strategy for dynamic flow scheduling. Besides, MicroTE [4] firstly separated the predictable and unpredictable flow and used ECMP for the unpredictable while used Bin-packing heuristic for predictable flow scheduling.

2.2 Reinforcement Learning

Reinforcement Learning (RL) [12] can make learning decisions dynamically based on the interaction between its learning agent and the environment. In every learning step, the agent observes the current state S_t and takes one action a_t from the action space $\mathbb{A}$. Note that there is a mapping π from the state space $\mathcal{S}$ to the action space $\mathbb{A}$ ($\pi : \mathcal{S} \rightarrow \mathbb{A}$). For each step it takes, the agent will get a reward r_t from the environment. The ultimate purpose of reinforcement learning is to output the optimal action, which maximizes the reward function $E[\sum_t \gamma^t r_t]$, where $\gamma > 0$ is called the discount factor. In MetisRL, we adopt a actor-critic [11] algorithm called DDPG [10], which is developed from the value-based algorithm that to supports temporal-difference update.

3 Problem Formulation

In this section, we define the input and the output for the reinforcement learning algorithm. For the input, we define a traffic matrix to describe the network communication demands. The traffic matrix $M = (m_{ij}), i, j \leq N$ is an $N \times N$ matrix; N represents the number of servers in the network. The $M[i][j]$ indicates the size of traffic flow from host i to host j. For example, consider there are 4 hosts named a, b, c, d in the network, then there are totally 4 flows in the network currently. The 0.3 in $M[1][3]$ indicates that there are 0.3G bps from host a will be sent to host c.

$$\begin{bmatrix} & a & b & c & d \\ a & \varnothing & \langle 0.3, 0.4, 0.2, 0.1 \rangle & \langle 0.3, 0.3, 0.3, 0.1 \rangle & \langle 0.4, 0.1, 0.4, 0.1 \rangle \\ b & \langle 0.5, 0.3, 0.1, 0.1 \rangle & \varnothing & \dots & \dots \\ c & \dots & \dots & \varnothing & \dots \\ d & \dots & \dots & \dots & \varnothing \end{bmatrix}$$

Fig. 1. The output of the reinforcement learning component

The output is an $N \times N$ matrix A, where $A[i][j]$ is a tuple representing the probability of forwarding network packets to different core-switches. For example, as shown in Fig. 1, the tuple $\langle 0.3, 0.4, 0.2, 0.1 \rangle$ in $A[1][2]$ represents that the flow from a to b will choose the path across switch $Core_0$ with a probability of 0.3, switch $Core_1$ with a probability of 0.4, etc. We employ the splitting scheme to implement the routing scheme, in which the network packets will be split and then forwarded among different paths. The probability from our output indicates the proportion of split packets that are transmitted through a certain path.

In the learning process, we firstly divide time into episodes with small intervals and represent the routing scheme in t as $R^{(t)}$. Based on the traffic matrix from episodes $t - m, t - k + 1, t - m + 2, ..., t - 1$ (m historical traffic matrices), we can adopt various objective functions as the reward in our reinforcement learning framework. In this way, the RL-based routing scheme is a highly flexible design.

4   System Overview and Model Design

4.1   MetisRL: Architecture

The architecture of MetisRL is shown in Fig. 2. MetisRL consists of three major components: the traffic monitoring component, the reinforcement learning component in the controller, and the routing implementation component which will allocate the routing decision to each switch. The network states are collected by the monitoring component and are used as the input for the reinforcement learning predicting component. Then the output of the reinforcement learning component will be implemented by the SDN controller.

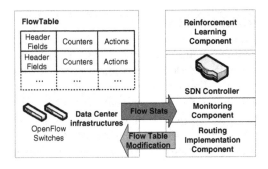

Fig. 2. Architecture of MetisRL

4.2   Reinforcement Learning Component

The reinforcement learning (RL) agent is implemented as part of the SDN controller. The RL agent consists of three parts, i.e., *States, Actions* and *Rewards*. An agent needs to interact with the environment through a sequence of states by performing actions to obtain rewards. The goal of the RL agent is to maximize its rewards by adjusting its policy based on the rewards. The SDN controller collects network information as the states in the RL model. We can elaborately design the reward function to realize different optimization objects.

States are a representation of the current environment of the task. In our model, the states are defined as a three-dimensional matrix S, which record the traffic status of the data center in the past period of time. For example, $S[m][i][j]$ means that in the mth time episode, the size of the traffic flow from host i to host j. By constructing the matrix S, the RL algorithm can be aware of the current status of the system.

Actions are some decisions an RL agent can do to change the states. As discussed in Sect. 3, our RL agent outputs a routing matrix A denoting the possible traffic flowing paths between any two hosts. Then the SDN controller will implement the optimal flow scheduling scheme according to this matrix as the corresponding actions. As for the action space, we can take full advantage of the Fat-tree topology. Fat-tree has 3 layers' structure so there is a core-path mapping in Fat-tree topology. This means that for each flow once we select a switch in the core layer to forward packets, we can define a unique path for each source-destination pair, which makes the action space simpler.

Rewards represent the feedback from the environment. In the MetisRL system, we aim to compute the scheduling scheme dynamically for each flow satisfying the bandwidth restriction and reduce the network latency via prediction. With a global view of the entire network, we are able to consider all the available paths for each flow and therefore achieve dynamic flow balancing in the whole data center. Corresponding, the optimization goal is to minimize $\max_{p \in P} \frac{\sum f(p)}{c(p)}$, where p represents the set of all network links connecting aggregation switches to core switches; $c(p)$ represents the capacity of these links and $\sum f(p)$ represents the sum of all flows running in this link currently. By minimizing the overall maximum link utilization rate, we are able to balance the data center's transmission workload, avoiding traffic collision and congestion. In addition, in order to avoid the condition that the sum of the flows in one link exceeds the capacity of this link, we set this restriction as a penalty in our reinforcement learning framework. The reward function is shown Eq. (1).

$$r = -\alpha \times \max_{p \in P} \frac{\sum f_a(p)}{c(p)} - \beta \times \frac{\sum f_b(p)}{c(p)} \tag{1}$$

where f_a means the legal flows in link p, and f_b means the flows that exceed the capacity of the link. α is a parameter used to normalize the reward function and β is a parameter used to control the penalty item. Correspondingly, the negative signs are added to Eq. (1), which aims to minimize the maximal link utilization and achieve load balance.

During the training, the monitor in the SDN controller collects information (e.g., flow size and link capacity) and calculates the reward based on Eq. (1). The reward will be fed back to the RL agent, which tries to get the higher reward by learning better actions. The purpose of RL is to incorporate downstream effects of current actions, yet future rewards may be uncorrelated. To this end, we finally introduce a discounter λ to define our optimization goal as below.

$$Q_t = E[r_t + \lambda \times r_{t+1} + \lambda^2 \times r_{t+2} + ... \lambda^h \times r_{t+h} + ...] \tag{2}$$

5   Experiment

5.1   Experiment Settings

We evaluate the performance of MetisRL on Mininet 2.2.1 and use the Ryu controller with Openflow v1.3. We implement our design on a 4-port Fat-tree data center topology, which is a representative and classical tree topology and can be easily transferred to other topologies [6]. We use DDPG [10] as our actor critic algorithm. DDPG being an actor-critic technique consists of two models: Actor and Critic. The network contains two fully-connected hidden layers with 40 and 4 neurons respectively, where 4 denotes the meaning of 4-port Fat-tree. The training episodes are set to 3000, and the activation function is the rectified linear unit (ReLU). The learning rate is set to 0.001 for the actor network, and 0.0001 for the critic network.

We compare MetisRL with **ECMP** (Equal Cost Multipath) [8] and **Global-first-fit** scheme in Hedera [2]. ECMP is one of the most famous static hash-based DCN flow scheduling schemes and Hedera is a dynamic scheduling scheme based on the heuristic algorithm. We define the network load with 20% of the maximum traffic matrix as the "light" traffic condition and network load with 80% of the maximum traffic matrix as the "heavy" traffic condition. We adopt the benchmark suite provided in [1], which contains the following traffic patterns:

- Random: A network host sends to any other host with uniform probability.
- Stride (i): A network host x only send to the host $(x + i)$ *mod* 16. In our experiment, we set stride i as $1, 2, 4, 8$

Table 1. Maximum link utilization under light traffic (20%)

Method	ECMP	Global-first	MetisRL
Random	25.4%	21.8%	19.4%
Stride (1)	20.6%	20.0%	18.1%
Stride (2)	22.1%	20.9%	18.4%
Stride (4)	19.7%	20.8%	16.9%
Stride (8)	20.1%	20.8%	17.0%

5.2   Experiment Results

Table 1 and Table 2 exhibits the maximum aggregation link utilization under light and heavy traffic. Maximum aggregation link utilization rate indicates the degree of traffic balance in the network. A flow scheduling scheme performs the best in balancing the load when the maximum aggregation link utilization rate is the minimum. As we can see, MetisRL performs best under all traffic conditions compared to other routing algorithms. The reason is that we flexibly inject the penalty attribute into the reward function, which can guarantee

Table 2. Maximum link utilization under heavy traffic (80%)

Method	ECMP	Global-first	MetisRL
Random	87.4%	83.8%	77.9%
Stride (1)	79.8%	79.3%	74.4%
Stride (2)	79.9%	80.1%	75.1%
Stride (4)	81.2%	80.8%	77.2%
Stride (8)	79.2%	79.8%	75.9%

minimal traffic congestion and collision. Equipped with the online reinforcement learning strategy, MetisRL obtains the highly dynamical routing ability by constantly interacting with the environment, which can avoid the NP-hard batched flows scheduling problem and improve the routing effectiveness. Besides, ECMP only selects a core switch randomly and ignores the network load balance. The Global-first-fit algorithm does not perform well in balancing the network load either since it lacks a global monitor of the entire network. No matter what traffic pattern we generate, MetisRL always performs the best among the three scheduling schemes.

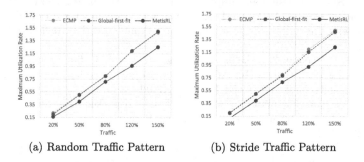

(a) Random Traffic Pattern   (b) Stride Traffic Pattern

Fig. 3. Maximum link utilization rate under different traffic load

Figure 3(a) and Fig. 3(b) reflect how the maximum link utilization grows as the traffic load grows. In both Random and Stride traffic pattern, the maximum link utilization rate grows with the network load linearly. As the figures mirror intuitively, MetisRL performs better than the other two flow scheduling types. Actually, one of the benefits of MetisRL is its flexible reward function. We can add new rewards to realized other network optimization goals (e.g. In Elastictree [7] which aims to economize and save energy, we can add network resource utilization in the reward function.). Besides, any network restrictions can be satisfied with a penalty in the reward function. This means we do not need to solve the NP-hard problems. Also, the neural network we used in the reinforcement learning process can also provide high scalability in the routing learning process.

6 Conclusion

In this paper, we present a flow scheduling approach combining reinforcement learning and SDN centralized control to achieve better load balance in the data center network. The data-driven method allows MetisRL to do traffic matrices predictions without knowing the current state. Besides, the reward function in the reinforcement framework allows us to design tailored routing schemes for various network optimization purposes. Different from other hashing-based static routing schemes and dynamic routing schemes using heuristic algorithms, MetisRL achieves the network optimization in a global view of the entire network. It not only reduces network collisions and congestions but also makes routing predictions and installs the flow table beforehand to reduce network latency.

References

1. Al-Fares, M., Loukissas, A., Vahdat, A.: A scalable, commodity data center network architecture. In: ACM SIGCOMM Computer Communication Review, pp. 63–74 (2008)
2. Al-Fares, M., Radhakrishnan, S., Raghavan, B., Huang, N., Vahdat, A.: Hedera: dynamic flow scheduling for data center networks. In: Usenix Symposium on Networked Systems Design and Implementation (NSDI), pp. 281–296 (2010)
3. Benson, T., Akella, A., Maltz, D.A.: Network traffic characteristics of data centers in the wild. In: ACM SIGCOMM Computer Communication Review, pp. 267–280 (2010)
4. Benson, T., Anand, A., Akella, A., Zhang, M.: MicroTE: fine grained traffic engineering for data centers. In: ACM Conference on Emerging Networking Experiments and Technologies (CoNEXT), p. 8 (2011)
5. Chen, T., Gao, X., Liao, T., Chen, G.: Pache: a packet management scheme of cache in data center networks. IEEE Trans. Parallel Distrib. Syst. (TPDS) **31**(2), 253–265 (2019)
6. Gao, X., Gao, Y., Zhu, Y., Chen, G.: U2-tree: a universal two-layer distributed indexing scheme for cloud storage system. IEEE/ACM Trans. Netw. (TON) **27**(1), 201–213 (2019)
7. Heller, B., et al.: Elastictree: saving energy in data center networks. In: Usenix Symposium on Networked Systems Design and Implementation (NSDI), vol. 10, pp. 249–264 (2010)
8. Hopps, C.: Analysis of an Equal-Cost Multi-Path Algorithm. RFC Editor (2000)
9. Li, Y., Pan, D.: Openflow based load balancing for fat-tree networks with multipath support. In: IEEE International Conference on Communications (ICC), pp. 1–5 (2013)
10. Lillicrap, T.P., et al.: Continuous control with deep reinforcement learning. arXiv preprint arXiv:1509.02971 (2015)
11. Peters, J., Schaal, S.: Natural actor-critic. Neurocomputing **71**(7–9), 1180–1190 (2008)
12. Sutton, R.S., Barto, A.G.: Reinforcement Learning: An Introduction, vol. 1. MIT Press Cambridge, Cambridge (1998)

CLZT: A Contrastive Learning Based Framework for Zero-Shot Text Classification

Kun Li[1,2], Meng Lin[1,2(✉)], Songlin Hu[1,2], and Ruixuan Li[2]

[1] School of Cyber Security, University of Chinese Academy of Sciences, Beijing, China
{likun2,linmeng,husonglin}@iie.ac.cn
[2] Institute of Information Engineering, Chinese Academy of Sciences, Beijing, China
liruixuan@iie.ac.cn

Abstract. Zero-shot text classification aims to predict classes which never been seen in training stage. The lack of annotated data and huge semantic gap between seen and unseen classes make this task extremely hard. Most of existing methods employ binary classifier-based framework, and regard it as a relatedness (yes/no) prediction problem between instances and every candidate class. However, these methods only consider the similarities between one instance and one class at a time, and ignore semantic relations between candidate classes. To alleviate this problem, we propose a novel **C**ontrastive **L**earning based **Z**ero-shot **T**ext classification framework (CLZT). With the contrastive optimized objects, we can capture the semantic relations between classes that need to be predicted and build more discriminative embeddings. Main experiment shows that our method achieves the best overall f1 score compared with baselines in three different datasets.

Keywords: Text classification · Zero-shot · Contrastive learning

1 Introduction

Traditional supervised classification task has been extensively studied and achieved great success due to the development of deep learning and rich data resources. However, supervised classification model requires massive human annotation and can only be used to predict the classes that have been seen in training stage [16]. Due to the high cost of data annotation and the dynamic change of classes, Zero-shot learning has recently attracted much attention from researchers. Zero-shot learning, known as **"train once, test anywhere"**, is trained on some labeled data belong to seen classes, and use transferable general knowledge to predict instances of other classes which have never been seen in training [9].

Most of existing zero-shot text classification methods adopt binary classifier-based framework, rather than traditional multi-class classification task [9,16,17].

Due to lacking of unseen data and their prototypes, models trained to find relatedness (yes/no) between one instance and one class independently. These methods utilized pre-trained language model through fine-tuning and achieve acceptable results. Since each class is treated independently in the predicting period, the neglect of feature relations between classes would lead to poor performance on detecting unseen data. For example, there are semantic relations between the classes "violence" and "crime", so distinguishing them requires that the model is able to compare all candidate classes, rather than predicting each one dependently.

Another typical framework, called projection-based method, has been widely employed in zero-shot image classification scenarios. Its basic idea is to projecting features of instances(texts, images) and all candidate classes prototypes (names or descriptions) into a common semantic space, then compare the distance between instances and each class separately [3,11,18]. However, In zero-shot text classification area, embeddings of instances and classes derived from pre-trained language model always **collapse** [1], which means the instances belonging to different classes would be mapped into very close space. This phenomenon makes it very difficult to apply directly to zero-shot text classification task.

Considering the above shortcomings, we propose a novel **C**ontrastive **L**earning based **Z**ero-shot **T**ext classification framework (**CLZT**). By optimizing the contrastive object of candidate classes, we can not only catch the semantic connections between candidate categories, but also alleviate the collapse problem derived from PLM. Specifically, all candidate classes and a instance are concatenated as one input, a BERT encoder and linear projecting layer are employed to produce embeddings which contain all candidate classes features. Then we optimize the Info Noise Contrastive Estimation(infoNCE) loss [8] and Supervised Contrastive Learning (SCL) loss [5] to build better representations. We conduct experiments on three benchmark datasets of different aspects (i.e. topic detection, emotion detection, and situation detection). Experimental results demonstrate the effectiveness of our model.

2 Related Word

Zero-shot classification drew a lot of attention in both image and text classification fields. In the early days [10] introduced a classical method called projection-based approach, which contain only two linear layers network, but the most important backbone in zero-shot image classification task. Projection-based model is kind of intuitive, but it is difficult to be directly used for zero-shot NLP tasks due to the representation collapse phenomenon of pretrained language model. [9] is the first work to regard zero-shot text classification as a relatedness (yes/no) prediction problem. This method trained three different deep neural networks to learn the matching probability between a instance and a class name. Then [16] proposed a text entailment based method and got state-of-the-arts results in three different zero-shot datasets. Recently, [7] and [15] studied the semi-supervised zero-shot setting (CTIT) which focused on how to pseudo-label the unlabeled data automatically, but they all need some external data resource.

Contrastive learning has gradually become the main domain of representing learning and has influenced many other machine learning fields. For NLP task, [13] employed multiple simple sentence-level augmentation strategies in order to learn a noise-invariant instance representation including word and span deletion, reordering, substitution. Due to the great success of pretrained language model, people also tried to improve its representation of instances through self-supervised contrastive method [14]. In addition to self-supervised scenarios, supervised contrastive learning has also been widely studied. [5] proposed a supervised contrastive learning (SCL) loss for the fine-tuning stage of pretraining language model and obtained significant improvements on multiple datasets of the GLUE benchmark in few-shot learning settings. Basically, we are inspired by [14] and [4] who use the contrastive learning method to avoid representation collapse derived from naive BERT-based pretrained language models. In our paper, we use the Info NCE loss proposed by [8] to build more discriminative representation of instances and classes, then add the SCL loss [5] to improve representation and classification result.

3   Methodology

In this section, we first formalize the zero-shot text classification problem. The classes covered by training instances are called seen classes, denoted as $C_s = \{c_i^s | i = 1, ..., N_s\}$, and the classes covered by testing instances are called unseen classes, denoted as $C_u = \{c_i^u | i = 1, ..., N_u\}$, where $C_s \cap C_u = \emptyset$. We aim to learn a zero-shot classification model $f(x, y; \theta)$ trained on the seen class instances $D^s = \{(x_i^s, y_i^s)\}_{i=1}^N$ to predict the testing instances $D^u = \{(x_i^u)\}_{i=1}^N$ of unseen classes.

We believe that all candidate class label words need to be input into the PLM model together with the instance. In this way, all classes can be taken into account when calculating the instance representation. And the representation of any class will be affected by other candidate classes through self-attention mechanism. Specifically, we use the hidden vector corresponding to $[CLS]$ in the final layer as the aggregate representation of whole instance, and for the class embedding, naturally, we use the word embedding of class to represent. Inspired by projection based method of zero-shot image classification, we add a shared linear projection layer to project instance and class embeddings produced by BERT encoder to the same feature space $\mathbb{R}^D$. So the instance embedding $e_s \in \mathbb{R}^H$ and class embedding $e_c^i \in \mathbb{R}^H$ can be formalized as below:

$$e_s = W_{proj}^T \cdot E\left(c^1...c^n; s\right)_{[CLS]} + b, \tag{1}$$

$$e_c^i = W_{proj}^T \cdot E\left(c^1...c^n; s\right)_{[c^i]} + b, \tag{2}$$

where E is the BERT encoder, the subscript symbol $[CLS]$ and c^i means to get the corresponding final hidden vector $h \in \mathbb{R}^H$.

After obtaining the representation of all candidate classes and instances through BERT encoder and projecting layer, we adopt the info noise contrastive estimation(infoNCE) loss as the contrastive objective, which forces the model to pull in the representations of instances and corresponding classes.

$$L_{nce} = -log \frac{exp(sim(e_c^k, e_s)/\tau)}{exp(\sum_{i=0}^{N} sim(e_c^i, e_s)/\tau)}, \tag{3}$$

where sim indicates the similarity function, d_s and d_c represent the final embedding of instance and class produced by BERT encoder, τ as a hyper-parameter controls the temperature.

In addition, we use the supervised contrastive loss (SCL) to cluster instances with same class and the representation of same class in different instances, which makes them more discriminative.

$$L_{scl} = \sum_{i=1}^{M} -\frac{1}{M-1} \sum_{j=1}^{M} 1_{i \neq j} 1_{y_i = y_j} log \frac{exp(sim(e^i, e^j)/\tau)}{\sum_{k=1}^{M} 1_{i \neq k} exp(sim(e^i, e^k)/\tau)}. \tag{4}$$

e represents the instance embedding and correct class embedding corresponding, M is the batch size, and sim is cosine similarity function, we use it in our paper instead of the original dot function in [5] to avoid calculating out of bounds. And also, τ as a hyper-parameter controls the temperature.

Finally, we optimize the L_{nce} and L_{scl} at the same time as below:

$$L = \lambda \cdot L_{nce} + (1 - \lambda) \cdot L_{scl}, \tag{5}$$

where L_{scl} is the average of instance and class L_{scl}, λ controls the weight of L_{nce} and L_{scl}. Figure 1 shows an overview of our framework for zero-shot text classification.

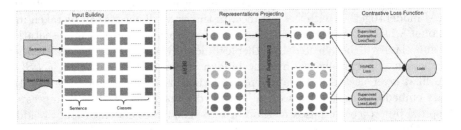

Fig. 1. Figure 1 shows the general framework of our method in training. We first combine all candidate classes and words of instance. The hidden embeddings are produced by the BERT encoder, then a shared linear projection layer project the BERT embedding to the same feature space. Finally, we can calculate the infoNCEL and SCL contrastive loss as optimized object.

4 Experiments

4.1 Datasets

We evaluate our proposed method on three standard zero-shot text classification datasets released by [16], including three different aspects: topic detection, emotion detection, and situation detection. For each dataset, there are two different separation versions have been adopted to avoid bias. The statistics of datasets are shown in Table 1.

Table 1. Statistics of three zero-shot text classification datasets, where version 1 and version 2 refer to different ways of partitions. Training data only contains seen classes and validating/testing data only contains unseen classes

	Partition version 1			Partition version 2		
	Train	Valid	Test	Train	Valid	Test
Topic	650000	30000	50000	650000	30000	50000
Emotion	20442	2457	4962	14110	4181	8802
Situation	2308	420	657	1682	709	1039

4.2 Implementation Details

In our experiment, we use the BERT-Base-Uncased [2] as our pretrained encoder. In practice, we implement the BERT encoder based on huggingface transformers module [12]. In training stage, we adopt Adam optimizer [6] and set the learning rate to be 5e−5, with $\beta 1$ 0.9,$\beta 2$ 0.999. We set the weight decay as 0.99 and use a linear learning rate warm-up over 10% of the training steps. The batch size is set to be 64 in all our experiments. The temperature τ of loss is set to be 1 for topic and situation classification task, 0.5 for emotion classification task. The loss control weight λ is set to be 0.9 for training with SCL loss, 1 for training without it in our experiment. These hyper-parameter are choiced through grid search in validation set and discussed in Sect. 4.5. Considering the different training dataset sizes, we set the training epoch size for topic, emotion, and situation to be 5, 5, 15 respectively. We set early stop criteria to avoid overfitting to seen class data.

4.3 Baseline Methods and Evaluation Metrics

We compare our model with baselines as following which can apply to same experiment setting: **Word2Vec**, **Label Similarity**, **FC**, **RNN**, **Binary-Bert** and **Entailment-Bert**. The approaches like [7] and [15], which need some unlabeled data potentially from unseen classes, will not be compared in our experiments. We use macro-F1 as evaluation metric in our experiments since datasets are not well balanced.

Table 2. Experimental results on three datasets, where v1 and v2 refer to two versions of partitions respectively. Most baselines are from [15], and we complete the Binary-Bert baseline as a supplement.

	Topic		Emotion		Situation	
	v1	v2	v1	v2	v1	v2
Word2vec	38.16	49.08	18.42	12.17	59.02	37.89
Label similarity	39.36	45.70	27.43	17.81	67.73	39.96
FC	20.93	29.29	33.76	12.98	38.47	34.15
RNN	31.09	28.63	33.05	19.47	32.98	25.61
Binary-BERT	58.29	55.86	35.30	19.16	74.57	54.83
Entail-BERT	67.73	60.20	29.31	11.96	75.08	51.48
Ours with infoNCE	68.22	63.41	43.64	19.52	77.61	**58.99**
Ours with infoNCE+SCL	**70.08**	**64.55**	**49.60**	**25.62**	**77.78**	58.95

4.4 Results and Discussion

Table 2 shows the experimental results on three zero-shot text classification datasets with two different partition versions. For baseline methods, The first two methods are totally unsupervised approaches which don't rely on any training data. As we can see, these static word embedding based methods have barely acceptable results because they can obtain some background knowledge from the pretrained datasets like WIKI. The next four methods are binary classification based models, and we can see the results of FC and RNN models are significantly worse than those of BERT-based models, we believe the reason is that these two methods have little background knowledge and the information available is barely from the training data.

For variants of our method, we can observe that both of them outperform all the other baselines. And we find that Our method with SCL loss can achieve better performance in average, especially on emotion dataset. But the performance on situation dataset is very close. We think it's because the size of the situation dataset is too small and the number of instances of each class is not uniformly distributed. For example, in situation v1 dataset, there are 798 "crime" instances and only 60 "regime change" instances. So on average, there are 20 "crime" instance and only 3 "regime change" instances in each batch, which is likely to produce unbalanced SCL loss in a batch. Moreover, all baselines and our methods work poorly in emotion detection task (especially on version 2). We suspect that the emotion classes are more semantically abstract, such as "love" "surprise", which makes the transfer of knowledge more difficult.

4.5 Visualized Analysis

One of the main purpose of our approach is to build better representations of instances and classes. In Fig. 2, we show the t-SNE plots of the learned

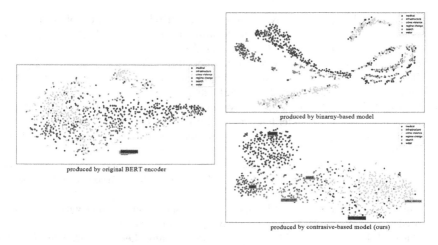

Fig. 2. Figure shows the representations of instance and class produced by original BERT model without training, binary BERT based model and ours contrasive learning based model.

embedding on situation **test** dataset. For instance points we directly use the embedded [CLS] vector. For class points, there is different embedding in each testing instance, so we use the average embedding which is correctly predicted. We compared the embedding produced by BERT without training, Binary-BERT model and our model. For raw BERT without training, instance and class embeddings are collapse, the embedding distribution is close to random. Binary-BERT model can distinguish some instances significantly, but still confuse on the whole. In our model, we can clearly see the contrastive object enforces more compact clustering of instances with same class. The class embedding is also very close to the cluster centers of their respective instances.

5 Conclusion

In this paper, we propose a contrastive-based model for zero-shot text classification. Experimental results show that our method achieves the best overall accuracy compared with baselines and recent approaches in the same zero-shot classification experiment setting. The experiment also shows that our model can capture the semantic relations between classes and build more discriminative representations of instances and classes by optimizing contrastive objects. In the future, we plan to build a two-phrase frame to resolve generalize setting, and also study how the latest contrastive method can be used in the zero-shot scenario.

References

1. Chen, X., He, K.: Exploring simple Siamese representation learning. In: Proceedings of the IEEE/CVF Conference on Computer Vision and Pattern Recognition, pp. 15750–15758 (2021)

2. Devlin, J., Chang, M., Lee, K., Toutanova, K.: BERT: pre-training of deep bidirectional transformers for language understanding. In: NAACL-HLT (1), pp. 4171–4186. Association for Computational Linguistics (2019)
3. Frome, A., et al.: DeViSE: a deep visual-semantic embedding model (2013)
4. Gao, T., Yao, X., Chen, D.: SimCSE: simple contrastive learning of sentence embeddings. arXiv preprint arXiv:2104.08821 (2021)
5. Gunel, B., Du, J., Conneau, A., Stoyanov, V.: Supervised contrastive learning for pre-trained language model fine-tuning. In: ICLR. OpenReview.net (2021)
6. Kingma, D.P., Ba, J.: Adam: a method for stochastic optimization. In: ICLR (Poster) (2015)
7. Meng, Y., et al.: Text classification using label names only: a language model self-training approach. In: EMNLP, vol. 1, pp. 9006–9017. Association for Computational Linguistics (2020)
8. van den Oord, A., Li, Y., Vinyals, O.: Representation learning with contrastive predictive coding. arXiv preprint arXiv:1807.03748 (2018)
9. Pushp, P.K., Srivastava, M.M.: Train once, test anywhere: zero-shot learning for text classification. arXiv preprint arXiv:1712.05972 (2017)
10. Romera-Paredes, B., Torr, P.: An embarrassingly simple approach to zero-shot learning. In: International Conference on Machine Learning, pp. 2152–2161. PMLR (2015)
11. Shigeto, Y., Suzuki, I., Hara, K., Shimbo, M., Matsumoto, Y.: Ridge regression, hubness, and zero-shot learning. In: Appice, A., Rodrigues, P.P., Santos Costa, V., Soares, C., Gama, J., Jorge, A. (eds.) ECML PKDD 2015. LNCS (LNAI), vol. 9284, pp. 135–151. Springer, Cham (2015). https://doi.org/10.1007/978-3-319-23528-8_9
12. Wolf, T., et al.: Transformers: state-of-the-art natural language processing. In: EMNLP (Demos), pp. 38–45. Association for Computational Linguistics (2020)
13. Wu, Z., Wang, S., Gu, J., Khabsa, M., Sun, F., Ma, H.: Clear: contrastive learning for sentence representation. arXiv preprint arXiv:2012.15466 (2020)
14. Yan, Y., Li, R., Wang, S., Zhang, F., Wu, W., Xu, W.: ConSERT: a contrastive framework for self-supervised sentence representation transfer. In: ACL/IJCNLP, vol. 1, pp. 5065–5075. Association for Computational Linguistics (2021)
15. Ye, Z., et al.: Zero-shot text classification via reinforced self-training. In: Proceedings of the 58th Annual Meeting of the Association for Computational Linguistics, pp. 3014–3024 (2020)
16. Yin, W., Hay, J., Roth, D.: Benchmarking zero-shot text classification: datasets, evaluation and entailment approach. In: EMNLP/IJCNLP, vol. 1, pp. 3912–3921. Association for Computational Linguistics (2019)
17. Zhang, J., Lertvittayakumjorn, P., Guo, Y.: Integrating semantic knowledge to tackle zero-shot text classification. In: NAACL-HLT, vol. 1, pp. 1031–1040. Association for Computational Linguistics (2019)
18. Zhang, L., Xiang, T., Gong, S.: Learning a deep embedding model for zero-shot learning. In: Proceedings of the IEEE Conference on Computer Vision and Pattern Recognition, pp. 2021–2030 (2017)

InDISP: An Interpretable Model for Dynamic Illness Severity Prediction

Xinyu Ma[1]([✉])[iD], Meng Wang[1][iD], Xing Liu[2][iD], Yifan Yang[3][iD],
Yefeng Zheng[4][iD], and Sen Wang[5][iD]

[1] Southeast University, Nanjing, China
{xinyu_ma,meng.wang}@seu.edu.cn
[2] Department of Anesthesiology, Third Xiangya Hospital,
Central South University, Changsha, China
xingxingmail@csu.edu.cn
[3] Tencent, Shenzhen, China
tobyfyang@tencent.com
[4] Tencent Jarvis Lab, Shenzhen, China
yefengzheng@tencent.com
[5] University of Queensland, Brisbane, Australia
sen.wang@uq.edu.au

Abstract. At present, there are a large number of methods for predicting the dynamic illness severity. However, these methods have two limitations: (1) they are not sensitive enough to abrupt changes in illness severity; (2) they are not comprehensive enough to explain the predicted results. To tackle these two challenges, we propose a novel Interpretable model for Dynamic Illness Severity Prediction (InDISP), effectively combining the patient status, structured medical knowledge, and drug usage to predict the trend of Sequential Organ Failure Assessment (SOFA) scores using a temporal convolutional network. The capture of drug usage events makes InDISP get better results with an abrupt change of the SOFA score. In addition, InDISP can explain the influence of features on the prediction results and explore the internal mechanism of their combined influence.

Keywords: Illness severity prediction · Interpretability · Knowledge graph · Deep learning model

1 Introduction

With the continuous development of database technology, hospitals gradually collect and store a large number of Electronic Medical Records (EMR). How to mine the knowledge from the data has gradually attracted the attention of researchers. Knowledge discovery and machine learning methods can be used not only to discover new patterns in patient data but also for classification and prediction purposes, such as the outcome or risk assessment [9]. For the caregivers in the Intensive Care Unit (ICU), real-time illness severity, which is the key to saving the lives of

A. Bhattacharya et al. (Eds.): DASFAA 2022, LNCS 13246, pp. 631–638, 2022.
https://doi.org/10.1007/978-3-031-00126-0_46

patients, is an important concern. It will be a great contribution to clinical practice if we can learn rich EMR information and provide strong support for ICU clinical decision-making. Since the early 1990s, the Sequential Organ Failure Assessment (SOFA) score [16] has been included in all aspects of critical care, and has been widely applied to the daily monitoring of acute incidence rates in critical care wards. The higher the score, the higher the severity of the illness. In the task of illness severity prediction, if the SOFA score trend of ICU patients can be dynamically predicted, it can help clinicians to better deal with the conditions of patients and make more appropriate clinical decisions.

At present, a large number of illness severity prediction models have been proposed based on EMR data mining. Adibi et al. [1] develop and validate a generalizable model to predict the individualized rate and severity of chronic obstructive pulmonary disease exacerbation. Chen et al. [4] propose an Attended Multi-Task Recurrence Neural Networks (AMRNN) for dynamic illness severity prediction. However, although these existing methods perform well in the task of illness severity prediction, there are still two challenges that remained to achieve good clinical practice: (1) evidence from a series of observational studies shows that even slight changes in SOFA scores are potentially related to significant changes in mortality prognosis [12]. However, existing methods show weaknesses in capturing unexpected SOFA. As shown in Fig. 1, the outputs of the Long Short-Term Memory (LSTM) model [8] differ a lot when predicting SOFA scores of two different ICU records. For ICU record 1, due to the stable trend of the SOFA scores, the model has a good output. However, for ICU record 2, it is obvious that the model reports a defect result to the point with an abrupt change of the SOFA scores, and (2) for clinical decision-makers in health care, the interpretability of model predictions is a priority for implementation and utilization [2]. However, most of the existing illness severity prediction methods only focus on the accuracy of prediction but do not give an explanation for prediction. Recently, several methods provide the interpretability with the attention mechanism of the model, but cannot leverage the medical domain knowledge to explain the predicted results. The lack of interpretability makes it difficult for doctors to accept the prediction results from the model.

To address the above two challenges, we propose a novel Interpretable model for Dynamic Illness Severity Prediction, InDISP for short. InDISP fuses patient status, drug usage, medical domain knowledge, and previous SOFA scores into a deep prediction model to achieve SOFA abrupt change prediction and provides better interpretability for each SOFA trend prediction result. Specifically, we first extract patient status and drug usage from EMRs. Secondly, we link drug usage to an existing knowledge graph and construct a drug-related subgraph, which will help to better incorporate the medical domain knowledge of the drug usage. Finally, we encode the drug-related subgraph, patient status, and SOFA scores to a Temporal Collaborative Network (TCN) [13]. For the predicted results, InDISP highlights the key patient status and drugs that contribute to SOFA score changes. Moreover, the paths between key drugs and SOFA scores in a drug-related subgraph are provided as evidence to explain why they cause the SOFA trend changes. We will release both code and testing datasets on Github upon acceptance.

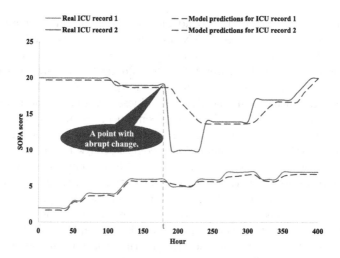

Fig. 1. Prediction results of an existing LSTM model under different SOFA score trends.

In summary, our contributions are threefold:

(1) We propose a novel model InDISP for the task of SOFA score trend prediction. To the best of our knowledge, InDISP is the first to fuse patient status, drug usage, medical domain knowledge, as well as SOFA scores, to make illness severity prediction. InDISP is able to capture the point with abrupt change of the SOFA scores and achieve more accurate prediction results.

(2) Our proposed model provides interpretable SOFA score trends by visualizing the key impacts of patient status and drug usage, and reveals the paths between drugs and SOFA scores in the medical knowledge graph. In this way, we explain how patient status and drug usage affect SOFA score changes respectively.

The remaining sections are arranged as follows. Section 2 introduces the related work. The method proposed by this paper is described in Sect. 3. Section ?? introduces the experiments. In Sect. 4, we summarize and provide future research directions of this work.

2 Related Work

At present, there are a large number of works to predict the illness severity [1,3,6], but there are few methods for dynamic illness severity prediction targeting a wide range of illnesses and clinical features. Chen et al. [4] propose an Attended Multi-Task Recurrence Neural Networks (AMRNN) for dynamic illness severity prediction, which learns the clinical features of each organ system by an LSTM as a special task, and a shared LSTM task is used to further improve the performance by taking advantage of the correlation between different learning tasks. Notably, AMRNN uses the attention mechanism to provide interpretability for 41 clinical features it used. But this interpretability is not comprehensive enough. Interpretability is a dominant feature of a successful predictive model [7]. However, the current interpretability in the medical field is still limited, especially in the dynamic illness severity prediction task. The existing methods usually introduce the attention mechanism to inform doctors of the variables that have outstanding contributions to the model prediction, but fail to fuse the medical domain knowledge. Faced with the challenge, there are two inspiring works. Zhang et al. [18] introduce a clinical knowledge graph into the heart failure prediction task, which can tackle the missing-data problem in the training set and guide the prediction of the model. Deng et al. [5] introduce a concept (named event) into the stock trend prediction task. The event embedding is obtained from the common knowledge graph as a factor influencing the stock price fluctuation, and then the internal relationship between multiple events with common influence is explored. Both works suggest that it may be a good direction to guide prediction and provide interpretability by integrating knowledge graphs in medical prediction tasks.

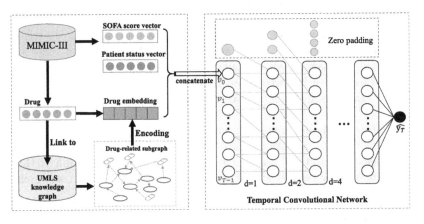

Fig. 2. Prediction results of an existing LSTM model under different SOFA score trends.

3 Methodology

The details of InDISP are presented in Fig. 2. Specifically, SOFA scores, patient status, and drug usage are first extracted from EMRs. Secondly, the drug usage is linked to an existing knowledge graph and a drug-related subgraph is constructed. Finally, the drug-related subgraph, patient status, and previous SOFA scores are encoded to a Temporal Convolutional Network (TCN) model.

3.1   Data Preprocessing

All clinical data were extracted from the Medical Information Mart for Intensive Care III (MIMIC-III) [11].

Data Cleaning. The MIMIC-III contains ICU records of multiple age groups. We mainly study the adult population, so the records of patients younger than 18 years old are ignored. Erroneous data is removed, such as ICU records of patients aged 300. Moreover, although most of the vital signs and other dynamic information representing the patient status are theoretically measured once an hour, there are still missing values in fact. We delete the ICU records without complete vital signs, including heart rate, systolic blood pressure, diastolic blood pressure, mean arterial pressure, respiratory rate, and temperature. For the drug usage sequence, the drug brands representing the same drug are combined under the guidance of professional doctors. In the remaining ICU records, we fill in the missing values.

Missing Value Filling. There are missing values in the patient status time series. Exploiting the forward-fill imputation strategy [14], we make further improvements for our data as follows: (1) In an ICU record, if a feature is vacant at a certain time, the nearest non-missing value of the feature before that time is utilized to fill in. (2) If all the features before the vacancy time are vacancies, the latest non-missing value after the vacancy time is utilized. (3) If an ICU record has not measured the feature at all times, the average value of the feature in all data is utilized.

3.2   Construction and Embedding of Drug Knowledge Graph

The purpose of drug embedding is to learn the low-dimensional dense vector representation of drugs in Knowledge Graph (KG). We link all drugs in the drug usage sequence to the medical knowledge graph and construct a drug-related subgraph G. A graph embedding model is finally implemented to train G and obtain the drug representation, which enables the model to leverage more useful information about the drug.

Entity Linking and KG Construction. The REST API of UMLS [15] is utilized to link drug names to UMLS standard terminology. Specifically, we use its endpoints to search and retrieve UMLS content. The first search result is chosen as the drug entity linking to the drug name when there are multiple search results. Based on these entities, we continue to implement the API to search their atomic information and use each atomic information to retrieve their relationship and corresponding medical entities. G is constructed by these entities and relationships, composed of triples, each of which is represented by (h, r, t), where h is the head entity, r is the relationship and t is the tail entity.

KG embedding is the process of encoding entities and relationships in G to a low-dimensional continuous vector space. For a triple (h, r, t), the entity and relation vector are first initialized randomly. Secondly, a score function $f(h, r, t)$ [17] is defined to judge the correctness. The larger $f(h, r, t)$ is, the probability of (h, r, t) being true is greater. By maximizing the score function, we can finally get the embedding e_i of the ith drug. With the drug embedding, the drug usage sequence D is translated into a drug embedding sequence D' as:

$$D = \{a'_0, a'_1, ..., a'_i, ..., a'_{T-1}\}, \tag{1}$$

$$a'_i = [d'_1, d'_2, ..., d'_j, ..., d'_{T-1}] = [e_1 d_1, e_2 d_2, ..., e_j d_j, ..., e_M d_M], d_i \in (0, 1), \tag{2}$$

where a'_i represents the drug embedding at the ith hour, e_j represents the embedding of the jth drug, and d_j represents whether the jth drug is used.

3.3 Temporal Convolutional Network

InDISP is the first model to utilize TCN, a variant of the convolutional neural network for processing sequence modeling tasks, in predicting the SOFA score trend. The distinguishing characteristics of TCNs are: (1) the convolutions in the architecture are causal, meaning that there is no information leakage from future to past, and (2) the architecture can take a sequence of any length and map it to an output sequence of the same length. Beyond this, we emphasize how to build effective history sizes using a combination of very deep networks and dilated convolutions. To grab longer dependencies, more layers are needed to stack sequentially. TCN proposes the dilated convolution, which allows input interval sampling during convolution. Specifically, for the input $\{v_0, v_1, ...v_i, ...v_{T-1}\}$ and a filter $f : \{0, 1, ..., k-1\} \rightarrow \mathbb{R}$, the dilated convolution operation F on i_{th} element of input sequence is defined as:

$$F(i) = \sum_{j=0}^{k-1} f(j) \cdot v_{i-d \cdot j}, \tag{3}$$

where d is the dilation factor, k is the filter size, and $i - d \cdot i$ accounts for the direction of the past. Dilation is thus equivalent to introducing a fixed step between every two adjacent filter taps. As shown in Fig. 2, $d = 1$ in the bottom layer indicates that every point is fed as input, $d = 2$ in the middle layer indicates that every two points are sampled as input. Using larger dilation factor enables an output at the top level to represent a wider range of inputs, thus effectively expanding the receptive field with fewer layers. Given a filter size k and dilation factor d, the effective history of one such layer is $(k-1)d$.

3.4 The Interpretability of InDISP

SOFA score trend prediction is treated as a classification task of 7 classes in InDISP, each representing a different degree of SOFA change. For patient status

S and drug usage D, two weight matrices are defined separately to capture the impact of different patient status and drug usage on the predicted class. Specifically, the patient status impact is defined as $W_S \in \mathbb{R}^{C*N}$, where C is the class number (i.e. $C = 7$), N is the number of patient status features, and θ_{ij} in W_S represents the impact of the jth patient status feature on the predicted class i. The impact of drug usage is defined as $W_D \in \mathbb{R}^{C*M}$, where M is the total number of drugs, and μ_{ij} in W_D represents the impact of the jth drug on the predicted class i. When the last class label $y_{T-1} = 2$, the patient status vector and drug embedding will be multiplied by the second line of θ and μ separately in the embedding layer. These two impact weights explain the influence of patient status and drug usage on the SOFA score trend. We highlight the patient status and drug with greater θ and μ as key patient status and key drug respectively.

Moreover, based on the key drug, we provide an evidence graph to show the paths between drugs and SOFA scores. Specifically, given a key drug as the begin node and a SOFA score node as the end node, all the paths between them in the drug-related subgraph are explored, which constitute the evidence graph.

4   Conclusions

In this paper, we proposed a novel model InDISP for the task of SOFA score trend prediction. InDISP fuses patient status, drug usage, medical domain knowledge, as well as SOFA scores, to conduct illness severity prediction. Compared to most existing models, InDISP is able to capture the abrupt change of the SOFA scores and make more accurate prediction results. Moreover, InDISP provides comprehensive interpretability by visualizing the key impact of patient status and drug usage on SOFA scores. The paths between key drugs and SOFA scores in the drug-related subgraph are provided as evidence to explain why they cause the SOFA trend changes. We identify several potential directions for this work: (1) medical domain knowledge is used in drug usage in this paper, whereas medical domain knowledge can also be introduced into the patient status in the future, (2) new MIMIC-IV data sets [10] can be used in the future, and more time-stamped information can be extracted from clinical data, such as procedures, and (3) our method can be used for drug interaction research in the future.

References

1. Adibi, A.: The Acute COPD Exacerbation Prediction Tool (ACCEPT): a modelling study. Lancet Respir. Med. **8**(10), 1013–1021 (2020)
2. Ahmad, M.A., Eckert, C., Teredesai, A.: Interpretable machine learning in healthcare. In: Proceedings of the ACM International Conference on Bioinformatics, Computational Biology, and Health Informatics, pp. 559–560 (2018)
3. Aşuroğlu, T., Oğul, H.: A deep learning approach for sepsis monitoring via severity score estimation. Comput. Meth. Programs Biomed. **198**, 105816 (2021)

4. Chen, W., Long, G., Yao, L., Sheng, Q.Z.: AMRNN: attended multi-task recurrent neural networks for dynamic illness severity prediction. World Wide Web **23**(5), 2753–2770 (2019). https://doi.org/10.1007/s11280-019-00720-x

5. Deng, S., Zhang, N., Zhang, W., Chen, J., Pan, J.Z., Chen, H.: Knowledge-driven stock trend prediction and explanation via temporal convolutional network. In: World Wide Web Conference, pp. 678–685 (2019)

6. Dervishi, A.: Fuzzy risk stratification and risk assessment model for clinical monitoring in the ICU. Comput. Biol. Med. **87**, 169–178 (2017)

7. Gilpin, L.H., Bau, D., Yuan, B.Z., Bajwa, A., Specter, M., Kagal, L.: Explaining explanations: an overview of interpretability of machine learning. In: 2018 IEEE 5th International Conference on Data Science and Advanced Analytics (DSAA), pp. 80–89. IEEE (2018)

8. Hochreiter, S., Schmidhuber, J.: Long short-term memory. Neural Comput. **9**(8), 1735–1780 (1997)

9. Jensen, P.B., Jensen, L.J., Brunak, S.: Mining electronic health records: towards better research applications and clinical care. Nat. Rev. Genet. **13**(6), 395–405 (2012)

10. Johnson, A., Bulgarelli, L., Pollard, T., Horng, S., Celi, L., Mark, R.: MIMIC-IV (version 0.4) (2020). https://doi.org/10.13026/a3wn-hq05

11. Johnson, A.E.: MIMIC-III, a freely accessible critical care database. Sci. Data **3**(1), 1–9 (2016)

12. Lambden, S., Laterre, P.F., Levy, M.M., Francois, B.: The SOFA score-development, utility and challenges of accurate assessment in clinical trials. Crit. Care **23**(1), 1–9 (2019)

13. Lea, C., Vidal, R., Reiter, A., Hager, G.D.: Temporal convolutional networks: a unified approach to action segmentation. In: Hua, G., Jégou, H. (eds.) ECCV 2016. LNCS, vol. 9915, pp. 47–54. Springer, Cham (2016). https://doi.org/10.1007/978-3-319-49409-8_7

14. Lipton, Z.C., Kale, D., Wetzel, R.: Directly modeling missing data in sequences with RNNs: improved classification of clinical time series. In: Machine Learning for Healthcare Conference, pp. 253–270. PMLR (2016)

15. McCray, A., Razi, A.: The UMLS knowledge source server. Int. J. Med. Inf. **8**, 144–147 (1995)

16. Vincent, J.L., et al.: The SOFA (Sepsis-related Organ Failure Assessment) score to describe organ dysfunction/failure (1996)

17. Wang, Q., Mao, Z., Wang, B., Guo, L.: Knowledge graph embedding: a survey of approaches and applications. IEEE Trans. Knowl. Data Eng. **29**(12), 2724–2743 (2017)

18. Zhang, X., Qian, B., Li, Y., Yin, C., Wang, X., Zheng, Q.: KnowRisk: an interpretable knowledge-guided model for disease risk prediction. In: IEEE International Conference on Data Mining, pp. 1492–1497. IEEE (2019)

Learning Evolving Concepts with Online Class Posterior Probability

Junming Shao[1,2], Kai Wang[1,2], Jianyun Lu[2,3], Zhili Qin[1,2],
Qiming Wangyang[1,2], and Qinli Yang[1,2(✉)]

[1] Yangtze Delta Region Institute (Huzhou), University of Electronic Science
and Technology of China, Chengdu, China
[2] Data Mining Lab, University of Electronic Science and Technology of China,
Chengdu, China
{junmshao,qinli.yang}@uestc.edu.cn
[3] School of Artificial Intelligence and Big Data, Chongqing College of Electronic
Engineering, Chongqing, China

Abstract. In this paper, we propose a new instance-aware concept drift detection and learning algorithm, called CPP, by dynamically capturing the evolving concepts via online Class Posterior Probability. Instead of comparing distributions of two adaptive windows of data or using prediction performance to infer concept drifts, we model and trace the concept drift by investigating the change of class posterior probability via ensemble of classifier chains directly. Building upon the intuitive concept drift modeling, CPP shows several attractive benefits: (a) It is capable of detecting and learning both gradual and abrupt concept drifts effectively, and thus supports accurate predictions. (b) CPP allows distinguishing real concept drifts from noisy instances or virtual concept drifts. (c) The time-changing concepts are captured and learned at the instance level, and this suggests that CPP lends itself to fast concept drift detection and learning. Empirical results show that our method allows effective and efficient concept drift detection and has good prediction performance compared to many baselines.

Keywords: Data stream · Concept drift · Class posterior probability · Classifier chain

1 Introduction

Learning evolving concepts on data streams has received growing attention in the data mining community, and many drift-aware adaptive learning algorithms have been proposed due to its wide emerging real-world problems such as target marketing, email filtering and network intrusion detection. Formally, concept drift is defined in online supervised learning scenarios where the conditional distribution of the target variable given the input data changes over time [6]. To identify concept drifts explicitly, there are mainly two strategies: the distribution-based detectors [3,11,16] and the error-rate based detectors [2,6,12,15].

A. Bhattacharya et al. (Eds.): DASFAA 2022, LNCS 13246, pp. 639–647, 2022.
https://doi.org/10.1007/978-3-031-00126-0_47

The distribution-based methods detect concept drifts by dynamically monitoring the change of distributions between two fixed or adaptive windows of data. Due to the evolving nature of data streams, it is often a non-trivial task to determine the appropriate window size. Moreover, window-depending approaches tend to detect and learn concept drifts in a relatively slower way. The error-based methods infer concept drifts building upon the assumption that prediction performance suffers in the presence of concept drifts. This type of methods allows handling concept drifts incrementally. However, characterizing concept drifts based on the error-rate is not always a good strategy. The reason is that the prediction performance usually not only depends on the time-changing concepts, but also heavily lies in the presence of noisy instances and the learning model itself. Another implicit technique to handle concept drifts is to focus on the recent data by using adaptive window or decay function [10]. Such forgetting mechanism is usually suitable for data streams with gradual concept drifts. In fact, for many established data stream learning algorithms, the implicit and explicit strategies are often used simultaneously. Although established approaches have already achieved some success, detecting and learning evolving concepts on data streams is still a big challenge.

In this paper, we present an intuitive way to capture the evolving concepts by dynamically investigating the change rate of class posterior probability, where a slow change of class posterior probability characterizes a gradual concept drift while an abrupt drift corresponds to a fast change of the class posterior probability. More precisely, the class posterior probability of each arriving example is estimated by the ensemble of classifier chains in an online setting. Relying on the change rate of class posterior probability, the gradual and sudden concept drifts are directly modeled, and more importantly, learned as soon as possible at the instance level.

2   Proposed Method

In this section, we present the CPP algorithm for concept drift detection and learning on evolving data streams.

2.1   Concept Drift Modelling with Class Posterior Probability

Concept drift refers to the change of the conditional distribution of the target variable given the input data over time [6]. Namely, a concept drift directly corresponds to a change of the conditional distribution $f(C|\mathbf{X})$. For the context of data stream learning, the target variable C is the class variable, and $f(C|\mathbf{X})$ is the online class posterior distribution, which can be estimated using Bayes' theorem. Building upon the class posterior probability, the concept drift is directly modeled. In the context of data streams, it is infeasible to keep track of all historical examples to estimate the posterior probability for each class. Therefore, to monitor the change of class posterior probability, an online estimation scheme is required. In this section, we introduce a new method to estimate the class posterior probability in the streaming environments via classifier chains. Classifier chain [14] has been originally used to characterize the label dependency in multi-label classification problem, where classifiers are linked along a chain where each classifier deals with the binary relevance problem associated with one label.

Classifier Chain. To estimate the posterior probability for each class C_i, according to the Bayes' theory, we have:

$$P(C_i|\mathbf{X}) = \frac{P(C_i)P(\mathbf{X}|C_i)}{P(\mathbf{X})} = \frac{P(C_i)P(\mathbf{X}^1,\cdots,\mathbf{X}^d|C_i)}{P(\mathbf{X})} \tag{1}$$

$$= \frac{P(C_i)\prod_{j=1}^d P(\mathbf{X}^j|\bigcap_{k=1}^{j-1}\mathbf{X}^k, C_i)}{\sum_{i=1}^l P(C_i)\prod_{j=1}^d P(\mathbf{X}^j|\bigcap_{k=1}^{j-1}\mathbf{X}^k, C_i)} \tag{2}$$

Now, given an infinite stream of data, we allow estimating the class posterior probability of each incoming example by employing classifier chains. The basic idea is to factor the conditional probability $P(\mathbf{X}^1,\cdots,\mathbf{X}^d|C_i)$ as d smaller parts (i.e. $\prod_{j=1}^d P(\mathbf{X}^j|\mathbf{X}^{j-1},\mathbf{X}^{j-2},\cdots,C_i)$), and each part is modeled as a classifier (i.e. $f_1,\cdots,f_d$). With the strategy, the dependencies among features are modeled. Namely, each classifier is used to compute the corresponding probability (i.e. $f_1(\mathbf{X}^1,C_i)$ for $P(\mathbf{X}^1|C_i)$, $f_2(\mathbf{X}^1,\mathbf{X}^2,C_i)$ for $P(\mathbf{X}^2|\mathbf{X}^1,C_i)$, and $f_j(\mathbf{X}^1,\cdots,\mathbf{X}^j,C_i)$ for $P(\mathbf{X}^j|\mathbf{X}^{j-1},\cdots,\mathbf{X}^1,C_i)$). Here, we use the Naïve Bayes classifier to compute the corresponding probability due to its simplicity and effectiveness.

Ensemble of Classifier Chains. In previous section, conditional probability is estimated by a single classifier chain. However, it only provides one way to compute the conditional probability. With different order of features, the result may vary (i.e. $P(\mathbf{X}^1,\cdots,\mathbf{X}^d|C_i)=\prod_{j=1}^d P(\mathbf{X}^{g(j)}|\bigcap_{k=1}^{j-1}\mathbf{X}^{g(k)}, C_i)$, where $g:\{1,\cdots,d\} \rightarrow \{1,\cdots,d\}$ is a bijective mapping). Hence, to increase robustness, a series of classifier chains need to be generated. The class posterior probability is estimated by each classifier chains separately, and all estimations are combined to yield the final result. Many studies have demonstrated that ensembles of a few ensemble members are sufficient to obtain a solution sufficiently close to the optimum [5,7]. Here, five classifier chains are used in this study. Moreover, to well ensemble a set of classifier chains, one intuitive way is to weight the classifier chains according to their performance instead of averaging.

2.2   Instance-Sensitive Concept Drift Detection and Learning

Gradual Concept Drift. As gradual drift corresponds to a slow change of class posterior probability, the key point is to adaptively capture the dynamic change. Instead of using traditional window-style investigation, we introduce the instance-aware concept drift detection approach. The key idea is to monitor the subtle change of class posterior probability for each arriving example dynamically.

Formally, given an instance X_j which belongs to class C_i, let $P(C_i|X_j)$ represent its estimated class posterior probability, and μ_{C_i} be the mean posterior probability of instances of class C_i. To characterize the change of posterior probability, we first define the **representative mean class posterior probability** $\mu_{C_i}^R$ for the class C_i as:

$$\mu_{C_i}^R = \frac{1}{m_i} \sum_{j=1}^{n_i} P(C_i|X_j) \cdot \delta(X_j), \tag{3}$$

where n_i is the number of instances of class C_i, and m_i is the number of corresponding instances which are correctly classified based on the our estimator (class posterior). $\delta(X_j)$ equals 1 if the label is correctly predicted, and 0 otherwise.

Here the representative average class posterior probability $\mu_{C_i}^R$ captures the concept which current estimator allows modeling (correct predictions), while μ_{C_i} represents the overall performance of the class C_i. Therefore, to present, the overall change of class posterior probabilities of instances caused by concept drift or noise is the difference ω between $\mu_{C_i}^R$ and μ_{C_i}.

$$\omega_{C_i} = \frac{\mu_{C_i}^R - \mu_{C_i}}{\mu_{C_i}^R} \tag{4}$$

Afterwards, the change class posterior probability $\Delta\omega_{C_i}(j)$ resulting from the instance X_j can be defined as:

$$\Delta\omega_{C_i}(j) = \omega_{C_i}(j) - \omega_{C_i}(j-1) \tag{5}$$

Building upon the instance-aware variation, the change rate of posterior probability for a given class C_i is heuristically estimated as:

$$\beta_{C_i}(j) = \beta_{C_i}(j-1) + \Delta\omega_{C_i}(j) \tag{6}$$

where $\beta_{C_i}(1)$ is 0 at the beginning, indicating there is no concept drift happened. Finally, the overall change rate of posterior probability resulting from all classes are computed as:

$$\beta(j) = \sum_{i=1}^{k} \frac{1}{\alpha_i} \beta_{C_i}(j) \tag{7}$$

where α_i is the prior probability for each class C_i.

Moreover, as the speed of concept drift corresponds to the change rate of class posterior probability, the prediction model (i.e. online class posterior probability estimator) needs to be adaptively updated to learn recent relevant instances (i.e. instances really characterizing current concept). Thus, we dynamically update the online class posterior probability estimator using a weighting scheme as follows.

$$w(j) = e^{\beta(j)} \cdot w(j-1) \tag{8}$$

Abrupt Concept Drift. To distinguish noisy instances from real abrupt concept drifts, an early-warning-late-confirmation mechanism is introduced. First, we mark an instance as a potential drifting point if its class posterior probability is significantly differ from its representative mean class posterior probability. Let X_j be an instance belongs to class C_i. It is viewed as a potential drifting point if

$$P(X_j) < \mu_{C_i}^R - k \cdot \sigma_{C_i} \qquad (9)$$

where σ_{C_i} is the standard deviation of the corresponding class posterior probabilities. k is a constant, which is used to control the sensitivity to potential drifting points. In this study, $k = 2$ is used. The larger of the k, less points will be marked as potential drifting points.

Once a potential drifting point is marked, we continuously monitor the following incoming data. Meanwhile, a tentative estimator (i.e. a new ensemble of classifier chains) is used to learn the incoming instances to compute their online posterior probabilities. If the estimates between two estimators (current estimator and tentative estimator) are different significantly, the abrupt drift is confirmed, and the tentative estimator replaces the current estimator. To characterize the difference of the two estimators on the data back to the first potential drifting point, we apply the *chi-square test* with Yates's correction [18].

3 Experimental Evaluation

3.1 Experiment Setup

Data Sets: For synthetic data streams with abrupt drifts, we use the well-known *Sine* [6], which generates the data with class boundary $y = sin(x)$. Moreover, we also evaluate our algorithm on five real-world data, which include *Nebraska Weather, Shuttle, Electricity, US Census Income*, and *Sensor*.

Selection of Comparison Methods. We compare CPP to the most widely used approaches, which include the distribution-based methods: ADWIN [3], PL [1], and STEPD [12], and the error-rate based methods: PHT [13], DDM [6], EDDM [2] and ECDD [15]. To further evaluate the benefit of CPP for data stream classification, we compare it to: *SingleClassifier* with DDM and EDDM, respectively, Naïve Bayes with PairedLearner (PL), HoeffdingAdaptiveTree [9] with ADWIN; *Weighted Ensemble* [17], *OzaBagAdwin* [4], and PASC [8].

3.2 Proof of Concept

We start the evaluation with synthetic data streams to demonstrate some properties of CPP.

Concept Drift Detection: To investigate the performance of sudden concept drift detection, we perform the experiments on the *Sine*-family data sets. Here we generate 10 synthetic data streams, where each data stream consists of 100,000 instances with **NINE** sudden drifts (i.e. abrupt concept drift occurs per 10,000

Table 1. The abrupt concept drift detection performance of different algorithms.

	FA	MD	WD	CD	DD
CPP	**0.037**	**0.000**	**0.000**	17.170	**9.333**
ADWIN	0.062	0.011	16.178	16.178	9.533
PL	0.931	**0.000**	8.670	8.670	130.800
PHT	0.988	0.985	255.281	309.622	10.567
STEPD	0.155	**0.000**	6.851	**8.115**	10.793
DDM	0.805	0.637	23.211	56.319	17.467
EDDM	0.930	0.507	28.552	85.148	64.267
ECDD	0.897	0.007	5.581	25.696	86.867

Table 2. The performance of different concept drift detection algorithms on evolving streams with different noise levels.

	Noise level (1%)					Noise level (10%)				
	FA	MD	WD	CD	DD	FA	MD	WD	CD	DD
CPP	**0.01**	**0.00**	**0.00**	18.3	**9.07**	**0.00**	**0.00**	**0.90**	20.9	**9.03**
ADWIN	0.06	0.02	16.3	16.3	9.43	0.60	0.59	25.5	25.5	9.13
PL	0.94	**0.00**	9.03	**9.03**	139.9	0.97	0.01	10.89	**10.9**	350.4
PHT	1.00	1.00	263.3	317.7	10.6	1.00	1.00	309.4	368.7	11.3
STEPD	0.14	0.00	7.03	8.36	10.6	0.04	0.00	9.98	12.49	9.45
DDM	0.83	0.66	25.1	62.5	17.9	0.91	0.84	39.4	98.3	16.0
EDDM	0.93	0.48	31.3	92.1	68.5	0.95	0.70	95.7	278.6	58.4
ECDD	0.90	0.03	14.6	72.9	87.6	0.90	0.01	9.67	26.0	88.9

instances). All algorithms work on the ten data streams and the results are averaged. Table 1 summaries the performances of the eight distinct concept drift detection algorithms in terms of the five measures (FA, MD, WD, CD and DD). From the table, we can observe that CPP outperforms comparing algorithms in terms of all measures expect for confirmation delay (CD). The confirmation delay of CPP is (sub)optimal time to exclude the noise effect.

Noise Handling: To assess the noise effect on different concept drift detection algorithms, like experiments for abrupt drift detection, we generate the *Sine*-family data streams with different noise levels. Table 2 gives the results with different levels of noise, including 1% and 10%. We can observe that CPP allows yielding good results in terms of the false alarm rate, the missing detection rate and the number of detected drifts compared to other algorithms.

3.3 Prediction Performance Analysis

To systematically demonstrate the benefit of CPP for data stream classification, we further compare it to seven representatives of data stream learning paradigms: SingleClassifier with DDM or EDDM, HoeffdingAdaptiveTree, OzaBagAdwin, WeightedEnsemble, PASC, and Naïve Bayes with Paired Learner, on the mentioned five real data streams. Table 3 summaries the prediction performance of different data stream classification algorithms in terms of different evaluation measures. For the five real data streams, CPP yields good results, and outperforms other algorithms in terms of accuracy and F1-measure.

Table 3. The performance of different data stream classification algorithms on real-world data sets.

Data	#Obj	#Dim	Methods	Acc.	Prec.	Rec.	F_1	Time (ms)
Weather	18,159	8	CPP	**0.757**	0.716	**0.708**	**0.712**	747
			DDM	0.707	0.675	0.694	0.684	643
			EDDM	0.730	0.685	0.679	0.682	**250**
			PL	0.731	0.686	0.677	0.681	1030
			HoeffAdaTree	0.735	0.691	0.684	0.687	742
			OzaBagAdwin	0.750	**0.719**	0.657	0.686	1077
			WEnsemble	0.703	0.672	0.691	0.681	2185
			PASC	0.705	0.664	0.672	0.668	1221
Shuttle	43,500	9	CPP	**0.993**	**0.656**	0.441	**0.527**	9731
			DDM	0.925	0.332	**0.463**	0.387	938
			EDDM	0.919	0.334	0.445	0.381	1182
			PL	0.929	0.337	0.341	0.339	922
			HoeffAdaTree	0.970	0.389	0.430	0.408	**731**
			OzaBagAdwin	0.976	0.410	0.407	0.408	4291
			WEnsemble	0.970	0.470	0.338	0.393	11670
			PASC	0.932	0.321	0.335	0.328	3712
Electri.	45,312		CPP	**0.882**	**0.877**	**0.877**	**0.877**	2481
			DDM	0.812	0.809	0.803	0.806	**949**
			EDDM	0.848	0.845	0.845	0.381	1037
			PL	0.871	0.868	0.869	0.869	2043
			HoeffAdaTree	0.834	0.832	0.826	0.829	2264
			OzaBagAdwin	0.843	0.845	0.833	0.839	4305
			WEnsemble	0.709	0.702	0.702	0.702	5914
			PASC	0.786	0.782	0.777	0.779	1981

(*continued*)

Table 3. (*continued*)

Data	#Obj	#Dim	Methods	Acc.	Prec.	Rec.	F_1	Time (ms)
Census	32,561	14	CPP	**0.855**	0.804	**0.788**	**0.796**	6173
			DDM	0.834	0.787	0.728	0.756	**660**
			EDDM	0.832	0.781	0.726	0.753	1227
			PL	0.762	0.667	0.648	0.657	905
			HoeffAdaTree	0.828	0.787	0.703	0.742	976
			OzaBagAdwin	0.845	**0.815**	0.728	0.769	2512
			WEnsemble	0.829	0.775	0.728	0.751	6132
			PASC	0.775	0.687	0.666	0.676	2466
Sensor	2,219,803	5	CPP	**0.875**	**0.860**	0.838	**0.849**	2102010
			DDM	0.808	0.790	0.780	0.785	**136374**
			EDDM	0.847	0.833	0.819	0.826	217781
			PL	0.375	0.422	0.398	0.410	238879
			HoeffAdaTree	0.650	0.644	0.641	0.643	316232
			OzaBagAdwin	0.873	0.850	**0.848**	**0.849**	1572264
			WEnsemble	0.675	0.792	0.681	0.732	2834150
			PASC	0.824	0.802	0.801	0.801	443028

4   Conclusion

In this paper, we introduce a new concept drift detection algorithm, CPP, to detect and learn time-changing concepts on evolving data streams. While existing approaches focus on identifying concept drift using distribution-based or prediction performance inferring methods, we use the class posterior probability to model the concept drift directly. In comprehensive experiments, we have shown that CPP outperforms state-of-the-art concept detection approaches and the associated classification algorithms.

Acknowledgments. This work is supported by the National Natural Science Foundation of China (61976044, 52079026), and Sichuan Science and Technology Program (2022YFG0260, 2020YFH0037).

References

1. Bach, S.H., Maloof, M.: Paired learners for concept drift. In: ICDM, pp. 23–32 (2008)
2. Baena-García, M., del Campo-Ávila, J., Fidalgo, R., Bifet, A., Gavaldà, R., Morales-Bueno, R.: Early drift detection method (2006)
3. Bifet, A., Gavalda, R.: Learning from time-changing data with adaptive windowing. In: SIAM SDM, vol. 7, p. 2007 (2007)

4. Bifet, A., Holmes, G., Pfahringer, B., Kirkby, R., Gavaldà, R.: New ensemble methods for evolving data streams. In: ACM SIGKDD, pp. 139–148 (2009)
5. Frank, E., Kramer, S.: Ensembles of nested dichotomies for multi-class problems. In: ICML, p. 39 (2004)
6. Gama, J., Medas, P., Castillo, G., Rodrigues, P.: Learning with drift detection. In: Bazzan, A.L.C., Labidi, S. (eds.) SBIA 2004. LNCS (LNAI), vol. 3171, pp. 286–295. Springer, Heidelberg (2004). https://doi.org/10.1007/978-3-540-28645-5_29
7. Geilke, M., Frank, E., Karwath, A., Kramer, S.: Online estimation of discrete densities. In: ICDM, pp. 191–200 (2013)
8. Hosseini, M.J., Ahmadi, Z., Beigy, H.: Using a classifier pool in accuracy based tracking of recurring concepts in data stream classification. Evol. Syst. 4(1), 43–60 (2013)
9. Hulten, G., Spencer, L., Domingos, P.: Mining time-changing data streams. In: ACM SIGKDD, pp. 97–106 (2001)
10. Koychev, I.: Tracking changing user interests through prior-learning of context. In: Adaptive Hypermedia and Adaptive Web-Based Systems, pp. 223–232 (2002)
11. Kuncheva, L.I.: Classifier ensembles for detecting concept change in streaming data: overview and perspectives. In: Proceedings of the 2nd Workshop SUEMA, vol. 2008, pp. 5–10 (2008)
12. Nishida, K., Yamauchi, K.: Detecting concept drift using statistical testing. In: Corruble, V., Takeda, M., Suzuki, E. (eds.) DS 2007. LNCS (LNAI), vol. 4755, pp. 264–269. Springer, Heidelberg (2007). https://doi.org/10.1007/978-3-540-75488-6_27
13. Page, E.: Continuous inspection schemes. Biometrika 41, 100–115 (1954)
14. Read, J., Pfahringer, B., Holmes, G., Frank, E.: Classifier chains for multi-label classification. Mach. Learn. 85(3), 333–359 (2011)
15. Ross, G.J., Adams, N.M., Tasoulis, D.K., Hand, D.J.: Exponentially weighted moving average charts for detecting concept drift. Pattern Recogn. Lett. 33(2), 191–198 (2012)
16. Vorburger, P., Bernstein, A.: Entropy-based concept shift detection. In: ICDM, pp. 1113–1118 (2006)
17. Wang, H., Fan, W., Yu, P.S., Han, J.: Mining concept-drifting data streams using ensemble classifiers. In: ACM SIGKDD, pp. 226–235 (2003)
18. Yates, F.: Contingency tables involving small numbers and the χ^2 test. Suppl. J. Roy. Stat. Soc. 1, 217–235 (1934)

Robust Dynamic Pricing in Online Markets with Reinforcement Learning

Bolei Zhang[1,2] and Fu Xiao[1]

[1] School of Computer, Nanjing University of Posts and Telecommunications, Nanjing, People's Republic of China
{bolei.zhang,xiaof}@njupt.edu.cn
[2] State Key Laboratory for Novel Software Technology, Nanjing University, Nanjing, People's Republic of China

Abstract. In online markets, reinforcement learning (RL) is a promising way for dynamic pricing, due to its ability in maximizing long-term cumulative return. However, directly optimizing RL policies in the online markets can be costly since RL requires trial-and-error with the environment, which may lead to drastic revenue loss. In this paper, we propose a robust dynamic pricing algorithm using RL. The main idea is to train the dynamic pricing policy in an adversarial simulation environment built with a generative adversarial framework. In this framework, the generator is trained to: 1) imitate real customers behaviors; 2) generate adversarial behaviors. The algorithm is proved to converge under certain assumptions. The experiment results show that our algorithm can be comparable with the algorithm directly trained in the real environment. Moreover, it outperforms other baseline significantly in different scenarios.

Keywords: Dynamic pricing · Robust reinforcement learning · GAN

1 Introduction

Dynamic pricing, which sets flexible prices for products or services at different periods, has been a common practice in a variety of commercial industries. As dynamic pricing requires making sequential decisions, reinforcement learning (RL), which aims to maximize the long-term cumulative return, has become a promising approach for optimizing the pricing policy. However, in the real-world applications such as dynamic pricing, it is still a critical challenge for RL algorithms to work: First, in model-free environments, RL uses trial-and-error in the environment to optimize the policy. While in the application of dynamic pricing, **arbitrary exploration** may lead to drastic revenue loss [2,3]; Second, when deploying the RL algorithm in real-world environments, there may be **malicious or even adversarial customers** (e.g. customers from a competing opponent) that behave against the retailer [8]. Therefore, a robust pricing policy is required to deal with the adversarial behaviors.

© The Author(s), under exclusive license to Springer Nature Switzerland AG 2022
A. Bhattacharya et al. (Eds.): DASFAA 2022, LNCS 13246, pp. 648–655, 2022.
https://doi.org/10.1007/978-3-031-00126-0_48

Considering the above challenges, in this paper, we propose a robust RL algorithm for dynamic pricing using only offline data set. The underlying intuition is that the pricing strategy can be robust to transfer to the real environment even when there are adversarial customers. We first adopt a generative adversarial framework, which generates "virtual customers" with two objectives: 1. The generated customers are indistinguishable from the real customers by the discriminator; 2. The generated customers behave adversarially against the retailer to minimize the revenue. Then, we train the dynamic pricing policy in the environment with the adversarial customers. The main contributions of this paper include:

- We model the robust pricing strategy as a sequential decision problem, where the aim is to maximize long term profit with adversarial customers.
- A novel generative adversarial framework is proposed, where the customers aim to imitate the real customers and behave against the retailer;
- Both theoretical and empirical analysis are conducted to validate the effectiveness of our algorithm.

2 Problem Formulation

In the dynamic pricing problem, suppose the time is discretized into finite horizon as $\{0, 1, ..., T\}$. We consider a simplistic but practical setting as follows:

Setting: At each step $t \in \{0, 1, ..., T\}$, the retailer first sets a price p_i^t for the provided product (or service) i. The customers as a whole would then decide the sale quantity q_i^t based the provided price. By observing the sale quantity in this step, the retailer will set the product price p_i^{t+1} for the next step. The aim of the retailer is to maximize the long-term cumulative profit of each product i.

In this setting, the decision process of both the retailer and the customers can be formulated as MDP (Markov Decision Process) quintuples. Moreover, the retailer and the customers are part of the environments of each other. For the retailer, the elements of the MDP $\langle \mathcal{S}, \mathcal{A}, \mathcal{T}, \mathcal{R}, \gamma \rangle$ can be described as follows:

- **State.** The state of the retailer consists of the product information, the action history of the retailer and customers, and the context information.
- **Action.** The action of the retailer corresponds to the price set at the given step p_i^t.
- **Reward.** The reward of the retailer can be naturally defined as the revenue at each step: $r_t = q_i^t(p_i^t - c_i)$, where c_i is the cost of the product i.

Similarly, the MDP of the customers can also be formulated as $\langle \mathcal{S}_c, \mathcal{A}_c, \mathcal{T}_c, \mathcal{R}_c, \gamma_c \rangle$:

- **State.** The state of the customers consists of four parts: The product information, the action history of the retailer, the context information, and the action of the retailer at this step: p_i^t;
- **Action.** The action of the customers corresponds to the sale quantity at the given step q_i^t.
- **Reward.** Compared to the retailer, the reward of the customers is unknown.

3   Robust Dynamic Pricing

3.1   The Framework

As formulated above, optimizing the pricing policy for the retailer requires inter-acting with the customers to get sale quantity. Since the customers are also making sequential decisions, it can be hard to simulate the sale quantity in the environment. In this section, we introduce a generative adversarial framework, in which the generator makes sequential prices and sale quantities to confuse the discriminator. The framework is illustrated in Fig. 1. As presented, the frame-work mainly consists of three components: a generator, a discriminator and a retailer network. The generator and discriminator are similar to that of GAIL [4,9], and the retailer network generates optimized pricing policy directly. Dif-ferent components of the framework are introduced as follows:

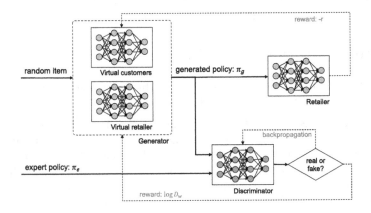

Fig. 1. The robust dynamic pricing framework.

Generator: The generator is composed of two networks: a virtual retailer net-work and a virtual customers network. The generator takes random item as input and generates price-quantity trajectories using the networks. The virtual retailer network first generates the price p_i^t, and the virtual customers network generates the sale quantity q_i^t by observing p_i^t. Both the networks use RL to optimize the parameters. Therefore, we define reward function to satisfy the fol-lowing objectives: 1) The networks can generate price-quantity trajectories that are indistinguishable from the true expert trajectories by the discriminator; 2) The virtual customers network generates adversarial behaviors that minimize the retailer's return.

Discriminator: The discriminator is a binary classifier that tries to distin-guish the expert trajectory and the generated trajectory. Similar to GAN, the discriminator can be trained using traditional supervised learning techniques by predicting whether the input is real or fake.

Retailer: The retailer network is used to optimize the dynamic pricing strategy. Meanwhile, the generator will observe the retailer's price and generate adversarial sale quantities that minimize the retailer's return.

3.2 Model Training

Suppose the **retailer policy network** π_r is parameterized by θ_r. The policy gradient of the retailer network can be written as:

$$\mathbb{E}[\nabla_{\theta_r} \log \pi_r(p|s) Q_r(s,p)] - \lambda \nabla_{\theta_r} H(\pi_r) \tag{1}$$

where $H(\pi_r)$ is $\gamma-$discounted causal entropy of the policy π_r, indicating that we prefer policies with high-entropy. The expectation $\mathbb{E}$ computes the empirical average over a finite batch of samples. The function $Q_r(s,p) = \mathbb{E}_{\pi_r}[R|s_0 = s, p_0 = p]$ represents the approximated state-action value of the generator network and can be updated by minimizing the temporal-difference error [7].

For the **discriminator**, we minimize cross entropy loss to distinguish the expert trajectories and generated trajectories. The gradient of the discriminator can be written as:

$$\mathbb{E}_{\tau_g}[\nabla_w \log(D_w(s_g, p_g, q_g))] + \mathbb{E}_{\tau_e}[\nabla_w \log(1 - D_w(s, p, q))] \tag{2}$$

where D_w represents the network of discriminator parameterized by w. τ_g and τ_e represent the state action batch samples generated by the generator π_g and the expert policy π_e respectively. s_g is a random state picked from the expert trajectories.

The **generator** can be updated as follows:

$$\mathbb{E}_{\tau_g}[\nabla_{\theta_g} \log(\pi_g(p_g, q_g|s_g)) Q_g(s_g, p_g, q_g)] - \lambda_g \nabla_{\theta_g} H(\pi_g) \tag{3}$$

where π_g is the joint policy of the virtual retailer and virtual customers, and $Q_g(s_g, p_g, q_g) = \mathbb{E}_{\tau_g}[R_g|s_0 = s_g, p_0 = p_g, q_0 = q_g]$ represents the critic network that estimates the state-action value of the generator (both the virtual customers and virtual retailers). The reward function can be written as:

$$r_g = (1 - \delta) \log D_w(s_g, p_g, q_g) - \delta r \tag{4}$$

where $\delta \in [0, 1]$ is a coefficient that controls the weight of the adversary. With a larger weight, the generated customers will focus more on being adversarial against the retailer. Otherwise, the generated customers will try to imitate the real customers' behaviors.

We present our algorithm RoboPricing (ROBust Offline Pricing) in Algorithm 1. In this algorithm, we first initialize the network parameters of different components. In the first iteration, we fix the retailer network, and update the generator and discriminator, where the reward function of the generator is simply $\log D_w(s_g, p_g, q_g)$. Then in the second iteration, by fixing the discriminator and the virtual retailer, we update the generator and retailer networks.

Algorithm 1: RoboPricing

---

1 Initialize retailer policy networks π and generator π_g;
2 Initialize discriminator as D_w;
   /* Fix retailer, update generator and discriminator   */
3 **while** *not converged* **do**
4 | Sample random state $s_g \in \tau_e$;
5 | Generated trajectories: $\tau_g \sim \pi_g(s_g)$;
6 | Sample expert trajectories: τ_e;
7 | Update the discriminator according to Eq. 2;
8 | Get the reward function as $r_g = \log D_w(s_g, p_g, q_g)$;
9 | Update the generator according to Eq. 3;
10 **end**
   /* Fix discriminator and virtual retailer, update virtual customers
     and retailer   */
11 **while** *not converged* **do**
12 | Update the retailer according to Eq. 1 with reward r;
13 | Get the reward function as $r_g = (1 - \delta) \log D_w(s, p_g, q_g) - \delta r$;
14 | Update the virtual customers according to Eq. 3 with reward r_g;
15 **end**

3.3 Convergence Analysis

The establishment of the convergence mainly depends on the corollary of Szepesvari and Littman [10], and can be proved by showing that the state-action value $Q = (Q_r, Q_{\hat{c}})$ is a fixed point with contraction mapping, where $Q_r, Q_{\hat{c}}$ are the state-action value of the retailer and virtual customers respectively. The mapping function and the convergence point can be defined as:

Definition 1: P_t is mapping function on the complete metric space $\mathbb{Q} \to \mathbb{Q}$, $P_t Q = (P_t Q_r, P_t Q_{\hat{c}})$, that satisfies

$$P_t Q_k(s, p_r, q_g) = r_k + \gamma Q_k(s', \pi_r(s'), \pi_{\hat{c}}(s')) \tag{5}$$

for $k \in \{r, \hat{c}\}$, where s' is the next state.

Definition 2: Q^* is a convergence point if it satisfies:

$$Q_k^*(s, p_r, q_g) = r_k + \gamma \sum_{s' \in S} p(s'|s, p_r, q_g) Q_k^*(s', \pi_r(s'), \pi_{\hat{c}}(s')), \text{ for } k \in \{r, \hat{c}\} \tag{6}$$

With the above definitions, we can prove that the P_t is a contraction mapping and Q^* is a convergence point:

Lemma 1: Q^* is a fixed pointed: $\mathbb{E}[P_t Q^*] = Q^*$.

Proof of Lemma 1:

$$Q_k^*(s, p_r, q_g) = r_k + \gamma \sum_{s'} p(s'|s, p_r, q_g) Q_k^*(s', \pi_r(s'), \pi_{\hat{c}}(s'))$$

$$= \mathbb{E}[P_t Q_k^*]$$

Lemma 2: P_t is a contraction mapping: $\|P_t Q - P_t Q^*\| \leq \gamma \|Q - Q^*\|$.

The proof of Lemma 2 can be directly inferred from Lemma 16 of Hu and Wellman [5], with the following assumptions:

Assumption 1: Every state $s \in S$ and action p_i, q_i are visited infinitely often.

Assumption 2: One of the following conditions holds during learning.

– **Condition A**. Every stage game $(Q_r(s), Q_{\hat{c}}(s))$ has a global optimal point, and agents' payoffs in this equilibrium are used to update their Q-functions.
– **Condition B**. Every stage game $(Q_r(s), Q_{\hat{c}}(s))$ has a saddle point, and agents' payoffs in this equilibrium are used to update their Q-functions.

Assumption 3: The critic learning rate α_t for the t-th transaction satisfies $\sum_{t=0}^{\infty} \alpha^t(s, p_t, q_t) = \infty$, $\sum_{t=0}^{\infty} [\alpha^t(s, p_t, q_t)]^2 < \infty$ hold uniformly and with probability 1.

With the above lemmas, we have the following main theorem:

Theorem 1: The sequence of $(Q_r^t, Q_{\hat{c}}^t)$ in Algorithm 1 can converge to a fixed value $Q^* = (Q_r^*, Q_{\hat{c}}^*)$ with probability 1.

4 Experiments

4.1 Experiment Setting

Dataset Description. We use a real-world data set extracted from JD,[1] where 100 items among 732 items with highest sale quantities are selected in the fridge category. At each day, we get the lowest price (if there are fluctuations in the price) and the number of sale quantities for each item. In addition, each item is associated with 15 attributes, including compressor type, volume, depth, width, size, brand, etc. The data is collected from 2017.01.01 to 2019.10.01. The cost of each item is assumed to be the historical lowest price.

We train and evaluate our algorithm in both **test (online) environment** and **simulation environment**. The test environment is built with an XGboost model which predicts the price and sale quantities. The simulation environment is built with the generator.

Baselines. We compare our algorithm with the following baselines:

– **SL**: We use supervised learning to predict the sale quantity, and find the optimal price by iterating over the model for the highest profit.
– **PPO-sim**: We train PPO in the simulation environment without adversarial customers.
– **PPO-real**: We train PPO directly in the real environment.
– **CQL**: CQL is an offline RL algorithm that is directly trained with offline data e with a simple Q-value regularizer [1].
– **BCQ**: BCQ is another offline RL algorithm which restricts the action space in order to force the agent towards behaving close to on-policy [6].

---

[1] JD is one of the largest retail company in China, www.jd.com.

4.2   Performance Analysis

Convergence of Training. First we train RoboPricing w.r.t different episode length and observe the convergence of training. During training, we evaluate the algorithms in the corresponding environment iteratively without exploration. The algorithms are trained 3 times and the shaded area represents one standard deviation. As presented in Fig. 2, all the algorithms can converge steadily during training, which validates our theorem in Sect. 3.3.

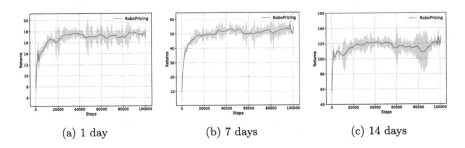

(a) 1 day   (b) 7 days   (c) 14 days

Fig. 2. The convergence of RoboPricing w.r.t different episode length.

Performance in the Test Environment. In Fig. 3a, we evaluate different algorithms without exploration in the test environment. The start date is chosen as 2019.09.15. As shown in the figure, our RoboPricing algorithm can achieve high performance across different settings. The profit of the retailer in RoboPricing is comparable to PPO-real, which is trained directly in this environment. PPO-sim's performance is limited mainly due to the gap between simulation environment and real environment. BCQ and CQL may face the problem of sample inefficiency and therefore have high variance.

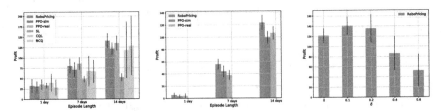

(a) Episode profit of differ- (b) Episode profit of differ- (c) Episode profit of Robo-
ent algorithms in the test ent algorithms in the adver- Pricing in the test environ-
environment.   sarial environment.   ment w.r.t different δ.

Fig. 3. Performance in the test/adversarial environment

Performance in the Adversarial Environment. We then evaluate the trained retailer's pricing strategy in the adversarial environment with the weight $\delta = 0.1$. As shown in Fig. 3b, in all the cases, our RoboPricing can achieve the best performance by considering the adversarial behaviors. We also validate different weights of robustness. In particular, we adjust the value of δ so that RoboPricing can handle different adversarial behaviors. Figure 3c is the result of RoboPricing with different values of δ in the test (real) environment. As presented, RoboPricing can achieve the highest profit when $\delta = 0.1$. With larger values of δ, the performance of RoboPricing would also decrease.

5 Conclusion

This paper proposes a novel robust pricing strategy for dynamic pricing. The main idea is to train the pricing strategy and adversarial customers model simultaneously to find the equilibrium. The trained policy can therefore be robust to adversarial behaviors of the customers. The results are validated both theoretically and empirically.

Acknowledgements. This research was supported by National Key Research and Development Program of China (No. 2019YFB2101704); Natural Science Foundation of Jiangsu Province (No. BK20200752); The NUPTSF (No. NY220080).

References

1. Fujimoto, S., Meger, D., Precup, D.: Off-policy deep reinforcement learning without exploration. In: International Conference on Machine Learning, pp. 2052–2062. PMLR (2019)
2. Garcia, J., Fernández, F.: Safe exploration of state and action spaces in reinforcement learning. J. Artif. Intell. Res. **45**, 515–564 (2012)
3. Hans, A., Schneegaß, D., Schäfer, A.M., Udluft, S.: Safe exploration for reinforcement learning. In: ESANN, pp. 143–148. Citeseer (2008)
4. Ho, J., Ermon, S.: Generative adversarial imitation learning. In: Advances in Neural Information Processing Systems, vol. 29, pp. 4565–4573 (2016)
5. Hu, J., Wellman, M.P.: Nash Q-learning for general-sum stochastic games. J. Mach. Learn. Res. 4(Nov), 1039–1069 (2003)
6. Kumar, A., Zhou, A., Tucker, G., Levine, S.: Conservative Q-learning for offline reinforcement learning. arXiv preprint arXiv:2006.04779 (2020)
7. Mnih, V., et al.: Playing Atari with deep reinforcement learning. arXiv preprint arXiv:1312.5602 (2013)
8. Pinto, L., Davidson, J., Sukthankar, R., Gupta, A.: Robust adversarial reinforcement learning. In: International Conference on Machine Learning, pp. 2817–2826. PMLR (2017)
9. Shi, J.C., Yu, Y., Da, Q., Chen, S.Y., Zeng, A.X.: Virtual-taobao: virtualizing real-world online retail environment for reinforcement learning. In: Proceedings of the AAAI Conference on Artificial Intelligence, vol. 33, pp. 4902–4909 (2019)
10. Szepesvári, C., Littman, M.L.: A unified analysis of value-function-based reinforcement-learning algorithms. Neural Comput. **11**(8), 2017–2060 (1999)

Multi-memory Enhanced Separation Network for Indoor Temperature Prediction

Zhewen Duan[1,2,3], Xiuwen Yi[2,3(✉)], Peng Li[2,3], Dekang Qi[2,3,4], Yexin Li[2,3], Haoran Xu[2,3], Yanyong Huang[5], Junbo Zhang[2,3], and Yu Zheng[1,2,3]

[1] Xidian University, Xi'an, China
[2] JD Intelligent Cities Research, Beijing, China
xiuwenyi@foxmail.com, {lipeng367,qidekang,liyexin}@jd.com
[3] JD iCity, JD Technology, Beijing, China
[4] Southwest Jiaotong University, Chengdu, China
[5] Southwestern University of Finance and Economics, Chengdu, China
huangyy@swufe.edu.cn

Abstract. Indoor temperature prediction is vital to predictive control on district heating systems. Due to the data collection in practice, there always exist residential areas with limited historical data. Transferring the knowledge from residential areas with sufficient data is of great help to address the data scarcity problem. However, it is still challenging as the data distribution shifts among residential areas and shifts over time. In this paper, we proposed a Multi-Memory enhanced Separation Network (**MMeSN**) to predict indoor temperature for residential areas with limited data. MMeSN is a parameter-based multi-source transfer learning method, mainly consisting of two components: *Source Knowledge Memorization* and *Memory-enhenced Aggregation*. Specifically, the former component jointly decouples the domain-independent & domain-specific information which separately memorize the specific historical patterns for each source. The latter component memorizes the historical relationships between the target and multiple sources and further aggregates the domain-specific & domain-independent information. We conduct extensive experiments on a real-world dataset, and the results demonstrate the advantages of our approach.

Keywords: Time series prediction · Transfer learning · Deep learning · Indoor temperature · Urban computing

1 Introduction

District heating system is widely used during winter, supplying heat to residents for keeping the house warm. For monitoring the performance of heating services, heat companies have deployed some temperature sensors in the house, generating real-time indoor temperature records. Accurate predicting future indoor temperature is important for the predictive control of heating systems. By controlling

A. Bhattacharya et al. (Eds.): DASFAA 2022, LNCS 13246, pp. 656–663, 2022.
https://doi.org/10.1007/978-3-031-00126-0_49

the heat supply intelligently, the indoor temperature can be sustained at a comfortable level while reducing energy consumption.

However, considering the real-world data collection mechanism, over-fitting problem often happened with traditional time-series prediction approaches when there is not sufficient training data. For example, we can only get limited amount of data for the residential area with newly-deployed temperature sensors due to the cold-start problem.

One feasible idea is to transfer knowledge from a data-rich domain to the data-poor domain [10]. Besides, multi-source transfer learning has become a hotspot since more source information will contribute to a robust model. In this paper, we propose a multi-memory enhanced separation network, named MMeSN, to predict residential indoor temperature with limited data. MMeSN is a multi-source parameter-based transfer learning method, which mainly contains source knowledge memorization and memory-enhanced aggregation. The former memorizes useful knowledge of multiple sources, and the latter adapts the source knowledge to the target domain. Our main contributions are summarized as follows:

- *Multiple memory modules.* We design multiple source memories to learn the specific patterns of each source and design a target-source memory to capture the correlations between target and source domains, which is robust for historical knowledge memorization with better transfer generalization.
- *Joint decomposition architecture.* We design the joint decomposition architecture for decoupling independent & specific information among all domains, which helps distill individual historical knowledge and alleviates the complexity of knowledge transfer.
- *Real evaluation.* We conduct extensive experiments on a real-world dataset with four residential areas and the results demonstrate the advantages of our approach over several state-of-the-art baselines.

2 Overview

2.1 Problem Definition

Formulation of Prediction Task. For a residential area, we specify $\mathbf{x}_i = (x_i^1, x_i^2, ..., x_i^m) \in \mathbb{R}^m$ as m different sensor readings at time interval i, including *indoor temperature, heating temperature, outdoor temperature, outdoor humidity, wind speed, and wind direction.* Besides, we specify y_i as the indoor temperature. The prediction task can be defined as predicting the next indoor temperature y_{L+1}, given the historical observations $\mathcal{X} = \{\mathbf{x}_i | i = 1, ..., L\}$.

Formulation of Transferring Task. Given n source residential areas $\{\mathcal{X}^s | s = 1, 2, ..., n\}$ with sufficient historical data and one target residential area $\mathcal{X}^t$ with limited data. The transferring task can be defined as predicting the next indoor temperature of the target residential area $y_{L^t+1}^t$, where L^t is the length of historical records of the target area.

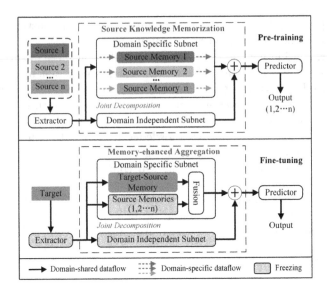

Fig. 1. Framework of our proposed MMeSN.

3   Methodology

Figure 1 illustrates the framework of MMeSN method, trained by two stages: 1) *source data pre-training*, which jointly trains the network by multiple source data to memorize the source knowledge; 2) *target data fine-tuning*, which re-trains the network with only target data to transfer the knowledge from multiple sources.

In the pre-training stage, we feed all source data into the network separately to jointly train the parameters. Firstly, the extractor learns the feature representation from the original data. Then, in source knowledge memorization component, we propose a joint decomposition architecture which contains a domain independent subnet and a domain specific subnet with multiple memories to learn the shared and specific historical information. Finally, the concatenation of these two outputs is fed into the predictor to forecast the next value.

In the fine-tuning stage, the network parameters are re-trained only using the target data. To utilize the source knowledge, we freeze the parameters of the extractor, domain independent subnet, and multiple source memories. Then, in memory-enhanced aggregation component, we reconstruct the domain specific subnet by target-source memory and fusion module for target specific knowledge learning. After that, the outputs are fed into a fusion module for knowledge aggregation. Likewise, the shared and specific knowledge are aggregated together to make the final prediction by the predictor.

MMeSN mainly consists of four components: extractor, source knowledge memorization, memory-enhanced aggregation, and predictor. Here, the extractor extracts the feature representations from the input data, capturing the interactions between multiple factors. We implement it with a feature embedding layer followed by a flatten layer and two fully connected layers with the activation function ReLU, which is denoted as $f_E(\cdot)$ for short. While the predictor receives the concatenation of the output from domain specific and independent subnet, then generates the prediction utilizing a linear transformation with Sigmoid function. In the following, we detail the key components: source knowledge memorization and knowledge-enhanced aggregation.

3.1 Source Knowledge Memorization

To learn the historical patterns of different residential areas, we design the source knowledge memorization component, consisting of a joint decomposition architecture and multiple source memories. The former architecture jointly decouples the domain-independent & domain-specific information for all sources, and the latter separately memorize the specific historical patterns for each source by the memory network.

Though the temperature changes differently among residential areas, it still obeys the same underlying heat exchange rules in the real-world scenario. For example, the indoor temperature rises with the increment of heating temperature and outdoor temperature. Motivated by such fact, we design the joint decomposition architecture, which consists of the domain independent subnet (DI) and domain specific subnet (DS) with n branches for independent and specific knowledge learning, thus alleviating the complexity of knowledge transfer.

In our task, given the source data $\mathbf{X}^s \in \mathbb{R}^{K \times m}$ with a fixed window size K, we specify the source input as the feature representations $f_E(\mathbf{X}^s)$ learned by the extractor. Then the extracted domain independent knowledge by DI and domain specific knowledge by DS-s are further concatenated as the output of source knowledge memorization. We implement the domain independent subnet with two fully connected layers followed with ReLU and batch normalization. The output of this module is denoted as $\mathbf{z}_{DI}^s = f_{DI}(f_E(\mathbf{X}^s))$.

As for the domain specific subnet, we design n branches for n different sources. Considering the distribution of each source data changes over time, we implement each branch with a memory network trained by the corresponding source data to memorize the historical information. Memory network is effective to model the sequential data, storing the long-term dependencies. The structure of memory network is illustrated in Fig. 2 (a). We construct a memory matrix $\mathbf{M}^s \in \mathbb{R}^{V \times d}$ to store the historical information for source s, which contains V memory representations with dimension d. Each row of the memory matrix can be regarded as one distribution pattern of historical data. For each input source data $\mathbf{X}^s$, we get the key vector $\mathbf{k}^s = f_E(\mathbf{X}^s) \in \mathbb{R}^d$, which is then utilized to calculate the similarity score p_j^s with each slice of memory representation $\mathbf{M}_j^s$. The output of domain specific subnet $\mathbf{z}_{DS}^s$ is calculated by the weighted sum of memory slices shown in Eq. 1.

$$p_j^s = \frac{exp(<\mathbf{k}^s, \mathbf{M}_j^s>)}{\sum_{i=1}^{V} exp(<\mathbf{k}^s, \mathbf{M}_i^s>)}, \quad \mathbf{z}_{DS}^s = \sum_{j=1}^{V} p_j^s * \mathbf{M}_j^s \tag{1}$$

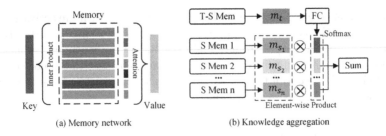

(a) Memory network   (b) Knowledge aggregation

Fig. 2. The process of knowledge aggregation

When retrieving the historical patterns by the memory network, it generates the most relevant memory representation by merging all memory slices with the similarity scores. Compared with the fully connected network learning the mapping function weights, the memory network is more general and robust in transferring scenarios as it learns the historical patterns stored in memory slices.

3.2 Memory-Enhanced Aggregation

After the source data pre-training stage, the shared and specific knowledge from multiple sources is memorized within the DI and DS modules. For better adapting the source knowledge to the target domain, in memory-enhanced aggregation, we freeze the parameters of domain independent subnet and multiple source memories. Then, we reconstruct the domain specific subnet, appending the target-source memory and fusion module to help transfer the source specific knowledge to the target domain.

Similar to the procedure of source knowledge memorization, in this component, the shared knowledge behind all residential areas is learned by the domain independent subnet with the output $z_{DI}^t = f_{DI}(f_E(\mathbf{X}^t))$. While the specific knowledge of the target domain can be viewed as the aggregation of all source knowledge, whose detail is shown in Fig. 2 (b). Given the target input data $\mathbf{X}^t \in \mathbb{R}^{K \times m}$, we get the key vector from the extractor $\mathbf{k}^t = f_E(\mathbf{X}^t) \in \mathbb{R}^d$ with window size K, which is then fed into the domain specific subnet to generate n source memory representations $\mathbf{v}_1, \mathbf{v}_2, ..., \mathbf{v}_n$. Besides, we utilize the target-source memory to help incorporate the source knowledge which shares the same structure of the source memory network with different parameters. Here, the key vector $\mathbf{k}^t$ is also fed into the target-source memory to generate the correlation representation $\mathbf{v}^t$. Then, a linear transformation is adopted to output the normalized similarity scores c with a Softmax function. After that, the specific knowledge of the target domain z_{DS}^t can be calculated by the weighted sum of aggregating source memory representations shown in Eq. 2.

$$\mathbf{c} = \text{Softmax}(\mathbf{W}_p \cdot \mathbf{v}^t + \mathbf{b}_p), \ \mathbf{z}_{DS}^t = \sum_{i=1}^{n} c_i * \mathbf{v}^i \qquad (2)$$

4   Experiments

4.1   Settings

Dataset. We conduct experiments on a real-world indoor temperature dataset with four residential areas, collected from 2018/12/15 to 2019/03/15 with hourly time intervals, including *indoor temperature, heating temperature, outdoor temperature, outdoor humidity, wind speed,* and *wind direction.* For evaluation, we use 4-fold cross-validation, where one residential area is regarded as the target area and the others as source areas.

Baselines. We compare MMeSN with 7 baselines. **No Transfer, MFSAN** [9], **CoDATS** [6], **DANN** [1], **TL-MLP** [2], **SHL-DNN** [4], **TL-SMI** [5]. In our setting, we adjust the implementations of MFSAN, DANN and CoDATS by changing the CNN layers to fully connected layers and the classification loss to regression loss. We also compare several model variants by removing some components (**only DI, only DS, one branch**) and replace the memory module with fully connected layers (**w/o Mem, w/o S Mem. w/o T-S Mem**).

Model Details. We use min-max normalization to normalize the continuous value to $[0,1]$. The extractor is implemented by a feature embedding layer with unit size 8 followed by a flatten layer and 2 fully connected layers with unit size $\{64, 16\}$. As for the source and target memory size, we set the source memory as 7×16 and target memory as 6×16. For domain independent subnet, we use two fully connected layers with sizes $\{16, 8\}$. The slide window size is set as 12.

4.2   Model Comparison

Comparison Among Different Baselines. As shown in Table 1, we compare MMeSN with multiple baselines on four areas. When directly trained with the observed target data, the DNN model (No Transfer) performs worst, showing the importance of transfer learning for the data scarcity problem. The next three approaches (MFSAN, CoDATS, and DANN) achieve a relatively higher MAE, as these approaches are designed to learn the domain-invariant features by aligning the distribution of source and target observed data. However, the future distribution is unforeseeable. The performance decreases when the distribution shift happens for the target domain. TL-MLP, SHL-DNN, and TL-SMI are parameter-based methods for time-series prediction, which achieve a smaller error margin. Note that TL-SMI achieves comparatively better performance since this approach is designed for thermal load prediction and selects the optimum source to pre-train the model and then fine-tunes the whole network with target data. Our proposed model achieves the best performance comparing to all baselines with the average 8.6% and 8.3% relative improvement beyond TL-SMI on MAE and MAPE, respectively. This is because MMeSN decouples the independent & specific knowledge for each domain and memorizes multiple intra- and inter-correlations between target and source domains, which improves the generation for the target prediction.

Table 1. Comparison with different baselines.

Methods	Area 1		Area 2		Area 3		Area 4		Average	
	MAE	MAPE	MAE	MAPE	MAE	MAPE	MAE	MAPE	MAE	MAPE
No transfer	0.625	3.319	0.720	3.923	0.636	2.723	1.171	6.321	0.788	4.071
MFSAN	0.341	1.798	0.474	1.589	0.167	0.715	0.577	2.876	0.390	1.744
CoDATS	0.250	1.316	0.241	1.258	0.190	0.815	0.567	2.685	0.312	1.518
DANN	0.260	1.358	0.264	1.421	0.229	0.985	0.576	2.886	0.333	1.662
TL-MLP	0.316	1.669	0.291	1.562	0.203	0.870	0.466	2.283	0.319	1.596
SHL-DNN	0.293	1.560	0.210	1.222	0.138	0.590	0.428	2.096	0.267	1.367
TL-SMI	0.259	1.371	0.230	1.232	0.122	0.514	0.413	2.021	0.256	1.284
MMeSN	**0.243**	**1.273**	**0.207**	**1.112**	**0.120**	**0.513**	**0.366**	**1.810**	**0.234**	**1.177**

Comparison Among Different Distributions. To verify the generalization for distribution shift problem, we compare MMeSN with different baselines in Fig. 3 (a) and different model variants in Fig. 3 (b). From the comparison results, our proposed MMeSN could achieve the best performance both among all baselines and variants, which demonstrates the effectiveness of our model to alleviate the impact of distribution shift problem.

5   Related Work

Existing transfer learning approaches mainly can be divided into two folds. One is the domain adaptation based methods by aligning the distribution between source and target [6,9]. The other is the parameter-based methods pre-trained by source data and fine-tuned by limited target data. For urban transfer learning, it has been widely applied in three-class applications. The first is the prediction problem, e.g. crowd flow prediction for a new city [7]. The second is the deployment problem, e.g. commercial store site recommendations [8]. The third is the detection problem which detects the objects of interest [3]. Our proposed MMeSN adopts the parameter-based methods with multiple memories to predict the indoor temperature with limited data, which is robust for memorizing historical knowledge with better transfer generalization.

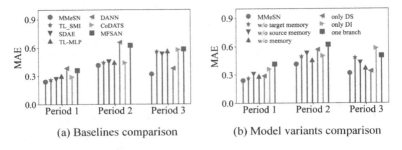

(a) Baselines comparison   (b) Model variants comparison

Fig. 3. Comparison on different time periods of area 4

6   Conclusion

In this paper, we propose a multi-memory enhanced separation network, MMeSN, to predict the residential indoor temperature with limited data. For transferring multi-source knowledge, we adopt a joint decomposition architecture to decouple the domain independent & specific information and utilize multiple source memories and target-source memory to learn the historical patterns. Experimental results demonstrate the advantage of our approach over several baselines.

Acknowledgments. This work was supported by Beijing Nova program (Z211 100002121119) and the Youth Fund Project of Humanities and Social Science Research of Ministry of Education (No. 21YJCZH045).

References

1. Ganin, Y., et al.: Domain-adversarial training of neural networks. J. Mach. Learn. Res. **17**, 59:1–59:35 (2016)
2. Gao, N., Shao, W., Rahaman, M.S., Zhai, J., David, K., Salim, F.D.: Transfer learning for thermal comfort prediction in multiple cities. Build. Environ. **195**, 107725 (2021)
3. He, T., et al.: What is the human mobility in a new city: transfer mobility knowledge across cities. In: Proceedings of The Web Conference 2020, pp. 1355–1365 (2020)
4. Hu, Q., Zhang, R., Zhou, Y.: Transfer learning for short-term wind speed prediction with deep neural networks. Renew. Energy **85**, 83–95 (2016)
5. Lu, Y., Tian, Z., Zhou, R., Liu, W.: A general transfer learning-based framework for thermal load prediction in regional energy system. Energy **217**, 119322 (2021)
6. Wilson, G., Doppa, J.R., Cook, D.: Multi-source deep domain adaptation with weak supervision for time-series sensor data. In: KDD, pp. 1768–1778 (2020)
7. Yao, H., Liu, Y., Wei, Y., Tang, X., Li, Z.: Learning from multiple cities: a meta-learning approach for spatial-temporal prediction. In: WWW, pp. 2181–2191 (2019)
8. Yao, Y., et al.: Fine-scale intra-and inter-city commercial store site recommendations using knowledge transfer. Trans. GIS **23**(5), 1029–1047 (2019)
9. Zhu, Y., Zhuang, F., Wang, D.: Aligning domain-specific distribution and classifier for cross-domain classification from multiple sources. In: AAAI, vol. 33, pp. 5989–5996 (2019)
10. Zhuang, F., et al.: A comprehensive survey on transfer learning. Proc. IEEE **109**, 43–76 (2021)

An Interpretable Time Series Classification Approach Based on Feature Clustering

Fan Qiao[1], Peng Wang[1(✉)], Wei Wang[1], and Binjie Wang[2]

[1] School of Computer Science, Fudan University, Shanghai, China
{fqiao20,pengwang5,weiwang1}@fudan.edu.cn
[2] School of Mechanical, Electronic and Control Engineering,
Beijing Jiaotong University, Beijing, China
bjwang2@bjtu.edu.cn

Abstract. The time series classification problem has been an important mining task and applied in many real-life applications. A large number of approaches have been proposed, including shape-based approaches, dictionary-based ones, ensemble-based ones and some deep-learning approaches. However, these approaches either suffer from low accuracy or need massive features which hinder the interpretability. To overcome these challenges, in this paper, we propose a novel approach, FCCA, based on the feature clustering. We first present the formal definition features of various types. Then we propose the approaches of feature candidates generation, feature filtering and feature clustering. With a small number of representative features, FCCA not only achieves high accuracy, but also improves the interpretability greatly. Extensive experiments are conducted on UCR benchmark to verify the effectiveness and efficiency of the proposed approach.

Keywords: Time series · Interpretablity · Classification · Feature

1 Introduction

The time series classification problem has been an important mining task that has been growing rapidly in the past few decades. It is useful in many real-life applications, including industrial 4.0, Internet of Things and human activity recognition, and so on. A large number of TSC algorithms have been proposed, including shape-based approaches [3,15], structure-based approaches [9,12] and interval-based ones [6].

The work is supported by the Ministry of Science and Technology of China, National Key Research and Development Program (No. 2020YFB1710001), NSFC (No. 52172397).

Shape-based approach builds models based on raw numerical values by defining similarity metrics. The typical model is 1NN with Euclidean distance or Dynamic Time Warping (DTW) distance. Elastic Ensemble (EE) [5] further increases the accuracy combining 11 NN classifiers. Shapelet-based approaches are more accurate, which builds models based distinguishing subsequences. The initial approach is proposed in [15], and has lots of extensions such as Fast Shapelet [3]. However, the interpretable nature of shapelet is not always satisfactory because most approaches generate plenty of shapelets. Although approaches like Fast Shapelet can generate fewer shapelets, the accuracy will suffer due to the missing of critical shapelets. Structure-based approach builds models based on features transformed from time series. Some features are statistical features like mean, standard deviation or autocorrelation coefficient [6]. Other features are approximate representations like Symbolic Aggregate approXimation (SAX) [4] and Symbolic Fourier Approximation (SFA) [12]. Still, structure-based approaches need a large quantity of features.

Two other types of approaches are ensemble-based approach and deep learning approach. The ensemble-based approaches integrate various features or classifiers to achieve higher accuracy, like TS-CHIEF [13] and HIVE-COTE [6]. As for the deep learning approaches, despite of the success in other fields such as LIME [10] and SHAP [7], the accuracy is not satisfactory and lower than some ensemble-based approaches. Moreover, the interpretability is low.

In general, the current TSC approaches are either not satisfactory in accuracy, or not interpretable due to the exploration of feature numbers. To address these issues, in this paper, we propose a novel approach, named Feature Clustering based Classification Approach (FCCA for short). FCCA integrates three types of features time domain, frequency domain and interval-based features, so that it can achieve high accuracy in various fields. FCCA first randomly generates a large number of feature candidates and selects high-quality ones based on the mechanism similar with the decision tree. Next, we propose a feature filtering strategy to prune sub-optimal features. Finally, we propose a clustering approach to aggregate similar and redundant features into representative ones, and based on them we build the classification model. A small number of representative features can not only increase the accuracy, but also improve the interpretability greatly. The contributions of our work are summarized as follows:

- We present the formal definition features of various types.
- We propose a tree-structure model to generate discriminative feature candidates, integrating time domain, frequency domain and interval-based features. A feature ranking strategy is utilized to filter the low-quality features.
- We propose a clustering approach to extract representative features, including techniques like definitions of feature distance, adaptively determining the number of clusters and so on.
- We conduct extensive experiments on UCR benchmark to verify the effectiveness and efficiency of FCCA.

2   Approach

2.1   Preliminary and Overview

A time series $T = (t_1, t_2, ..., t_n)$ is a sequence of ordered numeric values, where n is the length of T. A labelled dataset $\mathcal{D} = \{(T_1, L_1), (T_2, L_2), \cdots, (T_N, L_N)\}$ is a set of N labelled time series with $|C|$ class labels, $\{c_1, c_2, \cdots, c_{|C|}\}$. $L_i \in C$ is the class label of T_i.

Figure 1 shows an overview of our approach. In the first step, we generate feature candidates $\mathcal{F}_0$ integrating time domain features, frequency domain features and interval-based features through a set of trees. In the second step, we filter features in $\mathcal{F}_0$ to get the optimal feature group $\mathcal{F}$. In the third step, we cluster features in $\mathcal{F}$ to generate a small number of representative features.

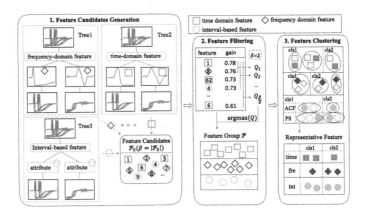

Fig. 1. Approach overview

2.2   Feature Definition

Definition 1 (Feature). *A feature f is a quaterple, $f = \langle R_f, P_f, \theta_f, L_f \rangle$. It consists of a feature representation R_f, transformation parameters P_f, similarity threshold θ_f and feature label L_f. The representation R_f is specific to different transformations, which can be a symbolic word or a continuous value. The threshold θ_f provides metrics that measure the similarity between time series and features. Feature label L_f is the class that can be recognized by feature f.*

The time domain feature captures the *shape* characteristics of the raw time series. We use SAX approximation [4], which transforms time series to a symbolic word, called as a SAX word. For time series T, we turn length-l subsequences into SAX words. Hence, the representation R_f is a length-w SAX word and transformation parameters P_f consists of subsequence length l, SAX word length w, and alphabet size Σ. We define the similarity between T and f as the smallest distance between f and all these SAX words.

The frequency domain feature is similar with the time domain feature, except that it transforms a subsequence in the frequency domain with SFA representation [12]. So the representation R_f is a SFA word. Note that the SFA word represents the *shape* in the frequency domain.

The interval-based feature is the statistic information of a fixed interval of time series. We utilize two transformations of $RISE$ [6], ACF and PS. P_f consist of (s, e, λ, τ), in which s and e are the starting point and ending point of the interval, $\lambda \in \{ACF, PS\}$ is the transformation type, and τ represents which coefficient is selected. R_f is the typical value of this coefficient of time series with label L_f.

2.3   Step One: Feature Generation

In step one, we generate a set of high quality features, denoted as $\mathcal{F}_0$. We split the labelled dataset $\mathcal{D}$ into training set $\mathcal{D}_{train}$ and validation set $\mathcal{D}_{valid}$ randomly. $\mathcal{F}_0$ is generated based on $\mathcal{D}_{train}$. We build K number of trees, named feature tree and denoted as $FTree$, which are similar with the decision tree. The tree construction starts from the root node and recursively builds the sub-trees. If a node N is not pure enough, we split it into $|C|$ child nodes, each of which corresponds to a class label.

We split a node based on $|C|$ features, which satisfy 1) they belong to the same type; 2) they have the same parameters P_f. We call these $|C|$ features as a splitting function, denoted as $SP = (f_1, f_2, \cdots, f_{|C|})$, in which the label of f_i is c_i. For each feature type, we randomly generate N_s number of splitting functions, compute their information gain respectively, and select one with the highest quality to split this node.

Given node N, for time domain SP, first we randomly generate the transformation parameters $P_f = (l, w, \Sigma)$. Next we randomly select a *pilot time series* T of class c_i from node N, and transform T based on P_f to obtain a set of distinct SAX words. Then, we select the most distinguishing one which can achieve the highest information gain as the representation of class c_i. Frequency domain splitting function is similar with that of the time domain. For the interval-based SP, we first randomly select an interval $[s, e]$. For each time series within N, we compute all coefficients of both ACF and PS transformation on interval $[s, e]$. To generate the feature for class c_i, we select the coefficient whose information gain is higher than those of other coefficients, and consider the corresponding transformation and the coefficient as the transformation parameters.

The quality of SP is the average information gain of all $|C|$ features, denoted as $Q(SP)$. Among all $3 \cdot N_s$ splitting functions, we select the one with highest $Q(SP)$ to split node N. Also, we add all features in it into $\mathcal{F}_0$.

2.4   Step Two: Feature Filtering

In step two, we first rank all features within $\mathcal{F}_0$ according to the information gain, that is, after ranking, $\mathcal{F}_0 = \{f_1, f_2, \cdots, f_{|\mathcal{F}_0|}\}$, in which $gain(f_i) \geq gain(f_{i+1})$. Then we select the optimal feature group, which is defined as below.

Definition 2 (Feature group). *Given ranked $\mathcal{F}_0$ and a parameter δ, feature group FG_i is a feature subset from $\mathcal{F}_0$, that is, $FG_i = \{f_1, f_2, \cdots, f_{i*\delta}\}$. Obviously, it holds that $FG_i \subset FG_{i+1}$.*

If FG_i is the optimal one, we filter all features outside FG_i. We evaluate the quality of each group FG_i, denoted as $Q(FG_i)$ $(1 \leq i \leq \lceil |\mathcal{F}_0|/\delta \rceil)$ on the validation set $\mathcal{D}_{valid}$ to avoid the overfitting.

Before defining $Q(FG_i)$, we first introduce a weighted voting strategy to classify a time series T based on feature group FG_i. The weight of feature $f \in FG_i$ is the information gain, $gain(f)$. If time series T satisfies feature f, we vote class L_c with positive weight $gain(f)$. Otherwise, we vote class L_c with negative weight $gain(f)$. Formally, for each label c, the overall weight voted to T by all features in FG is denoted as $score(T, c)$. We determine the label of T, $\hat{L}(T)$, as the label with largest $score(T, c)$.

For each feature group FG_i, we compute its quality $Q(FG_i)$ with Eq. 1.

$$Q(FG_i) = \frac{|\{L(T) = \hat{L}(T)|T \in \mathcal{D}_{valid}\}|}{|\mathcal{D}_{valid}|} \tag{1}$$

Finally, we select the one with highest quality as the optimal feature group, denoted as $\mathcal{F}$, which will participate in the next step of feature clustering.

2.5   Step Three: Feature Clustering

In this step, we cluster features and extract the most representative features. We only cluster features with both the same type and the same class label. So we first split features in $\mathcal{F}$ into some disjoint groups, each of which corresponds to a specific feature type and a class label. Each group is named as homogeneous feature group, denoted as $HFG_{i,\lambda}^{type}$, where *type*, i and λ represent the feature type, class label and transformation type respectively[1].

Distance Between Features. We first define the distance between features within an HFG. For time domain features, we measure the distance of SAX words by reverting a SAX word into a numeric subsequence and computing the Euclidean distance. For frequency domain features, we calculate the Euclidean distance between two SFA words with a multiplier factor $ln(m/m')$ to constraint that the length m of feature's original time series cannot differ greatly. For interval-based features, we define the distance between two features as the Euclidean distance of parameters s, e and τ.

Representative Feature of the Cluster. After we cluster group HFG into some clusters, within each cluster, we select one highest quality feature, named representative feature, which can vote when classifying time series.

---

[1] The subscript λ only works for the interval-based feature group.

The quality of feature f is measured by multiplying two components, 1) information gain within $\mathcal{D}_{train}$; 2) accuracy within $\mathcal{D}_{valid}$. The latter can be measured similarly with Eq. 1 by considering feature f as a feature group. The only difference is that we modify the weight of feature f as $gain(f) \cdot \ln(N_{cls})$ where N_{cls} is the number of features in cluster cls. In other words, the larger a cluster, the higher weight its representative feature.

Clustering Process. Finally, we discuss the cluster process. The naive strategy is to cluster features within each HFG individually. However, it may cause sub-optimal globally. So, we put clustering of each HFG in a global view. To ease the description, we denote all HFG's as a sequence, $\{HFG_1, HFG_2, \cdots, HFG_{|H|}\}$, where $|H|$ is the number of HFG's.

We cluster the features in a top-down fashion. Initially, each HFG is considered as a cluster $\mathcal{C}_0 = \{cls_1, cls_2, \cdots, cls_{|H|}\}$. Then we split the clusters iteratively until meeting the stop conditions. In each step, we select a cluster and split it into two sub-clusters, so that the increasing of the total accuracy is largest. We split a cluster with a greedy strategy. Given two clusters cls and cls', we define the distance between them as the maximal distance between any pair of features from cls and cls' respectively. The clustering process terminates when one of these two conditions is met, 1) the total accuracy decreases; 2) each HFG is split into at least 5 clusters.

3   Experimental Results

We conduct experiments on the UCR benchmark [1]. All experiments are conducted on Windows PC with Intel(R) Core(TM) i7-9700K CPU(8-core 3.00 GHz CPU), 16 GB memory and 2 TB disk space. We set the number of feature trees $K = 100$ and the number of features generated at each node for different types $N_s = 100$. For time domain and frequency domain features, sliding window length $l \in \{10...n\}$, word length $w \in \{6, 8, 10, 12, 14, 16\}$, and the alphabet size $\Sigma = 4$. The parameter δ of a feature group in feature filtering is set to 5. The source code and all detailed experimental results are available[2].

Baseline Approaches. We compare our approach with 8 baseline approaches of 5 different types. For Time-domain feature based approach, we select SAX-VFSEQL [9], Fast shapelet (FS) [3] and Triple Shapelet Network(TSN) [8]. BOSS [11] and RISE [6] are chosen for frequency-domain feature based approach and interval-based approach respectively. TS-CHIEF [13] is the state-of-the-art ensemble approach. For deep Learning approach, we compare FCCA with two deep learning approaches, Resnet [2] and mWDN [14].

---

[2] Our Github repository: https://github.com/EuphoriaFF/FCCA.

Accuracy Comparison. In the first experiment, we compare the accuracy between our approach FCCA with baselines, and the results can be found in our github repository. Moreover, in Fig. 2, we give the diagram which demonstrates the difference of approaches in rank. We test TS-CHIEF with tree number 100 and 500 respectively, since 100 is the default number of our approach and 500 is the default value in TS-CHIEF paper [13]. Detailed accuracy table is available on our website.

Our approach achieves the best results on 17 datasets, which is better than other approaches apparently. Among all baselines, TS-CHIEF is the most competitive one, because it integrates a large number of features of different types. However, the ensemble strategy makes it hard to understand the classification model. Another interesting phenomenon is that deep learning based approach cannot beat traditional TSC approaches. Finally, it can be seen from the table of accuracy that when TS-CHIEF builds the same number of trees as our approach, $K = 100$, its accuracy is far more inferior than our approach, which verifies the effectiveness of our feature clustering strategy.

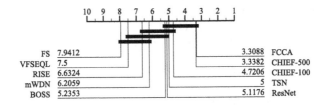

Fig. 2. A critical difference diagram for different approaches

Different Type of Features. In this experiment, we design ablation experiments to evaluate the contribution of each type of features. By disabling any feature type, we can get three variants of FCCA. Figure 3 compares the accuracy between variants using two types of features (horizontal) and FCCA (vertical) in the three scatter-plots. The parameter settings for variants are consistent with FCCA.

It can be seen that FCCA outperforms three variants apparently. The contributions of time domain features and frequency domain ones are larger than interval-based features. However, there are still two datasets whose accuracy benefits greatly from the interval-based features in Fig. 3(c), verifying the necessity for the existence of interval-based features.

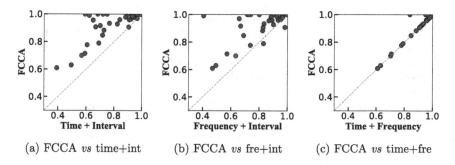

Fig. 3. Contribution of different feature types

4   Conclusion

In this paper, we propose a novel approach to extract multiple types of features based on feature clustering. In contrast to most ensemble approaches using all generated features to classify directly, feature clustering algorithm further selects the most representative features to reduce a large number of redundant features. An extensive experiments shows the superiority of our approach both in accuracy and interpretability with a small number of features.

References

1. Chen, Y., et al.: The UCR time series classification archive, July 2015. http://www.timeseriesclassification.com
2. Ismail Fawaz, H., Forestier, G., Weber, J., Idoumghar, L., Muller, P.-A.: Deep learning for time series classification: a review. Data Min. Knowl. Disc. **33**(4), 917–963 (2019). https://doi.org/10.1007/s10618-019-00619-1
3. Keogh, E.J., Rakthanmanon, T.: Fast Shapelets: a scalable algorithm for discovering time series Shapelets. In: SIAM 2013, pp. 668–676 (2013)
4. Lin, J., Keogh, E.J., Wei, L., Lonardi, S.: Experiencing SAX: a novel symbolic representation of time series. DMKD **15**(2), 107–144 (2007). https://doi.org/10.1007/s10618-007-0064-z
5. Lines, J., Bagnall, A.: Time series classification with ensembles of elastic distance measures. Data Min. Knowl. Disc. **29**(3), 565–592 (2014). https://doi.org/10.1007/s10618-014-0361-2
6. Lines, J., Taylor, S., Bagnall, A.J.: HIVE-COTE: the hierarchical vote collective of transformation-based ensembles for time series classification. In: ICDM 2016 (2016)
7. Lundberg, S.M., Lee, S.: A unified approach to interpreting model predictions. In: NeurIPS 2017, pp. 4765–4774 (2017)
8. Ma, Q., Zhuang, W., Cottrell, G.W.: Triple-Shapelet networks for time series classification. In: ICDM 2019, pp. 1246–1251 (2019)
9. Nguyen, T.L., Gsponer, S., Ifrim, G.: Time series classification by sequence learning in all-subsequence space. In: ICDE 2017, pp. 947–958 (2017)
10. Ribeiro, M.T., Singh, S., Guestrin, C.: "Why should I trust you?": explaining the predictions of any classifier. In: SIGKDD 2016, pp. 1135–1144 (2016)

11. Schäfer, P.: The BOSS is concerned with time series classification in the presence of noise. Data Min. Knowl. Disc. **29**(6), 1505–1530 (2014). https://doi.org/10.1007/s10618-014-0377-7

12. Schäfer, P., Högqvist, M.: SFA: a symbolic Fourier approximation and index for similarity search in high dimensional datasets. In: EDBT 2012, pp. 516–527 (2012)

13. Shifaz, A., Pelletier, C., Petitjean, F., Webb, G.I.: TS-CHIEF: a scalable and accurate forest algorithm for time series classification. Data Min. Knowl. Disc. **34**(3), 742–775 (2020). https://doi.org/10.1007/s10618-020-00679-8

14. Wang, J., Wang, Z., Li, J., Wu, J.: Multilevel wavelet decomposition network for interpretable time series analysis. In: SIGKDD 2018, pp. 2437–2446 (2018)

15. Ye, L., Keogh, E.: Time series Shapelets: a new primitive for data mining. In: SIGKDD 2009, pp. 947–956 (2009)

Generative Adversarial Imitation Learning to Search in Branch-and-Bound Algorithms

Qi Wang[1], Suzanne V. Blackley[2], and Chunlei Tang[2($\boxtimes$)]

[1] Fudan University, Shanghai 200438, China
[2] Harvard Medical School, Boston, MA 02120, USA
ctang5@partners.org

Abstract. Recent studies have shown that reinforcement learning (RL) can provide state-of-the-art performance at learning sophisticated heuristics by exploiting the shared internal structure combinatorial optimization instances in the data. However, existing RL-based methods require too much trial-and-error reliant on sophisticated reward engineering, which is laborious and inefficient for practical applications. This paper proposes a novel framework (RAIL) that combines RL and generative adversarial imitation learning (GAIL) to meet the challenge by searching in branch-and-bound algorithms. RAIL has a policy architecture with dual decoders, corresponding to the sequence decoding of RL and the edge decoding of GAIL, respectively. The two complement each other and restrict each other to improve the learned policy and reward function iteratively.

Keywords: Combinatorial optimization · Reinforcement learning · Branch-and-bound · Generative adversarial imitation learning

1 Introduction

Modeling combinatorial optimization (CO) [1] can be considered a constrained minimization problem. Given a mathematical expression, we can make decisions in the feasible region by constraining the range of variables to minimize the target value. Correspondingly, the decision variable can be linear, an integer (or 0/1), or a mixture of integer and linear, and we call them linear, integer, and mixed-integer programming, respectively.

Branch and bound (B&B) algorithm [2], as the driving force behind mixed-integer programming, consists of a systematic enumeration of candidate solutions, and finds the bounds of the cost function f given certain subsets. Solution solvers contain an exact algorithm similar to B&B algorithms to achieve high accuracy. Algorithms such as B&B can return the superior (lower) bound of the problem, that is, the limit value of the optimal global solution.

The selected policy can significantly affect the exact algorithms' running time when selecting branch nodes in branch and bound. At present, the common ones are depth-first, breadth-first, etc., but the effects of these methods are far from outstanding, so we can look to machine learning to get some new, better performing policies. Many existing methods of applying machine learning to solve CO problems focus on learning heuristics.

Reinforcement learning (RL) usually learns the optimal policy by calculating cumulative rewards, and the design of reward explains what agent's behavior is favorable for humans. However, the rewards it obtains in multi-step decision-making are often very sparse, and the use of cumulative reward-based RL faces a vast search space. More importantly, in areas where it is unclear how to design rewards, agents trained by RL algorithms usually get bad policies and perform worse than expected. Imitation learning (IL) [3] came into being in this situation, and it has solved multi-step decision-making problems very well, especially in robotics. One of IL's significant limitations is that it relies too much on demonstrations, but the acquisition cost of demonstrations may be relatively high. One solution to this challenge may be to combine IL and RL to learn from fewer or simpler demonstrations.

This paper expresses VRP as a mixed-integer programming problem and applies a GNN to model the exact B&B algorithm on bipartite graphs. We design two decoders, a sequence decoder and an edge classification decoder, which output the probability distribution of solutions according to the same policy. We propose a training framework with an adaptive reward function that bridges the DDPG algorithm [4] and GAIL [3]. We only need to give it a small set of labeled data to drive the agent to interact with the environment to form a decoding policy in the initial stage. Then we fed the decoded solution sequences to the GAIL discriminator as the expert trajectory to improve the edge decoder policy. In a word, the dual optimization methods of actor-critic and generator-decoder constitute an optimization system to improve the learned policies and reward functions iteratively.

To sum up, the main contributions of this paper are as follows:

- We propose a GAIL-based plug-and-play component to learn optimization policies and reward functions through imitating demonstrations.
- We design two decoders combining GAIL and DDPG to build a novel training framework for learning decoding policies (without targeted rewards and heuristics).
- We significantly reduce the interaction required to learn a good imitation policy by designing an off-policy learning procedure that relies on experience retention.

2   Related Work

To make this paper self-consistent, we introduce in this section the learning algorithms related to our method and related works for learning B&B and CO heuristics.

2.1   Imitation Learning

Past Imitation learning works have brought impressive research results. Still, they learned to imitate one skill and did not accelerate their learning to imitate the next skill. Compared with the existing IL methods, the recently proposed generative adversarial imitation learning (GAIL) [3] has significantly improved performance when simulating complex expert policies in a large, high-dimensional environment. GAIL promotes IRL by formulating the IL problem as a minimax optimization, which can be solved more efficiently and scalable with alternate gradient-based algorithms.

2.2 Learning the B&B Algorithm

He et al. [5] learned such a strategy (an adaptive searching order) selecting the branch node containing the optimal branching solution. The algorithm learns by continuously collecting expert behaviors through the entire branching process. They propose to learn a branching strategy evaluation function to speed up the branching process. Gasse et al. [2] proposed a graph convolutional neural network model based on the learning B&B variable selection policy for CO.

2.3 Learning for Combinatorial Optimization

Most current works apply attention and RNN or graph neural networks (GNNs) [1] to model CO instances. Bello et al. [6] took the lead in applying RL to learn CO algorithms. Nazari et al. [7] applied permutation invariant layers to encode nodes instead of sequentially encoding nodes in a pointer network. Kool et al. [8] applied a transformer to encode nodes and the pointer attention mechanism to decode solutions. Dai et al. [9] applied the deep Q-learning algorithm to train a GNN to learn the graph's topology. An extensive review of different methods at the junction of RL and CO is given in Wang et al. [1].

3 Methodology

3.1 Background

Markov Decision Process. We assimilate the sequential decisions made in B&B into a Markov decision process (MDP). We aim to simulate solution solvers and learn branching variable selections, so we consider the solvers as the environment and the branchers as agents. State $s \in S$ corresponds to a set of cities visited (B&B tree with all past branching decisions and the leaf node currently focused on). Action a is when the agent selects one of the unvisited cities according to the policy $\pi(a_t|s_t)$ and adds it to the set (state$s$) of cities that have already been visited. The probability of a trajectory $\tau = (s_0, a_0, \ldots, s_t, a_t, \ldots, s_T, a_T)$ depends on the branching policy π and the residual component of the solver, specifically:

$$p_\pi(\tau) = p(s_0) \prod_{t=0}^{T-1} \sum_a \pi(a_t|s_t)p(s_{t+1}|s_t, a_t) \tag{1}$$

3.2 Neural Network Architecture

Our framework (Fig. 1) runs on bipartite graphs, and the agent aims to select a node as its next destination. Given some initial node features, we perform a fixed number of iterations of graph neural network refinement on them and then select the node with the largest value as the target node of the agent.

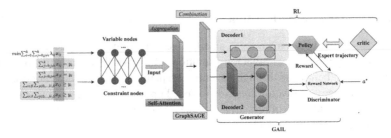

Fig. 1. Overview of our framework

Encoder. The encoder in previous works often only considers the node information. Still, the information on edges, including the actual distance, is essential in practical problems, so we simultaneously consider the node and edge information. Given the d_x dimension node feature x_i and the d_y dimension edge feature y_{ij}, the encoder first calculates the d_h dimension node and edge embedding:

$$\hat{e}_i^{(0)} = W_1 x_i + b_1 \tag{2}$$

$$\hat{e}_{e_{ij}}^{(0)} = W_2 y_{ij} + b_2 \tag{3}$$

Here W_1, W_2, b_1, b_2 are the linear mapping parameters. Then these embeddings pass through an L-layer graph convolution (GC) layer. Node embeddings after passing through the l layer is defined as follows:

$$\hat{e}_{N(v_i)}^l = ReLU\left(W^l AGG\left(\left\{\hat{e}_j^{l-1}, \forall j \in N(v_i)\right\}\right)\right) \tag{4}$$

$$\hat{e}_i^l = CONCAT\left(\hat{e}_i^{l-1}, \hat{e}_{N(v_i)}^l\right) \tag{5}$$

Here, $\hat{e}_{N(v_i)}^l$ represents the aggregation features of the neighbor nodes $N(v_i)$ of the node v_i. AGG is an aggregation function. W^l is a trainable matrix shared by all nodes on the l layer. $CONCAT$ combines its embeddings and the aggregation embeddings of neighbor nodes.

Decoders. We propose two decoders to handle node and edge embeddings. We apply RL to train one node sequence prediction decoder and GAIL to train the other edge prediction decoder. The solutions produced by the node sequence decoder learned by RL can serve as expert demonstrations of the edge decoder for GAIL to imitate. The reward function learned by GAIL can guide the RL agent to learn better policies.

Node Decoder. We apply GRU and context-based attention to map node embeddings into a sequence $\bar{s}$. Given input embeddings, the decoder needs to generate a sequence $\bar{s} = \{\bar{s}_t, t = 1, \ldots, T\}$ of length T. We concretize the branch policy in formula (2) into a decoding policy:

$$\pi(\bar{s}|s; \theta) = \prod_{t=0}^{T} \pi(\bar{s}_{t+1}|\psi(S, \theta_e), \bar{s}_t; \theta_d) \tag{6}$$

Here $\psi(S, \theta_e)$ is the encoder with any input S and trainable parameters θ_e, θ_d is a learnable parameter. The sequence decoder generates a node sequence $\bar{s}$ according to node embeddings generated by GCN and the GRU hidden state h_t. In the decoding time step t, the context weight c_{ti} of the node v_i is calculated as follows:

$$c_{ti} = \begin{cases} -\infty, \text{ if node } v_i \text{ should be masked}, \\ v_a^T tanh(W_D[v_i; h_t]), \text{ otherwise}, \end{cases} \tag{7}$$

where v_a^T, W_D are trainable parameters. Then we apply the softmax function to obtain the pointing distribution towards input nodes:

$$\pi(\bar{s}_t|h_t, \psi(S, \theta_e); \theta_d) = softmax(c_{ti}), i \in \{1, 2, \ldots, n\} \tag{8}$$

Edge Decoder. Since node and edge embeddings both contain the information of graphs and influence each other, when learning converges, the solutions derived from the sequence decoder and the edge decoder should be consistent with being logical. We regard the edge decoder as the generator of GAN, composed of a simple multi-layer perceptron (MLP). The edge sequence $\{e_{ij}\} \subseteq E$ can also be expressed as a solution of VRP:

$$\pi_{e_{ij}} = softmax(MLP(\hat{e}_{e_{ij}}^L)) \tag{9}$$

Here, the embedding $\hat{e}_{e_{ij}}^L$ passes through an MLP and then obtains a softmax distribution, which can serve as the probability of the appearance of the edge e_{ij}. We use solutions obtained by the sequence decoder as ground truth and feed them to the GAN discriminator. The discriminator strives to distinguish the solutions generated by the generator from the ground truth to improve the generated solutions' quality continuously.

3.3 Training

We incorporate GAIL into RL to omit the manual design of RL's reward functions and produce more exact and diverse solutions. In turn, RL provides GAIL with imitated expert trajectories and improves its sampling efficiency. Our training framework (Fig. 1) is built based on the DDPG algorithm [4] by relying on off-policy actor-critic architecture and adopting deterministic policies in the context of imitation learning. Rewards and policies are included in the adversarial training of GAN, while the policy and critic module are trained as actor-critic architecture. Therefore, the reward and critic can play similar roles that forge and maintain a signal that enables reward-seeking policies to adopt desired behaviors.

Imitation Learning with Pre-training and Teacher Forcing. We usually apply pre-training and teacher forcing to cope with the problem of unstable RL and GAN training.

We first employ an LKH3 solver to obtain a small amount of label data as ground truth a^* and feed them to GAIL for imitating. We then apply teacher forcing to pre-train the agent, forcing the agent to perform the same behavior as the ground-truth past trajectories at each step:

$$\ell_{pre}(\theta) = -E_\pi \left[\sum_t log\pi(a_t^*; s_t) \right] \tag{10}$$

Reward. We bring in a reward network with parameters ϕ as a GAIL discriminator, and the cross-entropy loss applied to train the reward network is:

$$E_{\mathcal{R}}\big[-\log(1 - D_\phi(s, a))\big] + E_{\pi_E}\big[-logD_\phi(s, a)\big] + \lambda_1 \mathfrak{R}_{GP}(\phi) \tag{11}$$

Here π_E is the expert policy obtained from the sequence decoder. $\mathfrak{R}_{GP}(\phi)$ is a penalty for the gradient of our Wasserstein GAN [10] discriminator, helping our method improve its stability further. We sample transitions from a replay buffer $\mathcal{R}$ collected during off-policy training instead of directly sampling the policy trajectory $\pi_{e_{ij}}$.

Critic. Our critic's loss $\ell(\varphi)$ contains three components: [11] (1) 1-step Bellman residual $\ell_1(\varphi)$, (2) n-step Bellman residual $\ell_n(\varphi)$, (3) a weight decay regulariser $\mathfrak{R}_{WD}(\varphi)$.

$$\ell(\varphi) = \ell_1(\varphi) + \ell_n(\varphi) + \lambda_2 \mathfrak{R}_{WD}(\varphi) \tag{12}$$

$$\tilde{Q}_\varphi^1(s_t, a_t) \cong r_\phi(s_t, a_t) + \gamma Q_\varphi(s_{t+1}, \pi_\theta(s_{t+1})) \tag{13}$$

$$\tilde{Q}_\varphi^n(s_t, a_t) \cong \sum_{k=0}^{n-1} \gamma^k r_\phi(s_{t+1}, a_{t+k}) + \gamma^n Q_\varphi(s_{t+n}, \pi_\theta(s_{t+n})) \tag{14}$$

Here λ_2 is a parameter. $\ell_1(\varphi)$ and $\ell_n(\varphi)$ are respectively defined by the 1-step and n-step lookahead versions of the Bellman equation.

Policy. We update the policy π_θ to maximize the performance objective (minimize its loss $\ell_{RL}(\theta)$), specifically:

$$\nabla_\theta J(\pi_\theta) \approx E_{\mathcal{R}}\big[\nabla_\theta Q_\varphi(s_t, \pi_\theta(s_t))\big] = E_{\mathcal{R}}[\nabla_\theta \pi_\theta(s_t)\nabla_a Q_\varphi(s_t, a)|_{a=\pi_\theta(s_t)}] \tag{15}$$

To combine supervised pre-training and RL, we apply a linear combination of their loss functions and treat the final losses as follows:

$$\ell(\theta) = \alpha \ell_{pre}(\theta) + (1 - \alpha)\ell_{RL}(\theta) \tag{16}$$

where $\alpha \in [0, 1]$ is a hyper-parameter to be tuned, and θ denotes the learned parameter vector. In our experiments, we set $\alpha = 0.2$.

4   Experiments

4.1   Comparative Study

As we can see from the comparison results in Table 1, learning-based methods perform significantly better than traditional heuristic algorithms. RAIL outperforms other learning-based methods in accuracy and time on TSP and CVRP instances (number of nodes 20, 50, and 100).

Table 1. RAIL vs. baselines. "Obj." indicates the shortest distance obtained by the different methods, "Gap" indicates the gap between the shortest distance obtained by the different methods and the optimal solutions, and "Time" indicates the time required for testing

Method	n = 20			n = 50			n = 100			Problem
	Obj.	Gap	Time	Obj.	Gap	Time	Obj.	Gap	Time	
LKH3	3.84	0.00%	18 s	5.7	0.00%	5 m	7.76	0.00%	21 m	TSP
Nearest Insertion	4.33	12.91%	1 s	6.78	19.03%	2 s	9.46	21.82%	6 s	
Ptr-RL [6]	–	–	–	5.75	0.95%	–	8	3.03%	–	
S2V-DQN [9]	3.89	1.42%	–	5.99	5.16%	–	8.31	7.03%	–	
AM [8]	3.84	0.08%	5 m	5.73	0.52%	24 m	7.94	2.26%	1 h	
RAIL	**3.84**	0.00%	1 s	**5.7**	0.00%	2 s	**7.76**	0.00%	4 s	
LKH3	6.14	0.58%	2 h	10.38	0.00%	7 h	15.65	0.00%	13 h	CVRP
PRL [7]	6.4	4.92%	–	11.15	7.46%	–	16.96	8.39%	–	
AM [8]	6.25	2.49%	6 m	10.62	2.40%	28 m	16.23	3.72%	2 h	
RAIL	**6.1**	0.00%	**1 s**	**10.38**	0.00%	**2 s**	**15.65**	0.00%	**5 s**	

4.2 Learning Curves

Figure 2(a) shows that when our GNN is trained by RL (only), IL (both GAIL and pre-trained IL), and RAIL, respectively, RAIL converges the fastest, while RL converges faster than IL. The combined performance of RL and GAIL is more robust than either of them, which illustrates the necessity and usefulness of the proposed joint training algorithm. Figure 2(b) shows that test curves on TSP100 fluctuate in a range and are not as stable as the learning curves during training, which is reasonable because the data distribution varies somewhat from instance to instance. The test curves gradually stabilize, proving RAIL's effectiveness and stability when reasoning.

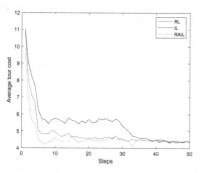 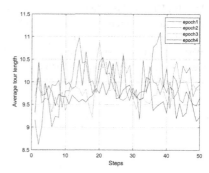

Fig. 2. Learning curves in training and testing. The left figure (a) shows RAIL convergence in three training modes. The right figure (b) shows the learning curves of RAIL trained on TSP100 and then tested on the new TSP100.

5 Conclusion

This paper uses a bipartite graph to convert CO into an integer linear programming problem. Modeling such an instance (e.g., a routing problem) can be transformed by learning to search in the branch and bound algorithm. We next created a novel training framework of RL and GAIL, which roughly correspond to the node decoder and the edge decoder. We use the edge decoder as the generator, the reward network as the discriminator, and the policy as the co-optimized target link of RL and GAIL. Feeding node sequences as expert trajectories to the discriminator can help determine whether solutions from the edge decoder are correct to improve the policy and solution continuously.

References

1. Wang, Q., Tang, C.: Deep reinforcement learning for transportation network combinatorial optimization: a survey. Knowl. Based Syst. (2021). https://doi.org/10.1016/j.knosys.2021.107526
2. Gasse, M., Chételat, D., Ferroni, N., Charlin, L., Lodi, A.: Exact combinatorial optimization with graph convolutional neural networks. Adv. Neural Inf. Process. Syst. **32** (2019)
3. Ho, J., Ermon, S.: Generative adversarial imitation learning. In: Proceedings of the 30th International Conference on Neural Information Processing Systems (NIPS 2016) (2016). https://doi.org/10.2307/j.ctv1dp0vwx.25
4. Lillicrap, T.P., et al.: Continuous control with deep reinforcement learning. In: Proceedings of the 4th International Conference on Learning Representations (ICLR 2016) - Conference Track Proceedings (2016)
5. He, H., Daumé, H., Eisner, J.: Learning to search in branch-and-bound algorithms. Adv. Neural Inf. Process. Syst. **4**, 3293–3301 (2014)
6. Bello, I., Pham, H., Le, Q.V., Norouzi, M., Bengio, S.: Neural combinatorial optimization with reinforcement learning. Proceedings of the 5th International Conference on Learning Representations (ICLR 2017) - Working Track Proceedings, pp. 1–15 (2019)
7. Nazari, M., Oroojlooy, A., Takáč, M., Snyder, L.V.: Reinforcement learning for solving the vehicle routing problem. Adv. Neural Inf. Process. Syst. **2018**, 9839–9849 (2018)
8. Kool, W., Van Hoof, H., Welling, M.: Attention, learn to solve routing problems! In: Proceedings of the 7th International Conference on Learning Representations (ICLR 2019), pp. 1–25 (2019)
9. Dai, H., Khalil, E.B., Zhang, Y., Dilkina, B., Song, L.: Learning combinatorial optimization algorithms over graphs. Adv. Neural Inf. Process. Syst. **2017**, 6349–6359 (2017)
10. Gulrajani, I., Ahmed, F., Arjovsky, M., Dumoulin, V., Courville, A.: Improved training of Wasserstein GANs. Adv. Neural Inf. Process. Syst. **2017**, 5768–5778 (2017)
11. Vecerik, M., et al.: Leveraging Demonstrations for Deep Reinforcement Learning on Robotics Problems with Sparse Rewards, pp. 1–10 (2017)

A Trace Ratio Maximization Method for Parameter Free Multiple Kernel Clustering

Yan Chen[1], Lei Wang[2], Liang Du[2,3(✉)], and Lei Duan[1(✉)]

[1] School of Computer Science, Sichuan University, Chengdu, China
chenyan557@stu.scu.edu.cn, leiduan@scu.edu.cn
[2] School of Computer and Information Technology, Shanxi Univiersity,
Taiyuan, China
duliang@sxu.edu.cn
[3] Institute of Big Data Science and Industry, Shanxi Univiersity, Taiyuan, China

Abstract. Multiple Kernel Clustering (MKC) is helpful to leverage complementary information from various contexts and alleviate the difficulty of kernel determination. However, the key weighting strategies for optimal kernel learning around individual kernels are not derived from their optimization problems but embedded in a plug-and-play manner and lead to sub-optimal objective function value. More seriously, the hyper-parameters, introduced by the additive balance of these two coupled sub-tasks, are hard to determine in unsupervised learning scenarios and lead to inconsistent and less satisfying results. To avoid the problems mentioned above, we present a novel parameter-free MKC method with the trace ratio criterion (TRMKC in short), which minimizes the approximation errors between consensus and base kernels using the corr-entropy induced metric and maximizes the mean similarities based on the consensus kernel. The trade-off between these two coupled sub-procedures can be automatically balanced, and the performance could be mutually reinforced. To solve the trace ratio criterion and the corr-entropy induced non-quadratic function optimization problem, we present an alternative strategy with monotonic convergence proof, which reformulates it into a series of sub-problems with trace difference and quadratic programming by utilizing the half-quadratic optimization technique. Extensive MKC experimental results well demonstrate the effectiveness of TRMKC.

Keywords: Multiple kernel clustering · Corr-entropy induced kernel weighting · Trace ratio criterion

This work was supported in part by the National Key Research and Development Program of China (2018YFB0704301-1), the National Natural Science Foundation of China (61972268, 61976129, 62176001), the Sichuan Science and Technology Program (2020YFG0034).

A. Bhattacharya et al. (Eds.): DASFAA 2022, LNCS 13246, pp. 681–688, 2022.
https://doi.org/10.1007/978-3-031-00126-0_52

1   Introduction

As a fundamental research field in machine learning and data mining, clustering has been widely used in various applications. The Multiple Kernel Clustering (MKC) methods are widely investigated to incorporate complementary information across kernels and avoid the kernel selection or design problem. The paradigm of MKC has received considerable attentions, and several methods have been proposed recently. In general, existing MKC algorithms can be roughly grouped into three categories. The first category takes the early fusion strategy, where one sub-task is to learn the consensus kernel from multiple candidate kernels, and another is to perform clustering on the single consensus kernel [5,7,8,10]. The second category algorithms also take the early fusion strategy via the paradigm of multiple graph clustering. One sub-task is to extract multiple affinity graphs from multiple kernels, and another one is to integrate these affinity graphs to get the final consensus graph. These two sub-tasks can be concatenated separately as in [6,15] or optimized jointly as in [14,16]. The third category utilizes the late fusion strategy, where multiple base partitions are first generated from individual kernels and then integrated to build the final consensus partition [9,11,17].

Based on the inherent connection between consensus kernel learning and clustering on consensus kernel, most existing MKC algorithms jointly optimize these procedures within a unified learning framework, where these targets are often manually balanced by additional hyper-parameters. These algorithms often perform well with suitable parameters detected by the grid search strategy, which uses the ground truth to determine the proper parameters. However, the limitation of such a strategy is also clear. On the one hand, the performance of MKC algorithms is less stable and largely dependent on the choice of hyper-parameters. On the other hand, it is less applicable and even infeasible to search for the best hyper-parameters via the so-called grid search strategy for the unsupervised MKC scenario, where no label information is available. Therefore, it is much preferable to develop MKC methods without parameter turning or even parameter-free.

To address the problems mentioned above on candidate kernel weighting and hyper-parameter sensitivity, we propose to learn the optimal neighborhood kernel called TRMKC by minimizing the approximation errors between consensus and base kernels using the corr-entropy induced metric, where the large errors caused by poor quality kernels could be largely suppressed. In contrast, good kernels corresponding to small errors could be carefully emphasized. It should be noted that the kernel weight can be estimated directly from the optimization problem, not like the plug-and-play strategy as [13,19]. TRMKC is proposed to integrate consensus kernel learning and clustering by maximizing the trace ratio criterion without introducing additional hyper-parameters, where the balance between these two coupled sub-procedures can be automatically balanced, and the performance could be mutually reinforced. To solve the optimization problem with the trace ratio criterion and the corr-entropy induced non-quadratic function, we present an alternative optimization strategy with monotonic convergence proof, which reformulates it into a series of sub-problems with trace difference and quadratic programming by utilizing the half-quadratic optimization techniques. Extensive

MKC experimental results on 11 benchmark datasets with ten recent MKC algorithms well demonstrate the effectiveness of TRMKC.

2   Design of TRMKC

Given multiple kernel grammatrices $\{\mathbf{K}^i\}_{i=1}^m$ on n samples, where $\mathbf{K}^i \in \mathcal{R}^{n \times n}$ is the i-th kernel matrix and it is PSD as required. The MKC algorithm aims to identify the c consensus cluster structure across these candidate kernels.

2.1   Consensus Kernel Learning via CIM

Instead of using a traditional quadratic function like [13,19], the approximation error between consensus kernel $\mathbf{K}$ and single individual kernel $\mathbf{K}^i$ is measured by the corr-entropy induced non-quadratic loss function, which leads to the following optimization problem for consensus kernel learning,

$$\min_{\mathbf{K}} \sum_{i=1}^m \ell(\mathbf{K}, \mathbf{K}^i) = \sum_{i=1}^m \left(1 - \exp(-\frac{||\mathbf{K} - \mathbf{K}^i||^2}{\delta^2})\right), \quad \text{s.t.} \quad \mathbf{K} \succeq 0, \quad (1)$$

where δ is the bandwidth of the Gaussian function, which is estimated automatically in this paper. From the perspective of the loss function, the benefits of the above non-quadratic loss function against the traditional quadratic one are two folds. Firstly, the CIM loss is upper bounded and change slowly for large errors. Therefore, the effect of a low-quality kernel with large error could be large suppressed. Secondly, the CIM loss changes quickly on small errors. As a result, similar high-quality kernels, which correspond to small errors, could be carefully distinguished. The different behavior of CIM function on large and small errors makes it is appropriate to characterize the expectation of consensus kernel across these candidate kernels.

From the perspective of kernel weighting, the implicit kernel weights can also be derived from the CIM loss function according to the half-quadratic optimization theory [1]. It assigns larger weights to good kernels with small errors and small weights to bad kernels with large errors. Moreover, we further introduce the explicit weight for each kernel by solving the following problem,

$$\min_{\mathbf{K}, \mathbf{s}} \sum_{i=1}^m \frac{1}{s_i} \left(1 - \exp(-\frac{||\mathbf{K} - \mathbf{K}^i||^2}{\delta^2})\right), \text{s.t.} \quad \mathbf{K} \succeq 0, \sum_{i=1}^m s_i = 1, s_i \geq 0, \forall i,$$
$$(2)$$

where $\mathbf{s} \in \mathcal{R}^{m \times 1}$ and $\frac{1}{s_i}$ can be seen as the explicit weight for the i-th kernel.

2.2   Clustering on Consensus Kernel

Spectral clustering and k-means are two major traditional clustering methods. Ratio cut aims to maximize the mean similarities between data points in the same cluster while k-means aim to minimize the mean distances between points in the same cluster. Recently, it has been shown that both k-means and ratio cut can be unified into the following framework [12],

$$\max_{\mathbf{H}} \quad \text{tr}(\mathbf{H}^T \mathbf{A} \mathbf{H}), \quad \text{s.t.} \quad \mathbf{H}^T \mathbf{H} = \mathbf{I}, \tag{3}$$

where $\mathbf{H} \in \mathcal{R}^{n \times c}$ is the partition matrix and $\mathbf{A} \in \mathcal{R}^{n \times n}$ is the similarity matrix. The mere difference between ratio-cut and k-means is that the former usually uses the Gaussian kernel to capture the pairwise similarity, while k-means adopts the linear kernel. Based on such a unified view of these two methods, the final clustering on consensus kernel can also be achieved by solving the optimization problem in Eq. (3), where we have $\mathbf{A} = \mathbf{K}$.

2.3 The Proposed TRMKC

We have two sub-tasks, i.e., minimizing Eq. (2) for consensus kernel learning and maximizing Eq. (3) for clustering on consensus kernel. Instead of introducing additional hyper-parameter to balance these targets, we take the trace ratio criterion to incorporate these two coupled procedures in a unified learning framework, which can be presented as follows,

$$\max_{s, \mathbf{H}, \mathbf{K}} \quad \frac{\text{tr}(\mathbf{H}^T \mathbf{K} \mathbf{H})}{\sum_{i=1}^{m} \frac{1}{s_i}(1 - \exp(-\frac{\|\mathbf{K} - \mathbf{K}^i\|_F^2}{\delta^2}))} \tag{4}$$

$$\text{s.t.} \quad \mathbf{H}^T \mathbf{H} = \mathbf{I}, \mathbf{K} \succeq 0, \sum_{i=1}^{m} s_i = 1, s_i \geq 0, \forall i.$$

Eq. (4) learns the consensus kernel around individual kernels by minimizing the weighted corr-entropy induced non-quadratic loss in the denominator and performs clustering on the consensus kernel by maximizing the quadratic term in the nominator. The consensus kernel $\mathbf{K}$ is involved both in the nominator and denominator. These two targets can be automatically balanced without the involvement of hyper-parameters. Based on these discussions, the optimization problem in Eq.(4) is suitable for MKC.

2.4 Optimization

We present an alternative learning algorithm to maximize Eq. (4).

Update H with fixed K Given s and $\mathbf{K}$, the sub-problem with respect to $\mathbf{H}$ can be simplified as

$$\max_{\mathbf{H}} \quad \text{tr}(\mathbf{H}^T \mathbf{K} \mathbf{H}), \quad \text{s.t.} \quad \mathbf{H}^T \mathbf{H} = \mathbf{I}. \tag{5}$$

Since the consensus kernel $\mathbf{K}$ is PSD, the optimal solution of $\mathbf{H}$ can be obtained by the c eigenvectors corresponding to the c largest eigenvalues.

Update s with fixed K and H Denote $h_i = 1 - \exp(-\frac{||\mathbf{K}-\mathbf{K}^i||_F^2}{\delta^2})$, the rest problem w.r.t. **H** can be rewritten as

$$\min_{\mathbf{s}} \quad \sum_{i=1}^{m} \frac{h_i}{s_i}, \quad \text{s.t.} \sum_{i=1}^{m} s_i = 1, s_i \geq 0, \forall i. \tag{6}$$

The optimal solution can be computed by $s_i = \frac{\sqrt{h_i}}{\sum_{i'=1}^{m} \sqrt{h_{i'}}}$.

Update K with fixed H and s Given **H** and **s**, the problem w.r.t. **K** can be written as,

$$\max_{\mathbf{K}} \quad \frac{\text{tr}(\mathbf{H}^T \mathbf{K} \mathbf{H})}{\sum_{i=1}^{m} \frac{1}{s_i}(1 - \exp(-\frac{||\mathbf{K}-\mathbf{K}^i||_F^2}{\delta^2}))}, \quad \text{s.t.} \quad \mathbf{K} \succeq 0. \tag{7}$$

The trace ratio problem in Eq. (7) can be solved by a series of trace difference problems [3]. Given **H**, **s** and $\mathbf{K}^{t-1}$ at the $t-1$-th iteration, we first introduce the following auxiliary variable at the t-th iteration,

$$\lambda^t = \frac{\text{tr}(\mathbf{H}^T \mathbf{K}^{t-1} \mathbf{H})}{\sum_{i=1}^{m} \frac{1}{s_i}(1 - \exp(-\frac{||\mathbf{K}^{t-1}-\mathbf{K}^i||_F^2}{\delta^2}))}, \tag{8}$$

where the index of $(t-1)$ on **H**, **s** is omitted. Equation (7) can be reformulated into the following trace difference problem,

$$\max_{\mathbf{K}} \quad \text{tr}(\mathbf{H}^T \mathbf{K} \mathbf{H}) - \lambda^t \sum_{i=1}^{m} \frac{1}{s_i}(1 - \exp(-\frac{||\mathbf{K} - \mathbf{K}^i||_F^2}{\delta^2})), \quad \text{s.t.} \quad \mathbf{K} \succeq 0. \tag{9}$$

According to the HQ theory, we introduce the auxiliary variable $\mathbf{g} \in \mathcal{R}^{m \times 1}$ with $g_i = \exp(-\frac{||\mathbf{K}^{t-1}-\mathbf{K}^i||^2}{\delta^2})$. Then Eq. (9) can be rewritten as,

$$\min_{\mathbf{K}} \quad \lambda^t \sum_{i=1}^{m} \frac{g_i}{s_i}||\mathbf{K} - \mathbf{K}^i||^2 - \text{tr}(\mathbf{K}^T(\mathbf{H}\mathbf{H}^T)), \quad \text{s.t.} \quad \mathbf{K} \succeq 0. \tag{10}$$

By introducing the auxiliary variable $\hat{\mathbf{K}} = \frac{1}{\sum_{i'=1}^{m} \frac{g_{i'}}{s_{i'}}} \sum_{i=1}^{m} \frac{g_i}{s_i}\mathbf{K}^i + \frac{1}{2\lambda^t \sum_{i'=1}^{m} \frac{g_{i'}}{s_{i'}}}\mathbf{H}\mathbf{H}^T$, the problem in Eq. (10) can be simplified as,

$$\min_{\mathbf{K}} \quad ||\mathbf{K} - \hat{\mathbf{K}}||^2, \quad \text{s.t.} \quad \mathbf{K} \succeq 0. \tag{11}$$

The above problem is the quadratic projection of $\hat{\mathbf{K}}$ within the positive semi-definite space. The optimal solution can be obtained by setting

$$\mathbf{K} = \mathbf{U}_{\hat{\mathbf{K}}} \mathbf{S}_{\hat{\mathbf{K}}}^+ \mathbf{V}_{\hat{\mathbf{K}}}^T, \tag{12}$$

where $\hat{\mathbf{K}} = \mathbf{U}\mathbf{S}\mathbf{V}^T$ is the singular value decomposition (SVD) of $\hat{\mathbf{K}}$ and $\mathbf{S}^+$ is the non-negative projection of $\mathbf{S}$, i.e., $\mathbf{S}^+ = \max(\mathbf{S}, \mathbf{0})$.

3   Experiments

We compare TRMKC[1] with 10 MKC algorithms to show the effectiveness and provide the convergence analysis of TRMKC on benchmark datasets.

3.1   Experiment Setup

We use 11 widely used data sets for multiple kernel clustering comparison [2,4]. We omit the details of these data sets due to the space limitation. We compare TRMKC with 10 recent developed MKC methods, including RMSC [18], RMKKM [2], MKCMR [8], LKAM [5], ONKC [10], JMKSC [19], MKCSS [20], ONALK [7], SPMKC [13], OPLF [9]. The codes are mostly provided by the authors in public repository or emails. We adopt clustering accuracy (ACC) to evaluate the clustering result.

Table 1. ACC(%) of 11 algorithms on 11 benchmark datasets

	RMSC	RMKKM	MKCMR	LKAM	ONKC	JMKSC	MKCSS	ONALK	SPMKC	OPLF	TRMKC
JAFFE	67.14	79.34	95.77	88.26	97.18	82.16	38.03	93.90	96.24	96.71	**98.59**
CSTR	42.74	70.11	64.84	65.26	65.26	65.26	62.53	45.89	65.89	65.05	**80.42**
TR45	37.83	61.45	69.86	67.10	72.03	68.99	65.07	25.94	74.20	68.70	**76.09**
TR41	53.08	55.47	64.92	49.43	61.50	60.82	53.19	45.44	57.18	59.34	**68.00**
MPeg7	46.29	49.00	53.21	47.86	53.93	49.36	52.50	48.21	55.71	52.29	**56.64**
BA	33.69	43.45	45.80	41.24	44.09	29.06	45.44	46.23	34.54	37.11	**52.28**
COIL	58.75	61.88	64.51	60.21	65.90	58.96	71.67	67.64	64.10	56.04	**78.96**
Wap	36.22	44.62	44.10	41.35	46.35	36.47	35.26	32.95	47.88	43.91	**59.81**
Digit	59.43	78.91	83.14	76.74	83.08	65.33	87.76	87.65	78.02	80.30	**88.48**
Palm	56.90	70.55	73.50	59.05	78.80	76.70	74.80	71.05	80.80	78.05	**88.15**
K1B	66.84	75.73	64.10	58.08	67.48	78.29	84.57	63.33	63.03	72.44	**84.79**
Avg	50.81	62.77	65.80	59.51	66.87	61.04	60.98	57.11	65.24	64.54	**75.65**

We follow the same strategy in [2,19] to build 12 base kernels to evaluate the clustering performance of the MKL methods. There are some parameters should be set in advance for fair. So, it is non-trivial to choose the proper multiple hyper-parameters corresponding to the best clustering performance for LKAM, ONKC, JMKSC, MKCSS, ONALK, and SPMKC across different datasets. As a result, some hyper-parameters of these algorithms are determined according to the suggestions from their papers, while others are determined by observing the corresponding sensitivity study results and choosing the stable value.

The concrete setting of hyper-parameters for all these algorithms are presented as follows. For all these MKC algorithms, the number of clusters is set to the true number of classes for all the data sets. For RMSC, the regularization parameter is set to $\lambda = 0.005$ as in [18]. For RMKKM, the parameter γ to control the kernel weight distribution is set to $\gamma = 0.3$ as in [2]. For MKCMR [8],

---

[1] The cold of TRMKC has been released at https://github.com/YanChenSCU/TRMKC-DASFAA-2022.

the regularization parameter is fixed to $\lambda = 2^{-2}$. For LKAM [5], the regularization parameter and the neighborhood size are set to $\lambda = 2^{-1}$ and $\tau = 0.05n$. For ONKC [10], the importance of the neighborhood kernel learning and the kernel diversity balancing term are set to $\rho = 2^{-4}$ and 2^{-7}. For JMKSC, the hyper-parameters are set to $\alpha = 10^{-1}, \beta = 20, \gamma = 5$ as suggested in [19]. For MKCSS [20], the constrained rank value, the importance of the Frobenius term, the kernel diversity balancing term and the number of neighbors are fixed as $0.1n$, $10^{-4}\|\mathbf{K}_{avg}\|_F$, 2^{-2} and $0.01n$. For ONALK [7], the kernel diversity balancing term, the importance of the neighborhood kernel learning and the threshold of neighborhood similarity are set to $\lambda = 2, \rho = 2^{-1}, \tau = 0.1$. For SPMKC [13], the hyper-parameters are set to $\lambda_1 = 4, \lambda_2 = 1, \lambda_3 = 100, \lambda_4 = 1000$. For OPLF, it is free of hyper-parameters [9]. For TRMKC, we use the neighborhood kernel as [20] and the local size is fixed to $5\log(n)$ on all the datasets without tuning. For all these methods except for OPLF, we run k-means 10 times and obtain the final result corresponding to the minimal value of k-means objective. The variance for all the compared algorithms is 0 in our experiments.

3.2 Clustering Results Analysis

Based on the clustering result in Table 1, we can see that

- TRMKC outperforms all these 10 MKC algorithms on 11 benchmark data sets in most cases. We can see that our method achieves 13.59% improvement against the second-best result on averages in terms of ACC.
- The results for MKCMR, LKAM, ONKC, JMKSC, MKCSS, ONALK and SPMKC degenerate in general and exhibit poor performance in many cases. These algorithms would achieve better results by tuning the parameters according to the grid-search strategy. However, it is infeasible in practice. Since we run these algorithms with fixed hyper-parameters, the ubiquitous degeneration indeed indicates that these algorithms are less stable without careful tuning.
- Considering the difficulty of hyper-parameter determination in the MKC scenario and the degeneration of exiting MKC algorithms on fixed parameters, these experimental results motivate us to develop a parameter-free method for the MKC task. Compared with baseline methods, the results of TRMKC demonstrate not only the superiority in performance but also the stableness across all data sets.

4   Conclusion

In this paper, we propose a novel parameter-free MKC method with the trace ratio criterion (TRMKC), which minimizes the approximation errors for optimal kernel learning and maximizes the mean similarities on the consensus one. An alternative algorithm has been developed to solve the optimization problem with trace ratio criterion and the non-quadratic CIM loss function. Comprehensive experimental results on 11 widely-used datasets show that exiting MKC methods with hyper-parameters are indeed less stable and exhibit poor results without proper parameter determination. Moreover, TRMKC consistently outperforms the state-of-the-art competitors in terms of clustering performance.

References

1. Du, L., Li, X., Shen, Y.D.: Robust nonnegative matrix factorization via half-quadratic minimization. In: ICDM, pp. 201–210 (2012)
2. Du, L., et al.: Robust multiple kernel k-means using l21-norm. In: IJCAI, pp. 3476–3482 (2015)
3. Jia, Y., Nie, F., Zhang, C.: Trace ratio problem revisited. IEEE Trans. Neural Netw. **20**(4), 729–735 (2009)
4. Kang, Z., Peng, C., Cheng, Q., Xu, Z.: Unified spectral clustering with optimal graph. In: AAAI, pp. 3366–3373 (2018)
5. Li, M., Liu, X., Wang, L., Dou, Y., Yin, J., Zhu, E.: Multiple kernel clustering with local kernel alignment maximization. In: IJCAI, pp. 1704–1710 (2016)
6. Li, X., Ren, Z., Lei, H., Huang, Y., Sun, Q.: Multiple kernel clustering with pure graph learning scheme. Neurocomputing **424**, 215–225 (2021)
7. Liu, J., et al.: Optimal neighborhood multiple kernel clustering with adaptive local kernels. In: TKDE, p. 1 (2021)
8. Liu, X., Dou, Y., Yin, J., Wang, L., Zhu, E.: Multiple kernel k-means clustering with matrix-induced regularization. In: AAAI, vol. 30 (2016)
9. Liu, X., et al.: One pass late fusion multi-view clustering. In: ICML, pp. 6850–6859 (2021)
10. Liu, X., et al.: Optimal neighborhood kernel clustering with multiple kernels. In: AAAI, pp. 2266–2272 (2017)
11. Liu, X., et al.: Late fusion incomplete multi-view clustering. TPAMI **41**(10), 2410–2423 (2019)
12. Pei, S., Nie, F., Wang, R., Li, X.: Efficient clustering based on a unified view of k-means and ratio-cut. In: NeurIPS (2020)
13. Ren, Z., Sun, Q.: Simultaneous global and local graph structure preserving for multiple kernel clustering. TNNLS **32**(5), 1839–1851 (2021)
14. Ren, Z., Sun, Q., Wei, D.: Multiple kernel clustering with kernel k-means coupled graph tensor learning. In: AAAI, pp. 9411–9418 (2021)
15. Ren, Z., Yang, S.X., Sun, Q., Wang, T.: Consensus affinity graph learning for multiple kernel clustering. IEEE Trans. Cybern. **51**(6), 3273–3284 (2020)
16. Sun, M., et al.: Projective multiple kernel subspace clustering. IEEE Trans. Multim. 1 (2021)
17. Wang, S., et al.: Multi-view clustering via late fusion alignment maximization. In: IJCAI, pp. 3778–3784 (2019)
18. Xia, R., Pan, Y., Du, L., Yin, J.: Robust multi-view spectral clustering via low-rank and sparse decomposition. In: AAAI, vol. 28 (2014)
19. Yang, C., Ren, Z., Sun, Q., Wu, M., Yin, M., Sun, Y.: Joint correntropy metric weighting and block diagonal regularizer for robust multiple kernel subspace clustering. Inf. Sci. **500**, 48–66 (2019)
20. Zhou, S., et al.: Multiple kernel clustering with neighbor-kernel subspace segmentation. TNNLS **31**(4), 1351–1362 (2020)

Supervised Multi-view Latent Space Learning by Jointly Preserving Similarities Across Views and Samples

Xiaoyang Li[1], Martin Pavlovski[2], Fang Zhou[1(✉)], Qiwen Dong[1],
Weining Qian[1], and Zoran Obradovic[2]

[1] School of Data Science and Engineering, East China Normal University,
Shanghai, China
51195100010@stu.ecnu.edu.cn, {fzhou,qwdong,wnqian}@dase.ecnu.edu.cn
[2] Temple University, Philadelphia, PA, USA
{martin.pavlovski,zoran.obradovic}@temple.edu

Abstract. In multi-view learning, leveraging features from various views in an optimal manner to improve the performance on predictive tasks is a challenging objective. For this purpose, a broad range of approaches have been proposed. However, existing works focus either on capturing (1) the common and complementary information across views, or (2) the underlying between-view relationships by exploiting view pair similarities. Besides, for the latter, we find that the obtained similarities cannot representatively reflect the differences among views. Towards addressing these issues, we propose a novel approach called MELTS (Multi-viEw LatenT space learning with Similarity preservation) for multi-view classification. MELTS first utilizes distance correlation to explore hidden between-view relationships. Furthermore, by assuming that different views share certain common information and each view carries its unique information, the method leverages both (1) the similarity information of different view pairs and (2) the label information of distinct sample pairs, to learn a latent representation among multiple views. The experimental results on both synthetic and real-world datasets demonstrate that MELTS considerably improves classification accuracy compared to other alternative methods.

Keywords: Multi-view classification · Latent representation learning · Distance correlation

1 Introduction

In the real world, an object can often be described by multiple different features. For example, in the vision domain, a video can be described by language, visual and audio features. These data are called multi-view data and are often collected from different data sources or measured by different methods. Multi-view classification widely exist in real-world applications, such as image classification [9] and disease diagnosis [11].

© The Author(s), under exclusive license to Springer Nature Switzerland AG 2022
A. Bhattacharya et al. (Eds.): DASFAA 2022, LNCS 13246, pp. 689–696, 2022.
https://doi.org/10.1007/978-3-031-00126-0_53

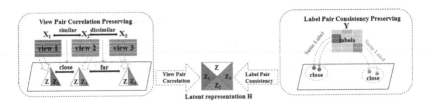

Fig. 1. An illustration of the constituent components of the MELTS framework.

Multi-view data have two important principles: consensus and complementary principles [4]. Since different views share information, the consensus principle suggests that distinct views should be in agreement. Existing methods focus primarily on extracting common information shared by all views (e.g. [3]). The main idea is to map the original views into a latent subspace and maximize the consensus among the learned representations from distinct views. This group of methods reduce feature redundancy, however, they fail to capture the complementary information contained in individual views.

On the other hand, the complementary principle assumes that each view contains some unique information. The recent methods [4,8] tend to learn representations that contain both the consistency information shared by all views and the complementary information of each individual view, simultaneously. To further make effective use of complementary information, a diversity constraint is added to enforce the learned representations from different views to be independent of one other [7]. Promoting sufficient diversity across different views essentially ignores the *true* underlying relationship among the views.

To explore the relationship among views, recently, [10] presented a method that utilizes Jensen Shannon divergence (JS-divergence) to calculate the similarities between view pairs. However, we observe that the view-pair similarities cannot be captured correctly when the number of samples and some between-sample distances are large, due to the manner in which JS-divergence is formulated.

To address the aforementioned issues, we propose a novel approach MELTS (Multi-viEw LatenT space learning with Similarity preservation) aimed at preserving similarities between view pairs as well as between sample pairs in the learned representations. Figure 1 illustrates the framework of the approach. MELTS learns a latent subspace that is composed of a shared component across all views and a specific component for each view. To capture the relationship among views, we propose to utilize distance correlation [6] to calculate the similarity between each view pair in the original data. We assume that if two views are less similar, then their specific components will contain more discriminative information. Otherwise, their specific components will be similar to each other. Such a latent representation *true* learning approach would tend to preserve the latent relationships between views. In addition, by taking label information into account, MELTS minimizes the distances between any sample pairs in the subspace if they are from the same class, which further enhances the separability of the learned features. After learning the representations, a linear classifier is applied for classification.

The contributions of this work are summarized as follows:

- We identified the cause of the problem of using JS-divergence to calculate the relationship between view pairs and proposed to utilize distance correlation as a remedy.
- We proposed a method MELTS that simultaneously leverages both (1) the similarity information for different view pairs and (2) the label consistency of sample pairs to learn a latent representation among multiple views.

2 Proposed Approach

Let the matrix $\mathbf{X}_i = \left[\mathbf{x}_1^i, \cdots, \mathbf{x}_n^i\right] \in \mathbb{R}^{d_i \times n}$ represent a set of data samples in the i-th view, where d_i denotes the feature dimension of the i-th ($i = 1, \cdots, k$) view and k denotes the number of views. Further, let $\mathbf{Y} \in \mathbb{R}^{n \times c}$ be the label matrix, where c is the number of classes and each row is a label vector in which the b-th entry is 1 and the rest are -1 if its corresponding sample falls in the b-th class. The goal is to classify each sample by learning the latent representations among k different views of the training samples. Since an individual sample is described by multiple views, we assume that the latent representations contain a shared component $\mathbf{Z} \in \mathbb{R}^{n \times d_s}$ among all views and a specific component $\mathbf{Z}_i \in \mathbb{R}^{n \times d}$ within each single view, where d_s and d are the dimensions of the shared and specific components, respectively.

2.1 View Pair Correlation Preserving

Before describing our model, we first introduce the view pair correlation preserving term, which intends to capture the discriminative information among multiple views by exploring correlation between view pairs. We hypothesize that if a view-pair correlation under the original representation is strong, then the view-specific components of two views should be similar. For example, if two views' original representations $\mathbf{X}_i$ and $\mathbf{X}_j$ are correlated, we expect that $\mathbf{Z}_i$ and $\mathbf{Z}_j$ are similar, which means $\mathbf{Z}_i$ and $\mathbf{Z}_j$ are close to each other in the projected space. By utilizing this term, the latent relationships between the views can be preserved. However, each view may contain different feature dimension, which brings a challenge to measure the between-view correlation. To address this issue, one recent work [10] applies JS-divergence to measure the similarity between view pairs. Specifically, they first calculate the distance between any two samples in each view. After computing the distance probability between all sample pairs in each view, they use JS-divergence to calculate the similarity between view pairs. When view pairs are more similar, JS-divergence is closer to zero.

Computing the difference of distance probability distributions in two views allows to measure the similarity between views with different feature dimensions. However, we found that when (1) the number of samples is large (or even when some values of sample pairs' distances are large) and (2) the difference of distance probability between two views is small; the view pair similarity calculated by JS-divergence *tends to be small*, which fails to capture the relationship among the views. The main reason is that larger values of the number of samples and sample pairs' distances make the distance probabilities smaller and cause negligible difference between the distance probability distributions of two views, which leads to the JS-divergence being close to zero.

In our model, we adopt the idea of distance correlation [6], which originally was designed to describe the dependence between two random variables. Distance correlation utilizes distance covariance to capture the relationship between views and avoids to calculate the distance probability, which can overcome the problem of using JS-divergence. Let $l_{a,b}^i$ represent the Euclidean distance between the samples a and b in the i-th view, and let $\mathbf{L}^i$ be the distance matrix containing all pairwise distances. After double centering of each distance, the distance correlation between views i and j is: $dCorr(\mathbf{X}_i, \mathbf{X}_j) = \frac{dCov(\mathbf{X}_i, \mathbf{X}_j)}{\sqrt{dVar(\mathbf{X}_i)dVar(\mathbf{X}_j)}}$, where $dCov(\mathbf{X}_i, \mathbf{X}_j) = \sqrt{\frac{1}{n^2} \sum_{a=1}^n \sum_{b=1}^n t_{a,b}^i t_{a,b}^j}$ and $dVar(\mathbf{X}_i) = dCov(\mathbf{X}_i, \mathbf{X}_i)$. The view pair correlation preserving is achieved by minimizing the following term:$\frac{1}{k(k-1)} \sum_{i<j} \|\mathbf{Z}_i - \mathbf{Z}_j\|_F^2 C_{i,j}$, where $C_{i,j} = dCorr(\mathbf{X}_i, \mathbf{X}_j)$ acts as a weight based on the correlation between the views i and j. When the correlation value is large, the term intends to pull the specific components of the two views closer in the projected space. Otherwise, it tries to pull them further apart.

2.2   Label Pair Consistency Preserving

To further leverage the label information, we introduce a term that enforces the samples from the same class to be closer to one another in the latent space, which enhance the separability of the learned representations. Specifically, by concatenating the latent representations of all views results in $\mathbf{H} = [\mathbf{Z}\, \mathbf{Z}_1 \cdots \mathbf{Z}_k] \in \mathbb{R}^{n \times (d_s + kd)}$, we obtain the samples that contain all views' information in the projected space. For samples q and l, the learned features are $\mathbf{H}^q \in \mathbb{R}^{(d_s + kd)}$ and $\mathbf{H}^l \in \mathbb{R}^{(d_s + kd)}$, respectively. The consistency weights are defined as follows: if q and l are from the same class, $s_{q,l} = 1$; otherwise, $s_{q,l} = 0$; where $q, l = 1, ..., n$. Therefore, the sample label pair consistency term is: $\frac{1}{n^2} \sum_{q=1}^n \sum_{l=1}^n \left\| \mathbf{H}^q - \mathbf{H}^l \right\|_2^2 s_{q,l} = \frac{2}{n^2} tr(\mathbf{H}^T \mathbf{L} \mathbf{H})$, where $\mathbf{L} = \mathbf{D} - \mathbf{S}$ is the Laplacian of the similarity matrix $\mathbf{S}$, and $\mathbf{D}$ is the diagonal matrix with $d_{q,q} = \sum_{l=1}^n s_{q,l}$. By minimizing the sample label pair consistency term, when two samples are from the same class, the distance between their latent representations is reduced.

2.3   Complete Objective Function

Recall that multi-view data follows the consensus and complementary principles. Our goal is to learn a latent representation for each view that preserves (1) the information shared among all views and (2) specific information coming from each single view. Following [8], this can be achieved by minimizing the following term:$\sum_{i=1}^{k} \left\| \mathbf{X}_i^{\mathrm{T}} \mathbf{P}_i + \mathbf{1b}_i^{\mathrm{T}} - [\mathbf{Z} \ \mathbf{Z}_i] \right\|_F^2$, where $\mathbf{P}_i \in \mathbb{R}^{d_i \times (d_s+d)}$ is the transformation matrix of the i-th view used to map the original representation $\mathbf{X}_i$ into the latent space; $\mathbf{b}_i \in \mathbb{R}^{(d_s+d)}$ is a bias term and $\mathbf{1} \in \mathbb{R}^n$ is an all-ones vector. The learned feature vector for the i-th view $[\mathbf{Z} \ \mathbf{Z}_i] \in \mathbb{R}^{n \times (d_s+d)}$ consists of the shared component $\mathbf{Z}$ across all views and the specific component $\mathbf{Z}_i$ of the i-th view. After learning the view specific and shared components, a linear classifier is designed for classification. Following [8], the corresponding loss function is formulated as: $\left\| \mathbf{HW} + \mathbf{1b}^{\mathrm{T}} - \mathbf{Y} \right\|_F^2$, where $\mathbf{W} \in \mathbb{R}^{(d_s+kd) \times c}$ is a transformation matrix that maps the projected features $\mathbf{H}$ into the label space and $\mathbf{b} \in \mathbb{R}^c$ is a bias term. By minimizing the loss function, the difference between the predicted outputs $\mathbf{HW} + \mathbf{1b}^{\mathrm{T}}$ and the ground-truth $\mathbf{Y}$ tends to reduce. Therefore, the complete objective function is defined as:

$$
\min_{\mathbf{W},\mathbf{P}_i,\mathbf{Z},\mathbf{Z}_i,\mathbf{b},\mathbf{b}_i} \sum_{i=1}^{k} \left\| \mathbf{X}_i^{\mathrm{T}} \mathbf{P}_i + \mathbf{1b}_i^{\mathrm{T}} - [\mathbf{Z} \ \mathbf{Z}_i] \right\|_F^2 + \frac{\alpha}{k(k-1)} \sum_{i<j} \left\| \mathbf{Z}_i - \mathbf{Z}_j \right\|_F^2 C_{i,j}
$$

$$
+ \frac{\beta}{n^2} \sum_{q=1}^{n} \sum_{l=1}^{n} \left\| \mathbf{H}^q - \mathbf{H}^l \right\|_2^2 s_{q,l} + \gamma \left\| \mathbf{HW} + \mathbf{1b}^{\mathrm{T}} - \mathbf{Y} \right\|_F^2 + \theta \left\| \mathbf{W} \right\|_F^2 , \quad (1)
$$

where $\left\| \mathbf{W} \right\|_F^2$ is a regularization term included to help prevent overfitting. Note that the first three terms aim to learn the latent representation among multiple views by utilizing the label information of the sample pairs and the similarity information of the view pairs. The last two terms constitute a regularized loss between the predicted outputs and the ground-truth.

2.4 Optimization Procedure

For the objective function from Eq. (1), we solve the problem by using an alternating optimization strategy and update $\mathbf{W}$, $\{\mathbf{P}_i\}_{i=1}^{k}$, $\mathbf{Z}$, $\{\mathbf{Z}_i\}_{i=1}^{k}$, $\mathbf{b}$, $\{\mathbf{b}_i\}_{i=1}^{k}$, iteratively to optimize these variables.

First, we fix $\mathbf{W}$, $\mathbf{Z}$, $\{\mathbf{Z}_i\}_{i=1}^{k}$, $\mathbf{b}$ and update $\{\mathbf{P}_i\}_{i=1}^{k}$, $\{\mathbf{b}_i\}_{i=1}^{k}$: by setting the derivative of Eq. (1) w.r.t. $\mathbf{P}_i$, $\mathbf{b}_i$ to zero, we get $\mathbf{P}_i = -(\mathbf{X}_i \mathbf{X}_i^{\mathrm{T}})^{-1} \mathbf{X}_i (\mathbf{1b}_i^{\mathrm{T}} - [\mathbf{Z} \ \mathbf{Z}_i])$ and $\mathbf{b}_i = \frac{1}{n}[\mathbf{Z} \ \mathbf{Z}_i]^{\mathrm{T}} \mathbf{1} - \frac{1}{n} \mathbf{P}_i^{\mathrm{T}} \mathbf{X}_i \mathbf{1}$.

Next, we fix $\{\mathbf{P}_i\}_{i=1}^{k}$, $\{\mathbf{b}_i\}_{i=1}^{k}$, $\mathbf{Z}$, $\{\mathbf{Z}_i\}_{i=1}^{k}$ and update $\mathbf{W}$, $\mathbf{b}$: by setting the derivative of the Eq. (1) w.r.t. $\mathbf{W}$, $\mathbf{b}$ to zero, yields $\mathbf{W} = -(\mathbf{H}^{\mathrm{T}} \mathbf{H} + \frac{\theta}{\gamma} \mathbf{I})^{-1} \mathbf{H}^{\mathrm{T}} (\mathbf{1b}^{\mathrm{T}} - \mathbf{Y})$ and $\mathbf{b} = \frac{1}{n} \mathbf{Y}^{\mathrm{T}} \mathbf{1} - \frac{1}{n} \mathbf{W}^{\mathrm{T}} \mathbf{H}^{\mathrm{T}} \mathbf{1}$.

Lastly, we fix $\{\mathbf{P}_i\}_{i=1}^{k}$, $\{\mathbf{b}_i\}_{i=1}^{k}$, $\mathbf{W}$, $\mathbf{b}$ and update $\mathbf{Z}$, $\{\mathbf{Z}_i\}_{i=1}^{k}$: Note that we update $\mathbf{Z}_i$ in a view-by-view manner. When updating $\mathbf{Z}_i$, $\mathbf{Z}$ and $\{\mathbf{Z}_j\}_{j \neq i}$ are fixed. Now, considering that $\mathbf{H} = [\mathbf{Z} \ \mathbf{Z}_1 \ \cdots \ \mathbf{Z}_k]$, $\frac{\beta}{n^2} \sum_{q=1}^{n} \sum_{l=1}^{n} \left\| \mathbf{H}^q - \mathbf{H}^l \right\|_2^2 s_{q,l}$ can be rewritten as $\frac{2\beta}{n^2} (tr(\mathbf{Z}^{\mathrm{T}} \mathbf{LZ}) + \cdots + tr(\mathbf{Z}_k^{\mathrm{T}} \mathbf{LZ}_k))$. By setting $\mathbf{B} = \mathbf{1b}^{\mathrm{T}} - \mathbf{Y}$ and $\mathbf{E}_1 = \mathbf{Zw} + \sum_{j \neq i}^{k} \mathbf{Z}_j \mathbf{w}_j + \mathbf{B}$, $\gamma \left\| \mathbf{HW} + \mathbf{1b}^{\mathrm{T}} - \mathbf{Y} \right\|_F^2$ can be rewritten as $\gamma \left\| \mathbf{Z}_i \mathbf{w}_i + \mathbf{E}_1 \right\|_F^2$, where $\mathbf{w}_i \in \mathbb{R}^{d \times c}$.

By setting $\mathbf{A}_i = \mathbf{X}_i^T\mathbf{P}_i + \mathbf{1b}_i^T$ and denoting $\mathbf{P}_i = [\mathbf{P}_{i1}\ \mathbf{P}_{i2}]$, $\mathbf{P}_{i1} \in \mathbb{R}^{d_i \times d_s}$, $\mathbf{P}_{i2} \in \mathbb{R}^{d_i \times d}$, $\mathbf{b}_i = [\mathbf{b}_{i1}\ \mathbf{b}_{i2}]$, $\mathbf{b}_{i1} \in \mathbb{R}^{d_s}$, and $\mathbf{b}_{i2} \in \mathbb{R}^d$, we get $\mathbf{A}_i = [\mathbf{A}_{i1}\ \mathbf{A}_{i2}] = [\mathbf{X}_i^T\mathbf{P}_{i1} + \mathbf{1b}_{i1}^T\ \mathbf{X}_i^T\mathbf{P}_{i2} + \mathbf{1b}_{i2}^T]$. Taking the derivative of Eq. (1) and setting it to zero yields $\frac{2\beta}{n^2}\mathbf{LZ}_i + \mathbf{Z}_i\left(\mathbf{I} + \frac{\alpha}{k(k-1)}(\sum_{j \neq i}C_{i,j})\mathbf{I} + \gamma\mathbf{w}_i\mathbf{w}_i^T\right) = \mathbf{A}_{i2} + \frac{\alpha}{k(k-1)}\sum_{j \neq i}C_{i,j}\mathbf{Z}_j - \gamma\mathbf{E}_1\mathbf{w}_i^T$, where $\mathbf{I} \in \mathbb{R}^{d \times d}$ is an identity matrix. The closed-form solution of the given term can be computed using the algorithm from [1].

When updating $\mathbf{Z}$, $\{\mathbf{Z}_i\}_{i=1}^k$ are fixed. By setting $\mathbf{E}_2 = \sum_{i=1}^k \mathbf{Z}_i\mathbf{w}_i + \mathbf{B}$ and the derivative of Eq. (1) w.r.t. $\mathbf{Z}$ to zero, we obtain $\frac{2\beta}{n^2}\mathbf{LZ} + \mathbf{Z}\left(k*\mathbf{I} + \gamma\mathbf{ww}^T\right) = \sum_{i=1}^k \mathbf{A}_{i1} - \gamma\mathbf{E}_2\mathbf{w}^T$, where $\mathbf{I} \in \mathbb{R}^{ds \times ds}$ is an identity matrix. The optimization problem of the given term can also be solved using the algorithm from [1].

3   Experiments

3.1   Experiment Setting

Three publicly available and widely used datasets, including image, text, or even multi-source data, were used in the experiments. **MSRC-V1** [12] is a scene image dataset. For a given image, the task is to predict the image's category. The dataset consists of six views, 7 classes and a total of 210 samples. **TweetFit** [2] consists of recordings from individual users' sensors and the data were collected from multiple social media. For a given user, the task is to predict the user's body mass index (BMI). We selected users with data available for all data sources, which consists of 8 classes, 205 samples and three views. **BBCSport**[1] is a sport news text dataset. For a given text, the task is to predict the text's category. The dataset consists of 116 samples, 5 classes and four views.

We compared the classification accuracy of MELTS with the following methods. **SVM** applied to the concatenation of multiple views. **MVDA** [5] determines a discriminant common space by learning linear transforms of each view. **MVCS** [8] learns a latent subspace from multiple views by simultaneously considering the correlated information across the views and the unique information within each single view. **WeReg** [9] adaptively assigns weights to distinct views to account for view importance.

In all experiments, standard fivefold cross-validation was utilized and the average accuracies along with their standard deviations on each dataset were reported. For each of the 5 trials, we randomly choose 75% of the data for training, and the rest for validation.

The parameters were fine-tuned based on validation performance. PCA was used on the original data to initialize the shared component $\mathbf{Z}$ and each specific component $\mathbf{Z}_i$. Let d_i^* denote the feature dimension of the i-th view representation obtained by PCA and d^* denote the minimum of $\{d_1^*, \cdots, d_k^*\}$. As for the hyperparameters of MELTS, $\alpha, \beta, \gamma, \theta$ were selected from $\{10^{-3}, 10^{-2}, \cdots, 10^2, 10^3\}$ based on optimal performance, while ds and d were selected from $\{1, \frac{1}{4}d^*, \frac{1}{2}d^*, \frac{3}{4}d^*, d^*\}$. The tradeoff parameter C, i.e. the inverse

---

[1] http://mlg.ucd.ie/datasets/segment.html.

regularization strength, for SVM was selected from $\{10^{-3}, 10^{-2}, \cdots, 10^2, 10^3\}$. For MVDA, a 1-Nearest Neighbor classifier was applied to the low-dimensional representations for classification. Since MELTS is built on the basis of MVCS, it was compared to MVCS under the same ds and d to show the influence of the added terms. In the case of WeReg, for fair comparison, a sample's label is predicted based on the maximum label probability instead of using the kNN-based prediction approach from [9].

3.2   Results on Synthetic Data and Real-World Datasets

We initially assess the performance of the proposed method on the synthetic data. We generated a dataset with three views containing 300 samples, each having 200 features. The samples were generated from six normal distributions with different parameters, thus defining six separate classes of samples. 50 out of the 200 features were generated to be similar to increase feature redundancy. To better understand the effect of view similarity, we generate similar parameter values for the first two views and generate the third view by using distinct parameter values. The pairwise between-view similarities based on distance correlation are following: $dCorr(\mathbf{X}_1, \mathbf{X}_2) = 0.99$, $dCorr(\mathbf{X}_1, \mathbf{X}_3) = dCorr(\mathbf{X}_2, \mathbf{X}_3) = 0.59$. It can be inferred that view 1 is more similar to view 2, than to view 3.

Table 1 shows the classification accuracies obtained by all models on the synthetic data. First, SVM and WeReg perform poorly since these three methods directly concatenate the features of all views and ignore the feature redundancy across multi-view representations. Compared with SVM and WeReg, MELTS attains an improvement of 39% and 35%, respectively. Second, MELTS achieves 34% higher accuracy than MVDA since the method ignores the hidden specific information of each single view. Furthermore, MELTS produces 2% higher accuracy than MVCS, which suggests that leveraging the sample pairs' label information and the view pairs' similarities can help learn more relevant and discriminative features.

The experimental results on the three real-world datasets are also reported in Table 1. Compared with MVCS, MELTS achieves considerable improvements on the three datasets. For example, MELTS achieves 7% higher accuracy than MVCS on MSRC-V1 and 14% on TweetFit. WeReg obtains lower performance than MELTS on the three datasets, as it ignores the latent relationships among the different views. The results obtained by MVDA are much lower than those of MELTS on all datasets. Moreover, on most datasets, MELTS obtains the smallest standard deviation, which demostrates the stability of our method.

Table 1. Average accuracy obtained by MELTS and the alternative methods.

Methods	MSRC-V1	TweetFit	BBCSport	Synthetic data
SVM	0.97 ± 0.01	0.34 ± 0.04	0.89 ± 0.05	0.19 ± 0.03
MVDA	0.93 ± 0.02	0.29 ± 0.04	0.80 ± 0.08	0.24 ± 0.07
MVCS	0.91 ± 0.07	0.24 ± 0.07	0.74 ± 0.07	0.56 ± 0.05
WeReg	0.96 ± 0.02	0.33 ± 0.05	0.94 ± 0.05	0.23 ± 0.07
MELTS	**0.98 ± 0.01**	**0.38 ± 0.01**	**0.95 ± 0.05**	**0.58 ± 0.07**

4   Conclusion

In this paper, we proposed MELTS, a novel multi-view learning approach. MELTS learns a latent subspace, containing the common information across all views and the individual information carried by each view. MELTS effectively models the between-view relationships and utilizes the label information to enhance the discriminability in the learned latent subspace. The results indicate that MELTS achieves better classification performance than other methods.

Acknowledgements. This research was supported in part by NSFC grant 61902127 and Natural Science Foundation of Shanghai 19ZR1415700.

References

1. Bartels, R.H., Stewart, G.W.: Solution of the matrix equation ax+ xb= c [f4]. Commun. ACM **15**(9), 820–826 (1972)
2. Farseev, A., Chua, T.S.: Tweetfit: fusing multiple social media and sensor data for wellness profile learning. In: AAAI, pp. 95–101 (2017)
3. Frome, A., Corrado, G., Shlens, J., et al.: Devise: a deep visual-semantic embedding model. In: NeurIPS, pp. 2121–2129 (2013)
4. Jia, X., Jing, X.Y., Zhu, X., Chen, S., et al.: Semi-supervised multi-view deep discriminant representation learning. IEEE TPAMI (2020)
5. Kan, M., Shan, S., Zhang, H., Lao, S., Chen, X.: Multi-view discriminant analysis. IEEE TPAMI **38**(1), 188–194 (2015)
6. Székely, G.J., Rizzo, M.L., Bakirov, N.K., et al.: Measuring and testing dependence by correlation of distances. Ann. Statist. **35**(6), 2769–2794 (2007)
7. Wu, F., Jing, X.Y., et al.: Semi-supervised multi-view individual and sharable feature learning for webpage classification. In: WWW, pp. 3349–3355 (2019)
8. Xue, X., Nie, F., Wang, S., Chang, X., Stantic, B., Yao, M.: Multi-view correlated feature learning by uncovering shared component. In: AAAI, pp. 2810–2816 (2017)
9. Yang, M., Deng, C., Nie, F.: Adaptive-weighting discriminative regression for multi-view classification. Pattern Recogn. **88**, 236–245 (2019)
10. Zhang, J., Zhang, P., Liu, L., et al.: Collaborative weighted multi-view feature extraction. Eng. Appl. Artif. Intell. **90**, 103527 (2020)
11. Zhang, M., Yang, Y., Shen, F., Zhang, H., Wang, Y.: Multi-view feature selection and classification for Alzheimer's disease diagnosis. Multim. Tools Appl. **76**(8), 10761–10775 (2017)
12. Zhou, T., Zhang, C., Gong, C., et al.: Multiview latent space learning with feature redundancy minimization. IEEE Trans. Cybern. **50**(4), 1655–1668 (2018)

Market-Aware Dynamic Person-Job Fit with Hierarchical Reinforcement Learning

Bin Fu[1], Hongzhi Liu[1(✉)], Hui Zhao[2], Yao Zhu[3], Yang Song[6], Tao Zhang[7], and Zhonghai Wu[4,5(✉)]

[1] School of Software and Microelectronics, Peking University, Beijing, China
{binfu,liuhz}@pku.edu.cn
[2] Wangxuan Institute of Computer Technology, Peking University, Beijing, China
1701212454@pku.edu.cn
[3] Center for Data Science, Peking University, Beijing, China
yao.zhu@pku.edu.cn
[4] National Engineering Center of Software Engineering,
Peking University, Beijing, China
[5] Key Lab of High Confidence Software Technologies (MOE),
Peking University, Beijing, China
wuzh@pku.edu.cn
[6] BOSS Zhipin NLP Center, Beijing, China
songyang@kanzhun.com
[7] BOSS Zhipin, Beijing, China
kylen.zhang@kanzhun.com

Abstract. Person-Job Fit (PJF) is the core of online recruitment. Several recent methods took PJF as a preference-learning problem, and tried to learn two-sided preferences from their historical behaviors. However, they ignored users' interactive feedbacks (*accepted* or *rejected*) received from the other side which may change users' preferences. In addition, they neglected the status of local market which may affect the final matching results of person-job pairs.

To solve these issues, we propose a market-aware dynamic PJF method with <u>hi</u>erarchical <u>re</u>inforcement learning (HIRE). We design a two-level hierarchy of reinforcement learning policies. Two low-level policies aim to learn the dynamic preferences of both persons and jobs with consideration of interactive feedbacks, and a high-level policy aims to learn the optimal dynamic matching strategy with consideration of local market state. Extensive experiments on two real-world datasets show the effectiveness of HIRE compared with the state-of-the-art.

Keywords: Person-Job Fit · Bilateral decision process · Hierarchical reinforcement learning · Local market state

1 Introduction

Due to the development of the Internet, online recruitment has become a popular mode in the recruitment field. Its core is Person-Job Fit (PJF), which aims to

A. Bhattacharya et al. (Eds.): DASFAA 2022, LNCS 13246, pp. 697–705, 2022.
https://doi.org/10.1007/978-3-031-00126-0_54

match right persons with right jobs by satisfying both the job requirements and the personal desires [1]. However, with a tremendous number of job postings (JPs) and curriculum vitaes (CVs) available on the Internet, both employees and employers suffer from the problem of information overloading. To overcome this issue, several early studies took PJF as a recommendation task [2–4], and designed recommendation models based on handcrafted features extracted from JPs and CVs. Recently, some studies regarded PJF as a semantic matching problem [5–7], and tried to learn the representations of CVs and JPs with deep learning techniques. Nevertheless, CVs and JPs do not always contain nor fully reflect users' preferences. To solve this problem, some methods [8–10] were proposed to learn two-sided implicit preferences from their historical behaviors.

However, prior studies ignore the effects of interactive feedbacks and local market state. Users' preferences are affected by the interactive feedbacks (e.g. *accepted* or *rejected*) received from the other party in the two-sided market. For example, a user may change his/her expectation after receiving rejection(s). The status of recruitment market (e.g. supply-demand situation), may also affect the matching results. For instance, with less *IOS Development* jobs available in the market, some job seekers are driven to match the related jobs like *Android Development*. Prior methods make recommendations from the individual perspective, which are prone to cause the fierce competition and matching failures.

To solve these issues, we propose to model two-sided personalized states and the local market state. Specifically, we model users' personalized states based on their previous behaviors and interactive feedbacks, and adopt the supply-demand situation as the local market state. Then, we proposed to utilize reinforcement learning (RL) to learn a market-aware matching policy based on these states. Since a bilateral matching decision can be decomposed into two unilateral decisions, i.e. person decision and job decision, we further propose to learn two unilateral policies, i.e. a person policy and a job policy, to learn the intrinsic preferences of both persons and jobs from their decision processes. In addition, we design a two-level hierarchy for these three policies to transfer the preferences from both person policy and job policy to the matching policy, and call this proposed method with **hi**erarchical **re**inforcement learning as HIRE.

The main contributions of this paper are summarized as follows: (1) we propose to consider the effects of interactive feedbacks and local market status; (2) we design a dynamic PJF method with a two-level hierarchy of RL policies; (3) we conduct extensive experiments on two real-world datasets, and show the effectiveness of HIRE compared with the state-of-the-art.

2   Related Work

Early studies took PJF as a recommendation task [2, 3, 11, 12] or a bilateral preference-matching task [4] based on the handcrafted features extracted from CVs and JPs. However, the manual design of features is of expensive cost and inefficient. With the surge of deep learning techniques, several studies considered PJF as a deep semantic matching problem between the textual CVs and JPs [5–7, 13]. Text-CNN [5] and hierarchical attentive BiLSTM [6] were proposed to learn better representations of both CVs and JPs. Le et al. [13] tried to incorporate two-sided intentions to learn the co-attentive interdependence between CVs and JPs. Adversarial learning [7] and domain transfer [14] were used to learn more expressive representations of CVs and JPs. Nevertheless, CVs and JPs may not contain or fully reflect the preferences of job seekers and recruiters. To solve this issue, Yan et al. [8] tried to learn users' preferences from their historical successful matchings. However, the successful matchings are extremely sparse, since users are likely to leave the online recruitment platform once achieving a successful matching. For this, a multi-view co-teaching mechanism [15], and a fusion of semantic entities and textual features [10], were proposed to enhance the representation learning. Fu et al. [9] tried to profile two-sided dynamic preferences from their historical multiple behaviors. However, to the best of our knowledge, prior studies ignored the effects of interactive feedbacks and the market status.

3   Problem Definition

Let P and J denote the person (i.e. job seeker) set and the job (i.e. recruiter) set, respectively. Each person $p \in P$ has a textual curriculum vitae C_p. Each job $j \in J$ has a textual job posting D_j. An occupation taxonomy is often used to organize CVs and JPs. Each user has an interaction history, including unilateral decisions (e.g. *not apply/invite, apply/invite, accept* and *reject*) and interaction feedbacks (e.g. *rejected* and *accepted*). Let $H_{p,t} = \{(j_k, b_k, k) \mid k < t\}$ and $H_{j,t} = \{(p_k, b_k, k) \mid k < t\}$ denote the historical interactions of job seeker p and recruiter j at time t, respectively. (j_k, b_k, k) in $H_{p,t}$ denotes that person p had an interaction b_k (i.e. made a unilateral decision or received an interactive feedback) with job j_k at time k. Let $x^{(t)}$ denote the local market state at time t.

The task of market-aware dynamic PJF is to find a function $f(\cdot)$, which can predict the matching degree of any person-job pair $<p, j>$ at any time t, given their textual profiles (C_p, D_j), historical interactions $(H_{p,t}, H_{j,t})$ and the market state $x^{(t)}$, i.e. $f(<p, j> \mid C_p, D_j, H_{p,t}, H_{j,t}, x^{(t)})$. Then, $f(\cdot)$ is used to find those potential matching person-job pairs and make recommendations.

4   The Proposed Approach

We formulate the sequential interaction process between persons/jobs and the recruitment system (matching agent) as a MDP. We design a two-level hierarchy of three policies to learn two-sided preferences and predict the matching results (Fig. 1).

Markov Decision Process (MDP). The MDP is defined by seven elements $(S, A, P, R_M, R_P, R_J, \gamma)$. **State S:** each state $s_t \in S$ consists of the two-sided personalized states $(s_{p,t}, s_{j,t})$ and the local market state $x^{(t)}$. **Action A:** an action $a_t \in A$ is to select a person $p_{j,t}$ (job $j_{p,t}$) from the available candidate set $A_{j,t}$ $(A_{p,t})$ and recommend it to the target job j (person p). i.e. $a_t = p_{j,t} \in A_{j,t}$ for job j or $a_t = j_{p,t} \in A_{p,t}$ for person p. **State Transition P:** each user's state (e.g.

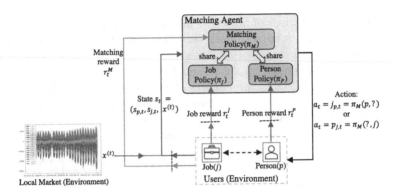

Fig. 1. The sequential interaction process as a MDP. At each time t, the matching agent observes state s_t and performs action a_t, e.g. recommends job j (i.e. $j_{p,t}$) to the target person p (i.e. $a_t = j_{p,t} = \pi_M(p,?)$), or recommends person p (i.e. $p_{j,t}$) to the target job j (i.e. $a_t = p_{j,t} = \pi_M(?,j)$). Then, the matching agent receives three rewards $\{r_t^M, r_t^P, r_t^J\}$ for three policies $\{\pi_M, \pi_P, \pi_J\}$ according to users' feedbacks and the matching result.

$s_{p,t}$) is updated according to his/her decision and interactive feedback, and the local market state $x^{(t)}$ is updated by the realtime supply-demand situation.

Reward Function $\mathcal{R}_M, \mathcal{R}_P, \mathcal{R}_J$: they measure the scalar rewards of each action for policies $\{\pi_M, \pi_P, \pi_J\}$, according to the corresponding users' feedbacks.

Discount Rate γ: it's the discount rate for future rewards, $\gamma \in [0,1]$. Our goal is to learn an optimal matching policy π_M^* which maximizes the cumulative matching reward (r_t^M), with consideration of the cumulative unilateral rewards of person policy π_P (r_t^P) and job policy π_J (r_t^J).

Personalized User State. We model users' personalized states based on their most recent N behaviors and interactive feedbacks, including the positive (*apply/invite* and *accept*, $\heartsuit$) and negative (*not apply/invite* and *reject*, -) behaviors, and the positive (*being-accepted* +) and negative (*being-rejected* $\Diamond$) interactive feedbacks. The textual CVs and JPs are encoded into vectors using text-CNN [5], then GRU is used to model each sequence. We concatenate the outputs of these GRUs with user's profile embedding to obtain user's state representation, e.g. $v_{p,t} = [\ e_p \parallel v_{p,t}^+ \parallel v_{p,t}^\heartsuit \parallel v_{p,t}^- \parallel v_{p,t}^\Diamond\ ]$.

Local Market State. We model the local market state from the supply-demand situation, and $x^{(t)} = <x_1^{(t)}, x_2^{(t)}, ..., x_l^{(t)}>$ where $x_i^{(t)}$ $(i \in [1,l])$ denotes the supply-demand difference in the i-th occupation category o_i at time t. We normalize $x^{(t)}$, i.e. $\bar{x}^{(t)} = x^{(t)}/\max_{i \in [1,l]} |x_i^{(t)}|$, and feed it into a linear layer.

Reward Functions $\mathcal{R}_P, \mathcal{R}_J, \mathcal{R}_M$. We hope to give a higher matching reward to those matched person-job pairs which reduce the supply-demand imbalance of the current local market. Specifically, we hope the matching person-job pair $< p, j >$ can reduce the talent surplus for category o_p (i.e. $x_{o_p}^{(t)} > 0$) and the talent gap for category o_j (i.e. $x_{o_j}^{(t)} < 0$), where o_p and o_j denote the occupation

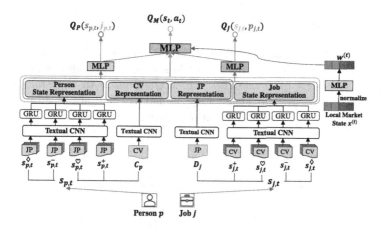

Fig. 2. Hierarchical architecture of Q functions $\{Q_M, Q_P, Q_J\}$ for policies $\{\pi_M, \pi_P, \pi_J\}$.

category of person p and job j, respectively. For any person-job pair $<p, j>$, according to their matching result and the supply-demand situation, we define the matching reward function: $r_t^M = \mathcal{R}_M(s_t, a_t) = 1 + \alpha \mathbb{I}(x_{o_p}^{(t)} > 0) * \mathbb{I}(x_{o_j}^{(t)} < 0)$ if $<p, j>$ *is matched*, else 0. $\alpha \in [0, 1]$ is a market reward coefficient. This two-fold reward helps the matching policy to identify those matching person-job pairs which are more fit to the current market. We formulate the unilateral feedbacks as unilateral rewards for the low-level policies π_P and π_J to learn users' preferences. Then, for any person-job pair $<p, j_{p,t}>$, according to the unilateral decision of person p to job $j_{p,t}$, the unilateral preference reward function $\mathcal{R}_P$ is defined: $r_t^P = \mathcal{R}_P(s_{p,t}, j_{p,t}) = 1$ if the unilateral decision is one in $\{apply/invite, accept\}$, else 0. $r_t^J = \mathcal{R}_J(s_{j,t}, p_{j,t})$ for job policy π_J is defined similarly.

Hierarchical Reinforcement Learning. To transfer the preference information from both person policy π_P and job policy π_J to the matching policy π_M, we design a two-level hierarchy of RL policies. Specifically, we adopt Q-learning [16] to optimize these policies $\{\pi_M, \pi_P, \pi_J\}$, and design a hierarchical sharing architecture of their Q functions $\{Q_M, Q_P, Q_J\}$ as shown in Fig. 2.

For any person-job pair $<p, j>$, $Q_M(\cdot)$ is defined based on two-sided states $(s_{p,t}, s_{j,t})$ and the local market state $x^{(t)}$. For each individual user, since the states of other users and the local market state are unseen, we assume his/her decision process is a Partially Observable Markov Decision Process (POMDP). Then, the Q functions $\{Q_M, Q_P, Q_J\}$ are defined as follows: $Q_M(s_t, a_t) = \text{MLP}_M([v_{p,t} \| v_{j,t} \| w^{(t)}])$, $Q_P(s_{p,t}, j_{p,t}) = \text{MLP}_P([v_{p,t} \| e_{j_{p,t}}])$ and $Q_J(s_{j,t}, p_{j,t}) = \text{MLP}_J([v_{j,t} \| e_{p_{j,t}}])$, where $\{e_{p_{j,t}}, e_{j_{p,t}}\} \in \mathbb{R}^{d_1}$ are the profile embeddings of person $p_{j,t}$ and job $j_{p,t}$, and $\{v_{p,t}, v_{j,t}\} \in \mathbb{R}^{d_2}$ are the state representations of person p and job j at time t. $w^{(t)} \in \mathbb{R}^{d_3}$ is the latent representation of local market state $x^{(t)}$. We adopt the off-policy temporal difference [17] to optimize the Q functions, and minimize their mean-square losses according to the Bellman equation [18].

5   Experiments

Experimental Settings. We collected two PJF datasets *Fina.* and *Tech.* from the largest online recruitment platform 'Boss Zhipin'[1] in China, which contain the textual CVs/JPs from one city and interactions with timestamps across four months in Finance and Technology industry. There are seven occupation categories in *Fina.* and twelve occupation categories in *Tech* (Table 1).

Table 1. Statistics of dataset *Fina.* and *Tech.*

Dataset	Fina	Tech
#Behaviors:		
#not apply/invite	134,574	483,588
#apply/invite	68,058	113,122
#reject	63,228	100,750
#accept (matching)	4830	12,372
#Profiles:		
#Curriculum Vitae (CV)	8240	18,817
#Job Posting (JP)	6507	26,668
#Ave. words of CVs	264	220
#Ave. words of JPs	368	230
#Occupation Category(l)	7	12
#Month Range	2019.3~6	2019.2~5

Several PJF methods are used as baselines, including PJFNN [5], APJFNN [6], JRMPM [8],[2] IPJF [13], MV-CoN [15][3] and DPJF-MBS [9].[4] The evaluation settings are following the prior work [9] but without negative sampling during testing. For HIRE, we set $d_1 = 16$ and $d_2 = 8$, and let $N = 3$. MLP_P and MLP_J are set with two hidden layers of $\{64, 32\}$ units (i.e. $64 = 4d_2 + 2d_1$), and MLP_M is set with two hidden layers of $\{168, 80\}$ units (i.e. $168 = 2(4d_2 + d_1) + 2 \times 32 + d_3$). For fair comparisons, we tune the hyperparameters of all the comparative methods on validation set, and adopt the same pre-trained word embeddings.

Comparison with Baselines. From Table 2, the major findings are summarized: (1) HIRE performs significantly better than all the baselines on four evaluation metrics, which confirms the effectiveness of HIRE. It considers both local market state and interactive feedbacks, and can better adapt to both users' shifting preferences and dynamic market state. (2) Among baselines, DPJF-MBS

---

[1] https://www.zhipin.com.
[2] https://github.com/leran95/JRMPM.
[3] https://github.com/RUCAIBox/Multi-View-Co-Teaching.
[4] https://github.com/BinFuPKU/DPJF-MBS.

Table 2. Ranking performance. * indicates $p \leq 0.01$ on the Wilcoxon signed rank test.

Dataset	Metric	Method							Impr.
		PJFNN	APJFNN	JRMPM	MV-CoN	IPJF	DPJF-MBS	HIRE	
Fina.	MAP	0.4245	0.4202	0.4392	0.4651	0.4831	0.4993	**0.5321***	6.6%
	MRR	0.4610	0.4550	0.4392	0.5129	0.5365	0.5492	**0.5818***	5.9%
	NDCG	0.5830	0.5793	0.6015	0.6161	0.6464	0.6677	**0.7023***	5.2%
	AUC	0.6151	0.6253	0.6354	0.6643	0.7030	0.7231	**0.7587***	4.9%
Tech.	MAP	0.5100	0.5060	0.5460	0.5790	0.5929	0.6288	**0.7066***	12.4%
	MRR	0.5433	0.5385	0.5957	0.6274	0.6341	0.6683	**0.7487***	12%
	NDCG	0.6491	0.6461	0.6846	0.7041	0.7143	0.7387	**0.8034***	8.8%
	AUC	0.7232	0.7198	0.7733	0.8002	0.8107	0.8289	**0.8708***	5.1%

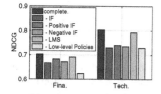

Fig. 3. Ablation study.

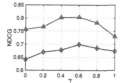
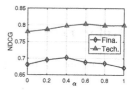

Fig. 4. Parameter sensitivity.

performs the best since it models the dynamics of two-sided preferences. {IPJF, JRMPM} utilize users' intentions/behaviors besides profiles, which outperform the pure profile-matching models {PJFNN, APJFNN}.

Ablation Study. We try to evaluate the effectiveness of interactive feedbacks $\{s^{+}, s^{\diamond}\}$, local market state $x^{(t)}$ and two low-level unilateral policies $\{\pi_P, \pi_J\}$ in HIRE, by removing each of them. Figure 3 show the results, and confirms the effectiveness of each component in HIRE. Moreover, we have some observations: (1) Removing negative interactive feedbacks (-Negative IF) can cause a larger performance drop than removing positive interactive feedbacks (-Positive IF), this may be because that positive interactive feedbacks are more sparse. (2) Negative and positive interactive feedbacks are mutual complementary to some extent. (3) Removing two low-level unilateral policies degrades the model performance seriously. (4) Local market state (LMS) makes HIRE aware of the dynamic market, which can make better matching predictions to be more fit to the market. (5) Overall, all the components of HIRE working together yield the best results.

Parameter Sensitivity. Figure 4 shows the impacts of hyper-parameters: discount factor γ and market reward coefficient α. We find that: (1) when increasing γ from 0 to 1, the performance first increases and then drops, and achieves the best when $\gamma = 0.6$. (2) The market reward enables HIRE to identify and boost those potential matching person-job pairs. It's suggested to set α to 0.4 or 0.6.

6   Conclusion

In this paper, we propose to consider both interactive feedbacks and market status for dynamic PJF, and design a hierarchical reinforcement learning method. Extensive experiments on two real-world datasets confirm that: (1) considering market status is beneficial for PJF, (2) interactive feedbacks affect users' preferences to some extent, (3) our proposed method provides an effective way to achieve the market-aware dynamic PJF. For future work, we think some components of the proposed model can be further improved, and also want to study the effects of interactive feedbacks on users' preferences in a fine-grained way.

Acknowledgements. This work was partially sponsored by National 863 Program of China (Grant No. 2015AA016009).

References

1. Edwards, J.R.: Person-job fit: a conceptual integration, literature review, and methodological critique. Int. Rev. Indust. Organ. Psychol. **6**, 283–357 (1991)
2. Lu, Y., Helou, S.E., Gillet, D.: A recommender system for job seeking and recruiting website. In: WWW (Companion Volume). International World Wide Web Conferences Steering Committee, pp. 963–966. ACM (2013)
3. Paparrizos, I., Cambazoglu, B.B., Gionis, A.: Machine learned job recommendation. In: Proceedings of the Fifth ACM Conference on Recommender Systems, Series (RecSys 2011), pp. 325–328. ACM, New York (2011)
4. Hong, W., Li, L., Li, T., Pan, W.: IHR: an online recruiting system for Xiamen talent service center. In: KDD, pp. 1177–1185. ACM (2013)
5. Zhu, C., et al.: Person-job fit: adapting the right talent for the right job with joint representation learning. ACM Trans. Manag. Inf. Syst. **9**(3), 1–17 (2018)
6. Qin, C., et al.: Enhancing person-job fit for talent recruitment: an ability-aware neural network approach. In: SIGIR, pp. 25–34. ACM (2018)
7. Luo, Y., Zhang, H., Wen, Y., Zhang, X.: Resumegan: an optimized deep representation learning framework for talent-job fit via adversarial learning. In: CIKM, pp. 1101–1110. ACM (2019)
8. Yan, R., Le, R., Song, Y., Zhang, T., Zhang, X., Zhao, D.: Interview choice reveals your preference on the market: to improve job-resume matching through profiling memories. In: KDD, pp. 914–922. ACM (2019)
9. Fu, B., Liu, H., Zhu, Y., Song, Y., Zhang, T., Wu, Z.: Beyond matching: modeling two-sided multi-behavioral sequences for dynamic person-job fit. In: Jensen, C.S., et al. (eds.) DASFAA 2021. LNCS, vol. 12682, pp. 359–375. Springer, Cham (2021). https://doi.org/10.1007/978-3-030-73197-7_24
10. Jiang, J., Ye, S., Wang, W., Xu, J., Luo, X.: Learning effective representations for person-job fit by feature fusion. In: CIKM, pp. 2549–2556. ACM (2020)
11. Gupta, A., Garg, D.: Applying data mining techniques in job recommender system for considering candidate job preferences. In: ICACCI, pp. 1458–1465. IEEE (2014)
12. Yang, S., Korayem, M., AlJadda, K., Grainger, T., Natarajan, S.: Combining content-based and collaborative filtering for job recommendation system: a cost-sensitive statistical relational learning approach. Knowl. Based Syst. **136**, 37–45 (2017)

13. Le, R., et al.: Towards effective and interpretable person-job fitting. In: CIKM, pp. 1883–1892. ACM (2019)
14. Bian, S., Zhao, W.X., Song, Y., Zhang, T., Wen, J.: Domain adaptation for person-job fit with transferable deep global match network. In: EMNLP/IJCNLP (1), pp. 4809–4819. ACL (2019)
15. Bian, S., et al.: Learning to match jobs with resumes from sparse interaction data using multi-view co-teaching network. In: CIKM, pp. 65–74. ACM (2020)
16. Watkins, C.J.C.H., Dayan, P.: Technical note q-learning. Mach. Learn. **8**, 279–292 (1992)
17. Precup, D., Sutton, R.S., Dasgupta, S.: Off-policy temporal difference learning with function approximation. In: ICML, pp. 417–424. Morgan Kaufmann (2001)
18. Sutton, R.S., Barto, A.G.: Reinforcement learning - an introduction, series. In: Adaptive Computation and Machine Learning. MIT Press (1998)

TEALED: A Multi-Step Workload Forecasting Approach Using Time-Sensitive EMD and Auto LSTM Encoder-Decoder

Xiuqi Huang, Yunlong Cheng, Xiaofeng Gao$^{(\boxtimes)}$, and Guihai Chen

MoE Key Lab of Artificial Intelligence, Department of Computer Science and
Engineering, Shanghai Jiao Tong University, Shanghai, China
{huangxiuqi,aweftr}@sjtu.edu.cn, {gao-xf,gchen}@cs.sjtu.edu.cn

Abstract. Many data-driven methods and machine learning techniques
are constantly being applied to the database management system
(DBMS), which are based on the judgment of future workloads to achieve
a better tuning result. We propose a novel multi-step workload forecasting approach named TEALED which applies time-sensitive empirical
mode decomposition and auto long short-term memory based encoder-
decoder to predict resource utilization and query arrival rates for DBMSs.
We first improve the empirical mode decomposition method by considering time translation and extending short series. Then we utilize the
encoder-decoder network to extract features from decomposed workloads and generate workload predictions. Moreover, we combine hyper-
parameter search technologies to guarantee performance under varying
workloads. The experiment results show the effectiveness and robustness
of TEALED, and indicate the ability of multi-step workload forecasting.

Keywords: Workload forecasting · Multi-step prediction · Self-driving
DBMS · Long short-term memory · Empirical mode decomposition

1 Introduction

With the development of self-driving DBMS [10,14], machine learning methods
have become active in the field of databases, including but not limited to, using
machine learning to tune configurations, optimize queries, select indexes and
choose materialized views. Many of these self-tuning models are data-driven and
query-based, such as the QTune [5], QueryBot [7] and Ortiz et al. [9], which
can all benefit from workload forecasting. A good prediction of short-term and
long-term workloads can fundamentally help these data-driven and query-based
methods to improve self-driving DBMS. Hence, workload forecasting is the first

This work was supported by the National Key R&D Program of China [2020YF
B1707903]; the National Natural Science Foundation of China [61872238, 61972254],
Shanghai Municipal Science and Technology Major Project [2021SHZDZX0102], and
the ByteDance Research Project [CT20211123001686].

and most critical step in self-driving DBMS to understand application work-loads and predict future trends for better performance. Obviously, numerous database optimizations can achieve better results based on forecasted workload, especially multi-step forecasting that can provide sufficient adjustment time. However, modeling and forecasting workloads are actually very difficult due to complex system functions, numerous configuration parameters, huge data scales, and diverse application requirements.

To meet the challenges of forecasting for self-driving DBMS, we propose TEALED, which uses time-sensitive empirical mode decomposition (EMD) and auto LSTM encoder-decoder to deal with multi-step database workload fore-casting. TEALED includes two main components: parser, and forecaster. In the parser, we use a novel time-sensitive EMD method to cope with complex database workloads. It decomposes the pre-process workload data extracted from DBMS logs into several sub-time series. In the forecaster, we utilize an auto long short-term memory based encoder-decoder and design a weighted loss function to extract features from decomposed workloads and generate multi-step workload forecasting. Besides, we combine a sequential model-based optimization method and a neural architecture search approach to tune hyper-parameters and neural network layers for different kinds of workloads.

The rest of this paper is organized as follows. We introduce related work and formulate DBMS workload forecasting problem in Sect. 2. In Sect. 3, we present the architecture overview and exhibit details about the parser and forecaster of TEALED. In Sect. 4, we evaluate TEALED and show the experiment results. Finally, we draw a conclusion of this paper in Sect. 5.

2 Related Work and Problem Statement

2.1 Related Work

For database workload forecasting, Holze et al. [3] put forward ideas about using Markov chains, Petri nets, neuronal nets, and fuzzy logic to identify and pre-dict significant workload shifts based on SQL workload monitoring. Pavlo et al. [11] present a Markov model-based approach to predict the probability distri-bution of transactions. DBSeer [8] provides resource and performance analysis and prediction for highly concurrent OLTP workloads by employing statistical models. Ma et al. [7] use clustering methods to deal with SQLs and propose a hybrid learning method which combines LR, KR and RNN to predict the arrival rate. Liu et al. [6] improve RNN to predict future workload for Database-as-a-Service. These researches use relatively simple methods, some only capture certain patterns in the workloads, and some focus on SQL processing, which are not enough to model short-term and long-term workload forecasting to support complex database self-tuning.

2.2 Problem Statement

Database workloads change in a random process and have a strong correla-tion with time. We categorize workloads of DBMSs into two parts: query and resource. Query forecasting helps the DBMS to optimize indexes and data partitions in advance to reduce query latency. Resource forecasting assists the DBMS to schedule computing resources to improve resource utilization.

To meet the optimization requirements of the self-driving DBMS, it is necessary to provide forecasting workloads online and give multi-step forecasting to allow time for adjustment. We define the problem as below.

Definition 1 (Online Multi-step Workload Forecasting). *Given a online database workload series $X = (x_1, x_2, \cdots, x_t)$ comprising t observations and the required prediction step length m, the goal of online multi-step workload forecasting is to forecast m workload values $\hat{\boldsymbol{y}}_t = \{\hat{y}_{t+1}, \hat{y}_{t+2}, \cdots, \hat{y}_{t+m}\}$ before the time $t + 1$. When time is $t + 1$, taking the real workload value y_{t+1} into observations.*

3   System Overview and Model Design

3.1   TEALED Architecture

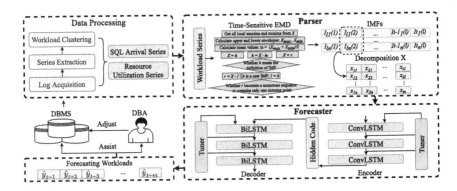

Fig. 1. TEALED architecture

Figure 1 shows a high-level architecture of TEALED that we propose for multi-step workload forecasting to help self-driving DBMS. TEALED contains two main parts: parser and forecaster. Starting from the workload series extracted from database system logs, TEALED parser decomposes the workload series into several sub-time series components of different frequencies with strong regularity. We propose a time-sensitive EMD algorithm, which uses a moving decomposition window and extends workload values at the boundaries to help decomposition. This algorithm is beneficial to handle database workloads affected by complex factors so that the sub-time series obtained have stronger regularity. Then, TEALED forecaster forecasts multi-step future workloads from workload sub-time series, which follows the encoder-decoder architecture. The flexibility of the encoder-decoder allows us to adapt to diverse workloads while ensuring long-term and short-term forecasting accuracy. Besides, we combine automatic machine learning methods as the tuner with long short-term memory based encoder-decoder and design a weighted loss function.

3.2 Parser

In the parser, we use the EMD based data processing method to decompose the workload series containing multiple factors into several sub-time series components of different frequencies with strong regularity. Empirical Mode Decomposition (EMD) is first proposed by Huang et al. [4], in which the core idea is to decompose a signal X into intrinsic mode functions (IMFs) and a residual in an iterative process. An IMF satisfies the following two properties: the number of extrema points and zero crossings are exactly equal or differ at most by one; and the upper and lower envelope are symmetric. However, the inability to reuse newly arrived data makes EMD useless for online multi-step workload forecasting. Hence, we propose the time-sensitive EMD algorithm which uses a moving decomposition window and boundary extension to deal with time-series movement and marginal effect.

First, we choose the length of the moving decomposition window, which usually depends on the trade-off between decomposition accuracy and decomposition speed. The longer the moving decomposition window, the less likely the decomposition will be affected by the marginal effect, but at the same time, the calculation cost such as cubic spline interpolation increases. And we use the traditional EMD algorithm to decompose the first w workload values and get the number of base iterations n. Then, we slice the subsequent workloads using a moving window of w length and decompose them under the given number of iterations. Besides, we found that the value of IMFs closer to the boundary of the window is more susceptible to marginal effects. So, in each move, we use a simple LSTM model to extend e workload values backward to help decomposition. The input of the parser is the database historical workload series $X = x_1, x_2, x_3, \cdots$. And the output is n time-sensitive IMFs that for each x_i there has $\boldsymbol{x}_i = \{x_{i_1}, x_{i_2}, \cdots, x_{i_n}\}$.

3.3 Forecaster

The proposed TEALED forecaster follows the seq2seq architecture [13]. This architecture consists of two components: encoder and decoder. The aim of encoder is to learn a representation, a code, for a set of input data. And decoder tries to decode the code into an output sequence. Our forecaster consists of several layers of autoencoders, where the output of each hidden layer is connected to the input of the successive hidden layer. We use ConvLSTM as encoder and Bi-directional LSTM as decoder. The number of layers is tuned in the ML tuner.

ConvLSTM Encoder. In the forecaster, the input at time T is actually a $n \times t$ matrix to be convoluted, which is $\mathcal{X}_T = \{\boldsymbol{x}_{T-t+1}, \boldsymbol{x}_{T-t+2}, \cdots, \boldsymbol{x}_T\}$. The inputs $\mathcal{X}_T$ will be encoded to the code, or the hidden state $\mathcal{H}_T \in \mathcal{R}^{n \times t}$ through several ConvLSTM layers following $(\mathcal{H}_T^j, \mathcal{C}_T^j) = f_c(\mathcal{X}_T^j)$, $\mathcal{X}_T^j = \mathcal{H}_T^{j-1}(j > 1)$, where $\mathcal{C}_T^j$ is the cell output and $j = 1, 2, \cdots, l_e$, l_e is the number of encoder layers. The key equations of f_c are shown in Eq. 1, where '$*$', '$\circ$' and 'σ' represent the convolution operator, Hadamard product and sigmoid function.

$$i_T = \sigma \left(W_{xi} * \mathcal{X}_T + W_{hi} * \mathcal{H}_{T-1} + W_{ci} \circ \mathcal{C}_{T-1} + b_i \right)$$
$$f_T = \sigma \left(W_{xf} * \mathcal{X}_T + W_{hf} * \mathcal{H}_{T-1} + W_{cf} \circ \mathcal{C}_{T-1} + b_f \right)$$
$$\mathcal{C}_T = f_T \circ \mathcal{C}_{T-1} + i_T \circ \tanh \left(W_{xc} * \mathcal{X}_T + W_{hc} * \mathcal{H}_{T-1} + b_c \right) \qquad (1)$$
$$o_T = \sigma \left(W_{xo} * \mathcal{X}_T + W_{ho} * \mathcal{H}_{T-1} + W_{co} \circ \mathcal{C}_T + b_o \right)$$
$$\mathcal{H}_T = o_T \circ \tanh \left(\mathcal{C}_T \right)$$

In the forecaster, the ConvLSTM layers are stacked to extract the temporal features hierarchically. We use a ML tuner to determine the number of stacked layers, in order to get the best performance when the real application varies. The number of the filters in ConvLSTM is also tuned by the ML tuner.

Bi-directional LSTM Decoder. In the forecaster, we use BiLSTM to be our decoder. We use $\mathcal{H}_T$ to represent the code of $\mathcal{X}_T$ encoded by ConvLSTM layers. The code will be repeated 10 times to form the inputs of the decoder. So the input $\mathcal{H}_T^d = \{\mathcal{H}_T^1, \mathcal{H}_T^2, \cdots, \mathcal{H}_T^{10}\}$. By doing this, the BiLSTM layer can be used to fully utilize the code. The inputs $\mathcal{H}_T^d$ will be mapped into the final result $\hat{y}_T$ through several BiLSTM layers following $h_{j+1}^i = f_b(h_j^i)$, $h_1^i = \mathcal{H}_T^d$, where $j = 1, 2, \cdots, l_d$, l_d represents the number of decoder layers. Rectified Linear Unit(RELU) activation is applied to the BiLSTM. The BiLSTM decoder is also stacked to decode the output from the ConvLSTM encoder, the number of decoder layers is also determined by the ML tuner. The output of the decoder can be a single step and multiple steps. The forecaster predicts multi-step workloads in order to obtain a more informative result and get a deeper insight into the future workload.

Modified Loss. The mean squared error (MSE) loss function is one of the most commonly used loss function for time series prediction tasks. However, it cannot fully capture the temporal information of multi-step forecasting with the same weights of each step. In fact, the predictions closer to the current time is more important, and the accuracy should be ensured as much as possible. Therefore, we proposed a modified loss which assign weights to the multi-step forecasting.

The output generate from the forecaster is $\hat{y}_T = \{\hat{y}_{T+1}, \cdots, \hat{y}_{T+m}\}$, where m denotes the number of forecasting steps. The modified loss follows $loss_m = \frac{\sum_{j=1}^m \alpha^{j-1}(\hat{y}_{T+j} - y_{T+j})^2}{m}$, where α represents the weight that shows the importance of predictions. α is usually smaller than 1 because the predictions closer to the current time is considered more important.

ML Tuner. In deep learning, the performance of the deep learning models depends on the pre-determined hyper-parameters heavily. However, the hyper-parameters cannot be learned during the process of model training. In order to get better performance for different kinds of database workloads, we use ML tuner to tune the hyper-parameters as well as the number of layers of encoder and decoder of the forecaster.

In our ML tuner, we use Tree-structured Parzen Estimator (TPE) to search hyper-parameters. It is a sequential model-based optimization (SMBO) approach. TPE models $P(x|y)$ and $P(y)$ where x is the parameters and y is the

Table 1. Tuned hyper-parameters

Hyper-parameter	Value	Hyper-parameter	Value
Observed workload length	20, 30, **60**, 90	Learning rate	1E-2, **1E-3**, 1E-4
ConvLSTM filter number	32, 64, **128**, 256	ConvLSTM filter size	**(1, 3)**, (1, 5), (1, 7)
ConvLSTM layer number	1, 2, 3	BiLSTM layer number	1, **2**, 3
BiLSTM hidden size	**64**, 128, 256, 512	batch size	16, 32, **64**

evaluation matric [2]. In addition, we apply ENAS [12] to tune the number of neural network layers of TEALED forecaster. ENAS uses parameter sharing between child models to accelerate the neural architecture search. In TEALED forecaster, 8 hyper-parameters are optimized in total. The optimal parameters and neural architecture search results of 5-step prediction as shown in Table 1. The bold parameters in the table represent the optimal values of the parameters after the ML tuner tuning.

4   Experimental Analysis

In this section, we prove the efficiency and robustness of TEALED's ability to forecasting multi-step workloads. First, we introduce the information of three evaluation metrics and two datasets. Then, we compare the prediction accuracy of TEALED with three baseline methods in different forecasting steps.

4.1   Evaluation Metrics and Datasets

Metrics. We use three evaluation metrics, MSE, MAE, and MDA, to compare the overall forecasting efficiency of different forecasting methods. MDA is the average accuracy of directions between the prediction and actual value, which is defined as: $\frac{1}{m} \sum_{i=T}^{m+T} count((\hat{y}_{i+1} - y_i)(y_{i+1} - y_i) > 0)$.

Datasets. We use two real-world datasets. Alibaba Cluster Trace [1] is a resource utilization dataset from a production cluster and we randomly choose one machine's CPU utilization. The unit time interval is 1 min and we use 10,000-time intervals. Bus Tracker Trace [7] are queries come from a BusTracker application and we use the arrival rates of the most common SQL template 1 after clustering. The unit time interval is 10 min and we use 8,000-time intervals. Both of them, we use the first 90% as the training set while the last 10% as the testing set.

4.2   Prediction Accuracy Evaluation

We compare the prediction accuracy of TEALED with three widely used forecasting methods, including ARIMA, Holt-Winter's, and LSTM.

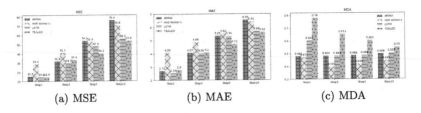

(a) MSE   (b) MAE   (c) MDA

Fig. 2. Prediction accuracy comparison in Alibaba-CPU

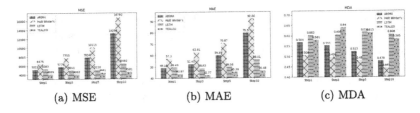

(a) MSE   (b) MAE   (c) MDA

Fig. 3. Prediction accuracy comparison in BUS-SQL1

Figures 2 and 3 shows the performance of the three baselines and TEALED in Alibaba-CPU and BUS-SQL1 with different values of steps. In Alibaba-CPU, the MSE and MAE of ARIMA, LSTM, and TEALED are close when the step is small. As the step increases, the performance of LSTM and TEALED is more stable and achieves better performance than the traditional methods, which have a poor ability to capture long-term patterns. In BUS-SQL1, the SQL arrival rates jitter very intensely over time. The MSE and MAE of TEALED are smaller than others, which means TEALED has better ability for short-term and long-term workload forecasting. It is worth noting the MDA, which characterizes the ability of the forecasting method to judge the trend. In Alibaba-CPU, TEALED is more stable and more capable for finding long-term trend than LSTM, while in BUS-SQL1, it is also comparable to LSTM facing of drastically workload changes.

5   Conclusion

In this paper, we propose TEALED: a novel multi-step workload forecasting approach for database systems using time-sensitive empirical mode decomposition and auto long short-term memory based encoder-decoder. Starting from workload series, TEALED parser uses the time-sensitive EMD algorithm to decompose workload series, in which adopts moving decomposition windows and extending workload values backward. Then, TEALED forecaster forecasts multi-step future workloads from workload sub-time series, which utilizes the encoder-decoder architecture. We combine a tree-structured parzen estimator and a neural architecture search approach with the long short-term memory based encoder-decoder and design a weighted loss function. The output of TEALED includes short-term and long-term forecasting series of database workloads, which can improve automatic database tuning. By conducting experiments based on two real-world datasets: Bus Tracker Trace and Alibaba Cluster Trace, we prove that TEALED performs well in multi-step workload forecasting than both traditional and neural network-based methods.

References

1. Alibaba. Alibaba cluster trace. In: Alibaba Cluster Traces v2018. https://github. com/alibaba/clusterdata (2018)
2. Bergstra, J.S., Bardenet, R., Bengio, Y., Kégl, B.: Algorithms for hyper-parameter optimization. In: Advances in Neural Information Processing Systems (NeurIPS), pp. 2546–2554 (2011)
3. Holze, M., Ritter, N.: Towards workload shift detection and prediction for autonomic databases. In: Proceedings of the ACM First Ph.D. Workshop in CIKM, pp. 109–116 (2007)
4. Huang, N.E., et al.: The empirical mode decomposition and the Hilbert spectrum for nonlinear and non-stationary time series analysis. Roy. Soc. Lond. Ser. A Math. Phys. Eng. Sci. **454**, 903–995 (1998)
5. Li, G., Zhou, X., Li, S., Gao, B.: Qtune: a query-aware database tuning system with deep reinforcement learning. In: Proceedings of the VLDB Endowment (PVLDB), vol. 12, pp. 2118–2130 (2019)
6. Liu, C., Mao, W., Gao, Y., Gao, X., Li, S., Chen, G.: Adaptive recollected RNN for workload forecasting in database-as-a-service. In: International Conference on Service-Oriented Computing (ICSOC), pp. 431–438 (2020)
7. Ma, L., Van Aken, D., Hefny, A., Mezerhane, G., Pavlo, A., Gordon, G.J.: Query-based workload forecasting for self-driving database management systems. In: ACM International Conference on Management of Data (SIGMOD), pp. 631–645 (2018)
8. Mozafari, B., Curino, C., Jindal, A., Madden, S.: Performance and resource modeling in highly-concurrent OLTP workloads. In: ACM International Conference on Management of Data (SIGMOD), pp. 301–312 (2013)
9. Ortiz, J., Balazinska, M., Gehrke, J., Keerthi, S.S.: Learning state representations for query optimization with deep reinforcement learning. In: Workshop on Data Management for End-to-End Machine Learning (DEEM) (2018)
10. Pavlo, A., et al.: Self-driving database management systems. In: Conference on Innovative Data Systems Research (CIDR), pp. 1–6 (2017)
11. Pavlo, A., Jones, E.P., Zdonik, S.: On predictive modeling for optimizing transaction execution in parallel OLTP systems. In: Proceedings of the VLDB Endowment (PVLDB), vol. 5, pp. 2150–8097 (2011)
12. Pham, H., Guan, M., Zoph, B., Le, Q., Dean, J.: Efficient neural architecture search via parameters sharing. In: International Conference on Machine Learning (ICML), pp. 4095–4104 (2018)
13. Sutskever, I., Vinyals, O., Le, Q.V.: Sequence to sequence learning with neural networks. In: Advances in Neural Information Processing Systems (NeurIPS), pp. 3104–3112 (2014)
14. Van Aken, D., Pavlo, A., Gordon, G.J., Zhang, B.: Automatic database management system tuning through large-scale machine learning. In: ACM International Conference on Management of Data (SIGMOD), pp. 1009–1024 (2017)

Author Index

Printed in the United States
by Baker & Taylor Publisher Services